纪念吴文俊先生诞辰100周年

国家出版基金项目
NATIONAL PUBLICATION FOUNDATION

The Complete Works of Wu Wen-Tsun

Topology IV

吴文俊全集・拓扑学卷 IV

吴文俊　著
李邦河　编订

科学出版社
龍門書局
北京

内 容 简 介

本卷收录了吴文俊在拓扑学领域发表的 56 篇学术论文，这些论文包含了吴文俊在示性类、示嵌类、示浸类、示痕类、能计算性与 I^*-量度等方面做出的一系列重要工作，蕴含了他在拓扑学领域的诸多原始思想.

本卷可作为数学或数学史研究人员、教师、研究生的参考文献，也可以作为拓扑学课程的参考书.

图书在版编目（CIP）数据

吴文俊全集. 拓扑学卷. IV/吴文俊著；李邦河编订. —北京：科学出版社，2019.5

国家出版基金项目

ISBN 978-7-5088-5549-3

Ⅰ. ①吴… Ⅱ. ①吴… ②李… Ⅲ. ①拓扑–文集 Ⅳ. ①O1-53

中国版本图书馆 CIP 数据核字(2019) 第 074594 号

责任编辑：李 欣 赵彦超 / 责任校对：彭珍珍
责任印制：肖 兴 / 封面设计：无极书装

科学出版社 出版
北京东黄城根北街 16 号
邮政编码：100717
http://www.sciencep.com

北京通州皇家印刷 印刷
科学出版社发行 各地新华书店经销
*
2019 年 5 月第 一 版 开本：720 × 1000 1/16
2019 年 5 月第一次印刷 印张：44
字数：890 000

定价：298.00 元

(如有印装质量问题，我社负责调换)

编 者 序

中国现代数学的崛起, 开始于 20 世纪初, 经历了几代人坚苦卓绝的努力. 在这百年奋战中涌现出来的数学家中, 吴文俊是最杰出的代表之一. 他早年留学法国, 留学期间就已在拓扑学方面做出了杰出贡献, 提出了后来以他的名字命名的 "吴公式" 和 "吴示性类". 回国后提出了 "吴示嵌类" 等拓扑不变量, 发展了统一的嵌入理论. 他关于示性类与示嵌类的研究, 已成为 20 世纪拓扑学的经典, 至今还在前沿研究中使用. 20 世纪 70 年代以来, 吴文俊院士在汲取中国古代数学精髓的基础上, 开创了崭新的现代数学领域 —— 数学机械化. 他发明的被国际上誉为 "吴方法" 的数学机械化方法, 改变了国际自动推理的面貌, 形成了自动推理的中国学派, 已使中国在数学机械化领域处于国际领先地位. 上述工作无疑属于 20 世纪中国数学赶超国际先进水平的标志性成果, 而吴文俊院士博大精深的科学研究, 除了拓扑学与数学机械化以外, 还跨越了代数几何、博弈论、中国数学史、计算图论、人工智能等众多领域, 并在每个领域都留下了这位多能数学家的重要贡献.

吴文俊先生是一位具有强烈爱国精神的数学家. 自 1950 年谢绝法国师友的挽留回到祖国后, 半个世纪如一日, 为在他深爱的中华故土发展数学事业而鞠躬尽瘁. 除了第一流的科研成果, 吴文俊先生长期身处中国数学界领导地位, 在团结带领整个中国数学界赶超世界先进水平方面, 也做出了不可磨灭的贡献. 特别是, 吴文俊先生在担任中国数学会理事长期间, 领导中国数学最终成功地加入了国际数学联盟, 此举大大提高了我国数学界的国际地位, 同时也为我国成功举办 2002 年国际数学家大会铺平了道路.

吴文俊治学严谨, 学术思想活跃, 无论获得多么高的声誉, 他总是勤奋地在科研第一线工作, 一生积极进取, 锲而不舍, 不断取得新的成就. 在开始从事机器证明时, 他已近花甲之年, 从零开始学习编写计算机程序, 每天十多个小时在机房连续工作, 终于在几何定理机器证明这一难题上取得成功.

吴文俊先生为中国现代数学的发展建立了丰功伟绩, 而他本人却始终淡泊、谦逊. 他处事公正豁达, 待人充满善意, 受过他帮助的人可以说不计其数. 正因如此, 这位有着崇高国际声望而平易近人的学者, 受到了每一个认识他的人格外的爱戴与尊敬.

2019 年 5 月 12 日是吴文俊先生百年诞辰. 为了纪念这个特殊的日子, 我们编辑出版了《吴文俊全集》, 通过系统地收录、整理吴文俊先生的学术著作和论文, 纪念吴先生的学术思想及学术成就. 全集共计 13 卷, 包括拓扑学 4 卷、数学机械化 5

卷以及数学史、博弈论与代数几何、数学思想各 1 卷；同时, 全集还设有附卷, 收录吴文俊先生的同事、学生和其他社会各界人士发表过的与吴先生有关的各类文献资料.

最后, 我们对在全集编辑中给予帮助的各位同事表示衷心感谢；感谢国家出版基金对于全集出版的资助；感谢科学出版社编辑人员在出版全集时认真细致的专业精神；感谢相关出版与新闻机构在版权方面提供的帮助.

李邦河　高小山　李文林

2019 年 3 月

前　　言

本卷收录了吴文俊先生在拓扑学领域发表的 56 篇论文. 按照时间顺序和工作重点, 我们大致可将吴先生的工作分成四个时期来介绍.

第一个时期的工作主要完成于 1953 年前, 包括他在中央研究院、法国以及回国初期的工作. 这一阶段的工作主要集中在示性类方面, 包括 Stiefel-Whithey 示性类、吴示性类、Pontrjagin 示性类等.

1948 年发表的论文 [2] 是吴文俊引起轰动的一篇论文. 该论文投稿于 1947 年 8 月, 此时他正在中央研究院工作. 该文给出了 Stiefel-Whitney 示性类的乘法公式的一个较为简短的证明. 此公式的证明最初由 Whitney 于 1940 年给出, 因其证明极为复杂, 没有全文刊出. 故在论文发表后, Whitney 仍不得不保留原稿. 陈省身先生十分欣赏吴的结果, 推荐到数学领域的顶级期刊 *Annals of Mathematics* 上发表. 据说该文章出来后, Whitney 认为不必再保留他的手稿了.

论文 [3-10] 则是吴文俊在法国留学和工作的时候完成的.

在法国期间, 吴文俊于斯特拉斯堡大学, 跟随 Ehresmann 学习. Ehresmann 是 E. Cartan 的学生, 他的博士论文是关于 Grassmann 流形的同调群的计算. 同时, 他还是纤维丛概念的创始人之一. 吴在其指导下继续进行纤维空间及示性类的研究, 完成了论文 [4, 6, 7], 这些论文所述定理及其完整证明后来整理到其博士学位论文中. 他的学位论文同 G. Reeb 的论文一起, 于 1952 年作为一本书出版. 作为副产品, 吴在 [3] 中还证明了 $4k$ 维球面没有近复结构, 这是第一个关于球面上的近复结构的非平凡的结论. 迄今为止, 有近复结构的 6 维球面上是否存在复结构仍然是一个著名的没有解决的问题.

此后, 吴文俊到巴黎, 在法国国家科学研究中心 (CNRS) 跟随 H. Cartan 继续拓扑学研究. 他于 1950 年发表的论文 [8], 研究了流形上 Stiefel-Whitney 示性类. 在该文中吴为研究 Stiefel-Whitney 示性类的计算问题, 定义了一类新的示性类, 并给出了一个重要公式. 这一公式后来被称为第一吴公式. 该公式使 Stiefel-Whithey 示性类的计算变得极其容易, 同时利用该公式还可以轻而易举地证明 Stiefel-Whithey 示性类不仅是拓扑不变的, 而且还是同伦不变的. 他在该文中提出来的新的示性类后来被称为吴示性类, 简称为吴类. 同年发表的论文 [9], 给出了 Grassmann 流形上的 Steenrod 幂运算的公式, 这就是著名的第二吴公式. 1956 年 Dold 证明, 这一公式给出了 Stiefel-Whithey 示性类之间所有可能的关系.

在法国期间完成的论文大都发表在《法国科学院通报》(*Comptes Rendus*) 上,

均是以简报的形式报道研究成果, 详细完整的证明则在后面的论文中给出, 如论文 [13, 14] 给出了 [8, 9] 中结果的详细证明.

在论文 [10] 中, 吴研究了 Smith 运算与 Steenrod 运算. 他发现此二组运算是等价的, 并确定二者互相决定的方式, 从而给出了 Steenrod 幂一个较自然的定义. 但是该文中的证明用到了 Thom 关于 Steenrod 幂的内在的公理系统理论. 后来, 在论文 [24] 中, 他给出了不借助于 Thom 工作的直接证明.

吴文俊自 1952 年回国后的大多数文章都发表在国内的期刊上. 论文 [12, 16-19] 是研究 Pontrjagin 示性类的一个系列文章, 他证明了该示性类模 3、模 4 的拓扑不变性, 还引领了对这一示性类的拓扑不变或非拓扑不变的进一步研究. Pontrjagin 示性类定义在微分流形上, 吴曾想证明其拓扑不变性, 后来有人发现这是不对的. Pontrjagin 示性类是有理不变的, 其示性类可变部分产生于其绕元素部分, 所以他关于该示性类模 3、模 4 的拓扑不变性是非常重要的工作. 他关于四维定向流形的 Pontrjagin 示性类是符号差的三倍的猜想, 对后世数学的发展影响非常深远, 成为 Hirzebruch 的符号差定理和 Atiyah-Singer 指标定理的源头.

第二个时期的工作大致在 1953 年至 1958 年间, 包括论文 [11, 20-23, 25-26, 28-30, 32-34, 41]. 这一阶段吴独创了关于复形的示嵌类、示浸类和示痕类等概念.

论文 [11] 中引入了关于复合形的一组非同伦不变的拓扑不变量. 由于熟知的大量拓扑不变量均是同伦不变的, 吴的这一发现受到了高度重视. 在这一发现的鼓舞下, 他在论文 [20] 中提出了示嵌类的概念, 在论文 [22] 中提出了示浸类, 并系统发展了复形在欧氏空间中的嵌入与浸入理论, 见 [21, 22, 25, 26]. 他独创地运用 Smith 周期变换定理于复形的 p 重约化积, 并于论文 [30, 33, 34] 中提出示痕类的概念. 运用这些类他给出了 n 维复形可嵌入于 $R^{2n}(n>2)$, 可浸入于 $R^{2n-1}(n>3)$ 的充分且必要条件, 以及 n 维复形在 $R^{2n+1}(n>1)$ 中的两个嵌入同痕的充要条件.

1957 年, 吴文俊将这一阶段成果整理成书, 在数学所油印成册. 直至 1965 年才将此油印本修订后, 总结成专著 *A Theory of Imbedding, Immersion, and Isotopy of Polytopes in an Euclidean Space*, 由科学出版社出版. 中译本《可剖形在欧氏空间中的实现问题》一直到 1978 年 5 月才正式出版. “文革” 期间, 吴对印刷电路与集成电路中的布线问题产生了兴趣, 这导致了他对 1 维复形在平面的可嵌入问题的思考. 吴完全解决了这一问题, 使著名的 Karatowski 不可嵌入定理成为其特例. 这一结果被作为附录收录到该中译本中.

第三个时期的工作可从 1958 年算起, 至 1973 年. 1958 年, 拓扑学被批判成“理论脱离实际”, 吴曾把主要精力放在更加联系实际的博弈论上. 但他对拓扑仍然保持兴趣, 也写了不少文章, 只是方向比较分散, 包括 [27, 35-40, 42-46].

可以看到 [27] 是一篇点集拓扑的论文, 论文 [36] 讨论了射影空间中某些实二次曲面的示性类, 论文 [38] 是一篇讨论欧氏空间中旋转的短文, [39, 40] 研究的是微分拓扑中的浸入问题, 论文 [45] 研究的是奇点理论, [46] 是一篇有关外微分形式的文章, 论文 [37] 证明了 $n>4$ 时可定向光滑流形能光滑嵌入于 R^{2n-1} 这个重要定理.

第四个时期的工作起始于 1973 年. 这一阶段, 吴文俊试图在代数拓扑方面构造性地统一处理同调群、同伦群、示性类、上同调运算等, 见 [47-55]. 他以 "能计算性" 的概念, 重新整理和改造 Sullivan 的极小模理论, 提出和解决了不少问题. 在早期的文章中, 吴文俊将这一新的理论命名为 I^* 函子, 见 [47-51], 后来发现这不是通常意义下的函子, 所以最后将其定名为 I^* 量度, 见 [52-54]. 这一时期, 吴在数学机械化和代数拓扑两条战线上同时作战. 在出版了这方面的专著 *Rational Homotopy Type: A Constructive Study via the Theory of the I*-Measure* 之后, 他便全力以赴于数学机械化方向了.

感谢李博副研究员对完成该前言的大力协助.

李邦河

目　录

1. Topologie——Note Surles Produits Essentiels Symétriques des Espaces Topologiques*

Note[1] de M. Wen-Tsün Wu, présentée par M. Élie Cartan

Dans un espace topologique donné E prenons un ensemble arbitraire de n points $\{x_1, \cdots, x_n\}$ qui ne sont nécessairement pas tous différents. Deux ensembles $\{x_1, \cdots, x_n\}$ et $\{y_1, \cdots, y_n\}$ sont considérés comme identiques si chaque x_i est égal à un certain y_j et chaque y_i est égal à un certain x_j. Nous définissons, d'après M M. Borsuk et Ulam①, dans cet espace une topologie d'une manière naturelle. L'espace topologique ainsi obtenu s'appelle produit essentiel symétrique de l'espaoe E et sera désigné par E(n). L'objet de cette Note est d'étudier des propriétés topologiques et en particulier de déterminer les groupes d'homologie des espaces I(n) et C(n), où I est un intervalle fermé et C est un cercle.

Considérons d'abord le cas de I(n). On remarque qu'un point de I(n) peut être représenté par un ensemble des valeurs $\{x_1, \cdots, x_n\}$ telles que $\mathrm{o} \leqslant x_1 \leqslant \cdots \leqslant x_n \leqslant \mathrm{I}$.

Prenons dans l'espace euclidien à n dimensions $\mathcal{R}^n$, $n + \mathrm{I}$ points

$$\begin{aligned} \mathrm{A}_0 &= (\mathrm{o}, \mathrm{o}, \cdots, \mathrm{o}, \mathrm{o}), \\ \mathrm{A}_1 &= (\mathrm{I}, \mathrm{I}, \cdots, \mathrm{I}, \mathrm{I}), \\ \mathrm{A}_2 &= (\mathrm{o}, \mathrm{I}, \cdots, \mathrm{I}, \mathrm{I}), \\ &\cdots\cdots \\ \mathrm{A}_n &= (\mathrm{o}, \mathrm{o}, \cdots, \mathrm{o}, \mathrm{I}). \end{aligned}$$

Désignons par $[x_0, \cdots, x_n]$ le point du simplexe $\sigma^n = \mathrm{A}_0\mathrm{A}_1 \cdots \mathrm{A}_n$ dont les coordonnées barycentriques sont $x_0, \cdots, x_n$. La transformation

$$f : [x_0, \cdots, x_n] \rightarrow \{x_1, x_1 + x_2, \cdots, x_1 + \cdots + x_n\}$$

* *C. R. Acad. Sci. Paris* 224, 1947: 1139-1141.

① *Bull. Amer. Math. Soc.*, 1931, 37: 875-882.

est une transformation continue de σ^n sur I(n). Deux points $[x_0, \cdots, x_n]$ et $[y_0, \cdots, y_n]$ ont la même image s'il existe des entiers $i_1 \geqslant$ I, $\cdots$, $i_k \geqslant$ I, $j_1 \geqslant$ I, $\cdots$, $j_k \geqslant$ I tels que l'on ait

$$\sum_{i=0}^{n} x_i = \sum_{i=0}^{n} y_i = \mathrm{I},$$
$$x_2 = \cdots = x_{i_1} = \mathrm{o}, \quad y_2 = \cdots = y_{j_1} = \mathrm{o},$$
$$x_{i_1+2} = \cdots = x_{i_1+i_2} = \mathrm{o}, \quad y_{j_1+2} = \cdots = y_{j_1+j_2} = \mathrm{o},$$
$$\vdots \qquad \vdots$$
$$x_{i_1+\cdots+i_{k-1}+2} = \cdots = x_{i_1+\cdots+i_k} = \mathrm{o}; \quad y_{j_1+\cdots+j_{k-1}+2} = \cdots = y_{j_1+\cdots+j_k} = \mathrm{o};$$
$$x_0 = y_0,$$
$$x_1 = y_1,$$
$$x_{i_1+1} = y_{j_1+1},$$
$$\vdots$$
$$x_{i_1+\cdots+i_{k-1}+1} = y_{j_1+\cdots+j_{k-1}+1}.$$

Il en résulte que l'espace déduit de σ^n par l'identification de toutes les paires de simplexes ordonnés

$$\mathrm{A}_0\mathrm{A}_1\mathrm{A}_{i_1+1} \quad \cdots \quad \mathrm{A}_{i_1+\cdots+i_{k-1}+1},$$
$$\mathrm{A}_0\mathrm{A}_1\mathrm{A}_{j_1+1} \quad \cdots \quad \mathrm{A}_{j_1+\cdots+j_{k-1}+1}$$

est homéomorphe à I(n). Nous avons done le théorème suivant :

Théorème I — *Soit* I$'(n)$ *le complexe déduit du simplexe* $\mathrm{A}_2 \cdots \mathrm{A}_n$ *par l'identification de tous les simplexes ordonnés de la table suivante* :

$$\mathrm{A}_3\mathrm{A}_4\mathrm{A}_5 \quad \cdots \quad \mathrm{A}_{n-1}\mathrm{A}_n,$$
$$\mathrm{A}_2\mathrm{A}_4\mathrm{A}_5 \quad \cdots \quad \mathrm{A}_{n-1}\mathrm{A}_n,$$
$$\mathrm{A}_2\mathrm{A}_3\mathrm{A}_5 \quad \cdots \quad \mathrm{A}_{n-1}\mathrm{A}_n,$$
$$\cdots\cdots$$
$$\mathrm{A}_2\mathrm{A}_3\mathrm{A}_4 \quad \cdots \quad \mathrm{A}_{n-2}\mathrm{A}_{n-1}.$$

Ators I(n) *est homéomorphe à l'umon de* I$'(n)$ *et d'un intervalle* $\mathrm{A}_0\mathrm{A}_1$.

Borsuk et Ulam ont démontré que I(n) ne peut pas être plongé dans $\mathcal{R}^n$ lorsque $n \geqslant 4$, Avec notre théorème il serait facile de voir que I(4) peut être plongé dans $\mathcal{R}^5$. En effet I'(4) est un triangle $\mathrm{A}_2\mathrm{A}_3\mathrm{A}_4$ avec l'identification des côtés $\mathrm{A}_2\mathrm{A}_3$, $\mathrm{A}_2\mathrm{A}_4$,

A_3A_4. Done $I'(4)$ peut être plongé dans $\mathcal{R}^3$. En prenant dans $\mathcal{R}^5$ un intervalle A_0A_1 sur une droite qui ne rencontre pas $\mathcal{R}^3$, on voit que l'union de A_0A_1 et $I'(4)$, c'est-à-dire $I(4)$, est plongé dans $\mathcal{R}^5$. Ce théorème a été démontré par M. Haratomi①.

L'espace $C(n)$ se déduit de $I(n)$ par l'identification des points frontières o et I de l'intervalle I. Cela correspond à l'identification des points

$$[y_1, o, y_2, \cdots, y_n] \quad \text{et} \quad [o, y_2, \cdots, y_n, y_1].$$

Cette observation conduit aux rèsultats suivants :

Théorème II — $C(n)$ *est homéomorphe au complexe déduit de* σ^n *par les identifications de la table suivante*:

$$\begin{cases} A_0A_1A_3A_4A_5 & \cdots & A_{n-1}A_n \\ A_0A_1A_2A_4A_5 & \cdots & A_{n-1}A_n \\ A_0A_1A_2A_3A_5 & \cdots & A_{n-1}A_n \\ & \vdots & \\ A_0A_1A_2A_3A_4 & \cdots & A_{n-2}A_{n-1}, \end{cases}$$

$$\begin{cases} A_0A_2A_3 & \cdots & A_n \\ A_nA_1A_2 & \cdots & A_{n-1}. \end{cases}$$

Théorème III — $C(n)$ *est une* H-*sphère lorsque* n *est impair. Pour* n *pair, les seuls groupes d'homologie non nuls sont* H^0 *et* H^{n-1}, *qui sont d'ailleurs isomorphes au groupe additif des entiers.*

Comme $I(n)$est un H-simplexe, loutes les transformations de $I(n)$en luimême ont an moins un point fixe. La situation est différente pour $C(n)$. Soient $o \leqslant \theta \leqslant 2\pi$ la coordonnée angulaire sur C et α une valeur inférieure à $2\pi/n$. La transformation

$$f : \{\theta_1, \cdots, \theta_n\} \to \{\theta_1 + \alpha, \cdots, \theta_n + \alpha\}$$

est une transformation de $C(n)$ en lui-même qui ne possède pas de points fixes.

①*Jap. Journal of Math.*, 1932, 9: 103-110.

2. On the Product of Sphere Bundles and the Duality Theorem Modulo Two*

Introduction①

Given two sphere bundles $\mathfrak{S}_1$ and $\mathfrak{S}_2$ over the base complexes K_1 and K_2 respectively, it is possible to define in a natural way a "product bundle" over the product complex $K_1 \times K_2$. When $K_1 = K_2 = K$(say), the part of the product bundle over the diagonal of the product complex $K \times K$ is the product bundle in the sense of Whitney.② We shall prove in the present paper that a certain duality theorem holds for the product bundle over $K_1 \times K_2$ and that Whitney's duality theorem for sphere bundles follows from this more general duality theorem as a consequence. (Throughout the paper coefficients mod 2 will be used.) The idea of this proof seems to be quite different from Whitney's original one, of which only a brief sketch is known.③

The paper is divided into three sections. In §1 some preliminary considerations and theorems on vector fields are given. A duality theorem for the product bundle over $K_1 \times K_2$ is then proved in §2. §3 is devoted to a proof of Whitney's duality theorem.

§1

1. We recall in this paragraph the definition of a bundle of linear spaces or more simply, a vector bundle.

A complex K with cells $\sigma_1, \sigma_2, \cdots$ and a v-dimensional vector space V are given. To each point p of K a v-dimensional vector space $V(p)$ is associated so that $V(p)$ and $V(q)$ are disjoint if p and q are distinct points of K. Suppose there are non-degenerate linear mappings $\xi_{\sigma_i p}$ of V on $V(p)$ for every σ_i of K and every point p of

* *Ann. of Math.*, 1948, 49(3): 641-653.

① The problems in this paper were suggested to me by Professor S. S. Chern, with whom I have many helpful discussions. To him are expressed here my thanks.

② Whitney, Lectures in Topology. Harvard Univ. , 1941: 131.

③ Whitney, Proc. , Nat. Acad. Sci. , 1940, 26: 143-148.

σ_i with the following condition satisfied: For p common to σ_i and σ_j, $\xi_{\sigma_i p}^{-1}\xi_{\sigma_j p}$ gives a continuous map of $\sigma_i \cap \sigma_j$ into the group of nondegenerate linear mappings of V on itself. Then we can make the union of all the spaces $V(p)$ into a single topological space $\mathfrak{B}$ in a natural way so that for every cell $\sigma_i \in K$ the topological product $V \times \sigma_i$ is homeomorphic to the union of all $V(p)$ for which $p \in \sigma_i$. This homeomorphism is in fact given by $\xi_{\sigma_i}(p, x) = \xi_{\sigma_i p}(x)$, where $x \in V$.

We shall introduce the following terminologies:

$\mathfrak{B}$, the vector bundle;

K, the base complex;

V, the director space;

$V(p)$, the vector space over p;

v, the bundle-dimension of $\mathfrak{B}$;

$\xi_{\sigma p}$, the coordinate system in σ.

2. Let $\mathfrak{B}$ be the vector bundle defined in paragraph 1. By a continuous mapping of K or a subcomplex of K into $\mathfrak{B}$ we shall always mean one that maps the points p of K or a subcomplex of K into the respective $V(p)$ over p. A set of continuous mappings $\varphi_1, \cdots, \varphi_m$ of a subcomplex L of K into $\mathfrak{B}$ is said to form an m-field $\Phi = \{\varphi_1, \cdots, \varphi_m\}$ over L. We say that Φ is continuous (or discontinuous) at a point $p \in L$ if $\varphi_1(p), \cdots, \varphi_m(p)$ are linearly independent (or linearly dependent) in $V(p)$ and that Φ is a continuous m-field over L if it is continuous at every point $p \in L$.

Let K^r be the r-dimensional skeleton of K. As is well known,① for $m \leqslant v$ continuous m-field $\{\varphi_1, \cdots, \varphi_m\}$ always exists over K^{v-m}. Now orient V and the cells of the complex arbitrarily and consider any $(v-m+1)$-dimensional oriented cell σ_i. For points p in $\partial\sigma_i$, $\xi_{\sigma_i p}^{-1}\varphi_1(p), \cdots, \xi_{\sigma_i p}^{-1}\varphi_m(p)$ together give a map φ_{σ_i} of $\partial\sigma_i$ into the Stiefel manifold② $V_{v,m}$ of all ordered sets of m linearly independent vectors in V. The characteristic ${}^4d(\sigma_i)$ of this mapping is either an integer or is defined only mod 2. In any way, the chain

$$w = \sum_i d(\sigma_i)\sigma_i$$

when reduced mod 2 if necessary, is a $(v-m+1)$-dimensional cocycle mod 2 of K the class of which is independent of the particular choice of the continuous m-field and the orientations of the cells of K. The classes thus obtained will be called the

① See for example Stiefel, Comm. Math. Helv., 1936, 8: 331.

② Stiefel, loc. cit., 310-323. The characteristic will be denoted sometimes by Char $\sigma_i\{\varphi_1, \cdots, \varphi_m\}$, Char $\varphi_{\sigma_i}\sigma_i$, etc.

characteristic classes and their cocycles characteristic cocycles. We denote them by W^r, $r = 1, 2, \cdots, v$.

For convenience we shall define W^0 to be the class containing the cocycle I, which is the sum of all vertices of K. We also put all $W^r = 0$, for $r > v$.

3. We shall put $[a - b] = a - b$ for $a \geqq b$ and $[a - b] = 0$ for $b > a$. We shall prove that in a vector bundle $\mathfrak{B}$ it is possible to construct on K an m-field $\Phi = \{\varphi_1, \cdots, \varphi_m\}$, ($m$ not necessarily $\leqslant v$) which satisfies the following conditions (C_r), $r = 0, 1, 2, \cdots$:

(C_r). Let p be any point in K^r. The conditions are:

CASE 1. If $m + r \leqslant v$, then $\varphi_1, \cdots, \varphi_m$ are linearly independent at p;

CASE 2. If $m + r > v$, then there is an integer $0 \leqslant i \leqslant m - [v - r]$ such that $\varphi_1(p), \cdots, \varphi_{[v-r]+i}(p)$ are linearly independent while

$$\varphi_{[v-r]+i+1}(p) = \cdots = \varphi_m(p) = 0.$$

Such a field will be called a canonical field. We shall construct it successively over K^0, K^1, $\cdots$ as follows:

1^0. Construction over $K^0, \cdots, K^{[v-m]}$.

If $m < v$, we take $\varphi_1(p), \cdots, \varphi_m(p)$ at a vertex p to be any m linearly independent vectors in $V(p)$ and then extend successively to K^{v-m}. If $m \geqq v$ and p a vertex, we take $\varphi_1(p), \cdots, \varphi_v(p)$ to be v linearly independent vectors in $V(p)$, while we set $\varphi_{v+1}(p) = \cdots = \varphi_m(p) = 0$. In this way $(C_0), \cdots, (C_{[v-m]})$ evidently hold.

2^0. Construction over K^r, $r > [v - m]$, assuming that $\varphi_1, \cdots, \varphi_m$ have been constructed over K^{r-1} with (C_{r-1}) satisfied.

Consider any r-dimensional cell σ^r. We can denote its points by tp, $0 \leqslant t \leqslant 1$, where p is a point on $\partial\sigma^r$ and $0 \cdot p = 0$ is a fixed interior point of σ.

$\{\varphi_1, \cdots, \varphi_{[v-r]}\}$ being defined and continuous on ∂^r we can extend it continuously into the interior of σ^r. Then $\bar{\varphi}_1^*(tp) = \xi_{\sigma p}^{-1}\varphi_1(tp), \cdots, \bar{\varphi}_{[v-r]}^*(tp) = \xi_{\sigma_p}^{-1}\varphi_{[v-r]}(tp)$ are $[v - r]$ continuous mappings of σ^r into V so that for each tp these are $[v - r]$ linearly independent vectors. We can find $v - [v - r]$ further mappings $\bar{\varphi}_{[v-r]+1}^*(tp), \cdots, \bar{\varphi}_v^*(tp)$ of σ^r into V so that for every tp in σ^r, $\bar{\varphi}_1^*(tp), \cdots, \bar{\varphi}_v^*(tp)$ are linearly independent and form a positive system in V, assuming that V has been oriented. Put $\xi_{\sigma p}\bar{\varphi}_i^*(tp) = \bar{\varphi}_i(tp)$, $i = 1, 2, \cdots, v$, we get a v-field $\{\bar{\varphi}_1, \cdots, \bar{\varphi}_v\}$ continuous over σ^r. We have moreover

$$\tilde{\varphi}_i(tp) = \varphi_i(tp), \qquad \text{for } i = 1, \cdots, [v - r].$$

Let

$$\varphi_i(p) = \sum_{j=1}^{v} a_j^{(i)}(p)\bar{\varphi}_j(p), \qquad i = [v-r]+1, \cdots, m,$$

where $a_j^{(i)}(p)$ are real numbers. We define

$$\varphi_i(tp) = \sum_{j=1}^{v} t a_j^{(i)}(p)\bar{\varphi}_j(tp), \qquad i = [v-r]+1, \cdots, m.$$

The field $\{\varphi_i, \cdots, \varphi_m\}$ is then extended over σ^r. Doing this for all σ^r, we get an m-field over K^r.

The only places in σ^r where discontinuity occurs are either 1) 0, or 2) tp, $t \neq 0$ for which $\{\varphi_1, \cdots, \varphi_m\}$ is discontinuous at p.

In case 1), $\{\varphi_1, \cdots, \varphi_{[v-r]}\}$ is continuous, while $\varphi_{[v-r]+1}(0) = \cdots = \varphi_m(0) = 0$. Condition (C_r) is thus satisfied with $i = 0$.

In case 2), there exists by induction an integer i so that $\{\varphi_1, \cdots, \varphi_{[v-r+1]+i}\}$ is continuous at p, while

$$\varphi_{[v-r+1]+i+1}(p) = \cdots = \varphi_m(p) = 0.$$

This is true when p is replaced by tp, $t \neq 0$. As

$$[v-r+1] = \begin{cases} [v-r]+1, & \text{for} \quad r < v+1, \\ [v-r], & \text{for} \quad r \geqq v+1, \end{cases}$$

we see that condition (C_r) is satisfied.

We remark in passing that in case $r \leqslant v$, not all of

$$a_{v-r+1}^{(v-r+i)}(p), \cdots, a_v^{(v-r+1)}(p), \qquad p \in \partial\sigma^r$$

are zero. For we have

$$\varphi_i(p) = \bar{\varphi}_i(p), \qquad i = 1, 2, \cdots, v-r,$$
$$\varphi_{v-r+1}(p) = \sum_{j=1}^{v} a_j^{(v-r+1)}(p) \cdot \bar{\varphi}_j(p).$$

As $\varphi_{v-r+1}(p)$ is linearly independent of $\varphi_i(p)$, $i = 1, 2, \cdots, v-r$, for $p \in \partial\sigma^r$, the matrix of their components with respect to the field $\{\bar{\varphi}_1, \cdots, \bar{\varphi}_v\}$ over σ^r, namely,

$$\begin{pmatrix} 1 & 0 & \cdots & 0 & 0 & \cdots & 0 \\ 0 & 1 & \cdots & 0 & 0 & \cdots & 0 \\ \vdots & \vdots & & \vdots & \vdots & & \vdots \\ 0 & 0 & \cdots & 1 & 0 & \cdots & 0 \\ a_1^{(v-r+1)} & a_2^{(v-r+1)} & \cdots & a_{v-r}^{(v-r+1)} & a_{v-r+1}^{(v-r+1)} & \cdots & a_v^{(v-r+1)} \end{pmatrix}$$

must be of rank $v-r+1$.

4. Given two vector bundles $\mathfrak{B}_1$, $\mathfrak{B}_2$, of which the base complexes and the other symbols are distinguished by the subscripts 1 and 2, we shall define a third bundle $\mathfrak{B}$ according to the following table:

Base complex: $K = K_1 \times K_2$;

Director space: $V = V_1 \oplus V_2$; ①

Vector space over $p_1 \times p_2$: $V(p_1 \times p_2) = V_1(p_1) \oplus V_2(p_2)$; ①

Bundle dimension: $v = v_1 + v_2$;

Coordinate system in $\sigma_1 \times \sigma_2$:

$$\xi_{\sigma_1\times\sigma_2, p_1\times p_2}(\mathfrak{r}_1+\mathfrak{r}_2) = \xi_{\sigma_1 p_1, 1}(\mathfrak{r}_1) + \xi_{\sigma_2 p_2, 2}(\mathfrak{r}_2), \quad \text{where } \mathfrak{r}_1 \in V_1, \mathfrak{r}_2 \in V_2.$$

This bundle $\mathfrak{B}$, as a topological space, is a topological product of $\mathfrak{B}_1$ and $\mathfrak{B}_2$. In fact, by means of the coordinate system $\xi_{\sigma_1\times\sigma_2, p\times p_2}(\mathfrak{r}_1+\mathfrak{r}_2)$ in $\sigma_1 \times \sigma_2$, we map the point $\xi_{\sigma_1\times\sigma_2, p_1\times p_2}(\mathfrak{r}_1+\mathfrak{r}_2)$ into the point $(\xi_{\sigma_1 p_1, 1}(\mathfrak{r}_1), \xi_{\sigma_2 p_2, 2}(\mathfrak{r}_2))$ of $\mathfrak{B}_1 \times \mathfrak{B}_2$. This mapping is clearly topological. We can therefore write $\mathfrak{B}_1 \times \mathfrak{B}_2$ for $\mathfrak{B}$ without confusion and shall call $\mathfrak{B}$ the product bundle of $\mathfrak{B}_1$ and $\mathfrak{B}_2$.

Let the characteristic classes of $\mathfrak{B}$, $\mathfrak{B}_1$ and $\mathfrak{B}_2$ be respectively denoted by W, W_1, W_2, with the convention made at the end of paragraph 2. Then our main theorem is the following

Theorem I *The characteristic classes of the product bundle $\mathfrak{B} = \mathfrak{B}_1 \times \mathfrak{B}_2$ are expressible in terms of those of $\mathfrak{B}_1$ and $\mathfrak{B}_2$. More precisely, we have the formula*

$$W^r = \sum_{i=0}^{r} W_1^i \times W_2^{r-i}, \quad r = 1, 2, \cdots, v. \tag{4.1}$$

The multiplication of cohomology classes occurred in this formula may be explained as follows:

① $V_1 \oplus V_2$ means the join of the two vector spaces. We assume that $V(p_1 \times p_2)$ and $V(q_1 \times q_2)$ are disjoint for $p_1 \times p_2 \neq q_1 \times q_2$ even if $p_1(p_2)$ may coincide with $q_1(q_2)$.

Let

$$C_1=\sum_i a_{i1}\sigma_{i1},\quad C_2=\sum_j a_{j2}\sigma_{j2}$$

be any two chains of K_1 respectively K_2, where a_{i1}, a_{j2} are elements of a coefficient ring(in the case considered above is the ring of residue classes mod 2). We define a product chain $C=C_1\times C_2$ of $K_1 \times K_2$ by

$$C=\sum_{i,j} a_{i1}a_{j2}(\sigma_{i1}\times\sigma_{j2}).$$

Since

$$\delta(\sigma_{i1}\times\sigma_{j2})=\delta\sigma_{i1}\times\sigma_{j2}+(-1)^p\sigma_{i1}\times\delta\sigma_{j2},\qquad p=\dim\sigma_{i1},$$

we have

$$\delta C=\delta C_1\times C_2+(-1)^p C_1\times\delta C_2,$$

if C_1 is a p-dimensional chain.

It follows readily that we can define the product class of two cohomology classes in a unique way. It is this product that we denote by $W_1\times W_2$ in (4.1).

5. If two vector bundles $\mathfrak{B}_1$, $\mathfrak{B}_2$ of which the symbols are again distinguished by subscripts 1 and 2 are defined on the same base complex K, a third bundle $\breve{\mathfrak{B}}$ over the same K can be defined according to the following table:

Base complex: K;

Director space: $\breve{V}=V_1\oplus V_2$; ①

Vector space over p : $\breve{V}(p)=V_1(p)\oplus V_2(p)$;

Bundle-dimension: $v=v_1+v_2$;

Coordinate system in σ:

$$\breve{\xi}_{\sigma p}(\mathfrak{r}_1+\mathfrak{r}_2)=\xi_{\sigma p,1}(\mathfrak{r}_1)+\xi_{\sigma p,2}(\mathfrak{r}_2),\quad \text{where } \mathfrak{r}_1\in V_1\cdot\mathfrak{r}_2\in V_2.$$

$\breve{\mathfrak{B}}$ will be called the span bundle of $\mathfrak{B}_1$ and $\mathfrak{B}_2$ and we shall write $\mathfrak{B}=\mathfrak{B}_1\smile\mathfrak{B}_2$. This notation is suggested by the following "duality theorem" of Whitney:

Theorem II *For the span bundle $\mathfrak{B}=\mathfrak{B}_1\smile\mathfrak{B}_2$ we have*

$$\breve{W}^r=\sum_{i=0}^r W_1^i\smile W_2^{r-i},\quad r=1,2,\cdots,v, \tag{5.1}$$

where $\breve{W}$, W_1, W_2 are the respective characteristic classes of $\breve{\mathfrak{B}}$, $\mathfrak{B}_1$, $\mathfrak{B}_2$ with the convention of paragraph 3.

① As in 5) the spaces $\breve{V}(p)$ are assumed to be disjoint from each other.

§2

6. Throughout §2 the notations of paragraph 4 will be used. Let $\Phi_1 = \{\varphi_{1,1}, \cdots, \varphi_{m,1}\}$ and $\Phi = \{\varphi_{1,2}, \cdots, \varphi_{m,2}\}$ be canonical m-fields on K_1, K_2 as defined in paragraph 3, $m \leqslant v$. We now construct an m-field $\Phi = \{\varphi_1, \cdots, \varphi_m\}$ on $K = K_1 \times K_2$ by setting

$$\varphi_i(p_1 \times p_2) = \varphi_{i,1}(p_1) + \varphi_{m-i+1,2}(p_2), \qquad i = 1, 2, \cdots, m.$$

Φ is continuous on the skeleton K^{v-m} of K.

Proof First consider the points $p_1 \times p_2$ in the cell $\sigma_1^{v_1-m_1} \times \sigma_2^{v_2-m_2}$, where $0 \leqslant m_1 \leqslant v_1$, $0 \leqslant m_2 \leqslant v_2$ and $m_1 + m_2 = m$.

Suppose that Φ is discontinuous at $p_1 \times p_2$. Then there would exist real numbers $a_1, \cdots, a_m$ not all zero such that

$$a_1\varphi_1(p_1 \times p_2) + \cdots + a_m\varphi_m(p_1 \times p_2) = 0, \tag{6.1}$$

that is,

$$a_1\varphi_{1,1}(p_1) + \cdots + a_m\varphi_{m,1}(p_1) = 0, \tag{6.2}$$

and

$$a_1\varphi_{m,2}(p_2) + \cdots + a_m\varphi_{1,2}(p_2) = 0.$$

Since the fields Φ_1 and Φ_2 are canonical, there are integers i_1 and i_2 such that

1) $\varphi_{1,1}(p_1), \cdots, \varphi_{m_1+i_1,1}(p_1)$ are linearly independent and $\varphi_{1,2}(p_2), \cdots, \varphi_{m_2+i_2,2}(p_2)$ are linearly independent.

2)
$$\varphi_{m_1+i_1+1,1}(p_1) = \cdots = \varphi_{m,1}(p_1) = 0,$$

$$\varphi_{m_2+i_2+1,2}(p_2) = \cdots = \varphi_{m,2}(p_2) = 0.$$

By 2), (6.2) becomes

$$a_1\varphi_{1,1}(p_1) + \cdots + a_{m_1+i_1}\varphi_{m_1+i_1,1}(p_1) = 0,$$
$$a_m\varphi_{1,2}(p_2) + \cdots + a_{m-m_2-i_2+1}\varphi_{m_2+i_2,2}(p_2) = 0.$$

By 1),
$$a_1 = \cdots = a_{m_1+i_1} = 0,$$

$$a_m = \cdots = a_{m-m_2-i_2+1} = 0.$$

As $m - m_2 = m_1$, it follows that all the a's are zero, which is a contradiction.

Consider next the points $p_1 \times p_2$ in a cell $\sigma_1^{v_1+m_1} \times \sigma_2^{v_2-m_2}$, where $0 \leqslant m_1$, $0 \leqslant m_2 \leqslant v_2$ and $m_2 - m_1 = m$.

We then have $m \leqslant m_2$ and Φ_2 is continuous on $\sigma_2^{v_2-m_2}$. Hence from the second equation of (6.2) we would have again $a_1 = \cdots = a_m = 0$, which is also a contradiction.

The other case is similar, and thus our assertion is proved.

7. Before evaluating the characteristics from the field constructed in the above paragraph, we shall prove in this section a lemma on the degree of mapping.

Let S_i', S_i be spheres of dimensions n_i which bound the cells V_i', V_i, $i = 1, 2$. Denote by S', S the joins of the pairs of spheres S_1', S_2' and S_1, S_2 respectively, i.e.,

$$\begin{aligned} S' &= S'_1 \times V'_2 + S'_2 \times V'_1, \\ S &= S_1 \times V_2 + S_2 \times V_1. \end{aligned} \tag{7.1}$$

The points of $S'(S)$ can be conveniently denoted by

$$t_1 x_1' \times t_2 x_2' \qquad (t_1 x_1 \times t_2 x_2)^{①},$$

where $x_i' \in S_i'(x_i \in S_i)$,

$$t_1 = 1 \text{ and } 0 \leqslant t_2 \leqslant 1 \text{ or } t_2 = 1 \text{ and } 0 \leqslant t_1 \leqslant 1.$$

Or, $0_i'(0_i)$ being the centers of $V_i'(V_i)$, $S'(S)$ is composed of the sets $x_1' \times 0_2' x_2'$ and $0_1' x_1' \times x_2'(x_1 \times 0_2 x_2$ and $0_1 x_1 \times x_2)$.

Let $S_i'(S_i)$ be oriented. Orient $V_i'(V_i)$ in coherence with $S_i'(S_i)$, then (7.1) defines an orientation of $S'(S)$.

Suppose we are given continuous mappings f_i of S_i' into S_i with degree d_i. Define a mapping f of S into S by

$$f(t_1 x_1' \times t_2 x_2') = t_1 f_1(x_1') \times t_2 f_2(x_2'). \tag{7.2}$$

Then we have the following

Lemma *The degree of the mapping f is given by*

$$d = d_1 d_2. \tag{7.3}$$

Proof Subdivide S_i', $S_i (i = 1, 2)$ into sufficiently fine simplexes and deform f_1, f_2 into simplicial approximations f_i^*. Next we subdivide S', S into cells so that the

①We can equally well denote these points by $t_1 x_1' + t_2 x_2'$ $(t_1 x_1 + t_2 x_2)$.

ground-cells of S, say, are of the form

$$(0_1\sigma_1)\times\sigma_2, \qquad \sigma_1\times(0_2\sigma_2),$$

where σ_i are ground-simplexes of the subdivisions of $S_i (i=1,2)$.

Deform f to f^* so that during the deformation relations analogous to (7.2) always hold, with the final result

$$f^*(t_1x_1'\times t_2x_2')=t_1f_1^*(x_1')\times t_2f_2^*(x_2').$$

We shall determine the degree of mapping of f^*.

For this purpose consider any oriented ground-cell of S, say

$$(0_1\sigma_1)\times\sigma_2.$$

For ground-simplexes τ_{1i}' of S_1' and τ_{2j}' of S_2', we have

$$f*[(0_1'\tau_{1i}')\times\tau_{2j}']=\pm(0_1\sigma_1)\times\sigma_2, \tag{7.4}$$

if and only if

$$f_1^*(\tau_{1i}')=\pm\sigma_1, \qquad f_2^*(\tau_{2j}')=\pm\sigma_2. \tag{7.5}$$

Let the number of simplexes τ_{1i}, τ_{2j} for which (7.5) hold with positive (negative) sign be P_1, P_2, (N_1,N_2) so that

$$P_1-N_1=d_1, \qquad P_2-N_2=d_2.$$

Then the number of cells $(0_1'\tau_{1i}')\times\tau_{2j}'$ for which (7.4) holds with positive (negative) sign is $P=P_1P_2+N_1N_2(N=P_1N_2+P_2N_1)$. Hence

$$\begin{aligned} d=P-N&=(P_1-N_1)(P_2-N_2)\\ &=d_1d_2. \end{aligned}$$

The lemma is thus proved.

8. We now come to the determination of characteristics for the field defined in paragraph 6. For this purpose let us consider a $(v-m+1)$-cell $\sigma=\sigma_1^{r1}\times\sigma_2^{r_2}$ of K, where $r_1+r_2=v-m+1$. We suppose first that $0<r_1\leqslant v_1$, $0<r_2\leqslant v_2$.

As shown in paragraph 3, there are continuous mappings $\bar\varphi_{i,1}$, $i=1,2,\cdots,v_1$ of $\sigma_1^{r_1}$ in $\mathfrak{B}_1$ and continuous mappings $\bar\varphi_{j,2}$, $j=1,2,\cdots,v_2$ of $\sigma_2^{r_2}$ in $\mathfrak{B}_2$ satisfying the following conditions:

1°. $\bar{\varphi}_{1,1}, \cdots, \bar{\varphi}_{v_1,1}$ are linearly independent at every point $tp_1 \in \sigma_1$ and $\bar{\varphi}_{1,2}, \cdots, \bar{\varphi}_{v_2,2}$ are linearly independent at every point $tp_2 \in \sigma_2$, where we denote as usual by p_1 a point in $\partial\sigma_1$, p_2 a point in $\partial\sigma_2$ and $0 \leqslant t \leqslant 1$.

2°. $\varphi_{i,1}(tp_1) = \bar{\varphi}_{i,1}(tp_1)$ for $i = 1, 2, \cdots, v_1 - r_1$,

$$\varphi_{j,2}(tp_2) = \bar{\varphi}_{j,2}(tp_2) \text{ for } j = 1, 2, \cdots, v_2 - r_2.$$

3°. If

$$\varphi_{i,1}(p_1) = a_{1,1}^{(i)}(p_1)\bar{\varphi}_{1,1}(p_1) + \cdots + a_{v_1,1}^{(i)}(p_1)\bar{\varphi}_{v_1,1}(p_1),$$
$$i = v_1 - r_1 + 1, \cdots, v_1,$$

$$\varphi_{j,2}(p_2) = a_{1,2}^{(j)}(p_2)\bar{\varphi}_{1,2}(p_v) + \cdots + a_{v_2,2}^{(j)}(p_v)\bar{\varphi}_{v_2,2}(p_v),$$
$$j = v_2 - r_2 + 1, \cdots, v_2,$$

then

$$\varphi_{i,1}(tp_1) = ta_{1,1}^{(i)}(p_1)\bar{\varphi}_{1,1}(tp_1) + \cdots + ta_{v_1,1}^{(i)}(p_1)\bar{\varphi}_{v_1,1}(tp_1),$$
$$i = v_1 - r_1 + 1, \cdots, v_1,$$

$$\varphi_{j,2}(tp_2) = ta_{1,2}^{(j)}(p_v)\bar{\varphi}_{1,2}(tp_v) + \cdots + ta_{v_2,2}^{(j)}(p_v)\bar{\varphi}_{v_2,2}(tp_v),$$
$$j = v_2 - r_2 + 1, \cdots, v_2.$$

4°. If V_1 and V_2 are definitely oriented, $\bar{\varphi}_{1,1}^*(tp_1), \cdots, \bar{\varphi}_{v_2,1}^*(tp_1)$ form a positive system in V_1 and $\bar{\varphi}_{1,2}^*(tp_2), \cdots, \bar{\varphi}_{v_2,2}^*(tp_2)$ form a positive system in V_2, where

$$\bar{\varphi}_{i,k}^*(tp_k) = \xi_{\sigma_k,tp_k,k}^{-1}\bar{\varphi}_{i,k}(tp_k), \quad i = 1, 2, \cdots, v_k; k = 1, 2.$$

Whence the vectors

$$\bar{\varphi}_{i,1}(t_1p_1), \bar{\varphi}_{j,2}(t_2p_2), \quad i = 1, 2, \cdots, v_1; j = 1, 2, \cdots, v_2,$$

form a basis in $V(t_1p_1 \times t_2p_2)$.

Write for simplicity

$$a_{i,k}^{(j)}(p_k) = a_{i,k}^{(j)}, \quad i = 1, \cdots, v_k; j = v_k - r_k + 1, \cdots, v_k; k = 1, 2.$$

Then we have for every point $t_1p_1 \times t_2p_2$ in $\partial\sigma$, where

$$t_1 = 1 \text{ and } 0 \leqslant t_2 \leqslant 1 \text{ or } t_2 = 1 \text{ and } 0 \leqslant t_1 \leqslant 1$$

$$(\Phi)\begin{cases}\varphi_i(t_1p_1\times t_2p_2)=\bar{\varphi}_{i,1}(t_1p_1)+t_2a_{1,2}^{(m+1-i)}\bar{\varphi}_{1,2}(t_2p_2)+\cdots+t_2a_{v_2,2}^{(m+1-i)}\bar{\varphi}_{v_2,2}(t_2p_2)\\ \hfill 1\leqslant i\leqslant v_1-r,\\ \varphi_i(t_1p_1\times t_2p_2)=t_1a_{1,1}^{(i)}\bar{\varphi}_{1,1}(t_1p_1)+\cdots+t_1a_{v_1,1}^{(i)}\bar{\varphi}_{v_1,1}(t_1p_1)+\bar{\varphi}_{m+1-i,2}(t_2p_2),\\ \hfill v_1-r_1+2\leqslant i\leqslant m_1\\ \varphi_{v_1-r_1+1}(t_1p_1\times t_2p_2)=t_1a_{1,1}^{(v_1-r_1+1)}\bar{\varphi}_{1,1}(t_1p_1)+\cdots+t_1a_{v_1,1}^{(v_1-r_1+1)}\bar{\varphi}_{v_1,1}(t_1p_1)\\ \qquad +t_2a_{1,2}^{(v_2-r_2+1)}\bar{\varphi}_{1,2}(t_2p_2)+\cdots+t_2a_{v_2,2}^{(v_2-r_2+1)}\bar{\varphi}_{v_2,2}(t_2p_2).\end{cases}$$

Now take arbitrary positive basis in V_1 and V_2, say $\mathfrak{r}_{1,1},\cdots,\mathfrak{r}_{v_1,1}$ and $\mathfrak{r}_{1,2},\cdots,\mathfrak{r}_{v_2,2}$. As $\bar{\varphi}_{i,1}^*,\bar{\varphi}_{j,2}^*$ are defined both on the boundary and in the interior of σ_1 and σ_2 respectively, we can deform continuously the set of mappings $\{\bar{\varphi}_{1,1}^*,\cdots,\bar{\varphi}_{v_1,1}^*,\bar{\varphi}_{1,2}^*,\cdots,\bar{\varphi}_{v_2,2}^*\}$ of $\partial\sigma$ in V into a set of constant mappings which orders to any point in $\partial\sigma$ the system of constant vectors $\mathfrak{r}_{1,1},\cdots,\mathfrak{r}_{v_1,1},\mathfrak{r}_{1,2},\cdots,\mathfrak{r}_{v_2,2}$. This deformation then induces a continuous deformation of the m-field $\Phi=\{\varphi_1,\cdots,\varphi_m\}$ over $\partial\sigma$ into a continuous m-field $\Theta=\{\theta_1,\cdots,\theta_m\}$ over $\partial\sigma$ given by the following set of equations:

$$(\Theta)\begin{cases}\theta_i(t_1p_1\times t_2p_2)=\xi_{\sigma,t_1p_1\times t_2p_2}\theta_i^*(t_1p_1\times t_2p_2),\quad i=1,2,\cdots,m\\ \theta_i^*(t_1p_1\times t_2p_2)=\mathfrak{r}_{i,1}+t_2a_{1,2}^{(m+1-i)}\mathfrak{r}_{1,2}+\cdots+t_2a_{v_2,2}^{(m+1-i)}\mathfrak{r}_{v_2,2},\quad 1\leqslant i\leqslant v_i-r_1\\ \theta_i^*(t_1p_1\times t_2p_2)=t_1a_{1,1}^{(i)}\mathfrak{r}_{1,1}+\cdots+t_1a_{v_1,1}^{(i)}\mathfrak{r}_{v_1,1}+\mathfrak{r}_{m+1-i,2},\quad v_1-r_1+2\leqslant i\leqslant m\\ \theta_{v_1-r_1+1}^*(t_1p_1\times t_2p_2)=t_1a_{1,1}^{(v_1-r_1+1)}\mathfrak{r}_{1,1}+\cdots+t_1a_{v_1,1}^{(v_1-r_1+1)}\mathfrak{r}_{v_1,1}\\ \qquad +t_2a_{1,2}^{(v_2-r_2+1)}\mathfrak{r}_{1,2}+\cdots+t_2a_{v_2,2}^{(v_2-r_1+1)}\mathfrak{r}_{v_2,2}.\end{cases}$$

It follows from the remark at the end of paragraph 3, that for points $t_1p_1\times t_2p_2\in\partial\sigma$, at least one of

$$t_1a_{v_1-r_1+1,1}^{(v_1-r_1+1)},\cdots,\quad t_1a_{v_1,1}^{(v_1-r_1+1)},\quad t_2a_{v_2-r_2+1,2}^{(v_2-r_2+1)},\cdots,t_2a_{v_2,2}^{(v_2-r_2+1)}$$

is not 0. Hence we can deform continuously the m-field Θ into a continuous m-field $\Psi=\{\psi_1,\cdots,\psi_m\}$ given by

$$(\Psi)\begin{cases}\psi_i(t_1p_1\times t_2p_2)=\xi_{\sigma,t_1p_1\times t_2p_2}\psi_i^*(t_1p_1\times t_2p_2),\quad i=1,2,\cdots,m\\ \psi_i^*(t_1p_1\times t_2p_2)=\mathfrak{r}_{i,1},\quad i=1,2,\cdots,v_1-r_1\\ \psi_i^*(t_1p_1\times t_2p_2)=\mathfrak{r}_{m+1-i,2},\quad i=v_1-r_1+2,\cdots,m\\ \psi_{v_1-r_1+1}^*(t_1p_1\times t_2p_2)=t_1a_{v_1-r_1+1,1}^{(v_1-r_1+1)}\mathfrak{r}_{v_1-r_1+1,1}+\cdots+t_1a_{v_1,1}^{(v_1-r_1+1)}\mathfrak{r}_{v_1,1}\\ \qquad +t_2a_{v_2-r_2+1,2}^{(v_2-r_2+1)}\mathfrak{r}_{v_2-r_2+1,2}+\cdots+t_2a_{v_2,2}^{(v_2-r_2+1)}\mathfrak{r}_{v_2,2}.\end{cases}$$

Moreover, there is no loss of generality in assuming that

$$[a_{v_1-r_1+1,1}^{(v_1-r_1+1)}]^2+\cdots+[a_{v_1,1}^{(v_1-r_1+1)}]^2=1,$$
$$[a_{v_2-r_2+1,2}^{(v_2-r_2+1)}]^2+\cdots+[a_{v_2,2}^{(v_2-r_2+1)}]^2=1,$$

for otherwise we can bring about this by a further deformation. (See the remark at the end of paragraph 3.)

For $p_1 \in \partial\sigma_1$, $p_2 \in \partial\sigma_2$, let us put

$$\begin{aligned}
&\psi_{i,1}^*(p_1) = \mathfrak{r}_{i,1}, \quad i = 1, 2, \cdots, v_1 - r_1,\\
&\psi_{v_1-r_1+1,1}^*(p_1) = a_{v_1-r_1,+1,1}^{(v_1-r_1+1)}\mathfrak{r}_{v_1-r_1+1,1} + \cdots + a_{v_1,1}^{(v_1-r_1+1)}\mathfrak{r}_{v_1,1},\\
&\psi_{j,2}^*(p_2) = \mathfrak{r}_{j,2}, \quad j = 1, 2, \cdots, v_2 - r_2,\\
&\psi_{v_2-r_2+1,2}^*(p_2) = a_{v_2-r_2+1,2}^{(v_2-r_2+1)}\mathfrak{r}_{v_2-r_2+1,2} + \cdots + a_{v_2,2}^{(v_2-r_2+1)}\mathfrak{r}_{v_2,2},\\
&\xi_{\sigma_1 p_1}\psi_{i,1}^*(p_1) = \psi_{i,1}(p_1), \quad i = 1, 2, \cdots, v_1 - r_1 + 1,\\
&\xi_{\sigma_2 p_2}\psi_{j,2}^*(p_2) = \psi_{j,2}(p_2), \quad j = 1, 2, \cdots, v_2 - r_2 + 1.
\end{aligned}$$

Then we can prove in a similar manner that $\Psi_1 = \{\psi_{1,1}, \cdots, \psi_{v_1-r_1+1,1}\}$ and $\Psi_2 = \{\psi_{1,2}, \cdots, \psi_{v_2-r_2+1,2}\}$ form continuous $(v_1 - r_1 + 1)$- and $(v_2 - r_2 + 1)$-fields on $\partial\sigma_1$ and $\partial\sigma_2$ and are respectively continuously deformable from the fields $\Phi_1 = \{\varphi_{1,1}, \cdots, \varphi_{v_1-r_1+1,1}\}$ and $\Phi_2 = \{\varphi_{1,2}, \cdots, \varphi_{v_2-r_2+1,2}\}$.

Put①

$$\text{Char.}_{\sigma_k}\Psi_k = d(\sigma_k), \qquad k = 1, 2$$

and

$$\text{Char.}_{\sigma}\Psi = d(\sigma),$$

then, from what has proved above, we have

$$\text{Char.}_{\sigma_k}\Phi_k = d(\sigma_k), \quad k = 1, 2; \quad \text{Char.}_{\sigma}\Phi = d(\sigma).$$

The vectors $\psi_{v_1-r_1+1,1}^*(p_1)$, $\psi_{v_2-r_2+1,2}^*(p_2)$ and $\psi_{v_1-r_1+1}^*$ now define respectively a map f_1 of $\partial\sigma_1$ in an $(r_1 - 1)$-dimensional sphere S_1, a map f_2 of $\partial\sigma_2$ in an $(r_2 - 1)$-dimensional sphere S_2, and a map f of $\partial\sigma$ in the join of S_1 and S_2, of which the degrees are respectively say d_1, d_2 and d. As these maps are connected by the relation

$$f(t_1p_1 \times t_2p_2) = t_1f_1(p_1) \times t_2f_2(p_2)\ ,$$

it follows from paragraph 7,

$$d = d_1d_2.$$

Since ψ_i^*, $\psi_{i,1}^*$, $\psi_{j,2}^*$ are constant vectors for $i \neq v_1 - r_1 + 1$, $j \neq v_2 - r_2 + 1$, d, d_1 and d_2 are respectively equal, or congruent mod 2, to the characteristics $d(\sigma)$,

① See note 4).

$d(\sigma_1)$ and $d(\sigma_2)$. Hence

$$d(\sigma) \equiv d(\sigma_1) \cdot d(\sigma_2) \mod 2. \tag{8.1}$$

Next consider a $(v-m+1)$-cell $\sigma = \sigma_1^{r_1} \times \sigma_2^{r_2}$ of K where $r_2 > v_2$, $r_1 > 0$. We must then have $v_1 - r_1 + 1 > m$. The field $\{\varphi_{1,1}, \cdots, \varphi_{m,1}\}$ is thus continuous at every point of $\sigma_1^{r_1}$. Using the same reasoning as before, we see that the given field $\{\varphi_1, \cdots, \varphi_m\}$ is deformable to a second one $\{\psi_1, \cdots, \psi_m\}$ so that

$$\xi^{-1}_{\sigma, t_1 p_1 \times t_2 p_2} \psi_i(t_1 p_1 \times t_2 p_2) = \mathfrak{r}_{i,1}, \quad i = 1, 2, \cdots, m.$$

Hence in this case

$$d(\sigma) = 0. \tag{8.2}$$

The case $r_1 > v_1$ is similar.

For the last case where $r_1 = 0$, $r_2 \leqslant v_2$ or $r_2 = 0$, $r_1 \leqslant v_1$ we can prove in the same way that

$$d(\sigma) \equiv d(\sigma_2) \bmod 2, \tag{8.3}$$

respectively

$$d(\sigma) \equiv d(\sigma_1) \bmod 2,$$

where $d(\sigma_i)$ is the characteristic of the field $\{\varphi_{1,i}, \cdots, \varphi_{v_i - r_i + 1, i}\}$ on $\partial\sigma_i$, and $d(\sigma)$ is that of the field $\{\varphi_1, \cdots, \varphi_m\}$ on $\partial\sigma$.

9. The proof of Theorem I is now immediate. The canonical fields on K_1, K_2 constructed in the preceding sections give chains

$$\begin{aligned} w_1^{r_1} &= \sum_k d(\sigma_{k,1}^{r_1}) \cdot \sigma_{k,1}^{r_1}, \qquad r_1 = 0, 1, \cdots, v_1, \\ w_2^{r_2} &= \sum_k d(\sigma_{k,2}^{r_2}) \cdot \sigma_{k,2}^{r_2}, \qquad r_2 = 0, 1, \cdots, v_2, \end{aligned} \tag{9.1}$$

in which $d(\sigma_{k,i}^{r_i})$ is the characteristic of the field $\{\varphi_{1,i}, \cdots, \varphi_{v_i - r_i + 1, i}\}$ on $\partial\sigma_{k,i}^{r_i}$, $\sigma_{k,i}^{r_i} \in K_i$. Also the m-field $\{\varphi_1, \cdots, \varphi_m\}$ defines a chain

$$w^r = \sum_k d(\sigma_k^r) \cdot \sigma_k^r, \qquad r = v - m + 1. \tag{9.2}$$

When reduced mod 2 if necessary, these chains give the characteristic cocyles of the respective bundles.

According to (8.1), (8.2) and (8.3), the coefficients of these chains are connected by the following relations:

$$\begin{aligned} &d(\sigma_{j,1}^{r_1} \times \sigma_{k,2}^{r_2}) \equiv d(\sigma_{j,1}^{r_1}) \cdot d(\sigma_{k,2}^{r_2}) \bmod 2, \qquad \text{for } r_1 \leqslant v_1 \text{ and } r_2 \leqslant v_2, \\ &d(\sigma_{j,1}^{r_1} \times \sigma_{k,2}^{r_2}) = 0, \qquad \text{for } r_1 > v_1, v_2 > 0 \quad \text{or} \quad r_2 > v_2, r_1 > 0, \\ &d(\sigma_{j,1}^{0} \times \sigma_{k,2}^{i_2}) \equiv d(\sigma_{k,2}^{r_2}) \bmod 2, \qquad \text{for } r_2 \leqslant v_2, \\ &d(\sigma_{j,1}^{r_1} \times \sigma_{k,2}^{0}) \equiv d(\sigma_{j,2}^{r_1}) \bmod 2, \qquad \text{for } r_1 \leqslant v_1. \end{aligned} \tag{9.3}$$

Let us put

$$w_i^{r_i} = \sum_k d(\sigma_{k,i}^{r_i}) \cdot \sigma_{k,i}^{r_i} = 0 \quad \text{for} \quad r_i > v_i, i = 1, 2$$

and

$$w_i^0 = \sum_k d(\sigma_{k,i}^0) \cdot \sigma_{k,i}^0 = \sum_k \sigma_{k,i}^0, \qquad i = 1, 2.$$

This is in agreement with the convention made in the end of paragraph 2. Then the equations (9.3) can be mingled into a single one:

$$d(\sigma_{j,1}^{r_1} \times \sigma_{k,2}^{r_2}) \equiv d(\sigma_{j,1}^{r_1}) \cdot d(\sigma_{k,2}^{r_2}) \quad \bmod 2.$$

It follows therefore from (9.1) and (9.2) that

$$w^r = \sum_{1=0}^{r} w_1^i \times w_2^{r-i} \qquad \bmod 2$$

i.e.,

$$W^r = \sum_{1=0}^{r} W_1^i \times W_2^{r-i}.$$

§3

10. This section will be devoted to a proof of Whitney's duality theorem mod 2. In preparing for the proof we shall make a few remarks.

Our first remark is concerned with the definition of the cup product in a complex K by means of our product defined in paragraph 4. This method of introducing the cup product is due to Lefschetz, ① but we shall summarize for our purpose the main result in a simplified form.

① Lefschetz. Algebraic Topology. Amer. Math. Soc. Colloquium publ., 1942: 173-181.

Let β, γ be two cohomology classes of K. Then $\beta \times \gamma$ is a cohomology class of $K \times K$. The diagonal mapping

$$d : K \to K \times K,$$

defined by

$$d : \sigma \to \sigma \times \sigma, \quad \sigma \in K,$$

induces a chain transformation of K into $K \times K$ and hence a homomorphism d^* of the cohomology groups of $K \times K$ into those of K. The theorem of Lefschetz asserts that $d^*(\beta \times \gamma) = \beta \cup \gamma$, where the latter is the cup product.

11. Our second remark is related to the notion of an induced sphere or vector bundle. Let a complex L be mapped simplicially into K by f. To a point $q \in L$ we associate the vector space $q \times V(f(q))$. The union of all these vector spaces can be made in a natural way to a vector bundle over L, the coordinate systems $\eta_{r_i q}$, for $q \in \tau_i \in L$, being defined by $\eta_{\tau_i q} = \xi_{f(r_i) f(q)}$. We shall call this vector bundle the bundle over L *induced* by the mapping f. It follows immediately from the definition of the induced vector bundle that $f^* W^r$, $r = 0, 1, \cdots$ are the characteristic cohomology classes of L, where W^r are the same of K and f^* the homomorphism of the cohomology groups of K into those of L induced by f.

12. With all the above preparations the proof of Theorem II follows immediately:

Let $\mathfrak{B}_1$, $\mathfrak{B}_2$ be bundles over K, and let $\mathfrak{B}_1 \times \mathfrak{B}_2$ be the product bundle over $K \times K$. Let $d : K \to K \times K$ be the diagonal map, and $\breve{\mathfrak{B}}$ the bundle over K induced by d. It is clear that $\breve{\mathfrak{B}}$ is the span bundle of $\mathfrak{B}_1$ and $\mathfrak{B}_2$. From paragraphs 10 and 11 it follows respectively that

$$d^*(W_1^i \times W_2^{r-i}) = W_1^i \cup W_2^{r-i}$$

and that

$$d^* W^r = \breve{W}^r.$$

Applying the homomorphism d^* to the formula (4.1) we are therefore led to (5.1). This proves Theorem II.

3. Topologie—Sur L'existence D'un Champ D'éléments de Contact Ou D'une Structure Complexe Sur Une Sphère*

Note (*) de M. Wu Wen Tsun, présentee par M. Élie Cartan

1. Champ d'éléments de contact. — Il est évident que l'existenee d'un champ de p vecteurs dans une variété implique l'existence d'un champ d'éléments de contact de dimension p dans cette variété. Mais la réciproque n'est pas vraie en général. Un contre-exemple est fourni par le produit de deux espaces projectifs P_2 et P_5 respectivement de dimension 2 et 5. En effet, désignons les classes caractéristiques de deux variétés quelconques M_1, M_2 et de leur produit $\mathrm{M}_1 \times \mathrm{M}_2$ par W_1^i, W_2^j et W^k respectivement, le groupe de coefficients étant Panneau des entiers mod. 2, on a① : $W^r = \sum\limits_{i=1} W_1^i \times W_2^{r-i}$, $r = 1, 2, \cdots$. Pour $\mathrm{M}_1 = \mathrm{P}_2$, $\mathrm{M}_2 = \mathrm{P}_5$, on en déduit que $W^0 \neq 0$. Ceci démontre la nonexistence existence du champ de 2 vecteurs dans la variété $P_2 \times P_5$, tandis que l'existence d'un champ de surfaces dans $\mathrm{P}_2 \times \mathrm{P}_5$ est bien évidente.

Néanmoins on a le théoréme suivant:

Théoréme I — *Dans une variété* V_n *de dimension* n*, dont*

$$H_i[V_n, \pi_{l-1}(\Gamma_p)] = 0, \quad 0 < i < n,$$

où Γ_p *est le groupe orthogonal d'une sphére* S_{p-1}*, et* H_i *sont les groupes de cohomologie, l'existence d'un champ d'éléments de contact de dimension* P *implique celle d'un champ de* P *vecteurs pourvu que* $2p \leqslant n$.

Démonstration② — Subdivisons V_n en un complexe K et désignons par K′

---

* *C. R. Acad. Sci. Paris* 226, 1948: 2117-2119.

① Wu Wen Tsün. *On the product of sphere bundles and the duality theorem modulo two* (à paraitre en *Annals of Math.*).

② M. Ehresmann m'a fait remarquer que pour la sphére ordinaire on peut démontrere théoréme plus simplemem comme suit : Il existe un champ de p ecteurs situés dans les éléments du champ donné et linéairement indépendants, sauf en un point A, de même unchampde q vecteurs situés dans les éléments orthogonaux et linéairement indépendants auf en un point $B \neq A$. On fait la somme de vecteurs correspondantscomme dans la démonstration précédente.

son dual. Par l'hypothése on peut définir un champ de p vecteurs $\mathrm{C}=\{v_1,\ldots,v_p\}$, tels que v_1, $\cdots$, v_p au point $x\in V_n$ soient siiués daps l'élément E_x du champ donné et soient linéairement indépendannts sauf aux sommets de K'. De même on peut définir un champ de $q=n-p$ vecteurs $C'=\{v_1',\cdots,v_q'\}$ dans V_n, telsque v_1', $\cdots$, v_q' au point x soient situés dans l'élément E_x' orthogonal à E_x et soient linéairement indépendants sa uf aux sommets de K. Le champ $C^\star=\{(v_1^\star,\ldots,v_p^\star\}$, *où* $v_i^*=v_i+v_i'$, $i=1,\ldots,p$, est alors eontinu en tous les points de V_{n^t} C.Q.F.D.

Corollaire I — *La sphére de dimension paire ne peut pas étre un espace feuilleté* ①. *Si la sphére* S_n *est un esppace feuilleté dons les feuilles sont de dimensionn* p, *on á* $p=1$ *ou* $n-1$, *si* $n=4k+1$; $p=1$, 2, 3, $n-3$, $n-2$ *ou* $n-1$, *si* $n=8k+3$②.

Corollaire II — *Dans une vvariété* V_n *dont* $\mathrm{H}_2=o$, *l'existence d'un champ* 《*orientable*》 *de surfaces implique celle d'un champ de deux vecteurs.*

2. *Structure complexe.* — Soient M une variété diférentielle réelle et orientable de dimension $2n(n>1)$, M$'$ la variété des vecteurs tangents de M, et M$''$ la variété des systémes de deux vecteurs tangents et orthogonaux de M. M$'$ et M$''$ sont orientables et M$''$ est un espace fibré sur la base M$'$ dont les fibres sont des sphéres de dimension $2n-2$. Pour que la variété M ait une structure presque complexe ③, il faut qu'il existe un champ de I-transformalions de M$'$ tels que (a) Iv soit orthogonal à v, $v\in M'$; (b) $\mathrm{I}^2=-1$. On peut démontrer que (b) est une conséquence de (a). Évidemment (a) est équivalent à l'existence d'une section dans la fibration de M$''$ sur M$'$. Les théorémes de M. Eckmann et de M. Gysin④ nous donnent alors les conditions nécessaires suivantes: $(1)_r$ $\pi_r(M'')\approx\pi_r(M')+\pi_r(S_{2n-2})$:$(2)_s$ $H^s(M'')\approx H^s(M'\times S_{2n-2})$. Si M est une sphére S_{2n}, M$'$ et M$''$ sont respectivement les variétés de Stiefel $\mathrm{V}_{2n+1,2}$ et $\mathrm{V}_{2n+1,3}$. Les groupes d'homotopie $\pi_r(\mathrm{V}_{n,m})$ ont été déterminés jusqu'à $r=n-m+2$⑤. En particulier, $\pi_{2n-1}(V_{2n+1,2})\approx G_4$ si n est pair, où G_4 est le groupe d'entiers mod 4. On voitdonc que $(1)_{2n-1}$ n'est pas vérifié si n est pair. Les autres conditions $(1)_r$, $(2)_s$ sont satisfaites pour tous les n, s et r où $r\leqslant 2n$ et $2n-1$ et par suite ne

① Pour la définition de variété feuilletée voir C. Ehresmann et G. Reeb, Comptes rendus, 1944, 218: 995; G. Reeb, Comptes rendus, 1947, 224: 1613.

② Cf. par exemple G. W. WHITEHEAD, Annals of Math., 1946, 47: 779-785; 1947, 48: 783-785.

③ C. Ehresmann. *Sur la théorie des espaces fibrés* (*Colloque de Topologie algébrique*, 1947): voir aussi H HOPF, *Zur Topologie der komplexen Mannigfaltigkeiten*, 1948.

④ *Comm. Math. Helv.* , 1941-1942, 14: 234-256 et 61-122.

⑤ J. H. C. Whitehead. *Proc. Lond. Math. Soc.* , 1944, 48: 243-291; 1947, 49: 479-481; cf. aussi B. ECKMANN. *Comm. Math. Helv.* , 1942-1943, 15: 318-339.

nous donnent rien. En résumé, on a

Théorème II — *Les sphères* S_{4r} *n'admettent pas la structure presque complexe, ni par conséquent la structure complexe.*

4. Topologie—Sur Les Classes Caractéristiques D'un Espace Fibré En Sphères*

Note (*) de M. Wu Wen-Tsun, transmise par M. Élie Cartan

1. Soit $\mathfrak{s}^{n-1}$ un espace fibré en sphères [abréviation. $(n-1)$-e.f.s.] sur la base K dont les fibres sont des sphères de dimension $n-1$. Entre les diverses classes caractéristiques (C. C.) $\mathrm{W}^r(\mathfrak{s})$, $r=1,\cdots,n$, il existe les relations suivantes①

$$\frac{1}{2}\delta\omega\,\mathrm{W}^{2k}=\mathrm{W}^{2k+1} \qquad (1)_k$$

En effet, considérons un champ de m vecteurs indépendants $\Phi_m=\{\varphi_1,\cdots,\varphi_m\}$ sur le squelette K^{n-m} et un champ de $(m+1)$ vecteurs indépendants

$$\Phi_{m+1}=\{\,\varphi_1,\cdots,\varphi_m,\ \varphi_{m+1}\,\}$$

sur le squelette K^{n-m-1}. La sphère $\mathrm{S}^{n-1}(p)$ sur un point $p\in\mathrm{K}^{n-m}$ est alors le joint de deux sphères $\mathrm{S}_1^{m-1}(p)$ et $\mathrm{S}_2^{n-m-1}(p)$, dont $\mathrm{S}_1(p)$ est déterminée par les vecteurs $\varphi_1(p),\cdots,\varphi_m(p)$ et $\mathrm{S}_2(p)$ est la grande sphère de $\mathrm{S}(p)$ complètement orthogonal à $\mathrm{S}_1(p)$. L'ensemble de sphères $\mathrm{S}_2(p)$, $p\in\mathrm{K}^{n-m}$ forme un $(n-m-1)$-e.f.s. $\mathfrak{s}_2$ sur K^{n-m} et un champ de vecteurs $\Phi'=\{\varphi_{m+1}\}$ a été défini sur K^{n-m-1} dans $\mathfrak{s}_2$. Les champs Φ_m et Φ_{m+1} dans $\mathfrak{s}$ définissent respectivement deux cocycles caractéristiques w^{n-m+t} et w^{n-m} de $\mathfrak{s}$ dans K, et le champ Φ' dans $\mathfrak{s}_2$ un cocycle caractéristique c^{n-m} de $\mathfrak{s}_2$ dans K^{n-m}. On démontre que, dans le cas $n-m=2k$, $w^{n-m}=(c^{n-m})_2$ et $\delta c^{n-m}=2w^{n-m+1}$. Done $1/2\ \delta\omega\ \mathrm{W}^{2k}=\mathrm{W}^{2k+1}$.

Nous étendrons la définition des C. C. en posant $\mathrm{W}^r(\mathfrak{s})=0$ pour $r>n$ et $\mathrm{W}^0(\mathfrak{s})=1$ =classe unité de K, coefficient $\in \mathrm{I}_2$ = groupe d'entiers mod2. La classe réduite de $\mathrm{W}_2^r(\mathfrak{s})$ par réduction mod2 sera désignée par $\mathrm{W}_2^r(\mathfrak{s})$. Les I_2-classes $\bar{\mathrm{W}}_2^r$ déterminées par récurrence par

---

* *C. R. Acad. Sci. Paris* 227, 1948: 582-584.

① H. Whitney. *Michigan Lectures*: 123. Remarquons que δ est tune opération locale au sens de M. N. E. Steenrod (*Annals of Math.*, 1942, 43: 116-131).

$$\sum_{i=0}^{r} \mathrm{W}_2^i(\mathfrak{s}) \cup \overline{\mathrm{W}}_2^{r-i}(\mathfrak{s}) = 0 \quad (r \geqslant 0) \tag{2}$$

seront appelées les C. C. duales (C. C. D.) de $\mathfrak{s}$. Introduisons une indéter-minée t et posons

$$\mathrm{W}(\mathfrak{s}) = \sum_{r=0}^{\infty} \mathrm{W}^r(\mathfrak{s})t^r, \quad \mathrm{W}_2(\mathfrak{s}) = \sum_{r=0}^{\infty} \mathrm{W}_2^r(\mathfrak{s})t^r, \quad \overline{\mathrm{W}}_2(\mathfrak{s}) = \sum_{r=0}^{\infty} \overline{\mathrm{W}}_2^r(\mathfrak{s})t^r.$$

Nous les appellerons respectivement le polynome caractéristique (P. C.), le P. C. réduit (P. C. R.) et le P. C. R. dual de $\mathfrak{s}$. On peut alors réunir les formules (2) dans une seule: $\mathrm{W}_2(\mathfrak{s}) \cup \overline{\mathrm{W}}_2(\mathfrak{s}) = 1$.

Soit K′ un autre complexe; une application continue f de K′ dans K induit un e.f.s. sur K′, désigné par $f^\star\mathfrak{s}$. Soit $f^\star$ aussi l'homomorphisme de l'anneau de cohomologie de K dans celui de K′, on aura

$$f^\star\mathrm{W}(\mathfrak{s}) = \mathrm{W}(f^\star\mathfrak{s}), \quad f^\star\mathrm{W}_2(\mathfrak{s}) = \mathrm{W}_2(f^\star\mathfrak{s}), \quad f^\star\overline{\mathrm{W}}(\mathfrak{s}) = \overline{\mathrm{W}}_2(f^\star\mathfrak{s}). \tag{3}$$

Soient $\mathfrak{s}_i (i = 1, 2)$ deux e.f.s. sur la même base K. On construit un troisième e.f.s. sur K, dont les sphères S(p) sont le joint de sphères $\mathrm{S}_i(p)$ de $\mathfrak{s}_i$. Nous l'appellerons le produit des $\mathfrak{s}_i$ et le désignerons par $\mathfrak{s}_1 \cup \mathfrak{s}_2$. On a

$$\mathrm{W}_2(\mathfrak{s}_1 \cup \mathfrak{s}_2) = \mathrm{W}_2(\mathfrak{s}_1) \cup \mathrm{W}_2(\mathfrak{s}_2), \quad \overline{\mathrm{W}}_2(\mathfrak{s}_1 \cup \mathfrak{s}_2) = \overline{\mathrm{W}}_2(\mathfrak{s}_1) \cup \overline{\mathrm{W}}_2(\mathfrak{s}_2). \tag{4}$$

Ces deux formules constituent ce qu'on appelle le théorème de dualité (mod$_2$) de M. Whitney①.

2. Soit $\mathfrak{s}^{n-1} (n > 1)$ un e.f.s. sur une *variété différentiable* M et f la projection de $\mathfrak{s}^{n-1}$ sur M; $\mathrm{M}^\star$ = l' e.f.s. tangent de M; $\mathfrak{s}^\star$ =l'e. f. s. tangent de $\mathfrak{s}^{n-1}$.

Théorème I —

$$\mathrm{W}_2(\mathfrak{s}^\star) = f^\star\mathrm{W}_2(\mathfrak{s}) \cup f^\star\mathrm{W}_2(\mathrm{M}^\star), \quad \overline{\mathrm{W}}_2(\mathfrak{s}^\star) = f^\star\overline{\mathrm{W}}_2(\mathfrak{s}) \cup f^\star\overline{\mathrm{W}}_2(\mathrm{M}^\star).$$

Démonstration —Soit T_x (N_x) le plan tangent (normal) de S(p) au point $x \in \mathrm{S}(p)$, sphère sur $p \in$ M. L' espace des directions dans T_x (N_x) issues de x engendre un e. f. s. $\mathfrak{J}(\mathfrak{A})$ dont la base est $\mathfrak{s}$. On peut considérer S(p) comme la sphère unité d'un espace vectoriel V(p) d'origine p. Soit L_x la droite passant par $x \in \mathrm{S}(p)$ et p. Les L_x définissent naturellement un o-e.f.s. $\mathfrak{S}$ sur la base $\mathfrak{s}$. On

① *Loc. cit.* : 132. Pour une démonstration, *voir* Wu Wen-Tsun. *Annals of Math.* , 1948, 49: 641-653.

voit que (A)$\mathfrak{S}$ est simple; (B) $\mathfrak{L}\cup\mathfrak{J}$ est isomorphe à $f^\star(\mathfrak{s})$; (C)$\mathfrak{N}$ est isomorphe à $f^\star \mathrm{M}^\star$; (D) $\mathfrak{s}^\star=\mathfrak{J}\cup\mathfrak{A}$. Done en utilisant (3) et (4), on a respectivement:

(A) $\mathrm{W}_2(\mathfrak{L}\cup\mathfrak{J})=\mathrm{W}_2(\mathfrak{J})$, $\quad\overline{\mathrm{W}}_2(\mathfrak{L}\cup\mathfrak{J})=\overline{\mathrm{W}}_2(\mathfrak{J})$;

(B) $\mathrm{W}_2(\mathfrak{L}\cup\mathfrak{J})=f^\star\mathrm{W}_2(\mathfrak{s})$, $\quad\overline{\mathrm{W}}_2(\mathfrak{L}\cup\mathfrak{J})=f^\star\overline{\mathrm{W}}_2(\mathfrak{s})$;

(C) $\mathrm{W}_2(\mathfrak{A})=f^\star\mathrm{W}_2(\mathrm{M}^\star)$, $\quad\overline{\mathrm{W}}_2(\mathfrak{A})=f^\star\overline{\mathrm{W}}_2(\mathrm{M}^\star)$;

(D) $\mathrm{W}_2(\mathfrak{s}^\star)=\mathrm{W}_2(\mathfrak{J})\cup\mathrm{W}_2(\mathfrak{A})$, $\quad\overline{\mathrm{W}}_2(\mathfrak{s}^\star)=\overline{\mathrm{W}}_2(\mathfrak{J})\cup\overline{\mathrm{W}}_2(\mathfrak{A})$.

Le théorème I est une consequence des formules (A) - (D).

Comme une conséquence immédiate du théorème I on a:

Théorème II —*Pour qu'une variété* M^n *puisse être fibrée en sphères à dimension d, il est nécessaire que* $\mathrm{W}_2^r(\mathrm{M}^\star)=0$, *et* $\bar{\mathrm{W}}_2^r(\mathrm{M}^\star)=0$, *pour* $r>n-d$.

Le théorème I nous permet aussi de déterminer toutes lesC. C. des variétés de Stiefel. En effet, on a $\mathrm{W}_2^r(\mathrm{V}_{n,1}^\star)=\mathrm{W}_2^r(\mathrm{S}^{n-1})=0$, $r>0$, c'est-à-dire $\mathrm{W}_2(\mathrm{V}_{n,2}^\star)=1$. Supposons que $\mathrm{W}_2(\mathrm{V}_{n,m}^\star)=1$ et considérons $\mathrm{V}_{n,m+1}^\star$ comme un e.f.s. $\mathfrak{s}$ sur $\mathrm{V}_{n,m}$ avec projection f.

Le théorème I nous donne $\mathrm{W}_2(\mathrm{V}_{n,m+1}^\star)=f^\star\mathrm{W}(\mathfrak{s})\cup f^\star\mathrm{W}_2(\mathrm{V}_{n,m}^\star)$. Done $\mathrm{W}_2(\mathrm{V}_{n,m+1}^\star)=f^\star\mathrm{W}_2(\mathfrak{s})$. Néanmoins on a $\mathrm{W}_2^r(\mathfrak{s})=0$ pour $r>n-m$, d'après la définition des C. C. ; pour $0<r<n-m$, puisque $\mathrm{H}_r(\mathrm{V}_{n,m})=0$; et $f^\star\mathrm{W}_2^{n-m}(\mathfrak{s})=0$, d'après un théorème de S. Chern①. Donc $\mathrm{W}_2(\mathrm{V}_{n,m+1}^\star)=1$. En utilisantles formules (1) et en remarquant que $\chi(\mathrm{V}_{n,m+1})=\chi(\mathrm{V}_{n,m})\chi(\mathrm{S}^{n-m-1})=0$, χ étant la caractéristique d'Euler-Poincaré, on aura $\mathrm{W}\,(\mathrm{V}_{n,m})=1$. Donc:

Théorème III —*Toutes les classes caractéristiques de dimension* $r>0$ *des variétés de Stie fel* $\mathrm{V}_{n,m}(m>1)$ *sont nulles.*

Remarque — $\mathrm{V}_{n,m}(m>1)$ est en général non parallélisable. Ex. $\mathrm{V}_{4,2}$ et $\mathrm{V}_{8,2}$.

① *Annals of Math.*, 1946, 47: 85-121. (Théorème 8.)

5. Topologie——Sur Le Second Obstacle D'un Champ D'éléments de Contact Dans Une Structure Fibrée Sphérique*

Note de M. Wu Wen-Tsun, présentée Par M. Élie Cartan

1. Considérons une structure fibrée $\mathfrak{F}$ sur la base K et de fibre F. Supposons que les deux premiers groupes d'homotopie non nuls de F soient π_{h-1} (F) et $\pi_{k-1}(\mathrm{F})$ $k > h > 1$. Définissons un *champ* (section) $\{\varphi\}$ sur le squelette K^n et étendons-le, pas à pas aux squelettes K^1, K^2, $\cdots$, jusqu'à K^{h-1}. On rencontre alors une difficulté et l'on obtient un cocycle h-dimensionel z^h à coefficients $\in \pi_{h-1}(\mathrm{F})$. La classe ζ^h de z^h, appelée le *premier obstacle*, est indépendant du champ $\{\varphi\}$ particulier et sa nullité est une condition nécessaire et suffisante pour l'existence d'un certain champ sur K^h. Soit alors $\zeta^h = 0$ et $\{\varphi\}$ un tel champ sur K^h. On peut étendre $\{\varphi\}$ aux K^{h+1}, $\cdots$, K^{k-1} et l'on obtient un cocycle z^k à coefficients $\in \pi_{k-1}(\mathrm{F})$. La classe ζ^k de z^k s'appelle le *deuxième obstacle* associé au champ (φ). Cette classe dépend du champ $\{\varphi\}$ défini et le problème de déterminer les relations entre les divers deuxièmes obstacles a fait l'objet de recherches récentes de topologie algébrique, mais n'est résolu que dans des cas particuliers①. Nous nous proposons d'étudier dans cette Note le deuxième obstacle d'un champ de m-éléments (ou éléments de contact de dimension m) dans une structure fibrée sphérique $\mathfrak{s}^{(n-1)}$ de fibre S^{n-1}, ou, ce qui revient au même, d'un champ dans une structure fibrée associée dont la fibre est une variété grassmannienne. Notre méthode s'appuie sur un théorème de H. Whitney, à savoir le suivant②:

Soient $\mathfrak{s}_1^{(n_1-1)}$ et $\mathfrak{s}_2^{(n_2-1)}$ deux structures fibrées sphériques sur la même base K et $\mathfrak{s}_0^{(n_0-1)} = \mathfrak{s}_1^{(n_1-1)} \cup \mathfrak{s}_2^{(n_2-1)} (n_0 = n_1 + n_2)$ leur *produit* défini de manière que la sphère $\mathrm{S}_0(p)$ dans $\mathfrak{s}_0$ sur $p \in \mathrm{K}$ soit le joint des sphères $\mathrm{S}_1(p)$ et $\mathrm{S}_2(p)$ dans $\mathfrak{s}_1$ et $\mathfrak{s}_2$. Soient $\mathrm{W}_2(\mathfrak{s}_i) = 1 + \mathrm{W}_2^1(\mathfrak{s}_i) \cdot t + \cdots + \mathrm{W}_2^{n_i}(\mathfrak{s}_i) \cdot t^{n_i} (i = 0, 1, 2)$ le polynome caractéristique de $\mathfrak{s}_i$, où $\mathrm{W}_2^k(\mathfrak{s}_i)$ sont les classes caractéristiques réduites mod 2 de $\mathfrak{s}_i$ au sens de

* *C. R. Acad. Sci. Paris* 227, 1948: 815-817.

① N. E. Steenrod. *Annals of Math.*, 1947, 48: 290-320; H. HOPF, *Colloque de Topologie algébrique*, Paris, 1947. Voir aussi Note(1).

② *Michigan Lectures*: 132.

Stiefel-Whitney et 1 la I_2-classe unitée o-dimen sionnelle (I_m = groupe d'entiers mod m). Alors

$$W_2(\mathfrak{s}_0) = W_2(\mathfrak{s}_1) \cup W_2(\mathfrak{s}_2).$$

2. Étudions d'abord le cas d'éléments linéaires non orientés① dans $\mathfrak{s}^{(n-1)}$ sur $K(n > 2)$. Le premier obstacle est nul et le. deuxième obstacle est donc défini. Soit ζ^n (àcoefficients $\in I_0$) le. deuxième obstacle associé à un champ $\{\varphi\}$ d'éléments linéaires non orientés sur K^{n-1}. Soit $\{\psi\}$ un autre champ déduit d'un champ de vecteurs. L'obstacle de *coincidence* de $\{\varphi\}$ et $\{\psi\}$ est une I_2-classe ξ_2^1 de dimension 1 qui ne dépend, en réalité, que de $\{\varphi\}$. La classe ζ_2^n, déduite de ζ^n par réduction mod 2, s'exprime alors en fonction de ξ_2^1 et des classes caractéristiques $W_2^k = W_2^k(\mathfrak{s}^{(n-1)})$ par

$$\zeta_2^n = (\xi_2^1)^n + W_2^1\,(\xi_2^1)^{n-1} + \cdots + W_2^{n-1}\xi_2^1 + W_2^n\,, \tag{1}$$

la multiplication et la puissance étant le $\bigcup$ produit.

Pour le démontrer, considérons une structure fibrée sphérique $\mathfrak{s}^{(0)}$ *simple* sur la base K et formons le produit: $\mathfrak{s}^{(n)} = \mathfrak{s}^{(n-1)} \cup \mathfrak{s}^{(0)}$. Le champ $\{\varphi\}$ donné n'est défini que sur K^{n-1}, mais, considéré comme un champ dans $\mathfrak{s}^{(n)}$, s'étend sur le squelette K^n qui définit alors d'une manière évidente une structure fibrée sphérique $\mathfrak{s}_1^{(0)}$ sur la base K^n. Soit $\mathfrak{s}_1^{(n-1)}$ la structure sur K^n définie par un champ de n-éléments dans $\mathfrak{s}^{(n)}$ complètement orthogonaux à $\{\varphi\}$. Alors $\mathfrak{s}^{(n)}/K = \mathfrak{s}_1^{(0)} \cup \mathfrak{s}_1^{(n-1)}$.

En appliquant le théorème de Whitney et en remarquant que $W_2(\mathfrak{s}^{(0)}) = 1$, on aura:

$$W_2(\mathfrak{s}^{(n-1)}) = W_2(\mathfrak{s}^{(n)}) = W_2(\mathfrak{s}_1^{(0)}) \cup W_2(\mathfrak{s}_1^{(n-1)}),$$

ou

$$I + W_2^1 t + \cdots + W_2^n t^n = (I + \xi_2^1 t)(I + \zeta_2^1 t + \cdots + \zeta_2^n t^n), \tag{2}$$

dont ξ_2^1 et ζ_2^n sont les mèmes que dans la formule (1). (1) est alors une conséquence de (2).

Dans le cas d'un champ d'éléments linéaires complexes dans une structure fibrée sphérique complexe la formule (1) devient②

$$\zeta^{2n} = (\xi^2)^n \;+\; C^2(\xi^2)^{n-1} + \ldots + \; C^{2n},$$

① Le problème d'éléments linéaires non orientés m'a été posé par M. T. H. Kiang.

② La présente Note est inspirée par cette formule, due à M. Kundert, d'après une communication orale de M. H. Hopf.

où C^{2k} sont les classes caractéristiques définies par S. Chern①. On le démontre par la même méthode, en utilisant une généralisation du théorème de Whitney.

3. Pour un champ de 2 éléments orientés dans $\mathfrak{s}^{(n-1)}(n > 4)$, le deuxième obstacle ζ^{n-1}, qui est une I_0- ou I_2-classe selon que n est pair ou impair, est donné par les formules suivantes :

$$\zeta_2^{n-1} = \begin{cases} (\xi_2^2)^{\frac{n-1}{2}} + W_2^2(\xi_2^2)^{\frac{n-3}{2}} + \cdots + W_2^{n-1}, & n \text{ impair}, \\ W_2^1(\xi_2^2)^{\frac{n-2}{2}} + W_2^3(\xi_2^2)^{\frac{n-4}{2}} + \cdots + W_2^{n-1}, & n \text{ pair}, \end{cases}$$

où ξ_2^2 est une I_2-classe quelconque, déduite d'une I_0-classe par réduction mod 2. Dans le cas d'un champ de 2-éléments non orientés, le *troisième obstacle* ζ^{n-1}, réduit mod 2, s'exprime en termes de $W_2^k(\mathfrak{s}^{(n-1)})$ et deux indéterminées ξ_2^1 et ξ_2^2, mais les formules sont compliquées.

La méthode précédente s'applique de même aux champs de m-éléments $(m > 2)$ dans quelques cas particuliers : par exemple, l'obstacle ζ^6 pour m-éléments $(m > 5)$ dans $\mathfrak{s}^{(n-1)}(n = m+5)$ et l'obstacle ζ^7 pour m-éléments $(2 < m \neq 6)$ dans $\mathfrak{s}^{(n-1)}(n = m + 6)$. De même pour les champs d'éléments composés (figures constituées d'un système d'éléments deux à deux orthogonaux).

① *Annals of Math.* , 1946, 47: 85-121.

6. Topologie——Sur La Structure Presque Complexe D'une Variété Différentiable Réelle de Dimension 4*

Note (*) de M. Wu Wen-Tsun, présentée par M. Elie Cartan

1. Soit $\Re_{2n,4}$ la variété grassmannienne des éléments *orientés* de dimension 4 dans un espace euclidien E de dimension $2n+4$ $(n>2)$ et passant par l'origine de E. Il existe① dans $\Re_1$, un I_2-cycle $\mathfrak{r}^{8n-2}$ et deux I_0-cycles $\mathfrak{r}_1^{8n-4}, \mathfrak{r}_2^{8n-4}$ qui sont représentés respectivement par les symboles $[n-1, n-1, n, n]$, $[n-2, n-2, n, n]$ et $[n-1, n-1, n-1, n-1]$, avec les notations analogues à celles de S.Chern②. Nous désignerons par $\wedge^2$, $\wedge_1^4$, $\wedge_2^4$ les classes de cohomologie correspondantes.

Soit M^4 une variété différentiable et orientable de dimension 4.La structure fibrée sphérique Langente de M *orientée* est induite par une certaine application $f: M \to \Re_{2n,4}$ dont le type d'homotopie est bien déterminé③ Les classes $f^\star\wedge^2 = W^2$, $f^\star\wedge_i^4 = W_i^4 (i=1,2)$ sont des invariants de M④ Nous établirons dans cette Note le théorème suivant:

Pour que la variété orientée M^4 *admette une structure presque complexe*⑤, *il est néccssaire et suffisant qu'il existe une* e_0*-classe* C^2 *telle que*

$$(C^2)_2 = W^2 \qquad \text{et} \qquad C^2 \cup C^2 = W_1^4 + 2W_2^4. \tag{1}$$

où $(\ )_2$ est l'opération de réduction $\bmod_2$. Plus précisément, le système des structures presque complexes est en (1-1)-correspondance au système de I_0-classes C^2 satisfaisant à la condition (1).

---

* *C. R. Acad. Sci Paris* 227, 1948: 1076-1078.

① Cf.L.PONTRJARGIN. *C.R.U.R.S.S.*, 35, 1942, p.34-37.

② Cf.par exemple, *Annals of Math.*, 1948, 49: 362-372. I_m = groupe d'entiers mod. m.

③ L.PONTRJARGIN. *C. R. U. R. S. S.*, 1945, 47: 242-245.

④ W^2 est en effet une classe caractéristique de Stiefel-Whitney, et W_2^4, M^2 la caractéristique d'Euler-Poincaré. D'après Pontrjargin, W_1^4, M^2 est un autre invariant numérique de M^4 qui change le signe quand onchange l'orientation de M. f^A est l'homomorphisme inverse d'une application f.

⑤ CH. EHRESMANN, *Sur la théorie des espaces fibrés* (*Collogue de Topologie algebrique*, Paris, 1947).

IT peut arriver done qu'une variété M^4 admette une infinité de structures presque complexes non équivalentes. Exemple: $S^1 \times S^1 \times S^1 \times S^1$; Le théorème nous donne aussi un critère permettant de déterminer si la variété M,admettant une structure presque complexe correspondant à une de. ses orientations, admettra aussi une structure presque complexe correspondant à l'autre①.

La démonstration de notre théorème est basée sur une méthode de L. Pontrjargin②.

2. Soit $\mathfrak{C}_{n,2}$ la variété grassmanienne des plans complexes dans un espace unitaire U de dimension complexe $n+2 (n>2)$ et passant par l'origine de U. Soient $\mathcal{E}^{4n-2}$, $\mathcal{E}_1^{4n-4}$, $\mathcal{E}_2^{4n-4}$, $\mathcal{E}^2$, $\mathcal{E}_1^2, \mathcal{E}_2^2$ les I_0-cycles correspondant aux symboles $[n-1,n]$, $[n-2,n]$, $[n-1,n-1]$, $[0,1]$, $[0,2]$ et $[1,1]$ respectivement①. Les classes de cohomologie Γ^2, Γ_1^4 et Γ_2^4 duales aux $\mathcal{E}^2$, et $\mathcal{E}_i^4$ sont liees par la relation suivante

$$\Gamma^2 \cup \Gamma^2 = \Gamma_1^4 + \Gamma_2^4.$$

Les groupes d'homotopie de $\mathfrak{C}_{n,2}$ sont: $\pi_1 = \pi_3 = 0$, $\pi_2 \approx \pi_4 \approx I_0$. Soient S_0^2, S_0^4 les deux représentants des générateurs de π_2 et π_4. On aura les indices d'intersection:

$$S_0^2 \circ \xi^{4n-2} = 1, \quad S_0^4 \circ \xi_1^{4n-4} = 1, \quad S_0^4 \circ \xi_2^{4n-4} = -1. \tag{3}$$

On déduit de (2) et (3) les propositions suivantes:

a. A chaque système de I_0-classes de cohomologie C^2, C_1^4, C_2^4 dans un complexe K de dimension 4 tels que $C^2 \bigcup C^2 = C_1^4 + C_2^4$ il existe une application $g: K \to \mathfrak{C}_{n,2}$ tels que

$$g^\star \Gamma^2 = C^2, \quad g^\star \Gamma_1^4 = C_1^4, \quad g^\star \Gamma_2^4 = C_2^4. \tag{4}$$

b. Deux applications $g_i: K^4 \to \mathfrak{C}_{n,2} (i=1,2)$ sont homotopes si et seulement si $g_1{}^\star \Gamma^2 = g_2^\star, \Gamma^2, g_1^\star \Gamma_i^4 = g_2^\star \Gamma_i^4 (i=1,2)$.

Identifions U et E et considérons $\mathfrak{C}_{n,2}$ comme une sous-variété de $\mathfrak{A}_{2n,4}$.

Soit φ l'application identique de $\mathfrak{C}_{n,2}$ dans $\mathfrak{R}_{2n,4}$. On aura

$$\mathfrak{r}^{sn-2} \circ [\varphi(\xi^2)]_2 = 1 \tag{5}$$

$$\mathfrak{r}_1 - \circ \varphi(\xi_1^s) = +1, \quad \mathfrak{r}_1^{sn-1} \circ \varphi(\xi_2^s) = -1, \quad \mathfrak{r}_2^{sn-4} \circ \varphi(\xi_1^s) = 0, \quad \mathfrak{r}_2^{sn-1} \circ (\xi_2^1) = +1, \quad (5)'$$

① Cf. H. Hopf. *Zur Topologie der komplexen Mannigfaltigkeiten.* 11, Courant Anniversary volume, 1948: 167-185.

② *C.R.U.R.S.S.*, 1945, 47: 322-325.

d'ou

$$\varphi^{*}\Lambda^{2}=(\Gamma^{2})_{2},\quad \varphi^{*}\Lambda_{1}^{1}=\Gamma_{1}^{1}-\Gamma_{2}^{1},\quad \varphi^{*}\Lambda_{2}^{4}=\Gamma_{2}^{4}. \tag{6}$$

Pour les groupes d'homotopie de $\mathfrak{R}_{2n,4}$ on a $\pi_1=\pi_3=0$, $\pi_2\approx I_2, \pi_4\approx I_0+I_0$. Soient S^2, S_1^4 et S_2^4 les représentants de ses générateurs.Alors

$$S^{2}\circ\mathfrak{r}^{\delta n-2}=1,\quad S_{1}^{4}\circ\mathfrak{r}_{1}^{\delta n-1}=4,\quad S_{1}^{4}\circ\mathfrak{r}_{2}^{\delta n-1}=0,\quad S_{2}^{4}\circ\mathfrak{r}_{2}^{\delta n-1}=1. \tag{7}$$

3. Soit $f: M^4\to\mathfrak{R}_{2n,4}$l'application qui induit la structure fibrée sphérique tangente de M.La condition nécessaire et suffisante pour que M^4 admette une structure presque complexe est que f soit homotope à une application de M dans $\mathfrak{C}_{n,2}$. La nécessité de notre théorème est done évidente, voir (6). Supposons que la condition (1) soit vérifiée.Posons

$$C_{1}^{4}=W_{1}^{4}+W_{2}^{4}\quad \text{et}\quad C_{2}^{4}=W_{2}^{4}, \tag{8}$$

tel que $C^2\cup C^2= C_1^4+C_2^4$. D'après a, il existe une application $g: M\to\mathfrak{C}_{n,2}$. vérifiant (4). Ilrésulte de (1), (4) et (8) que

$$f^{\star}\wedge^{2}=(\varphi g)^{\star}\wedge^{2},\quad f^{\star 0}\wedge_{i}^{4}=(\varphi g)^{\star}\wedge_{i}^{4}\quad (i=1,2). \tag{9}$$

En utilisant (7) et en remarquant que M n'admette pas de co-torsion de dimension 4, on déduit de (g) que $f\simeq\varphi g$, ce qui démontre que la condition de notre théorème est suffisante.

7. Topologie——Sur La Structure Presque Complexe D'une Variété Différentiable Réelle*

Note (*) de M. Wu Wen-Tsün, présentée par M. Élie Cartan

1. Soit $G_{n,m}$① la variété des sous-espaces *orientés* X^m de dimension m d'un espace vectoriel R^{n+m} de dimension $n+m$. L'ensemble des $X^m \in G_{n,m}$ coupant un sous-espace fixe de dimension $n+2k-2$ de R^{n+m} suivant un sous-espace de dimension $\geqslant 2k$ est un cycle entier de dimension $mn-4k$, $o \leqslant 2k \leqslant m$. Leurs classes de cohomologie duales de dimension $4k$ seront désignées par P^{4k}. L'ensemble des $X^m \in G_{n,m}$ coupant un sous-espace fixex de dimension $n+k-1$ de R^{n+m} suivant un sous-espace de dimension $\geqslant k$ est un cycle mod 2 dont les classes de cohomologie duales de dimension k, aussi mod 2, seront désignées par W^k, $0 \leqslant k < m$. De même, la classe de cohomologie entière duale au cycle entier constitué par les sous-espaces $X^m \in G_{n,m}$ contenus dans un sous-espace fixe de dimension $n+m-\mathrm{I}$ de R^{n+m} sera désignée par P_0^m. On pose $W^m = (P_0^m)_2$, où $(_2)$ est la réduction mod 2.

Soit M^m une variété différentiable orientale de dimension m. Sa structure fibrée tangente, dont les fibres sont définitivement orientées correspondant à une orientation fixe de M^m, est engendrée par une certaine application $f: M^m \to G_{n,m}$ pour n assez grand. Les classes $f^\star P_0^m$, $f^\star P^{4k}$, $o \leqslant 4k \leqslant m$, et $f^\star W^k$, $o \leqslant k \leqslant m$ sont alors des invariants de la variété M^m que nous désignerons respectivement par $P_0^m(M)$, $P^{4k}(M^m)$ et $W^k(M^m)$. Nous remarquerons que pour la variété M^m orientée en sens inverse, ou - M^m, on a $P_0^m(M^m) = -P_0^m(-M^m)$, $P^{4k}(M^m) = P^{4k}(-M^m)$ et $W^k(M^m) = W^k(-M^m)$.

Nous introduirons une indéterminée t et poserons

$$P(M^m, t) = \sum_k (-1)^k P^{4k}(M^m) t^{2k},$$

---

* *C. R. Acad. Sci. Paris* 228, 1949: 972-973.

① Cf. L. Pontrjargin. *Mat. Sbornik*, 1947, 21(63): 233-284.

$$W(M^m,t)=\sum_k W^k(M^m)t^k.$$

2. Soient n, m des nombres pairs① et $\mathrm{C}_{n,m}$ la variété des sous-espaces complexes U^m de dimension complexe $m/2$ d'un espace unitaire U^{n+m} de dimension complexe $(n+m)/2$. L'ensemble des $\mathrm{U}^m \in \mathrm{C}_{n,m}$ coupant un sous-espace complexe fixe de dimension complexe $n/2+k-1$ de U^{n+m} suivant un espace complexe de dimension complexe $\geqslant k$ est un cycle entier de dimension 2 k dont la classe de cohomologie duale sera désignée par C^{2k}, $0 \leqslant 2k \leqslant m$. Les classes mod 2, déduites de C^{2k} par réduction mod 2, seront désignées par C_2^{2k}.

Une structure fibrée sphérique presque complexe σ définie sur la base K et à fibre sphérique de dimension $m-1$, m paire, peut être engendrée par une certaine application $f:\mathrm{K}\to\mathrm{C}_{n,m}$, pour n assez grand. Les classes $f^\star\,\mathrm{C}^{2k}$ et $f^\star\,C_2^{2k}$ sont alors des invariants de cette structure que nous désignerons par $\mathrm{C}^{2k}(\sigma)$ et $\mathrm{C}_2^{2k}(\sigma)$ respectivement. Ce sont des invariants de Chern. Nous poserons

$$\mathrm{C}(\sigma,t)=\sum_k \mathrm{C}^{2k}(\sigma)t^k, \qquad \mathrm{C}_2(\sigma,t)=\sum_k \mathrm{C}_2^{2k}(\sigma)t^k.$$

3. Supposons que n, m soient pairs. En identifiant les espaces U^{n+m} et R^{n+m} et en considérant un espace complexe de dimension complexe $m/2$ comme un sous-espace de dimension m naturellement orienté, on obtient une application canonique $\delta:\mathrm{C}_{n,m}\to\mathrm{G}_{n,m}$. La considération de son type d'homologie nous donne le théorème suivant:

Théorème — *Si la variété orientée* M^m *(m pair). admet une structure presque complexe* σ*, on a.*

$$\begin{aligned}
&\mathrm{P}_0^m(\mathrm{M}^m)=\mathrm{C}^m(\sigma),\\
&\mathrm{W}(\mathrm{M}^m,t)=\mathrm{C}_2(\sigma,t),\\
&\mathrm{P}(M^m,t)=\mathrm{C}(\sigma,t)\cup\mathrm{C}(\sigma,-t).
\end{aligned}$$

Ce théorème donne quelques conditions nécessaires pour qu'une variété orientée admette une structure presque complexe. Ces conditions sont aussi suffisantes dans le cas $m=4$②.

① Cf. S. Chern. *Annals of Math.*, 1946, 47: 85-121.
② Cf. *Comptes rendus*, 1948, 227: 1076-1078.

8. Topologie Algébrique——Classes Caractéristiques Et I-carrés D'une Variété*

Note (*) de M. Wu Wen-Tsün, présentée par M. Élie Cartan

1. Soit M un espace topologique vérifiant les conditions suivantes:

a. Le groupe $H^n(M)$① est de rang 1 dont la base est X_1^n;

b. On a $H^p(M) \approx \text{Hom}\,[H^{n-p}(M), Z_2]$ dont l'isomorphisme est établi par le cup produit $X^p(Y^{n-p})X_1^n = X^p \cup Y^{n-p}$, $X^p \in H^p(M)$, $Y^{n-p} \in H^{n-p}(M)$.

Pa exemple, une variété compacte de dimension n est un tel espace. Dans un tel espace on peut définir un système de classes $U^p \in H^p(M)$, $0 \leqslant 2p \leqslant n$, par les équations suivantes:

$$U^p \cup Y^{n-p} = Sq^p Y^{n-p} \quad [\text{pour } Y^{n-p} \text{ quelconque de } H^{n-p}(M)]. \tag{1}$$

②

Nous les appellerons les *classes canoniques* on les U-*classes* de M. Les classes W^i, $0 \leqslant i \leqslant n$, definies par

$$W^i = \sum_p Sq^{i-p} U^p \tag{2}$$

seront alors appelées les *classes caractéristiques* on les W-*classes* de M. On a, par exemple, $W^0 = U^0 = 1$, $W^1 = U^1$, $W^2 = U^2 + U^1 \cup U^1$, etc.

Le nom des classes caractéristiques est justifié par le théorème suivant:

Théorème — *Pour une variété compacte* M *les* W-*classes ainsi définies s'identifient aux classes caracteristiques de Sliefel-Whitney de cette variété.*

2. La démonstration de ce théorème s'appuie sur un théorème de Thom③ , et le lemme suivant, démontré par H. Cartan①:

---

* *C. R. Acad. Sci. Paris* 230, 1950: 508-511.

① $H^*(M)[H^p(M)]$ = le groupe de cohomologie (de dimension p) de l'espace M. Le groupe des coefficients sera exclusivement le groupe Z_2 des entiers mod 2 sauf, mention du contraire. La classe unite de $H^0(M)$ est désignée par 1.

② Nous adoptons ici la nouvelle notation de Steenrod pour les *i-carrés*: $Sq^p X^q = Sq_{q-p} X^q$. *Cf.* Steenrod, *Annals of Math.*, 1947, 48: 290-319.

③ *Voir* la Note précédente de Thom sur les *variétés plongées* et *i-carrés*(même numéro des *Comptes rendus*) et la Note précédente de H. Cartan sur *une théorie axiomatique des i-carrés* (*Comptes rendus*, 1950, 230: 425).

Lemme – *Dans un espace-produit* M × M′ *on a*

$$Sq^i(X \otimes Y) = \sum_j Sq^j X \otimes Sq^{i-j} Y, \quad X \in H^\star(M), \quad Y \in H^\star(M').$$

On en déduit que, dans un même espace M, *on a*

$$Sq^i(X \cup Y) = \sum_j Sq^j X \cup Sq^{i-j} Y, \quad X, Y \in H^\star(M).$$

Prenons maintenant une base $\{X^p_\alpha\}$ do $H^\star(M)$ dans la variété M, supposée de dimension n, telle que $X^p_\alpha \cup X^{n-p}_\beta = \delta_{\alpha\beta} X^n_1$. La classe $\Delta^n \in H^n(M \times M)$ correspondant à la diagonale de l'espace-produit M×M s'exprime alors par $\Delta^n = \sum_{\alpha,p} (X^p_\alpha \otimes X^{n-p}_\alpha)$. D'après le lemme précédent, on a donc

$$Sq_i \Delta^n = \sum_{\alpha,p,j} Sq^{i-j} X^p_\alpha \otimes Sq^j X^{n-p}_\alpha.$$

D'autre part, soit $\sum_\alpha a^t_\alpha X^i_\alpha$ la classe caractéristique de Stiefel-Whitney de dimension i de la variété M qui est aussi la classe de Stiefel-Whitney de la structure normale de M par rapport à M × M. On a, d'après la formule (6) de Thom[①], $Sq^i \Delta^n = \sum_\alpha \psi^\star \alpha^i_\alpha X^i_\alpha$, où $\psi^\star$ applique $H^i(M)$ dans $H^{n+i}(M \times M)$. On en déduit

$$Sq^i \Delta^n = \sum_{\alpha,\mu,q} \alpha^i_\alpha X^q_\mu \otimes (X^i_\alpha \cup X^{n-q}_\mu).$$

En considérant les termes de la forme $X^i_\mu \otimes X^n_1$ dans les deux expressions de $Sq^i \Delta^n$, on trouve que

$$\sum_\mu a^i_\mu X^i_\mu = \sum_p Sq^{i-p} U^p,$$

c'est-à-dire la classe de Stiefel-Whitney de dimension i coincide avec la classe W^i, définie par (1) et (2). C. Q. F. D.

3. Le théorème précédent montre que les classes caractéristiques de Stiefel-Whitney d'une variété compacte de dimension n sont complètement déterminées par les classes canoniques U^p, $0 \leqslant 2p \leqslant n$, et par conséquent par la structure des cup produits et les i-carrés de cette variété. On peut en déduire d'autres propriétés concernant les classes de Stiefel-Whitney, ainsi qu'il suit[①]:

a. *Les classes* W^i *pour* $2\ i > n$ *sont complètement determinées par les classes* W^i *pour* $o \leqslant 2i \leqslant n$, *et par les opérations de carrés.*

b. $W^n = 0$ pour n impair; $W^n = Sq^k U^k = U^k \cup U^k$ pour $n = 2k$ pair; $W^2 = W^1 \cup W^1$ pour $n = 3$; $W^1 \cup W^1 \cup W^2 = o$ pour $n = 4$(on peut même démontrer $W^1 \cup W^2 = o$ pour $n \leqslant 5$).

① Cf. H. Whitney. *Michigan Lectures*, 1941: 101-141.

c. Pour M *orientable* et $n = 2k$ pair, U^k est une classe de première espèce, c'est-à-dire, U^k est déduite d'une classe aux coefficients entiers par réduction mod 2. Pour $n = 4$ la classe $W^2 = U^2 + U^1 \cup U^1 = U^2$ est alors de première espèce et par conséquent la troisieme classe de Stiefel-Whitney (aux coefficients entiers) est nulle; ce n'est pas le cas en général pour $n > 4$, comme le montre la variété orientable de dimension 5 construite de la facon suivante: M^5 est le produit topologique d'un plan projectif complexe P et d'un segment $I = [0,1]$ avec l'identification $(x, y, z) \times (0) = (\bar{x}, \bar{y}, \bar{z}) \times (1)$,où x, y, z sont des nombres complexes, coordonnées homogènes de P, et $\bar{x}$, $\bar{y}$, $\bar{z}$ leurs complexes conjugués.

d. Définissons un autre système de classe U^p (ici $0 \leqslant p \leqslant n$) par récurrence par les équations $\bar{U}^0 = U^0 = 1$ et $\sum_i \bar{U}^i \cup U^{p-i} = 0$, pour $p > 0$. Les classes $\bar{W}$ définies par $\bar{W}^i = \sum_p Sq^{i-p}\bar{U}^p (0 \leqslant i \leqslant n)$ satisfont alors aux équations $\bar{W}^0 = 1$ et $\sum_i \bar{W}^i \cup W^{p-i} = 0$ pour $p > 0$. Cela veut dire que les classes $\bar{W}^i$ ne sont autres que les classes caractéristiques duales de M introduites par Whitney. On a $\bar{W}_n = \sum_p Sq^{n-p}\bar{U}^p = \sum_p U^{n-p} \cup \bar{U}^p = 0$, d'après (1).

e. D'après H. Cartan, $U^p = 0$ pour p impair et M orientable. On en déduit que $W^{n-1} = 0$ pour M orientable et $n = 4k + 2$, ce qui est aussi une conséquence de *c*.

(Extrait des *Comptes rendus des séances de l'Académie des Sciences*,t. 230, p. 508-511, séance du 6 février 1950.)

9. Topologie Algébrique——Les *i*-carrés Dans Une Variété Grassmannienne*

Note de M. Wu Wen-Tsün, présentée par M. Élie Cartan

1. L'anneau de coefficients de l'anneau de cohomologie $H^*(M)$ d'un espace M sera dans ce qui suit exclusivement l'anneau des entiers mod 2.

Soient W^i, $i \geqslant 0$ quelconque, les W-classes (classes caractéristiques de Stiefel-Whitney) d'une s.f.s. (structure fibrée sphérique) avec la convention $W^0 = 1$ (classe unité de la base), et $W^i = 0$ si $i > m$, $m - 1$ étant la dimension de la fibre sphère. Nous allons démontrer la formule suivante :

$$\mathrm{Sq}^r W^s = \sum_t C^t_{s-r+t-1} W^{r-t} W^{s+t} \quad (s \geqslant r > 0), \tag{1}$$

où C^q_p = coefficient binomial pour $p \geqslant q > 0$, $= 0$ pour $p < q > 0$, et $= 1$ pour $p = -1$ et $q = 0$(tous sont réduits mod 2).

Signalons d'abord quelques conséquences de cette formule : définissons, dans la base, un système de classes $U^p (p \geqslant 0$ quelconque) par les équations suivantes:

$$W^i = \sum_p \mathrm{Sq}^{i-p} U^p, \quad i \geqslant 0 \text{ quelconque;} \tag{2}$$

nous les appellerons les *classes canoniques* de la structure considérée. Si la s.f.s. est en particulier la structure tangente associée à une variété différentiable M de dimension m, on voit, en comparant avec les équations (1) et (2) d'une Note précédente ①, que le nom de classes canoniques est justifié; de plus, parmi toutes les s.f.s. (aux fibres S^{m-1}) sur la variété M comme base, la structure tangente de M possède la propriété remarquable suivante :

$$U^p = 0 \quad \text{pour} \quad 2p > m, \tag{3}$$

De (1) et (3) on déduit:

a. Pour une structure orientable on a $U^{2k+1} = 0$, k quelconque, ce qui généralise un thèorème de H. Carlan①.

---

* *C. R. Acad. Sci. Paris* 230, 1950: 918-920.

① *Comptes rendus*, 1950, 230: 508-511.

b. Pour la structure tangente d'une variété différentiable de dimension m, on a $W^1W^{m-2} = 0$ si $m = 4k$; $W^1W^{m-3} = 0$; $W^1W^{m-1} = 0$ si $m = 4k+1$; $W^m = W^1W^{m-1}$ si $m = 4k+2$; $W^1W^{m-1} = 0$, $W^{m-1} = W^1W^{m-2}$si $m = 4k+3$.

2. Soit $G_{n,m}$ la variété grassmannienne des m-éléments linéaires dans un espace euclidien R^{n+m} de dimension $n+m$ passant par l'origine de R^{n+m}. On sait ① que l'*anneau* $H^\star(G_{n,m})$ est engendré par les classes W^i de la s.f.s. $\mathscr{J}_{n,m}$. (fibres S^{m-1}) de base $G_{n,m}$ canoniquement associée à $G_{n,m}$.De plus, comme m'a fait remarquer H. Cartan :

Lemme 1 – *Soit* $\varphi_p(W^i)$ *un polynome non identiquement nul en* $W^1, \cdots, W^m$ *telque pourchaque terme* $W^{i_1} \ldots W^{i_k}$ *de ce polynome on ait* $i_1+\cdots+i_k = p \leqslant n$.*Alors* $\varphi_p(W^i)$ *est un élément non nul de* $H^\star(G_{n,m})$.

Supposons alors que R^{n+m} soit le produit de deux espaces euclidiens $R_j^{n_j+m}$ de dimension $n_j+m_j (j=1,2)$. Soient $G_{n_j,m_j} (j=1,2)$ les variétés grassmanniennes définies respectivement dans $R_j^{n_j+m_j}$. Pour $X_j \in G_{n_j,m_j}$ soil $X \in G_{n,m}$ le joint de X_1 et X_2, on a alors une application canonique

$$f: \quad G_{n_1,m_1} \times G_{n_2,m_2} \to G_{n,m}$$

définie par $f(X_1 \times X_2) = X$. En désignant par $W_j^i (j=1,2)$ respectivement les W-classes des structures $\mathscr{J}_{n_j,m_j}$ on a:

Lemme 2 – *Le type d'homologie mod 2 de f est déterminé par* ② :

$$f^\star W^i = \sum_j W_1^j \otimes W_2^{i-j} \quad (i \geqslant 0 \text{ quelconque}) .$$

Comme conséquence des lemmes 1 et 2, en conservant les notations, on a :

Lemme 3 – *Pour* $p \leqslant n_1$ et n_2, $\varphi_p(W^i)$ *est un élément non nut de* $H^\star(G_{n,m})$ *si et seulement si* $f^\star\varphi_p(W^i)$ *est un élément non nul de* $H^\star(G_{n_1,m_1} \times G_{n_2,m_2})$.

3. *Démonstration de* (1). – Nous poserons

$$\varphi_{r,s}(W^i) = Sq^r W^s + \sum_t C_{s-r+t-1}^t W^{r-t} W^{s+t}.$$

La formule (1), ou, ce qui revient au même, la formule $\varphi_{r,s}(W_j^i) = 0$, étant évidente pour $m=1$, nous supposerons par induction qu'elle est exacte pour les structures dont les fibres sphéres ont une dimension $< m-1$, où $m > 1$. Soient maintenant

① S. CHERN, *Annals of Math.*, 1948, 49: 362-372.

② Nous remarquons que le théoréme de Whitney sur le produit de deux structures fibrées sphériques est une conséquence de ce lemme dont la démonstration est donnée dans ma Thése, Strasbourg, 1949.

W^i, W_j^i respectivement les W-classes des structures $\mathscr{I}_{n,m}$ et $\mathscr{I}_{n_j,m_j}(j = 1,2)$ où $n = n_1 + n_2$, $n_j \geqslant r + s$ $m_1 = m - 1$, $m_2 = 1$. De la formule $f^\star \mathrm{Sq}^i = \mathrm{Sq}^i f^\star$, d'un théoréme de H. Cartan ①, et du lemme 2 du paragraphe 2, on déduit

$$f^\star \varphi_{r,s}(W^i) = \varphi_{r,s}(W_1^2) \otimes I + \varphi_{r,s-1}(W_1^i) \otimes W_2^1 + \varphi_{r-1,s-1}(W_1^i) \otimes (W_2^1)^2.$$

D'aprés l'hypothése d'induction on a donc $f^* \varphi_{r,s}(W^i) = 0$ et par conséquent $\varphi_{r,s}(W^i) = 0$ d'aprés le lemme 3. La structure $\mathscr{I}_{n,m}$étant universelle pour n assez grand, on a $\varphi_{r,s}(W^i) = 0$ pour une s. f. s. quelconque. La formule (1) est ainsi-démontrée par induction.

Soient en particulier W^i les W-classes de la structure $\mathscr{I}_{n,m}$ sur la base $G_{n,m}$. L'anneau $H^*(G_{n,m})$ étant engendré par les classes W^i, on voit que la formule (1) détermine complètement les i-carrés dans $G_{n,m}$ en les exprimant comme des polynômes en W^i.

(Extrait des *Comptes rendus des séances de l'Académie des Sciences*, t. 230, p. 918-920, séance du 6 mars 1950.)

① *Comptes rendus*, 1950, 230: 425-427.

10. Sur les Puissances de Steenrod*

§1. Complexes à transformations périodiques

Soient $\hat{K}$ un complexe simplicial fini et t une transformation simpliciale de $\hat{K}$ tels que 1° $t^p =$ identité 1, où p est un nombre premier; 2° les simplexes σ de $\hat{K}$ pour lesquels $\sigma = t\ \sigma$ forment un sous-complexe *fermé* $\hat{L}$ de $\hat{K}$. Posons

$$d = 1 - t, \quad s = 1 + t + \cdots + t^{p-1}$$

et désignons par les mêmes symboles les transformations duales à t, d et s. Soient K et L des complexes-quotient de $\hat{K}$ et $\hat{L}$ par t, (c'est-à-dire par la relation d'équivalence σ équivalent à $t\ \sigma$), où L est isomorphe à $\hat{L}$. Supposons dans la suite que $\hat{L}$ soit nul. La projection $\pi : \hat{K} \to K$ induit alors une cochaíne-transformation (c'es-à-dire commutant avec δ)

$$\pi : c^n(\hat{K}, G) \to c^n(K, G)$$

de la façon suivante: soit $x \in c^n(\hat{K}, G)$. Pour $\tau \in K$ soient $\tau_i, i = 1, \cdots, p$ les simplexes de $\hat{K}$ tel que $\pi\tau_i = \tau$. On définit alors πx par $\pi x \tau = \sum\limits_i x_i \tau_i$.

Définissons maintenant ($z = z_0$ ou z_p)

$$r^n : H^n(K, z) \to H^{n+2}(K, Z)$$

et

$$v : H^n(K, z) \to H^{n+1}(K, Z_p)$$

de la manière suivante:

Soit $X \in H^n(K, z)$. Prenons $x \in X$ et $x_1 \in C^n(\hat{K}, z)$ tel que $\pi x_1 = x$. On aura alors pour certaines cochaînes $x_2 \in C^{n+1}(\hat{K}, z), x_3 \in C^{n+2}(\hat{K}, z)$,

$$\pi^* x = 5x_1, \quad \delta x_1 = dx_2, \quad \delta x_2 = 5x_3. \tag{1}$$

---

* *Colloque de Topologie de Strasbourg*, 1951, no. IX, 9pp. La Bibliothéque Nationale et Universitaire de Strasbourg, 1952. Nous donnons dans ce qui suit quelques résultats fragmentaires sur les puissances de Steenrod.

Par définition, vx sera la classe, réduite mod p, de πx_2 et kx la classe de πx_3.

Théorème: Soient 1 la classe-unité de K et $U = p + 1 \in H^2(K, z), v = v_1 \in H^1(K, z_p)$.

Alors: en définissant les v-produits par **I**'sccouplement naturel des groupes de coefficients.

Démonstration. Prenons les cochaínes u_0, u_1 et u_2, tels que $\pi u_0 = 1, \pi^* 1 = \delta u_1, \delta u_0 = du_1, \delta u_1 = su_2$.

On a alors $\pi k_2 \in U, \pi u_1 \in V$ et (cf. (1)),

$$\pi(u_0 \cup sx_1) = \pi u_0 \cup \pi x_1 = x,$$

lesv-produits étant définis par rapport à certains ordres de sommets dans $\hat{}K$ et K tels que π soit une application préservant I′ordre. Π en résulte:

$$\begin{aligned}
&\pi^* x = s(u_0 \cup sx_1),\\
&\delta(u_0 \cup sx_1) = \delta u_0 \cup sx_1 = du_1 \cup sx_1 = d(u_1 \cup sx_1),\\
&\delta(u_1 \cup sx_1) = \delta u_1 \cup sx_1 = su_1 \cup sx_1 = s(u_2 \cup sx_1),\\
&\pi(u_1 \cup sx_1) \ \ \mathrm{mod} p = \pi u_1 \cup \pi x_1 \ \mathrm{mod} p \in V \cup x,\\
&\pi(u_2 \cup sx_1) = \pi u_2 \cup \pi x_1 \in U \cup x.
\end{aligned}$$

C′es-à-dire $vx = V \cup X$ et $rX = U \cup X$.

Π va sans dire que les classes U et V peuvent être définies pour des espaces génĕraux dans lesquels opère une transformation périodique sans points fixes.

§2. Les puissances de Steenrod

Soit K un espace quelconque. Steenrod a découvert pour chaque $p \geqslant \lambda$ des opérations topologiques

$$\mathcal{D}_j^p : H^n(K, R) \to H^{n-i}(K, R'), \quad \begin{pmatrix} R\text{kn anneau} \\ \text{commutatif} \end{pmatrix}$$

où $R' = R$, R/pR ou R/zR selon les parités de j, p et r. Nous écrirons dans tout ce qui suit $st^i_{(p)}$ ou tout simplement st^i au lieu de $D^p_{(p-1)n-j}$ tel que

$$st^i : H^n(K, R) \to H^{n+j}(K, R')$$

augmentent toujours par j la dimension de la classe sur laquelle st^i opère. D′après Steenrod, ces *puissances*

st^i jouissent des propriétés suivantes:

1° st^i sont des homomprphismes pour p premier:

2° Pour $X \in H^n(K, R), st^j X = 0$ pour $j < 0$ et $j > (p-1)n$.

3° Pour $X \in H^n(K, Z_1), st^{(p-1)n} X = X \cup \cdots \cup X$ (p fois), réduit mod p ou non selon que $(p-1)q$ est pair ou impair;

4° Pour $X \in H^n(K, z_p)$,

$$st^\circ X = \begin{cases} x, & \text{pour } p = 2. \\ \lambda_{p,n} X, & \text{pour } p = 2k+1, \text{premier} \end{cases}$$

impair, où $\lambda_{p,n} = (-1)^{kn(n-1)/2}(R!)^n$ est non nul mod p;

5° Pour $\lambda_i \in H^{n_i}(K_i, z_p), i = 1, 2$, on a dans $K_1 \times K_2$ la formule suivante de Cartan-Steenrod:

$$st^i(X_1 \otimes X_2) = \sum_{j=0}^{k} st^j X_1 \otimes st^{k-j} X_2, \quad \text{pour } p = 2,$$

$$st^{2i}(X_1 \otimes X_2) = \pm \sum_{j=0}^{k} st^{2j} X_1 \otimes st^{2k-2j} X_2, \quad \text{pour } pi \text{ mpair;}$$

6° Pour une application $f : K_1 \to K_2$ on a $t^* st^j = st^i t^*$. On peut aussi facilement démontrer que.

7° $st^{2k+1} = \lambda st^{2k}$, où

$$\lambda : H^n(K, z_p) \to H^{n+1}(K, z_p)$$

est I'homomorphisms de Whitney-Bockstein.

§3. Les puissances des classes U et V

Déterminons maintenant les puissances des classes U et V définies dans §1 pour un système (K̂, t) avec $t^p = 1$ et $\hat{L} = 0$. Le groupe de coefficients sera exclusivement z_p. Pour cela considérons une sphère S de dimension $zn - 1$: $\sum_{i=1}^{n} z_i \bar{z}_i = 1$ et la transformation t : $z'_k = \exp(2\pi i/p) z_k, k = 1, \cdots, n$. L'espace-quotient de S par t est alors un espace lenticulaire L^{2n-1}. Soient U et V les classes $n-1$ et $V1$ de L^{2n-1} définies comme dans §1. On sait que $H^{2n}(L^{2n-1})$ et $H^{2n+1}(L^{2n-1}), n = 0, 1, \cdots, K$ sont engendrés respectivement par $(U)^n$ et $(U)^n V$. Par conséquent on a nécessairement

(1) $st^{2n}V = a_{2n}(U)^nV$, $st^{2n+1}V = a_{2n+1}(U)^{n+1}$,

(1)′ $st^{2n}U = b_{2n}(U)^{n+1}$, $st^{2n+1}U = b_{2n+1}(U)^{n+1}V$.

Or L^{2n-1} (n assez grand) est un espace universel pour les espaces fibrés de groupe structural z_p, c′est-à-dire, pour un système $(\hat{K}, t)$ avec $t' = 1, c = 0$ et espa-ce-quotient $K = \hat{K}/t$, il existe une application $\hat{f} : \hat{K} \to S$ qui se projette sur une application $f : K \to L^{2n-1}$. Les classes U et V dans K sont alors les images inverses des classes U et V de L^{2n-1} et, d′après 6° §2n les formules (1) et (1)′ sont donc aussi exactes pour les classes U et V d′un système quelconque $(\hat{K}, t)$ avec les mêmes a, $\hat{b}$ pour dime K quelconque.

Cela étant, considérons deux tels systèmes quelconques $(\hat{K}_i; t_i)$ avec complexes-quotient $\hat{K}_\varepsilon = \hat{K}_i/t$. Soient $\hat{K}$ le complexe-quotient de $\hat{K}_1 \times \hat{K}_2$ par $t_1 \times t_2$ et t la chaîne-transformation définie par $tw(a_1 \times a_2) = w(t_1a_1 \times a_2)$, où $a_i \in \hat{K}_i, i = 1, 2$ et où ω est la projection de $\hat{K}_1 \times \hat{K}_2$ sur $\hat{K}$. Le couple $(\hat{K}, t)$ forme alors un système dont le complexe-quotient de $\hat{K}$ par t est $K_1 \times K_2$. Socent U_iV_i les classes r^1 et $v1$ dans $\hat{K}_i$. On peut alors vérifier que

$$V = V_1 \otimes 1 - 1 \otimes V_2, \quad U = U_1 \otimes 1 - 1 \otimes U_2. \tag{2}$$

D′après $1°, 2°, 4°$ et $5°$ de §2, on a donc pour p premier impair,

$$st^{2n}V = c(st^{2n}V_1 \otimes 1 - 1 \otimes st^{2n}V_2), \quad c \not\equiv 0 \text{ mod p}.$$

Si a_{2n} n′est pas nul, ceci donne

$$(U)^nV = c[(U_1)^nV_1 \otimes 1 - 1 \otimes (U_2)^nV_2],$$

ce qui est impossible pour $p > n > 1$, d′après (1), (2) et 1° de §2. En tenant compte de 2° de §2, on a donc pour la classe $V = vl$ d′un système $(\hat{K}, t)$ quelconque:

$$st^nV = 0, \quad n > 1 \quad (p \text{ premier impair}) \tag{3}$$

De même on aura

$$st^nU = 0, \quad n \neq 0 \text{ ou } 2(p-1) \quad (p \text{ premier impair}). \tag{3)'}$$

Dans les autres cas st^nV et st^nU sont complètement déterminés d′après 3° et 7° de §2 (p premier).

Rezarquons que les formules (3) et (3)′, etc. sont déduites des propriétés formelles $1° - 7°$ de §2, sans utiliser une réalisation conrète de st^n.

§4. Deux autres définitions des puissances de Steenrod

Dans ce qui suit les groupes de coefficients seront exclusivement le groupe z_p, p étant *un nombre premier*, sauf mention du contraire.

Considérons de nouveau un couple $(\hat{K}, t)$ avec sous-complexe invariant $\hat{L}$ et complexe-quotient K et L. D'après les procédés de Smith on peut alors définir les groupes spéciaux d'homologie $H_n^5(\hat{K})$ et de cohomologie $H_s^n(\hat{K}, L)$ où $s = d$ ou s. On peut aussi définir des homomorphisme s canoniques

$$\left.\begin{array}{l}\xi_s : H_s^n(\hat{K}, \hat{L}) \to H_\xi^{n+1}(\hat{K}, \hat{L}) \\ \eta_s : H_n^s(\hat{K}) \to H_{n-1}^{\bar{s}}(\hat{K})\end{array}\right\} (s, \bar{s}) = (d, s) \text{ ou } (s, d)$$

$$\left.\begin{array}{l}\theta_s : H^n(\hat{L}) \to H_s^{n+1}(\hat{K}, \hat{L}) \\ \varphi_s : H_n^s(\hat{K}) \to H_s(\hat{L})\end{array}\right\} s = d \text{ ou } s$$

Remarquons que $H_d^n(\hat{K}, \hat{L}) \approx H^n(K, L)$ d'après Thom, et on désignant cet isomorphisme par p, on a $\psi\xi_d\psi^{-1} = v$ et $\psi\xi_s\xi_d\psi^{-1} = p, p_s\mu p$ et vétant définis comme dans §1.

Soient maintenant $\hat{K}$ le produit de p exemplaires d'un même complexe $\hat{L}$ et t la chaîne-transformation définie par $t(a_1 \otimes a_2 \otimes \cdots \otimes a_p) = (-1)^{pn}(a_p \otimes a_1 \otimes \cdots \otimes q_{p-1}))$, où $a_i \in \hat{\mathrm{L}}, q = \dim\ a_p$, et $\wedge = \dim(a_1 \otimes \cdots \otimes a_{p-1})$, qui est induite par la transformation cyclique de $\hat{K}$ si $\hat{K}$ est un complexe géométrique. Désignons la diagonale de $\hat{K}$ par le même symbol $\hat{L}$ et supposons que $\hat{K}$ a été subdivisé d'une façon invariante par rapport à t et contenant ainsi $\hat{L}$ comme souscomplexe. Par une étude profonde de la théorie de Smith-Richardson et en utilisant les résultats de §1 et §3; surtout la double interprétation des homomorphismes y et p_3^n Thom a établi une théorie "intrinsèque" des puissances de Steenrod. Sans une forme plus restrictive maix suffisante pour notre but, on peut énoncer son résultat prinoipal comme suit:

Theorème (THOM). Pour une classe $U \in H^n(\hat{L})$ soit $T^{(p-1)n}\ U$ la classe $(U)^p$. Il existe alors un système de classes $T^iU \in H^{n+j}(\hat{L})$ uniquement déterminé par l'équation suivante:

$$\sum_{j=0}^{(p-1)n} \xi^{(p-1)n-j}\theta_i T^j U = 0,$$

où

$$\theta_j = \begin{cases} \theta_s, & (p-1)n - j \quad \text{pair}, \\ \theta_d, & (p-1)n - j \quad \text{impair}, \end{cases}$$

et $\xi^{(p-1)n-j}$ est un produit alterné de ξ, et ξ_d, à ($(p-1)n-j$ facteurs). De plus, les classes T^jU ne diffèrent per per aes coefficients des puissances st^jU de Steenrod que par des coefficients $\neq 0$. D'autre part, pour une classe $X \in H_n(L)$ et un cycle $z \in X$ le cycle $x^p = x \otimes \cdots \otimes x$, après subdivision, détermine une classe X^p de $H^d_{pn}(K)$ qui ne dépend que de X. Posons

$$\varphi_s \eta^{(p-1)n+k} X^p = S_k X \in H_{n-k}(\hat{L}),$$

où

$$\varphi_s = \begin{cases} \varphi_d, & (p-1)_n + k \quad \text{pair}, \\ \varphi_s, & (p-1)n + k \quad \text{cmpair}, \end{cases}$$

et $\eta^{(p-1)n+k}$ est un produit alterné de η_s et η_d, à $(p-1)_n = j$ facteurs. Il est facile de voir que $s_k = 0$ pour $k < 0$, que s_e ne diffère de I'identité que par un coefficient non nul, et que les s_k sont des homomorphismes. z_p étent un corps, on peut définir I'homomor-phisme dual

$$s^k : H_n(\hat{L}) \to H_{n+n}(\hat{L}).$$

on a alors le théorème suivant:

Théorème. On peut exprimer les homomorphismes s^k en fonction de T^j et récipvoquement ($k > 0$):

$$\begin{cases} \displaystyle\sum_{j=0}^{n} s^{n-j} T^j = 0, & \text{pour } p{+}2 \text{ ou } p^k \text{ impair}, \\ \displaystyle\sum_{j=0}^{n} s^{(k-2)} T^{2j} = 0, & \text{pour } p > 2. \end{cases}$$

Ces deux théorèmes fournissent deux définitions intrinsèques, duales I'un et I'autre, des puissances de Steenrod. La dernière définition, bien que la moins puissante, est certainement la plus naturelle.

On peut définir S_k pour les coefficients entiers. Mais cela ne nous donne rien d'essentiellement nouveau.

§5. Les puissances de Steenrod dans les variétés grassmanniennes

Soient $G_{n,m}$ la variété grassmannienne des méléments dans un espace Euclidien de dimension $n+m$ et $G_{n,m}$ la variété grassmannienne des éléments complexes de dimension complexe m dans un espace unitaire de dimension complexe $n+m$. On sait que l'anneau de cohomologie mod 2 de $G_{n,m}$ est engendré par un système de classes de Stiefel Whitney $W^i, i=1,2,\cdots,m$ et que l'anneau de cohomologie entière de $C_{h,m}$ par un système de classes de Chen $c^{ni}, i=1,\cdots,m$. Nous avons esquissé une méthode about issant à des formules explicites de st^2_{i+1} dans $G_{n,m}$ qui donne en particulier:

$$st^i_{(2)}W^m = w^i w^m. \tag{1}$$

Par une méthode analogue on peut démontrer que, dans $C_{n,m}$ et pour p premier impair,

$$st^{2n}_{(p)}c^{2s} = 0 \quad \text{mod } p \text{ pour} x \not\equiv 0 \text{ mod}(p-1).$$

Ceci m'a conduit à la conjecture suivante: dans un espace quelconque les puissances $st_\phi X$ d'une classe X mod p du moins de prémière espèce sont toutes triviales sauf dans les dimensions $i=0$ ou 1 mod $2(p-1)$. Cette conjecture a été confirmée depuis par Thom en donnant une démonstration complète dans toute sa généralité.

En proftant d'une simplification, donnée par H. Cartan, de ma méthode pour déterminer $st^i_{(p)}$ dans $G_{n,m}$, on peut aussi déterminer $st^i_{(p)}$ dans $C_{n,m}$, p étant un premier quelconque. Je ne citerai que le résultat suivant:

Théorème. Considérons C^{2i} comme des indéterminés et soient $x_1, i=1,\cdots,m$ les racines de l'équation $f(x,C) \equiv x^m - C^2x^{m+1} + \cdots + C^{2m} = 0$. Pour $n \leqslant s, St^{2n(p-1)}C^{ns}$ est alors le polynome symétrique en x_i du terme typique $p_{n,k}(x_1 \cdots x_n)p_{x_{n+1}\cdots x_3}$ où

$$\begin{cases} p_{m,m} = 1, p_{n,1} = \lambda_{n,2}p_{n,s-1} & \text{pour } p \text{ impair} \\ p_{n,1} = 1 & \text{pour } p\text{=2}. \end{cases}$$

En particulier, soit

$$y^{m(p-1)} - R^2y^{m(p-1)-1} + \cdots \pm R^{2m(p-1)} = 0$$

l'équation obtenue à partir de $f(x,c)=0$ et de $y=x^{p-1}$ en éliminant x. On a alors

$$st^{2n(p-1)}_{(p)}C^{3m} = p_{n,m}R^{2n(p-1)}C^{2m}\text{mod}p. \tag{2}$$

§6. Applications aux classes caractéristiques d'une espace fibré sphérique

Soient E un espace fibré par S^{m-1}, K le complexe-base, A I'espace fibré par la m-boule fermée associée à E et A' le complémentaire de E dans A. A E on peut alors associer des suites exactes (coefficients entiers mixtes, soient tordus, soient non-tordus) de la façon suivante:

$$\begin{array}{ccccccccc} \to H^{n-1}(E) & \to & H^2(A') & \xrightarrow{\beta} & H^2(n) & \to & H^2(E) \to & & \text{(suite exacte de Thom)} \\ \| & & \uparrow \varphi^* & & \uparrow j & & \| & & \\ \to H^{n-1}(E) & \to & H^{n-k}(K) & \xrightarrow{\psi} & H^2(K) & \to & H^2(E) \to & & \text{(suite exacte de-Spanier)} \end{array}$$

D'autre part, on sait qu'à I'espace fibré E on peut associer des systèmes d'invariants homologiques visà-vis de sa structure fibrée, à savoir le système des classes de Stiefel (Whitney $W^i, i=1,\cdots,m$, aux coefficients locaux mod 2 ou entiers, et le système des classes de Pontrjagin $p^i, \psi_i \leqslant m$) aux coefficients entiers. D'aprèsSpanier, ψ est alors le cup-produit avec W^m où $\psi = j^1\rho\varphi^*, j$ et φ^* étant des isomorphismes sur, introduits par Thom.

Thom a pu démontrer que, si E est en particulier I'espace tangent associé à une structure différentiable d'une variété différentiable M de dimension m, les classes W^i de E dépendent uniquement de la structure topologique de M, mais non de la structure différentiable considérée. Les formules explicites de W^i en termes de cohomologie de M á l'aide de $st_{(2)}$ ont été aussi déterminées. Le succès est due uniquement à la formule suivante de Thom:

$$st^i_{(2)}\varphi * 1 = \varphi^* W^i. \tag{1}$$

Pour éviter des complications concernant les coefficients, nous allons considérer cette formule comme une formule mod 2, ce qui est suffisant pour notre but. Cela étant, nous esquissons ici une autre démonstration de cette formule:

De (1) de §5 on déduit:

$$st^i_{(2)}\psi_1 = \psi W: \quad \text{mod 2, d'où}$$

$$\beta st^i_{(2)}\varphi^* 1 = \beta\varphi^* W^i \quad \text{mod 2.} \tag{2}$$

Supposons que E est I'espace fibré canonique sur $K = G_{n,m}$. Pour n assez grand $\psi/H^n(k), n \leqslant km$, est alors biunivoque. Il en est ainsi pour β et (2) devient donc précisément (1). La formule (1), étant valable pour les structures canoniques sur $G_{n,m}$, est alors aussi valable pour nne structure quelconque.

Cette démonstration a I'avantage de pouvoir être étendue aux classes de Pontrjagin. Soit $m * 2m'$ et E une structure presque complexe avec classes de Chem C^{2i}. On a alors $C^{2m'} * (-1)^{m'} W^{2m'}$ et on déduit par la méthode précédente, en partant de la formule (2) de §5, que:

$$st_{(p)}^{2n(p-1)} \varphi * 1 = P_{n,m} \varphi^* R^{2(p-1)} \quad \bmod p, \tag{3}$$

p étant un premier quelconque.

Soit maintenant E une structure quelconque et Π_2 (E) le produit de Whittney de E par lui-même qui admet une structure presque complexe canonique. Pour les classes de Chem C^{2i} de Π_2 (E) et les classes de Pontrjar-gin $P^{4)}$ de E on a alors

$$C^{2i} = \begin{cases} 0, & \text{impair}, \\ (-1)^j p + j & i = 2j. \end{cases}$$

Il en résulte que les classes $R^{n(p-1)}$ dans la formile (3) correspondant à E sont des polynomes en P^j de E. En particulier, $R^j = p^j$ mod 3 pour p=3.

Considérons maintenant une variété différentiable compacte M de dimension m et le produit $M \times M \times M = M^3$ avec diagonale Δ. En appliquant (3) à I'espace normal (isomorphe à T_2 (E), E étant I'espace tangent de M) de Δ dans M^3 on obtient:

$$st_{(p)}^{2n(p-1)} \psi^* 1 = p_{n,m} \psi^* R^{2m(p-1)} \quad \bmod p,$$

où $\psi^* : H^n(M) \to H^{n+2m}(M^3)$ est le prolongement de φ^*. Cette formule donne I'invariance topologique de $R^{2n(p-1)}$ mod p, en particulier de P^j mod 3 pour p=3.

Si M est en particulier de dimension 4, on trouve à I'aide de la formule de Cartan-Steenrod et du théorème de Thom sur la trivialité de $st_{(p)}^i$, que $p^* = 0$ mod 3. Ce théorème confirme la conjecture suivante, que j'ai formulée, il y a quelques années:

Soit M une variété différentiable, compacte, orientée, et de dimension 4. Il existe une base $x_i, y_j, j = 1, \cdots, \bar{\tau}$, de $H_2(M)$ aux coefficients réels tel qu, on ait pour les indices d'intersection $x_i \cdot x_{i'} = +\delta_{ii}\xi, y_j \cdot y_{j'} = -\delta_i j_j$ et $x_i \cdot y_j = 0$. Posons $\tau = \bar{\tau} - \bar{\tau}$ et $\pi = p^u$ (M). La conjecture s'exprime alors $\pi = 3\tau$.

Bibliographie

[1] STEENROD N E. Reduced powers of cohomology classes. Conférences du Collège de France, Paris, 1951.

[2] RICHARDSON M et SMITH P A. Periodic transformation of. complexes. Annals of Math., 1938, 39: 611-633.

[3] THOM R. Une théorie intrinsèque des puissances de Steenrod. Collègue de Topologie de Strasbourg, 1951.

11. 有限可剖分空间的新拓扑不变量*

导　言

我们所知道的拓扑空间的许多拓扑不变量，如同伦群、下同调群、上同调环，以及在这种群或环上可以定义的种种运算如 Whitehead 积、Steenrod 平方和幂、Понтрягин平方之类，不但是这个空间的拓扑不变量，而且也是它的同伦不变量. 直到现在，我们只知道一些孤立的并非是同伦不变的拓扑不变量，例如空间的维数、一维可剖分空间 (polyhedron) 的分枝点数，和直到最近才证明是拓扑不变但非同伦不变的某种特殊三维流形的 Reidemeister 组合不变量之类 [1]. 可是，这种拓扑不变但非同伦不变的量的搜求是必要的，因为它们的发现可以使我们有希望去讨论空间的拓扑分类问题，也就是古典拓扑上的主要问题，而不致再局限于空间的同伦分类问题 (参阅 [2]，§1).

本文可视为这方面的一个尝试. 我们证明了这样的一条定理，任何一个有限的可剖分空间 P 可配以许多可剖分空间的组，这种空间组的同伦型乃是 P 的拓扑不变量. 因之，这些空间组的任一同伦不变量都是 P 的拓扑不变量，而且，只要原来那个同伦不变量是可以计算的，那么由此导出的拓扑不变量也是可以计算的. 简单的例子说明，这样所得来的拓扑不变量一般说来都不是同伦不变的. 这使我们得到了一个普遍的方法，可以从空间或空间组的同伦不变量以导出有限可剖分空间的一组拓扑不变而非同伦不变的量来.

这些新不变量中最简单的一种是由 Euler-Poincaré 示性数 (以后简称示性数) 这个同伦不变量所导出的，乃是与大于 1 的整数 n 有关的一组整数值不变量. 在本文中详细地讨论了这种不变量的计算方法，证明它们一般说来是非同伦不变的，甚至可据以区分两个可以缩成一点的空间的拓扑型 (topological type). 但另一方面，在可剖分空间是一个复合的闭流形 (combinational closed manifold) 时，它们完全由这个流形的示性数所定.

此外，我们又证明一个有限可剖分空间的 Betti 数，可由这种新不变量，即拓扑不变而非同伦不变的量表示出来，反过来自然不可能. 这说明那些新不变量至少要比一部分古典不变量如 Betti 数之类要更基本.

目前的一个重要问题，乃是决定如果我们把所讨论的空间范围限制到闭流形时，这些新不变量中是否还有非同伦不变者存在，但于此我们尚不能作确切的

*Acta Math. Sinica, 1953, 3: 261-290.

回答.

§1　一些名词与记号的解释

因为拓扑学上许多最基本的概念如复合形，可剖分空间之类的意义在流行的书刊中颇不一致，因此在本节中，我们将先明确一下在以后常要用到而用法有歧义的一些名词 (和记号) 的意义.

在欧氏空间中的凸集、凸胞腔、凸胞腔的边和维数、单纯形和它们的有关名词将如通常那样定义.

在同一欧氏空间中的任两凸集 A 和 B 可确定一个既包含 A 又包含 B 的最小凸集 X. 假设 A, B 不相交而 X 具有下列性质: 对于任一 $x \in X-A-B$，有一个唯一的点 $a \in A$，和一个唯一的点 $b \in B$，使 x 在线段 ab 之上，换言之，x 可唯一地表成 $x=(a,b,r)=ra+(1-r)b$ 的形状，其中 $a \in A$, $b \in B$，$0<r<1$，$1-r:r$ 为线段 ax 与 xb 之比. 这时我们就说 X 是 A，B 两凸集的“联合”(join)，记以 $X=A\circ B$.

我们将抽象地引进一个 -1 维的凸胞腔 ε，并作下列规约: $\varepsilon\circ\varepsilon=\varepsilon$，又对于任一凸集 A，$A\circ\varepsilon=\varepsilon\circ A=A$.

若 σ, τ 是任两凸胞腔而 $\sigma\circ\tau$ 有定义，则显然

$$\dim(\sigma\circ\tau)=\dim\sigma+\dim\tau+1,$$

此处 σ, τ 亦可为 ε.

在本文中所讨论的“复合形”和“可剖分空间”都是有限的，因之“有限”两字将一概省去.

在以后所谓“欧氏 (有限) 复合形”，是指一个在某一欧氏空间中的“有限”个凸胞腔的非空集合 K，满足这样的条件: 这集合中任一凸胞腔的维数 $\geqq 0$ 的边也是这个集合的元素，且这集合中任两凸胞腔的交在非空集合的时候，必同时是这两凸胞腔的边，因之也属于这个集合. K 中所有凸胞腔的和集成一空间，我们记作 $\bar{K}$. 我们叫 $\bar{K}$ 为 K 的“空间”，K 为 $\bar{K}$ 的一个“剖分”，在 K 中的凸胞腔都是单纯形时，K 就叫做一个“欧氏单纯复合形”，而是 $\bar{K}$ 的一个“单纯剖分”.

假定一个欧氏空间中的子空间 P 可以看作是其中一个欧氏复合形的空间时，也就是 P 可以有剖分时，我们就说 P 是一个“欧氏可剖分空间”.

有时，我们把 -1 维胞腔 ε 也加入欧氏复合形 K 那个集合里去，并把它看作 K 中任一凸胞腔的边. 这时，我们就说 K 是被“扩大”了，而把这个扩大了的欧氏复合形记作 K_ε

以后所谓“复合形”，(或“单纯复合形”)，是指一个拓扑空间 Q，一个欧氏复合形 (或欧氏单纯复合形)K，以及一个从 $\bar{K}$ 到 Q 之上的拓扑映像 T 的总和而言. 这

个复合形将记作 ${}_TK$，Q 叫做复合形 ${}_TK$ 的“空间”：$Q \equiv {}_T\bar{K}$，而 ${}_TK$ 叫做 Q 的一个“剖分”(或“单纯剖分”)

假定一个拓扑空间可以看作是一个复合形的空间时，也就是可以有剖分时，就叫做一个“可剖分空间”. 如所久知，一个可剖分空间必然有单纯剖分.

假定 K，L 是某同一欧氏空间中的两个欧氏复合形，而 L 中每一凸胞腔同时也是 K 中的一个凸胞控，那么我们就说 L 是 K 的一个“子复合形”，记作 $L \subset K$，此时显然应有 $\bar{L} \subset \bar{K}$.

假定 K，L 是某同一欧氏空间中的欧氏复合形而 $L \subset K$，T 是 $\bar{K}$ 到一拓扑空间 Q 上的拓扑映像，因而 ${}_TK$ 是一个复合形，以 Q 为其空间，那么 ${}_TL$ 也是一个复合形，以 ${}_T(\bar{L}) = P(\subset Q)$ 为其空间. 这时，我们就说 ${}_TL$ 是 ${}_TK$ 的一个“子复合形”，记作 ${}_TL \subset {}_TK$.

若 K_1，K_2 都是同一欧氏空间中的两个欧氏复合形，而对于每一 $\sigma_1 \in K_1$，$\sigma_2 \in K_2$，$\sigma_1 \circ \sigma_2$ 恒有定义，那么一切凸胞腔 $\sigma_1 \circ \sigma_2, \sigma_1 \circ \varepsilon$ 或 $\varepsilon \circ \sigma_2(\sigma_1 \in K_1, \sigma_2 \in K_2)$ 成一欧氏复合形. 我们将把它叫做 K_1 和 K_2 的“联合复合形”(join complex)，记作 $K_1 \circ K_2$.

§2 正则的复合形偶与可剖分空间偶

假定 ${}_TK$，${}_TL$ 是两个复合形而 ${}_TL \subset {}_TK$，此处 K, L 是某同一欧氏空间中的欧氏复合形 $(L \subset K)$，那么在下面的条件甲能满足时，我们就说 $({}_TK, {}_TL)$ 成一“正则的复合形偶”，或说 ${}_TL$ 正则地浸没于 ${}_TK$ 中，记作 ${}_TL \Subset {}_TK$. 条件甲如次:

甲. K 中任意一个既有 $\bar{L}$ 的点也有不在 $\bar{L}$ 中的点的凸胞腔 σ 必是两个凸胞腔 σ_1，σ_0 的联合: $\sigma = \sigma_1 \circ \sigma_0$，此处 $\sigma_0 \in L$，$\sigma_1 \in K$ 但不 $\in L$，且 $\sigma_0 = \sigma \cap \bar{L}$.

在特别的情形，假定 ${}_TK$，${}_TL({}_TL \subset {}_TK)$ 都是单纯复合形，即 K，$L(L \subset K)$ 都是欧氏单纯复合形时，很容易证明条件甲就与下面的条件乙等值.

乙. K 中任意一个单纯形 σ 的顶点若都属于 L 则单纯形 σ 也属于 L.

今设 ${}_TL \Subset {}_TK$. 我们将引入下面的一些记号.

K 中一切与 $\bar{L}$ 无公共交点的凸胞腔成一 K 的子复合形，记作 $K_1^{(+)}(K, L)$ 或 $K_1^{(+)}$.

K 中一切与 $\bar{L}$ 有公共交点的凸胞腔以及所有它们的边成一 K 的子复合形，记作 $K_1^{(-)}(K, L)$ 或 $K_1^{(-)}$.

K 中一切既在 $K_1^{(+)}$ 内又在 $K_1^{(-)}$ 内的凸胞腔成一 K 的子复合形，即 $K_1^{(+)}$ 和 $K_1^{(-)}$ 之交，记作 $K_1^{(0)}(K, L)$ 或 $K_1^{(0)}$.

假定 x 是 $\bar{K}_1^{(-)}$ 中的一点，但不 $\in \bar{L}$ 且不 $\in \bar{K}_1^{(0)}$，那么 x 必是 K 中某一凸胞腔 σ 的点，而 σ 既有 $\bar{L}$ 中的点也有不在 $\bar{L}$ 中的点. 由条件甲，应有 $\sigma = \sigma_1 \circ \sigma_0$，

此处 $\sigma_0 \in L$，$\sigma_0 = \sigma \cap \bar{L}$，$\sigma_1 \in K$ 但不 $\in L$，显然 $\sigma_1 = \sigma \cap \bar{K}_1^{(0)} \in K_1^{(0)}$. 由§1 开首所述，我们可表 x 成下列形状 $x = (x_1, x_0, r) = rx_1 + (1-r)x_0$，此处 $x_1 \in \sigma_1$，$x_0 \in \sigma_0$，$0 < r < 1$. 要是 x 在另一 $\sigma' \in K$ 内，而 σ' 也既有 $\bar{L}$ 中的点也有不在 $\bar{L}$ 中的点，且 $\sigma' = \sigma_1' \circ \sigma_0'$, $\sigma_0' = \sigma' \cap \bar{L} \in L$, $\sigma' = \sigma' \cap \bar{K}_1^{(0)} \in K_1^{(0)}$. 若表 x 为 $x = (x_1', x_0', r')$, $x_1' \in \sigma_1'$, $x_0' \in \sigma_0'$, $0 < r' < 1$, 那么由于这种表示的几何意义，可以知道这两种表示方法必然一致. 我们应有 $x_1' = x_1 \in \sigma_1 \cap \sigma_1' (\neq \varnothing)$, $x_0' = x_0 \in \sigma_0 \cap \sigma_0' (\neq \varnothing)$, $r' = r > 0$ 且 < 1. 此外，我们将作下列记号上的规定. 若 $x_1 \in \bar{K}_1^{(0)}$ 而 $x_0 \in \bar{L}$, 我们将以 $(x_1, x_0, 0)$ 表 x_0(x_1 任意)，而以 $(x_1, x_0, 1)$，表 x_1(x_0 任意). 这样，由前所述，我们就可得到下面的一些结论：

1°. 在 $\bar{K}_1^{(-)}$ 中的任一点 x 必可表作形状 $x = (x_1 \cdot x_0, r)$，此处 $x_1 \in \bar{K}_1^{(0)}$，$x_0 \in \bar{L}$, $0 \leqslant r \leqslant 1$.

2°. 在 $x \in \bar{K}_1^{(-)} - \bar{L} - \bar{K}_1^{(0)}$ 时，这个表示是唯一的，且 $0 < r < 1$. 在 $x \in \bar{L}$ 时，$x_0 = x$ 而 x_1 任意，$r = 0$. 在 $x \in \bar{K}_1^{(0)}$ 时，$x_1 = x$ 而 x_0 任意，$r = 1$.

3°. 在这种表示之下，映像 $x \to r$ 是连续的. 而在 x 不 $\in \bar{L}$ 时，映像 $x \to x_1$ 也是连续的，在 x 不 $\in \bar{K}_1^{(0)}$ 时，则映像 $x \to x_0$ 是连续的.

以下暂设 r 是一个任意的固定实数, $\geqq 0$ 且 $\leqslant 1$.

所有或 $\in \bar{K}_1^{(+)}$ 或 $\in \bar{K}_1^{(-)}$ 而在表示 $x = (x_1, x_0, r')$ 中 $r' \geqq r$ 的一切点 x 成一空间记作 $\bar{K}_r^{(+)}(K, L)$ 或

所有 $\in \bar{K}_1^{(-)}$ 而在表示 $x = (x_1, x_C, r')$ 中 $r' \leqslant r$ 的一切点 x 成一空间记作 $\bar{K}_r^{(-)}(K, L)$ 或 $\bar{K}_r^{(-)}$.

所有 $\in \bar{K}_1^{(-)}$ 而在表示 $x = (x_1, x_0, r')$ 中 $r' = r$ 的一切点 x 成一空间记作 $\bar{K}_r^{(0)}(K, L)$ 或 $\bar{K}_r^{(0)}$.

在 $r = 1$ 时，这里所定义的 $\bar{K}_r^{(+)}$, $\bar{K}_r^{(-)}$, $\bar{K}_r^{(0)}$ 等显然与以前所定义的 $\bar{K}_1^{(+)}$, $\bar{K}_1^{(-)}$, $\bar{K}_1^{(0)}$ 一致，又在 $r = 0$ 时，$\bar{K}_0^{(+)} = \bar{K}$, $\bar{K}_0^{(-)} = \bar{L}$, $\bar{K}^{(0)} = \bar{L}$.

假定 σ 是 K 中任意一个既有 $\bar{L}$ 的点也有不在 $\bar{L}$ 中的点的凸胞腔. 不论 r 如何 $(0 \leqslant r \leqslant 1)$，命 $\sigma_r^{(+)} = \bar{K}_r^{(+)} \cap \sigma$, $\sigma_r^{(-)} = \bar{K}_r^{(-)} \cap \sigma$, $\sigma_r^{(0)} = \bar{K}_r^{(0)} \cap \sigma$, 那么 $\sigma_r^{(+)}$, $\sigma_r^{(-)}$ 和 $\sigma_r^{(0)}$ 都是凸胞腔。我们容易看出 $\bar{K}_r^{(+)}$, $\bar{K}_r^{(-)}$, $\bar{K}_r^{(0)}$ 都是可剖分空间. 各有一个如下所定义的剖分 $K_r^{(+)}$, $K_r^{(-)}$ 和 $K_r^{(0)}$.

$K_r^{(+)}$ 是一切 $\in K_1^{(+)}$ 以及一切 $\sigma_r^{(+)}$ 和它们的边所成的复合形 $(K_0^{(+)} = K)$.

$K_r^{(-)}$ 是一切 $\in L$ 以及一切 $\sigma_r^{(-)}$ 和它们的边所成的复合形 $(K_0^{(-)} = L)$.

$K_r^{(0)}$ 是一切 $\sigma_r^{(0)}$ 所成的复合形 $(K_0^{(0)} = L)$.

我们还不难证实 $K_r^{(0)}$ 是复合形 $K_r^{(+)}$ 与 $K_r^{(-)}$ 的“交”：

$$K_r^{(0)} = K_r^{(+)} \cap K_r^{(-)} \quad (0 \leqslant r \leqslant 1)\,. \tag{1}$$

我们又有

$$\bar{K} = \bar{K}_r^{(+)} \cup \bar{K}_r^{(-)}, \quad \bar{K}_r^{(0)} = \bar{K}_r^{(+)} \cap \bar{K}_r^{(-)} \quad (0 \leqslant r \leqslant 1)\,, \tag{2}$$

$$_T\bar{K} = {}_T\bar{K}_r^{(+)} \cup {}_T\bar{K}_r^{(-)}, \quad {}_T\bar{K}_r^{(0)} = {}_T\bar{K}_r^{(+)} \cap {}_T\bar{K}_r^{(-)} \quad (0 \leqslant r \leqslant 1)\,. \tag{2$'$}$$

从 $\bar{K}_r^{(+)}$ 等的定义，又可得下面两项事实，证明因很简单从略：

1. 对于任两 >0 且 <1 的 r, r', $\bar{K}_r^{(+)}$, $\bar{K}_r^{(-)}$, $\bar{K}_r^{(0)}$(或 ${}_T\bar{K}_r^{(+)}$, ${}_T\bar{K}_r^{(-)}$, ${}_T\bar{K}_r^{(0)}$), 各与 $\bar{K}_{r'}^{(+)}$, $\bar{K}_{r'}^{(-)}$, $\bar{K}_{r'}^{(0)}$(或 ${}_T\bar{K}_{r'}^{(+)}$, ${}_T\bar{K}_{r'}^{(-)}$, ${}_T\bar{K}_{r'}^{(0)}$) 有相同的拓扑型，或简称“同拓”(homomorphic).

2. 对于任一 >0 且 <1 的 r, $\bar{K}_1^{(+)}$(或 ${}_T\bar{K}_1^{(+)}$) 是 $\bar{K}_r^{(+)}$(或 ${}_T\bar{K}_r^{(+)}$) 也是 $\bar{K}-\bar{L}$(或 ${}_T\bar{K} - {}_T\bar{L}$) 的一个变状收缩核 (deformation retract)，$\bar{K}_0^{(-)} = \bar{L}$(或 ${}_T\bar{K}_0^{(-)} = {}_T\bar{L}$) 是 $\bar{K}_r^{(-)}$(或 ${}_T\bar{K}_r^{(-)}$) 的一个变状收缩核. 因之，${}_T\bar{K}_r^{(+)}$ 与 ${}_T\bar{K}_1^{(+)}$ 或 ${}_T\bar{K} - {}_T\bar{L}$ 有相同的一切同伦同调性质，${}_T\bar{K}_r^{(-)}$ 与 ${}_T\bar{L}$ 亦然.

今设 P，Q 是两个可剖分空间，其中 P 是 Q 的子空间. 假定 P, Q 各有剖分 ${}_TL$ 与 ${}_TK$ 存在，使 K，L 在同一欧氏空间中而 $L \Subset K$，那么我们就说 (Q, P) 成一“正则的可剖分空间偶”，或说 P 正则地浸没于 Q，记作 $P \Subset Q$. 我们又说 $({}_TK, {}_TL)$ 是 (Q, P) 的一个“正则剖分”.

§3 主要定理的叙述

在以下，所提到的拓扑空间都假定属于某一个固定的类型，例如正常空间、可剖分空间之类.

假定有有限个拓扑空间 $Q_1, \cdots, Q_m (m \geqq 1)$，具有这样的性质：其中任意两个空间 Q_i，Q_j 有公共点时，$Q_{ij} = Q_i \cap Q_j$ 不论作为 Q_i 或是 Q_j 的子空间看来，乃是相同的拓扑空间. 这时我们就说这一组 (有一定次序的) 空间 $(Q_1, \cdots, Q_m)$ 成一 (对于某一固定类型的) 拓扑空间组或简称“空间组”.

假定 $\mathbf{Q} = (Q_1, \cdots, Q_m)$，$\mathbf{R} = (R_1, \cdots, R_m)$ 是两个 (属于同一固定类型的) 具有同样多个数的空间，且只要 $Q_i (i = 1, \cdots, m)$ 的和与交之间有某种包含或相等的关系，与之相当的 $R_i (i = 1, \cdots, m)$ 之间也有相同的包含或相等的关系，那么我们说 $\mathbf{Q}$ 和 $\mathbf{R}$ 是“相似”的.

在以下，$\mathbf{P} = (P_1, \cdots, P_m)$，$\mathbf{Q} = (Q_1, \cdots, Q_m)$ 和 $\mathbf{R} = (R_1, \cdots, R_m)$ 是指 (属于同一类型的) 相似空间组. 假定 $f_i : Q_i \to R_i (i = 1, \cdots, m)$ 是一组连续映像，具有这样的性质，任两空间 Q_i, Q_j 只要有公共点时，对于任一 $x \in Q_{ij} = Q_i \cap Q_j$ 有 $f_i(x) = f_j(x) \in R_{ij} = R_i \cap R_j (\neq \varnothing)$. 这时，我们就说 $\mathbf{f} = (f_1, \cdots, f_m)$ 是 $\mathbf{Q}$ 到 $\mathbf{R}$ 中的一个连续映像，记作 $\mathbf{f} : \mathbf{Q} \to \mathbf{R}$. 这种连续映像的存在，自然须预先假定 $\mathbf{Q}$, $\mathbf{R}$ 两空间组具有这样的性质，只要 $Q_i \cap Q_j \neq \varnothing$，一定也有 $R_i \cap R_j \neq \varnothing$.

在特别情形，每一 R_i 与 $Q_i(i=1,\cdots,m)$ 相合，而每一 $f_i:Q_i\to R_i$ 都是恒同变换即 $f_i=1(i=1,\cdots,m)$ 时,$1=(1,\cdots,1)$ 是一个连续映像. 又若 $f_i:P_i\to Q_i$，$g_i:Q_i\to R_i(i=1,\cdots,m)$ 组成连续映像 $\mathbf{f}:\mathbf{P}\to\mathbf{Q}$，$\mathbf{g}:\mathbf{Q}\to\mathbf{R}$，那么 $h_i=g_if_i:P_i\to R_i(i=1,\cdots,m)$ 显然也组成一个连续续映像 $\mathbf{h}:\mathbf{P}\to\mathbf{R}$.

今以 T 表闭线段 $[0,1]$，那么 $Q\times T=(Q_1\times T,\cdots,Q_m\times T)$ 显然也是一个空间组 (我们假定 $Q_i\times T$ 仍属于原来那个空间的类型). 假定对于每一 $t\in T$，有一连续变换 $\mathbf{D}_t:\mathbf{Q}\to\mathbf{R}$，此处 $\mathbf{D}_t=(D_{t,1},\cdots,D_{t,m})$, $D_{t,i}:Q_i\to R_i(i=1,\cdots,m)$, 满足下面的条件：定义 $D_i:Q_i\times T\to R_i$ 为 $D_i(x_i,t)=D_{t,i}(x_i)$，$x_i\in Q_i,t\in T$ 时，$\mathbf{D}=(D_1,\cdots,D_m)$:$\mathbf{Q}\times T\to\mathbf{R}$ 是一个连续映像. 这时，我们就说 $\mathbf{D}_t,t\in T$，是 $\mathbf{D}_0$ 与 $\mathbf{D_1}$ 之间的一个"变状". 如果对于两个连续映像 $\mathbf{f},\mathbf{g}:\mathbf{Q}\to\mathbf{R}$ 有一变状 $\mathbf{D}_t:\mathbf{Q}\to\mathbf{R},t\in T$ 存在使 $\mathbf{D}_0=\mathbf{f},\mathbf{D}_1=\mathbf{g}$，那么我们说 $\mathbf{f}$ 与 $\mathbf{g}$"同伦"，记作 $\mathbf{f}\simeq\mathbf{g}$.

假定 $\mathbf{f}:\mathbf{Q}\to\mathbf{R},\mathbf{g}:\mathbf{R}\to\mathbf{Q}$ 都是连续变换，而有

$$\mathbf{fg}\simeq\mathbf{1},\quad \mathbf{gf}\simeq\mathbf{1},$$

那么我们说空间组 $\mathbf{Q}$ 与 $\mathbf{R}$ 有相同的"同伦型"，或简称"同伦"，记作 $\mathbf{Q}\simeq\mathbf{R}$. 显然，假定 $\mathbf{Q}'=(Q_{i_1},\cdots,Q_{i_S})$，$\mathbf{R}'=(R_{i_1},\cdots,R_{i_S})$，$1\leqslant i_1$，$\cdots$，$i_s\leqslant m$，那么 $\mathbf{Q}'$，$\mathbf{R}'$ 也是空间组, 而从 $\mathbf{Q}\simeq\mathbf{R}$ 可得 $\mathbf{Q}'\simeq\mathbf{R}'$.

空间组间的同伦关系，显然是一个同值观念，因此，我们可依常法定义所谓空间组的同伦不变性或同伦不变量等.

在 $m=1$ 时，空间组只有一个空间，这时上面所引进的一些观念与关于空间间的相当观念符合.

今将空间的类型限制为 (有限的) 可剖分空间. 下面是这种空间组的一个同伦不变量的例子.

假定 Q_1,Q_2 都是 Q 的子空间而 $Q_{12}=Q_1\cap Q_2$. 设 $j_{i*}:H_k(Q_{12})\to H_k(Q_i)$, $i=1,2(k\geqq 0)$，是由恒同变换 $j_i:Q_{12}\to Q_i$, $i=1,2$ 所引起的同构变换，此处 H_k 是指对于以某同一固定数域为系数群的下同调群. 命 $N_k=j_{1*}^{-1}(0)\cap j_{2*}^{-1}(0)\subset H_k(Q_{12})$ 而以 q^k 表 N_k 的秩 (rank)，则 q^k 是空间组 $\mathbf{Q}=(Q,Q_1,Q_2,Q_{12})$ 的一个同伦不变量. 证之如次：

设 $\mathbf{Q}\prime=(Q',Q_1',Q_2',Q_{12}')$ 是与 $\mathbf{Q}$ 相似的一个空间组. 与 $\mathbf{Q}$ 同样，可定义 j_i', j_{i*}', N_k', q'^k 等. 假定 $\mathbf{Q}\simeq\mathbf{Q}\prime$, 而 $\mathbf{f}=(f,f_1,f_2,f_{12})$ $\mathbf{Q}\to\mathbf{Q}'$, $\mathbf{f}'=(f',f_1',f_2',f_{12}')$: $\mathbf{Q}'\to\mathbf{Q}$ 是实现这个同伦关系的连续映像，即 $\mathbf{ff}'\simeq\mathbf{1}$, $\mathbf{f}'\mathbf{f}\simeq\mathbf{1}$. 从后二者可知 $f_if_i'\simeq 1$, $f_i'f_i\simeq 1$, $i=1,2$, 以及 $f_{12}f_{12}'\simeq 1$, $f_{12}'f_{12}\simeq 1$, 因之 $f_{i*}:H_k(Q_i)\approx H_k(Q_i')$, f_{i*}':$H_k(Q_i')\approx H_k(Q_i)$, $i=1,2$, 且 $f_{12*}:H_k(Q_{12})\approx H_k(Q_{12}')$, $f_{12*}':H_k(Q_{12}')\approx H_k(Q_{12})$. 又因 $f_{12}\equiv f_i/Q_{12}$, $f_{12}'\equiv f_i'/Q_{12}'$, $i=1,2$, 故 $j_i'f_{12}=f_ij_i$, $j_if_{12}'=f_i'j_i'$,

$i=1$, 2. 由此知 $f_{12*}j_{i*}^{-1}(0)=j_{i*}^{\prime -1}$ (0), $f'_{12*}j_{i*}^{\prime -1}=(0)=j_{i*}^{-1}(0)$, $i=1$, 2, 即 $f_{12*}N_k=N'_k$, $f'_{12*}N'_k=N_k$. 因之 $N_k\approx N'_k$ 而 $q^k=q'^k$, 即所欲证.

今设 (Q,P) 是一个正则可剖分空间偶，$({}_TK,{}_TL)$ 是 (Q,P) 的任意一个正则剖分. 依§2，对于任一 $\geqq 0$ 及 $\leqslant 1$ 的 r，我们可从 $({}_TK,{}_TL)$ 作出一个空间组 $(Q,{}_T\bar{K}_r^{(+)},{}_T\bar{K}_r^{(-)},{}_T\bar{K}_r^{[0]},P)$. 本文中的主要定理乃是:

定理 1 假定 $({}_TK,{}_TL)$ 是正则可剖分空间偶 (Q,P) 的一个正则剖分，而 $0<r<1$，那么从 $({}_TK,{}_TL)$ 作出的空间组 $(Q,{}_T\bar{K}_r^{(+)},{}_T\bar{K}_r^{(-)},{}_T\bar{K}_r^{(0)},P)$ 的同伦型与所选择的正则剖分 $({}_TK,{}_TL)$ 无关. 换言之，这个空间组的同伦型乃是正则可剖分空间偶 (Q,P) 的拓扑不变量.

从正则可剖分空间偶 (Q,P) 的同一个正则剖分 $({}_TK,{}_TL)$ 所作出的不同空间组 $(Q,{}_T\bar{K}_r^{(+)},{}_T\bar{K}_r^{(-)},{}_T\bar{K}_r^{(0)},P)$ 的同伦型之与 r $(0<r<1)$ 无关，可以像下面那样来证明它.

设 $0<r_1<r_2<1$. 由§2，在 ${}_T\bar{K}_1^{(-)}$ 中的每一个点 x 都可表成 $x=T(x_1,x_0,r)$ 的形状，此处 $x_1\in\bar{K}_1^{(0)}$, $x_0\in\bar{L}$, $0\leqslant r\leqslant 1$, 而在 x 不 $\in{}_T\bar{K}_1^{(0)}$ 或 P 时，这种表示是唯一的，且 $0<r<1$. 今定义 $f:Q\to Q$ 如次 $({}_T(x_1,x_0,r)\in{}_T\bar{K}_1^{(-)})$:

$$f:\begin{cases}{}_T(x_1,x_0,r)\to{}_T\left(x_1,x_0,\dfrac{1-r_2}{1-r_1}\cdot r+\dfrac{r_2-r_1}{1-r_1}\right), & 1\geqq r\geqq r_1,\\ {}_T(x_1,x_0,r)\to{}_T\left(x_1,x_0,\dfrac{r_2}{r_1}\cdot r\right), & r_1\geqq r\geqq 0,\\ x\to x,\quad x\in{}_T\bar{K}_1^{(+)}.\end{cases}$$

显然 f 是 Q 到 Q 上的一个拓扑映像，而且

$$f({}_T\bar{K}_{r_1}^{(+)})={}_T\bar{K}_{r_2}^{(+)},\quad f({}_T\bar{K}_{r_1}^{(-)})={}_T\bar{K}_{r_2}^{(-)},\quad f({}_T\bar{K}_{r_1}^{(0)})={}_T\bar{K}_{r_2}^{(0)},\quad f(P)=P.$$

因之在 f 之下空间组 $(Q,{}_T\bar{K}_{r_1}^{(+)},{}_T\bar{K}_{r_1}^{(-)},{}_T\bar{K}_{r_1}^{(0)},P)$ 的各个空间与空间组 $(Q,{}_T\bar{K}_{r_2}^{(+)},{}_T\bar{K}_{r_2}^{(-)},{}_T\bar{K}_{r_2}^{(0)},P)$ 的各相当空间同拓. 自然更应有

$$(Q,{}_T\bar{K}_{r_1}^{(+)},{}_T\bar{K}_{r_1}^{(-)},{}_T\bar{K}_{r_1}^{(0)},P)\simeq(Q,{}_T\bar{K}_{r_2}^{(+)},{}_T\bar{K}_{r_2}^{(-)},{}_T\bar{K}_{r_2}^{(0)},P),$$

即所欲证.

在一般的情形，即空间组系从不同的正则剖分所作出的时候，定理 1 的证明较为复杂. 这个证明，我们将移至§4 中来完成它，目前先列举一些从定理 1 直接就可获得的推论.

首先，定理 1 包括了下面的

定理 2 在定理 1 的假定下，${}_T\bar{K}_r^{(0)}(0<r<1)$ 的同伦型是 (Q,P) 的拓扑不变量.

因为 ${}_T\bar{K}_r^{(-)}(0<r<1)$ 以 P 为它的一个变状收缩核，而 ${}_T\bar{K}_r^{(+)}(0<r<1)$ 与 $Q-P$ 都以 ${}_T\bar{K}_1^{(+)}$ 为变状收缩核，故 ${}_T\bar{K}_r^{(-)}\simeq P$, ${}_T\bar{K}_r^{(+)}\simeq Q-P(0<r<1)$, 因而 ${}_T\bar{K}_r^{(+)}$ 和 ${}_T\bar{K}_r^{(-)}$ 的同伦型在 $0<r<1$ 时都是 (Q,P) 的拓扑不变量，甚为明显，但如定理 2 所述 ${}_T\bar{K}_r^{(0)}$ 的同伦型之为拓扑不变量，则并不显然.

从上面的一些定理可知：

定理 3　在定理 1 的假定下，空间组 $(Q,{}_T\bar{K}_r^{(+)},{}_T\bar{K}_r^{(-)},{}_T\bar{K}_r^{(0)},P), 0<r<1$ 的任一个同伦不变量必是 (Q,P) 的一个拓扑不变量. 特别，${}_T\bar{K}_r^{(+)}$ 或 ${}_T\bar{K}_r^{(0)}(0<r<1)$ 的任一同伦不变量必是 (Q,P) 的一个拓扑不变量.

特别，在 P 是可剖分空间 Q 的一个点时，(Q,P) 必是一个正则的可剖分空间偶. 这时的定理 2 是一个很早就知道的定理，可参阅 [3] 的第五章. ${}_T\bar{K}_r^{(0)}$ 的下同调群是 ${}_T\bar{K}_r^{(0)}$ 的同伦不变量，因之是 (Q,P) 的拓扑不变量，这就是所谓 Q“在 P 点的下同调群”.

§4　主要定理的证明

像§3 中那样，假定 (Q,P) 是一个正则的可剖分空间偶，$({}_TK,\ {}_TL)$ 和 $({}_{T'}K',{}_{T'}L')$ 是它的任意两个正则剖分. 对于任一 $\geqq 0$ 而 $\leqslant 1$ 的 s，从 $({}_TK,{}_TL)$ 可以作出一个空间组 $(Q,{}_T\bar{K}_s^{(+)},{}_T\bar{K}_s^{(-)},{}_T\bar{K}_s^{(0)},P)$. 同样，从 $({}_{T'}K',\ {}_{T'}L')$ 也可以作出一个空间组 $(Q,{}_{T'}K'^{(+)}_s,T'\bar{K}'^{(-)}_s,{}_{T'}\bar{K}'^{(0)}_s,P)$. 那么，定理 1 无非是说在 $0<s<1$ 时，

$$(Q,{}_T\bar{K}_s^{(+)},{}_T\bar{K}_s^{(-)},{}_T\bar{K}_s^{(0)},P)\simeq(Q,{}_{T'}\bar{K}'^{(+)}_s,{}_{T'}\bar{K}'^{(-)}_s,{}_{T'}\bar{K}'^{(0)}_s,P). \tag{I}$$

定理 2 是说在 $0<s<1$ 时，

$${}_T\bar{K}_s^{(0)}\simeq{}_{T'}\bar{K}'^{(0)}_s. \tag{II}$$

虽然定理 2 只是定理 1 的推论，但因为它的证明比较简单，所以我们先来证明定理 2，也就是证明 (II) 式.

令

$$N={}_T\bar{K}_1^{(-)}-{}_T\bar{K}_1^{(0)}-P,\quad N'={}_{T'}\bar{K}'^{(-)}_1-{}_{T'}\bar{K}'^{(0)}_1-P, \tag{1}$$

则 $N\cup P$ 和 $N'\cup P$ 都是 P 在 Q 中的一个邻域. 因为 K, L, K', L' 都是“有限的”复合形，故必有数 r_0', r_1', r_0 使

$$0<r_0'<r_1'<1,\quad 0<r_0<1; \tag{2}$$

$${}_{T'}\bar{K}'^{(-)}_{r'_0}\subset{}_T\bar{K}^{(-)}_{r_0}\subset{}_{T'}\bar{K}'^{(-)}_{r'_1}\subset N\cup P. \tag{2$'$}$$

由§2，N 中的点可唯一地表成形状 ${}_T(x_1,x_0,r)$，此处 $0<r<1$, $x_1\in\bar{K}_1^{(0)}$, $x_0\in\bar{L}$. 同样 N' 中的点也可唯一地表成形状 ${}_{T'}(x_1',x_0',r')$，此处 $0<r'<1$, $x_1'\in\bar{K}_1'^{(0)}$, $x_0'\in\bar{L}'$.

今定义映像 ($t\in{}_T=[0,1]$)

$$F_t:N\to N,\quad F_t':N'\to N'$$

为 (${}_T(x_1,x_0,r)\in N$, ${}'_T(x_1',x_0',r')\in N'$)

$$\begin{cases}F_{tT}(x_1,x_0,r)={}_T(x_1,x_0,tr_0+\overline{1-t}\cdot r),\\ F'_{tT'}(x'_1,x'_0r')={}_{T'}(x'_1,x'_0,tr'_0+\overline{1-t}\cdot r').\end{cases}\tag{3}$$

像§2 所指出的那样，F_t，F_t' 都是连续映像且连续地依赖于 $t\,(\in T)$. 由 (3) 知

$$F_0=1,\quad F_1/{}_T\bar{K}_{r_0}^{(0)}=1,\quad F_1(N)\subset{}_T\bar{K}_{r_0}^{(0)},$$
$$F'_0=1,\quad F'_1/{}_{T'}\bar{K}'^{(0)}_{r'_0}=1,\quad F'_1(N')\subset{}_{T'}\bar{K}'^{(0)}_{r'_0}.$$

又由 (1)，(2)，(2)′ 可知

$$F_t({}_{T'}\bar{K}'^{(0)}_{r'_0})\subset F_t({}_{T'}\bar{K}^{(-)}_{r_0}-P)\subset{}_T\bar{K}^{(-)}_{r_0}-P\subset{}_{T'}\bar{K}'^{(-)}_{r'_1}-P\subset N',$$
$$F'_t({}_T\bar{K}^{(0)}_{r_0})\subset F'_t({}_{T'}\bar{K}'^{(-)}_{r'_1}-P)\subset{}_{T'}\bar{K}'^{(-)}_{r'_1}-P\subset N.$$

因此，如果我们如下定义一些映像 ($x\in{}_T\bar{K}^{(0)}_{r_0}, x'\in{}_{T'}\bar{K}'^{(0)}_{r'_0}$)

$$f(x)=F_1'(x),\quad f'(x')=F_1(x'),$$

$$D_t\,(x)=\begin{cases}F_1F'_{2t}\,(x), & 0\leqslant t\leqslant\dfrac{1}{2},\\ F_1F'_1\,(x), & \dfrac{1}{2}\leqslant t\leqslant 1;\end{cases}$$

$$D'_t\,(x')=\begin{cases}F'_1F_{2t}\,(x'), & 0\leqslant t\leqslant\dfrac{1}{2},\\ F'_1F_1\,(x'), & \dfrac{1}{2}\leqslant t\leqslant 1;\end{cases}$$

那么

$$f:{}_T\bar{K}^{(0)}_{r_0}\to{}_{T'}\bar{K}'^{(0)}_{r'_0}, f':{}_{T'}\bar{K}'^{(0)}_{r'_0}\to{}_T\bar{K}^{(0)}_{r_0},$$
$$D_t:{}_T\bar{K}^{(0)}_{r_0}\to{}_T\bar{K}^{(0)}_{r_0}, D'_t:{}_{T'}\bar{K}'^{(0)}_{r'_0}\to{}_{T'}\bar{K}'^{(0)}_{r'_0}$$

都是连续映像且 D_t，D_t' 连续地依赖于 $t(\in T)$. 显然

$$D_0=1,\quad D_1=f'f,$$
$$D'_0=1,\quad D'_1=ff',$$

故

$$f'f' \simeq 1, \quad ff' \simeq 1,$$

亦即

$${}_T\bar{K}^{(0)}_{r_0} \simeq {}_{T'}\bar{K}'^{(0)}_{r'_0}.$$

如§2 末所指出，从此式即得 (II) 式，而定理 2 已证毕.

在证定理 1 时，我们将先取数 r_i, r'_i $(i=0,1,2)$ 使

$$1 > r_0 > r_1 > r_2 > 0, \quad 1 > r'_0 > r'_1 > r'_2 > 0, \tag{4}$$

$${}_T\bar{K}^{(+)}_1 \subset {}_{T'}\bar{K}'^{(+)}_{r'_0} \subset {}_T\bar{K}^{(+)}_{r_0} \subset {}_{T'}\bar{K}'^{(+)}_{r'_1} \subset {}_T\bar{K}^{(+)}_{r_1} \subset {}_{T'}\bar{K}'^{(+)}_{r'_2} \subset {}_T\bar{K}^{(+)}_{r_2}, \tag{5}$$

或即

$${}_T\bar{K}^{(-)}_1 \supset {}_{T'}\bar{K}'^{(-)}_{r'_0} \supset {}_T\bar{K}^{(-)}_{r_0} \supset {}_{T'}\bar{K}'^{(-)}_{r'_1} \supset {}_T\bar{K}^{(-)}_{r_1} \supset {}_{T'}\bar{K}'^{(-)}_{r'_2} \supset {}_T\bar{K}^{(-)}_{r_2}. \tag{5$'$}$$

因为我们的复合形都是“有限的”，故这样的 r_i，r'_i 必然存在.

对于 $t \in T$ 及任意 $i,j=0,1,2$ 而 $i \neq j$ 定义

$$F_{ij,t}, F'_{ij,t} : Q \to Q$$

如次 $({}_T(x_1,x_0,r) \in {}_T\bar{K}^{(-)}_1$, ${}_{T'}(x'_1,x'_0,r') \in {}_{T'}\bar{K}'^{(-)}_1)$:

1°. $r_i > r_j$ 因而 $i<j$ 时，

$$F_{ij,t}: \begin{cases} {}_T(x_1,x_0,r) \to {}_T\left(x_1,x_0,(1-t)r + t\left(\dfrac{1-r_j}{1-r_i}\cdot r - \frac{r_i-r_j}{1-r_i}\right)\right), & r_i \leqslant r \leqslant 1, \\ {}_T(x_1,x_0,r) \to {}_T(x_1,x_0,(1-t)r+tr_j), & r_j \leqslant r \leqslant r_i, \\ x \to x, \quad x \in {}_T\bar{K}^{(+)}_1 \text{或} {}_T\bar{K}^{(-)}_{r_j}. \end{cases}$$

2°. $r_i < r_j$ 因而 $i>j$ 时，

$$F_{ij,t}: \begin{cases} {}_T(x_1,x_0,r) \to {}_T\left(x_1,x_0,(1-t)r + t\cdot\dfrac{r_j}{r_i}\cdot r\right), & 0 \leqslant r \leqslant r_i, \\ {}_T(x_1,x_0,r) \to {}_T(x_1,x_0,(1-t)r+tr_j), & r_i \leqslant r \leqslant r_j, \\ x \to x, \quad x \in {}_T\bar{K}^{(+)}_{r_j}. \end{cases}$$

3°. $r'_i > r'_j$ 因而 $i<j$ 时，

$$F'_{ij,t}: \begin{cases} {}_{T'}(x'_1,x'_0,r') \to {}_{T'}(x'_1,x'_0,(1-t)r' + t(\frac{1-r'_j}{1-r'_i}\cdot r' - \frac{r'_i-r'_j}{1-r'_i})), & r'_i \leqslant r' \leqslant 1, \\ {}_{T'}(x'_1,x'_0,r') \to {}_{T'}(x'_1,x'_0,(1-t)r'+tr'_j), & r'_j \leqslant r' \leqslant r'_i, \\ x' = x', \quad x' \in {}_{T'}\bar{K}'^{(+)}_1 \text{或} {}_{T'}\bar{K}'^{(-)}_{r'_j}. \end{cases}$$

4°. $r'_i < r'_j$ 因而 $i > j$ 时，

$$F'_{ij,t}:\begin{cases} {}_{T'}(x'_1, x'_0, r') \to {}_{T'}(x'_1, x'_0, (1-t)r' + t\cdot\frac{r'_j}{r'_i}\cdot r'), & 0 \leqslant r' \leqslant r'_i, \\ {}_{T'}(x'_1, x'_0, r') \to {}_{T'}(x'_1, x'_0, (1-t)r' + tr'_j), & r'_i \leqslant r' \leqslant r'_j, \\ x' \to x', \quad x' \in {}_{T'}\bar{K}'^{(+)}_{r'_j}. \end{cases}$$

上面这些映像，容易知道都是连续的而且连续地依赖于 $t(\in T)$.

其次，我们定义一些映像 $(t \in T)$

$$f, f', D_t, D'_t : Q \to Q$$

如次：

$$f = F'_{12,1}F_{01,1}, \quad f' = F_{10,1}F'_{21,1},$$

$$D_t = \begin{cases} F_{20,6t}, & 0 \leqslant t \leqslant \frac{1}{6}, \\ F_{20,1}F_{01,6t-1}, & \frac{1}{6} \leqslant t \leqslant \frac{1}{3}, \\ F_{20,1}F'_{12,6t-2}F_{01,1}, & \frac{1}{3} \leqslant t \leqslant \frac{1}{2}, \\ F_{20,1}F'_{21,6t-3}f, & \frac{1}{2} \leqslant t \leqslant \frac{2}{3}, \\ F_{20,1}F_{10,6t-4}F'_{21,1}f, & \frac{2}{3} \leqslant t \leqslant \frac{5}{6}, \\ F_{20,6-6t}f'f & \frac{5}{6} \leqslant t \leqslant 1; \end{cases}$$

$$D'_t = \begin{cases} F'_{02,6t}, & 0 \leqslant t \leqslant \frac{1}{6}, \\ F'_{02,1}F'_{21,6t-1}, & \frac{1}{6} \leqslant t \leqslant \frac{1}{3}, \\ F'_{02,1}F_{10,6t-2}F'_{21,1}, & \frac{1}{3} \leqslant t \leqslant \frac{1}{2}, \\ F'_{02,1}F_{01,6t-3}f', & \frac{1}{2} \leqslant t \leqslant \frac{2}{3}, \\ F'_{02,1}F'_{12,6t-4}F_{01,1}f', & \frac{2}{3} \leqslant t \leqslant \frac{5}{6}, \\ F'_{02,6-6t}ff', & \frac{5}{6} \leqslant t \leqslant 1. \end{cases}$$

这些映像，也都是连续的而且 D_t, D'_t 都连续地依赖于 $t(\in T)$.

由 (4), (5) 或 $(5)'$ 以及 $F_{ij,t}$, $F'_{ij,t}$ 的定义，我们不难验算出下面的一些关系：

$$f({}_T\bar{K}^{(+)}_{r_0}) \subset {}_{T'}\bar{K}'^{(+)}_{r'_2}, \quad f({}_T\bar{K}^{(-)}_{r_0}) \subset {}_{T'}\bar{K}'^{(-)}_{r'_2}, \quad f({}_T\bar{K}^{(0)}_{r_0}) \subset {}_{T'}\bar{K}'^{(0)}_{r'_2}, \quad f(P) \subset P.$$

因此，f 定义了下面的空间组间的连续映像，

$$\mathbf{f}:(Q,{}_T\bar{K}_{r_0}^{(+)},{}_T\bar{K}_{r_0}^{(-)},{}_T\bar{K}_{r_0}^{(0)},P)\rightarrow(Q,{}_{T'}\bar{K}'^{(+)}_{r'_2},{}_{T'}\bar{K}'^{(-)}_{r'_2},{}_{T'}\bar{K}'^{(0)}_{r'_2},P).$$

同样，我们经过一些繁长的验算可以知道 f'，D_t，D'_t 也定义了某些空间组间的连续映像如次:

$$\begin{aligned}
&\mathbf{f}:(Q,{}_{T'}\bar{K}'^{(+)}_{r'_2},{}_{T'}\bar{K}'^{(-)}_{r'_2},{}_{T'}\bar{K}'^{(0)}_{r'_2},P)\rightarrow(Q,{}_T\bar{K}_{r_0}^{(+)},{}_T\bar{K}_{r_0}^{(-)},{}_T\bar{K}_{r_0}^{(0)},P),\\
&\mathbf{D}_t:(Q,{}_T\bar{K}'^{(+)}_{r_0},{}_T\bar{K}_{r_0}^{(-)},{}_T\bar{K}_{r_0}^{(0)},P)\rightarrow(Q,{}_T\bar{K}_{r_0}^{(+)},{}_T\bar{K}_{r_0}^{(-)},{}_T\bar{K}_{r_0}^{(0)},P),\\
&\mathbf{D'}_t:(Q,{}_{T'}\bar{K}'^{(+)}_{r'_2},{}_{T'}\bar{K}'^{(-)}_{r'_2},{}_{T'}\bar{K}'^{(0)}_{r'_2},P)\rightarrow(Q,{}_{T'}\bar{K}'^{(+)}_{r'_2},{}_{T'}\bar{K}'^{(-)}_{r'_2},{}_{T'}\bar{K}'^{(0)}_{r'_2},P).
\end{aligned}$$

此外，由 f，f'，D_t，D'_t 的定义得

$$\begin{aligned}
&\mathbf{D}_0=1,\quad \mathbf{D}_1=\mathbf{f}'\mathbf{f},\\
&\mathbf{D'}_0=1,\quad \mathbf{D'}_1=\mathbf{f}\mathbf{f}',
\end{aligned}$$

因此

$$\mathbf{f}\mathbf{f}'\simeq\mathbf{f}'\mathbf{f},$$

亦即

$$(Q,{}_T\bar{K}_{r_0}^{(+)},{}_T\bar{K}_{r_0}^{(-)},{}_T\bar{K}_{r_0}^{(0)},P)\simeq(Q,{}_{T'}\bar{K}'^{(+)}_{r'_2},{}_{T'}\bar{K}'^{(-)}_{r'_2},{}_{T'}\bar{K}'^{(0)}_{r'_2},P).$$

依§3 关于定理 1 的讨论，这一式的两边各与 (I) 式的两边同伦，因而 (I) 式亦即定理 1 已完全证明.

§5　可剖分空间的拓扑不变量

假定 P 是一个可剖分空间，n 是 >1 的正整数. P 的 n 次自乘的拓扑积将表以 $Q=P^n$，它的对角空间即一切作 $(x_1,\cdots,x_n)$ 形状而 $x_1=\cdots=x_n\in P$ 的点所成的子空间与 P 同拓，我们将仍用 P 来表示. 我们有

定理 4　(Q,P) 是一个正则的可剖分空间偶.

在证明这一定理之先，我们将先作一准备.

假定 L 是在 m 维欧氏空间 R^m 中的一个单纯复合形，那么在 mn 维欧氏空间 $(R^m)^n=R^{mn}$ 中有一个复合形 $K'=L^n$，即 L 自乘 n 次的积，它的胞腔都是 $\sigma_1\times\cdots\times\sigma_n(\sigma_i\in L)$ 形状的凸集. $\bar{K}'=\bar{L}^n$ 的对角空间与 $\bar{L}$ 同拓，因之有一与 L 同构的剖分，这个对角空间与它的剖分将仍用 $\bar{L}$ 和 L 来表示. 在以下我们将定义一个以 L 为子复合形的 $\bar{K}'$ 的“标准”剖分 K.

先设 $\sigma_1, \cdots, \sigma_n$ 是 L 中任意 n 个单纯形，$\sigma_i (i = 1, \cdots, n)$ 恰有 $r+1$ 个 $(r \geqq 0)$ 顶点 $a_0, \cdots, a_r$ 公共：

$$\sigma_j = (a_0, \cdots, a_r, b_1^{(j)}, \cdots, b_{kj}^{(j)}), \quad j = 1, \cdots, n. \tag{1}$$

令

$$(1)' \qquad I = \{0, 1, \cdots, r\}, \quad J = \{1, \cdots, n\}, \quad K_j = \{1, \cdots, k_j\}, \quad j \in J.$$

对于任一 $I' \subset I$，令 $\sigma_{I'}$ = 由 $i \in I'$ 的一切 a_i 所成的单纯形，特别在 $I' = \varnothing$ 时，$\sigma_{I'} = \varepsilon$. 又令

$$\sigma'_j = (b_1^{(j)}, \cdots, b_{kj}^{(j)}), \quad j \in J, \tag{1$''$}$$

特别在 $K_j = \varnothing$ 时，$\sigma'_j = \varepsilon$. 于是

$$\sigma_j = \sigma_I \circ \sigma'_j, \quad j \in J.$$

假定对于每一 $j \in J$ 有一 $I_j \subset I$ 和一 $K'_j \subset K_j$ 而 $I_1 \cap \cdots \cap I_n = \varnothing$ 且对于任一 $j \in J$, I_j 与 K'_j 不同时为 $\varnothing$，则 (有次序的) 集合组 $(I_1, \cdots, I_n; K'_1, \cdots, K'_n)$ 将称为一个“分组.” 对于每一分组 $(I_1, \cdots, I_n; K'_1, \cdots, K'_n)$ 在 R^{mn} 中有一凸胞腔

$$\xi = (\sigma_{I_1} \circ \sigma'_{K'_1}) \times \cdots \times (\sigma_{I_n} \circ \sigma'_{K'_n}) (\subset \sigma_1 \times \cdots \times \sigma_n). \tag{2}$$

所有这些胞腔 ξ 显然成一复合形 E. 又令 $\delta_{I'} (I' \subset I, I' \neq \varnothing)$ 表一切作 $(x_1, \cdots, x_n)$ 形状而 $x_1 = \cdots = x_n \in \sigma_{I'}$，的点所成的凸胞腔，则所有 $\delta_{I'}$ 显然组成一 δ_I 的复合形 D. 我们有

预备定理 1 对于每一作 (2) 形状的 ξ 以及每一 $\subset I$ 而 $\neq \varnothing$ 的 I'，$\delta_{I'} \circ \xi$ 有意义，因而 D 和 E 的联合复合形 $F = D \circ E$ 有定义且是 $\sigma_1 \times \cdots \times \sigma_n$ 的一个剖分.

证 很容易知道我们只须证明下面的三点即可：

1°. 对于每一 $\bar{D}$ 的点 x 和 $\overline{E}$ 的点 y，以及任一 $\geqq 0$ 且 $\leqslant 1$ 的实数 t，$(1-t)x + ty \in \sigma_1 \times \cdots \times \sigma_n$.

因为 $\sigma_1 \times \cdots \times \sigma_n$ 是一个凸集，故 1° 甚显然.

2°. 对于每一在 $\sigma_1 \times \cdots \times \sigma_n$ 内的点 z，必可找到一点 $x \in \bar{D}$ 和一点 $y \in \bar{E}$ 以及一数 $t \geqq 0$ 且 $\leqslant 1$，使 $z = (1-t)x + ty$.

3°. 对于在 $\sigma_1 \times \cdots \times \sigma_n$ 内但不在 $\bar{D}$ 或 $\bar{E}$ 内的点 z，2° 中所说的 x, y 和 t 乃是唯一地决定的，且 $0 < t < 1$.

兹先证明 2°.

假定所与的点是

$$z = (z_1, \cdots, z_n) \in \sigma_1 \times \cdots \times \sigma_n, \tag{3}$$

此处对于每一 $j \in J$，z_j 依重心坐标的表示是

$$z_j = \sum_{i \in I} \alpha_i^{(j)} a_i + \sum_{k \in K_j} \beta_k^{(j)} b_k^{(j)} \in \sigma_j, \tag{3)'$$

$$\begin{cases} \sum_{i \in I} \alpha_i^{(j)} + \sum_{k \in K_j} \beta_k^{(j)} = 1, \\ 1 \geqq \alpha_i^{(j)}, \beta_k^{(j)} \geqq 0, \quad i \in I, k \in K_j. \end{cases} \tag{4}$$

令

$$\alpha_i^* = \min_{j \in J} \alpha_i^{(j)}, \tag{5}$$

并如下定义 $I_j \subset I (j \in J)$:

$$\begin{cases} i \in I_j \text{与} \alpha_i^{(j)} > \alpha_i^* \quad \text{同义}, \\ i \text{不} \in I_j \text{与} \alpha_i^{(j)} = \alpha_i^* \quad \text{同义}. \end{cases} \tag{6}$$

则

$$I_1 \cap \cdots \cap I_n = \varnothing \tag{7}$$

对于任两指数 $j, l \in J$，有

$$\sum_{i \in I} \alpha_i^{(j)} + \sum_{k \in K_j} \beta_k^{(j)} = \sum_{i \in I} \alpha_i^{(l)} + \sum_{k \in K_l} \beta_k^{(l)} (= 1),$$

从此可得

$$\sum_{i \in I_j} (\alpha_i^{(j)} - \alpha_i^*) + \sum_{k \in K_j} \beta_k^{(j)} = \sum_{i \in I_l} (\alpha_i^{(l)} - \alpha_i^*) + \sum_{k \in I_l} \beta_k^{(l)} .$$

今以 t 表这个与 $j \in J$ 无关的公共值:

$$t = \sum_{i \in I_j} (\alpha_i^{(j)} - \alpha_i^*) + \sum_{k \in K_j} \beta_k^{(j)}, \quad j \in J. \tag{8}$$

由 (4) 得

$$0 \leqslant t \leqslant 1. \tag{9}$$

情形一. $t = 0$.

$t = 0$ 成立的充要条件是

$$I_j = \varnothing, j \in J; \quad \beta_k^{(j)} = 0, j \in J, k \in K_j.$$

从前者得

$$\alpha_i^{(j)} = \alpha_i^*, \quad i \in I, j \in J ,$$

故

$$z_1 = \cdots = z_n = \sum_{i \in I} \alpha_i^* a_i \in \sigma_I = (a_0 \cdots a_r),$$

$$z = (z_1, \cdots, z_n) \in \bar{D},$$

因之 2° 成立.

情形二. $t = 1$.

比较 (4) 与 (8)，可知 $t = 1$ 的充要条件是

$$\alpha_i^{(j)} = \begin{cases} 0, & i \notin I_j, \\ a_i^{(j)} - \alpha_i^*, & i \in I_j. \end{cases}$$

这时 z_j 变为

$$z_j = \sum_{i \in I_j} (\alpha_i^{(j)} - \alpha_i^*) a_i + \sum_{k \in K_j} \beta_k^{(j)} b_k^{(j)} \in \sigma_{I_j} \circ \sigma'{}_j, \quad j \in J,$$

且由此可知 I_j, K_j 不能同时为 $\varnothing$ 而 $(I_1, \cdots, I_n; K_1, \cdots, K_n)$ 为一分组. 故

$$z = (z_1, \cdots, z_n) \in (\sigma_{I_1} \circ \sigma_1') \times \cdots \times (\sigma_{I_n} \circ \sigma_n') \subset \bar{E},$$

而 2° 成立.

情形三. $0 < t < 1$.

这时可用以下各式定义 $\lambda_i^{(j)}, \mu_k^{(j)}, \delta_i$:

$$t\lambda_i^{(j)} = \alpha_i^{(j)} - \alpha_i^*, \quad i \in I_j, j \in J, \tag{10_1}$$

$$t\mu_k^{(j)} = \beta_k^{(j)}, \quad k \in K_j, \quad j \in J, \tag{$10)_2$}$$

$$(1-t)\delta_i = \alpha_i^*, \quad i \in I. \tag{$10)_3$}$$

由此可知

$$0 \leqslant \mu_k^{(j)}, \delta_i, \quad i \in I, j \in J, k \in K_j, \tag{$11)_1$}$$

$$\lambda_i^{(j)} > 0, \quad i \in I_j, i \in J. \tag{$11)_2$}$$

又由 (8) 与 $(10)_1$—$(10)_3$，得

$$t = \sum_{i \in I_j} (\alpha_i^{(j)} - \alpha_i^*) + \sum_{k \in K_j} \beta_k^{(j)} = t \sum_{i \in I_j} \lambda_i^{(j)} + t \sum_{k \in K_j} \mu_k^{(j)}, \quad j \in J,$$

故

$$\sum_{i \in I_j} \lambda_i^{(j)} + \sum_{k \in K_j} \mu_k^{(j)} = 1, \quad j \in J. \tag{12}$$

$$\lambda_i^{(j)}, \mu_k^{(j)} \leqslant 1, \quad j \in J, k \in K_j \tag{11$_3$}$$

特别 (12) 式说明 $I_j;K_j(j \in J)$ 不能同时为 $\varnothing$ 而由 (7)，$(I_1, \cdots, I_n; K_1, \cdots, K_n)$ 是一分组. 由 (11)，(12) 又知

$$y_j = \sum_{i \in I_j} \lambda_i^{(j)} a_i + \sum_{k \in K_j} \mu_k^{(j)} b_k^{(j)} \in \sigma_{I_j} \circ \sigma'_j, \quad j \in J, \tag{13}$$

$$y = (y_1 \cdots, y_n) \in (\sigma_{I_1} \circ \sigma'_1) \times \cdots \times (\sigma_{I_n} \circ \sigma'_n) \subset \bar{E}. \tag{13$'$}$$

由 $(10)_1, (10)_3$，(6)，

$$\begin{cases} t\lambda_i^{(j)} + (1-t)\delta_i = \alpha_i^{(j)}, & i \in I_j, j \in J \\ (1-t)\delta_i = \alpha_i^{(j)}, & i \notin I_j, j \in J. \end{cases} \tag{14}$$

将 (14) 各式相加，并由 (4)，$(10)_2$ 及 (12)，对于 $j \in J$，得

$$\begin{aligned}(1-t)\sum_{i \in I} \delta_i &= \sum_{i \in I} \alpha_i^{(j)} - t \sum_{i \in I_j} \lambda_i^{(j)} \\ &= 1 - \sum_{k \in K_j} \beta_k^{(j)} - t \sum_{i \in I_j} \lambda_i^{(j)} \\ &= 1 - t \sum_{k \in K_j} \mu_k^{(j)} - t \sum_{i \in I_j} \lambda_i^{(j)} \\ &= 1 - t,\end{aligned}$$

因之

$$\sum_{i \in I} \delta_i = 1, \tag{15}$$

$$\delta_i \leqslant 1, \quad i \in I, \tag{11$_4$}$$

$$x_j = \sum_{i \in I} \delta_i a_i \in (a_0, \cdots, a_r), \tag{16}$$

$$x = (x_1, \cdots, x_n) \in \bar{D}. \tag{16$'$}$$

由 (3)，$(3)'$，$(10)_2$，(13)，$(13)'$，(14)，(16)，$(16)'$ 诸式，得

$$z = (1-t)x + ty, \tag{17}$$

此处 $x \in \tilde{D}$, $y \in \bar{E}$, $0 < t < 1$, 因之 2° 成立.

由情形一至三，2° 已完全证明. 今证 3° 如次:

假定 z 如 (3)，$(3)'$ 及 (4) 所示而可表成形状 (17)，其中 $x \in \bar{D}$ 由 (16) 与 $(16)'$ 所定，而 $y \in \bar{E}$ 则由 (13)，$(13)'$ 所定. 我们并假定 z 不 $\in \bar{D}$ 也不 $\in \bar{E}$，

因而 $0 < t < 1$. 此外 $\mu_k^{(j)}$, δ_i 等满足 $(11)_1, (11)_3, (11)_4$，(12)，(15) 等式，且可假定 $(11)_2$ 也成立，而无损于一般性，因为若有不满足于 $(11)_2$ 的 $\lambda_i^{(j)}$，我们可把使 $\lambda_i^{(j)} = 0$ 的 i 从 I_j 中取消，如果这种缩小的 $I_j(j \in J)$ 仍用 I_j 来表示，那么 $(I_1, \cdots, I_n, K_1, \cdots, K_n)$ 仍是一个分组. 这时，(7) 式成立.

对于每一 $i \in I$，用 (5) 式来定义 α_i^*. 由 (17) 可得 (14). 若 $i \in I$ 已定，那么由于 (7) 式，将至少有一个 $j \in J$ 使 i 不 $\in I_j$，而对于任意这样的 j 由 (14) 可得

$$(1-t)\delta_i = \alpha_i^{(j)}, \quad i \notin I_j. \tag{18}$$

另外由 (14) 与 $(11)_2$ 又有

$$(1-t)\delta_i < \alpha_i^{(j)}, \quad i \in I_j. \tag{19}$$

因之

$$(1-t)\delta = \min_{j \in J} \alpha_i^{(j)} = \alpha_i^*, \quad j \in I, \tag{20}$$

而 (18)，(19) 变为

$$\alpha_i^{(j)} \begin{cases} > \alpha_i^*, & i \in I_j, \\ = \alpha_i^*, & i \notin I_j. \end{cases} \tag{21}$$

(21) 式说明 $I_j, j \in J$ 乃是由所与的点 $z = (z_1, \cdots, z_n)$ 所唯一地决定的

由 (14)，(20) 可得 $(10)_1, (10)_3$. 由 (13)，(13)′，(16)，(16)′ 和 (17) 可得 $(10)_2$，又由 $(10_1, (10)_2$ 和 (12) 可得 (8)，而 (8) 式说明 t 也是由所与的的点 z 所唯一地决定的. 而由 (10)，(13)，(13)′，(16)，(16)′ 诸式 $x \in \bar{D}, y \in \bar{E}$ 也由 z 所完全决定.

因此，形状如 (17) 的分解乃是唯一的，而 3° 得证. 这样，预备定理 1 已完全证明.

预备定理 2 $\bar{K}'$ 有一剖分 K，以对角复合形 L 为一子复合形.

证 令 $K_1'^{(+)}$ 是 K' 中一切作 $\sigma_1 \times \cdots \times \sigma_n$ 形状的凸胞腔所成的复合形，此处 $\sigma_j \in L, (j = 1, \cdots, n)$，而 $\sigma_1, \cdots, \sigma_n$ 无公共顶点. 另一面，对于 L 中具有公共顶点的任意 n 个 (依照一定次序的) 单纯形 $\sigma_1, \cdots, \sigma_n$，可依预备定理 1 作一 $\sigma_1 \times \cdots \times \sigma_n$ 的剖分 $F = D \circ E$，记之为 $F(\sigma_1, \cdots, \sigma_n)$，$D$ 和 E 也各记之为 $D(\sigma_1, \cdots, \sigma_n)$ 和 $E(\sigma_1, \cdots, \sigma_n)$. 要证明预备定理 2，只须证明 $K_1'^{(+)}$ 与所有的 $F(\sigma_1, \cdots, \sigma_n)$ 合成为一个 $\bar{K}'$ 的剖分即可.

首先，对于一组有公共顶点的 $\sigma_1, \cdots, \sigma_n \in L$ 所作成的 $\sigma_1 \times \cdots \times \sigma_n$ 的剖分 $F = D \circ E$，与 $K_1'^{(+)}$ 的公共部分，就是属于 E 的那些凸胞腔，而且 $\bar{E} = \bar{K}_I'^{(+)} \cap (\sigma_1 \times \cdots \times \sigma_n)$.

其次，假定 $\sigma_1 \times \cdots \times \sigma_n \in K'$ 而 σ_j 如 (1) 所示，又 $\tau_1 \times \cdots \times \tau_n \in K'$ 而

$$\tau_j=(c_0\cdots c_s d_1^{(j)}\cdots d_{l_j}^{(j)}),\quad j\in J,$$
$$\tau_1\cap\cdots\cap\tau_n=(c_0\cdots c_s).$$

此处 $s\geqq 0$，则 $\sigma_1\times\cdots\times\sigma_n$ 与 $\tau_1\times\cdots\times\tau_n$ 有不属于 $\bar{K}_1'^{(+)}$ 的公共点的充要条件乃是 $(a_0,\cdots,a_r)$ 与 $(c_0,\cdots,c_s)$ 须有公共的点. 设 $(a_0,\cdots,a_r)$ 与 $(c_0,\cdots,c_s)$ 的公共边为 $(c_0,\cdots,e_r)$，$t\geqq 0$，而 σ_j 与 τ_j 的公共边为

$$\zeta_j=(e_0,\cdots,e_t f_1^{(j)},\cdots,f_{mj}^{(j)}),\quad j\in J.$$

那么 $\zeta_1,\cdots,\zeta_n$ 的公共顶点就是 $c_0,\cdots,e_t$, 而

$$(\sigma_1\times\cdots\times\sigma_n)\cap(\tau_1\times\cdots\times\tau_n)=(\sigma_1\cap\tau_1)\times\cdots\times(\sigma_n\cap\tau_n)=\zeta_1\times\cdots\times\zeta_n,$$

我们在下面将证明 $F(\zeta_1,\cdots,\zeta_n)$ 的每一胞腔，必同时也是 $F(\sigma_1,\cdots,\sigma_n)$ 和 $F(\tau_1,\cdots,\tau_n)$ 的胞腔反之亦然，即 $F(\zeta_1,\cdots,\zeta_n)$ 是复合形 $F(\sigma_1,\cdots,\sigma_n)$ 与 $F(\tau_1,\cdots,\tau_n)$ 的交. 为这个目的，假定 $(I_1',\cdots,I_n';M_1',\cdots,M_n')$ 是对于 $I'=\{0,1,\cdots,t\}$ 以及 $M_j=\{1,\cdots,m_j\}$ 所作的一个分组 $\zeta_j'=(f_1^{(j)}\cdots f_{m_j}^{(j)})$, 而

$$\zeta=(\zeta_{I'_1}\circ\zeta'_{M'_1})\times\cdots\times(\zeta_{I'_n}\circ\zeta'_{M'_n}),\quad (I_j'\subset I',M_j'\subset M_j)$$

是一个像 (2) 那样的 $E(\zeta_1,\cdots,\zeta_n)$ 中的一个胞腔. 对于每一 $j\in J$, $\zeta_{I'_j}\circ\zeta'_{M'_j}$ 必作 $\sigma_{T_j}\circ\sigma'_{K'_j}$ 的形状，此处 $K_j'\subset K_j$，$I_j\subset I$，I，K_j 等依 (1) 那样定义：如果有一个 $i\in I_1\cap\cdots\cap I_n$，那么 a_i 将是每一 $\zeta_{I'_j}\circ\zeta_j'$ 的顶点，因而也是每一 $\zeta_{I'_j}$ 的顶点，由 $(I_1',\cdots,I_n';M_1',\cdots,M_n')$ 是对于 I' 和 $M_j(i\in J)$ 的分组的假定，这不可能. 因此 $I_1\cap\cdots\cap I_n=\varnothing$，而对于每一 j，因 I_j'，M_j' 不同时为 $\varnothing$，故 I_j，K_j' 也不能同时为 $\varnothing$，由此知 $(I_1,\cdots,I_n;K_1',\cdots,K_n')$ 是对于 I 和 $K_j(j\in J)$ 的一个分组，而 $\xi=(\sigma_{I_1}\circ\sigma'_{K'_1})\times\cdots\times(\sigma_{I_n}\circ\sigma'_{M'_n})$ 乃是 $E(\sigma_1,\cdots,\sigma_n)$ 的一个胞腔. 这样, 每个 $E(\zeta_1,\cdots,\zeta_n)$ 的胞腔也是 $E(\sigma_1,\cdots,\sigma_n)$ 的胞腔，同样也是 $E(\tau_1,\cdots,\tau_n)$ 的胞腔，显然 $D(\zeta_1,\cdots,\zeta_n)$ 的胞腔必同时是 $D(\sigma_1,\cdots,\sigma_n)$ 和 $D(\tau_1,\cdots,\tau_n)$ 的胞腔. 因此 $F(\zeta_1,\cdots,\zeta_n)=D(\zeta_1,\cdots,\zeta_n)\circ E(\zeta_1,\cdots,\zeta_n)$ 的胞腔也必然同时是 $F(\sigma_1,\cdots,\sigma_n)=D(\sigma_1,\cdots,\sigma_n)\circ E(\sigma_1,\cdots,\sigma_n)$ 和 $F(\tau_1,\cdots,\tau_n)=D(\tau_1,\cdots,\tau_n)\circ E(\tau_1,\cdots,\tau_n)$ 的胞腔. 它的反面也容易看出. 这证明了上面所说的话.

从上面所说的两点，就得到了预备定理 2.

今如定理 2, 设 P 是一个有限可剖分空间，${}_TL$ 是 P 的一个单纯剖分，$Q=P^n$ 而 $K'=L^n$. 定义 $T:\bar{K}'\to Q$ 如 $T(z_1,\cdots,z_n)=(Tz_1,\cdots,Tz_n)$, $z_j\in\bar{L}$. 从预备定理 2 中所说到的剖分 K 的定义，显然 K 和它的子复合形 L 成一个正则的复合形偶而 $({}_TK,{}_TL)$ 为 (Q,P) 的一个正则剖分，因此我们得到了定理 4. 剖分 ${}_TK$ 或 (K) 以后将称为 $Q=P^n$ 的一个“标准剖分”.

今设 G_n 是 n 个整数 $1,\cdots,n$ 上的置换群，对于每一 $\omega\in G_n,(z_1,\cdots,z_n)\to(z_{\omega(1)},\cdots,z_{\omega(n)})$, $z_j\in P$, 定义了一个 $Q=P^n$ 的拓扑变换 g_ω，而且 g_ω 也引起了一个变换 (仍记以 g_ω) 把标准剖分 K 的胞腔变为 K 的胞腔. 因之，对于 G_n 的任一子群 Γ, Q 有一个"法空间"Q_Γ(即把 Q 中凡可用 Γ 的元素互相变换而得的点叠合成一点所得到的空间), K 也有一个"法复合形" K_Γ(即把 K 中凡可用 Γ 的元素所引起的胞腔变换互相变换而得的胞腔叠合成一个胞腔所得到的复合形)，而且 $_T:\bar{K}\to Q$ 引起了一个拓扑映像 $_{T\Gamma}:\bar{K}_\Gamma\to Q_\Gamma$ 使 $_{T\Gamma}K_\Gamma$ 是 Q_Γ 的一个剖分. 在自然地定义的投影 $\pi:Q\to Q_\Gamma$(或 $K\to K_\Gamma$) 之下，Q(或 K) 的对角空间 P(或对角复合形 L) 被投影而为一个与 P(或 L) 同拓 (或同构) 的子空间 (或子复合形)，兹仍记之为 P(或 L). 那么定理 4 可作下列补充.

定理 4′ 对于任一 G_n 的子群 $\Gamma,(Q_\Gamma,P)$ 是一个正则的可剖分空间，以 $(_{T\Gamma}K_\Gamma,{}_TL)$ 为一正则剖分，此处 $_{T\Gamma}:\bar{K}_\Gamma\to Q_\Gamma$ 定义如 $_{T\Gamma}\pi(z_1,\cdots,z_n)=(_Tz_1,\cdots,{}_Tz_n)$，式中 π 为 $_T\bar{K}\to T_\Gamma\bar{K}_\Gamma$ 即 $Q\to Q_\Gamma$ 的投影.

在 $\Gamma=1$ 时，$_{T\Gamma}$ 将简记为 T，此处 $_T:\bar{K}\to Q$ 定义如 $_T(z_1,\cdots,z_n)=(z_1,\cdots,z_n)$, $z_j\in\bar{L}$.

根据§4, 从 (Q_Γ,P) 所作的空间组 $(Q_\Gamma,{}_{T\Gamma}\bar{K}^{(+)}_{\Gamma,r},{}_{T\Gamma}\bar{K}^{(-)}_{\Gamma,r},{}_{T\Gamma}\bar{K}^{(0)}_{\Gamma,r},P)$ $(0<r<1)$, 的任一同伦不变量, 一定是 (Q_Γ,P) 的拓扑不变量, 因之也就是 P 的拓扑不变量下面的简单的例，可以说明这种拓扑不变量，一般并非是同伦的不变量.

例: 设 L 是一个一维单纯复合形，有四个顶点 o,a,b,c，三个一维胞腔 (oa)，(ob) 和 (oc). 取 $n=2$，K 是 $K'=L^2=L\times L$ 的标准剖分，则对于正则复合形偶 (K,L) 而言，$K_1^{(+)}$ 是一个一维的复合形有顶点 12，一维胞腔 12. $K_1^{(+)}$ 的一维下同调群是一个无限的循环群. 下面的下闭链所定的下同调类作成这个一维同调群的一个母素 (一维胞腔都已定向, 如记号所示):

$$\begin{aligned}z=&a\times(bo)+a\times(oc)+b\times(co)+b\times(oa)+c\times(ao)+c\times(ob)\\&+(bo)\times a+(oc)\times a+(co)\times b+(oa)\times b+(ao)\times c+(ob)\times c.\end{aligned}$$

其次. 假定 L 是一个一维的单纯复合形，由一个一维单纯形和它的两个顶点所成，取 $n=2$，K 是 $K'=L^2=L\times L$ 的标准剖分. 对于正则复合形偶 (K,L) 而言，$K_1^{(+)}$ 是一个 o 维的复合形，有顶点 2，而 $K_1^{(+)}$ 的一维下同调群显然是 o.

由此可见. 上面的两个可剖分空间 $\overline{L}$ 虽有同样的同伦型，甚至都可收缩成一个点，两个 $\overline{K_1^{(+)}}$ 的一维同调群并不一一同构.

但据本节以及§4 所论，若 L 是一个单纯复合形，K 是 $L^2=L\times L$ 的标准剖分，则 $K_1^{(+)}$ 的一切同伦不变量特别如下同调群乃是可剖分空间 $\overline{L}$ 的拓扑不变量. 上面的例说明，这样的拓扑不变量通常并非同伦不变量.

§6 可剖分空间的 n 乘 Euler-Poincaré 示性数

假定 P 是一个可剖分空间，整数 $n \geqq 1$ 而 $Q = P^n$. ${}_TL$ 是 P 的一个单纯剖分，则由§5，Q 有一个标准剖分 ${}_TK$，以 ${}_TL$ 为子复合形而 ${}_TL \Subset_T K$，此处 $T : \bar{K} \to Q$ 定义如 ${}_T(z_1, \cdots, z_n) = ({}_Tz_1, \cdots, {}_Tz_n)$, $z_j \in \bar{L}$ 对于任一 > 0 且 < 1 的 r，从正则可剖分空间偶 (Q, P) 所得出的空间 ${}_T\bar{K}_r^{(+)}$, ${}_T\bar{K}_r^{(-)}$, ${}_T\bar{K}_r^{(0)}$ 的 Euler-Poincaré 示性数 (以下简称示性数) 都是 (Q, P), 因之也是 P 的拓扑不变量，我们将它们依次叫做 P 的 “n 乘 Euler-Poincaré 外、内和中示性数”，或简称 P 的“n 乘外、内和中示性数”，并记之为 $\chi_n^{(+)}(P), \chi_n^{(-)}(P)$ 和 $\chi_n^{(0)}(P)$. 本节的目的，在获得一些计算这些不变量的约化公式.

首先，${}_T\bar{K}_r^{(-)}$ 以 ${}_T\bar{L}$ 为它的一个变状收缩核，因之 ${}_T\bar{K}_1^{(-)}$ 和 ${}_T\bar{L}$ 即 P 的示性数相同，即

$$\chi_n^{(-)}(P) = \chi(P), \tag{I}$$

此处 χ 表 Euler-Poincaré 示性数.

其次，由§2 的 (1)，(2)，(2)′ 诸式，可得

$$\chi({}_T\bar{K}_r^{(+)}) + \chi({}_T\bar{K}_r^{(-)}) - \chi({}_T\bar{K}_r^{(0)}) = \chi({}_T\bar{K}).$$

因 ${}_T\bar{K} = Q = P^n$, $\chi({}_T\bar{K}) = \chi(P^n) = [\chi(P)]^n$, 故上式可由 (I) 书为

$$\chi_n^{(+)}(P) - \chi_n^{(0)}(P) = [\chi(P)]^n - \chi(P). \tag{II}$$

为了求得 $\chi_n^{(+)}(P)$ 和 $\chi_n^{(0)}(P)$ 的约化公式，我们先定义任意一个单纯复合形 L 以及它的扩大复合形 L_ε 的 n 乘外示性数如次：首先，我们将定义 $\chi_n^{(+)}(L)$ 为 $\chi_n^{(+)}(\bar{L})$. 如前令 K 为 L^n 的标准剖分，$\bar{K}_r^{(+)}$, $\bar{K}_r^{(-)}$, $\bar{K}_r^{(0)}$ 也像前面那样定义. 因为 $\bar{K}_r^{(+)}(0 < r < 1)$ 以 $\overline{K}_1^{(+)}$ 为它的一个变状收缩核，故 $\chi_n^{(+)}(L) = \chi_n^{(+)}(\bar{L}) = \chi(\bar{K}_r^{(+)}) = \chi(\bar{K}_1^{(+)})$. 由§5，复合形 $K_1^{(+)}$ 系由一切作 $\sigma_1 \times \cdots \times \sigma_n$ 形状的胞腔所组成，此处 $\sigma_j \in L$，而 $\sigma_1, \cdots, \sigma_n$ 无公共顶点. 因此，$\chi_n^{(+)}(L)$ 也可以用下面的方法来定义它. 令一切 L(或 L_ε) 中 n 个单纯形 $\sigma_1, \cdots, \sigma_n$ 所成的 (有次序的) 组的集合为 $\mathfrak{R}(L)$(或 $\mathfrak{R}(L_\varepsilon)$). 在 $\mathfrak{R}(L)$(或 $\mathfrak{R}(L_\varepsilon)$) 中使 $\dim\sigma_1 + \cdots + \dim\sigma_n = i$ 的组 $(\sigma_1, \cdots, \sigma_n)$ 的子集记为 $\mathfrak{R}_i(L)$(或 $\mathfrak{R}_i(L_\varepsilon)$)，在 $\mathfrak{R}_i(L)$ 或 $\mathfrak{R}_i(L_\varepsilon)$) 中一切使 $\sigma_1, \cdots, \sigma_n$ 无公共顶点的那些组 $(\sigma_1, \cdots, \sigma_n)$ 的个数为 $h_{n,i}(L)$(或 $h_{n,i}(L_\varepsilon)$)，那么 $\chi_n^{(+)}(L)$ 显然可定义为

$$\chi_n^{(+)}(L) = \sum_{i \geqq 0} (-1)^i h_{n,i}(L). \tag{1}$$

同样，我们用类似的方法来定义 $\chi_n^{(+)}(L_\varepsilon)$, 即

$$\chi_n^{(+)}(L_\varepsilon)=\sum_{i\geqq -n}(-1)^i h_{n,i}(L_\varepsilon). \tag{2}$$

对于任意单纯复合形 L，有

$$\chi_n^{(+)}(L_\varepsilon)=\chi_n^{(+)}(L)+[\chi(L)-1]^n-[\chi(L)]^n. \tag{III}$$

证 命 $\mathfrak{R}_i(L)$ 中的个数为 $h'_{n,i}(L)$. 则 $\mathfrak{R}_i(L_\varepsilon)$ 中使 $\sigma_1,\cdots,\sigma_n$ 无公共顶点且其中恰有 j 个为 ε 的组 $(\sigma_1,\cdots,\sigma_n)$ 的个数显然在 $j\geqq 1$ 时为 $\binom{n}{j}h'_{n-j,i+j}(L)$, 在 $j=0$ 时为 $h_{n,i}(L)$. 因之，

$$h_{n,i}(L_\varepsilon)=h_{n,i}(L)+\sum_{j\geqq 1}\tbinom{n}{j}h'_{n-j,i+j}(L). \tag{3}$$

由 (1)，(2) 及 (3)，得

$$\begin{aligned}\chi_n^{(+)}(L_\varepsilon)&=\sum_{i\geqq 0}(-1)^i h_{n,i}(L)+\sum_{j\geqq 1}\Big[(-1)^j\tbinom{n}{j}\sum_{i\geqq -n}(-1)^{i+j}h'_{n-j,i+j}(L)\Big]\\&=\chi_n^{(+)}(L)+\sum_{j\geqq 1}\Big[(-1)^j\tbinom{n}{j}\sum_{i\geqq 0}(-1)^i h'_{n-j,i}(L)\Big]\\&=\chi_n^{(+)}(L)+\sum_{j\geqq 1}(-1)^j\tbinom{n}{j}\chi(L^{n-j})\\&=\chi_n^{(+)}(L)+\sum_{j\geqq 1}(-1)^j\tbinom{n}{j}[\chi(L)]^{n-j}\\&=\chi_n^{(+)}(L)+[\chi(L)-1]^n-[\chi(L)]^n,\end{aligned}$$

即 (III) 式.

今设 P，$Q=P^n$，以及 ${}_TK$，${}_TL$ 等如本节开首所示. 对于 L 中的每一单纯形 σ，命 L_σ 系由 L 中所有这样的单纯形 σ' 所组成的子复合形，σ' 与 σ 无公共顶点，但 $\sigma\circ\sigma'$ 有意义且 $\in L$. 我们又以 $L_{\sigma\varepsilon}$ 表 L_σ 被扩大了的复合形，则有 ①

$$\chi_n^{(+)}(P)=[\chi(P)]^n-\sum_{\sigma\in L}(-1)^{n(\dim\sigma+1)}\chi_n^{(+)}(L_{\sigma\varepsilon}), \tag{IV}$$

$$\chi_n^{(0)}(P)=\chi(P)-\sum_{\sigma\in L}(-1)^{n(\dim\sigma+1)}\chi_n^{(+)}(L_{\sigma\varepsilon}). \tag{V}$$

①若 σ 不是 L 中任一维数更高的单纯形的边，即 L_σ 不存在时，$L_{\sigma\varepsilon}$ 指只含有一个 -1 维单纯形 ε 的复合形.

证　由 (II) 式，我们只需证明 (IV) 式即可.

显然，在 $\mathfrak{R}_i(L)$ 中使 $\sigma_1,\cdots,\sigma_n$ 恰有一个 r 维的、边 σ 公共的一切组 $(\sigma_1,\cdots,\sigma_n)$ 的个数是 $h_{n,i-n(r+1)}(L_{\sigma\varepsilon})$. 因之 L^n 中 i 维胞腔的个数是

$$h_{n,i}(L)+\sum_{\sigma\in L}h_{n,i-n(\dim\sigma+1)}(L_{\sigma\varepsilon}),$$

故

$$\chi(L^n)=\sum_{i\geqq 0}(-1)^i h_{n,i}(L)+\sum_{\sigma\in L}\sum_{i\geqq 0}(-1)^i h_{n,i-n(\dim\sigma+1)}(L_{\sigma\varepsilon})$$

或

$$[\chi(L)]^n=\chi_n^{(+)}(L)+\sum_{\sigma\in L}(-1)^{n(\dim\sigma+1)}\sum_{i\geqq -n}(-1)^i h_{n,i}(L_{\sigma\varepsilon}),$$

由 (2)，上式即为 (IV) 式.

在 (IV)，(V) 中出现的 $\chi_n^{(+)}(L_{\sigma\varepsilon})$ 可由 (III) 式化为 $\chi_n^{(+)}(L_\sigma)$ 去求，而 L_σ 的维数显较原来的 L 的维数为低，因之 (IV)，(V) 可视为一组求 $\chi_n^{(+)}(P)$ 和 $\chi_n^{(0)}(P)$ 的约化公式. 下节中将举例说明这些公式的用法.

§7　杂　　例

例一　一维复合形.

假定 L 是一个一维的单纯复合形，它的一维单纯形的个数是 a'，恰为 p 个一维单纯形之端的顶点 (以下简称 p 枝点) 的个数是 $\alpha_p^0(p\geqq 0)$. 今求 $\chi_n^{(+)}(\bar{L})$ 与 $\chi_n^{(0)}(\bar{L})$ 如次:

对于每一一维的单纯形 σ, $L_{\sigma\varepsilon}$ 只有一个胞腔，即 ε，故 $\chi_n^{(+)}(L_{\sigma\varepsilon})=(-1)^n$.

对于每一 $p\geqq 1$ 的 p 枝点 σ，L_σ 系由 p 个孤立的点所组成，因之 $\chi(L_\sigma)=p$，$\chi_n^{(+)}(L_\sigma)=p^n-p$, 而由§6(III),

$$\chi_n^{(+)}(L_{\sigma\varepsilon})=\chi_n^{(+)}(L_\sigma)+[\chi(L_\sigma)-1]^n-[\chi(L_\sigma)]^n=(p^n-p)+(p-1)^n-p^n,$$

或

$$\chi_n^{(+)}(L_{\sigma\varepsilon})=(p-1)^n-p.$$

此式在 $p=0$ 时 (亦即 σ 是 L 中孤立的顶点时) 也显然成立.

应用§6(V)，得

$$\chi_n^{(0)}(\bar{L})=\chi(\bar{L})-\sum_{p\geqq 0}(-1)^n[(p-1)^n-p]\alpha_p^0-(-1)^n\alpha^1.$$

因 $\chi(\bar{L}) = -\alpha^1 + \sum\limits_{p \geqq 0} \alpha_p^0$，故上式亦可改书为

$$\chi_n^{(0)}(\bar{L}) = [1 + (-1)^n]\chi(\bar{L}) + (-1)^{n+1} \sum_{p \neq 2} (p-1)[(p-1)^{n-1} - 1]\alpha_p^0, \tag{1}$$

又由§6(IV) 或 (II) 得

$$\chi_n^{(+)}(\bar{L}) = [\chi(\bar{L})]^n + (-1)^n \chi(\bar{L}) + (-1)^{n+1} \sum_{p \neq 2} (p-1)[(p-1)^{n-1} - 1]\alpha_p^0. \tag{2}$$

根据[3]，第五章，在 $p \neq 2$ 时，α_p^0 都是 $\bar{L}$ 的拓扑不变量. (1)，(2) 两式说明一个一维可剖分空间的任一乘示性数，可由它的示性数以及它的一切 $p \neq 2$ 和 1 的不变量 α_p^0 所完全决定. 反过来，我们也可以证明任一 $p \neq 2$ 和 1 的不变量 α_p^0, 系由一切 n 乘中 (或外) 示性数和示性数及 α_1^0 所完全决定. 证之如下：

设 L 和 L' 是两个一维单纯复合形，有相同的示性数，相同的 1 枝点的个数，而所有的 n 乘示性数也相同. 假定有一整数 $q > 2$ 使 $p > q$ 时 $\alpha_p^0(\bar{L}) = \alpha_p^0(\bar{L}')$ 而 $\alpha_q^0(\bar{L}) > \alpha_q^0(\bar{L}') \geqq 0$，因为在 $p < q$ 时，$\lim\limits_{n \to \infty} \left(\frac{p-1}{q-1}\right)^n = 0$，故可取 n 充分大使

$$\begin{aligned}&\left|\sum_{2 \neq p < q} (p-1)[(p-1)^{n-1} - 1][\alpha_p^0(\bar{L}) - \alpha_p^0(\bar{L}')]\right| \\ <&\left|(q-1)[(q-1)^{n-1} - 1][\alpha_q^0(\bar{L}) - \alpha_q^0(\bar{L}')]\right|.\end{aligned}$$

因为我们已假定 $\chi(\bar{L}) = \chi(\bar{L}')$，故对于这样的 n，据 (1) 式及 (2) 式将有 $\chi_n^{(0)}(\bar{L}) \neq \chi_n^{(0)}(\bar{L}')$ 及 $\chi_n^{(+)}(\bar{L}) \neq \chi_n^{(+)}(\bar{L}')$，与假设相反. 因之对于每一 $p > 2$，应有 $\alpha_p^0(\bar{L}) = \alpha_n^0(\bar{L}')$，再从 (1) 和 (2) 又可得 $\alpha_0^0(\bar{L}) = \alpha_0^0(\bar{L}')$ 而证明完毕.

以上的结果说明一切 $p \neq 2$ 的不变量 α_p^0 以及示性数 χ 的集合与一切 $n > 1$ 的 $\chi_n^{(+)}$(或 $\chi_n^{(0)}$) 及 χ 与 α_1^0 等不变量所组成的集合"同值". 在这方面可参看§8 末所举的例.

例二 P_m 是一个 m 维的单纯形，则

$$\chi_n^{(0)}(P_m) = \chi_n^{(+)}(P_m) = 1 + (-1)^{m(n+1)+1}. \tag{3}$$

证 在 $m = 0$ 时此式显然成立，又由例 1，在 $m = 1$ 时此式也成立. 今假定上式在 $m < r$ 时成立.

取 P_r 的一个剖分 $_TL$，此处 L 即由一 r 维的欧氏单纯形及其边所组成. 若 σ 为 L 中一个 s 维的单纯形 $(r \geqq s \geqq 0)$，则 $\bar{L}_\sigma$ 为一 $r-s-1$ 维的单纯形，由归纳假定得 $(r > s \geqq 0)$

$$\chi_n^{(+)}(L_\sigma) = \chi_n^{(+)}(\bar{L}_\sigma) = 1 + (-1)^{(r-s-1)(n+1)+1},$$

又由§6(III)，

$$\begin{aligned}\chi_n^{(+)}(L_{\sigma\varepsilon}) &= \chi_n^{(+)}(L_\sigma) + [\chi(L_\sigma) - 1]^n - [\chi(L_\sigma)]^n \\ &= (-1)^{(r-s-1)(n+1)+1}.\end{aligned}$$

而此式不特在 $r > s \geqq 0$ 时，即在 $s = r$ 时也显然成立. 在 L 中 s 维单纯形的个数是 $\begin{pmatrix} r+1 \\ s+1 \end{pmatrix}$，故由§6(V)，

$$\begin{aligned}\chi_n^{(0)}(P_r) &= \chi(P_r) - \sum_{\sigma\in L}(-1)^{n(\dim\sigma+1)}\chi_n^{(+)}(L_{\sigma\varepsilon}) \\ &= 1 - \sum_{s=0}^{r}(-1)^{r(n+1)+s}\begin{pmatrix} r+1 \\ s+1 \end{pmatrix} \\ &= 1 + (-1)^{r(n+1)}\left[\sum_{i=0}^{r+1}(-1)^i\begin{pmatrix} r+1 \\ i \end{pmatrix} - 1\right] \\ &= 1 + (-1)^{r(n+1)+1},\end{aligned}$$

又因 $\chi(P_r) = 1$，由§6(II)，

$$\chi_n^{(+)}(P_r) = \chi_n^{(0)}(P_r).$$

故由归纳法得知 (3) 式普遍成立.

例三 P_m 是一个 m 维球面，则

$$\chi_n^{(0)}(P_m) = 0, \tag{4}$$

$$\chi_n^{(+)}(P_m) = [1 + (-1)^m]^n + (-1)^{mn+1}[1 + (-1)^m]. \tag{5}$$

证 在 $m = 0$ 或 1(例 1) 时，上两式易见其成立. 现假定在 $m < r$ 时 (4)，(5) 已成立.

取 P_r 的一个剖分 ${}_TL$，此处 L 为一 $r+1$ 维欧氏单纯形的维数 $\leqslant r$ 的边所成的复合形. 对于 L 中任一 s 维的复合形 $\sigma(r \geqq s \geqq 0)$，$\bar{L}_\sigma$ 是一个 $r-s-1$ 维的球面，故由归纳假定 $(r > s \geqq 0)$

$$\chi_n^{(+)}(L_\sigma) = \chi_n^{(+)}(\bar{L}_\sigma) = [1 + (-1)^{r-s-1}]^n + (-1)^{(r-s-1)n+1}[1 + (-1)^{r-s-1}],$$

此外 $\chi(L_\sigma) = 1 + (-1)^{r-s-1}$，故由§6(III) 得

$$\chi_n^{(+)}(L_{\sigma\varepsilon}) = (-1)^{(r-s-1)n+r-s}.$$

而此式不特在 $r > s \geqq 0$ 时，即在 $r = s$ 时也成立，因 L 中 s 维单纯形的个数是

$\binom{r+2}{s+1}$, $r \geqq s \geqq 0$, 故由§6(V),

$$\begin{aligned}\chi_n^{(0)}(P_r) =&\chi(P_r) - \sum_{s=0}^{r}(-1)^{n(s+1)}\binom{r+2}{s+1}\cdot(-1)^{(r-s-1)n+r-s}\\ =&1+(-1)^r+(-1)^{r(n+1)}\sum_{s=0}^{r}(-1)^{s+1}\binom{r+2}{s+1}\\ =&[1+(-1)^r][1+(-1)^{r(n+1)+1}].\end{aligned}$$

若 r 为奇数，则 $1+(-1)^r=0$，若 r 为偶数，则 $1+(-1)^{r(n+1)+1}=0$，故 $\chi_n^{(0)}(P_r)$ 恒为 0，而 (4) 式成立. 由 (4) 及§6(II) 可得 (5) 式.

例四 假定 M 是一个 m 维的闭流形，具有单纯剖分 ${}_TL$ 使对于 L 中任一 r 维的单纯形 $\sigma(m>r\geqq 0)$, $\bar{L}_\sigma$ 是一个 $m-r-1$ 维的球面，则

$$\chi_n^{(0)}(M)=0, \tag{6}$$

$$\chi_n^{(+)}(M)=[\chi(M)]^n-\chi(M). \tag{7}$$

证 设 L 中 r 维单纯形的个数为 a_r. 对于每一这样的 r 维单纯形 $\sigma,(r<m)\bar{L}_\sigma$ 为一 $m-r-1$ 维球面，故由§6(III) 及例 3，得

$$\chi_n^{(+)}(L_{\sigma\varepsilon})=(-1)^{(m+r+1)(n+1)+1}.$$

而此式即在 $r=m$ 也显然真确. 由§6(V)，

$$\begin{aligned}\chi_n^{(0)}(M) &= \chi(M) - \sum_{r=0}^{m}(-1)^{(r+1)n}\cdot(-1)^{(m+r+1)(n+1)+1}a_r\\ &= \chi(M)-(-1)^{m(n+1)}\sum_{r=0}^{m}(-1)^r a_r\\ &= \chi(M)+(-1)^{m(n+1)+1}\chi(M).\end{aligned}$$

在 m 为单数时，$\chi(M)=0$，m 为偶数时，$1+(-1)^{m(n+1)+1}=0$，故 $\chi_n^{(0)}(M)$ 恒 $=0$，而 (6) 式得证. 由§6(II) 及 (6) 式可得 (7) 式.

我们不难看出，(5) 式亦可书为 $\chi_n^{(+)}(P_m)=[\chi(P_m)]^n-\chi(P_m)$, 而为 (7) 式的特例.

§8 古典不变量 Betti 数与新不变量的关系

在本节中，所有下同调群的系数群是一个固定的域.

假定 P 是一个可剖分空间，$Q = P^n(n > 1)$，r 是一个 > 0 而 < 1 的实数. 像§5 那样，设 $P = {}_T\bar{L}$，$Q = {}_T\bar{K}$，此处 K 是被 L 所定的标准剖分 (L 是 P 的单纯剖分). 令 $Q_1 = {}_T\bar{K}_r^{(+)}$, $Q_2 = {}_T\bar{K}_r^{(-)}$, $Q_{12} = {}_T\bar{K}_r^{(0)}$, 此处

$$Q = Q_1 \cup Q_2, \quad Q_{12} = Q_1 \cap Q_2, \tag{1}$$

则据§3 中所论，空间组 (Q, Q_1, Q_2, Q_{12}) 有一组同伦不变量 $q^k(k \geqq 0)$，因之，照§4 的主要定理，q^k 是 (Q, P) 也是 P 的拓扑不变量.

应用 Mayer-Vietoris 定理于 (1) 式，得 $(q^{-1} = 0)$

$$p^k(Q) = p^k(Q_1) + p^k(Q_2) - p^k(Q_{12}) + q^k + q^{k-1}, \quad k \geqq 0. \tag{2}$$

令

$$p^k({}_T\bar{K}_r^{(+)}) = p_{(+)}^{k,n}, \quad p^k({}_T\bar{K}_r^{(-)}) = p_{(-)}^{k,n}, \quad p^k({}_T\bar{K}_r^{(0)}) = p_{(0)}^{k,n}, p^k(P) = p^k,$$

又改书 q^k 为 $q^{k,n}$，则 $p_{(-)}^{k,n} = p^k$. 又由 Künneth 公式得

$$p^{k(P^n)} = \sum_{(k)} p^{k_1} p^{k_2} \cdots p^{k_n}, \quad k \geqq 0,$$

此处和号 $\sum\limits_{(k)}$ 展开在所有 $\geqq 0$ 的 (有一定次序的) 整数组 $k_1, \cdots, k_n$ 上面，而 $k_1 + \cdots + k_n = k$. 因之，(2) 式亦可书为 $(q^{-1,n} = 0)$

$$\sum_{(k)} p^{k_1} p^{k_2} \cdots p^{k_n} = p_{(+)}^{k,n} + p^k - p_{(0)}^{k,n} + q^{k,n} + q^{k-1,n}, \quad k \geqq 0. \tag{3}$$

今取 $n = 2$，则从上式可得

$$(2p^0 - 1)p^k = p_{(+)}^{k,2} - p_{(0)}^{k,2} + q^{k,2} + q^{k-1,2} - \sum_{i=1}^{k-1} p^i p^{k-i}, \quad k \geqq 0. \tag{4}$$

因 $2p^0 - 1 \geqq 1$，故由对 k 的归纳可将 p^k 用这些新不变量 $p_{(+)}^{i,21}, p_{(0)}^{i,2}$ 和 $q^{i,2}(i \geqq 0)$ 表示出来，得

$$p^k = P_k(p_{(+)}^{i,2}, p_{(0)}^{i,2}, q^{i,2}), \quad k \geqq 0, \tag{5}$$

其中 P_k 在 $p^0 = 1$ 时是一个多项式.

从下面的例可以知道 $p_{(+)}^{i,n}$, $p_{(-)}^{i,n}$, $q^{i,n}$ 一般说来并不是 P 的同伦不变量，而 p^k 则不但是 P 的拓扑不变量，而且也是 P 的同伦不变量，因之那些新不变量不能由那些古典的不变量如 p^k 之类表示出来. 因此，(3) 的意义是说，新不变量要比那些古典不变量至少是 Betti 数 (系数群是任意一个域) 要来得更“基本”. 例如下：

令 P_1, P_2 是两个一维的可剖分空间，如附图所示.

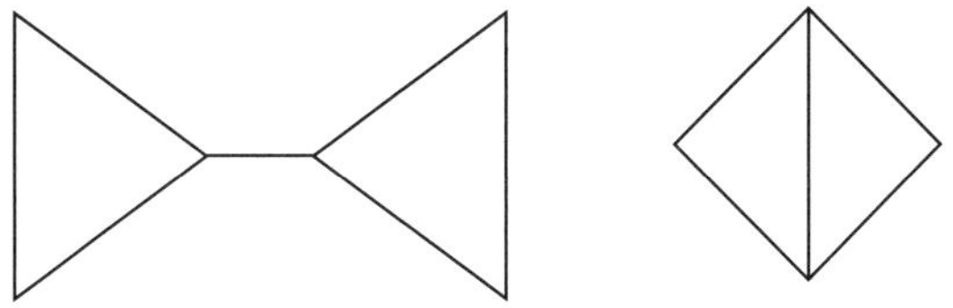

(P_1) (P_2)

由直接计算，可得

$$p_{(+)}^{2,2}(P_1)=2,\quad p_{(+)}^{2,2}(P_2)=0,$$
$$p_{(+)}^{1,2}(P_1)=7,\quad p_{(+)}^{1,2}(P_2)=5,$$
$$p_{(+)}^{0,2}(P_1)=1,\quad p_{(+)}^{0,2}(P_2)=1,$$
$$p_{(0)}^{1,2}(P_1)=7,\quad p_{(0)}^{1,2}(P_2)=7,$$
$$p_{(0)}^{0,2}(P_1)=1,\quad p_{(0)}^{0,2}(P_2)=1,$$
$$q^{1,2}(P_1)=2,\quad q^{1,2}(P_2)=4.$$

这两个空间的 $p_{(+)}^{i,2}$, $p_{(-)}^{i,2}$, $q^{i,2}$ 等并不一致. 但 P_1, P_2 有相同的一维的 Betti 数，因之有同样的同伦型，虽然它们的拓扑型并不相同. 这说明了这些新不变量都不是同伦不变量.

另一面，P_1, P_2 的三枝点数同样是 2，而其余的 p 枝点数 $(p\neq 2)$ 都是 0，因之仅由 p 枝点数 $(p\neq 2)$ 这些已知拓扑不变量不能区分 P_1, P_2 的拓扑型，而新不变量可能. 这说明在一维可剖分空间的情形，新的不变量也要比已知的那些不变量强 (参阅§7 例 1).

参考文献

[1] Moise. Affine Structures in 3-manifolds. *V.Annals of Mathematics*, 1952, 56: 96-114.

[2] Eilenberg S. On Problems of Topology. *ibid.*, 1949, 50: 247-260.

[3] Seifert H, and Threlfall W. Lebrbuch der Topologie.

Topological Invariants of New Type of Finite Polyhedrons

Abstract We describe in this paper a general method of deducing from homotopy invariants of spaces or systems of spaces invariants of a finite polyhedron which are invariants of the topological type of the polyhedron but are in general not invariants of homotopy type. It is shown that the betti numbers of a finite polyhedron may be expressed in terms of such new type invariants but, as is evident, not vice versa. It follows that these new type invariants are of more fundamental character than at least those classical invariants like betti numbers.

12. On Pontrjagin Classes I*

A sphere bundle $\mathcal{B}$(or a vector bundle in the sense of [1]) with orthogonal group or general linear group as structural group has as invariants a system of integral cohomology classes in the base space, besides the Stiefel-Whitney classes which are essentially mod 2, a fact which was first discovered by Pontrjagin in 1942[2]. Among these integral classes some of dimensions multiples of 4 are of particular importance which we have called Pontrjagin classes, denoted by $P^{4k}(\mathcal{B})$, of the given bundle. We know little about these classes and the fundamental problem about the topological invariance of these classes in the case of a differentiable manifold is not yet settled, in spite of the fact that the same problem about the Stiefel-Whitney classes has been solved completely in quite a neat manner[3, 4].

According to the original definition of Pontrjagin these classes are defined only through the intervention of grassmannian manifolds and are therefore not intrinsically defined in the bundle. The present paper furnishes two different intrinsic definitions of these classes which will serve as the starting point of our investigations about their topological invariance in the case of a differentiable manifold.

We will adopt the following notation. For any integral homology or cohomology class A, let A_0 and A_p(p an integer > 1) denote respectively the real class and the mod p class obtained from A by reductions and the reductions themselves will be denoted by $\rho_0 : A \to A_0$ and $\rho_p : A \to A_p$ respectively. We denote also by I, I_0, I_p the (additive) groups of integers, real numbers, and integers mod p respectively and $\rho_0 : I \to I_0$, $\rho_p : I \to I_p$ the corresponding reductions.

§1 The (real) grassmannian manifold $R_{n,m}$

Let $R_{n,m}$ be the grassmannian manifold of all m-elements in R^{n+m}. Let $\Omega_{n,m}$ be the set of all Schubert symbols a$= (a_1, \cdots, a_m)$, where $0 \leqslant a_1 \leqslant \cdots \leqslant a_m \leqslant n$. For such an a we will put $\mathrm{a}_{(i)} = (a_1, \cdots, a_{i-1}, a_i - 1, a_{i+1}, \cdots, a_m)$ if $a_i > a_{i-1}$ and $\mathrm{a}^{(i)} = (a_1, \cdots, a_{i-1}, a_i + 1, a_{i+1}, \cdots, a_m)$, if $a_i < a_{i+1}$, otherwise $\mathrm{a}_{(i)}$, $\mathrm{a}^{(i)}$ will be

*First published in Chinese in *Acta Mathcmatica Sinica*, 1953, 3(4): 291-315. *Sci. Sinica*, 1954, 3: 353-367; *Amer. Math. Soc. Translations, Ser.*, 1970, 2(92): 49-62.

undefined. We call $D(\mathrm{a}) = \sum_{i=1}^{m} a_i$ the dimension of a. Suppose a system of coordinates in R^{n+m} is given, then a certain canonical subdivision of $R_{n,m}$ may be so defined that to each $\mathrm{a} \in \Omega_{n,m}$ there corresponds in the subdivision an open cell $[\mathrm{a}]$ of dimension $D(\mathrm{a})$. With these cells oriented in a definite manner we have the following boundary formulas

$$\partial[\mathrm{a}] = \sum \varepsilon_i (1 + (-1)^{a_i+i+m})[\mathrm{a}_{(i)}], \quad \varepsilon_i = (-1)^{a_1+\cdots+a_i}, \tag{1}$$

the summation being over all i for which $\mathrm{a}_{(i)}$ is defined. Let $\{\mathrm{a}\}$ be the cochain taking value 1 on $[\mathrm{a}]$ but 0 on all other cells with same dimension, then dually we have

$$\delta\{\mathrm{a}\} = \sum \eta_i (1 + (-1)^{a_i+i+m+1})\{\mathrm{a}^{(i)}\}, \quad \eta_i = (-1)^{a_1+\cdots+a_i-1}, \tag{1$'$}$$

the summation being over all i for which $\mathrm{a}^{(i)}$ is defined.

Let

$$p_{4k} = (\underbrace{0, \cdots, 0}_{m-2k}, \underbrace{2, \cdots, 2}_{2k}), \quad 0 \leqslant 2k \leqslant m, \tag{2}$$

$$\bar{p}_{4k} = (\underbrace{0, \cdots, 0}_{m-2}, 2k, 2k), \quad 0 \leqslant 2k \leqslant n. \tag{3}$$

Then $\{p_{4k}\}$ and $\{\bar{p}_{4k}\}$ are integral cocycles of dimension $4k$ whose classes will be denoted by P^{4k} and $\bar{p}^{4k}$ respectively. These classes are independent of coordinates chosen in R^{n+m} and will be called P-classes and $\bar{P}$-classes in $R_{n,m}$ respectively.

Chains of the form $[\mathrm{a}]$, $\mathrm{a} \in \Omega_{n,m}$ will be called elementary chains. The set of all elementary chains of same dimension will be arranged in an order $<$ such that

$$[a_1, \cdots, a_m] < [b_1, \cdots, b_m]$$

if and only if for some index i we have $a_m = b_m$, $a_{m-1} = b_{m-1}$, $\cdots$, $a_{i+1}=b_{i+1}$, $a_i < b_i$. Any chain r may then be expressed in the "normal" form

$$\underline{x} = \sum_{i=1}^{k} \lambda_i \underline{x}_i, \quad \lambda_i \neq 0$$

in which all ξ_i are elementary chains and $\xi_1 < \cdots < \xi_k$ of which ξ_1 will be called the leading term of ε. The set of all r-dimensional chains will be arranged in an order $<$ such that for any two such chains $\xi = \sum_{i=1}^{k} \lambda_i \xi_i$, $\xi' = \sum_{i=1}^{k'} \lambda'{}_i \xi'{}_i$, in normal forms we have $\xi < \xi'$ if and only if for some index i we have $\xi_1 = \xi'_2$, $\cdots$, $\xi_{i-1} = \xi'_{i-2}$ and either $\xi_i < \xi'_i$ or $i - 1 = k < k'$.

Let us express an elementary chain ξ in the form

$$[0, \cdots, 0, \underbrace{b_s, \cdots, b_s}_{c_s}, \cdots, \underbrace{b_2, \cdots, b_2}_{c_2}, \underbrace{b_1, \cdots, b_1}_{c_1}], \tag{4}$$

where $0 < b_s < \cdots < b_2 < b_1$ and divide the set Ω^r of all r-dimensional elementary chains into three subsets Ω_i^r, $i = 0, 1, 2$ as follows: We put ξ in Ω_0^r if all b, c are even, while $\xi \in \Omega_1^r$ or Ω_2^r according as the first odd integer in the sequence $(b_1, c_1, b_2, c_2, \cdots, b_s, c_s)$ happens among $b's$ or $c's$. We say that a chain adheres to Ω_i^r if in the normal form all its terms are elementary chains of $\Omega_i^r (i = 0, 1, 2)$. It is easy to see that

(i) all chains adhering to Ω_0^r are integral cycles.

(ii) no chain adhering to Ω_2^r is an integral cycle, and the leading term of its boundary belongs to Ω_1^{r-1}.

Theorem 1 *If $r < n$ or $r = n =$ even, $C_r(R_{n,m})$ has a basis consisting of chains of the forms ξ', $\frac{1}{2}\partial\xi''$, ξ''' where ξ', ξ'', ξ''' are elementary chains of Ω_0^r, Ω_2^{r+1} and Ω_2^r respectively. If $r = n =$ odd, besides chains of the forms ξ', $\frac{1}{2}\partial\xi''$, ξ''' we should add the integral cycle $\xi_0 = [0, \cdots, 0, n]$ to form a basis of $C_r(R_{n,m})$.*

Proof Let $\xi = \sum_{i=1}^{k} \lambda_i \quad \xi_i \in C_r(R_{n,m})$, $r \leqslant n$, be in the normal form. If $r = n =$ odd and $\xi_1 = [0, \cdots, 0, n] = \xi_0$, then $\xi = \xi_0 \varepsilon \Omega_1^r$. Suppose that this is not so. If $\xi_1 \in \Omega_0^r$ or Ω_2^r, then $\xi - \xi_1$ is either 0 or $> \xi$. If $\xi_1 \in \Omega_1^r$, and ξ_1 be of the form (2) with b_j as the first odd integer in the sequence $(b_1, c_1, \cdots, b_s, c_s)$, we will take

$$\xi_1'' = [0, \cdots, 0, \underbrace{b_s, \cdots, b_s}_{c_s}, \cdots, \underbrace{b_j, \cdots, b_j}_{c_j - 1}, b_j + 1, \underbrace{b_{j-1}, \cdots, b_{j-1}}_{c_{j-1}}, \cdots, \underbrace{b_1, \cdots, b_1}_{c_1}] \in \Omega_2^{r+1}.$$

We have then in normal form $\partial\xi_1'' = 2\sum_{i=1}^{l} \mu_i \eta_i$, with $\mu_i = \pm 1$ and $\eta_1 = \xi_1$. Hence $\xi - \mu_1 \cdot \frac{1}{2}\partial\xi_1''$ is either 0 or $> \xi$. It follows that by subtracting successively from ξ chains of the form ξ', $\frac{1}{2}\partial\xi''$, ξ''' as stated in the theorem we get finally either 0 or ξ_0, in the case $r = n =$ odd. This proves our theorem.

From theorem 1 and B we get easily the following

Corollary 1 *If $r < n$, then $Z_r(R_{n,m})$ has a basis consisting of cycles of the form ξ', $\frac{1}{2}\partial\xi''$, where $\xi' \in \Omega_0^r$, $\xi'' \in \Omega_2^{r+1}$ for which ξ' are of infinite order while $\frac{1}{2}\partial\xi''$ are of order 2.*

Corollary 2 *If $r < n$ then $H_r(R_{n,m}, I_p)$, $p =$ odd or 0, has a basis with the elementary chains of Ω_0^r reduced mod p as the representative cycles. In particular, $H_r(R_{n,m}, I_p) = 0$ for $r<n$ and $r \not\equiv 0$ mod 4.*

§2 A certain canonical mapping l: $R_{n,m} \to C_{n,m}$

Let U^{n+m} be a unitary space of complex dimension $n+m$. The space $C_{n,m}$ of all complex m-elements in U^{n+m} is the (complex) grassmannian manifold defined in U^{n+m}. It is known that[5] the ring $H^*(C_{n,m})$ is generated by a system of classes $C^{2i} \in H^{2i}(C_{n,m})$, $0 \leqslant i \leqslant m$, which we will call C-classes of $C_{n,m}$. The C-classes $\bar{C}^{2i}$, $0 \leqslant i \leqslant n$, are then defined by the equations $\sum\limits_{i+j=k} C^{2i} \cup \bar{C}^{2j} = 0$ or unit class 1, according as $k > 0$ or $k = 0$.

In the unitary space U^{n+m} and the Euclidean space R^{n+m}, let us choose systems of complex and real coordinates $(z_1, \cdots, z_{n+m})$ and $(x_1, \cdots, x_{n+m})$. respectively. The equations $z_i = x_i$, $i = 1, \cdots, n+m$, then define a linear mapping

$$\vec{l}: \begin{cases} R^{n+m} \to U^{n+m}, \\ (x_1, \cdots, x_{n+m}) \to (z_1, \cdots, z_{n+m}), \end{cases}$$

which induce naturally a canonical mapping

$$l: \quad R_{n,m} \to C_{n,m}.$$

Though the definition of l depends on the choice of coordinates in R^{n+m} and U^{n+m}, it is easy to see that the homotopy class of l is independent of such choices.

Theorem 2 *Let $r = 0$ or an odd prime p_v then*

$$l^* C_r^{2k} = \begin{cases} 0, & k \text{ odd}, \\ (-1)^{k/2} P_r^{2k}, & k \text{ even}, \end{cases} \quad (0 \leqslant k \leqslant m), \tag{1}$$

$$l^* \bar{C}_r^{2k} = \begin{cases} 0, & k \quad \text{odd}, \\ \bar{P}_r^{2k}, & k \quad \text{even}, \end{cases} \quad (0 \leqslant k \leqslant n). \tag{1$'$}$$

Proof The case k odd being evident, we will suppose in what follows that k is even. We suppose also that $n > 2k$.

Let $\delta_k^* \subset C_{n,m}$ be defined by the set of complex m-elements $\{Z\}$ satisfying

$$\text{Comp. dim}(Z \cap U^{n+k-1}) \geqslant k, \tag{2}$$

where U^{n+k-1} is an arbitrarily chosen $(n+k-1)$-element in U^{n+m}. Let δ_k be the cycle represented by the naturally oriented δ_k^* and ξ_k the cycle $[\underbrace{0,\cdots,0}_{m-k},\underbrace{2,\cdots,2}_{k}]$ defined in the canonical subdivision with respect to the given coordinate system $(x_1,\cdots,x_{n+m})$ of R^{n+m}. By Cor.2 of §1, the second equation of (1) is equivalent to the following intersection formulas (I_r denotes the real or mod p index of intersection with respect to the natural orientation of $C_{n,m}$):

$$I_r(\delta_k, l_\xi) = 0, \qquad \xi \in \Omega_0^{2k}, \quad \xi \neq \xi_k; \tag{3}$$

$$I_r(\delta_k, l\xi_k) = (-1)^{k/2}. \tag{3$'$}$$

For the proof of(3), let us consider a cycle $\xi = [a_1,\cdots,a_m] \in \Omega_0^{2k}$, $\xi \neq \xi_k$. Since $\sum\limits_{i=1}^{m} a_i = 2k$ and each a_i is even, we must have $a_1 = \cdots = a_{m-k+1} = 0$. We know that ξ may be geometrically represented by the pseudomanifold ξ^* of which any m-element X should satisfy the condition

$$X \supset R^{m-k+1}, \tag{4}$$

where R^s is the element $x_{s+1} = \cdots = x_{n+m} = 0$. Since $\bar{l}(R^{m-k+1})$ is a complex $(m-k+1)$- element of U^{n+m}, we may choose in U^{n+m} a complex $(n+k-1)$- element U^{n+k-1} which intersects $l(\bar{R}^{m-k+1})$ in the origin of U^{n+m} only and determines δ_k^* by (2). Comparing (2) and (4), we see easily that δ_k^* and $l(\xi^*)$ can have no common points and (3) is thus proved.

To prove (3)$'$, let us remark that the cycle ξ_k is geometrically represented by the pseudomanifold ξ_k^* which is defined by

$$R^{m-k} \subset X \subset R^{m+2}.$$

Any such X has equations of the form

$$\left.\begin{aligned} &x_{m+3} = \cdots = x_{n+m} = 0, \\ &\sum_{i=m-k+1}^{m+2} a_{ji}x_i = 0, \quad j = 1,2 \end{aligned}\right\} \tag{5}$$

in which a_{ji} are real numbers and the last two equations are linearly independent.

Define now δ_k^* by (2) with U^{n+k-1} as follows:

$$z_1 = \cdots = z_{m-k} = 0, \qquad z_{m-k+1} = \mathrm{i}z_{m-k+2}. \tag{6}$$

It is easy to see that there is only one element $X_0 \in \xi_k^*$ for which $l(x_0) = Z_0$ is common to δ_k^* and $l(\xi_k^*)$, namely the element given by the following equations

$$x_{m-k+1} = x_{m-k+2} = 0, \qquad x_{m+3} = \cdots = x_{n+m} = 0. \tag{7}$$

To determine the index of intersection of $l\xi_k^*$ and δ_k^* at Z_0 we take in U^{n+m} a complex coordinate system $(z_1', \cdots, z_{n+m}')$ related to the previous system $(z_1, \cdots, z_{n+m})$ in the following manner:

$$\left.\begin{aligned}
&z'_i = z_{m+i+2}, \quad i = 1, \cdots, n-2;\\
&z'_{n-1} = z_{m-k+2};\\
&z'_i = z_{-n+m+i-k+3}, \quad i = n, \cdots, n+k-1;\\
&z'_i = z_{i-n-k}, \quad i = n+k+1, \cdots, n+m;\\
&z'_{n+k} = z_{m-k+1} - \mathrm{i} z_{m-k+2}.
\end{aligned}\right\} \tag{8}$$

By (6) and (7), U^{n+k-1} and $l(X_0) = Z_0$ are then given respectively by the following systems of equations:

$$U^{n+k-1} : z'_{n+k} = \cdots = z'_{n+m} = 0, \tag{9}$$

$$l(X_0) = Z_0 : z'_1 = \cdots = z'_{n-1} = 0, z'_{n+k} = 0. \tag{10}$$

In $C_{n,m}$ any complex m-element neighbouring Z_0 may be represented by the system of equations:

$$z'_j = \sum_{i=n}^{n+k-1} \zeta_{ji} z'_i + \sum_{i=n+k+1}^{n+m} \zeta_{ji} z'_i, \quad j = 1, \cdots, n-1, n+k,$$

in which $\zeta_{ji} = \zeta'_{ji} + i\,\zeta''_{ji}$, with ζ', ζ'' real numbers. Such elements form an open cell L of dimension $2nm$, of which ζ_{ji} may be regarded as a system of complex coordinates. The cell L becomes then a unitary space with a natural orientation concordant with that of $C_{n,m}$. By(2), (9), $L \cap \delta_k^* = L_1$ is a complex element of L with equations

$$\zeta_{n+k,n} = \cdots = \zeta_{n+k,n+k-1} = 0. \tag{11}$$

On the other hand, by (5) and (7), the m-elements $X \in \xi_k^*$ in the neighbourhood of X_0 may be represented by the following system of equations:

$$\left.\begin{aligned}
&x_j = \sum_{i=m-k+3}^{m+2} \xi_{ji} x_i, \quad j = m-k+1, m-k+2,\\
&x_j = 0, \qquad\qquad\qquad j = m+3, \cdots, n+m,
\end{aligned}\right\} \tag{12}$$

where ξ_{ji} are real numbers. By (8), the equations of $l(X)$ for such X are

$$\left.\begin{aligned} &{z'}_i = 0, \quad j = 1, \cdots, n-2;\\ &{z'}_{n-1} = \xi_{m-k+2,m-k+3}{z'}_n + \cdots + \xi_{m-k+2,m+2}{z'}_{n+k-1};\\ &{z'}_{n+k} = \mathrm{i}{z'}_{n-1} = \xi_{m-k+1,m-k+3}{z'}_n + \cdots + \xi_{m-k+1,m+2}{z'}_{n+k-1}. \end{aligned}\right\} \tag{13}$$

Such $l(X)$ form a complex element $L_2 = l(\xi_k^*) \cap L$ in the unitary space L. The elements L_1 and L_2 meet only in the origin $l(X_0) = Z_0$ of L at which the index of intersection is easily seen to be $(-1)^{k/2}$. This proves $(3)'$ in the case $n > 2k$. As the restriction $n > 2k$ is easily removed, (1) is completely proved.

$(1)'$ may be proved similarly or deduced from (1) by considering the canonical mappings $R_{n,m} \to R_{m,n}$ and $C_{n,m} \to C_{m,n}$.

Theorem 3 *Let $W^i(\bar{W}^i)$ be the (mod 2) W-classes($\bar{W}$-classes) of $R_{n,m}$ and $C_2^{2i}(\bar{C}_2^{2i})$ the* mod 2 *C-classes ($\bar{C}$-classes) of $C_{n,m}$. If $l : R_{n,m} \to C_{n,m}$ is the canonical mapping mentioned before, then*

$$l^* C_2^{2i} = (W^i)^2, \quad l^* \bar{C}_2^{2i} = (\bar{W}^i)^2. \tag{14}$$

Proof Using some notations as before, let R^{2n+2m} be the Euclidean space determined by U^{n+m}, with a(real)coordinate system $(x'_1, \cdots, x'_{n+m}, x''_1, \cdots, x''_{n+m})$ such that $z_i = x'_i + \mathrm{i}x''_i$, $i = 1, \cdots, n+m$. Consider R^{2n+2m} as a direct sum $R_1^{n+m} \oplus R_2^{n+m}$, where R_k^{n+m} is determined by $(x_1^{(k)}, \cdots, x_{n+m}^{(k)})$. Let $R_{n,m}^{(k)}$ and $R_{2n,2m}$ be the grassmannian manifolds defined in R_k^{n+m} and R^{2n+2m} respectively. Between the various grassmannian manifolds we have the following canonical mappings.

$$\begin{aligned} l: &\quad R_{n,m} \to C_{n,m},\\ k: &\quad C_{n,m} \to R_{2n,2m},\\ h: &\quad {R'}_{n,m} \times {R''}_{n,m} \to R_{2n,2m}, \end{aligned}$$

defined respectively as follows: for $Z \in C_{n,m}$, $X^{(k)} \in R_{n,m}^{(k)}$, $k(Z)$ is the $2m$-element in R^{2n+2m} determined by Z, $h(X', X'') = X' \oplus X''$ in R^{2n+2m}, and l is as before. Besides, the equations $x_i^{(k)} = x_i$, $i = 1, \cdots, n+m$; $k = 1, 2$ determine a linear mapping

$$\bar{j}_k = \begin{cases} R^{n+m} \to R_k^{n+m},\\ (x_1, \cdots, x_{n+m}) \to (x_1^{(k)}, \cdots, x_{n+m}^{(k)}), \end{cases}$$

which induces a mapping

$$j_k : \quad R_{n,m} \to R_{n,m}^{(k)} \quad (\bar{k} = 1, 2).$$

For $X \in R_{n,m}$ let $j(X) = j_1(X) \times j_2(X)$, then we have a mapping

$$j: \quad R_{n,m} \to R'_{n,m} \times R''_{n,m}.$$

Let the W-classes(mod 2)in $R^{(k)}_{n,m}$, $R_{2n,2m}$ be respectively $W^i_k (k = 1, 2)$, W'^i, then

$$j^*(W_1^{i_1} \otimes W_2^{i_2}) = j_1^* W_1^{i_1} \cup j_2^* W_2^{i_2} = W^{i_1} W^{i_2}. \tag{15}$$

Since we have known[6] that

$$\begin{aligned} hj &= kl, \\ k^* W'^{2i} &= C_2^{2i}, \\ h^* W'^{2i} &= \sum_{i_1+i_2=2i} W_1^{i_1} \otimes W_2^{i_2}, \end{aligned}$$

it follows easily from $l^* C_2^{2i} = l^* k^* W'^{2i} = j^* h^* W'^{2i}$ that

$$l^* C_2^{2i} = (W^i)^2, \tag{16}$$

which is the first formula of (14). Similarly we may prove the second formula of (14). This concludes Theorem 3.

Theorem 4

$$l^* C^{2k} = \begin{cases} \dfrac{1}{2}\delta(S_q^{k-1} W^k), & k \text{ odd}, \\ (-1)^{k/2} P^{2k} + \dfrac{1}{2}\delta(W^{k-1} W^k), & k \text{ even}. \end{cases} \quad (k \leqslant m) \tag{17}$$

$$\left. l^* \bar{C}^{2k} = \begin{cases} \dfrac{1}{2}\delta(S_q^{k-1} \bar{W}^k), & k \text{ odd}, \\ \bar{P}^{2k} + \dfrac{1}{2}\delta(\bar{W}^{k-1} \bar{W}^k), & k \text{ even}. \end{cases} \quad (k \leqslant n) \right\} \tag{17$'$}$$

Proof Let $\mathrm{a}_i = (0, \cdots, 0, i) \in \Omega_{n,m}$ and $\mathrm{a}_{j,i} = (0, \cdots, 0, j, i) \in \Omega_{n,m}$ where $j \leqslant i$. By $(1)'$ of § 1, we have

$$\begin{aligned} &\delta\{\mathrm{a}_{2j,4i-2j-1}\} = \pm 2\{\mathrm{a}_{2j,4i-2j}\} \pm 2\{\mathrm{a}_{2j+1,4i-2j-1}\}, \\ &\delta\{\mathrm{a}_{2j-1,4i-2j}\} = 0. \end{aligned}$$

On the other hand, we know that[5]

$$\{\mathrm{a}_{2i-1}\} \cup \{\mathrm{a}_{2i}\} \sim \sum_{j=0}^{2i-1} \{\mathrm{a}_{j,4i-j-1}\} \bmod 2,$$

$$\{\mathrm{a}_{2i}\} \cup \{\mathrm{a}_{2i}\} \sim \sum_{j=1}^{2i} \{\mathrm{a}_{j,4i-j}\} \bmod 2.$$

It follows that $\rho_2 \cdot \frac{1}{2}\delta(\bar{W}^{2i-1} \quad \bar{W}^{2i})$ contains a cocycle of the form

$$\begin{aligned}&\rho_2 \sum_{j=0}^{i-1} (\{a_{2j,4i-2j}\} + \{a_{2j+1,4i-2j-1}\}) \\ =&\rho_2 \sum_{j=0}^{2i-1} \{a_{j,4i-j}\} = \rho_2 \sum_{j=0}^{2i} \{a_{j,4i-j}\} + \rho_2\{a_{2i,2i}\},\end{aligned}$$

which belongs to $\bar{W}^{2i} \cup \bar{W}^{2i} + \rho_2 \bar{P}^{4i}$. Hence

$$\rho_2 \left[\bar{P}^{4i} + \frac{1}{2}\delta(\bar{W}^{2i-1}\bar{W}^{2i})\right] = \bar{W}^{2i} \cup \bar{W}^{2i}.$$

By Theorem 3, we have then

$$\rho_2 \left[l^*\bar{C}^{4i} - \bar{P}^{4i} - \frac{1}{2}\delta(\bar{W}^{2i-1}\bar{W}^{2i})\right] = 0. \tag{18}$$

By Theorem 2, we have also

$$\rho_0 \left[l^*\bar{C}^{4i} - \bar{P}^{4i} - \frac{1}{2}\delta(\bar{W}^{2i-1}\bar{W}^{2i})\right] = 0. \tag{19}$$

As $R_{n,m}$ has only torsion number 2, we have from (18) and (19)

$$l^*\bar{C}^{4i} - \bar{P}^{4i} - \frac{1}{2}\delta(\bar{W}^{2i-1}\bar{W}^{2i}) = 0, \tag{20}$$

which is the 2nd formula of (17)′. The other formulas may be proved in the similar manner.

§3 An intrinsic definition of pontrjagin classes

For any m-vector bundle $\mathcal{B}$ over a (finite) complex K we may form the Whitney product $\mathcal{B} \cup \mathcal{B}$ over the same base space K and turn $\mathcal{B} \cup \mathcal{B}$ into a "unitary bundle" as follows. Let the director space of $\mathcal{B}$ be R^m, the fiber in $\mathcal{B}$ over $p \in K$ be R_p^m, and the coordinate system in a coordinate neighborhood σ of K be $\xi_{op}: \ R^m \to R_p^m (p \in \sigma)$. Take two copies R_i^m of R^m and $R_{p,i}^m (i = 1,\ 2)$ of each R_p^m, realized by $\omega_i: \ R^m \to R_i^m, \ R_p^m \to R_{p,i}^m$. According to the definition of Whitney product, the director space of $\mathcal{B} \cup \mathcal{B}$ and the fiber over $p \in K$ in $\mathcal{B} \cup \mathcal{B}$ are respectively $R^{2m} = R_1^m \oplus R_2^m$ and $R_p^{2m} = R_{p,1}^m \oplus R_{p,2}^m$ with coordinate system $\varsigma_{\sigma p}: \ R^{2m} \to R_p^{2m}$, $(p \in \sigma)$ in σ defined by

$$\zeta_{op}(\omega_1\xi_1 + \omega_2\xi_2) = \omega_1\xi_{op}(\xi_1) + \omega_2\xi_{op}(\xi_2), \quad \xi_1, \xi_2 \in R^m, p \in \sigma. \tag{1}$$

Now for each complex number $C = a + ib$ (a and b are real) and any $\xi_1, \xi_2 \in R^m$, let us define a multiplication as follows:

$$
\begin{aligned}
c(\omega_1\xi_1 + \omega_2\xi_2) &= a(\omega_1\xi_1 + \omega_2\xi_2) + ib(\omega_1\xi_1 + \omega_2\xi_2),\\
a(\omega_1\xi_1 + \omega_2\xi_2) &= a\omega_1\xi_1 + a\omega_2\xi_2,\\
\mathrm{i}(\omega_1\xi_1 + \omega_2\xi_2) &= -\omega_1\xi_2 + \omega_2\xi_1.
\end{aligned}
$$

Under this multiplication $R^{2m} = R_1^m \oplus R_2^m$ becomes canonically a unitary space of complex dimension m, which will be denoted by U^m. Similarly, for each $p \in K$, we may endow R_p^{2m} with a natural complex structure to turn it into a unitary space U_p^m of complex dimension m. For any two overlapping coordinate neighborhoods σ, τ of K and any point $p \in \sigma \cap \tau$, we have by (1),

$$\mathrm{i}\zeta_{\tau p}^{-1}\zeta_{\sigma p}(\omega_1\xi_1 + \omega_2\xi_2) = \zeta_{\tau p}^{-1}\zeta_{\sigma p}\mathrm{i}(\omega_1\xi_1 + \omega_2\xi_2), \quad \xi_1, \xi_2 \in R^m.$$

It follows that, with respect to any complex coordinate system in U^m, $\zeta_{\tau p}^{-1}\zeta_{\sigma p}$ is a non-singular linear transformation on m complex variables and thus under $\zeta_{\sigma p}$, $\mathcal{B} \cup \mathcal{B}$ becomes an m-unitary bundle $\mathcal{B}^*$, Evidently this bundle $\mathcal{B}^*$ depends intrinsically on $\mathcal{B}$ and will be called the canonical unitary bundle associated to $\mathcal{B}$.

Now let $\Re_{n,m}$ and $\mathscr{D}_{n,m}$ be respectively the universal m-vector bundle and m-unitary bundle canonically defined over the grassmannian manifolds $R_{n,m}$ and $C_{n,m}$. Suppose that $\mathcal{B} = \alpha^*\Re_{n,m}$ where $\alpha : K \to R_{n,m}$ is a continuous mapping and n is sufficiently large. Then it is easy to see that $\mathcal{B}^*$ is induced from $\mathscr{D}_{n,m}$ by $l\ \alpha = \beta : K \to C_{n,m}$ where $l : R_{n,m} \to C_{n,m}$ is defined in § II, i.e. $\mathcal{B}^* \sim \beta^*\mathscr{D}_{n,m}$. By definition, the Pontrjagin classes $P^{4k}(\mathcal{B})$ of the vector bundle $\mathcal{B}$ and the Chern classes $C^{2i}(\mathcal{B}^*)$ of the unitary bundle $\mathcal{B}^*$ are respectively given by

$$P^{4k}(\mathcal{B}) = \alpha^* P^{4k}, \tag{2}$$

$$C^{2i}(\mathcal{B}^*) = \beta^* C^{2i}, \tag{3}$$

where P^{4k} and C^{2i} are respectively the P-classes of $R_{n,m}$ and C-classes of $C_{n,m}$. Similarly for the dual Pontrjagin classes and the dual Chern classes:

$$\bar{P}^{4k}(\mathcal{B}) = \alpha^* \bar{P}^{4k}, \tag{2$'$}$$

$$\bar{C}^{2i}(\mathcal{B}) = \beta^* \bar{C}^{2i}. \tag{3$'$}$$

The definitions of these classes are not intrinsic owing to the intervention of grassmannian manifolds. For the Chern classes we may define them intrinsically as the obstruction classes in Steenrod's sense of certain associated bundles of the given unitary bundle. The Pontrjagin classes may be defined intrinsically in a similar manner, as shown by Chern and Rokhlin, but with some complication. The purpose of this section is to give some relatively simple intrinsic definitions of Pontrjagin classes by reducing them to Chern classes as follows. By Theorems 2 and 4 of § II and the formulas (2), (3) above, we get easily

$$P_r^{4k}(\mathcal{B}) = (-1)^k C_r^{4k}(\mathcal{B}^*), \quad r = 0 \quad \text{or an odd prime}, \quad 0 \leqslant 2k \leqslant m, \tag{*(4)}$$

$$P^{4k}(\mathcal{B}) = (-1)^k C^{4k}(\mathcal{B}^*) + \frac{1}{2}\delta[W^{2k-1}(\mathcal{B}) \cup W^{2k}(\mathcal{B})], \quad 0 \leqslant 2k \leqslant m. \tag{*(5)}$$

Similar expressions may be obtained for $\bar{P}^{4k}(\mathcal{B})$ and $\bar{P}_r^{4k}(\mathcal{B})$. As $\mathcal{B}^*$, $C^{2i}(\mathcal{B}^*)$ and $W^i(\mathcal{B})$ are all intrinsically defined from $\mathcal{B}$, (4) and (5) furnish us the following

Intrinsic definition of Pontrjagin classes. To a given m-vector bundle $\mathcal{B}$, we may endow naturally the Whitney product $\mathcal{B} \cup \mathcal{B}$ of $\mathcal{B}$ by itself with a unitary structure to turn it into a unitary bundle $\mathcal{B}^*$. The Pontrjagin classes of $\mathcal{B}$ are then intrinsically defined in terms of the Chern classes $C^{2i}(\mathcal{B}^*)$ of $\mathcal{B}^*$ and the Stiefcl-Whitney classes $W^i(\mathcal{B})$ of $\mathcal{B}$ by the formulas (4) *and* (5).

§4 Certain formulas about characteristic classes

Let A be the bundle space of an m-unitary bundle σ, with base space a finite complex K and projection $\omega : A \to K$. In each fiber U_p^m over $p \in K$ in σ may be defined a (complex) grassmannian manifold $C_{m-1,1}(p)$. All of these manifolds form naturally a space B which is the bundle space of a fiberbundle $\mathfrak{R}$ associated with σ, the base being K, the fiber over $p \in K$, $C_{m-1,1}(p)$, and projection, $\pi : B \to K$. Let A_1 be the subspace of $A \times B$ formed by all points (a, b) such that $\omega a = \pi b (= p)$, and $a \in b \in C_{m-1,1}(p)$, then A_1 is the bundle space of a certain complex line bundle (or 1-unitary bundle) $\mathcal{I}$ with B as base.

Theorem 5 *In the space B we have*

$$\sum_{i=0}^{m} (-1)^i \pi^* C^{2i}(\sigma) \cup [C^2(\mathcal{I})]^{m-i} = 0. \tag{1}$$

Moreover, the classes $C^{2i}(\sigma)$ are uniquely determined by (1) *in the sense that if*

$$\sum_{i=0}^{m} (-1)^i \pi^* C'^{2i} \cup [C^2(\mathcal{I})]^{m-i} = 0,$$

where $C'^0 = C^0(\sigma) = 1$ *is the unit class in* K, *then* $C'^{2i} = C^{2i}(\sigma)$ *for* $i = 0, 1, \cdots, m$.

Proof Let us consider the m-unitary bundle $\pi^*\sigma$ over the base B. The bundle space Γ of this bundle is a subspace of $A \times B$, consisting of all points $(a, b) \in A \times B$ such that $\omega a = \pi b$, the projection $\tilde{\pi} : \Gamma \to B$ being defined by $\tilde{\pi}(a, b) = b$. In $\pi^*\sigma$ the fiber over $b \in B$ is an unitary space U_b^m of complex dimension m whose points (a, b) satisfy $a \in U_p^m$ and $p = \pi(b)$. Now b represents a complex line of U_p^m through its origin. The complex $(m-1)$-element in U_p^m completely orthogonal to b will be denoted by $\bar{b}$. In Γ the set of all points (a, b), where $a \in \bar{b}$ is naturally the bundle space, A_2, of an $(m-1)$-unitary bundle $\bar{\mathcal{I}}$ with B as base space. As $b \oplus \bar{b} = U_p^m(p = \pi(b))$, it turns out that $\pi^*\sigma$ is the Whitney product of $\mathcal{I}$ and $\bar{\mathcal{I}}$, i.e. $\pi^*\sigma \sim \mathcal{I} \cup \bar{\mathcal{I}}$. By duality theorem for product bundles[6] we have

$$\begin{aligned}&1 + C^2(\pi^*\sigma)t + \cdots + C^{2m}(\pi^*\sigma)t^m \\ =&[1 + C^2(\mathcal{I})] \cup [1 + C^2(\bar{\mathcal{I}})t + \cdots + C^{2m-2}(\bar{\mathcal{I}})t^{m-1}],\end{aligned}$$

t being an indeterminate. Multiplying both sides by $\sum_{i=0}^{\infty}(-1)^i[C^2(\mathcal{I})]^i t^i$, we get

$$\sum_{i=0}^{m} C^{2i}(\pi^*\sigma)t^i \cdot \sum_{i=0}^{\infty}(-1)^i[C^2(\mathcal{I})]^i\, t^i = \sum_{i=0}^{m-1} C^{2i}(\mathcal{I})t^i.$$

The comparison of coefficients of t^m gives then

$$\sum_{i=0}^{m} C^{2i}(\pi^*\sigma) \cup (-1)^{m-i}[C^2(\mathcal{I})]^{m-i} = 0.$$

Since $C^{2i}(\pi^*\sigma) = \pi^*C^{2i}(\sigma)$ we get(1).

We will give $C^2(\mathcal{I})$ in (1) another interpretation as follows. Take a fixed point $p \in K$ and consider the fiber $C_{m-1,1}(p)$ in $\Re$ as a complex projective space K_{m-1} of complex dimension $m-1$. Then $H^*(C_{m-1,1}(p))$ is generated by $D^2 \in H^2(C_{m-1,1}(p))$ which takes value + 1 on a complex line of K_{m-1}, with a single relation $(D^2)^m = 0$. On the other hand, $C_{m-1,1}(p)$ is a grassmannian manifold defined in U_p^m with a C-class C^2. It is easy to see that $C^2 = D^2$. Next, let $\partial_{m-1,1}(P)$ be the canonical 1-unitary bundle over $C_{m-1,1}(p)$, then $C^2 = C^2(\mathscr{D}_{m-1,1}(p))$, while the points of the bundle space $\tilde{C}_{m-1,1}(p)$ of $\mathscr{D}_{m-1,1}(p)$ are of the form (a, b) where $b \in C_{m-1,1}(p)$, $a \in b \subset U_p^m$. If $(a, b) \in \tilde{C}_{m-1,1}(p)$, then $\pi b = \omega a = p$, which means $(a, b) \in A_1$. Hence we may define identical mappings $\tilde{i} : \tilde{C}_{m-1,i}(p) \subset A_1$ and $i : C_{m-1,1}(p) \subset B$. Evidently $\tau_{\tilde{i}} = i\tau_p$, where τ_p and τ are projections of $\mathscr{D}_{m-1,1}(p)$ and $\mathcal{I}$ respectively.

Hence we have $\mathscr{D}_{m-1,1}(p) \sim i^*\mathcal{I}$, whence

$$i^*C^2(\mathcal{I}) = C^2(i^*\mathcal{I}) = C^2(\mathscr{D}_{m-1,1}(p)) = C^2 = D^2.$$

It follows that the bundle $\mathfrak{R}$ over K has the property that its fiber is "totally non-homologous in the bundle space B". The uniqueness of $C^{2i}(\sigma)$ occurring in (1) follows then from a theorem of Leray-Hirsch(Cf. e. g. [9]).

Let us remark that, the bundle $\mathcal{I}$ being intrisically associated with the given bundle σ, the above theorem gives an intrinsic definition of $C^{2i}(\sigma)$. Now combining with (4) or (5) of § III, we get an alternative intrinsic definition of Pontrjagin classes given by

Theorem 6 *To each m-vector bundle $\mathcal{B}$ over a finite complex let us form the Whitney product $\mathcal{B} \cup \mathcal{B}$ of $\mathcal{B}$ with itself and endow it with a natural unitary structure to turn it into a unitary bundle $\mathcal{B}^*$. Let B be the bundle space of the $C_{m-1,1}$- bundle $\mathfrak{R}$ associated with $\mathcal{B}^*$ and $\mathcal{I}$, the complex line bundle naturally defined over B as base space. Then the Pontrjagin classes $P^{4k}(\mathcal{B})$ are uniquely determined by the following formulas in the space B:*

$$\sum_{k=0}^{m'} (-1)^k \pi^* P_r^{4k}(\mathcal{B}) \cup [C_r^2(\mathcal{I})]^{m-2k} = 0, \tag{2}$$

$$\sum_{k=0}^{m'} \pi^*[(-1)^k P^{4k}(\mathcal{B}) + \frac{1}{2}\delta(W^{2k-1}(\mathcal{B}) \cup W^{2k}(\mathcal{B}))] \cup [C^2(\mathcal{I})]^{m-2k} = 0, \tag{3}$$

in which m'denotes the greatest integer $\leqslant m/2$, r is either 0 or an odd prime, and π is the projection of $\mathfrak{R}$.

In the case of Stiefel-Whitney classes we have a theorem analogous to theorem 5 as follows.

Theorem 5′ *Let $\mathcal{B}$ be an m-vector bundle over a finite complex K. Let β be a bundle associated with $\mathcal{B}$ with m-dimensional projective space as fibers. Over the bundle space B of β we may naturally define an 1-vector bundle $\mathcal{I}$. Then the (reduced) Stiefel-Whitney classes $W^i(\mathcal{B})$ of $(\mathcal{B})$ are uniquely determined by the following formula:*

$$\sum_{i=0}^{m'} \pi^* W^i(\mathcal{B}) \cup [W^1(\mathcal{I})]^{m-i} = 0, \tag{4}$$

π being the projection of the bundle β.

The proof of Theorem $5'$ is quite analogous to that of Theorem 5 and is therefore omitted. We remark only that both these two theorems are first discovered by G. Hirsch [9], a little before the author in 1950.

References

[1] Wu Wen-Tsün. On the Product of Sphere Bundles and the Duality Theorem Mod. Two. *Annals of Math.*, 1948, **49**: 641-653.

[2] Л. Понтрягин. Characteristic Cycles on Manifolds. *C. R. USSR*, 1942, **35**: 34-37.

[3] Thom R. Les Espaces Fibrés en Sphéres et Carreés de Steenrod. *Ann. Ec. Norm. Sup.*, 1952, (3)**69**: 110-182.

[4] Wu Wen-Tsün. Classes Caractéristiques et *i*-canés d'une variété. C. R. Paris, 1950, **230**: 508-511.

[5] Chern S. On the Multiplication in the Characteristic Ring of Sphere Bundle. *Annals of Math.*, 1948, **49**: 362-372.

[6] Wu Wen-Tsün. Sur les Classes Caractéristiques des Structures Fibrés Sphériques. *Act. Sc. Ind.* , No. 1183. Hermann, Paris, 1952.

[7] Ролин. Внутреннееопредениехарактерисгическихциклои Понтрягин. Дан., 1952, **84**: 449—452.

[8] Spanier E H. Homology Theory of Fiber Bundles. *Proceedings of International Congress of Math.*, 1950, II: 390-396.

[9] Hirsch G. Sur la Structure Multiplicative de l'Anneau de Cohomologie d'un Espace Fibré. C. R. Paris, 1950, **230**: 46-48.

13. On Squares in Grassmannian Manifolds*

Introduction

The present paper gives detailed proofs of the results announced in a note published in the *Comptes Rendus* of 1950[1]. The author wishes to express his gratitude to M. H. Cartan for the help and encouragement given him during his researches.

As is well known, any cohomology class on any coefficients in the Grassmannian manifold $G_{n,m}(n > m)$ corresponds to a certain invariant of an $(m-1)$ -sphere bundle which may be called a *characteristic* class of the bundle. In the case of coefficients mod 2, a theorem of Chern[2] shows that any class in $G_{n,m}$ may be expressed by means of cup-products in terms of a certain number of them, namely the classes W^i of dimension j, $0 < i \leqslant m$, which corresponds to the Stiefel-Whitney classes (reduced mod 2) of an $(m-l)$ -sphere bundle. In this paper, we shall give a complete determination of Steenrod's square operations[3] in $G_{n,m}$. It follows as a consequence that any class mod 2 in G may be expressed by means of cup-products and squares in terms of the classes w^{2^k} of dimension $2^k \leqslant m$. The study of characteristic classes mod 2 of a sphere bundle is thus reduced to the study of its reduced Stiefel-Whitney classes of dimensions 2^k.

I The additive structure of $H^*(G_{n,m})$[4]

Let R^{n+m} be a Euclidean space of dimension $n + m$. A linear subspace of dimension m of R^{n+m} passing through the origin of R^{n+m} will be called an m-clement of R^{n+m}. The Grassmannian manifold of all m-elements of R^{n+m} will be said to be defined in R^{n+m} and will be denoted by $G_{n,m}$.

Let R^i, $i = 1, 2,\cdots, n+m-1$ be a sequence of i-elements in R^{n+m}:

$$R^1 \subset R^2 \subset \cdots \subset R^{n+m-1}. \qquad (S)$$

Let $\Omega_{n,m}$ be the set of all sequences $\mathfrak{a}$ of integers $a_1, \cdots, a_m$ with $0 \leqslant a_1 \leqslant \cdots$

* First published in Chinese in *The Chinese Journal of Mathematics*, 1953, 2(4): 203-230. *Acta Sci. Sinica*, 1953, 2: 91-115; *Amer. Math. Soc. Translations, Ser.*, 1964, 2(38): 235-258.

$\leqslant a_m \leqslant n$. For any $\mathfrak{a} = (a_1, \cdots, a_m) \in \Omega_{n,m}$ let $\mathfrak{a}_{(i)}$ be the sequence $(a_1, \cdots, a_{i-1}, a_t - 1, a_{i+1}, \cdots, a_m) \in \Omega_{n,m}$ in the case $a_{i-1} \leqslant a_i - 1, 1 \leqslant i \leqslant m (a_o = 0)$; otherwise $\mathfrak{a}_{(i)}$ is undefined. Denote by $[\mathfrak{a}]^*$ the set of all elements X of $G_{n,m}$ such that

$$\mathrm{Dim}(X \cap R^{a_i+i}) \geqslant i, \quad j = 1, \cdots, m. \tag{1}$$

Then the set

$$|\mathfrak{a}| = [\mathfrak{a}]^* - \Sigma[\mathfrak{a}_{(i)}]^*,$$

where the summation is extended to all i for which $\mathfrak{a}_{(i)}$ is defined is by Ehresmann, an open cell of dimension

$$D\,(\mathfrak{a}) = \sum_{i=1}^{m} \mathfrak{a}_i.$$

Moreover, the set of all cells $[\mathfrak{a}]$ corresponding to all $\mathfrak{a} \in \Omega_{n,m}$ form a cell subdivision of $G_{n,m}$ and, when each cell is conveniently oriented, we would have

$$\partial[\mathfrak{a}] = \Sigma\eta_i[\mathfrak{a}_{(i)}], \quad \eta_i = 0, +2 \text{ or } -2,$$

the summation being again extended to all i for which $[\mathfrak{a}_{(i)}]$ is defined.

In what follows we shall be only interested in the case for which the coefficient group is the group of integers mod 2 with two elements 0, 1 and this will be tacitly assumed henceforth. For any $\mathfrak{a} \in \Omega_{n,m}$, $[\mathfrak{a}]$ is then always a (mod 2) cycle of which the homology class will be denoted by $Z[\mathfrak{a}]$. Remark that $[\mathfrak{a}]^*$ is a pseudomanifold and that the cycle carried by it is an element of $Z[\mathfrak{a}]$. Moreover, if $\mathfrak{a} = (a_1, \cdots, a_m)$ and $i_1 < \cdots < i_k$ be the indices such that $a_i < a_{i+1} (a_{m+1} = n)$, then the system of equations (1) is equivalent to the following system of equations

$$\mathrm{Dim}\,(X \cap R^{a_l+l}) \geqslant i, \quad i = i_1, \cdots, i_k. \tag{1$'$}$$

The pseudomanifold $[\mathfrak{a}]^*$ is then completely determined by the elements R^{a_i+i}, $i = i_1, \cdots, i_k$, but is independent of the choice of other $R^{i'}s$.

Now for any $\mathfrak{a} \in \Omega_{n,m}$ let u be the cochain for which $u[\mathfrak{a}] = 1$ and $u[\mathfrak{a}'] = 0$ for $\mathfrak{a}' \in \Omega_{n,m}, D(\mathfrak{a}') = D(\mathfrak{a})$, $\mathfrak{a}' \neq \mathfrak{a}$ is then a (mod 2) cocyclc of dimension $D(\mathfrak{a})$ whose cohomology class will be denoted by $(\mathfrak{a})$. Then we have

Theorem 1(Ehresmann) *The set of all classes $\{\mathfrak{a}\}$, $\mathfrak{a} \in \Omega_{n,m}$ forms an additive basis of generators of the cohomology ring $H^*(G_{n,m})$.*

Let us remark that the classes $\{\mathfrak{a}\}$ are in reality, as is easily seen, independent of the particular sequence (S) chosen and hence represent classes invariantly defined in $G_{n,m}$. Remark also that the class $\{\mathfrak{a}\}$ where $\mathfrak{a} = (0, \cdots, 0)$ is the unit class 1 of $H^*(G_{n,m})$.

The following classes are of particular interest:

$$W^i = \{\mathfrak{a}_i\}, \quad 0 \leqslant i \leqslant m,$$
$$\bar{W}^j = \{\bar{\mathfrak{a}}_j\}, \quad 0 \leqslant j \leqslant n,$$

where $\mathfrak{a}_i$ consists of $m-i$ 0's and i 1's, and $\mathfrak{a}_j$ consists of $m-1$ 0's and a single j. Remark that $W^0 = \overline{W}^0 = 1$. We shall extend the definition of W^i and $\bar{W}^j$ to all dimensions by setting

$$W^i = 0 \quad \text{for } i > m, \quad \overline{W^j} = 0 \quad \text{for } j > n.$$

These classes will be called respectively the *W-classes* and the *$\bar{W}$-classes* of $G_{n,m}$.

Now let $G_{n',m}$ be the Grassmannian manifold defined in a Euclidean space $R^{n'+m}$ where $n' > n$ and $R^{n'+m} \supset R^{n+m}$. Let us denote the classes, *etc.* , in $G_{n',m}$ by the same symbols as the corresponding classes, *etc.* , of $G_{n,m}$ by adding a prime. It is evident that $G_{n,m}$ is a submanifold of $G_{n',m}$. For the identical mapping

$$f : G_{n,m} \to G_{n',m},$$

we have

Lemma 1

$$f^*\{\mathfrak{a}\}' = \{\mathfrak{a}\}, \quad \text{for} \quad \mathfrak{a} \in Q_{n,m},$$
$$= 0, \quad \text{for } \mathfrak{a} \in \Omega_{n',m}, \quad \text{but } \notin \Omega_{n,m},$$

In particular, we have

$$f * W'^k = W^k, \quad f * \overline{W}'^k = \bar{W}^k$$

for all $k \geqslant 0$, with the above conventions.

Proof In $R^{n'+m}$ let us take an increasing sequence of i-elemerits R^i containing R^{n+m} as one of them:

$$R^1 \subset R^2 \cdots \subset R^{n+m} \subset \cdots \subset R^{n'+m}. \qquad (S)'$$

With respect to (S') let us define the cell subdivision of $G_{n',m}$ as before. As (S') contains (S) as a subsequence we have a corresponding cell subdivision of $G_{n,m}$ with respect to (S). For any $\mathfrak{a} \in \Omega_{n,m}$ the conditions (1) are then the same for either (S)

or (S'). It follows that with the above subdivisions $G_{n,m}$ is a subcomplex of $G_{n',m}$ and Lemma 1 follows immediately.

Consider now the Grassmannian manifold $G_{m,n}$ of n-elements in R^{n+m}. The classes, *etc.* of $G_{m,n}$ will again be denoted by the same symbols as those of $G_{n,m}$ by adding a prime. Define a natural mapping

$$g : G_{n.m} \to G_{m,n}$$

by $g(X) = \overline{X}$ where $\overline{X}$ is the n-element of $G_{m,n}$ completely orthogonal to the m-element X of $G_{n,m}$. Then we have

Lemma 2 $g^*W'^k = \bar{W}^k$, $g^*\bar{W}'^k = W^k$, $k \geqslant 0$.

Proof Let us make at first some preliminaries. For $\mathfrak{a} = (a_1, \cdots, a_m) \in \Omega_{n,m}$ define $\theta\mathfrak{a} = \mathfrak{b} = (b_1, \cdots, b_n) \in \Omega_{m,n}$ by $b_j = m - i$, for $0 \leqslant i \leqslant m$, $n - a_{i+1} < j \leqslant n - a_i$ $(a_0 = 0,\ a_{m+1} = n)$

$$\mathfrak{b} = (\underbrace{0, \cdots, 0}_{n-a_m}, \underbrace{1, \cdots, 1}_{a_m - a_{m-1}}, \cdots, \underbrace{m-i, \cdots, m-i}_{a_{i+1}-a_j}, \cdots, \underbrace{m, \cdots, m}_{a_1}). \qquad \text{i.e.}$$

Let $i(1 \leqslant i \leqslant m)$ be an index with $a_i > a_{i-1}(a_0 = 0)$ so that

$$\mathfrak{a}_{(i)} = (a_1, \cdots, a_{i-1}, a_l - 1, a_{j+1}, \cdots, a_m) \in \Omega_{n,m}$$

is defined. Then

$$\theta\mathfrak{a}_{(i)} = (\underbrace{0, \cdots, 0}_{n-a_m}, \cdots, \underbrace{m-i, \cdots, m+i}_{a_{i+1}-a_i+1}, \underbrace{m-i+1, \cdots, m-i+1}_{a_i - a_{i-1}-1}, \cdots, \underbrace{m, \cdots, m}_{a_1})$$

is easily seen to be $\mathfrak{b}_{(n-a_i+1)} \in \Omega_{m,n}$. Conversely, if $\mathfrak{b}_{(j)} \in \Omega_{m,n}$ is defined, we must have $j = n - a_i + 1$ for some i with $1 \leqslant i \leqslant m$ and $a_i > a_{i-1}$; then $\mathfrak{b}_{(j)} = \theta\, \mathfrak{a}_{(i)}$ where $\mathfrak{a}_{(i)} \in \Omega_{n,m}$ is defined.

The mapping $\theta : \Omega_{n,m} \to \Omega_{m,n}$ thus defined establishes evidently a 1-1-correspondence between $\Omega_{n,m}$ and $\Omega_{m,n}$. It has moreover a dual character, *i.e.* if $\theta : \Omega_{m,n} \to \Omega_{n,m}$ is similarly defined, then $\theta\mathfrak{b} = \mathfrak{a}$ if $\theta\mathfrak{a} = \mathfrak{b}$. The proof is also easy.

These preliminaries being made, let

$$R^1 \subset \cdots \subset R^{n+m-1} \qquad (S)$$

and

$$\overline{R^1} \subset \cdots \subset \bar{R}^{n+m-1} \qquad (S')$$

be two increasing sequences of i-elements in R^{n+m} such that R^i and $\bar{R}^{n+m-i}$ are completely orthogonal for $i = 1, \cdots, n+m-1$. We shall define cell subdivisions of $G_{n,m}$ and $G_{m,n}$ with respect to the sequences (S) and (S') respectively. For $\mathfrak{a} = (a_1, \cdots, a_m) \in \Omega_{n,m}$ the subset $[\mathfrak{a}]^*$ of $G_{n,m}$ is then the set of all m-elements X for which

$$\mathrm{Dim}(X \cap R^{a_1+\prime}) \geqslant i, \quad i = 1, \cdots, m. \tag{1}$$

When X is an m-element verifying the i-th equation of (1), $g(X) = \overline{X}$ and $\bar{R}^{n+m-a_i-i}$ will be contained in an element of dimension $\leqslant n+m-i$, namely the element completely orthogonal to the element $X \cap R^{a_i+1}$ of dimension at least i. It follows that

$$\mathrm{Dim}(X{\cap}R^{n+m-a_i-i}) \geqslant n + (n+m-a_1-i) - (n+m-i) = n - a_1.$$

The converse is easily seen to be also true. Consequently the set (1) is equivalent to the following set of conditions:

$$\mathrm{Dim}(\bar{X}{\cap}\bar{R}^{n+m-a_1-1}) \geqslant n - a_1, \quad i = 1, \cdots, m. \tag{2}$$

As is easily seen, (2) is equivalent in turn to the set of conditions

$$\mathrm{Dim}(X{\cap}R^{b_1+l}) \geqslant j, \quad j = 1, \cdots, n, \tag{3}$$

where $\mathfrak{b} = (b_1, \cdots, b_n) = \theta\mathfrak{a} \in \Omega_{m,n}$. As the elements $\bar{X}$ verifying (3) form the subset $[\mathfrak{b}]^*$ of $G_{m,n}$ and as g is evidently a homeomorphism of $G_{n,m}$ On $G_{m,n}$, g defines a homeomorphism of $[\mathfrak{a}]^*$ on $[\mathfrak{b}]^*(\mathfrak{b} = \theta\mathfrak{a})$. As we have already remarked, $\theta\mathfrak{a}_{(i)} - \mathfrak{b}_{(j)}$, where i runs over all indices for which $\mathfrak{a}_{(s)}$is defined and $j(j = n - a_i + 1)$ runs over all indices for which $\mathfrak{b}_{(j)}$ is defined, consequently g is also a homeomorphism of the open cell

$$[\mathfrak{a}] = [\mathfrak{a}]^* - \Sigma[\mathfrak{a}_{(i)}]^*,$$

on the open cell

$$[\mathfrak{b}] = [\mathfrak{b}]^* - \Sigma[\mathfrak{b}_{(j)}]^*.$$

This proves that g is a cell mapping of $G_{n,m}$ on $G_{m,n}$ with the above defined cell subdivisions. Algebraically (mod 2, as always), we have then for the homeomorphisms $g^{\#}$ of the chain groups induced by g:

$$g * [\mathfrak{a}] = [\mathfrak{b}], \quad \text{where } \mathfrak{b} = \theta\mathfrak{a}, \mathfrak{a} \in \Omega_{n,m}.$$

It follows that

$$g^*\{\mathfrak{b}\} = \{\mathfrak{a}\}, \tag{4}$$

where $\mathfrak{a} = \theta_1 \mathfrak{b} \in \Omega_{m,n}$. This determines completely the (mod 2) homology type of the mapping g. In particular, when $\mathfrak{b} = (\underbrace{0, \cdots, 0}_{n-k}, \underbrace{1, \cdots, 1}_{k})$ or $(\underbrace{0, \cdots, 0}_{n-1}, k)$ of $\Omega_{m,n}\theta\mathfrak{b} = \mathfrak{a}$ is given respectively by $\mathfrak{a} = (\underbrace{0, \cdots, 0}_{m-1}, k)$ or $(\underbrace{0, \cdots, 0}_{m-k}, \underbrace{1, \cdots, 1}_{k})$ of $\Omega_{n,m}$. Our lemma follows therefore from (4).

II The multiplicative structure of $H^*(G_{n,m})$[2]

By theorem 1 of Ehresmann as stated in §I, the cohomology ring $H^*(G_{n,m})$ of the grassmannian manifold $G_{n,m}$ has an additive basis consisting of all classes $\{\mathfrak{a}\}$, $\mathfrak{a} \in \Omega_{n,m}$. For the multiplicative structure of $H^*(G_{n,m})$ we have, first, the following result of Ehresmann in terms of intersections. For any $\mathfrak{a} = (a_1, \cdots, a_m) \in \Omega_{n,m}$ let $\mathfrak{a}' = (n - a_m, \cdots, n - a_1) \in \Omega_{n,m}$, then

$$\begin{cases} Z(\mathfrak{a}).Z(\mathfrak{a}') = 1, \\ Z(\mathfrak{a}).Z(\mathfrak{b}) = 0, \quad \mathfrak{b} \neq \mathfrak{a}', \quad D(\mathfrak{a}) + D(\mathfrak{b}) = nm. \end{cases} \tag{1}$$

Now in any r-dimensional closed manifold M there is a canonical isomorphism We will say that two elements of $H^i(M)$ and $H_{r-i}(M)$ respectively are dual to each other if they correspond under d. From (1) we see that $\{\mathfrak{a}\}$ and $Z(\mathfrak{a}')$ are dual elements. In particular, $W^i (0 < i \leqslant m)$ and $\bar{W}^i (0 < i \leqslant n)$ are dual respectively to $Z(\mathfrak{a}_i')$ and $Z(\bar{\mathfrak{a}}_i')$, where $\mathfrak{a}_i'$ has i $n{-}1$'s and $m - i$ n's, while $\bar{\mathfrak{a}'}_i$ has $1 n - i$ and $m - 1$ n's.

For the general multiplicative structure of the cohomology ring $H^*(G_{n,m})$, we have, by Chern,

$$\{\mathfrak{a}\} \cup \overline{W}^k = \Sigma\{\mathfrak{b}\}, \quad 0 \leqslant k \leqslant n, \tag{2}$$

where $\mathfrak{a} = (a_1, \cdots, a_m)$ and the summation is extended over all $\mathfrak{b} = (b_1, \cdots, b_m) \in \Omega_{n,m}$ such that

$$\begin{cases} a_i \leqslant b_i \leqslant a_{i+1} (a_{m+1} = n), \quad i = 1, \cdots, m, \\ D(\mathfrak{a}) + k = D(\mathfrak{b}). \end{cases} \tag{3}$$

From these formulas Chern deduces the following

Theorem 2(Chern) *The cohomology ring $H^*(G_{n,m})$ is generated by the classes W^k, $0 < k \leqslant m$ (respectively the classes $\bar{W}^k$, $0 < k \leqslant n$), so that each element of*

$H^*(G_{n,m})$ may be expressed as a polynomial of these $m-1$ classes (respectively $n-1$ classes) by means of cup-products.

We shall determine a little more precisely the multiplicative structure of the ring $H^*(G_{n,m})$. For this let x_i be "weighted" indeterminates of "weight" i, $i = 1, \cdots, s$ and let us call a polynomial (coefficients in the field of integers mod 2) *isobaric* of weight k if each of its terms $(x_1)^{i_1}(x_1)^{i_2}\cdots(x_s)^{i_s}$ is of weight k, i.e. if it satisfies the condition $i_1 + 2i_2 + \cdots + si_s = k$. The terms of the same weight k will be arranged in a lexicographic order $\ll$ as follows: We have

$$(x_1)^{i_2}(x_2)^{i_2}\cdots(x_s)^{i_3} \ll (x_1)^{j_1}(x_2)^{j_2}\cdots(x_3)^{j_3} \quad (\Sigma ri_r = \Sigma rj_1 = k),$$

if and only if either $\sum\limits_{r=1}^{s} i_r < \sum\limits_{r=1}^{s} j_r$, or $\sum\limits_{r=1}^{s} i_r = \sum\limits_{r=1}^{s} j_r$, and for some r, $0 \leqslant r < s$, we have

$$i_1 = j_1, \cdots, i_r = j_r, \quad i_{r+1} > j_{r+1}.$$

These being granted, we have then

Lemma 3 *Let $\varphi_k(x_i)$ be a non-identically zero polynomial in the weighted indeterminates x_i, $i = 1, \cdots, n$, which is isobaric of weight $k < 0$. If k is $\leqslant m$, then $\varphi_k(\overline{W}^i)$, with cup-products as multiplications, is a non-zero element of $H^*(G_{n,m})$. In other words, for $k \leqslant m$, the set of all k-dimensional classes of the form $(\overline{W}^1)^{k_1}(\overline{W}^2)^{k_2}\cdots(\overline{W}^n)^{k_n}$ are linearly independent in the ring $H^*(G_{n,m})$.*

Proof Let us arrange the classes of same dimension k in $H^*(G_{n,m})$ in a certain partial order $\prec$ as follows: Let $\mathfrak{a} = (a_1, \cdots, a_m)$, $\mathfrak{b} = (b_1, \cdots, b_m) \in \Omega_{n,m}$ with $D(\mathfrak{a}) = D(\mathfrak{b}) = k$. Then we define $\{\mathfrak{a}\} \prec \{\mathfrak{b}\}$ if and only if for somc r, $0 \leqslant r < m$, we have

$$a_1 = b_1, \cdots, a_r = b_r, \quad a_{r+1} < b_{r+1}.$$

By theorem 1 of §I, every k-dimensional class Z of $H^*(G_{n,m})$ may be written in the form $Z = \sum\limits_{i=1}^{r} \{\mathfrak{a}_i\}$ with $\mathfrak{a}_i \in \Omega_{n,m}$, $D(\mathfrak{a}_i) = k$, and

$$\{\mathfrak{a}_1\} \prec \{\mathfrak{a}_2\} \prec \cdots \prec \{\mathfrak{a}_r\}.$$

The class $(\mathfrak{a}_r)$ will then be called the *leading term* of the class Z. For any two k-dimensional classes Z and Z' we shall then define $Z \prec Z'$ if and only if for their leading terms $\{\mathfrak{a}\}$ and $\{\mathfrak{a}'\}$ we have $\{\mathfrak{a}\} \prec \{\mathfrak{a}'\}$.

From (2) and (3) we see that the leading term of $\{0, \cdots, 0, a_{j+1}, \cdots, a_m) \cup W^{a_j}$ with $a_j \leqslant a_{j+1}$, $j > 0$, when expanded into a linear sum of classes $[\mathfrak{b}]$, $\mathfrak{b} \in \Omega_{n,m}$, is the class $\{0, \cdots, 0, a_j, a_{j+1}, \cdots, a_m\}$. By induction it follows then easily that the leading term of a class $(\overline{W}^1)^{k_1}(\overline{W}^2)^{k_2}\cdots(\overline{W}^n)^{k_n}$, when expanded by formulas (2) and (3), is the class $\{0,\ldots,0,1\underbrace{,\ldots,1,}_{k_1}2\underbrace{,\ldots,2,}_{k_2}\cdots,n\underbrace{,\ldots,}_{k_n}n\}$ whenever $\sum\limits_{i=1}^{n} k_i \leqslant m$, in particular when the class is of dimension $k=\sum\limits_{j=1}^{n} rk_r \leqslant m$.

It follows that for any two k-dimensional classes $(\overline{W}^1)^{i_1}(\overline{W}^2)^{i_2}\cdots(\overline{W}^n)^{i_n}$ and $(\overline{W}^1)^{j_1}(\overline{W}^2)^{j_2}\cdots(\overline{W}^n)^{j_n}$, where $k\leqslant m$, we would have

$$(\overline{W}^1)^{i_1},(\overline{W}^2)^{i_2}\cdots(\overline{W}^n)^{i_n} \prec (\overline{W}^1)^{j_1}(\overline{W}^2)^{j_2}\cdots(\overline{W}^n)^{j_n} \tag{4}$$

if and only if $\{0,\cdots,0,\underbrace{1,\cdots,1,}_{i_1}\underbrace{2,\cdots,2,}_{i_2}\underbrace{n,\cdots,n}_{i_n}\} \prec \{0,\cdots,0,\underbrace{1,\cdots,1,}_{j_1}\underbrace{2,\cdots,2,}_{j_2}\underbrace{n,\cdots,n}_{j_n}\}$, i.e., if and only if either $\sum i_r < \sum j_r$ or $\sum i_r = \sum j_r$ and r exists with $0\leqslant r<n$, $i_1 = j_1, \cdots, i_r = j_r, i_{r+1} > j_{r+1}$. Therefore when $k\leqslant m$, (4) is eqnivalent to

$$(x_1)^{i_1}(x_2)^{i_2}\cdots(x_n)^{i_n} \ll (x_1)^{j_1}(x_2)^{j_2}\cdots(x_n)^{j_n},$$

where x_i, $i=1,\cdots,n$, are indeterminates of weight i.

Consider now a polynomial $\varphi_k(x_i)$ which is not identically zero and isobaric of weight $k\leqslant m$ in the indeterminates $x_1,\cdots,x_n$. We may write $\varphi_k(x_1)$ in the form

$$\varphi_k(x_i)=\sum_{r=1}^{s} z_r,$$

where each z_r is a term of the form $(x_1)^{r_1}(x_2)^{r_2}\cdots(x_n)^{r_n}$ and where $z_1\ll z_2\ll\cdots\ll z_s$. Then we have

$$\varphi_k(W^1,)=\sum_{r=1}^{s} Z_1,$$

where each Z_r is of the form $(\overline{W}^1)^{r_1}(\overline{W}^2)^{r_2}\cdots(\overline{W}^n)^{r_n}$ with exponents same as those of z_r. By what precedes we should have

$$Z_1<Z_2<\cdots<Z_3.$$

It follows that the leading term of Z_s is also the leading term of $\sum\limits_{r=1}^{s} Z_r$ and cannot be cancelled in the expansion of $\varphi_k(\overline{W}^i)$. By theorem 1, we have therefore $\varphi_k(\overline{W}^i)\neq 0$ and our lemma is proved.

Lemma 4 *Let $\varphi_k(x_i)$ be a polynomial not identically zero and isobaric of weight k in the weighted indeterminates. x_i, $i = 1, \cdots, m$. If $k \leqslant n$. then $\varphi_k(W^i)$ is a non-zero element of $H^*(G_{n,m})$.*

Proof Let its consider besides $G_{n,m}$ also the manifold $G_{m,n}$ and the mapping $g: G_{n,m} \to G_{m,n}$ defined in §I. The classes of $G_{m,n}$ will be, as before, distinguished from those of $G_{n,m}$ by adding primes. By Lemma 3, the class $\varphi_k(\overline{W}'^i)$ is not zero in $H^*(G_{m,n})$ when $k \leqslant n$. By Lemma 2, we have $g * \overline{W}' = W'$, $i \geqslant 0$. As g is a homeomorphism and g^* is an isomorphisin, we have also

$$\varphi_k(W') = \varphi_k(g^*\overline{W}'') = g^*\varphi_k(\overline{W}'') \neq 0$$

in $H^*(G_{n,m}) = g^*H^*(G_{m,n})$ when $k \leqslant n$. This proves Lemma 4.

The following much simpler proof of Lemma 4 is due to H. Cartan:

Let us consider the set of all k-dimensional classes of $H^*(G_{n,m})$ of the form

$$(W^1)^{k_1}(W^2)^{k_2}\cdots(W^m)^{k_m}. \tag{5}$$

The number α_k of such classes is equal to the number of sets of integers $(k_1, \cdots, k_m)$ such that

$$k_1 + 2k_2 + \cdots + mk_m = k, \quad k_1 \geqslant 0, \quad i = 1, \cdots, m. \tag{6}$$

Let us put

$$k_m = a_1, k_{m-1} = a_2 - a_1, \cdots, k_1 = a_m - a_{m-1}. \tag{7}$$

As $k_i \geqslant 0$, we have $0 \leqslant a_1 \leqslant a_2 \leqslant \cdots \leqslant a_m$. When $k \leqslant m$, we would have $a_m \leqslant \sum k_i \leqslant \sum ik_i \leqslant m$ and hence $\mathfrak{a} = (a_1, \cdots, a_m)$ is in $\Omega_{n,m}$ with $D(\mathfrak{a}) = \sum a_i = \sum ik_i = k$.

Conversely, for any such $\mathfrak{a} \in \Omega_{n,m}$ with $D(\mathfrak{a}) = k$ the equations (7) determine a system of values k_i satisfying (6) and give rise to a k-dimensional class of the form (5). This correspondence is evidently one-to-one and reciprocal. It follows that a_k, when $k \leqslant m$, is equal to the number of sets $\mathfrak{a} \in \Omega_{n,m}$ with $D(\mathfrak{a}) = k$, i.e., equal to the k-th betti number β_k of $G_{n,m}$, according to Theorem 1. By Theorem 2, every k-dimensional class of $H^*(G_{n,m})$ must be a sum of classes of the form (5). It follows that the β_k classes $(k \leqslant m)$ of the form (5) must be linearly independent and Lemma 4 follows immediately.

III　A canonical mapping $h: G_{n_1,m_1} \times G_{n_2,m_2} \to G_{n,m}$

Let R^{n+m} be the product of two Euclidean spaces $R_j^{n_j+m_j}$ of dimension n_j+m_j, $j=1, 2$, where $n=n_1+n_2$, $m=m_1+m_2$. Let $G_{n_j,m_j}, j=1, 2$ and $G_{n,m}$ be the grassmannian manifolds defined in $R_j^{n_j+m_j}$, $j=1, 2$, and R^{n+m} respectively. For any two m_j-elements X_i of $R_j^{n_j+m_j}$, $j=1, 2$, the product of X_1 and X_2 is an m-element X of R^{n+m}, hence $h(X_1 \times X_2)=X$ defines a continuous mapping

$$h: \quad G_{n_1,m_1} \times G_{n_2,m_2} \to G_m.$$

Now let $Wj'(\overline{W}j')$, $j=1, 2$ and $W'(\overline{W}^i)$, $i \geqslant 0$, be the W-classes ($\overline{W}$-classes) of G_{n_j,m_j}, $j=1, 2$ and $G_{n,m}$ respectively. By Theorem 2, the cohomology ring of $G_{n,m}$, is generated by $W^i(\overline{W}^i)$. Hence the homology type of h^* will be completely determined by $h^*W^i(h^*\overline{W}^i)$ for which we have

Lemma 5

$$h^*W^i = \sum_{i_1+i_2=i} W_1^{i_1} \otimes W_2^{i_2}, \qquad i \geqslant 0, \tag{1}$$

$$h^*\overline{W}^i = \sum_{i_1+i_2=i} \overline{W}_1^{i_1} \otimes W_2^{i_2}, \qquad i \geqslant 0. \tag{2}$$

Proof　The formula (2) is trivial for $i=0$ or $i>n$. Hence, to prove (2), we may assume that $0<i \leqslant n$.

Let $\mathfrak{a}_i \in \Omega_{n,m}$ be the sequence having $1\ n-i$ and $m-1\ n$'s and let $\mathfrak{a}_{i_j}^{(j)} \in \Omega_{n_j,m_j} j=1, 2$ be the sequence having m_i-1 0's and 1 i_i. In each of Ω_{n_j,m_j}, $j=1$, 2, take a sequence $\mathfrak{b}^{(j)}=(b_1^{(j)}, \cdots, b_{m_j}^{(j)})$ such that $D(\mathfrak{b}')+D(\mathfrak{b}'')=i$, then there are two possibilities, according as

$$b'_{m_1}+b''_{m_2}<i \tag{3}$$

or $=i$. In the former case at least one of the two $\mathfrak{b}^{(j)}$, $j=1, 2$, is not any $\mathfrak{a}_{i_j}^{(j)}$, in the latter case there will exist i_1, i_2 such that $i=i_1+i_2$, $\mathfrak{b}^{(j)}=\mathfrak{a}_{i_j}^{(j)}$, $j=1, 2$.

By (1) of §II it is evident that (2) is equivalent to the following intersection formulas:

$$h(Z(\mathfrak{b}') \times Z(\mathfrak{b}'')).Z(\mathfrak{a}_i)=\begin{cases} 0, & \text{if } b'_{m_1}+b''_{m_2}<i, \\ 1, & \text{if } \mathfrak{b}^{(j)}=\mathfrak{a}_{i_j}^{(j)}, j=1,2 \text{ and } i=i_1+i_2, \end{cases} \tag{4}$$

in which $\mathfrak{b}^{(j)}=(b_1^{(j)}, \cdots, b_{m_j}^{(j)}) \in \Omega_{n_j,m_j}$, $D(\mathfrak{b}')+D(\mathfrak{b}'')=i$.

Consider first the case that (3) is true. In that case we may take in each of $R_j^{n_j+m_j}$ a $b_{m_j}^{(j)}+m_j$-element R_j, $j=1, 2$, and in R^{n+m} an $n-i+1$-element R_0 such that, denoting by $R_1\times R_2$ the topological product of R_i, considered as a linear subspace of R^{n+m},

$$\mathrm{Dim}\ (R_0\cap(R_1\times R_2))=0. \tag{5}$$

The existence of such elements R_0, R_j, $j=1, 2$ is assured by (3).

Let $[\mathfrak{a}_i]^*$ be the set of m-elements X in $G_{n,m}$ satisfying the condition:

$$\mathrm{Dim}\ (X\cap R_0)\geqslant 1 \tag{6}$$

and $[\mathfrak{b}^{(j)}]^*$, $j=1, 2$ be the set of m_j-elements X_i in G_{n_j,m_j} satisfying respectively the conditions $\mathrm{Dim}\ (X_i\cap R^{b_i^{(j)}+1})\geqslant i$, $i=1,\cdots,m_j$, in particular the condition $X_j\subset R_j$. Then $[\mathfrak{a}_i]^*$, $[\mathfrak{b}^{(j)}]^*$ are all pseudomanifolds representing cycles in the homology classes $Z(\mathfrak{a}_i)$ and $Z(\mathfrak{b}^{(j)})$ respectively.

If $X_j\in[\mathfrak{b}^{(j)}]^*$, $j=1, 2$, so that $X_j\subset R_j$, then $h(X_1\times X_2)=X$ is $\subset R_1\times R_2$. By (5), X cannot satisfy (6), or $X\bar{\in}[\mathfrak{a}_i]^*$. Hence h $([\mathfrak{b}']^*\times[\mathfrak{b}'']^*)$ and $[\mathfrak{a}_i]^*$ are disjoint sets and we have

$$h(Z(\mathfrak{b}')\times Z(\mathfrak{b}'')).Z(\mathfrak{a}_i)=0.$$

This proves the upper part of (4).

Next, let $\mathfrak{b}^{(j)}=\mathfrak{a}_{i_j}^{(j)}$, $j=1, 2$ and $i_1+i_2=i$. Let us take in each of $R_j^{n_j+m_j}$, $j=1, 2$ a basis of vectors $(e_k^{(j)})=(e_1^{(j)},\cdots,e_{n_j+m_j}^{(j)})$. Let R_j^s be the linear subspace of $R_j^{n_j+m_j}$, $j=1, 2$, determined by the vectors $e_1^{(j)}$, $\cdots$, $e_s^{(j)}$, and let R^{n-i+1} he the linear subspace of R^{n+m} determined by vectors

$$e'_{m_1+i_1+1},\cdots,e'_{n_1+m_1};e''_{m_2+i_2+1},\cdots,e''_{n_2+m_2};e'_{m_1}+e''_{m_2}.$$

The set $[\mathfrak{b}^{(j)}]^*$, $j=1, 2$, of G_{n_j,m_j} formed by all m_j-elements X_j such that

$$R_j^{m_1-1}\subset X_1\subset R_j^{m_j+i_j}$$

and the set $[\mathfrak{a}_i]^*$ of $G_{n,m}$ formed by all m-elements X such that

$$\mathrm{Dim}\ (X\cap R^{n-i+1})\geqslant 1$$

are all pseudomanifolds, representing cycles belonging to $Z(\mathfrak{b}^{(j)})$, $j=1, 2$, and $Z(\mathfrak{a}_j)$ respectively. We see easily that $h([\mathfrak{b}']^*\times[\mathfrak{b}'']^*)$ and $[\mathfrak{a}_i]^*$ have one and only one element X_0 in common, where X_0 is determined by the vectors

$$e'_1,\cdots,e'_{m_1};e''_1,\cdots,e''_{m_2}.$$

It is easy to verify the intersection index of $h([\mathfrak{b}']^* \times [\mathfrak{b}'']^*)$ and $[\mathfrak{a}_i]^*$ to be 1, so that

$$h(Z(\mathfrak{b}') \times Z(\mathfrak{b}''), \quad Z(\mathfrak{a}_i) = 1.$$

This proves the lower part of (4).

(4) and hence (2) is thus proved.

We may use the same method to prove (1). But in what follows we will deduce (1) from (2) by means of Lemma 2.

Let us consider the grassmannian manifolds $G_{m,n}$ and G_{m_j,n_j} in R^{n+m} and $R_j^{n_j+m_j}$, $j=1$, 2, respectively. As in §1 we have canonical mappings

$$g: G_{n,m} \to G_{m,n}$$

and

$$g_j: G_{n_j,m_j} \to G_{m_j,n_j}, \quad j=1,2.$$

We have also, similar to h, a canonical mapping

$$h': G_{m_1,n_1} \times G_{m_2,n_2} \to G_{m,n}.$$

Let

$$g': G_{n_1,m_1} \times G_{n_2,m_2} \to G_{m_1,n_1} \times G_{m_2,n_2}$$

be the mapping detained by $g'(X_1 \times X_2) = g_1(X_1) \times g_2(X_2)$, $X_j \in G_{n_j,m_j}$. Then we have evidently $h'g' = gh$. Consequently

$$g'^* h'^* = h^* g^*. \tag{7}$$

Denote the classes of $G_{m,n}$ and G_{m_j,n_j} by the same symbols as, adding however a prime to distinguish from, the corresponding classes of $G_{n,m}$ and G_{n_j,m_j}, then we have, by Lemma 2 of §I,

$$g_j^* W'^{i_j}_j = W_j^{i_j}, \quad i_j \geqslant 0, j=1,2, \tag{8}$$

$$g^* \overline{W}'^i = W^i, \qquad i \geqslant 0. \tag{9}$$

We have also

$$g'^*(\overline{W}'^{i_1}_1 \otimes \overline{W}'^{i_2}_2) = g_1^* W'^{i_1}_1 \otimes g_2^* \overline{W}'^{i_2}_2. \tag{10}$$

Finally, by applying (2) on $G_{m,n}$, G_{m_j,n_j}, $j=1$, 2, and h', (2) becomes

$$h'^* W'^i = \sum_{i_1+i_2=1} \overline{W}'^{i_1}_1 \otimes \overline{W}'^{i_2}_2. \tag{11}$$

From (7)—(11), we get

$$
\begin{aligned}
h^*W' &= h^*g^*\overline{W}'^i \\
&= g'^*h'^*\overline{W}'^i \\
&= \sum_{i_1+i_2=1} g'^*(\overline{W}_1'^{i_1} \otimes \overline{W}_2'^{i_2}) \\
&= \sum_{i_1+i_2=i} (g_1^*\overline{W}_1'^{i_1} \otimes g_2^*\overline{W}_2'^{i_2}) \\
&= \sum_{i_1+i_2=i} \overline{W}_1^{i_1} \otimes \overline{W}_2^{i_2},
\end{aligned}
$$

this proves (1).

In an appendix an alternative proof of Lemma 5 will be given.

Lemma 6 *Let $\varphi_k(x_i)$ be a polynomial isobaric of weight k in the weighted indeterminates x_i, $0 < i \leqslant m$. Then for $k \leqslant n_1$ and n_2, $\varphi_k(W^i)$ is a non-zero element of $H^*(G_{n,m})$ if and only if $h^*\varphi_k(W^i)$ is a non-zero element of $H^*(G_{n_1,m_1} \times G_{n_2,m_2})$, where $h : G_{n_1,m_1} \times G_{n_2,m_2} \to G_{n,m}$ is the canonical mapping defined above.*

Before proving this lemma let us first establish the following

Lemma 7 Let $x_i^{(j)}$, $0 < i \leqslant m_j$, $j = 1, 2$, $m_1 + m_2 = m$, *be indeterminates such that* (Coefficients in the field of integers mod 2)

$$
x_i = \sum_{i_1+i_2=i} x'_{i_1}x''_{i_2}, \quad 0 < i \leqslant m \quad (x'_0 = 1, x''_0 = 1) \tag{12}
$$

i.e. ,

$$
\begin{cases}
x_1 = x'_1 + x''_1, \\
x_2 = x'_2 + x'_1x''_1 + x''_2, \\
\quad \vdots \\
x_m = x'_{m_1}x''_{m_2}.
\end{cases} \tag{12$'$}
$$

If the polynomial $\varphi(x_i)$ in x_i is not identically zero (mod 2), so is the polynomial $\varphi(\sum x'_{i_2}x''_{i_2})$ in the indeterminates $x_i^{(j)}$.

Proof Let F be a field of characteristic 2 containing $x_i^{(j)}$ such that the equations

$$
\zeta^{m_j} + x_1^{(j)}\zeta^{m_j-1} + \cdots + x_i^{(j)}\zeta^{m_j-i} + \cdots + x_{m_j}^{(j)} = 0, \quad j = 1,2 \tag{13}
$$

have roots $z_1^{(j)}$, $\cdots$, $z_{m_j}^{(j)}$. Then $x_i^{(j)}$ are elementary symmetric functions $p_i^{(j)}$ of $z_1^{(j)}$, $\cdots$, $z_{m_j}^{(j)}$, $0 < i \leqslant m$, $j = 1, 2$. Multiplying together $(13)_1$ and $(13)_2$, we get an equation which reduces, by (12) and (12)′, to the following equation:

$$\zeta^m + x_1\zeta^{m-1} + \cdots + x_i\zeta^{m-i} + \cdots + x_m = 0. \tag{14}$$

In the field F, (14) has roots z_1', $\cdots$, z_{m_1}', z_1'', $\cdots$, z_{m_2}'', and x_1 are elementary symmetric functions p_i of them. It is easily seen that

$$p_i = \sum_{i_1+i_2=i} p_{i_1}' p_{i_2}'', \quad 0 < i \leqslant m \quad (p_0' = 1, p_0'' = 1) \tag{15}$$

i. e.

$$\begin{cases} p_1 = p_1' + p_1'', \\ p_2 = p_2' + p_1'p_1'' + p_2'', \\ \quad\vdots \\ p_m = p_{m_1}'p_{m_2}''. \end{cases} \tag{15$'$}$$

Suppose now that $\varphi(x_i)$ is a polynomial not identically zero in x_i. By a theorem of algebra, $\varphi(p_i)$ is then a polynomial not identically zero in the indeterminates $z_i^{(j)}$, $0 < i \leqslant m_j$, $j = 1$, 2. By (15) or (15)′, this means that the polynomial $\varphi(\sum p_{i_1}' p_{i_2}'')$ is not identically zero in the indeterminates $z_1^{(j)}$. It follows that the polynomial $\varphi(\sum x_{i_1}' x_{i_2}'')$ cannot be identically zero in the indeterminates $x_i^{(j)} 0 < i \leqslant m$, $j = 1$, 2. This proves our lemma.

Proof of Lemma 6　Let $\varphi_K(x_i)$ be a polynomial isobaric of weight k in x_i, $i = 1$, 2, $\cdots$, m and suppose that $\varphi_k(W^i)$ is a non-zero element of $H^*(G_{n,m})$. This implies that the polynomial $\varphi_k(x_i)$ is not identically zero and hence by Lemma 7 $\varphi_k(\sum x_{i_1}' x_{i_2}'')$ is also non-identically zero in the indeterminates $x_i^{(j)}$, $0 < i \leqslant m$, $j = 1$, 2. By Lemma 4 the classes of any dimension i of the form $(W_j^1)^{i_1} \cdots (W_j^{m_j})^{i_{m_j}}$, where $i_1 + 2i_2 + \cdots + m_j i_{m_j} = i$, are linearly independent in the cohomology ring $H^*(G_{n_j,m_j})$, $j = 1$, 2, inasmuch as $i \leqslant n_j$. It follows that the classes of dimension k of the form

$$(W_1^1)^{r_1'}(W_1^2)^{r_2'} \cdots (W_1^{m_1})^{r'_{m_1}} \otimes (W_2^1)^{\rho''_1}(W_2^2)^{r''_2} \cdots (W_2^{m_2})^{r''_{m_2}},$$

where $i_1' + 2i_2' + \cdots + m_1 i_{m_1}' + i_1'' + 2i_2'' + \cdots + m_2 i_{m_2}'' = k$, are linearly independent in the cohonmology ring $H^*(G_{n_1,m_1} \times G_{n_2,m_2})$ inasmuch as $k \leqslant n_1$ and n_2. Consequently for $k \leqslant n_1$ and n_2, $\varphi_k(\sum W_1^{i_1} \otimes W_2^{i_2})$ is a non zero element of $H^*(G_{n_1,m_1} \times G_{n_2,m_2})$, since $\varphi_k(\sum x_{i_1}' x_{i_2}'')$ is not identically zero. By Lemma 4,

$$\varphi_k\left(\sum W_1^{i_1} \otimes W_2^{i_2}\right) = \varphi_k(h^* W^i) = h^* \varphi_k(W').$$

Hence $h^*\varphi_k(W^i)$ is non-zero in $H^*(G_{n_1,m_1}\times G_{n_2,m_2})$ for $k\leqslant n_1$ and n_2.

As it is evident that $\varphi_k(W^i)\neq 0$ in $H^*(G_{n,m})$ if $h^*\varphi(W^i)\neq 0$ in $H^*(G_{n_1,m_1}G_{n_2,m_2})$, our lemma is proved.

IV The squares in $G_{n,m}$

For the determination of square operations in $G_{n,m}$, we are in need of the following theorems, due to H. Cartan[5]:

Lemma 8 *If U_i, $i=1, 2$, are cohomology classes of topological spaces E_i, then*

$$Sq^r(U_1\otimes U_2)-\sum_{r_1+r_2=r} Sq^{r_1}U_1\otimes Sq^{r_2}U_2.$$

Lemma 9 *If U_i, $i=1,2$, are cohomology classes of same topological space E, then*

$$Sq^r(U_1\cup U_2)=\sum_{r_1+r_2=r} Sq^{r_1}U_1\cup Sq^{r_2}U_2.$$

In view of Lemma 9 and the Theorem 2 of Chern, it is sufficient to detertnine the squares of the classes W^i, $0<i\leqslant m$ (or W^i, $0<i\leqslant n$) of $G_{n,m}$ in order to determine the squares in $G_{n,m}$. The answer is given by the following theorem:

Theorem 3 *Let $\binom{p}{q}$ be binomial coefficients* (mod 2) *for $p\geqslant q>0$, with the convention $\binom{p}{q}=0$ for $p<q<0$ and $\binom{p}{o}=1$ for $p\geqslant -1$ (p, q are integers), then*

$$\text{(I)}\qquad Sq^rW^s=\sum_{t=0}^{r}\binom{s-r+t-1}{t}W^{r-t}W^{s+t},$$

$$\text{(II)}\qquad Sq^r\bar{W}^s=\sum_{t=0}^{r}\binom{s-r+t-1}{t}\overline{W}^{r-t}\overline{W}^{s+t}.\qquad (s\geqslant r\geqslant 0).$$

In particular, we have, e.g.

$$Sq^1W^s=W^1W^s+(s-1)W^{s+1},\tag{1}$$

$$Sq^2W^s=W^2W^s+(s-2)W^1W^{s+1}+\binom{s-1}{2}W^{s+2},\tag{2}$$

$$Sq^rW^m=W^rW^m.\tag{3}$$

Proof of (I)$_3$ Let us put

$$\varphi_{r,s}(W^i)=Sq^rW^s+\sum_{t=0}^{r}\binom{s-r+t-1}{t}W^{r-t}W^{s+t}.$$

The formula (I) is then equivalent to the following one:

$$\varphi_{r,s}(W^i) = 0, \quad s \geqslant r \geqslant 0. \tag{(I)$'$}$$

For the extreme cases $r = 0$ or $r = s$, we have respectively, taking into account our conventions:

$$\begin{aligned}
\varphi_{0,s}(W^i) &= Sq^0 W^s + \binom{s-1}{0} W^0 W^s = 0,\\
\varphi_{s,s}(W^i) &= Sq^s W^s + \sum_{t=0}^{s} \binom{t-1}{t} W^{s-t} W^{s+t}\\
&= (W^s)^2 + \binom{-1}{0} (W^s)^2\\
&= 0.
\end{aligned}$$

The formula (I)$'$ or (I) is thus trivially true in these cases. As $\varphi_{r,s}(W^i)$ is evidently 0 for $s > m$, we may therefore restrict henceforth our consideration to the case $m \geqslant s > r > 0$.

Let us prove now (I)$'$ by induction on m. For $m = 1$, we must have either $s > m$ or $r = s = 1$ or $r = 0$. All these have been seen to be trivially true. Suppose therefore that (I)$'$ has already been proved for $G_{n,k}$ where $k < m$ and let us consider (I)$'$ for $G_{n,m}$ where $m > 1$, with the restriction $m \geqslant s > r > 0$.

Let us consider first the case $n \geqslant 2(r+s)$. In this case we may take integers n_j, $j = 1, 2$ with $n = n_1 + n_2$, $n_j \geqslant r + s$. Put $m_1 = m - 1$, $m_2 = 1$ and define, as in §III, the canonical mapping

$$h : G_{n_1,m_1} \times G_{n_2,m_2} \to G_{n,m}.$$

Denote the W-classes of G_{n_j,m_j} by W_j^i, then, as W_2^i=0 for $i > 1$, Lemma 4 gives

$$h^* W^i = W_1^i \otimes 1 + W_1^{i-1} \otimes W_2^1, \quad i \geqslant 0$$

(with the convention $W_1^{-1} = 0$). By Lemma 8, we have also

$$Sq^r (W_1^s \otimes 1) = Sq^r W_1^s \otimes 1,$$

$$\begin{aligned}
Sq^r (W_1^{s-1} \otimes W_2^1) &= Sq^r W_1^{s-1} \otimes W_2^1 + Sq^{r-1} W_1^{s-1} \otimes Sq^1 W_2^1\\
&= Sq^r W_1^{s-1} \otimes W_2^1 + Sq^{r-1} W_1^{s-1} \otimes (W_2^1)^2.
\end{aligned}$$

It follows that

$$h^*\varphi_{r,s}(W^i) = Sq^r h^* W^s + \sum_{t=0}^{r} \binom{s-r+t-1}{t} h^* W^{r-t} \cdot h^* W^{s+t}$$

$$\begin{aligned}
=&Sq^r W_1^s \otimes 1 + Sq^r W_1^{s-1} \otimes W_2^1 + Sq^{r-1} W_1^{s-1} \otimes (W_2^1)^2 \\
&+\sum_{t=0}^{r} \binom{s-r+t-1}{t} W_1^{r-t} W_1^{s+t} \otimes 1 \\
&+\sum_{t=0}^{r} \binom{s-r+t-1}{t} W_1^{r-t-1} W_1^{s+t-1} \otimes (W_2^1)^2 \\
&+\sum_{t=0}^{r} \binom{s-r+t-1}{t} (W_1^{r-t-1} W^{s+t} + W_1^{r-t} W_1^{s+t-1}) \otimes W_2^1,
\end{aligned}$$

or

$$h^*\varphi_{r,s}(W^i) = \varphi_1 \otimes 1 + \varphi_2 \otimes W_2^1 + \varphi_3 \otimes (W_2^1)^2, \tag{4}$$

where we have put

$$\begin{aligned}
\varphi_1 =&Sq^r W_1^s + \sum_{t=0}^{r} \binom{s-r+t-1}{t} W_1^{r-t} W_1^{s+t} = \varphi_{r,s}(W_1^i), \\
\varphi_3 =& Sq^{r-1} W_1^{s-1} + \sum_{t=0}^{r} \binom{s-r+t-1}{t} W_1^{r-t-1} W_1^{s+t-1} \\
=& Sq^{r-1} W_1^{s-1} + \sum_{t=0}^{s-1} \binom{s-r+t-1}{t} W_1^{r-t-1} W_1^{s+t-1} \\
=& \varphi_{r-1,s-1}(W_1^i),
\end{aligned}$$

and

$$\begin{aligned}
\varphi_2 =&Sq^r W_1^{s-1} + \sum_{t=0}^{r} \binom{s-r+t-1}{t} W_1^{r-t-1} W_1^{s+t} \\
&+\sum_{t=0}^{r} \binom{s-r+t-1}{t} W_1^{r-t} W_1^{s+t-1} \\
=&Sq^r W_1^{s-1} + \sum_{r=1}^{s} \binom{s-r+t-2}{t-1} W_1^{r-t} W_1^{s+t-1} \\
&+\sum_{t=0}^{r} \binom{s-r+t-1}{t} W_1^{r-t} W_1^{s+t-1} \\
=&Sq^r W_1^{s-1} + W_1^r W_1^{s-1} + \sum_{t=1}^{r} \left[\binom{s-t+t-2}{t-1} \right.
\end{aligned}$$

$$+\binom{s-r+t-1}{t}\Big]W_1^{r-t}W_1^{s+t-1}.$$

Now $s>r>0$, as we have assumed. Hence for $1\leqslant t\leqslant r$ so that $s-r+t-1\geqslant t>0$, we have (mod 2)

$$\binom{s-r+t-2}{t-1}+\binom{s-r+t-1}{t}=\binom{s-r+t-2}{t}.$$

We get therefore

$$\begin{aligned}\varphi_2&=Sq^rW_1^{s-1}+W_1^rW_1^{s-1}+\sum_{t=1}^{r}\binom{s-r+t-2}{t}W_1^{r-t}W_1^{s+t-1}\\&=Sq^rW_1^{s-1}+\sum_{t=0}^{r}\binom{s-r+t-2}{t}W_1^{s-t}W_1^{s+t-1}\\&=\varphi_{r,s-1}(W_1^i).\end{aligned}$$

By our induction hypothesis we have

$$\varphi_{r,s}(W_1^i)=0,\varphi_{r,s-1}(W_1^i)=0,\varphi_{r-1,s-1}(W_1^i)=0,$$

i.e. $\varphi_1=0$, $\varphi_2=0$, $\varphi_3=0$ and therefore (4) becomes

$$h*\varphi_{r,s}(W^i)=0.$$

Now by Theorem 2 of Chern, Sq $^rW^s$ must be a polynomial in W^i by means of cup-products and hence $\varphi_{r,s}(W^i)$ is also a certain polynomial in W^i, say $\varphi_{r+s}(W^i)$. The form of the polynomial $\varphi_{r+s}(W^i)$ is unique since the dimension of $\varphi_{r,s}(W^i)$ is $r+s\leqslant n$ (cf. Lemma 4 of §II). Hence $\varphi_{r+s}(x_i)$ is a polynomial isobaric of weight $r+s$ in the indeterminates x_i, $0<i\leqslant m$. As $r+s\leqslant n_1$ and n_2 and $h^*\varphi_{r+s}(W^i)=h^*\varphi_{r,s}(W^i)$ is zero in $H^*(G_{n_1,m_1}\times G_{n_2,m_2})$, we have by Lemma 6 of §III,

$$\varphi_{r+s}(W^i)=\varphi_{r,s}(W^r)=0 \text{ in } H^*(G_{n,m}).$$

The formula (I) is thus proved by induction for the case $n\geqslant 2(r+s)$.

Suppose now $n<2(r+s)$. Let us take $n'<2(r+s)>n$ and consider the canonical mapping

$$f:G_{n,m}\to G_{n',m}$$

defined in §I. By what we have proved we have

$$\varphi_{r,s}(W'^i)=0,$$

where W'^i are the W-classes of $G_{n',m}$. By Lemma 1 we have then

$$\varphi_{r,s}(W^i) = \varphi_{r,s}(f^* W'^s) = f^* \varphi_{r,s}(W'^i) = 0.$$

The tormula $(1)'$ or (1) is thus conmpletely proced.

Proof of (II) Let us consider the canonical mapping

$$g : G_{n,m} \to G_{m,n}$$

defined in §1. Let W'^i be the W-classes of $G_{n,m}$, then, by (I), we have

$$Sq^r W'^s = \sum_{t=0}^{r} \binom{s-r+t-1}{t} W'^{r-t} W'^{s+t}, s \geqslant r \geqslant 0.$$

It follows by Lemma 2 that

$$\begin{aligned} Sq'\overline{W}^s &= Sq^r g^* W'^s = g^* Sq^r W'^s \\ &= \sum_{t=0}^{r} \binom{s-r+t-1}{t} g^* W'^{r-t} g^* W'^{s+t} \\ &= \sum_{s=0}^{r} \binom{s-r+t-1}{t} \overline{W}^{r-s} \overline{W}^{s+t}. \end{aligned}$$

This proves (II).

V The complete cohomology ring of $G_{n,m}$

The formula (I) of §III may be written in the form:

$$\binom{s-1}{r} W^{r+s} = Sq^r W^s + \sum_{t=0}^{r-1} \binom{s-r+t-1}{t} W^{r-t} W^{s+t}, \quad s \geqslant r > 0.$$

When $r = 2^k l$, $s = 2^k(l+1)$, we have (mod 2)

$$\binom{s-1}{r} = \binom{2^k(l+1)-1}{2^k l} = \binom{2^k l + 2^k - 1}{2^k - 1} = 1.$$

Hence, when i is of the form $2^k(2l+1)$ with $l > 0$, we would have

$$W^i = Sq^r W^s + \sum_{t=0}^{r=1} \binom{s-r+t-1}{t} W^{r-t} W^{s+t},$$

in taking $r = 2^k l > 0$ and $s = 2^k(l+1)$. This shows that inasmuch as i is not a power of 2, W^i may be expressed in terms of W' with $j < i$ by means of cup-products and squares. By successive reductions we may therefore express any class W^i in terms of the classes W^{2^k} of dimensions $2^k \leqslant m$ by means of cup-products and squares. For example,

$$\begin{aligned}
W^3 &= W^1W^2 + Sq^1W^2,\\
W^5 &= W^1W^4 + Sq^1W^4,\\
W^6 &= W^2W^4 + Sq^2W^4,\\
W^7 &= W^1W^2W^4 + W^1Sq^2W^4 + Sq^1(W^2W^4) + Sq^1Sq^2W^4,
\end{aligned}$$

etc.

Now for a topological space E let us call the totality of the cohomology ring (mod 2) of E as well as the square operations the *complete* cohomology ring (mod 2) of E. Then the above result, taking into account the Theorem 2 of Chern, may be formulated as follows:

Theorem 4 *The complete cohomology ring mod 2 of the Grassmannian manifold $G_{n,m}$ is generated by the classes W^{2^k}, where $0 < 2^k \leqslant m$.*

We have also the following

Theorem 5 *Any class W^{2^k}, where $0 < 2^k \leqslant m$ is independent of the other elements W' in the complcte cohomology ring mod 2 of $G_{n,m}$, when $n \geqslant m$. In other words, any such class W^{2^k} $(0 < 2^k \leqslant m)$ cannot be expressed in terms of other classes W' with $j \neq 2^k$ by means of cup-products and squares, when $n \geqslant m$.*

Proof For any r, s with $s \geqslant r > 0$, $r + s = 2^k(0 < 2^k \leqslant m)$ we have (mod 2)

$$\binom{s-1}{r} = \binom{2^k-1-r}{r} = \binom{2r}{r} = 0.$$

Consequently the formula (I) of §IV shows that in the expansion of Sq^rW^s there is no single term W^{2^k}.

Consider now an element of $H^*(G_{n,m})$ which is represented by a certain expression $\varphi(W^i)$ involving the classes W^i connected by means of cup-products and squares. By means of Lemma 9 of Cartan in §IV, we may express $\varphi(W^i)$ as a polynomial $P(W^i)$ of W^i with only cup-products as operations. By what precedes we see that $P(W^i)$ would not contain the single term W^{2^k} $(0 < 2^k \leqslant m)$ if W^{2^k} is not originally contained in the expression $\varphi(W^i)$. As $2^k \leqslant m \leqslant n$, we have by Lemma 4 of §II that $W^{2^k} \neq \varphi(W^i)$ if φ does not contain W^{2^k}. This proves our theorem.

The Theorems 5 and 6 combined together means that, when $n \geqslant m$, the set of elements W^{2^k}, $0 < 2^k \leqslant m$, form a set of *irreducible generators* of the complete cohomology ring mod 2 of $G_{n,m}$.

In either starting from the formula (II) of § IV or the Lemma 2 of § I, we may get the same results for the classes $\overline{W}^i$: When $n \leqslant m$, the set of elements $\overline{W}^{2^k}$, $0 < 2^k \leqslant n$, form a set of *irreducible generators* of the complete cohomology ring mod 2 of $G_{n,m}$.

Appendix
An alternative proof of Lemma 5

A fiber bundle $\mathfrak{B}$ for which the fiber is an m-dimensional linear vector space and the structural group is the group of all non-singular linear transformations of m variables L_m will be called an *m-vector bundle.* When the base space K of the m-vector bundle $\mathfrak{B}$ is a polyhedron, there exists a system of invariants $W_0^i(\mathfrak{B})$, $i = 1, 2, \cdots, m$, of $\mathfrak{B}$ which are cohomology classes of dimension i of K with local coefficients in $\pi_{i-1}(V_{m,m-i+1}) \approx z_0$ or z_2, where $V_{m,k}$ are Stiefel manifolds, and $z_0(z_2)$ is the group of integers (integers mod 2). These classes are the so-called Stiefel-Whitney classes of the bundle $\mathfrak{B}$ The classes mod 2 obtained from them by the natural map $\rho : \pi_{i-1}(V_{m,m-i+1}) \to z_2$ of coefficient groups will be called the *reduced Stiefel-Whitney classes* of the bundle $\mathfrak{B}$ and will be denoted in what follows by the symbols $W^i(\mathfrak{B})$, $i = 1, \cdots, m$. The definition of these reduced classes will also be extended over all non-negative integers i by setting $W^0(\mathfrak{B}) = 1$, the unit class mod 2 in the base K, and $W^i(\mathfrak{B}) = 0$ for $i > m$.

If $\mathfrak{B}$ is an m-vector bundle over a polyhedron K and h a continuous map of a polyhedron K' in K, the induced bundle of $\mathfrak{B}$ by h will be denoted by $h^*\mathfrak{B}$. We have then for their Stiefel-Whitney classes:

$$\begin{cases} h^* W_0^i(\mathfrak{B}) = W_0^i(h^*\mathfrak{B}), \\ h^* W^i(\mathfrak{B}) = W^i(h^*\mathfrak{B}). \end{cases} \tag{1}$$

If $\mathfrak{B}_i$, $i = 1, 2$ are m_i-vector bundles over polyhedrons K_i, we may define[6] a certain "product" bundle $\mathfrak{B}_1 \otimes \mathfrak{B}_2$ over the product space $K_1 \times K_2$, which is an m-vector bundle with $m = m_1 + m_2$. For their reduced Stiefel-Whitney classes we have then [6]

$$W^i(\mathfrak{B}_1 \otimes \mathfrak{B}_2) = \sum_{i_1+i_2=r} W^{i_1}(\mathfrak{B}_1) \otimes W^{i_2}(\mathfrak{B}_2). \tag{2}$$

Let us consider now the Grassmannian manifold $G_{n,m}$ defined in a Euclidean space R^{n+m} of dimension $n+m$. Let $\tilde{G}_{n,m}$ be the space of all pairs (x, X) of which X is an m-element of R^{n+m} and x is a vector in X. It is clear that $\tilde{G}_{n,m}$ is the bundle space of an m-vector bundle $\mathfrak{S}_{n,m}$ over $G_{n,m}$, as base with m-dimensional linear vector spaces X as fibers. The projection $\pi : \tilde{G}_{n,m} \to G_{n,m}$ is defined by the natural map $\pi(x, X) = X$. A system of coordinate neighborhoods and coordinate functions may be introduced as follows. Let V be a fixed nl-dimensional linear vector space with a fixed basis $v_1, \cdots, v_m$. For any m-element X of R^{n+m} the set V_X of all m-elements X' of R^{n+m} which has a non-degenerate orthogonal projection on X is an open set of $G_{n,m}$ containing X and will be one of our coordinate neighborhoods of the bundle $\mathfrak{S}_{n,m}$. Let us take in each $X \in G_{n,m}$ a system of m linearly independent vectors $x_1, \cdots, x_m$ and for any $X' \in V_x$ let $x'_1, \cdots, x'_m$ be the orthogonal projections of $x_1, \cdots, x_m$ on X'. Then the coordinate function $\varphi_X : V_X \times V \to \pi^{-1}(V_X)$ will be defined by $\varphi_X(X', \sum a_i v_i) = (\sum a_i x'_i, x')$, for any real numbers $a_1, \cdots, a_m$.

The above defined bundle $\mathfrak{S}_{n,m}$ will be called the *canonical* m-vector bundle over $G_{n,m}$. For its Stiefel-Whitney classes we have the following[7]

Theorem(Понтрягин) *The reduced Stiefel-Whitney classes of the canonical m-vector bundle $\mathfrak{S}_{n,m}$ over $G_{n,m}$, coincide with the W-classes W^i of $G_{n,m}$ defined in* §1 *for all i,* i. e.

$$W^i(\mathfrak{S}_{n,m}) = W^i, \quad i \geqslant 0. \tag{3}$$

Proof of Lemma 5 We will use same notations as in §III. Let $\mathfrak{S}_{n,m}$ be the canonical m-vector bundle over $G_{n,m}$ with bundle space $\tilde{G}_{n,m}$ and let $\mathfrak{S}_{n_j,m_j}$, $j=1$, 2 be the canonical m_j-vector bundle over G_{n_j,m_j} with bundle space $\tilde{G}_{n_j,m_j}$. By the definition of "product" bundle, it is readily seen that the bundle space of the product bundle

$$\mathfrak{S}'_{n,m} = \mathfrak{S}_{n_1,m_1} \otimes \mathfrak{S}_{n_2,m_2}, \tag{4}$$

which consists of all points $(x_1 + x_2, X_1 \times X_2)$ where $(x_j, X_j) \in G_{n_j,m_j}$, is a subspace of the bundle space $\tilde{G}_{n,m}$ of $\mathfrak{S}_{n,m}$. Moreover, if $\tilde{h}$ be the identical mapping of this space in $\tilde{G}_{n,m}$, we see that

$$\pi\tilde{h} = h\pi',$$

where π, π' are respectively the projections of the bundles $\mathfrak{S}_{n,m}$ and $\mathfrak{S}'_{n,m}$. It follows that $\tilde{h}$ is a bundle map of $\mathfrak{S}'_{n,m}$ in $\mathfrak{S}_{n,m}$ and $\mathfrak{S}'_{n,m}$ is equivalent to the induced bundle of $\mathfrak{S}_{n,m}$ by h:

$$\mathfrak{S}'_{n,m} \approx h^*\mathfrak{S}_{n,m}. \tag{5}$$

Therefore we have

$$W^i(\mathfrak{S}'_{n,m}) = W^i(h^*\mathfrak{S}_{n,m}), \quad i \geqslant 0.$$

By (1), (2) and (4), we have then

$$h^*W^i(\mathfrak{S}_{n,m}) = \sum_{i_1+i_2=i} W^{i_1}(\mathfrak{S}_{n_1,m_1}) \otimes W^{i_2}(\mathfrak{S}_{n_2,m_2}).$$

By the theorem of понтрягин (cf. (3)), this reduces to

$$h^*W^i = \sum_{i_1+i_2=i} W_1^{i_1} \otimes W_2^{i_2}.$$

This proves our lemma.

References

[1] Wu Wen-tsün. *Comptes Rendus*, 1950, **230**: 918-920.

[2] Chern S. *Annals of Mathematics*, 1948, **49**: 362-372.

[3] Steenrod N E. *Annals of Mathematics*, 1947, **48**: 290-320.

[4] Ehresmann Ch. Journal de Mathématiques, 1939, **104**.

[5] Cartan H. *Comptes Rendus*, 1950, **230**: 425-427.

[6] Wu Wen-tsün. *Annals of Mathematics*, 1948, **49**: 641-653.

[7] Понтрягин ,Л. ДАН , 1942, **35**: 34-37.

14.“格拉斯曼”流形中的平方运算*

引　言

本文及以下一文是作者在 1950—1951 年间关于球纤维丛示性类 (characteristic class①) 所得结果的综合报告, 其中有些结果已发表而无详细证明 [1,2] ②，有些则尚未发表过.

我们知道 $m-1$ 维球纤维丛 ③的理论 [3] 可归结到“格拉斯曼”流形 $G_{n,m}(n>m)$ 的研究. $G_{n,m}$ 中的任一上同调类 (系数群任意) 引出 $m-1$ 维球纤维丛的一个不变类, 叫做这个纤维丛的示性类. 在系数群是法 2 的整数加法群时，陈省身氏 [4] 曾证明 $G_{n,m}$ 的任一上同调类可以其中的 m 个维数是 i 的类 W^i，$0<i\leqslant m$ 用上积表示出来. 由这些类所引出的 $m-1$ 维球纤维丛的不变类即通常所说 (已约化为法 2) 的 Stietel-Whitney 类. 本文的目的，是在完全地决定 $G_{n,m}$ 中的 Steenrod 平方运算 [5]. 由此得到下面的结果: $G_{n,m}$ 中一切法 2 的上同调类都可以由维数是 2^k 形的那几个类 $W^{2^k}(0<2^k\leqslant m)$ 用上积及平方运算来表示. 因此，关于一个球纤维丛的法 2 示性类的问题，完全归结到那几个维数是 2^k 的法 2 Stietel-Whitney 类的研究了.

1950 年在作上述研究时，曾得到 H.Cartan 先生的一些协助和关心，作者谨在此志谢.

§1　$H^*(G_{n,m})$ 的加法构造 [6,7]

设 R^{n+m} 是一个 $n+m$ 维的欧氏空间. 为简单起见, 我们把一个经过 R^{n+m} 的原点的 m 维平直空间叫做 R^{n+m} 的一个 m 面. R^{n+m} 的一切 m 面很自然地成一个 nm 维的拓扑空间，即通常所谓“格拉斯曼”空间，我们把它记作 $G_{n,m}$，并说 $G_{n,m}$ 是定义于 R^{n+m} 中的.

---

* *J. Chinese Math. Soc.*, 1953, 2: 205-230.

① 关于拓扑学方面的中文名称， 大致依照江泽涵先生所译*Seitert-Threltall* 的拓扑学一书. 唯 Chain，cycle，homology group 等改称下链、下闭链、下同调群等以与上链 (cochain)、上闭链 (cocycle)、上同调环 (cohomology ring) 等区别. Cup product 则译作上积.

② 方括弧内数字是指示文末参考文献.

③我们假定球纤维丛的构造群是线性群或正交群.

设 R^i，$i=1$，2，$\cdots$，$n+m-1$ 是 R^{n+m} 中的 i 面，满足：

$$R^1 \subset R^2 \subset \cdots \subset R^{n+m-1}, \tag{S}$$

又设 $\Omega_{n,m}$ 是满足条件 $0 \leqslant a_1 \leqslant a_2 \leqslant \cdots \leqslant a_m \leqslant n$ 的一切整数序列 $\mathrm{a}=(a_1,\cdots,a_m)$ 的集合. 对于任一 $\mathrm{a}=(a_1,\cdots,a_m)\in\Omega_{n,m}$，和满足 $i\leqslant i\leqslant m$ 和 $a_{i-1}\leqslant a_i-1(a_0=0)$ 的任一正整数 i，令 $\mathrm{a}_{(i)}=(a_1,\cdots,a_{-1},a_j-1.a_{i+1},\cdots,a_m)\in\Omega_{n,m}$；在别种情形，$\mathrm{a}_{(i)}$ 无意义. 今以 $[\mathrm{a}]^*$ 表示 $G_{n,m}$ 中满足条件

$$\mathrm{Dim}(X\cap R^{a_i+i})\geqq i,\quad i=1,\cdots,m \tag{1}$$

的一切 m 面 X 的集合，那么集合

$$[\mathrm{a}]=[\mathrm{a}]^*-\sum[\mathrm{a}_{(i)}]^*$$

是一个维数是 $D(\mathrm{a})=\sum\limits_{i=1}^{m}a_i$ 的开胞腔，上式中的和号 $\sum$ 是开展到所有 $\mathrm{a}_{(i)}$ 有意义的 i 之上的. 而且，与一切 $\mathrm{a}\in\Omega_{n,m}$ 相当的胞腔 $[\mathrm{a}]$ 成一个 $G_{n,m}$ 的胞腔剖分，在各胞腔适当的定向后，应有

$$\partial[\mathrm{a}]=\sum\eta_i[\mathrm{a}_{(i)}],\quad \eta=0,+2 \text{ 或 } -2,$$

此处和号 $\sum$ 也开展到所有 $\mathrm{a}_{(i)}$ 有意义的 i 之上. 在以下我们只预备讨论系数群是只有两个元素 0，1 的法 2 整数加法群的情形，除非特别声明，以后总假定如此. 此时，对于任一 $\mathrm{a}\in\Omega_{n,m}$，$[\mathrm{a}]$ 总是一个下闭链，它的下同调类我们将用 $z(\mathrm{a})$ 来表示. 注意 $[\mathrm{a}]^*$ 是一个假流形，它所代表的下闭链是 $z(\mathrm{a})$ 的一个元素. 而且，假设 $\mathrm{a}=(a_1,\cdots,a_m)$ 而 1，$\cdots$，m 中使 $a_i<a_{i+1}(a_{m+1}=n)$ 的 i 是 i_1，$\cdots$，i_k，那么 (1) 与下面的 $(1)'$ 同值：

$$\mathrm{Dim}(X\cap R^{a_i+i})\geqq i,\quad i=i_1,\cdots,i_k, \tag{1$'$}$$

假流形 $[\mathrm{a}]^*$ 和它的类 $z(\mathrm{a})$ 也就由 R^{a_i+i}，$i=i_1$，$\cdots$，i_k 所完全决定，而与其他的 R^i 无涉.

今设 u 是一个上链，它使 $u[\mathrm{a}]=1$，而在 $\mathrm{a}'\in\Omega_{n,m}$，$D(\mathrm{a}')=D(\mathrm{a})$，$\mathrm{a}'\neq\mathrm{a}$ 时，$u[\mathrm{a}']=0$. 那么 u 是一个维数是 $D(a)$ 的上闭链. 它的上同调类将用 $\{\mathrm{a}\}$ 来表示. 我们有

定理 1(Enresmann) 所有 $\mathrm{a}\in\Omega_{n,m}$ 的类 $\{\mathrm{a}\}$ 成为上同调环 $H^*(G_{n,m})$ 的母素的一个加法基 (additive basis).

在此我们须注意这些类 $\{\mathrm{a}\}$ 与所选择的特殊序列 (S) 无关，因而是不变地定义于 $G_{n,m}$ 中的类，类 $z(a)$ 亦然. 又注意在 $\mathrm{a}=(0,\cdots,0)$ 时，类 $\{\mathrm{a}\}$ 就是 $H^*(G_{n,m})$ 的单位类 I (unit class) .

下面的类特别重要:

$$W^i=\{\mathrm{a}_i\},\qquad 0\leqslant i\leqslant m;$$
$$\overline{W}^i=\{\bar{\mathrm{a}}_i\},\qquad 0\leqslant i\leqslant n,$$

此处 a_j 有 $m-i$ 个 0 和 i 个 1, $\bar{\mathrm{a}}_i$ 则有 $m-1$ 个 0 和一个 i. 注意 $W^0=\overline{W}^0=I$, 而 W^i, $\overline{W}^i$ 的维数都是 i. 我们并拟把 W^i 和 $\overline{W}^i$ 的定义推广到任意维数 i 如下:

$$i>m \text{ 时 } W^i=0;\qquad i>n \text{ 时 } \overline{W}^i=0.$$

这些类在以后将叫做 $G_{n,m}$ 的"W 类"和"$\overline{W}$ 类".

今设 $G_{n',m}(n'>n)$ 是定义在一个包含 R^{n+m} 的欧氏空间 $R^{n'+m}$ 中的"格拉斯曼"空间. 在 $G_{n',m}$ 中的类、链等等我们采取与 $G_{n,m}$ 中相当的类、链等等同样的记号, 只在右上角加一撇以示区别. 很显然 $G_{n,m}$ 是 $G_{n',m}$ 的一个子流形. 对于恒同变换

$$f:\quad G_{n,m}\to G_{n',m}$$

我们有

预备定理 1

$$\begin{aligned} f^*\{\mathrm{a}\}'&=\{\mathrm{a}\},\quad \mathrm{a}\in\Omega_{n,m}\\ &=0,\qquad \mathrm{a}\in\Omega_{n',m}\text{但不}\in\Omega_{n,m}.\end{aligned}$$

特别对于任意整数 $i\geqq 0$ 有

$$f^*W'^i=W^i,\qquad f^*\overline{W}'^i=\overline{W}^i.$$

证 在 $R^{n'+m}$ 中取一个含 R^{n+m} 在其内的 i 面 R^i 的序列如次:

$$R^1\subset R^2\subset\cdots\subset R^{n+m}\subset\cdots\subset R^{n'+m-1}\qquad (S')$$

(S') 中在 R^{n+m} 前的一部分成为 R^{n+m} 中的一个序列 (S). 照前面所说, 从序列 (S) 可得 $G_{n,m}$ 的一个胞腔剖分. 同样, 从 (S') 也可以得到 $G_{n',m}$ 的一个胞腔剖分. 对于任一 $\mathrm{a}\in\Omega_{n,m}$, 条件 (1) 对于序列 (S) 或 (S') 都是一样的. 因之可知在上面的胞腔剖分下, $G_{n,m}$ 是 $G_{n',m}$ 的一个子复合形. 而且对于 $\mathrm{a}\in\Omega_{n,m}$, $f[\mathrm{a}]=[\mathrm{a}]'$. 由此立得预备定理 1.

试再讨论 R^{n+m} 中一切 n 面所成的"格拉斯曼"流形 $G_{m,n}$. 在 $G_{m,n}$ 中的类等等, 我们也用 $G_{n,m}$ 中类似的记号来表示. 只在右上角加一撇以示区别. 对于 $G_{n,m}$ 中任一 m 面 X, 令 $\tilde{X}$ 是与 X 完全垂直的 $G_{m,n}$ 中的 n 面. 那么 $g(X)=\overline{X}$ 很自然地定义了一个变换

$$g:G_{n,m}\to G_{m,n}.$$

很显然 $G_{n,m}$ 与 $G_{m,n}$ 同胚而 g 是一个拓扑变换. 我们并有

预备定理 2

$$g^*W'^i=\overline{W}^i,\quad g^*\overline{W}'^i=W^i,\quad i\geqq 0.$$

证 试先作若干预备如下:

对于任一 $a=(a_1,\cdots,a_m)\in\Omega_{n,m}$，令 $\theta\mathrm{a}=\mathrm{b}=(b_1,\cdots,b_n)\in\Omega_{m,n}$ 定义如下式:

$$n-a_{i+1}<j\leqslant n-a_i(a_0=0,a_{m+1}=n)\text{ 时},\quad b_j=m-i.$$

此处 $i=0$，1，$\cdots$, m. 换言之,

$$b=(\underbrace{0,\cdots,0,}_{n-a_m}\ \underbrace{1,\cdots,1,}_{a_m-a_{m-1}}\cdots,\underbrace{m-i,\cdots,m-i,}_{a_{i+1}-a_i}\cdots,\underbrace{m,\cdots,m}_{a_1}).$$

设指数 $i(1\leqslant i\leqslant m)$ 使 $a_i>a_{i-1}(a_0=0)$. 因而

$$\mathrm{a}(i)=(a_1,\cdots,a_{i-1},a_i-1,a_{i+1},\cdots,a_m)\in\Omega_{n,m}$$

有意义，则

$$\theta\mathrm{a}_{(i)}=(\underbrace{0,\cdots,0,}_{n-a_m}\cdots\underbrace{m-i,\cdots,m-i,}_{a_{i+1}-a_i+1}\underbrace{m-i+1,\cdots,m-i+1,}_{a_i-a_{i-1}-1}\cdots,\underbrace{m,\cdots,m)}_{a_1}$$

易见即为 $\mathrm{b}_{(n-a_i+1)}\in\Omega_{m,n}$. 反之，若 $\mathrm{b}_{(j)}\in\Omega_{m,n}$ 有意义，我们应有一指数 $i(1\leqslant i\leqslant m)$ 使 $j=n-a_i+1$ 而 $a_i>a_{i-1}$. 于是 $\mathrm{a}_{(i)}\in\Omega_{n,m}$ 有意义且 $\mathrm{b}_{(j)}=\theta\mathrm{a}_{(i)}$.

上面所定义的变换 $\theta:\Omega_{n,m}\to\Omega_{m,n}$ 显然规定 $\Omega_{n,m}$ 与 $\Omega_{m,n}$ 之间的一个一一对应，而且我们也容易证明若同样定义了 $\theta:\Omega_{m,n}\to\Omega_{n,m}$，那么从 $\theta_{\mathrm{a}}=\mathrm{b}$ 可得 $\theta\mathrm{b}=\mathrm{a}$.

有了这些准备，我们可在 R^{n+m} 中取两个序列

$$R^1\subset\cdots\subset R^{n+m-1}\tag{S}$$

和

$$\bar{R}^1\subset\cdots\subset\bar{R}^{n+m-1}\tag{$\bar{S}$}$$

使对于 $i=1$，$\cdots$, $n+m-1$，R^i 和 $\overline{R}^{n+m-i}$ 是完全垂直的一个 i 面和一个 $n+m-i$ 面. 对于 (S) 和 $(\overline{S})$ 依以前所说的方法各作 G_{nm} 和 $G_{m,n}$ 的胞腔剖分. 对于任一 $\mathrm{a}=(a_1,\cdots,a_m)\in\Omega_{n,m}$，$G_{n,m}$ 的子集合 $[\mathrm{a}]^*$ 亦即是满足条件

$$\mathrm{Dim}(X\cap R^{a_i+i})\geqq i,\quad i=1,\cdots,m\tag{1}$$

的一切 m 面的集合. 若 X 是满足 (1) 中第 i 个条件的 m 面那么与 $X\cap R^{a_i+i}$ 完全垂直的面 Y 的维数将至脆是 $n+m-i$，而 $g(X)=X$ 和 R^{n+m-a_i-i} 显然都在 Y 之内. 因之，

$$\mathrm{Dim}(\bar{X}\cap\bar{R}^{n+m-a_i-i})\geqq n+(n+m-a-i)-(n+m-i)=n-a.$$

反之，从最后一式也可以得到 (1) 的第 i 式. 因之条件 (1) 与下面的一组条件同值：

$$\mathrm{Dim}(\bar{X}\cap\bar{R}^{n+m-a_i-i})\geqq n-a_i,\quad i=1,\cdots,m, \tag{2}$$

而后者又易知与下面的一组条件同值：

$$\mathrm{Dim}(\bar{X}\cap\bar{R}^{b_j+j})\geqq j,\quad j=1,\cdots,n, \tag{3}$$

此处 $\mathrm{b}=(b_1,\cdots,b_n)=\theta\mathrm{a}\in\Omega_{m,n}$. 因为满足条件 (3) 的面 $\bar{X}$ 作成 $G_{m,n}$ 的子集合 $[\mathrm{b}]^*$ 而 g 是 $G_{n,m}$ 到 $G_{m,n}$ 上的拓扑变换, g 也定义了一个从 $[\mathrm{a}]^*$ 到 $[\mathrm{b}]^*$ 上的拓扑变换. 我们又说过从 $\theta\mathrm{a}=\mathrm{b}$ 可得 $\theta\mathrm{a}_{(i)}=\mathrm{b}(j)$，在此式中当 i 取使 $\mathrm{a}_{(i)}$ 有意义的一切指数时，$j(=n-a_i+1)$ 也适取尽了使 $\mathrm{b}_{(j)}$ 有意义的一切指数. 因之 g 也是开胞腔

$$[\mathrm{a}]=[\mathrm{a}]^*-\sum[\mathrm{a}_{(i)}]^*$$

到开胞腔

$$[\mathrm{b}]=[\mathrm{b}]^*-\sum[\mathrm{b}_{(j)}]^*$$

上的一个拓扑变换. 这证明了在上面的胞腔部分下，g 是 $G_{n,m}$ 到 $G_{m,n}$ 上的一个胞腔变换，由此知 g 所诱导的链群的同构变换 $g^{\#}$ 可由下式来决定：

$$g^{\#}[\mathrm{a}]=[\mathrm{b}],$$

此处 $h=\theta\mathrm{a}$，$\mathrm{a}\in\Omega_{n,m}$. 由此得

$$g^*\{\mathrm{b}\}=\{\mathrm{a}\}. \tag{4}$$

此处 $\mathrm{a}=\theta\mathrm{b}$，$\mathrm{b}\in\Omega_{m,n}$. 公式 (4) 完全决定了变换 g 的同调型 (homology type) . 在特例当 $\mathrm{b}\in\Omega_{m,n}$ 而 b 有 $n-i$ 个 0 和 i 个 1，或 $n-1$ 个 0 和一个 i 时，$\theta\mathrm{b}=\mathrm{a}\in\Omega_{n,m}$ 而 a 各有 $m-1$ 个 0 和一个 i 或 $m-i$ 个 0 或 i 个 1. 此时的公式 (4) 也就是本预备定理所要证明的公式.

§2 $H^*(G_{n,m})$ 的乘法构造 [4,6]

从§ 1 中所引述的定理 1，“格拉斯曼”流形 $G_{n,m}$ 的上同调环 $H^*(G_{n,m})$ 有一个加法的基底系由一切 $\mathrm{a}\in\Omega_{n,m}$ 的类 $\{\mathrm{a}\}$ 所组成. 关于 $H^*(G_{n,m})$ 的乘法构造, 亦即 $G_{n,m}$ 的交截环 (intersection ring), 则 Ehresmann 首先得有部分结果如下：假设 $\mathrm{a}=(a_1,\cdots,a_m)\in\Omega_{n,m}$ 而 $\mathrm{a}'=(n-a_m,\cdots,n-a_1)\in\Omega_{n,m}$，那么有下面的相交公式：

$$\begin{cases} z(\mathrm{a}), \quad z(\mathrm{a}')=1, \\ z(\mathrm{a}), \quad z(\mathrm{b})=0, \text{若 } \mathrm{b}\neq\mathrm{a}', D(\mathrm{a})+D(\mathrm{b})=nm. \end{cases} \tag{1}$$

我们知道在任意一个 r 维的流形 M 中有一个确定的一一同构：

$$d: \quad H^i(M)\approx H_{r-i}(M).$$

我们说在一一同构 d 之下的 $H^i(M)$ 与 $H_{r-i}(M)$ 的两个元素在 M 中是“对偶的”. 从 (1) 式可知 $\{\mathrm{a}\}$ 与 $z(\mathrm{a}')$ 是对偶的元素. 特例 W^i，$0<i\leqslant m$ 与 $\overline{W}^i$，$0<i\leqslant n$ 各与类 $z(\mathrm{a}_i)$ 与 $z(\bar{\mathrm{a}}_i)$ 对偶，此处 a_i 有 i 个 $n-1$ 和 $m-i$ 个 n，而 $\bar{\mathrm{a}}_i$ 则有一个 $n-i$ 和 $m-1$ 个 n.

关于 $H^*(G_{n,m})$ 的乘法构造, 陈省身证明了下面的一般公式:

$$\{\mathrm{a}\}\cup\overline{W}^i=\sum\{\mathrm{b}\}, \qquad 0\leqslant i\leqslant n, \tag{2}$$

此处 $\mathrm{a}=(a_1,\cdots,a_m)\in\Omega_{n,m}$ 而和号 $\sum$ 则开展到一切满足下面公式 (3) 的 $\mathrm{b}=(b_1,\cdots,b_m)\in\Omega_{n,m}$ 之上:

$$\begin{cases} a_j\leqslant b_j\leqslant a_{j+1} \quad (a_{m+1}=n), \quad j=1,\cdots,m, \\ D(\mathrm{a})+i=D(\mathrm{b}). \end{cases} \tag{3}$$

从这些公式陈省身得到了下面的

定理 2 上同调环 $H^*(G_{n,m})$ 可从 $G_{n,m}$ 的 W 类 H^i，$0<i\leqslant m$ 所产生. 换言之，$H^*(G_{n,m})$ 的任一元素都可表示为这 $m-1$ 个类的一个多项式 ①，其中的乘法即通常的上积，同样，$H^*(G_{n,m})$ 也可以由 $n-1$ 个 $\overline{W}$ 类 W^i，$0<i\leqslant n$，所产生，即它的任一元素可以写作 $\overline{W}^i$ 的一个多项式.

我们将进一步讨论环 $H^*(G_{n,m})$ 的乘法构造. 对此设 x_i 是具有“重量” (weight) i 的不定量，$i=1,\cdots,s$. 我们说 $(x_1)^{i_1}(x_2)^{i_2}\cdots(x_s)^{i_s}$ 形式的积以 $i_1+2i_2+\cdots+si_s$ 为其重量，并把一切有同样重量 k 的积排成辞典式的次序＜如次：

$$(x_1)^{i_1}(x_{i2})^{i_2}\cdots(x_s)^{i_s}<(x_1)^{j_1}(x_2)^{j_2}\cdots(x_s)^{j_s} \quad \left(\sum ri_r=\sum rj_r=k\right)$$

① 在本文内若非特别声明，多项式的系数都假定是在只有 0, 1 二元素的域之内.

的充要条件是或则 $\sum\limits_{r=1}^{s} i_r < \sum\limits_{r=1}^{s} j_r$，或则 $\sum\limits_{r=1}^{s} i_r = \sum\limits_{r=1}^{s} j_r$，而有某一 r 存在. $0 \leqslant r < s$，使

$$i_1 = j_1, \cdots, i_r = j_r, \quad i_{r+1} > j_{r+1}.$$

预备定理 3 设 $\varphi_k(x_i)$ 是一个不恒等于 0 的 x $(1 \leqslant i \leqslant n)$ 的多项式，其中各项的重量同是 $k > 0$. 若 $k \leqslant m$，那么多项式 $\varphi_k(\overline{W}^i)$ 以上积为乘法时，代表一个 $H^*(G_{n,m})$ 的非 0 元素. 换言之，在 $k \leqslant m$ 时，一切 $(\overline{W}^1)^{k_1}(\overline{W}^2)^{k_2}\cdots(\overline{W}^n)^{k_r}$ 形式的 k 维的类在环 $H^*(G_{n,m})$ 中是线性无关的.

证 我们把 $H^*(G_{n,m})$ 中有同一维数 k 的类排成一部分次序 $<$ 如次. 设 $\mathrm{a} = (a_1, \cdots, a_m)$，$\mathrm{b} = (b_1, \cdots, b_m) \in \Omega_{n,m}$ 且 $D(\mathrm{a}) = D(\mathrm{b}) = k$. 我们规定

$$\{\mathrm{a}\} < \{\mathrm{b}\}$$

的充要条件是有一 r 存在，$0 \leqslant r < m$，使

$$a_1 = b_1, \cdots, a_r = b_r, \quad a_{r+1} < b_{r+1}.$$

由§ 1 的定理 1，$H^*(G_{n,m})$ 的任一 k 维的类 z 可以写作下列形式:

$$z = \sum_{i=1}^{r} \{\mathrm{a}_i\},$$

此处 $\mathrm{a}_i \in \Omega_{n,m}$，$D(\mathrm{a}_i) = k$，而

$$\{\mathrm{a}_1\} < \{\mathrm{a}_2\} < \cdots < \{\mathrm{a}_r\},$$

此时我们把类 $\{\mathrm{a}_r\}$ 叫做 z 的领导项. 假设 z，z' 是任二 k 维的类，各以 $\{\mathrm{a}\}$，$\{\mathrm{a}'\}$ 为其领导项，那么我们规定 $z < z'$ 的充要条件是 $\{\mathrm{a}\} < \{\mathrm{a}'\}$. 在 $\{\mathrm{a}\} = \{\mathrm{a}'\}$ 时，我们不作任意次序上的规定.

从 (2) 与 (3) 我们可以看出 $a_j \leqslant a_{j+1}$ 而 $j > 0$ 时，$\{0, \cdots, 0, a_{j+1}, \cdots, a_m\} \cup \overline{W}^{a_j}$ 在展开成类 $\{\mathrm{b}\}$，$(\mathrm{b} \in \Omega_{n,m})$ 的和以后，它的领导项是类 $\{0, \cdots, 0, a_j, a_{j+1}, \cdots, a_m\}$. 由归纳易知在 $\sum\limits_{i=1}^{n} k_i \leqslant m$，特别在 $\sum\limits_{i=1}^{n} ik_i \leqslant m$ 时，类 $(\overline{W}^1)^{k_1}(\overline{W}^2)^{k_2}\cdots(\overline{W}^n)^{k_n}$ 用公式 (2) 与 (3) 展开后的领导项是类 $\{\mathrm{a}\}$，此处 a 有 $m - \sum\limits_{i=1}^{n} k_i$ 个 0，k_1 个 1，k_2 个 2，以迄 k_n 个 n. 由此知对于任意两个 k 维 $(k \leqslant m)$ 的类 $(\overline{W}^1)^{i_1}(\overline{W}^2)^{i_2}\cdots(\overline{W}^n)^{i_n}$ 和 $(\overline{W}^1)^{j_1}(\overline{W}^2)^{j_2}\cdots(\overline{W}^n)^{j_n}$，我们有

$$(\overline{W}^1)^{i_1}(\overline{W}^2)^{i_2}\cdots(\overline{W}^n)^{i_n} < (\overline{W}^1)^{j_1}(\overline{W}^2)^{j_2}\cdots(\overline{W}^n)^{j_n} \tag{4}$$

的充要条件是

$$\{0,\cdots,0,\underbrace{1,\cdots,1}_{i_1},\underbrace{2,\cdots,2}_{i_2},\cdots,\underbrace{n,\cdots,n}_{i_n}\}<\{0,\cdots,0,\underbrace{1,\cdots,1}_{j_1},\underbrace{2,\cdots,2}_{j_2},\cdots,\underbrace{n,\cdots,n}_{j_n}\}.$$

亦即或则 $\sum i_r<\sum j_r$，或则 $\sum i_r=\sum j_r$ 而有 r 存在 $0\leqslant r<n$ 使 $i_1=j_1$，$\cdots$，$i_r=j_r$，$i_{r+1}>j_{r+1}$. 因之当 $k\leqslant m$ 时，(4) 与下式同值：

$$(x_1)^{i_1}(x_2)^{i_2}\cdots(x_n)^{i_n}<(x_1)^{j_1}(x_2)^{j_2}\cdots(x_n)^{j_n},$$

此处 x_i，$i=1$，$\cdots$, n 是重量是 i 的不定量.

今设 $\varphi_k(x_l)$ 是一个不恒等于 0 且各项都有同样重量 $k\leqslant m$ 的多项式. 我们可把 $\varphi_k(x_i)$ 写作下列形式：

$$\varphi_k(x_i)=\sum_{r=1}^{s}z_r,$$

其中每一 z_r 是一个 $(x_1)^{r_1}(x_2)^{r_2}\cdots(x_n)^{r_n}$ 形状的项且

$$z_1<z_2<\cdots<z_s.$$

此时我们有

$$\varphi_k(\bar{W}^i)=\sum_{r=1}^{s}Z_r,$$

此处 $Z_r=(\overline{W}^1)^{r_1}(\overline{W}^2)^{r_2}\cdots(\overline{W}^n)^{r_n}$ 中的幂数与与之相当的项 $Z_r=(x_1)^{r_1}(x_2)^{r_2}\cdots(x_n)^{r_n}$ 中的幂数相同. 由前所述我们应该有

$$Z_1<Z_2<\cdots<Z_s,$$

因之 Z_s 的领导项也是 $\sum\limits_{r=1}^{s}Z_r$ 中的领导项而不能在 $\varphi_k(\overline{W}^i)$ 中被消去. 由定理 1，我们应有 $\varphi_k(\overline{W}^i)\neq 0$, 而定理得证.

预备定理 4 设 x_i，$i=1$，$\cdots$, m 是重量为 i 的不定量且 $\varphi_x(x_l)$ 是一个不恒等于 0 且每项都有同样重量 k 的 x_i 的多项式. 若 $k\leqslant n$，则 $\varphi_k(W^i)$ 是环 $H^*(G_{n,m})$ 中的一个非 0 元素.

证 除 $G_{n,m}$ 外, 试同时讨论流形 $G_{m,n}$ 并考察在§ 1 中所定义的变换 $g:G_{n,m}\to G_{m,n}$. 与前同样，$G_{m,n}$ 中的类将在右上角加撇以与 $G_{n,m}$ 中的类区别，由预备定理 3，在 $k\leqslant n$ 时，类 $\varphi_k(\overline{W}'^i)$ 是 $H^*(G_{m,n})$ 的非 0 元素. 由预备定理 2，我们有 $g^*\overline{W}'^i=W^i$，$i\geqq 0$. 因为 g 是一个拓扑变换而 g^* 是一个一一同构的变换，当 $k\leqslant n$ 时，我们应有

$$\varphi_k(W^i)=\varphi_k(g^*\overline{W}'^i)=g^*\varphi_k(\overline{W}'^i)\neq 0,$$

而定理得证.

下面的简单证明，系 H.Cartan 所示.

在 $H^*(G_{n,m})$ 中试考察所有作

$$(W^1)^{k_1}(W^2)^{k_2}\cdots(W^m)^{k_m} \tag{5}$$

形状的 k 维的类. 此种类的个数 a_k 显然等于满足条件

$$k_1+2k_2+\cdots+mk_m=k,\quad k_i\geqslant 0,\quad i=1,\cdots,m \tag{6}$$

的一切整数组 $(k_1,\cdots,k_m)$ 的个数. 令

$$k_m=a_1,k_{m-1}=a_2-a_1,\cdots,k_1=a_m-a_{m-1}. \tag{7}$$

因为 $k_i\geqq 0$，我们有 $0\leqslant a_1\leqslant\cdots\leqslant a_m$. 当 $k\leqslant m$ 时，又有 $a_m=\sum k_i\leqslant\sum ik_j\leqslant m$. 故此时 $\mathrm{a}=(a_1,\cdots,a_m)\in\Omega_{n,m}$ 且 $D(\mathrm{a})=\sum a_i=\sum ik_i=k$. 反之，对于任一 $D(\mathrm{a})=k$ 的 $\mathrm{a}\in\Omega_{n,m}$ 公式 (4) 定出一组满足 (6) 的整数组 $(k_1,\cdots,k_m)$ 因而得到一个形状如 (5) 的 k 维的类. 这样的对应显然是一对一且可逆的. 因之，当 $k\leqslant m$ 时，α_k 等于 $\Omega_{n,m}$ 中使 $D(\mathrm{a})=k$ 的 a 的个数，而由定理 1，后者就是 $G_{n,m}$ 的第 k 个 Betti 数. 由定理 2，$H^*(G_{n,m})$ 的任一 k 维的类一定是形状是 (5) 的若干个 k 维的类之和. 因之，当 $k\leqslant m$ 时，形状如 (5) 的 α_k 个类在 $H^*(G_{n,m})$ 中一定是线性无关的. 从这就立刻证明了预备定理 4.

§3　一个变换 h

假设 R^{n+m} 是两个 n_j+m_j 维欧氏空间 $R_j^{n_j+m_j}$，$j=1$，2 的积. $n=n_1+n_2$，$m=m_1+m_2$. 又设 $G_{n_j,m_j},j=1$，2，是定义于 $R_j^{n_j+m_j}$ 中的“格拉斯曼”空间，而 $G_{n,m}$ 是定义于 R^{n+m} 中的“格拉斯曼”空间. 对于 $R_j^{n_j+m_j}$，$j=1$，2，中的任两个 m_j 面 X_j，X_1 与 X_2 的积是 R^{n+m} 中的一个 m 面 X. 因之 $h(X_1\times X_2)=X$ 定义了一个连续变换

$$h:\quad G_{n_1,m_1}\times G_{n_2,m_2}\to G_{n,m},$$

今设 W_j^i (或 $\overline{W}_j^i$),$j=1$，2；$i\geqq 0$. 是 G_{n_j,m_j} 的 W 类 (或 $\overline{W}$ 类), W^i (或 $\overline{W}^i$), $i\geqq 0$, 是 $G_{n,m}$ 的 W 类 (或 $\overline{W}$ 类). 因为由定理 2，$G_{n,m}$ 的上同调环是由 W^i (或 $\overline{W}^i$) 所产生，h^* 的同调型可被 h^*W^i (或 $h^*\overline{W}^i$) 所完全决定，对此我们有

预备定理 5

$$h^*W^i=\sum_{i_1+i_2=i}W_1^{i_1}\otimes W_2^{i_2},\quad i\geqq 0, \tag{1}$$

$$h^*\overline{W}^i = \sum_{i_1+i_2=i} \overline{W}_1^{i_1} \otimes \overline{W}_2^{i_2}, \quad i \geqq 0. \tag{2}$$

证 公式 (1) 在 $i=0$ 或 $i>m$ 及公式 (2) 在 $i=0$ 或 $i>n$ 的情形都是很明显地成立的. 因之，欲证明例如公式 (2)，我们可假定 $0<i\leqslant n$.

假设 a_i 是 $\Omega_{n,m}$ 中含有一个 $n-i$ 和 $m-1$ 个 n 的序列而令 $\mathrm{a}_{ij}^{(j)}$，$j=1$，2，是 Ω_{n_j,m_j} 中含有 m_j-1 个 0 和一个 i_j 的序列. 在 Ω_{n_j,m_j}，$j=1$，2，中各任取一个序列 $\mathrm{b}^{(j)}=(b_1^{(j)},\cdots,b_{n_j}^{(j)})$ 使 $D(\mathrm{b}')+D(\mathrm{b}'')=i$，那么有两种可能的情形: 第一，

$$b'_{m_1}+b''_{m_2}<i, \tag{3}$$

这时两个 $\mathrm{b}^{(j)}$，$j=1$，2，不可能是任何 $\mathrm{a}_{ij}^{(j)}$ 使 $i_1+i_2=i$. 第二，$b'_{m_1}+b''_{m_2}=i$，这时应有 i_1，i_2 存在使 $i=i_1+i_2$，而 $\mathrm{b}^{(j)}=\mathrm{a}_{i_j}^{(j)}$，$j=1$，2.

从上面以及§ 2 开首所论，公式 (2) 由对偶定律显然与下面的相交公式同值:

$$h(z(\mathrm{b}')\times z(\mathrm{b}'')),\quad z(\mathrm{a}_i)=\begin{cases}0, & 在\ b'_{m_1}+b''_{m_2}<i\ 时,\\ 1, & 在\ \mathrm{b}^{(j)}=\mathrm{a}_{i_j}^{(j)}, j=1,2, 而\ i_1+i_2=i\ 时,\end{cases} \tag{4}$$

式中 $\mathrm{b}^{(j)}=(b_1^{(j)},\cdots,b_{m_j}^{(j)})\in\Omega_{n_j,m_j}$，$D(\mathrm{b}')+D(\mathrm{b}'')=i$.

先讨论 (3) 成立的情形. 这时，可在 $R_j^{n_j+m_j}$，$j=1$，2，中各取一序列 $(S_j)R_j^1\subset\cdots\subset R_j^{n_j+m_j-1}$，其中的 $b_{m_j}^{(j)}+m_j$ 面简记作 R_j，又在 R^{n+m} 中取一 $n-i+1$ 面 R_0，使

$$\mathrm{Dim}(R_0\cap(R_1\oplus R_2))=0. \tag{5}$$

由于 (3)，这样的 R_0 一定存在.

令 $[\mathrm{a}_i]^*$ 是 $G_{n,m}$ 中满足

$$\mathrm{Dim}(X\cap R_0)\geqq 1 \tag{6}$$

的一切 m 面 X 的集合. 同样令 $[\mathrm{b}^{(j)}]^*$，$j=1$，2，各是 G_{n_j,m_j} 中满足条件 $\mathrm{Dim}(X_j\cap R_j^{b_i^{(j)}+i})\geqq b_i^{(j)}$，$i=1$，$\cdots$，$m_j$ 的一切 m_j 面的集合. 那么 $[\mathrm{a}_i]^*$，$[\mathrm{b}^{(j)}]^*$ 都是假流形，而其所代表的下闭链各在下同调类 $z(\mathrm{a}_i)$ 和 $z(\mathrm{b}^{(j)})$ 内.

若 $x_j\in[\mathrm{b}^{(j)}]^*$，$j=1$，2，因而 $X_j\subset R_j$，那么 $h(X_1\times X_2)=X$ 将 $\subset R_1\cup R_2$，故由 (5)，X 不能满足 (6)，或 $X\bar{\in}[\mathrm{a}_i]^*$，因之 $h([\mathrm{b}']^*\times[\mathrm{b}'']^*)$ 与 $[\mathrm{a}_i]^*$ 无公共元素而有

$$h(z(\mathrm{b}')\times z(\mathrm{b}''))\cdot z(\mathrm{a}_i)=0,$$

此即 (4) 式的上面的一部分.

其次，设 $\mathrm{b}^{(j)}=\mathrm{a}_{i_j}^{(j)}$，$j=1$，2，而 $i_1+i_2=i$. 在 $R_j^{n_j+m_j}$，$j=1$，2 中各取一组矢量的基底 $(e_k^{(j)})=(e_1^{(j)},\cdots,e_{n_j+m_j}^{(j)})$，并设 R_j^s 是 $R_j^{n_j+m_j}$ 中由矢量 $e_1^{(j)}$，$\cdots$，$e_s^{(j)}$ 所定的平直空间. 又设 R^{n-i+1} 是 R^{n+m} 中由矢量

$$e'_{m_1+i_1+1},\cdots,e'_{n_1+m_1};\quad e''_{m_2+i_2+1},\cdots,e''_{n_2+m_2};\quad e'_{m_1}+e''_{m_2}$$

所定的平直空间，令 $[b^{(j)}]^*$，$i=1$，2，各是 G_{n_j,m_j} 中满足条件

$$R_j^{m_j-1}\subset X_j\subset R_j^{m_j+i_j}$$

的一切 m_j 面 X_j 的集合. 又令 $[\mathrm{a}_i]^*$ 是 $G_{n,m}$ 中满足条件

$$\mathrm{Dim}(X\cap R^{n-i+1})\geqq 1$$

的一切 m 面 X 的集合. 那么 $[\mathrm{b}^{(j)}]^*$ 和 $[\mathrm{a}_i]^*$ 都是假流形，其所代表的下闭链各属于 $z(\mathrm{b}^{(j)})$ 和 $z(\mathrm{a}_i)$. 我们容易看出 $h([\mathrm{b}']^*\times[\mathrm{b}'']^*)$ 与 $[\mathrm{a}_i]^*$ 有一个且只有一个元素 X_0 公共，这个元素 X_0 即是由矢量

$$e'_1,\cdots,e'_{m_1};\quad e''_1,\cdots,e''_{m_2}$$

所定的 m 面. 我们容易证明 $G_{n,m}$ 在 X_0 的附近，我们可以引入适当的坐标使 $h([\mathrm{b}']^*\times[\mathrm{b}'']^*)$ 与 $[\mathrm{a}_i]^*$ 在此坐标下面都是线性的面，且二者在一般的位置. 因之，二者的相交系数是 1，换言之，

$$h(z(\mathrm{b}')\times z(\mathrm{b}''))\cdot z(\mathrm{a}_i)=1,$$

此即 (4) 式下面的一部分.

由此 (4) 式已完全证明，亦即 (2) 已得证明.

我们也可以用同样的方法来证明 (1) 式. 但下面我们拟应用预备定理 2 从 (2) 式以证明 (1) 式.

我们试同时考察各在 R^{n+m} 和 $R_j^{n_j+m_j}$，$j=1$，2 中定义的“格拉斯曼”空间 $G_{m,n}$ 和 G_{m_j,n_j}. 如§ 1 中所示我们有一个变换

$$g:G_{n,m}\to G_{m,n}.$$

同样，我们也有相似的变换

$$g_j:G_{n_j,m_j}\to G_{m_j,n_j},\quad j=1,2.$$

此外，与 h 相似，我们也有一个变换

$$h':G_{m_1,n_1}\times G_{m_2,n_2}\to G_{m,n}.$$

令

$$g' : G_{n_1,m_1} \times G_{n_2,m_2} \to G_{m_1,n_1} \times G_{m_2,n_2}$$

是由 $g'(X_1 \times X_2) = g_1(X_1) \times g_2(X_2)$，$X_j \in G_{n_j,m_j}$ 所定义的一个变换. 那么我们显然有 $h'g' = gh$，因之

$$g'h'^* = h^*g^*. \tag{7}$$

我们若把 $G_{m,n}$ 和 G_{m_j,n_j} 中的类在右上角加撇以示区别，那么由§ 1 的预备定理 2 得

$$g_j^*\overline{W}'^{i_j}_j = W_j^{i_j}, \quad i_j \geqq 0, \quad j = 1, 2, \tag{8}$$

$$g^*\overline{W}'^i = W^i, \quad i \geqq 0. \tag{9}$$

此外又显然有

$$g'^*(\overline{W}'^{i_1}_1 \otimes W'^{i_1}_2) = g_1^* W'^{i_1}_1 \otimes g_2^* W'^{i_2}_2. \tag{10}$$

若应用 (2) 式于 $G_{m,n}$，G_{m_j,n_j} 及 h'，则 (2) 变为

$$h'^*\overline{W}'^i = \sum_{i_1+i_2=i} \overline{W}'^{i_1}_1 \otimes \overline{W}'^{i_2}_2. \tag{11}$$

从 (7) 至 (11) 各式，即得

$$\begin{aligned} h^*W^i &= h^*g^*\overline{W}'^i = g'^*h'^*\overline{W}'^i \\ &= \sum_{i_1+i_2=i} g'^*(\overline{W}'^{i_1}_1 \otimes \overline{W}'^{i_2}_2) \\ &= \sum_{i_1+i_2=i} (g_1^*\overline{W}'^{i}_1 \otimes g_2^*\overline{W}'^{i_2}_2) \\ &= \sum_{i_1+i_2=i} W_1^{i_1} \otimes W_2^{i_2}. \end{aligned}$$

而 (1) 式得证.

在附录里，我们将用另一方法来证明这一条预备定理 5.

预备定理 6 设 x_i，$0 < i \geqq m$ 是重量是 i 的不定量而 $\varphi_k(x_i)$ 是一个 x_i 的多项式，其中每一项的重量都是 k. 那么在 $k \leqslant n_1$ 和 n_2 时，$\varphi_k(W^i)$ 是 $H^*(G_{n,m})$ 的非 0 元素的充要条件是 $h * \varphi_k(W^i)$ 是 $H^*(G_{n_1,m_1} \times G_{n_2,m_2})$ 的非 0 元素，此处 $h : G_{n_1,m_1} \times G_{n_2,m_2} \to G_{n,m}(n = n_1 + n_2, m = m_1 + m_2)$ 是前面所定义的变换.

在证这一定理之先，我们先证明下面的

预备定理 7 设 $x_i^{(j)}$，$0 < i \leqslant m_j$，$j = 1$，2，$m_1 + m_2 = m$ 是不定量而

$$x_i = \sum_{i_1+i_2=i} x'_{i_1} x''_{i_2}, \quad 0 < i \leqslant m \quad (x'_0 = 1, x''_0 = 1), \tag{12}$$

即

$$\begin{cases} x_1 = x'_1 + x''_1, \\ x_2 = x'_2 + x'_1 x''_1 + x''_2, \\ \quad \vdots \\ x_m = x'_{m_1} x''_{m_2}. \end{cases} \tag{12$'$}$$

若 x_i 的多项式 $\varphi(x_i)$ 不恒等于 0，则，$\varphi(\sum x'_{i_1} x''_{i_2})$ 看作 $x_i^{(j)}$ 的多项式时，也不恒等于 0.

证　设 F 是一个示性数 (characteristic) 是 2 且含有 $x_i^{(j)}$，$j=1$，2，$i=1$，$\cdots$，m_j 的域，而使下面的方程式能完全分解 ①:

$$\zeta^{m_j} + x_1^{(j)} \zeta^{m_j-1} + \cdots + x_i^{(j)} \zeta^{m_j-i} + \cdots + x_{m_j}^{(j)} = 0, \quad j = 1, 2, \tag{13}$$

各恰有 m_j 个解. 设 (13) 的解是 $z_1^{(j)}$，$\cdots$，$z_{m_j}^{(j)}$，那么 $x_i^{[j]}$，$0 < i \leqslant m_j$，$j = 1$，2，就是 $z_1^{(j)}$，$\cdots$，$z_{m_j}^{(j)}$ 的初等对称函数 $p_i^{(j)}$. 将 (13) 的两个方程式相乘并利用 (12) 和 (12)$'$，得一方程式如下：

$$\zeta^m + x_1 \zeta^{m-1} + \cdots + x_i \zeta^{m-i} + \cdots + x_m = 0 \tag{14}$$

在场 F 中 (14) 的根是 z'_1，$\cdots$，z'_{m_1}，z''_1，$\cdots$，z''_{m_2}，而 x_i 是它们的初等对称函数 p_i. 我们易见

$$p_i = \sum_{i_1+i_2=i} p'_{i_1} p''_{i_2}, \quad 0 < i \leqslant m \quad (p'_0 = 1, p''_0 = 1), \tag{15}$$

即

$$\begin{cases} p_1 = p'_1 + p''_1, \\ p_2 = p'_2 + p'_1 p''_1 + p''_2, \\ \quad \vdots \\ p_m = p'_{m_1} p''_{m_2}. \end{cases} \tag{15$'$}$$

今设 $\varphi(x_i)$ 是不恒等于 0 的 x_i 的多项式. 由一代数的定理 ②，$\varphi(p_i)$ 视作不定量 $z_i^{(j)}$，$0 < i \leqslant m_j$，$j = 1$，2 的多项式时，也不恒等于 0. 由 (15) 与 (15)$'$，$\varphi(\sum p'_{i_1} p''_{i_2})$ 视作 $z_i^{(j)}$ 的多项式时且不恒等于 0，因之视作 $p_{i_j}^{(j)}$ 的多项式时更不会恒等于 0. 故 $\varphi(\sum x'_{i_1} x''_{i_2})$ 是不定量 $x_i^{(j)}$，$0 < i \leqslant m$，$j = 1$，2 的一个不恒等于 0 的多项式，而定理得证.

① 参阅 Van der Waerden，Moderne Algebra，26，27，29 诸节.

② 参阅 Van der Waerden，Moderne Algebra，Vol.1，§ 24，p.83.

预备定理 6 的证明: 设 $\varphi_k(x_i)$ 是 x_i, $i=1, 2, \cdots, m$ 的一个多项式, 其每项的重量都是 k, 并设 $\varphi_k(W^i)$ 在 $H^*(G_{n,m})$ 中不为 0. 由此假设知 $\varphi_k(x_i)$ 不恒为 0, 因而由预备定理 7 知 $\varphi_k(\sum x'_{i_1}x''_{i_2})$ 是不定量 $x_i^{(j)}$, $0<i\leqslant m_j$, $j=1, 2$ 的不恒为 0 的多项式. 由预备定理 4 一切作 $(W_j')^{i_1}\cdots(W_j^{m_j})^{i_{m_j}}$ 形状而 $i_1+2i_2+\cdots+m_ji_{m_j}=i$, $j=1, 2$ 的 i 维类在上同调环 $H^*(G_{n_j,m_j})$ 中只要 $i\leqslant n_j$ 就线性无关. 由此知一切作

$$(W_1^1)^{i_1'}(W_1^2)^{i_2'}\cdots(W_1^{m_1})^{i'_{m_1}}\otimes(W_2^1)^{i_1''}(W_2^2)^{i_2''}\cdots(W_2^{m_2})^{i''_{m_2}}$$

形状而 $i_1'+2i_2'+\cdots+m_1i_{m_1}'+i_1''+2i_2''+\cdots+m_2i_{m_2}''=k$ 的 k 维类在上同调环 $H^*(G_{n_1,m_1}\times G_{n_2,m_2})$ 中只要 $k\leqslant n_1$ 且 n_2 时就线性无关. 因此, 既然 $\varphi_k(\sum x'_{i_1}x''_{i_2})$ 不恒等于 0, 在 $k\leqslant n_1$ 及 n_2 时, $\varphi_k(\sum W_1^{i_1}\otimes W_2^{i_2})$ 在 $H^*(G_{n_1,m_1}\times G_{n_2,m_2})$ 中也就不等于 0. 又由预备定理 5,

$$\varphi_k(\sum W_1^{i_1}\otimes W_2^{i_2})=\varphi_k(h^*W^i)=h^*\varphi_k(W^i),$$

故 $h*\varphi_k(W^i)$ 在 $k\leqslant n_1$ 及 n_2 时, 是 $H^*(G_{n_1,m_1}\times G_{n_2,m_2})$ 中的非 0 元素. 由此知定理中所说的条件是必要的.

因为从 $h^*\varphi_k(W^i)$ 在 $H^*(G_{n_1,m_1}\times G_{n_2,m_2})$ 中 $\neq 0$ 时显然亦应有 $\varphi_k(W^i)$ 在 $H^*(G_{n,m})$ 中 $\neq 0$. 因此这个条件也是充分的而定理完全得证.

§4 $G_{n,m}$ 中的平方运算 [1]

在定 $G_{n,m}$ 中的平方运算时, 我们需要用到下面 H.Cartan 的两条定理 [8].

预备定理 8 若 U_i, $i=1, 2$, 是拓扑空间 E_i 的上同调类, 则

$$S_q{}^r(U_1\otimes U_2)=\sum_{r_1+r_2=r}S_q{}^{r_1}U_1\otimes S_q{}^{r_2}U_2.$$

预备定理 9 若 U_i, $i=1, 2$, 是同一拓扑空间 E 的上同调类, 则

$$S_q{}^r(U_1\cup U_2)=\sum_{r_1+r_2=r}S_q{}^{r_1}U_1\cup S_q{}^{r_2}U_2.$$

由预备定理 9 和§ 2 中的定理 2, 可知要定 $G_{n,m}$ 中的平方运算, 只要定 W^i, $0<i\leqslant m$ (或 $\overline{W}^i$, $0<i\leqslant n$) 的平方就够了, 结果见下面的

定理 3 设 $\binom{p}{q}$ 在 $p\geqq q>0$ 时是 p 物中取 q 物的组合数 (已约化为法 2 的数) 在 $p<q>0$ 时令 $\binom{p}{q}=0$ 而 $p\geqq -1$ 时令 $\binom{p}{0}=1$ (p, q 皆为整数), 则

$$\text{(I)}\qquad S_q{}^rW^s=\sum_{t=0}^{r}\binom{s-r+t-1}{t}W^{r-t}W^{s+t},$$

$$\text{(II)}\qquad S_q{}^r\bar{W}^s=\sum_{t=0}^{r}\binom{s-r+t-1}{t}\overline{W}^{r-t}\overline{W}^{s+t}.\qquad (s\geqq r\geqq 0).$$

例如：

$$
\begin{aligned}
&{S_q}^1 W^s = W^1 W^s + (s-1) W^{s+1},\\
&{S_q}^2 W^s = W^2 W^s + (s-2) W^1 W^{s+1} + \binom{s-1}{2} W^{s+2},\\
&{S_q}^r W^m = W^r W^m.
\end{aligned}
$$

(I) 的证明　令

$$
\varphi_{r,s}(W^i) = {S_q}^r W^s + \sum_{t=0}^{r} \binom{s-r+t-1}{t} W^{r-t} W^{s+t},
$$

则公式 (I) 与下式同值：

$$
\varphi_{r,s}(W^i) = 0, \quad s \geqq r \geqq 0. \tag{I$'$}
$$

在极端的情形 $r=0$ 或 $r=s$ 时我们各有

$$
\begin{aligned}
\varphi_{0,s}(W^i) &= {S_q}^0 W^s + \binom{s-1}{0} W^0 W^s = 0,\\
\varphi_{s,s}(W^i) &= {S_q}^s W^s + \sum_{t=0}^{s} \binom{t-1}{t} W^{s-t} W^{s+t}\\
&= (W^s)^2 + \binom{-1}{0} (W^s)^2\\
&= 0.
\end{aligned}
$$

因之公式 (I)$'$ 或 (I) 在此时是明显的真确的. 又因在 $s>m$ 时 $\varphi_{r,s}(W^i)$ 显然是 0，我们以后可以把讨论限制于 $m \geqq s > r > 0$ 的情形.

现在对 m 行归纳以证 (I)$'$. 在 $m=1$ 时我们必然有 $s>m$ 或 $r=s=1$ 或 $r=0$. 这时 (I)$'$ 都已知道是真确的. 假定对于 $G_{n,k}$ 而 $k<m>1$ 时的 (I)$'$ 式已真确，我们试来证明 $G_{n,m}$ 的 (I)$'$ 式. 自然，我们在以下假定 $m \geqq s > r > 0$.

试先考虑 $n \geqq 2(r+s)$ 的情形. 在此时我们可取整数 n_j，$j=1$，2，使 $n=n_1+n_2$，$n_j \geqq r+s$. 令 $m_1=m-1$，$m_2=1$ 而依§ 3 定义变换

$$
h: \quad G_{n_1,m_1} \times G_{n_2,m_2} \to G_{n,m}.
$$

今以 W_j^i，$j=1$，2，表 G_{n_j,m_j} 的 W 类，则因 $i>1$ 时 $W_2^i=0$，从预备定理 4 得

$$
h^* W^i = W_1^i \otimes 1 + W_1^{i-1} \otimes W_2^1, \qquad i \geqq 0,
$$

此处规定 $W_1^{-1}=0$. 由预备定理 8，我们又有

$$
\begin{aligned}
{S_q}^r(W_1^s\otimes 1)&={S_q}^rW_1^s\otimes 1,\\
{S_q}^r(W_1^{s-1}\otimes W_2^1)&={S_q}^rW_1^{s-1}\otimes W_2^1+{S_q}^{r-1}W_1^{s-1}\otimes {S_q}^1W_2^1\\
&={S_q}^rW_1^{s-1}\otimes W_2^1+{S_q}^{r-1}W_1^{s-1}\otimes (W_2^1)^2.
\end{aligned}
$$

由此得

$$
\begin{aligned}
h^*\varphi_{r,s}(W^i)=&{S_q}^rh^*W^s+\sum_{t=0}^{r}\binom{s-r+t-1}{t}h^*W^{r-t}\cdot h^*W^{s+t}\\
=&{S_q}^rW_1^s\otimes 1+{S_q}^rW_1^{s-1}\otimes W_2^1+{S_q}^{r-1}W_1^{s-1}\otimes (W_2^1)^2\\
&+\sum_{t=0}^{r}\binom{s-r+t-1}{t}W_1^{r-t}W_1^{s+t}\otimes 1\\
&+\sum_{t=0}^{r}\binom{s-r+t-1}{t}W_1^{r-t-1}W_1^{s+t-1}\otimes (W_2^1)^2\\
&+\sum_{t=0}^{r}\binom{s-r+t-1}{t}(W_1^{r-t-1}W_1^{s+t}+W_1^{r-t}W_1^{s+t-1})\otimes W_2^1,
\end{aligned}
$$

或

$$
h^*\varphi_{r,s}(W^i)=\varphi_1\otimes 1+\varphi_2\otimes W_2^1+\varphi_3\otimes (W_2^1)^2, \tag{1}
$$

此处

$$
\begin{aligned}
\varphi_1&={S_q}^rW_1^s+\sum_{t=0}^{r}\binom{s-r+t-1}{t}W_1^{r-t}W_1^{s+t}=\varphi_{r,s}(W_1^i),\\
\varphi_3&={S_q}^{r-1}W_1^{s-1}+\sum_{t=0}^{r}\binom{s-r+t-1}{t}W_1^{r-t-1}W_1^{s+t-1}\\
&={S_q}^{r-1}W_1^{s-1}+\sum_{t=0}^{r-1}\binom{s-r+t-1}{t}W_1^{r-t-1}W_1^{s+t-1}\\
&=\varphi_{r-1,s-1}(W_1^i).
\end{aligned}
$$

而

$$
\begin{aligned}
\varphi_2=&{S_q}^rW_1^{s-1}+\sum_{t=0}^{r}\binom{s-r+t-1}{t}W_1^{r-t-1}W_1^{s+t}\\
&+\sum_{t=0}^{r}\binom{s-r+t-1}{t}W_1^{r-t}W_1^{s+t-1}
\end{aligned}
$$

$$
\begin{aligned}
&= S_q{}^r W_1^{s-1} + \sum_{t=1}^{r} \binom{s-r+t-2}{t-1} W_1^{r-t} W_1^{s+t-1} \\
&\quad + \sum_{t=0}^{r} \binom{s-r+t-1}{t} W_1^{r-t} W_1^{s+t-1} \\
&= S_q{}^r W_1^{s-1} + W_1^r W_1^{s-1} \\
&\quad + \sum_{t=1}^{r} \left[\binom{s-r+t-2}{t-1} + \binom{s-r+t-1}{t} \right] W_1^{r-t} W_1^{s+t-1}.
\end{aligned}
$$

我们已经假定 $s>r>0$，因之在 $1 \leqslant t \leqslant r$ 时，$s-r+t-1 \geqq t>0$，而有

$$
\binom{s-r+t-2}{t-1} + \binom{s-r+t-1}{t} = \binom{s-r+t-2}{t}. \tag{法 2}
$$

所以

$$
\begin{aligned}
\varphi_2 &= S_q{}^r W_1^{s-1} + W_1^r W_1^{s-1} + \sum_{t=1}^{r} \binom{s-r+t-2}{t} W_1^{r-t} W_1^{s+t-1} \\
&= S_q{}^r W_1^{s-1} + \sum_{t=0}^{r} \binom{s-r+t-2}{t} W_1^{r-t} W_1^{s+t-1} \\
&= \varphi_{r,s-1}(W_1^i).
\end{aligned}
$$

由归纳假设我们有

$$
\varphi_{r,s}(W_1^i) = 0, \quad \varphi_{r,s-1}(W_1^i) = 0, \quad \varphi_{r-1,s-1}(W_1^i) = 0.
$$

即

$$
\varphi_1 = 0, \quad \varphi_2 = 0, \quad \varphi_3 = 0.
$$

故 (1) 变为

$$
h^* \varphi_{r,s}(W^i) = 0.
$$

由§ 2 的定理 2，$S_q{}^r W^e$ 一定是 W^i 的一个多项式，以上积为乘法. 因之 $\varphi_{r,s}(W^i)$ 也是 W^i 的一个多项式，设为 $\varphi_{r+s}(W^i)$. 因为类 $\varphi_{r,s}(W^i)$ 的维数是 $r+s \leqslant n$，做由§ 2 的预备定理 4，这个多项式 $\varphi_{r+s}(W^i)$ 的形状是唯一地由 $\varphi_{r,s}(W^i)$ 所次定的. 因之，若 x_i，$0<i \leqslant m$ 是重量为 i 的不定量，那么 $\varphi_{r+s}(x_i)$ 是一个确定的 x_i 的多项式，其每项的重量都是 $r+s$. 因 $r+s \leqslant n_1$ 和 n_2 且在 $H^*(G_{n_1,m_1} \times G_{n_2,m_2})$ 中 $h^* \varphi_{r+s}(W^i) = h^* \varphi_{r,s}(W^i) = 0$，故由§ 3 的预备定理 6 在 $H^*(G_{n,m})$ 中有

$$
\varphi_{r+s}(W^i) = \varphi_{r,s}(W^i) = 0.
$$

由此在 $n \geqq 2(r+s)$ 时已用归纳证明了公式 (I)$'$ 因而公式 (I).

今设 $n < 2(r+s)$. 取一整数 n' 使 $n' > 2(r+s) > n$ 且讨论如§ 1 中所定义的变换

$$f: G_{n,m} \to G_{n',m}.$$

若以 W'^i 表 $G_{n',m}$ 的 W 类，那么由适才所证，

$$\varphi_{r,s}(W'^i) = 0.$$

由§ 1 的预备定理 1 即得

$$\varphi_{r,s}(W^i) = \varphi_{r,s}(f^* W'^i) = f^* \varphi_{r,s}(W'^i) = 0.$$

故公式 (I)$'$ 或公式 (I) 已完全证明.

公式 (II) 的证明 设

$$g: \quad G_{n,m} \to G_{m,n}$$

是§ 1 中所定义的变换. 设 W'^i 是 $G_{m,n}$ 的 W 类，那么，由 (I)，我们有

$$Sq^r W'^s = \sum_{t=0}^{r} \binom{s-r+t-1}{t} W'^{r-t} W'^{s+t}, \quad s \geqq r \geqq 0.$$

由预备定理 2 得

$$\begin{aligned} Sq^r \bar{W}^s &= Sq^r g^* W'^s = g^* Sq^r W'^s \\ &= \sum_{t=0}^{r} \binom{s-r+t-1}{t} g^* \bar{W}'^{r-t} g^* \bar{W}'^{s+t} \\ &= \sum_{t=0}^{r} \binom{s-r+t-1}{t} \bar{W}^{r-t} \bar{W}^{s+t}. \end{aligned}$$

这证明了公式 (II) .

§5 $G_{n,m}$ 中的完全上同调环

§ 4 中的公式 (I) 也可以写作下列形式:

$$\binom{s-1}{r} W^{r+s} = Sq^r W^s + \sum_{t=0}^{r-1} \binom{s-r+t-1}{t} W^{r-t} W^{s+t}, \qquad s \geqq r > 0.$$

在 $r = 2^k l$，$s = 2^k(l+1)$ 时，

$$\binom{s-1}{r} = \binom{2^k(l+1)-1}{2^k l} = \binom{2^k l + 2^k - 1}{2^k - 1} = 1. \qquad \text{(法 2)}$$

故当 i 是 $2^k(2l+1)$ 形状的整数而 $l>0$ 时，

$$W^i = {S_q}^r W^s + \sum_{t=0}^{r-1} \binom{s-r+t-1}{t} W^{r-t} W^{s+t},$$

其中 $r=2^k l>0$ 而 $s=2^k(l+1)$. 这证明了只要 i 不是 2 的乘幂，那么 W^i 可以 $j<i$ 的 W^j 用上积和平方来表示. 这样的继续把每一维数 j 不是 2 的乘幂的 W^j 约化下去，可将任一类 W^i 用上积和平方以维数是 $2^k \leqslant m$ 的类 W^{2^k} 来表示. 例如:

$$\begin{aligned}
&W^3 = W^1W^2 + {S_q}^1W^2,\\
&W_5 = W^1W^4 + {S_q}^1W^4,\\
&W^6 = W^2W^4 + {S_q}^2W^4,\\
&W^7 = W^1W^2W^4 + W^1{S_q}^2W^4 + {S_q}^1(W^2W^4) + {S_q}^1{S_q}^2W^4 \text{ 等.}
\end{aligned}$$

今引入一新的名称如下：对于任一拓扑空间 E. 我们叫 E 的 (法 2) 上同调环和其中平方运算的总和作 E 的 (法 2) "完全上同调环". 于是，从上面所述和§ 2 的定理 2，我们所得的结果可以写成下面的

定理 4　"格拉斯曼"流形 $G_{n,m}$ 的 (法 2) 完全上同调环系由维数是 2 的乘幂的 W 类 W^{2^k}，$0<2^k \leqslant m$ 所产生.

此外，我们又有

定理 5　若 $n \geqq m$，则在 $G_{n,m}$ 中任一维数是 2 的乘幂的 W 类 W^{2^k}，$0<2^k \leqslant m$，在 $G_{n,m}$ 法 2 完全上同调环中必不依赖于其他的 W 类 W^j，$j \neq 2^k$. 换言之，在 $n \geqq m$ 时，任一这样的类 W^{2^k}，$0<2^k \leqslant m$ 不能用上积和平方以维数 $j \neq 2^k$ 的 W 类 W^j 来表示.

证　若 r，s 是任意整数使 $s \geqq r>0$. $r+s=2^k(0<2^k \leqslant m)$ 则

$$\binom{s-1}{r} = \binom{2^k-1-r}{r} = \binom{2r}{r} = 0. \qquad (\text{法 } 2)$$

因之§ 4 的公式 (I) 说明在 ${S_q}^r W^s$ 的展开式中不可能有 W^{2^k} 这样的项.

今设 $\varphi(W^i)$ 是一个用上积和平方连起来的一个 W^i 的式子，代表一个 $H^*(G_{n,m})$ 中 r 维的类，此处 $r=2^k \leqslant m$. 由§ 4 的预备定理 9 和公式 (I)，我们可以把 $\varphi(W^i)$ 表作一个 W^i 的多项式 $P(W^i)$ 仅以上积连结各 W^i. 又因 $n \geqq m$，由预备定理 4，$P(W^i)$ 的形式是唯一的. 由前所述，只要 $W^{2^k}(0<2^k \leqslant m)$ 不出现在原来的 $\varphi(W^i)$ 内，那么 $P(W^i)$ 就不可能含有 W^{2^k}. 因之，当 $\varphi(W^i)$ 不含有 W^{2^k} 时，$W^{2^k} \neq \varphi(W^i)$ 而定理得证.

定理 4 与 5 合起来的意义是说，在 $n \geqq m$ 时，类 W^{2^k}，$0 < 2^k \leqslant m$ 是 $G_{n,m}$ 的 (法 2) 完全上同调环的一组不可约母素.

同样，不论是从§ 4 的公式 (II) 或§ 1 的预备定理 2 出发，我们可以得到关于 $\bar{W}^i$ 的类似的结果：在 $n \leqslant m$ 时，类 $\bar{W}^{2^k}$，$0 < 2^k \leqslant n$，是 $G_{n,m}$ 的 (法 2) 完全上同调环的一组不可约母素.

附录　预备定理 5 别证

在以下所谓一个 m 矢纤维丛 (vector bundle)，意思是指一个纤维丛 (fiber bundle)，它的纤维是 m 维的矢量空间而其构造群 (structural group) 则是 m 个变量的线性群. 在 m 矢纤维丛 $\Re$ 的底空间 (base) K 是一个多面体时，$\Re$ 有一组所谓 Stiefel-Whitney 类的不变量 $W_0^i(\Re)$，$i = 1, 2, \cdots, m$. 它们是 K 中以 $\pi_{i-1}(V_{m,m-i+1})$ 为局部系数群 (local coefficient group) 的 i 维上同调类. 此处 $V_{m,m-i+1}$ 是 Stietel 流形，而 $\pi_{i-1}(V_{m,m-i+1})$ 与整数加法群 z_0 或法 2 整数加法群 z_2 一一同构. 因之，把系数法 2 约化后从 $W_0^i(\Re)$ 可得一组法 2 的类 $W^i(\Re)$，$i = 1, \cdots, m$，我们叫它们作纤维丛 $\Re$ 的 Stiefel-Whitney “约化类”. 我们并推广它们的定义到一切非负整数的维数 i 如下：

$$W^0(\Re) = 1, \quad W^i(\Re) = 0, \quad i > m,$$

此处 1 是 K 中的法 2 单位类.

若 $\Re$ 是多面体 K 上的一个 m 矢纤维丛而 h 是从多面体 K' 到 K 中的一个连续变换，则 h 从 $\Re$ 引导出一个 m 矢纤维丛，记作 $h^*\Re$. 对于 Stiefel-Whitney 类我们有

$$\begin{cases} h^* W_0^i(\Re) = W_0^i(h^*\Re), \\ h^* W^i(\Re) = W^i(h^*\Re). \end{cases} \tag{1}$$

若 $\Re_i$，$i = 1, 2$，各是多面体 K_i，$i = 1, 2$ 上的 m_i 矢纤维丛，那么我们可以在积空间 $K_1 \times K_2$ 上定义一个 $m_1 + m_2$ 矢纤维丛，叫做 $\Re_1$，$\Re_2$ 的“积”，记作 $\Re_1 \otimes \Re_2$[9]. 这个积的 Stiefel-Whitney 约化类可以从 $\Re_i$ 的 Stiefel-Whitney 约化类用下式定出：

$$W^i(\Re_1 \otimes \Re_2) = \sum_{i_1 + i_2 = i} W^{i_1}(\Re_1) \otimes W^{i_2}(\Re_2). \tag{2}$$

现在试讨论在一个 $n + m$ 维欧氏空间 R^{n+m} 中定义的“格拉斯曼”流形 $G_{n,m}$. 设 X 是 R^{n+m} 的任一 m 面而 x 是 X 中的任一矢量，那么一切对偶 (x, X) 成一空间 $\tilde{G}_{n,m}$. 显然 $\tilde{G}_{n,m}$ 是一个以 $G_{n,m}$ 为底的 m 矢纤维丛 $g_{n,m}$ 的纤维空间 (bundle space)，其纤维是 m 维的矢量空间 X. 投影 $\pi : \tilde{G}_{n,m} \to G_{n,m}$ 即是自然

的变换 $\pi(x, X)=X$. 对于这个纤维丛 $g_{n,m}$ 我们可引入一组坐标邻域 (coordinate eighborhoods) 和坐标函数 (coordinate functions) 如次，设 V 是一个固定的 m 维矢量空间在 V 中有一组固定的基底 $v_1, \cdots, v_m$. 对于 R^{n+m} 的任一 m 面 X，在 R^{n+m} 中一切在 X 上的垂直投影不蜕化的 m 面 X' 成一个 $G_{n,m}$ 中含 X 的开集 V_X，我们把它作为纤维丛 $g_{n,m}$ 的一个坐标邻域. 其次在每一 $X \in G_{n,m}$ 中取 m 个线性无关的矢量 $x_1, \cdots, x_m$，且对于任一 $X' \in V_X$ 令 $x'_1, \cdots, x'_m$ 是 $x_1, \cdots, x_m$ 到 X' 上的垂直投影. 那么坐标函数 $\varphi_X: V_X \times V \to \pi^{-1}(V_X)$ 可由下式

$$\varphi_X\left(X', \sum a_i v_i\right)=\left(\sum a_i x'_i, X'\right)$$

来定义，其中 $a_1, \cdots, a_m$ 是任意实数，

我们叫上面所定义的纤维丛 $g_{n,m}$ 为 $G_{n,m}$ 上的“模范矢纤维丛”. 据 Pontrjagin 的一条定理 [10]，$g_{n,m}$ 的 Stiefel-Whitney 约化类就是§ 1 中所定义的 $G_{n,m}$ 的 W 类 W^i，即

$$W^i(g_{n,m})=W^i, \quad i \geqq 0. \tag{3}$$

今设 R^{n+m} 是两个欧氏空间 $R_j^{n_j+m_j}$, $j=1,2$ 的积，$n=n_1+n_2$，$m=m_1+m_2$. 又设 G_{n_j,m_j}，$j=1$，2 是定义于 $R_j^{n_j+m_j}$ 中的“格拉斯曼”空间，而

$$h: \quad G_{n_1,m_1} \times G_{n_2,m_2} \to G_{n,m}$$

如§ 3 所定，则预备定理 5 可证明如次.

设 $g_{n,m}$ 是 $G_{n,m}$ 上的模范矢纤维丛，以 $\tilde{G}_{n,m}$ 为其纤维空间，又设 $g_{n_j m_j}$，$j=1$，2 各是 G_{n_j,m_j} 上的模范 m_j 矢纤维丛，以 $\tilde{G}_{n_j,m_j}$ 为其纤维空间. 由纤维丛之积的定义，易知积

$$g'_{n,m}=g_{n_1,m_1} \otimes g_{n_2,m_2} \tag{4}$$

的纤维空间 $\tilde{G}'_{n,m}$ 中的点都作下形 $(x_1+x_2, X_1 \oplus X_2)$，此处 $(x_j, X_j) \in \tilde{G}_{n_j m_j}$，因 $X_1 \oplus X_2 \in G_{n,m}$ 而 $x_1+x_2 \in X_1 \oplus X_2$，故 $\tilde{G}'_{n,m}$ 是 $g_{n,m}$ 的纤维空间 $\tilde{G}_{n,m}$ 的一个子空间. 而且，若 $\tilde{h}$ 是 $\tilde{G}'_{n,m}$ 到 $\tilde{G}_{n,m}$ 中的恒同变换，则易见

$$\pi \tilde{h}=h \pi',$$

此处 π，π' 各为纤维丛 $g_{n,m}$ 和 $g'_{n,m}$ 的投影. 由此知 $\tilde{h}$ 是 $g'_{n,m}$ 到 $g_{n,m}$ 的丛变换 (bundle map)，而 $g'_{n,m}$ 与 h 从 $g_{n,m}$ 所引导出来的纤维丛同构 (isomorphic)：

$$g'_{n,m} \approx h^* g_{n,m}, \tag{5}$$

因之

$$W^i(g'_{n,m})=W^i(h^* g_{n,m}), \quad i \geqq 0. \tag{6}$$

由 (1)，(2)，(4) 及 (6) 可得

$$h^*W^i(g_{n,m}) = \sum_{i_1+i_2=i} W^{i_1}(g_{n_1,m_1}) \otimes W^{i_2}(g_{n_2,m_2}).$$

又由 (3)，此式可写作

$$h^*W^i = \sum_{i_1+i_2=i} W_1^{i_1} \otimes W_2^{i_2},$$

即预备定理 (5).

参考文献

[1] Wu Wen-tsün. C.R.Paris，1950, 230: 918-920.

[2] ———. C.R.Paris，1950, 230: 508-511.

[3] Steenrod N E. Topology of Sphere Bundles，1951.

[4] Chern S. Annals of Mathematics，1948, 49: 362-372.

[5] Steenrod N E. Annals of Mathematics，1947, 48: 290-320.

[6] Ehresmann Ch. Journal de Mathématiques，1937, 16: 69-104.

[7] Pontrjagin L. Mat.Sbornik，21 (63): 233-284.

[8] Cartan H, C R Paris. 1950, 230: 425-427.

[9] Wu Wen-tsün. Annals of Mathematics, 1948, 49: 641-653.

[10] Pontrjagin L. ДАН, 1942, 35: 35-39.

On squares in grassmann manifolds

Abstract This paper gives detailed proofs of results announced in C.R.230 (1950). We determine completely the squares in grassmannian manifolds.As a consequence we prove that for an $(m-1)$-sphere bundle the mod 2 Stiefel-Whitney classes are all determined，with the aid of cup products and squares，by those of dimensions a power of 2.

15. 一个 H.Hopf 推测的证明*

前　　言

假设一个有限复合形 K 上的定向 S^2 丛 $\mathfrak{S}$ 在 K^3 上有截面，那么 Hopf 从它的第二阻碍公式获得了一个丛不变量 [1, 2]

$$\Delta^4(\mathfrak{S}) \in H^4(K).$$

Hopf 曾经猜测过这个丛不变量与丛的 4 维 Понтрягин 示性类有关. 本文的目的在证明这个推测是对的，更明确言之，应有

$$\Delta^4(\mathfrak{S}) = -P^4(\mathfrak{S}).$$

此处 $P^4(\mathfrak{S})$ 是定向丛 $\mathfrak{S}$ 的 Понтрягин 示性类，其定义可参阅 [3] 中第二章.

一般说来，一个球丛的丛不变量有两种普遍的方法可以获得. 第一种是作原来的丛的相配丛，再取 Steenrod 意义下的截面示性类. 第二种方法系 Понтрягин 所倡用，把原来的丛视为将底空间映像到适当格拉斯曼流形中去所引导而得，于是格拉斯曼流形中任一上同调类的逆影都是原来这个丛的丛不变量，Hopf 的方法提供了一个新的原则，在丛满足适当条件之下，从第二阻碍的考虑也可以获得一些丛不变量. 但本文的结果证明，在本文所讨论的那种情形 (事实上是最重要的情形)，Hopf 的方法并不能给我们以新的不变量，而只是对 4 维 Понтрягин 示性类给予了一个新的有趣的定义而已.

§1　Hopf 的推测

设 $\mathfrak{S} = (\widetilde{K}, K, \pi, S^2, R_3)$ 是有限复合形 K 上的一个可定向球丛 (R_m 指 m 个变量上的旋转群). 在以后我们恒将假定 $\mathfrak{S}$ 中的纤维都已协合地定向而称之为一定向丛. 若将其纤维都取相反定向，则所得的定向丛将记为一 $\mathfrak{S}$.

假设 $\mathfrak{S}$ 的 (未约化) 的三维 Stiefel-Whitney 示性类

$$W^3(\mathfrak{S}) = 0. \tag{1}$$

---

* *Acta Math. Sinica*, 1954: 491-500.

于是在 K^3 上有截面 $f: K^3 \to \widetilde{K}$ 使 $\pi f = 1$ 者存在. 任一这样的截面决定一阻碍上闭链

$$\bar{\gamma}_f^4 \in \bar{\Gamma}_f^4(\mathfrak{S}) \in H^4(K, \pi_3(S^2)).$$

将 $\pi_3(S^2)$ 的元素与其所定 Hopf 不变量相对应, 得一确定的同构变换 $\pi_3(S^2) \approx I$(I 为整数加法群), 由此引起 $i: H^4(K, \pi_3(S^2)) \approx H^4(K)$ 等.

令

$$i\,\bar{\gamma}_f^4 = \gamma_f^4, \quad i\,\bar{\Gamma}_f^4(\mathfrak{S}) = \Gamma_f^4(\mathfrak{S}) \in H^4(K).$$

其次对于任二截面 $f, g: K^3 \to \widetilde{K}$, 可作保纤伦移至 f', g' 使 f', g' 尽量分离, 以致在 K^1 上恒有 $f'(p) \neq g'(p)$, $P \in K^1$. 于是可得一二维整数上闭链 $\omega_{f',g}^2$, 如次: 任取一 $\sigma^2 \in K$, 则有拓扑对应 $\varphi_{\sigma P}: \pi^{-1}(P) \equiv S^2 (P \in \sigma^2)$, 使 $\varphi_\sigma: \pi^{-1}(\sigma^2) \equiv \sigma^2 \times S^2$ 为一拓扑对应, 此处 $\varphi_\sigma(q) = (p, \varphi_{\sigma p}(q))$, $P = \pi(q)$. 因 $\mathfrak{S}$ 已定向, 故 S^2 因之 $\pi^{-1}(\sigma^2)$ 也有了定向而成为一有边定向流形. 因 $f'(\sigma^2)$, $g'(\sigma^2)$ 在边上无公共交点, 故对定向的 $\pi^{-1}(\sigma^2)$ 而言, 二者有一确定的交点数, 定义之为 $\omega_{f',g'}^2(\sigma^2)$. 如此决定的上链 $\omega_{f',g'}^2$ 为一上闭链, 其上类 $n_{f,g}^2(\mathfrak{S})$ 与所择伦移无关:

$$\omega_{f',g'}^2 \in \Omega_{f,g}^2(\mathfrak{S}) \in H^2(K).$$

Hopf[1,2] 与 Болтянский [4] 都曾证明: 若已知 K^3 上的一个截面 g, 则与任一 K^3 上其他截面 f 相当的 Γ_f 的一般形式为

$$\Gamma_f^4(\mathfrak{S}) = \xi^2\xi^2 + \Omega_{g,g}^2(\mathfrak{S})\xi^2 + \Gamma_g^4(\mathfrak{S}), \tag{2}$$

其中以上积为乘法, 而 $\xi^2 \in H^2(K)$ 为由 f, g 所定的上类. 由 (2) 式 Hopf 证明

$$\Delta_f^4(\mathfrak{S}) = 4\Gamma_f^4(\mathfrak{S}) - \Omega_{f,f}^2(\mathfrak{S})\Omega_{f,f}^2(\mathfrak{S}) \in H^4(K) \tag{3}$$

与截面 f 的选择无关, 而为定向丛 $\mathfrak{S}$ 的不变量, 因之可表为 $\Delta^4(\mathfrak{S})$. 显见

$$\Gamma_f^4(\mathfrak{S}) = \Gamma_f^4(-\mathfrak{S}), \quad \Omega_{f,f}^2(\mathfrak{S}) = -\Omega_{f,f}^2(-\mathfrak{S}),$$

因之

$$\Delta^4(-\mathfrak{S}) = \Delta^4(\mathfrak{S}).$$

Hopf 曾推测 $\Delta^4(\mathfrak{S})$ 与 $\mathfrak{S}$ 的 4 维 Понтрягин 示性类有关, 明确言之:

$$\Delta^4(\mathfrak{S}) = -P^4(\mathfrak{S}). \tag{4}$$

本文的主要目的, 即在证明此一推测, 并因之得一 $\Delta^4(\mathfrak{S})$ 为不变量的直接证明, 而与公式 (2) 的建立无关. 这一结果说明在 (1) 的假定之下, Hopf 的方法并不能产生新的不变量来.

以下各节将保持本节的记号, 其中 $\mathfrak{S}$ 为定向丛, 满足条件 (1), f 为 K^3 上一个给定截面. 我们并假定 K 的维数是 4, 显然这并不损害其一般性.

§2 $\Omega^2_{f,f}(\mathfrak{S})$ 的一个解释

本节中将给予 $\Omega^2_{f,f}$ 以另一解释.

如前，f 仍为 K^3 上的一个截面. 对于 $p\in K^3$，令 S^0_p 为由 $f(p)$ 及 S^2_p 上与 $f(p)$ 直径相对的点 $\overline{f}(p)$ 所定的 0 维球，S^1_p 为 S^2_p 中与 $f(p)$ 完全垂直的大圆，则一切 S^0_p 与 $S^1_p(p\in K^3)$ 各决定了一个 K^3 上的 S^0 丛 σ_0 与 S^1 丛 σ_1，而 $\mathfrak{S}$ 在 K^3 上的部分即 $\mathfrak{S}_0$ 与 $\mathfrak{S}_1$ 的 Whitney 积: $\mathfrak{S}/K^3\sim\mathfrak{S}_0\cup\mathfrak{S}_1$. 显然 $\mathfrak{S}_1$ 是可定向丛，我们将定 $\mathfrak{S}_1$ 中各纤维 S^1_p 的向，使由 $f(p)$ 及 S^1_p 依此次序所定 S^2_p 的向即定向丛 $\mathfrak{S}$ 中原来所定 S^2_p 的向.

在 $\mathfrak{S}_1$ 中，有一定义于 K^1 上的截面 $g:g(p)\in S^1_p$，$p\in K^1$. 此截面 g 定义了一个阻碍上闭链 $w^2_g\in W^2(\sigma_1)\in H^2(K,\pi_1(S^1))$，此处 W^2 指 $\mathfrak{S}_1$ 在 Steenrod 意义下的截面示性类. 因 $\mathfrak{S}_1$ 已定向，故 $\pi_1(S^1)$ 确定地同构于 I，而可视 w^2_g，$W^2(\mathfrak{S}_1)$ 为整系数上闭链与整系数上类.

今取任一 $\sigma^2\in K^2$ 视之. σ^2 的点将表为 $p=(p_1,p_2)$，$|p|^2=p_1^2+p_2^2\leqslant 1$. 在 $\mathfrak{S}$ 中，设定向 S^2_p 中含有 $f(p)$ 的定向半球为 E^2_p，则取§1 中的记号，一切 $p\in\sigma^2$ 时，E^2_p 所成集合可在 φ_σ 之下恒同为一 (定向)4 维闭胞腔 $E^4=\sigma^2\times E^2\subset\sigma^2\times S^2$，此处 $E^2\subset S^2$ 的点可表为 $q=(q_1,q_2)$，$|q|^2=q_1^2+q_2^2\leqslant 1$，使

$$f(p)=(p,0),\quad p\in\sigma^2,\quad 0=(0,0)\in E^2,$$

而 S^1_p 为一切使 $|q|^2=1$ 的点 (p,q) 所成的集合. 今推广 σ^2 边界 $\dot\sigma^2$ 上的截面 g 至 σ^2 上，仍记之为 g 如次:

$$g(p)=(p,|p|\varphi_{\sigma,p/|p|}g(p/|p|)),\quad p\neq 0\in\sigma^2,\quad g(0)=(0,0),$$

于是 $f(\sigma^2)$，$g(\sigma^2)$ 只相交于一点 (0, 0). 若对于每一 σ^2 都定义 $g(\sigma^2)$ 如上述，则得一 K^2 上的截面 g. 显然 g 可自 f/K^2 依保纤伦移得来，故对任一 σ^2，有边定向流形 E^4 中 $f(\sigma^2)$ 与 $g(\sigma^2)$ 在 (0, 0) 的交点数即 $\omega^2_{f,g}(\sigma^2)$，而 $\omega^2_{f,g}\in\Omega^2_{f,f}(\mathfrak{S})$.

另一面，定义

$$\bar g:\quad \dot\sigma^2\to S^1_0$$

为

$$\bar g(p)=(0,\varphi_{\sigma p}g(p)),\quad p\in\dot\sigma^2,$$

则 $\bar g$ 所定之元素 $a\in\pi_1(S^1_0)\approx I$ 即 $w^2_g(\sigma^2)$, 但此 a 易见即为 $f(\sigma^2)$ 与 $g(\sigma^2)$ 在 (0, 0) 的交点数，因之

$$w^2_g(\sigma^2)=\omega^2_{f,g}(\sigma^2).$$

因此式对每一 $\sigma^2 \in K$ 皆然，故有 $w_g^2 = \omega_{f,g}^2$，而由此得

$$\Omega_{f,f}^2(\mathfrak{S}) = W^2(\mathfrak{S}_1). \tag{1}$$

换言之，$\Omega_{f,f}^2(\mathfrak{S})$ 可视为由 f 所定定向 S^1 从 $\mathfrak{S}_1$ 的 Steenrod 截面示性类，此即本节的目的.

其次，设有映像 $k : K' \to K$. 令 $k * \mathfrak{S} = \mathfrak{S}'$ 为从 $\mathfrak{S}$ 引导而得的定向丛，其丛空间为 $\tilde{K}'$，投影为 π'. 对于 $\mathfrak{S}$ 中 K^3 上的截面 f，显然引导得一 $\mathfrak{S}'$ 中 K'^3 上的截面 f' 使 $\tilde{k}\prime f\prime = fk\prime$，此处 $k\prime \simeq k$ 且 $k'(K'^3) \subset K^3$, $\widetilde{k}' : \widetilde{K}' \to \widetilde{K}$ 而 $k'\pi' = \pi\tilde{k}'$. 易见

$$\Gamma_{f'}^4(\mathfrak{S}') = k'^* \Gamma_f^4(\mathfrak{S}) = k^* \Gamma_f^4(\mathfrak{S}),$$
$$\Omega_{f',f'}^2(\mathfrak{S}') = k'^* \Omega_{f,f}^2(\mathfrak{S}) = k^* \Omega_{f,f}^2(\mathfrak{S}),$$

因之

$$\Delta_{f'}^4(k^*\mathfrak{S}) = k^* \Delta_f^4(\mathfrak{S}). \tag{2}$$

§3　$K = S^4$ 时的特殊情形

令 $R_{\pi,m}^{\wedge}$ 为欧氏空间 R^{n+m} 中一切"定向"m 面所成的格拉斯曼流形. 我们知道

$$\pi_4(R_{\widehat{n,3}}) \approx I \quad (n \geqslant 5), \tag{1}$$

其中一个母素 $\alpha \in \pi_4(R_{\widehat{\pi,3}})$ 可定义之如次：取球面 $S^3 : y_1^2 + \cdots + y_4^2 = 1$，及其所界闭胞腔 σ^4，表 σ^4 的点为 (y,t)，$y \in S^3$，$0 \leqslant t \leqslant 1$，而 $(y,0)$ 为 σ^4 的中心. 在 R^{n+3} 中取一组直角坐标 $(x_1, \cdots, x_{n+3})$ 及其相当矢量基 $(e_1, \cdots, e_{\pi+3})$. 对于任一 $y = (y_1, y_2, y_3, y_4) \in S^3$，令

$$\left.\begin{aligned} v_1(y,1) &= (y_4^2 - y_3^2 - y_2^2 + y_1^2)e_1 + 2(y_1y_2 - y_3y_4)e_2 + 2(y_1y_3 + y_2y_4)e_3, \\ v_2(y,1) &= 2(y_1y_2 + y_3y_4)e_1 + (y_4^2 - y_3^2 + y_2^2 - y_1^2)e_2 + 2(y_2y_3 - y_1y_4)e_3, \\ v_3(y,1) &= 2(y_1y_3 - y_2y_4)e_1 + 2(y_1y_4 + y_2y_3)e_2 + (y_4^2 + y_3^2 - y_2^2 - y_1^2)e_3, \end{aligned}\right\} \tag{2}$$

$$v_i(y,t) = v_i(y,1)\sin\frac{\pi t}{2} + e_{i+3}\cos\frac{\pi t}{2}, \quad i = 1,2,3,$$
$$X(y,t) = [v_1(y,t), v_2(y,t), v_3(y,t)].$$

此处 $[v_1, v_2, v_3]$ 表 v_i，$i = 1,2,3$，所定的定向 3 面 $\in R_{\widehat{n,3}}$. 将 σ^4 的边界恒同成一点得球 S^4，视 $t < 1$ 或 $t = 1$，表 S^4 中与 σ^4 中 (y,t) 相当的点为 (y,t) 或 y_0. 因不论

$y \in S^3$ 如何，$X(y,1)=[e_1,e_2,e_3]$，故 $(y,t)\to X(y,t)$ 定义了一个映像 $h:S^4\to R_{\widehat{n,3}}$，此 h 所定的同伦类即可取为 $\pi_4(R_{\widehat{n,3}})$ 的母素 a(因 $\pi_1(R_{\widehat{n,3}})=0$，故指标点无关紧要).

今考察 S^4 上的定向 S^2 丛 $\mathfrak{S}_0=h^*\Re_{\widehat{n,3}}$，此处 $\Re_{\widehat{n,3}}$ 为 $R_{\widehat{n,3}}$ 上的模范定向 S^2 丛，$\mathfrak{S}_0$ 的丛空间将记为 $\tilde{K}_0$. 显然

$$(y,t)\to v_1(y,t),\qquad t<1$$

确定了 $\mathfrak{S}_0$ 中 $S^4-(y_0)$ 上的一个截面 f. 令 S^2 为 $[e_1,e_2,e_3]$ 中的定向单位球，则 $-\gamma_f^4(S^4)$(S^4 自然定向) 显即由映像

$$f':\quad y\to v_1(y,1)$$

所定的 $\pi_3(S^2)\approx I$ 中的元素. 因 f' 的 Hopf 不变量为 -1，故

$$\gamma_f^4(S^4)=+1.$$

因 $\Omega_{f,f}^2(\mathfrak{S}_0)\in H^2(s^4)=0$，故得

$$\Delta_f^4(\mathfrak{S}_0)=4\Gamma_f^4(\mathfrak{S}_0)=+4\Sigma,\tag{3}$$

此处 $\Sigma\in H^4(S^4)$ 使 $\Sigma(S^4)=1$.

另一面，据Понтрягин，$R_{n,3}^{\wedge}$ 中有一 $3n-4$ 维下闭链 z，使

$$I(h\,S^4,z)=-4,$$

而 z 对偶于 $R_{\widehat{n,3}}$ 中的 4 维 P 类 P^4(参阅 [3] 第三章)，因之

$$P^4(\mathfrak{S}_0)(S^4)=h^*P^4(S^4)=-4,$$

或

$$P^4(\mathfrak{S}_0)=-4\Sigma.\tag{4}$$

从 (3)，(4) 得

$$\Delta_f^4(\mathfrak{S}_0)=-P^4(\mathfrak{S}_0).\tag{5}$$

因任一 S^4 上的定向 S^2 丛 $\mathfrak{S}$ 可从一映像 $S^4\to R_{\widehat{n,3}}$ 引导而得，且同伦的映像引导得同构的丛，故由 (1)，有映像 $k:S^4\to S^4$ 使

$$\mathfrak{S}\sim k^*\mathfrak{S}_0.$$

由§ 2 的 (2)，有 (f' 系从 f 引导而得)

$$\Delta^4_{f'}(\mathfrak{S}) = k^*\Delta^4_f(\mathfrak{S}_0). \tag{6}$$

又我们知道

$$P^4(\mathfrak{S}) = k^*P^4(\mathfrak{S}_0). \tag{7}$$

因之从 (5)，(6)，(7) 得

$$\Delta^4_{f'}(\mathfrak{S}) = -P^4(\mathfrak{S}). \tag{8}$$

今设底空间 $K = S^4$ 被赤道球分成的两个半球为 σ^4，σ'^4，且设 $K^3 \subset \sigma'^4$. 显然我们在上面可取 $k: S^4 \to S^4$ 使 f' 在 σ'^4 上有定义. 因在 K^3 上任两这样在 σ'^4 上有定义的 f'，g' 可依保纤伦移互相求得，故易见 $\Delta^4_{f'}(\mathfrak{S}) = \Delta^4_{g'}(\mathfrak{S})$. 因之在 (8) 式中若假定 f' 为任一可定义于 σ'^4 上的截面，(8) 式依然成立. 换言之，若 $K = S^4$ 及 f' 如上述，则与此相当的 Hopf 推测成立.

§4　在 K^4 上有截面时的特殊情形

设 $R_{\widehat{n,2}}$, $R_{\widehat{n,3}}$ 各定义于 R^{n+2} 与 R^{n+3} 中而 $R^{n+2} \subset R^{n+3}$, $R^{n+3} = R^{n+2} \oplus R^1$. 对于 $L \in R_{\widehat{n,2}}$, 令 $i(L) = R^1 \oplus L$, 且 $i(L)$ 依 (R^1, L) 的次序而定向 (R^1 亦定向), 则得一自然映像

$$i: \quad R_{\widehat{n,2}} \to R_{\widehat{n,3}}.$$

令 P^4，$p'4$ 各为 $R_{\widehat{n,3}}$，$R_{\widehat{n,2}}$ 中的 4 维 p 类，则

$$i^*P^4 = P'^4. \tag{1}$$

在 $R_{\widehat{n,2}}$ 中我们知有一 2 维整系数上类

$$X^2 = W^2(\Re_{\widehat{n,2}}), \tag{2}$$

此处 W^2 指 Steenrod 意义下 S^1 从 $\Re_{\widehat{n,2}}$ 的截面示性类，亦即 $\Re_{\widehat{n,2}}$ 的未约化 2 维 (整系数)Stiefel-Whitney 示性类. 不难证明在 $R_{\widehat{n,2}}$ 中

$$X^2 \cup X^2 = P'^4. \tag{3}$$

今设 $\mathfrak{S}$ 在 $K = K^4$ 上有截面 f，于是如§1，$\mathfrak{S}$ 可视为 $K = K^4$ 上一个定向 S^0 从 $\mathfrak{S}_0$ 与一定向 S^1 从 $\mathfrak{S}_1$ 的 Whitney 积：$\mathfrak{S}=\mathfrak{S}_0 \cup \mathfrak{S}_1$，其中 $\mathfrak{S}_0$ 在 K^4 上有截面 f，故为简单的. 若设

$$\mathfrak{S}_1 \sim h^*\Re_{\widehat{n,2}},$$

此处 $h: K \to R_{\widehat{n,2}}$，则显有

$$\mathfrak{S} \sim (ih)^* \mathfrak{R}_{\widehat{n,3}}. \tag{4}$$

由§ 2 的 (2) 及上面 (2) 式，得

$$\Omega^2_{f,f}(\mathfrak{S}) = W^2(\mathfrak{S}_1) = W^2(h^* \mathfrak{R}_{\widehat{n,2}}) = h^* W^2(\mathfrak{R}_{\widehat{n,2}}) = h^* X^2,$$

又由 (1)，(4)，得

$$P^4(\mathfrak{S}) = P^4(h^* i^* \mathfrak{R}_{\widehat{n,3}}) = h^* i^* P^4(\mathfrak{R}_{\widehat{n,3}}) = h^* i^* P^4 = h^* P'^4.$$

因之从 (3) 得

$$P^4(\mathfrak{S}) = \Omega^2_{f,f}(\mathfrak{S}) \cup \Omega^2_{f,f}(\mathfrak{S}). \tag{5}$$

因 Γ^4_f 显为 0，故 (5) 亦即

$$P^4(\mathfrak{S}) = -\Delta^4_f(\mathfrak{S}). \tag{6}$$

换言之，在 K^4 上有截面存在时，Hopf 的推测正确.

§5　一般情形时 Hopf 推测的证明

今设 $\mathfrak{S}$ 任意，并设 f 是 K^3 上的任一截面. 设 K^4 中一切 4 维定向胞腔的集合为 $\{\sigma^4_i\}$，$j \in J$. 另取一组 4 维定向胞腔 (σ'^4_j)，$j \in J$，并对每一 $j \in J$ 将 σ^4_j 与 σ'^4_j 的边黏合使成一 4 维球 S^4_i，且使

$$\partial(\sigma^4_j - {\sigma'}^4_j) = 0. \tag{1}$$

令黏合后所得的复合形为 K^*，其中, 由 σ'^4_j 及其边所成的子复合形为 $K' : K^* = K + K'$.

如§ 2，f/K^3 决定了一个 K^3 上的定向 S^0 丛 $\mathfrak{S}_0$，并使 $\mathfrak{S}$ 在 K^3 上的部分丛分解为 $\mathfrak{S}_0$ 与一个定向 S^1 丛 $\mathfrak{S}_1$ 的 Whitney 积. 因任一 S^3 上的球义都是简单的，因之对于每一 $j \in J$，$\mathfrak{S}_0$ 与 $\mathfrak{S}_1$ 在 $\dot{\sigma}'^4_j$ 上的部分丛都可扩充到 σ'^4_i 上，由此可扩充 $\mathfrak{S}_0$，$\mathfrak{S}_1$ 至 K'^4 上，扩充后所得的定向丛将记之为 $\mathfrak{S}'_0$，$\mathfrak{S}'_1$，而在 K'^4 上的 Whitney 积 $\mathfrak{S}'_0 \cup {\mathfrak{S}'}_1$ 将记为 $\mathfrak{S}'$，于是在 K^* 上，有一定向 S^2 丛 $\mathfrak{S}*$，系由 K^4 上的 $\mathfrak{S}$ 及 K'^4 上的 $\mathfrak{S}'$ 所合成. 显然有映像 $h: K* \to R^{\wedge}_{n,3}$，$h' : K'^4 \to R_{\widehat{n,2}}$，使

$$\mathfrak{S}^* \sim h^* \mathfrak{R}_{\widehat{n,3}}, \tag{2}$$

$$h/K^3 \equiv i\, h'/K^3, \tag{3}$$

此处 $i: R_{\widehat{n,2}} \to R_{\widehat{n,3}}$ 定义如 §4. 且 f 可扩张到每一 $\sigma_j^{\prime 4}$ 上，令由此所得的截面为 $f*: K^3 + K'^4 \to \widetilde{K}*.$，此处 $\tilde{K}^*$ 为 $\mathfrak{S}^*$ 的丛空间，则显然有

$$\gamma_{f^*}^4(\sigma'^4_j) = 0, \quad \gamma_{f^*}^4(\sigma_j^4) = \gamma_f^4(\sigma_j^4), \quad j \in J, \tag{4}$$

$$\Omega_{f^*,f^*}^2 \mathfrak{S} = \Omega_{f,f}^2 \mathfrak{S}^*. \tag{5}$$

今任取 $p^4 \in P^4(\mathfrak{S}^*)$，则应用 §3 于 S_j^4，$j \in J$，及 f^*, 得

$$4\gamma_{f^*}^4(\sigma_j^4 - \sigma'^4_j) = -p^4(\sigma_j^4 - \sigma'^4_j), \quad j \in J. \tag{6}$$

因 K' 上的 $\mathfrak{S}'$ 合于§4 的特殊情形，而 P^4 在 K' 上的限制显属于 $P^4(\mathfrak{S}')$，故有 $(\omega_{f^*,f^*}^2 \in \Omega_{f^*,f^*}^2(\mathfrak{S}))$

$$p^4 \sim \omega_{f^*,f^*}^2 \cup \omega_{f^*,f^*}^2 \quad (\text{在 } K' \text{ 中}). \tag{7}$$

即有 $a \in C^3(K^*)$，使

$$p^4 = \omega_{f^*,f^*}^2 \cup \omega_{f^*,f^*}^2 + \delta a \quad (\text{在 } K' \text{ 中}). \tag{8}$$

因之，

$$p^4(\sigma'^4_j) = (\omega_{f^*,f^*}^2 \cup \omega_{f^*,f^*}^2)(\sigma'^4_j) + a(\partial\sigma'^4_j), \quad j \in J. \tag{9}$$

又因在 S_j^4 中 $\omega_{f^*,f^*}^2 \sim 0$，故由 (1) 有

$$(\omega_{f^*,f^*}^2 \cup \omega_{f^*,f^*}^2)(\sigma'^4_j) = (\omega_{f^*,f^*}^2 \cup \omega_{f^*,f^*}^2)(\sigma_j^4), \quad j \in J. \tag{10}$$

由 (1)，(4)，(5)，(6)，(9)，(10) 诸式得 $(\omega_{f,f}^2 = \omega_{f^*,f^*}^2)$

$$\begin{aligned}
4\gamma_f^4(\sigma_j^4) &= 4\gamma_{f^*}^4(\sigma_j^4 - \sigma'^4_j) = -p^4(\sigma_j^4 - \sigma'^4_j) \\
&= -p^4(\sigma_j^4) + (\omega_{f^*,f^*}^2 \cup \omega_{f^*,f^*}^2)(\sigma'^4_j) + a(\partial\sigma'^4_j) \\
&= -p^4(\sigma_j^4) + (\omega_{f,f}^2 \cup \omega_{f,f}^2)(\sigma_j^4) + a(\partial\sigma_j^4),
\end{aligned}$$

或即

$$4\gamma_f^4(\sigma_j^4) - (\omega_{f,f}^2 \cup \omega_{f,f}^2)(\sigma_j^4) = -p^4(\sigma_j^4) + \delta a(\sigma_j^4).$$

因此式对任意 $i \in J$ 皆然，故得

$$4\gamma_f^4 - \omega_{f,f}^2 \cup \omega_{f,f}^2 \sim -p^4 \quad (\text{在 } K \text{ 中}).$$

因 p^4 在 K 中的限制显属于 $P^4(\mathfrak{S})$，故得

$$\Delta_f^4(\mathfrak{S}) = -P^4(\mathfrak{S}). \tag{11}$$

此即 Hopf 的推测. 因 $P^4(\mathfrak{S})$ 为丛不变量，故 $\Delta^4_f(\mathfrak{S})$ 亦然，而与 f 无关.

注意在上面的证明中，我们并不需要预先知道 Hopf 与 Болтянский 的第二阻碍公式 (§1 的 (2) 式) 与 $\Delta^4_f(\mathfrak{S})$ 的丛不变性，而相反，$\Delta^4_f(\mathfrak{S})$ 的丛不变性为 (11) 的推论.

参考文献

[1] Hopf H. Sur une formule de la théorie des espaces fibrés. *Colloque de Topologie à Bruxelles*, 1950: 117-121.

[2] Hopf H. Sur les champs d'éléments de surface dans les variétés à 4 dimensions. *Colloque International de Topologie Algébrique à Paris*, 1947: 55-59.

[3] Wu Wen-Tsün, Sur les classes caractéristiques des espaces fibrés sphériques. *Act.Sc.Ind.*, no.1183, Paris, 1952.

[4] В Болгянский. Векторные полянa многообразий. ДАН, 1951, 80: 305-307.

Proof of a certain conjecture of H.Hopf

Abstract For an oriented S^2-bundle $\mathfrak{S}$ with trivial 3-dim.Stiefel-Whitney class H.Hopf has deduced from his formula about second obstruction in this bundle acertain bundle invariant $\Delta^4(\mathfrak{S})$.[1,2] He has conjectured that this invariant coincides essentially with the 4-dim.Pontrjagin class of $\mathfrak{S}$. Without assuming a priori the invariance of $\Delta^4(\mathfrak{S})$, we prove directly that Hopf's conjecture is true.

16. 论ПОНТРЯГИН示性类 II*

前　　言

本文是［1］§§ 5，6 中所述结果的详细证明.

假定 M 是一个 m 维的可微分闭流形，$\mathfrak{D}$ 是 M 的一个微分构造. 对于 $\mathfrak{D}$ 而言，在 M 的各点的切面很自然地决定一个 m 欧氏空间丛 $\mathfrak{T}(\mathfrak{D})$. 如果从定义看，$\mathfrak{T}_{\mathfrak{D}}$ 的一切示性类都不但倚赖于 M，而且与微分构造 $\mathfrak{D}$ 有关. 可是就 Stiefel-Whitney 示性类而论，在 1949 年末，作者首先指出 $W^i(\mathfrak{T}(\mathfrak{D}))$ 与 M 中的平方运算有关，并证明了下列公式：

$$[W^2(\mathfrak{T})\mathfrak{D})) + W^1(\mathfrak{T}(\mathfrak{D})) \cup W^1(\mathfrak{T}(\mathfrak{D}))] \cup X = Sq^2 X, \quad X \in H^{m-2}(M, \mathrm{mod}\, 2).$$

由此知道 $W^2(\mathfrak{T}(\mathfrak{D}))$ 可由 M 的同调性质完全决定，因之实际上 $W^2(\mathfrak{T}(\mathfrak{D}))$ 只倚赖于 M 的拓扑构造，而与 M 上的微分构造 $\mathfrak{D}$ 无关①. 换言之，$W^2(\mathfrak{T}(\mathfrak{D}))$ 是 M 的拓扑不变量.

Thom 知道上述结果后，旋即证明所有 Stiefel-Whitney 示性类 $W^i(\mathfrak{T}(\mathfrak{D}))$，$i \geqslant 1$ 的拓扑不变性 [2,3,4]. 作者并得到了用 M 中上积与平方运算表示这些示性类的明确公式 [5].

此后一个自然发生的问题，自然是上述结论是否也适用于Понтрягин 示性类. 关于此，作者在 1950 —1951 年证明了一个可定向闭微分流形的Понтрягин示性类至少在法 3 约化后是拓扑不变的 [1]. 且若令 p 为一任意奇质数，则在 $p > 3$ 时，至少法 p 约化后Понтрягин 示性类应用上积所得的某种组合是拓扑不变的②. 这些结果的证明就是本文的主要内容.

在 Stiefel-Whitney 示性类的情形，拓扑不变性的证明是倚靠了下面的 Thom 公式：

$$Sq^i \tau^* 1 = \tau^* W^i(\mathfrak{B})$$

---

* *Acta Math. Sinica*, 1954, 4: 171-199; *Amer. Math. Soc. Translations, Ser.*, 1970, 2(92): 63-92.

① $W^1(\mathfrak{T}(\mathfrak{D}))$ 决定于 M 的可定向性，$W^m(\mathfrak{T}(\mathfrak{D}))$ 由 M 的 Euler-Poincaré 示性数所决定，因之在 $k=1$ 或 m 时，$W^k(\mathfrak{T}(\mathfrak{D}))$ 的拓扑不变性是显然的. 但在 $1 < k < m$ 时，问题即不简单.

② 由此并得下述结果：一个四维可定向微分闭流形的四维Понтрягин 示性类在此定向后的流形上所取的值必是 3 的倍数.

(说明见§ 4). 但 Thom 原来的证明，不易推广之于Понтрягин示性类. 在本文中，我们给予了这个公式以一个新的证明. 同样的证法可引得一个关于Понтрягин示性类的类示公式，由此得到了前述关于微分流形上Понтрягин的部分拓扑不变性的结果.

至于可微分流形上Понтрягин示性类的拓扑不变性的普遍问题，目前尚无可以解决的迹象. 有一些现象可以说明，这个问题之要完全解决，大概需要一些拓扑学中尚未发现的崭新工具. 不管如何，我们的工作可以说明，Понтрягин示性类的本质要比 Stiefel-Whitney 示性类的本质深刻得多. 它们的本质的了解对于拓扑学的发展必具有重大意义.

本文须用到前一文［6］的结果.

§1　局部紧空间上的同调群 [7—10]

本节中将简述以后所需的同调系统及其性质. 在以下，所有同调群的系数群是一个固定的可交换群，不明写.

对于任一紧空间偶 (X,A)，即一个紧空间 X 和 X 的一个闭子空间 A，$H_+^k(X,A)$ 将表 (X,A) 的 k 维 Cêch 上同调群.

今设 (X,A) 是一局部紧空间偶，即一个局部紧空间 X 和 X 的一个闭子空间 A. 取不属于 X 的一点 ∞，命 X^+ 为将 ∞ 加至 X 上依Александров方法紧化后所得的空间，又命 $A^+=A\cup\infty$. 定义

$$H^k(X,A)=H_+^k(X^+,A^+). \tag{1}$$

在此定义之下，$H^k(X,A)$ 中与 $U\in H_+^k(X^+,A^+)$ 相应的元素将以 U^- 表之，而 $H_+^k(X^+,A^+)$ 中与 $U\in H^k(X,A)$ 相应的元素则将以 U^+ 表之. 特别在 (X,A) 原为紧空间偶时，X^+，A^+ 各为 X 与 ∞，A 与 ∞ 的和空间，故此时有 $H^k(X,A)\approx H_+^k(X,A)$.

在以后，(X,A) 为紧空间偶或局部紧空间偶时，将各记作 $(X,A)\in\mathfrak{D}$ 与 $(X,A)\in\mathfrak{LC}$，又 ∞ 将取为不属于任何在讨论中的局部紧空间的点.

若 (X,A)，$(Y,B)\in\mathfrak{LC}$，映像 $f:(X,A)\to(Y,B)$ 具有下述性质：Y 中任一紧集的逆影是 X 的一个紧集，则 f 将叫做一个可允许的映像，记作 $f\in\mathfrak{A}$. 此时，可推广 f 为 $f^+:(X^+,A^+)\to(Y^+,B^+)$ 使 $f^+/X\equiv f$，$f^+(\infty)=\infty$. 我们定义 $f^*:H^k(Y,B)\to H^k(X,A)$ 为 $f^{+*}:H_+^k(Y^+,B^+)\to H_+^k(X^+,A^+)$. 明确言之：

$$f^*U=(f^{+*}U^+)^-, U\in H^k(Y,B). \tag{2}$$

上面的同调系统 H 具有下述诸性质：

$H1°$. 设 $(X,A)\in\mathfrak{LC}$. 定义 $i^+:(X^+,A^+)\to((X-A)^+,\phi^+)$ 为 $i^+/(X-A)=$ 恒同映像，$i^+(A^+)=\phi^+(=\infty)$，则 i^+ 连续且 $i^{+*}:H_+^k(((X-A)^+,\phi^+)\approx H_+^k(X^+,A^+)$，并且由此定义了一个确定的同构变换 $i^*:H^k(X-A)\approx H^k(X,A)$，此处

$$i^*U=(i^{+*}U^+)^-,\quad U\in H^k(X-A). \tag{3}$$

一般说来，若 (X,A)，$(Y,B)\in\mathfrak{LC}$，而 $f:(X,A)\to(Y,B)$ 是可允许的，且 $f/(X-A)$ 是到 $Y-B$ 上的同拓映像 (后一性质简称：f 是相对同拓映像)，则 $f^*:H^k(Y,\ B)\approx H^k(X,A)$.

$H2°$. 设 $X'\in\mathfrak{LC}$ 而 X 是 X' 的开集. 定义 $k^+:(X'^+,\phi^+)\to(X^+,\phi^+)$ 为 $k^+/X\equiv$ 恒同映像，而 $k^+((X'-X)\cup\infty)=\infty$，则 $k^{+*}:H_+^k(X^+,\phi^+)\to H_+^k(X'^+,\phi^+)$ 定义了一个确定的准同构变换 $k^*:H^k(X)\to H^k(X')$ 如次：

$$k^*U=(k^{+*}U^+)^-,\quad U\in H^k(X). \tag{4}$$

若命 $j:X'\subset(X',X'-X)$，则显见 $k^*=j^*i^*$，此处 $i^*:H^k(X)\approx H^k(X',X'-X)$ 如 $H1°$ 所示.

$H3°$. 对于任意 (X,A) 有一正确序列

$$\cdots\to H^k(X,A)\xrightarrow{j_k^*}H^k(X)\xrightarrow{i_k^*}H^k(A)\xrightarrow{\delta_k^*}H^{k+1}(X,A)\to\cdots$$

存在，其中诸变换各与下关于 (X^+,A^+,ϕ^+) 的正确序列的相当变换一致：

$$\cdots\to H_+^k(X^+,A^+)\xrightarrow{j_k^{+*}}H_+^k(X^+,\phi^+)\xrightarrow{i_k^{+*}}H_+^k(A^+,\phi^+)\xrightarrow{\delta_k^{+*}}H_+^{k+1}(X^+,A^+)\to\cdots$$

且 j_k^*，i_k^* 可各视为由包含关系 $j:X\subset(X,A)$ 与 $i:A\subset X$ 所引起.

$H4°$. H 满足 Eilenberg-Steenrod 的同调群公理系统，其中挖出 (excision) 公理与正确序列公理已如 1°，3° 所述.

$H5°$. 若 R^m 是一个 m 维的欧氏空间，则 (G 是系数群)

$$H^k(R^m)\approx\begin{cases}0, & k\neq m;\\ G, & k=m.\end{cases}$$

$H6°$. 若 X 有一胞腔剖分 ①K，则 $H^k(X)$ 同构于由 K 中有限上链所导出的上同调群 $H^k(K)$.

在以下，将简述 $H^*(X,\ A)=\sum\limits_k H^k(X,A)$ 中的环结构，此时系数群将假定是一个环.

① 这里的胞腔部分可取较广的定义，详见 [7].

假设 (X_1,A_1), $(X_2,A_2)\in\mathfrak{C}$, 则 $(X,A)=(X_1,A_1)\times(X_2,A_2)=(X_1\times X_2,A_1\times X_2\cup X_1\times A_2)\in\mathfrak{C}$. 任取 (X_1,A_1), (X_2,A_2) 的两个有限开遮盖 $\alpha_1=\{v_{j_1},j_1\in J_1\}$, $\alpha_2=\{v_{j_2},j_2\in J_2\}$, 并作 (X,A) 的有限开遮盖 $\alpha=\alpha_1\times\alpha_2=\{v_j,j\in J_1\times J_2\}$, 此处 $v_j=v_{j_1}\times v_{j_2}$, $j=(j_1,j_2)$. 易见这样形式的 α 在一切 (X,A) 的有限开遮盖所成的正向集系 (directed systems) 中成一同尾 (cofinal) 的子系. 命 α_1, α_2, α 的神经各为 $(K_{\alpha_1},L_{\alpha_1})$, $(K_{\alpha_2},L_{\alpha_2})$, (K_α,L_α). 对于任一单纯形 $\sigma\in(K_\alpha,L_\alpha)$ 设 σ 的顶点为 $v_{j_{10}}\times v_{j_{20}},\cdots,v_{j_{1k}}\times v_{j_{2k}}(j_{1i}\in J_1,j_{2i}\in J_2,i=0,1,\cdots,k)$, 则 $v_{j_{10}},\cdots,v_{j_{1k}}$ 决定了 K_{α_1} 中的一个单纯形 σ_1, $v_{j_{20}},\cdots,v_{j_{2k}}$ 决定了 K_{α_2} 中的一个单纯形 σ_2 而

$$\sigma\to\sigma_1\times\sigma_2$$

定义了一个胞腔对应 $i_\alpha:(K_\alpha,L_\alpha)\to(K_{\alpha_1},L_{\alpha_1})\times(K_{\alpha_2},L_{\alpha_2})=(K_{\alpha_1}\times K_{\alpha_2},L_{\alpha_1}\times K_{\alpha_2}+K_{\alpha_1}\times L_{\alpha_2})$. 因为 $\sigma_1\times\sigma_2$ 的闭包具有简单的同调性, 故 i_α 引起了一个确定的准同构变换 $i_\alpha^*:H^*((K_{\alpha_1},L_{\alpha_1})\times(K_{\alpha_2},L_{\alpha_2}))\to H^*(K_\alpha,L_\alpha)$.

今设 $U_1^+\in H_+^*(X_1,A_1)$, $U_2^+\in H_+^*(X_2,A_2)$. 取 (X_1,A_1), (X_2,A_2) 的有限开遮盖 α_1, α_2 使 U_1^+, U_2^+ 在 $H^*(K_{\alpha_1},L_{\alpha_1})$, $H^*(K_{\alpha_2},L_{\alpha_2})$ 中各有代表 $u_{a_1}^+$, $u_{a_2}^+$. 命 $\alpha=\alpha_1\times\alpha_2$, $u_a^+=i_a^*(u_{a_1}^+\otimes u_{a_2}^+)\in H^*(K_a,L_a)$, 而 $U^+\in H_+^*(X,A)$ 为 u_a^+ 所代表的元素. 易证 U^+ 与 α_1, α_2 的选择无关, 因而由 U_1^+, U_2^+ 所唯一地确定. 我们定义 $U^+=U_1^+\otimes U_2^+$.

假设 $(X_1,A_1),(X_2,A_2)\in\mathfrak{LC}$, 则 $(X,A)=(X_1,A_1)\times(X_2,A_2)=(X_1,\times X_2,A_1\times X_2\cup X_1\times A_1)\in\mathfrak{LC}$. 定义

$$f:(X_1^+,A_1^+)\times(X_2^+,A_2^+)=(X_1^+\times X_2^+,A_1^+\times X_2^+\cup X_1^+\times A_2^+)\to(X^+,A^+)$$

为 $f/X\equiv$ 恒同映像, $f(\infty\times X_2^+\cup X_1^+\times\infty)=\infty$, 则 f 是连续的, 且是相对同拓映像. 依 $H1^\circ$, $f^*:H^*(X^+,A^+)\approx H^*((X_1^+,A_1^+)\times(X_2^+,A_2^+))$. 对于任意 $U_1\in H^*(X_1,A_1)$, $U_2\in H^*(X_2,A_2)$, 我们定义 $U_1\otimes U_2\in H^*(X,A)$ 为与 $H^*(X^+,A^+)$ 中 $f^{*-1}(U_1^+\otimes U_2^+)$ 相当的元素:

$$U_1\otimes U_2=[f^{*-1}(U_1^+\otimes U_2^+)]^-. \tag{5}$$

假设 (X,A), (X_1,A_1), $(X_2,H_2)\in\mathfrak{LC}$, 且诸空间都是某同一空间的子空间, 而 $X=X_1\cap X_2$, $A=A_1\cup A_2$. 命 $d:(X,A)\to(X_1,A_1)\times(X_2,A_2)$ 为对角映像: $d(x)=x\times x$, $x\in X$, 如果 d 是可允许的, 则对于任意 $U_1\in H^*(X_1,A_1)$, $U_2\in H^*(X_2,A_2)$, 我们将定义 $U_1\cup U_2\in H^*(X,A)$ 为

$$U_1\cup U_2=d^*(U_1\otimes U_2). \tag{6}$$

特别在 $X=X_1$ 或 $X=X_2$ 时, d 必是可允许的, 因之 $U_1\cup U_2$ 恒有定义.

上面所定义的张量积 $\otimes$ 与上积 $\cup$ 除具有熟知的一些关于它们的性质外，又有以下诸性质：

$P1°$. 设 (X_1, A_1)，(X_2, A_2)，(Y_1, B_1)，$(Y_2, B_2) \in \mathfrak{LC}$，且

$$f_1 : (X_1, A_1) \to (Y_1, B_1), \quad f_2 : (X_2, A_2) \to (Y_2, B_2)$$

都是可允许的. 命 $(X, A) = (X_1, A_1) \times (X_2, A_2)$，$(Y, B) = (Y_1, B_1) \times (Y_2, B_2)$，且定义 $f = f_1 \times f_2 : (X, A) \to (Y, B)$ 为 $f(x_1, x_2) = f_1(x_1) \times f_2(x_2)$，$x_1 \in X_1$，$x_2 \in X_2$，则 f 也是可允许的. 对于任意 $U_1 \in H^*(Y_1, B_1)$，$U_2 \in H^*(Y_2, B_2)$，就有

$$f^*(U_1 \otimes U_2) = (f_1 \times f_2)^*(U_1 \otimes U_2) = f_1^* U_1 \otimes f_2^* U_2. \tag{7}$$

$P2°$. 设 (X, A)，(Y_1, B_1)，$(Y_2, B_2) \in \mathfrak{LC}$，且

$$f_1 : (X, A) \to (Y_1, B_1), \quad f_2 : (X, A) \to (Y_2, B_2)$$

都是可允许的. 定义 $f = f_1 \cup f_2 : (x, A) \to (Y_1, B_1) \times (Y_2, B_2)$ 为 $f(x) = f_1(x) \times f_2(x)$，$x \in X$，则 f 也是可允许的. 对于任意 $U_1 \in H^*(Y_1, B_1)$，$U_2 \in H^*(Y_2, B_2)$，就有

$$f^*(U_1 \otimes U_2) = (f_1 \cup f_2)^*(U_1 \otimes U_2) = f_1^* U_1 \cup f_2^* U_2.$$

$P3°$. 设 $X \in \mathfrak{LC}$ 且 A，A' 都是 X 的开集而 $A \subset A'$. 如 $H2°$ 定义 k^*: $H^*(A) \to H^*(A')$，则对于任意 $U \in H^*(A)$，$V \in H^*(X)$，有

$$k^*(U \cup V) = k^* U \cup V. \tag{8}$$

证 因 $d : A \to A \times X$，$d' : A' \to A' \times X$ 都是可允许的，故可推广 d，d' 为 $d^+ : (A^+, \phi^+) \to ((A \times X)^+, \phi^+)$，$d'^+ : (A'^+, \phi^+) \to ((A' \times X)^+, \phi^+)$. 因 A，$A \times X$ 各是 A' 和 $A' \times X$ 的开集，故如 $H2°$ 可定义 $k^+ : (A'^+, \phi^+) \to (A^+, \phi^+)$，$k_1^+ : ((A' \times X)^+, \phi^+) \to ((A \times X)^+, \phi^+)$. 又定义 $f : (A^+, \phi^+) \times (X^+, \phi^+) \to ((A \times X)^+, \phi^+)$，$f' : (A'^+, \phi^+) \times (X^+, \phi^+) \to ((A' \times X)^+, \phi^+)$ 如次. $f / A \times X =$ 恒同映像，$f(\infty \times X^+ \cup A^+ \times \infty) = \infty$，$f' / A' \times X =$ 恒同映像，$f'(\infty \times X \cup A' \times \infty) = \infty$. 于是由张量积的定义：

$$\begin{aligned}
&(U \otimes V)^+ = f^{*-1}(U^+ \otimes V^+), \\
&(k^* U \otimes V)^+ = f'^{*-1}((k^* U)^+ \otimes V^+) = f'^{*-1}(k^{+*} U^+ \otimes V^+).
\end{aligned}$$

又易见 $d^+ k^+ = k_1^+ d'^+$，$k_1^+ f' = f(k^+ \times 1)$，此处 $1 : (X^+, \phi^+) \to (X^+, \phi)$ 为恒同映像. 故由上积的定义：

$$(U \cup V)^+ = [d^*(U \otimes V)]^+ = d^{+*}(U \otimes V)^+ = d^{+*} f^{*-1}(U^+ \otimes V^+),$$

$$
\begin{aligned}
(k^*U\cup V)^+ &= [d'^*(k^*U\otimes V)]^+ = d'^{+*}(k^*U\otimes V)^+ = d'^{+*}f'^{*-1}(k^{+*}U^+\otimes V^+)\\
&= d'^{+*}f'^{*-1}(k^+\times 1)^*(U^+\otimes V^+) = d'^{+*}k_1^{+*}f^{*-1}(U^+\otimes V^+)\\
&= k^{+*}d^{+*}f^{*-1}(U^+\otimes V^+).
\end{aligned}
$$

两者相较，即得 (8) 式.

$P4°$. 设 A，X，$X'\in\mathfrak{LC}$，且 A 是 X' 的开集，X 是 X' 的闭集，$A\subset X$ 而 $\bar{k}: X\subset X'$，则对于任意 $U\in H^*(A)$，$V'\in H^*(X')$，有

$$
U\ \cup\ \bar{k}^*V\prime = U\ \cup V'. \tag{9}
$$

证 命 $d: A\to A\times X$，$d': A\to A\times X'$ 都是对角映像，$1: A\to A$ 是恒同映像，则 d，d'，$1\times\bar{k}\in\mathfrak{A}$ 且 $(1\times\bar{k})d=d'$. 故由 (6)，(7)，得

$$
U\cup\bar{k}^*V' = d^*(U\otimes\bar{k}^*V') = d*\overline{(1\times\bar{k})}^*(U\otimes V') = d\prime^*(U\otimes V') = U\cup V'.
$$

在以下，系数群将假定为法 p 整数域 I_p，此处 p 为一奇质数. 对于任意 (几何的) 复合形偶 (K,L)，Steenrod 证明有一组准同构变换

$$
St_p^i: H^q(K,L)\to H^{q+i}(K,L)
$$

存在，此处依 Steenrod 原来记号 $St_p^i = D^p_{k(p-1)-i}$. 在 $(X,A)\in\mathfrak{D}$ 时，对于任意 $U\in H_+^*(X,A)$，取 (X,A) 的一个有限开遮盖 α，以 (K,L) 为神经，使 U 在 $H^*(K,L)$ 中有代表 u，于是 St_p^iu 所代表的 (X,A) 的类与 α 的选择无关而我们将定义之为 $St_p^iU\in H_+^{k+i}(X,A)$. 在 $(X,A)\in\mathfrak{LC}$ 而 $U\in H^*(X,A)$ 时，我们将定义

$$
St_p^iU = (St_p^iU^+)^-. \tag{10}
$$

这样所定义的 Steenrod 幂 St_p^i，具有以下诸性质 ((X,A)，(X',A') 等 $\in\mathfrak{LC}$):

$S1°$. 若 $f:(X',A')\to(X,A)$ 是可允许的，则

$$
f^*St_p^i = St_p^if^*.
$$

$S2°$. $St_p^iU=0$, $i<0$ 或 $i>k(p-1)$, $U\in H^k(X,A)$.

$S3°$. $St_p^{k(p-1)}U = (U)^p$, $U\in H^k(X,A)$.

$S4°$. $St_p^0\,U = \lambda_kU$, $U\in H^k(X,A)$,

$$
\lambda_k = (-1)^{qk(k-1)/2}(q!)^k, q=(p-1)/2.
$$

$S5°$. 若 $U\in H^k(A)$, $\delta^*: H^k(A)\to H^{k+1}(X,A)$ 定义如 $H3°$, 则

$$
St_p^i\delta^*U = (-1)^{qk+i}q!\delta^*St_p^iU.
$$

$S6^\circ$. $St_p^{2i+1} = \left(\frac{1}{p}\delta\right) St_p^{2i}$.

$S7^\circ$. 若 $U_1 \in H^*(X_1)$, $U_2 \in H^*(X_2)$, 则

$$St_p^{2i}(U_1 \otimes U_2) = \sum_{i_1+i_2=i} St_p^{2i_1}U_1 \otimes St_p^{2i_2}U_2.$$

$S8^\circ$. 若 $U_1, U_2 \in H^*(X)$，则

$$St_p^{2i}(U_1 \cup U_2) = \sum_{i_1+i_2=i} St_p^{2i_1}U_1 \cup St_p^{2i_2}U_2.$$

$S9^\circ$. 若 $X \subset X'$ 而是 X' 的开集，$k^* : H^*(X) \to H^*(X')$ 定义如 $H2^\circ$，则

$$k^* St_p^i = St_p^i k^*.$$

试证 $S9^\circ$ 如次:

设 $U \in H^*(X)$，如 $H2^\circ$ 可定义 $k^+ : (X'^+, \phi^+) \to (X^+, \phi^+)$. 由 k^*，St_p^i 的定义，即 (4)，(10) 与 $S1^\circ$，得

$$\begin{aligned} k^* St_p^i U &= k^*(St_p^i U^+)^- = (k^{+*} st_p^i U^+)^- \\ &= (St_p^i k^{+*} U^+)^- = [St_p^i (k^* U)^+]^- \\ &= St_p^i k^* U, \end{aligned}$$

故 $S9^\circ$ 成立.

与 $p=2$ 相当的平方运算也有与以上类似的性质，但 $S4^\circ$，7°, 8° 须略加修改.

§2 格拉斯曼流形中的 Steenrod 幂 [1,11]

在本节中, p 是一个固定的奇质数，所提到的多项式和方程，其系数都假定在域 I_p 内.

设 $C_{n,m}$ 是一个复格拉斯曼流形，它的整系数 C 类是 C^{2i}，$0 < i \leqslant m$，在将系数法 p 约化后所得的法 pC 类是 C_p^{2i}，$0 < i \leqslant m$. 我们将设 C^0 和 C_p^0 各为整系数和法 p 单位类，$C^{2i}=0$，$C_p^{2i}=0$，$i > m$.

我们知道 $C_{n,m}$ 无挠系数，且它的整系数 (因之法 p) 上同调环的构造，由 $C^{2i}(C_p^{2i})$ 所产生，与实格拉斯曼流形 $R_{n,m}$ 的法 2 上同调环构造之由 W^i 所产生者相似. 与 [12] 中定理 2、预备定理 4，5，6 同样，我们有下面的两条预备定理:

预备定理 1 上同调环 $H^*(C_{n,m}, I_p)$ 可从 $C_{n,m}$ 的法 pC 类 C_p^{2i}，$0 < i \leqslant m$ 所产生，且在这环中维数 $\leqslant 2n$ 的部分中，这些 C 类是这部分的乘法基. 换言之，$C_{n,m}$

中任一 $2k$ 维的法 p 上同调类都可表为 C_p^{2i}，$0<i\leqslant m$，的一个多项式 (乘法是上积)，而 $k\leqslant n$ 时，这种表法是唯一的.

预备定理 2 设 $n_1+n_2=n$，$m_1+m_2=m$，则有一自然地定义的映像

$$h: C_{n_1,m_1}\times C_{n_2,m_2}\to C_{n,m}. \tag{1}$$

命 C_{n_j,m_j} 中的法 pC 类为 $C^{(j)2i}$ 等，则

$$h^*C_p^{2i}=\sum_{i_1+i_2=i} C_p^{\prime 2i_1}\otimes C_p^{\prime\prime 2i_2}, \tag{2}$$

且在 $k\leqslant n_1$ 和 n_2 时，

$$h^*: H^{2k}(C_{n,m},I_p)\subset H^{2k}(C_{n_1,m_1}\times C_{n_2,m_2},T_p).$$

今设 $n\geqslant mp$. 由预备定理 1，$St_p^{2r}C_p^{2s}$，$0<s\leqslant m$，$n\geqslant mp\geqslant r+s$，乃是 C_p^{2i}，$0<i\leqslant m$ 的一个确定的多项式，以上积为乘法，命之为

$$St_p^{2r}C_p^{2s}=\varphi_{r,s,m,n}(C_p^{2i}),\quad 0<s\leqslant m, n\geqslant mp\geqslant r+s, \tag{3}$$

此处 $\varphi_{r,s,m,n}(c_i)$ 乃是重量为 i 的不定量 c_i，$i=1,2,\cdots,m$ 的一个多项式，其每项的重量都是 $r+s$. 若 $n'>n\geqslant mp$，而 $C_p^{\prime 2i}$ 是 $C_{n',m}$ 的法 pC 类，则对于自然映像 $f: C_{n,m}\to C_{n',m}$ 而言，有 $f^*C_p^{\prime 2i}=C_p^{2i}$，故

$$\begin{aligned}\varphi_{r,s,m,n'}(C_p^{2i})&=\varphi_{r,s,m,n'}(f^*C_p^{\prime 2i})=f^*\varphi_{r,s,m,n'}(C_p^{\prime 2i})=f^*St_p^{2r}C_p^{\prime 2s}\\&=St_p^{2r}f^*C_p^{\prime 2s}=St_p^{2r}C_p^{2s}=\varphi_{r,s,m,n}(C_p^{2i}).\end{aligned}$$

因而只须 n 相当大 ($\geqslant mp$) 时，$\varphi_{r,s,m,n}(c_i)$ 的形状实际上与 n 无关. 我们定义

$$\varphi_{r,s,m}(c_i)=\varphi_{r,s,m,n}(c_i),\quad 0<s\leqslant m,\quad n\geqslant mp\geqslant r+s\,; \tag{4}$$

$$\varphi_{r,s,m}(c_i)=0,\quad s>m \text{ 或 } mp<r+s\,; \tag{4$'$}$$

$$\varphi_{r,o,m}(c_i)=\begin{cases}1, & r=0,\\ 0, & r>0.\end{cases} \tag{4$''$}$$

则只须 n 相当大，即有

$$St_p^{2r}C_p^{2s}=\varphi_{r,s,m}(C_p^{2i}),\quad r,s\geqslant 0. \tag{3$'$}$$

应用自然映像 $f: C_{n,m}\to C_{n',m}(n'>n)$，可见不论 n 如何，上式仍普遍成立.

今取正整数 n_1，n_2，m_1，m_2 使 $n_1+n_2=n$，$m_1+m_2=m$，$n_i\geqslant mp$，则对于自然映像 h 而言 (见 (1))，由预备定理 2，得 $(r+s\leqslant mp)$

$$\begin{aligned} h^*St_p^{2r}C_p^{2s} &=St_p^{2r}h^*C_p^{2r}=St_p^{2r}\left(\sum_{s_1+s_2=s}C_p'^{2s_1}\otimes C_p''^{2s_2}\right)\\ &=\sum_{s_1+s_2=s}St_p^{2r}(C'^{2s_1}_p\otimes C''^{2r_2}_p)=\sum_{\substack{s_1+s_2=s\\ r_1+r_2=r}}St_p^{2r_1}C'^{2s_1}_p\otimes St_p^{2r_2}C''^{2s_2}_p, \end{aligned}$$

或

$$\varphi_{r,s,m}(h^*C_p^{2i})=h^*\varphi_{r,s,m}(C_p^{2i})=\sum_{\substack{r_1+r_2=r\\ s_1+s_2=s}}\varphi_{r_1,s_1,m_1}(C_p'^{2i})\otimes\varphi_{r_2,s_2,m_2}(C_p''^{2i}),\tag{5}$$

即

$$\varphi_{r,s,m}(\sum_{i_1+i_2=i}C_p'^{2i_1}\otimes C_p''^{2i_2})=\sum_{\substack{r_1+r_2=r\\ s_1+s_2=s}}\varphi_{r_1,s_1,m_1}(C_p'^{2i_1})\otimes\varphi_{r_2,s_2,m_2}(C_p''^{2i_2}).\tag{5$'$}$$

由假设 $n_i\geqslant mp$，$i=1,2$，故由预备定理 1，在 $C_{n_1,m_1}\times C_{n_2,m_2}$ 中，任一 $2k$ 维的法 p 上同调类都可表成 $C_p'^{2i_1}\otimes C_p''^{2i_2}$，$0<i_1$，$i_2\leqslant m_j$ 的多项式 (乘法是上积)，且在 $k\leqslant n_1$ 和 n_2 时，这种表法是唯一的. 因之由 $(5)'$ 可知

$$\varphi_{r,s,m}\left(\sum_{i_1+i_2=i}c'_{i_1}c''_{i_2}\right)=\sum_{\substack{r_1+r_2=r\\ s_1+s_2=s}}\varphi_{r_1,s_1,m_1}(c'_{i_1})\varphi_{r_s,s_2,m_2}(c''_{i_2}),\tag{6}$$

此处 $c'_{i_1}(i_1=1,2,\cdots,m_1)$，$c''_{i_{2x}}(i_2=1,2,\cdots,m_2)$ 各是重量为 i_1 或 i_2 的不定量.

预备定理 3

$$\varphi_{r,1,m}(c_i)=\begin{cases}0, & r\neq 0,\quad p-1;\\ \lambda_2c_1, & r=0,\quad \lambda_2=(-1)^q(q!)^2, q=(p-1)/2;\\ (c_1)^p, & r=p-1.\end{cases}\tag{7$_m$}$$

证 在 $s=1$ 时，$(5)'$ 变为 (见 $(4)''$)

$$\varphi_{r,1,m}\left(\sum_{i_1+i_2=i}C_p'^{2i_1}\otimes C_p''^{2i_2}\right)=\varphi_{r,1,m_1}(C_p'^{2i})\otimes 1+1\otimes\varphi_{r,1,m_2}(C_p''^{2i}).\tag{8}$$

今将 $C_{n,m}$ 中一切类 $C^{2r_1}\cdots C^{2r_k}(0<r_1\leqslant\cdots\leqslant r_k\leqslant m)$ 依辞典排列. 暂设 m 相当大，$m>r+1$，并取 $m_1=1$. 设 $\varphi_{r,1,m}$ 不恒等于 0，可命 $C^{2i_1}\cdots C^{2i_k}$，i_1+

$\cdots + i_k = r+1$，为 $\varphi_{r,1,m}(C^{2i})$ 中次序最高而系数 $a \neq 0 (a \in I_p)$ 的项，则在 $\varphi_{r,1,m}(\sum C_p'^{2i_1} \otimes C_p''^{2i_2})$ 的展开式中，有一项 $C_p'^2 \otimes C_p''^{2i_1-2} C_p''^{2i_2} \cdots C_p''^{2i_k}$，而此项的系数等于 at，此处 $i_1 = \cdots = i_t \neq i_{t+1}$. 在 $0 < r < p-1$ 时，$t \leqslant ti_1 \leqslant r+1 < p$，故 $at \neq 0$ $(at \in I_p)$. 但 (8) 式右端不能有这样的项，故知

$$0 < r < p-1\text{时}, \quad \varphi_{r,1,m}(C_p^{2i}) = 0.$$

在 $m \leqslant r+1$ 时，由常法知上式仍成立.

由§ 2 Steenrod 幂的性质，知

$$\varphi_{r,1,m}(C_p^{2i}) = \begin{cases} 0, & r \neq 0, p-1; \\ \lambda_2 C_p^2, & r = 0 (\lambda_2\text{见}\S 1, S4^\circ); \\ (C_p^2)^p, & r = p-1. \end{cases}$$

由此即得 (7) 式.

从前述我们知道对于任意整数 $m > 0$，r，$s \geqslant 0$，在 I_p 中有一每项重量都是 $r+s$ 的 $c_i (0 < i \leqslant m)$ 的多项式 $\varphi_{r,s,m}(c_i)$ 存在，其中 c_i 是重量为 i 的不定量，具备有下面三个性质：

1°. (4)′，(4)″ 式成立.

2°. (7) 式成立.

3°. (6) 式成立. 换言之，若 m_1，$m_2 > 0$ 而

$$m_1 + m_2 = m, \tag{9}$$

又 $c_i'(0 < i \leqslant m_1)$，$c_i''(0 < i \leqslant m_2)$ 是另两组重量为 i 的不定量，而有

$$c^i = \sum_{i_1+i_2=i} c'_{i_1} c''_{i_2} \qquad (c'_0 = 1, c''_0 = 1), \tag{10}$$

则

$$\varphi_{r,s,m}(c_i) = \sum_{\substack{r_1+r_2=r \\ s_1+s_2=s}} \varphi_{r_1,s_1,m_1}(c_i') \varphi_{r_2,s_2,m_2}(c_i''). \tag{11}$$

预备定理 4　在 1°–3° 假定之下，$\varphi_{r,s,m}(c_i)$ 必是由下法所定的多项式. 扩张域 I_p 为 F_p 使包含 c_i 且在此域中方程

$$x^m - c_1 x^{m-1} + \cdots + (-1)^m c_m = 0 \tag{12}$$

有解 $x_1, \cdots, x_m$，则

$$\varphi_{r,s,m}(c_i) = \begin{cases} 0, & \text{若}r \neq 0 \mod (p-1); \\ (\lambda_2)^{s-t} \cdot S((x_1 \cdots x_t)^p x_{t+1} \cdots x_s), & \text{若}r = t(p-1), \end{cases} \tag{13$_m$}$$

其中 $S(x_1^{\lambda_1}\cdots,x_m^{\lambda_m})$ 表对于 $x_1,\cdots,x_m$ 而言，以 $x_1^{\lambda_1},\cdots,x_m^{\lambda_m}$ 为标准项的对称函数①而 λ_2 见 $(7)_m$：

$$\lambda_2 = (-1)^q\,(q!)^2, \quad q=(p-1)/2. \tag{13$'$}$$

证 设 c_i'，c_i'' 如 3°. 扩张域 F_p 使包含 c_i'，c_i'' 且在此域中方程

$$x^{m_j}-c_1^{(j)}x^{m_j-1}+\cdots+(-1)^{m_j}c_{m_j}^{(j)}=0, \quad j=1,2$$

都各有解 $x_1^{(j)},\cdots,x_{m_j}^{(j)}(j=1,2)$. 由 (9)，(10)，此两方程相乘所得的结果即为方程 (12)，故 $x_1,\cdots,x_m$ 可视为 $x_1',\cdots,x_{m_1}',x_1'',\cdots,x_{m_2}''$ 的一个排列，而有 ①

$$\begin{aligned}S(x_1\cdots x_t)^p x_{t+1}\cdots x_s = \sum_{\substack{s_1+s_2=s\\ t_1+t_2=t}}&[S'((x_1'\cdots x_{t_1}')^p x_{t_1+1}'\cdots x_{s_1}')\\ &\cdot S''((x_1''\cdots x_{t_2}'')^p x_{t_2+1}''\cdots x_{s_2}'')],\end{aligned} \tag{14}$$

其中 S'，S'' 各指对于 x_i'，x_i'' 而言的对称函数.

今对 m 施行归纳法以证 (13) 式. 在 $m=1$ 时，此时 $(13)_1$ 由 1°，2° 已证明. 今假设 $(13)_1$，$\cdots$，$(13)_{m-1}$ 已证明. 任取 m_1，m_2，c_i'，c_i'' 如前，则 m_1，$m_2<m$. 在 $r\not\equiv 0 \bmod (p-1)$ 时，对于任意使 $r_1+r_2=r$ 的 r_1, r_2 中，必至少有一 $\not\equiv 0 \bmod (p-1)$，因而由归纳假设，对于任意使 $s_1+s_1=s$ 的 s_1，s_2，$\varphi_{r_1,\,s_1,\,m_1}(c_i')$，$\varphi_{r_2,s_2,m_2}(c_i'')$ 中必至少有一 $=0$，故由 (11) 式得 $\varphi_{r,s,m}(c_i)=0$，此即 $(13)_m$ 的第一部分. 次设 $r=t(p-1)$，则由归纳假设与 (11)，(14) 得

$$\begin{aligned}\varphi_{t(p-1),s,m,}(c_i) &= \sum_{\substack{t_1+t_2=t\\ s_1+s_2=s}}\varphi_{t_1(p-1),s_1,m_1}(c_i'),\varphi_{t_2(p-1),s_2,m_2}(c_i'')\\ &=\sum_{\substack{t_1+t_2=t\\ s_1+s_2=s}}(\lambda_2)^{s_1-t_1}S'((x_1'\cdots x_{t_1}')^p x_{t_1+1}'\cdots x_{s_1}')\\ &\quad\cdot(\lambda_2)^{s_2-t_2}S''((x_1''\cdots x_{t2}'')^p x_{t_2+1}''\cdots x_{s_2}'')\\ &=(\lambda_2)^{s-t}S((x_1\cdots x_t)^p x_{t+1}\cdots x_s),\end{aligned}$$

此即 $(13)_m$ 的第二部分，而定理证毕.

在 $s=m$ 时，$\varphi_{r,m,m}$ 可明确地表示出如下. 命 c_i，x_i 如前，此处 c_i 是 x_i 的初等对称函数，又命

$$s_k=x_1^k+\cdots+x_m^k=S(x_1^k),$$

则由 Waring 公式，

$$(-1)^k\frac{s_k}{k}=\sum_{(a)}(-1)^{a_1+\cdots+a_m}\cdot\frac{(a_1+\cdots+a_m-1)!}{a_1!\cdots a_m!}\cdot c_1^{a_1}\cdots c_m^{a_m}$$

① 在 $s=0$ 时，$S((x_1\cdots x_t)^p x_{t+1}\cdots x_s)=0$.

$$=\sum_i{}^{(k)}\frac{(-1)^i}{i}(c_1+\cdots+c_m)^i, \tag{15}$$

其中 $\sum\limits_{(a)}$ 展开于一切使 $\alpha_i\geqslant 0$，$\alpha_1+2\alpha_2+\cdots+m\alpha_m=k$ 的整数组 $(\alpha_1,\cdots,\alpha_m)$ 上，$\sum^{(k)}$ 指只取展开式中各项重量都是 k 的部分和. 今以 $x_1^{p-1},\cdots,x_m^{p-1}$ 为根作方程

$$y^m-d_1y^{m-1}+\cdots+(-1)^m d_m=0,$$

并以 σ_k 表此方程之根的 k 幂和，则 $\sigma_k=s_k(p-1)$，而由另一经典公式有

$$d_r=S(x_1^{p-1}\cdots x_r^{p-1})=\sum_{(\beta)}(-1)^{r+\beta_1+\cdots+\beta_m}\cdot\frac{1}{\beta_1!\cdots\beta_m!}\cdot\left(\frac{\sigma_1}{1}\right)^{\beta_1}\left(\frac{\sigma_2}{2}\right)^{\beta_2}\cdots\left(\frac{\sigma_m}{m}\right)^{\beta_m}$$

$$=\sum_{(\beta)}(-1)^{\beta_1+\cdots+\beta_m}\cdot\frac{1}{\beta_1\cdots\beta_m!}\left(-\frac{s_{p-1}}{1}\right)^{\beta_1}\left(+\frac{s_{2(p-1)}}{2}\right)^{\beta_2}\cdots\left((-1)^m\cdot\frac{s_{m(p-1)}}{m}\right)^{\beta_m},$$

或

$$d_r=\sum_j{}^{(r)}\frac{(-1)^j}{j!}\left(-\frac{s_{p-1}}{1}+\frac{s_2(p-1)}{2}-\cdots+(-1)^m\cdot\frac{s_{m(p-1)}}{m}\right)^j, \tag{16}$$

其中 $\sum\limits_{(\beta)}$ 展开于一切使 $\beta_i\geqslant 0$，$\beta_1+2\beta_2+\cdots+m\beta_m=r$ 的整数组 $(\beta_1,\cdots,\beta_m)$ 上，而 $\sum^{(r)}$ 指只取展开式中各项重量都是 $r(p-1)$ 的部分和 (s_k 的重量是 k). 特别在 $p=3$ 时，

$$d_r=(-1)^r\sum_{i+j=2r}(-1)^{ij}c_ic_j\quad(p=3). \tag{17$_3$}$$

由 (15)，(16)，得

$$d_r=Sx_1^{p-1}\cdots x_r^{p-1}=\sum_j{}^{(r)}\frac{(-1)^j}{j!}\left(\sum_i{}^{[p-1]}\frac{(-1)^i}{i}(c_1+\cdots+c_m)^i\right)^j, \tag{17}$$

其中 $\sum^{(r)}$ 如前，$\sum^{[p-1]}$ 系指只取展开式中各项重量都可被 $p-1$ 除尽的部分和. 由 $(13)_m$，$r=t(p-1)$ 时，

$$\varphi_{r,m,m}(c_i)=S(x_1\cdots x_t)^px_{t+1}\cdots x_m=x_1\cdots x_m\cdot S(x_1^{p-1}\cdots x_t^{p-1})=c_md_t,$$

或

$$\varphi_{t(p-1),m,m}(c_i)=c_m\cdot\sum_j{}^{(t)}\frac{(-1)^j}{j!}\left(\sum_j{}^{[p-1]}\frac{(-1)^i}{i}(c_1+\cdots+c_m)^i\right)^j. \tag{18}$$

今定义 $D_p^{2r} \in H^{2r}(C_{n,m}, I_p), 0 < r \leqslant m(p-1)$ 为

$$D_p^{2r} = \begin{cases} 0, & r \not\equiv 0 \mod (p-1); \\ \sum_j^{(r)} \frac{(-1)^j}{j!} \left(\sum_i^{[p-1]} \frac{(-1)^i}{i} (C_p^2 + \cdots + C_p^{2m})^i \right)^j, & r = t(p-1); \end{cases} \tag{19}$$

其中 $\sum^{[p-1]}$ 指只取所展开式中维数是 $2(p-1)$ 倍数之项的部分和，$\sum^{(r)}$ 指只取展开式中维数是 $2r$ 的项的部分和. 于是由 (18)，$(3)'$，得

$$St_p^{2r} C_p^{2m} = \begin{cases} 0, & r \not\equiv 0 \mod (p-1); \\ C_p^{2m} D_p^{2r}, & r = t(p-1), \quad 0 < t \leqslant m. \end{cases} \tag{20}$$

在 $p = 3$ 时，D_p^{2r} 由 $(17)_3$ 为

$$D_3^{2r} = \begin{cases} 0, & r \not\equiv 0 \mod 2; \\ (-1)^k \sum\limits_{i+j=2k} (-1)^{ij} C_3^{2i} C_3^{2j}, & r = 2k. \end{cases} \tag{21_3}$$

§3 欧氏空间丛的 Gysin 正确序列 [4,15,16]

设 $\mathfrak{B} = (A, \pi, K, R^m, G)$ 是一个 m 欧氏空间丛，其中 A 是丛空间，π 是投影，K 是底空间，构造群 G 是正交群 O_m 或其子群. 与 $\mathfrak{B}$ 相配在 K 上有一个 m 维闭胞腔丛 $\bar{\mathfrak{B}}$ 和一个 $m-1$ 维球丛 $\mathfrak{S}$，命其丛空间各为 $\bar{A}$，E，投影各为 $\bar{\pi}$，ω. 我们不妨假定 $\bar{A}$ 是 ω 的映像柱 (mapping cylinder)，$A = \bar{A} - E$，$\omega = \bar{\pi}/E$，$\pi = \bar{\pi}/A$，且对于任一 $p \in K$，$\omega^{-1}(p)$ 是 $\bar{\pi}^{-1}(p)$ 的边，而 $\pi^{-1}(p) = \bar{\pi}^{-1}(p) - \omega^{-1}(p)$.

我们将假定 K 是一个有限的可剖分空间，则 A，$\bar{A}$，E 亦然. 任取 K 的胞腔剖分，仍记之为 K，则 A，亦有一剖分，仍记之为 A，如次. 命 σ 为 K 的任意胞腔，则 A 由一切 $\tilde{\sigma} = \pi^{-1}(\sigma)$ 形状的胞腔所组成. 显然 $\dim \tilde{\sigma} = \dim \sigma + m$.

在以下，H^k 指如§ 1 所述的 k 维上同调群，系数群将略去不再明写. 为避免不必要的麻烦起见，我们只讨论下面两种情形：

第一, 系数群是 I_2，构造群 $G = O_m$.

第二, 系数群是 I_0，构造群 G 是旋转群. 换言之，$\mathfrak{B}$，$\bar{\mathfrak{B}}$，$\mathfrak{S}$ 都是可定向的丛，此时我们并将假定诸纤维都已定向，而成有定向的丛，且对于任一 $p \in K$，$\pi^{-1}(p)$，$\bar{\pi}^{-1}(p)$ 和 $\omega^{-1}(p)$ 的定向是互相协合的.

对于空间偶 $(\bar{A}, E)$，依§ 1 的 $H1°$ 和 $H3°$ 有一正确序列

$$\cdots \to H^{r-1}(E) \xrightarrow{\alpha^*} H^r(A) \xrightarrow{\beta^*} H^r(\bar{A}) \xrightarrow{r^*} H^r(E) \to \cdots \tag{1}$$

且 β^* 即视 A 为 $\bar{A}$ 的开集时，与§ 1$H2°$ 中 k^* 相当的准同构变换.

因 $K \subset \bar{A}$ 且 K 是 $\bar{A}$ 的变状收缩核，故命 $c: K \to \bar{A}$ 为恒同映像，则

$$\begin{cases} \bar{\pi}^*: H^*(K) \approx H^*(\bar{A}), \\ c^*: H^*(\bar{A}) \approx H^*(K), \\ c^* = \bar{\pi}^{*-1}. \end{cases} \tag{2}$$

此外，我们将定义同构变换

$$\tau^*: H^r(K) \approx H^{r+m}(A) \tag{3}$$

如次. 设 $u \in C^r(K)$. 定义 $\tau^\# u \in C^{r+m}(A)$ 为

$$\tau^\# u\,(\tilde{\sigma}) = u(\sigma). \tag{4}$$

在第二情形时，因 $\tilde{\sigma}$ 与 $R^m \times \sigma$ 同拓，$\tilde{\sigma}$ 的定向将依前面所已选定的纤维 R^m 的定向和 σ 的定向依此次序所合成. 于是 $\tau^\#: C^r(K) \approx C^{r+m}(A)$ 是一个确定的同构变换. 显然 $\tau^\# \delta = \delta \tau^\#$，故 $\tau^\#$ 引出 (3).

τ^* 系 Thom 所引进，故将称之为 Thom 变换. 此变换具有以下诸性质:

1°. 设 K' 是另一有限可剖分空间，$f: K' \to K$ 是连续映像，$\mathfrak{B}' = f^*\mathfrak{B}$，$\bar{\mathfrak{B}}' = f^*\bar{\mathfrak{B}}$，$\mathfrak{S} = f^*\mathfrak{S}$，各以 A'，$\bar{A}'$，E' 为丛空间，余类推. 命 $\tilde{f}: A' \to A$ 为丛映像 (bundle map) 满足 $\pi\tilde{f} = f\pi'$. 取各丛的定向使 f 保持纤维的定向并命 τ^*，τ'^* 为与此等定向相当 (对于 K，K' 的适当剖分而言) 的 Thom 变换，则

$$\tilde{f}^*\tau^* = \tau'^* f^*. \tag{5}$$

证　$\tilde{f}$ 之为可允许的映像，容易看出取 K，K' 的剖分仍记之为 K，K' 且设 f 是胞腔映像，则 $\tilde{f}$ 亦然，而由 (4) 可得 $\tilde{f}^\#\tau^\# = \tau^\# f^\#$，由此得 (5). 在一般情形，可取 K' 的一个适当小剖分 K'_1，变状 f 为一 K'_1 到 K 的胞腔映像 f_1. 由纤维丛的同伦遮盖定理知有一丛映像 $\tilde{f}_1: A' \to A$ 使 $\tilde{f}_1 \simeq \tilde{f}$，$\pi\tilde{f}_1 = f_1\pi'$. 于是有 $\tilde{f}_1^*\tau^* = \tau'_1 * f_1^* = \tau'^* f_1^*$，亦即 (5) 式.

由 1° 可知

2°. Thom 变换 τ^* 与 K 的剖分无关.

3°. 对于任意 U_1，$U_2 \in H^*(K)$，有

$$\tau^*(U_1 \cup U_2) = \tau^* U_1 \cup \bar{\pi}^* U_2. \tag{6}$$

证　定义

$$\pi_1: \quad A \times K \to K \times K$$

为 $\pi_1(p,q)=(\pi(p),q)$，$p\in A$，$q\in K$. 命 $d:K\to K\times K$ 为对角映像. 又定义

$$\tilde{d}:\quad A\to A\times K$$

为 $\tilde{d}(p)=(p,\pi(p))$, $p\in A$，则

$$\pi_1\tilde{d}=d\pi. \tag{7}$$

易见 $\tilde{d}$ 是可允许的. 又 $A\times K$ 显然是一个 $K\times K$ 上的 m 欧氏空间从 $\mathfrak{B}_1$ 的从空间，$\mathfrak{B}_1$ 以 π_1 为从投影，$(p,\ q)\in K\times K$ 上的纤维为 $\pi^{-1}(p)\times q$，且 $\tilde{d}$ 是 $\mathfrak{B}$ 到 $\mathfrak{B}_1$ 的从映像，因之由 (7)，

$$d^*\mathfrak{B}_1\sim\mathfrak{B}. \tag{8}$$

对于 K 的任意胞腔 σ_1，σ_2，$\pi_1^{-1}(\sigma_1\times\sigma_2)=\pi^{-1}(\sigma_1)\times\sigma_2=\tilde{\sigma}_1\times\sigma_2$ 而所有此种胞腔成一 $A\times K$ 的剖分. 定向 $\mathfrak{B}_1$ 使 $\tilde{d}$ 保持相当纤维的定向，则每一胞腔 $\tilde{\sigma}_1\times\sigma_2$ 的定向即依前面已定的 $\tilde{\sigma}_1$，σ_2 的定向所定. 命 $\tau_1^{\#}$，τ_1^* 为对此定向从 $\mathfrak{B}_1$ 而言的 Thom 变换，则对于任意 u_1，$u_2\in C^*(K)$，从 (4) 式可得

$$\begin{aligned}\tau_1{}^{\#}(u_1\otimes u_2)(\tilde{\sigma}_1\times\sigma_2)&=(u_1\otimes u_2)(\sigma_1\times\sigma_2)=u_1(\sigma_1)\cdot u_2(\sigma_2)\\&=\tau^{\#}u_1(\tilde{\sigma}_1)\cdot u_2(\sigma_2)=(\tau^{\#}u_1\otimes u_2)(\tilde{\sigma}_1\times\sigma_2),\end{aligned}$$

故有 $\tau_1^{\#}(u_1\otimes u_2)=\tau^{\#}u_1\otimes u_2$，因之

$$\tau_1{}^*(U_1\otimes U_2)=\tau^*U_1\otimes U_2,\quad U_1,U_2\in H^*(K). \tag{9}$$

由 (5)，(7)，(8)，(9) 得 $(U_1,U_2\in H^*(K))$

$$\begin{aligned}\tau^*(U_1\cup U_2)=&\tau^*d^*(U_1\otimes U_2)=\tilde{a}^*\tau_1{}^*(U_1\otimes U_2)\\=&\tilde{d}^*(\tau^*U_1\otimes U_2).\end{aligned}$$

今命 $\tilde{d}':A\to A\times\bar{A}$ 为对角映像，则 $\tilde{d}=(1\times\bar{\pi})\tilde{d}'$，此处 $1:A\to A$ 是恒同映像，故 $\tilde{d}^*=\tilde{d}'^*(1\times\bar{\pi})^*$，而由§ 1 的 (6)，(7) 从上式得

$$\begin{aligned}\tau^*(U_1\cup U_2)&=\tilde{d}^*(\tau^*U_1\otimes U_2)=\tilde{d}'^*(1\times\bar{\pi})^*(\tau^*U_1\otimes U_2)\\&=\tilde{d}'^*(\tau^*U_1\otimes\bar{\pi}^*U_2)=\tau^*U_1\cup\bar{\pi}^*U_2,\end{aligned}$$

即所欲证.

今定义

$$\varphi^*=\tau^{*-1}\alpha^*, \tag{10}$$

$$\psi^*=\bar{\pi}^{*-1}\beta^*\tau^*=c^*\beta^*\tau^*, \tag{11}$$

则因 $\omega^* = \gamma^*\pi^*$，(1) 式变为下所谓 Gysin 正确序列:

$$\cdots \to H^{r-1}(E) \xrightarrow{\varphi^*} H^{r-m}(K) \xrightarrow{\psi^*} H^r(K) \xrightarrow{\omega^*} H^r(E) \to \cdots \tag{11$'$}$$

由 (6) 式可得 ($1=$ 单位类，$U \in H^*(K)$)

$$\psi^*U = c^*\beta^*\tau^*U = c^*\beta^*\tau^*(1 \cup U) = c^*\beta^*(\tau^*1 \cup \bar{\pi}^*U).$$

应用§ 1 的 $P3^\circ$ 于 $A \subset \bar{A} \subset \bar{A}$，应有

$$\beta^*(\tau^*1 \cup \bar{\pi}^*U) = \beta^*\tau^*1 \cup \bar{\pi}^*U,$$

故前式变为

$$\psi^*U = c^*(\beta^*\tau^*1 \cup \bar{\pi}^*U) = c^*\beta^*\tau^*1 \cup c^*\bar{\pi}^*U,$$

即

$$\psi^*U = \psi^*1 \cup U. \tag{12}$$

§4　Thom 定理别证及其推广 [1,4]

设 $\mathfrak{B} = (A, \pi, K, R^m, G)$, $\bar{\mathfrak{B}} = (\bar{A}, \bar{\pi}, K, \bar{R}^m, G)$, $\mathfrak{S} = \{E, \omega, K, S^{m-1}, G\}$ 如§3 开首所示，其中 $\bar{R}^m$ 是 m 维的闭胞腔. 在§ 3 所述第二情形时，各丛都已定向，如§ 3 命 $W_2^i(\mathfrak{B})$ 是它们的法 $2W$ 示性类，则 Thom 曾证明 [4]

$$Sq^i\tau^*1 = \tau^*W_2^i(\mathfrak{B}), \tag{1}$$

此处 $\tau^*: H^i(K, I_2) \approx H^{i+m}(A, I_2)$ 是 Thom 变换，1 是法 2 单位类. 此外有

$$\pi^{*-1}\beta^*\tau^*1 = \psi^*1 = -c^m(\mathfrak{S}), \tag{2}$$

此处在第二情形时，$c^m(\mathfrak{S}) \in H^m(K, \pi_{m-1}(S^{m-1})) \approx H^m(K, I_0)$ 为 Steenrod 定义下的截面示性类，其同构变换 $\pi_{m-1}(S^{m-1}) \approx I_0$ 系由定向丛 $\mathfrak{S}$ 中纤维的已有定向所确定，1 是整系数单位类. 在第一情形时，(2) 中各类系指法 2 约化后所得的类.

今将公式 (1) 作一别证如次.

先设 $K = R_{n,m}$ 为格拉斯曼流形，$\mathfrak{B} = \mathfrak{R}_{n,m}$ 为 $R_{n,m}$ 上的模范 m 欧氏空间丛. 命 W_2^i 为 $R_{n,m}$ 中的法 $2W$ 类. 对于 $\mathfrak{R}_{n,m}$ 而言，命与 τ^* 相当的变换为 $\tau_{n,m}^*$，与 A 相当的空间为 $A_{n,m}$，余类推. 由 [11] 的§ 4，有

$$Sq^iW_2^m = W_2^iW_2^m. \tag{3}$$

由 (2) 与§ 3 的 (12)，有

$$W_2^m = \psi_{n,m}^* 1_{n,m}, \tag{4}$$

$$XW_2^m = \psi_{n,m}^* X, \quad X \in H^*(K, I_2), \tag{5}$$

此处 $\psi_{n,m}^* = \pi_{n,m}^{*-1}\beta_{n,m}^*\tau_{n,m}^*$，$1_{n,m}$ 为 $R_{n,m}$ 中，的法 2 单位类. 因 (参阅§ 1 的 $S9°$ 及§ 3 的记号)

$$\bar{\pi}_{n,m}^{*-1} Sq^i = c_{n,m}^* Sq^i = Sq^i c_{n,m}^* = Sq^i \bar{\pi}_{n,m}^{*-1},$$
$$\beta_{n,m}^* Sq^i = Sq^i \beta_{n,m}^*,$$

故从 (3)，(4)，(5) 得

$$\bar{\pi}_{\pi,m}^{*-1}\, \beta_{n,m}^*\, Sq^i\, \tau_{n,m}^*\, 1_{n,m} = \bar{\pi}_{n,m}^{*-1}\, \beta_{n,m}^*\, \tau_{n,m}^*\, W_2^i. \tag{6}$$

由 [11] 的预备定理 4 及 (5) 式可知

$$\psi_{n,m}^* : H^k(R_{n,m}, I_2) \subset H^{k+m}(R_{n,m}, I_2), \quad k+m \leqslant n,$$

由此得

$$\beta_{n,m}^* : H^k(A_{n,m}, I_2) \subset H^k(\bar{A}_{n,m}, I_2), \quad k \leqslant n.$$

因之在 n 充分大 ($n \geqslant 2m$) 时，从 (6) 式可得

$$Sq^i \tau_{n,m}^* 1 = \tau_{n,m}^* W_2^i. \tag{7}$$

今设任意丛 $\mathfrak{B}$，$\bar{\mathfrak{B}}$，$\mathfrak{S}$ 如前，则有 $\lambda: K \to R_{n,m}$，使 $\mathfrak{S} = \lambda * \mathfrak{R}_{n,m}$($n$ 充分大). 命 $\tilde{\lambda}: A \to A_{n,m}$ 为其丛映像，则由§ 3 的 (5)，得

$$\begin{aligned} Sq^i\tau^* 1 &= Sq^i\tau^*\lambda^* 1_{n,m} = Sq^i\tilde{\lambda}^*\tau_{n,m}^* 1_{n,m} = \tilde{\lambda}^* Sq^i \tau_{n,m}^* 1_{n,m} \\ &= \tilde{\lambda}^*\tau_{n,m}^* W_2^i = \tau^*\lambda^* W_2^i = \tau^* W_2^i(\mathfrak{B}). \end{aligned}$$

此即 (1) 式.

上述证明可推广之以得关于Понтрягин示性类的一个类似于 (1) 的公式如次.

先取特例 $C_{n,m}$ 上的模范 m 酉空间丛 $\mathfrak{C}_{n,m}$ 视之，据§ 2，

$$St_p^{2t(p-1)} C_p^{2m} = C_p^{2m} D_p^{2t(p-1)} \quad (0 < t \leqslant m). \tag{8}$$

由 [13]，[14] 及 (2) 式又各有

$$C^{2m} = c^{2m}(\mathfrak{C}_{n,m}), \tag{9}$$

$$c^{2m}(\mathfrak{C}_{n,m}) = (-1)^m c^{2m}(\mathfrak{C}'_{n,m}), \tag{10}$$

$$\psi^*_{n,m} 1_{n,m} = -c^{2m}(\mathfrak{C}'_{n,m}), \tag{11}$$

其中 c^{2m} 系指 Steenrod 定义下相当丛的截面示性类，$\mathfrak{C}'_{n,m}$ 指将 $\mathfrak{C}_{n,m}$ 系视为 $2m$ 欧氏空间丛时由各纤维的自然定向所定的定向丛，$c^{2m}(\mathfrak{C}_{n,m})$，$c^{2m}(\mathfrak{C}'_{n,m}) \in H^{2m}(C_{n,m}, \pi_{2m-1}(S^{2m-1})) \approx H^{2m}(C_{n,m}, I_0)$，其中系数群 $\pi_{2m-1}(S^{2m-1}) \approx I_0$ 系由纤维 S^{2m-1} 的自然定向所定，$\psi_{n,m}$ 为 $\mathfrak{C}'_{n,m}$ 中与 ψ 相当的准同构变换，$1_{n,m}$ 为 $C_{n,m}$ 中的整系数单位类，余类推. 将此诸式法 p 约化后从§ 3 的 (12) 可得

$$\psi^*_{n,m} X = (-1)^{m+1} C_p^{2m} X, X \in H^{2k}(C_{n,m}, I_p). \tag{12}$$

代入 (8) 式，即得

$$St_p^{2t(p-1)} \psi^*_{n,m} 1_{n,m} = \psi^*_{n,m} D_p^{2t(p-1)}. \tag{13}$$

由§ 2 的预备定理 1，若 n 充分大 $(n \geqslant k+m)$，

$$\psi^*_{n,m} = \bar{\pi}^{*-1}_{n,m} \beta^*_{n,m} \tau^*_{n,m} : H^{2k}(C_{n,m}, I_p) \subset H^{2k+2m}(C_{n,m}, I_p),$$

因之

$$\beta^*_{n,m} : H^{2k}(A_{n,m}, I_p) \subset H^{2k}(\bar{A}_{n,m}, I_p), \quad n \geqslant k,$$

此处 $A_{n,m}$，$\bar{A}_{n,m}$ 为 $\mathfrak{C}'_{n,m}$ 中与 A，$\bar{A}$ 相当的空间. 于是与证 (1) 式时同样，从 (13) 可得

$$St_p^{2t(p-1)} \tau^*_{n,m} 1_{n,m} = \tau^*_{n,m} D_p^{2t(p-1)}, \quad n \geqslant 2mp. \tag{14}$$

今设 K 上任意 m 欧氏空间向从 $\mathfrak{B} = \lambda^* \mathfrak{R}_{n,m}$，$\lambda : K \to R_{n,m}$($n$ 充分大). 命 $\mathfrak{B} \cup \mathfrak{B}$ 所自然地决定的 m 酉空间丛为 $\mathfrak{B}^*$，则由［6］的§ 5，

$$\mathfrak{B}^* \sim (l\lambda)^* \mathfrak{C}_{n,m},$$

此处 $l : R_{n,m} \to C_{n,m}$ 为自然映像. 由此得

$$\mathfrak{B} \cup \mathfrak{B} \sim (l\lambda)^* \mathfrak{C}'_{n,m}.$$

于是应用§ 4 的 (5)，命 τ^* 为对于 $\mathfrak{B} \cup \mathfrak{B}$ 而言的 Thom 变换，$\tilde{\lambda}$ 为 $\mathfrak{B} \cup \mathfrak{B}$ 到 $(l\lambda)^* \mathfrak{C}'_{n,m}$ 的丛映像，从 (14) 可得

$$\begin{aligned} St_p^{2t(p-1)} \tau^* 1 &= St_p^{2t(p-1)} \tau^* (l\lambda)^* 1_{n,m} = St_p^{2t(p-1)} \tilde{\lambda}^* \tau^*_{n,m} 1_{n,m} \\ &= \tilde{\lambda}^* St_p^{2t(p-1)} \tau^*_{n,m} 1_{n,m} = \tilde{\lambda}^* \tau^*_{n,m} D_p^{2t(p-1)} \\ &= \tau^* (l\lambda)^* D_p^{2t(p-1)} = \tau^* \lambda^* l^* D_p^{2t(p-1)}. \end{aligned}$$

定义

$$l^* D_p^{2t(p-1)} = Q_p^{2t(p-1)}, \tag{15}$$

$$Q_p^{2t(p-1)}(\mathfrak{B}) = \lambda^* Q_p^{2t(p-1)}. \tag{15$'$}$$

则从前式得

$$St_p^{2t(p-1)} \tau^* 1 = \tau^* Q_p^{2t(p-1)}(\mathfrak{B}). \tag{16}$$

由 [6] 的§ 5，有

$$l^* C_p^{2i} = \begin{cases} 0, & i \text{ 是奇数;} \\ (-1)^{i/2} P_p^{2i}, & i \text{ 是偶数;} \end{cases} \tag{17}$$

其中 P_p^{4k} 是 $R_{n,m}$ 中的法 pP 类. 由§ 2 的 (19)，(21)，可得 $(q=(p-1)/2)$

$$Q_p^{2t(p-1)}(\mathfrak{B}) = \sum_j{}^{(t(p-1))} \frac{(-1)^j}{j!} \left[\sum_i{}^{[p-1]} \frac{(-1)^i}{i} \left(-P_p^4(\mathfrak{B}) + P_p^8(\mathfrak{B}) - \cdots\right)^i \right]^j \quad (0 < t \leqslant m), \tag{18}$$

$$Q_3^{4k}(\mathfrak{B}) = \sum_{i+j=k} P_3^{4i}(\mathfrak{B}) P_3^{4j}(\mathfrak{B}), \tag{18$_3$}$$

其中记号 $\sum^{(\,)}$ 及 $\sum^{[\,]}$ 的说明见§ 2.

§5 微分流形的ПОНТРЯГИН示性类

设 M 是一个 m 维可微分流形. 任取 M 的一个微分构造 $\mathfrak{D}$，此处 $\mathfrak{D}$ 由一组坐标邻域 V_j，$j \in J$，与一组坐标函数 $\varphi_j : R^m \to V_j$ 所定 (φ_j 是拓扑映像)，我们记作 $\mathfrak{D} = \{M, V_j, \varphi_j, j \in J\}$. 对于 $\mathfrak{D}$ 而言，M 的切矢自然地决定一 m 欧氏空间丛 $\mathfrak{T}(\mathfrak{D})$，称为 M 在 $\mathfrak{D}$ 下的切丛 (tangent bundle). $\mathfrak{T}(\mathfrak{D})$ 的构造群是一般线性群，但由 Ehresmann 的一条定理，有一 m 欧氏空间丛从属 (Subordonne) 于 $\mathfrak{T}(\mathfrak{D})$，以正交群 O_m 为构造群，且对 O_m 而言，其丛同构类完全由 $\mathfrak{T}(\mathfrak{D})$ 所确定. 因此，我们不妨假定 $\mathfrak{T}(\mathfrak{D})$ 的构造群即是 O_m.

今对任意整数 $n > 1$ 作拓扑积 $M^n = M_1 \times \cdots \times M_n$，此处 M_i 是 M 的 n 个模型，并命 $\omega_i : M^n \to M_i$ 为 M^n 到第 i 个因子的投影. 于是 $\mathfrak{D}$ 在 M^n 上引起一个微分构造 $\mathfrak{D}^n = (M^n, V_{(j_1,\cdots,j_n)}, \varphi_{(j_1,\cdots,j_n)}, (j_1,\cdots,j_n) \in J^n\}$，此处 $V_{(j_1,\cdots,j_n)} = V_{j_1} \times \cdots \times V_{j_n}, \varphi_{(j_1,\cdots,j_n)} : R^{mn} = \underbrace{R^m \oplus \cdots \oplus R^m}_{n} \to V_{(j_1,\cdots,j_n)}$ 定义如

$\varphi_{(j_1,\cdots,j_n)}(x_1,\cdots,x_n)=(\varphi_{j_1}(x_1),\cdots,\varphi_{j_n}(x_n))$. 显然 $\mathfrak{D}^n$ 是与 $\mathfrak{D}$ 同级的一个 M^n 上的微分构造.

我们知道 $\mathfrak{D}$ 必有一与之相容的 Riemann 计量 ds^2. 命 ds_i^2 为 ds^2 在 M_i 中的相当计量，则 $d\tilde{s}^2=ds_1^2+\cdots+ds_n^2$ 是 M^n 上与 $\mathfrak{D}^n$ 相容的一个 Riemann 计量. 命 $d:M\to M^n$ 为对角映像，则对于 $d\tilde{s}^2$ 而言，在 $d(M)$ 各点而与 $d(M)$ 垂直的 $\mathfrak{D}^n$ 的一切切矢自然地决定一 $m(n-1)$ 欧氏空间从 $\mathfrak{R}(\mathfrak{D})$，以 M 为底空间. 与前同样，我们不妨假定 $\mathfrak{R}(\mathfrak{D})$ 的构造群是正交群 $O_{m(n-1)}$. 我们有

预备定理 5 $\mathfrak{R}(\mathfrak{D})\sim\underbrace{\mathfrak{T}(\mathfrak{D})\cup\cdots\cup\mathfrak{T}(\mathfrak{D})}_{n-1}$, 其中 $\cup$ 指欧氏空间从的 Whitney 积.

证 设 $\tilde{p}=d(p)$, $p\in M$. 命 $T^n(\tilde{p},\mathfrak{D}^n)$ 是 M^n 在 $\tilde{p}$ 点对 $\mathfrak{D}^n$ 而言的切面，又命 $T(p,\mathfrak{D})$ 为 M 在 p 点对于 $\mathfrak{D}$ 而言的切面，则 $\omega_i:M^n\to M$ 引起了线性映像 $\bar{\omega}_i:T^n(\tilde{p},\mathfrak{D}^n)\to T(p,\mathfrak{D})$, $i=1,\cdots,n$. 设 $\mathrm{a}\in T^n(\tilde{p},\mathfrak{D}^n)$, $\bar{\omega}_i(\mathrm{a})=\mathrm{a}_i\in T(p,\mathfrak{D})$, $p=d(\tilde{p})$，则 $\mathrm{a}\leftrightarrow(\mathrm{a}_1,\cdots,\mathrm{a}_n)$ 显然建立了 $T^n(\tilde{p},\mathfrak{D}^n)$ 与 $\underbrace{T(p,\mathfrak{D})\oplus\cdots\oplus T(p,\mathfrak{D})}_{n}$ 间的一个线性拓扑对应. 在此对应之下，我们容易证明:

1°. a 与 $d(M)$ 相切的充要条件是

$$\mathrm{a}_1=\cdots=\mathrm{a}_n.$$

2°. a 与 $d(M)$ 对 $d\tilde{s}^2$ 而言相垂直的充要条件是

$$\mathrm{a}_1+\cdots+\mathrm{a}_n=0.$$

命 $N^n(p)$ 为 M^n 在 $\tilde{p}\in d(M)$ 对于 $d\tilde{s}^2$ 而言的垂面. 在 $N^n(\tilde{p})$ 中，命 $N_k^n(\tilde{p})$, $k=2,\cdots,n$，为一切满足下述条件 (C_k) 的切矢 $\mathrm{a}=(\mathrm{a}_1,\cdots,\mathrm{a}_n)\in N^n(\tilde{p})$ 所成的 m 面:

$$(C_k)\qquad \mathrm{a}_{k+1}=\cdots=\mathrm{a}_n=0,\quad -(k-1)\mathrm{a}_1=\mathrm{a}_2=\mathrm{a}_3=\cdots=\mathrm{a}_k.$$

显然 $N^n(\tilde{p})=N_2^n(\tilde{p})\oplus\cdots\oplus N_n^n(\tilde{p})$. 集合 $\bigcup\limits_{\tilde{p}\in d(M)}N_k^n(\tilde{p})=N_k^n$ 自然地成一拓扑空间. 命 $\lambda_k:N_k^n\to M$ 定义如 $\lambda_k(\mathrm{a})=p$，此处 $\mathrm{a}\in N_k^n(\tilde{p})$, $\tilde{p}=d(p)$. 于是 N_k^n 成为一个在 M 上的 m 欧氏空间从 $\mathfrak{R}_k^n$ 的从空间，以 λ_k 为投影，且

$$\mathfrak{R}(\mathfrak{D})\sim\mathfrak{R}_2^n\cup\cdots\cup\mathfrak{R}_n^n.\tag{1}$$

今对于 $\mathrm{a}_p\in T(p,\mathfrak{D})$, $p\in M$，命

$$h_k(\mathrm{a}_p)=((k-1)\mathrm{a}_p,-\underbrace{\mathrm{a}_p,\cdots,-\mathrm{a}_p}_{k-1},\underbrace{0,\cdots,0}_{n-k})\in N_k^n(p),\quad \tilde{p}=d(p),$$

则 h_k 显然是 $\mathfrak{T}(\mathfrak{D})$ 到 $\mathfrak{N}_k^n$ 的一个丛映像，因而 $\mathfrak{N}_k^n \sim \mathfrak{T}(\mathfrak{D})$. 因之从 (1) 式即得定理.

今设 M 是一个可定向闭流形，并设 $\mathfrak{D}$ 的级 $\geqslant 3$，则 M^n 也是可定向闭流形，且 $\mathfrak{D}^n$ 的级也 $\geqslant 3$. 对于任一 $\mathrm{a} \in N^n(\tilde{p})$，$\mathrm{a} \neq 0$，$\tilde{p} = d(p)$，$p \in M$，命 $g(\mathrm{a})$ 为对 $d\tilde{s}^2$ 而言，自 $\tilde{p}$ 出发而切于 a 的最短线，$q(\mathrm{a}, \varepsilon)$ 为在 $g(\mathrm{a})$ 上使 $\tilde{p}$ 与 $q(\mathrm{a}, \varepsilon)$ 的最短弧长等于 ε 的点. 因 M，M^n 都是闭的，故必有一 $\eta < 0$ 存在使只须 $\mathrm{a} \neq \mathrm{a}'(\mathrm{a} \in N^n(\tilde{p})$，$\mathrm{a}' \in N^n(\tilde{p}')$，$\mathrm{a}, \mathrm{a}' \neq 0)$ 或 $\varepsilon \neq \varepsilon'(\eta \geqslant \varepsilon, \varepsilon' \geqslant 0)$，即有 $q(\mathrm{a}, \varepsilon) \neq q(\mathrm{a}', \varepsilon\prime)$，且一切使 $a \in N^n(\tilde{p})$，$\tilde{p} = d(p)$，$p \in M$，$\eta \geqslant \varepsilon \geqslant 0$ 的点 $q(\mathrm{a}, \varepsilon)$ 成一 $d(M)$ 在 M^n 中的邻域，命之为 A，而一切使 $\eta \geqslant \varepsilon \geqslant 0$ 的点 $q(\mathrm{a}, \varepsilon)$ 成 A 在 M^n 中的闭包，命之为 $\bar{A}$. 于是 A 即可视为丛 $\mathfrak{N}(\mathfrak{D})$ 的丛空间，其投影为 $\pi(q(\mathrm{a}, \varepsilon)) = p$，此处 $\mathrm{a} \in N^n(\tilde{p})$，$\tilde{p} = d(p)$.$\bar{A}$ 可视为 $\mathfrak{N}(\mathfrak{D})$ 的相配闭胞腔丛的丛空间，其投影为 $\bar{\pi}(q(\mathrm{a}, \varepsilon)) = p$, a, $\tilde{p}$, p 如前，而 $\bar{\pi}/A \equiv \pi$.

任取 M 的一个定向，由此 $M^n = \underbrace{M \times \cdots \times M}_{n}$，因之 A 也都确定了定向. 其次对于每一 $p \in M$，取 $\pi^{-1}(p) \subset A$ 的定向使在 p 的附近，$\pi^{-1}(p)$ 和 M 依此次序所决定的 A 的定向与前相同. 命

$$\tau^* : H^k(M) \approx H^{k+m(n-1)}(A)$$

是对此定向丛 $\mathfrak{N}(\mathfrak{D})$ 而言的 Thom 变换，①其中系数群为 I_p, p 为奇质数，不明写，下同. 因 A 是 M^n 的开集，故依§ 1 的 $H2^\circ$ 有变换

$$k^* : H^k(A) \to H^k(M^n).$$

定义

$$\bar{\tau}^* = k^* \tau^* : H^k(M) \to H^{k+m(n-1)}(M^n). \tag{2}$$

容易看出 $\bar{\tau}^*$ 与 η 的大小及 ds^2 的选择无关，但我们将进一步证明下面的

定理 1 $\bar{\tau}^*$ 与 M 的微分构造 $\mathfrak{D}$ 的选择无关. 换言之，$\bar{\tau}^*$ 是拓扑不变地定义于 M 和 M^n 之间的变换. 此外，

$$\bar{\tau}^* : H^*(M) \subset H^*(M^n).$$

证 设 $\mathfrak{D}'$ 是 M 上另一级 $\geqslant 3$ 的微分构造. 凡从 $\mathfrak{D}'$ 所得的种种将与 $\mathfrak{D}$ 所得的相当事物以同样记号表示之，但于右上角加一撇以示区别. 此外，在定义 A' 与

---

① 由经典的定理，知道 M 必有与 $\mathfrak{D}$ 相容的胞腔剖分，故 τ^* 必可定义，且依§ 3 的 2° 知 τ^* 与所选择的胞腔剖分无关.

$\bar{A}'$ 时，我们将选择 η' 充分小使 $\bar{A}' \subset A$. 因 M 是闭的，故此事必可能. 依§ 1 的 $H2^\circ$，有变换

$$j^* : H^*(A') \to H^*(A)$$

存在，且

$$k^* j^* = k'^* : H^*(A') \to H^*(M^n). \tag{3}$$

我们知道 ①$H^m(M)$，$H^{mn}(M^n)$，$H^{mn}(A)$，$H^{mn}(A')$ 都 $\approx I_p$. 与以前所决定的 M, M^n，A，A' 的定向相应，这些群都有一个确定的母素，各以 M_0，M_0^n，A_0，A_0' 表之，则

$$k^*(A_0) = M_0^n, \quad k'^*(A_0') = M_0^n, \quad j^*(A_0') = A_0. \tag{4}$$

任取 M 的一个剖分，仍记之为 M，依§ 3 可作 A 的一个相应剖分，仍记之为 A，则依 $\tau^\#$ 的定义和 M，A 定向的选择，易见

$$\tau^*(M_0) = A_0, \tag{5}$$

或依§ 3 的 (6)，

$$\tau^* 1 \cup \bar{\pi}^* M_0 = A_0. \tag{6}$$

同样有

$$\tau'^* 1 \cup \bar{\pi}'^* M_0 = A_0'. \tag*{(6)$'$}$$

命 $\bar{j} : \bar{A}' \subset \bar{A}$, $c : M \to \bar{A}$, $c' : M \to \bar{A}'$ 定义如对角映像 d，则 $\bar{j}c' = c$，又 $\bar{\pi}^* = c^{*-1}$，$\bar{\pi}'^* = c'^{*-1}$，故

$$\bar{j}^* \bar{\pi}^* = \bar{\pi}'^*. \tag{7}$$

应用§ 1 的 $P3^\circ$，$P4^\circ$ 于是得

$$\begin{aligned} j^*(\tau'^* 1 \cup \bar{\pi}'^* M_0) &= j^*(\tau'^* 1 \cup \bar{j}^* \bar{\pi}^* M_0) = j^*(\tau'^* 1 \cup \bar{\pi}^* M_0) \\ &= j^* \tau'^* 1 \cup \bar{\pi}^* M_0. \end{aligned}$$

而由 (4)，(6)$'$，得

$$j^* \tau'^* 1 \cup \bar{\pi}^* M_0 = A_0. \tag{8}$$

因 $H^{m(n-1)}(A) \approx H^0(M) \approx I_p$, 故 $\tau^* 1$, $j^* \tau'^* 1 \in H^{m(n-1)}(A)$ 二者只能相差一法 p 整数倍数，而比较 (6)，(8) 两式可知必须有

$$\tau^* 1 = j^* \tau'^* 1. \tag{9}$$

① 参阅 [7]，特别§ 7.

由 (2)，(3)，(9) 得

$$\bar{\tau}^*1 = \bar{\tau}'^*1. \tag{10}$$

换言之，$\bar{\tau}^*1$ 为不变地定义于 M^n 中的类，而与 $\mathfrak{D}$ 的选择无关.

其次，设 $U \in H^*(M)$ 任意，则由 § 3 的 (6)，本节的 (7)，(9) 以及 § 1 的 $P3^\circ$，$P4^\circ$，得

$$\begin{aligned} j^*\tau'^*U &= j^*(\tau'^*1 \cup \bar{\pi}'^*U) = j^*(\tau'^*1 \cup \bar{j}^*\bar{\pi}^*U) = j^*(\tau'^*1 \cup \bar{\pi}^*U) \\ &= j^*\tau'^*1 \cup \bar{\pi}^*U = \tau^*1 \cup \bar{\pi}^*U = \tau^*U. \end{aligned}$$

故

$$j^*\tau'^* = \tau^*. \tag{11}$$

由 (3)，(11) 与 (2) 得 $\bar{\tau}'^* = k'^*\tau'^* = k^*j^*\tau'^* = k^*\tau^* = \tau^*$, 或

$$\bar{\tau}'^* = \bar{\tau}^*. \tag{12}$$

换言之，$\bar{\tau}^*$ 与 M 的微分构造 $\mathfrak{D}$ 无关. 这证明了定理的第一部分.

今设 $V \in H^*(M^n)$. 命 $\bar{k}: \bar{A} \subset M^n$, 则 $\bar{k}c = d$, $c^*\bar{k}^* = d^*$, $\bar{k}^* = c^{*-1}d^* = \bar{\pi}^*d^*$, 故由 (2)，§ 1 的 $P3^\circ$，$P4^\circ$ 和§ 3 的 (6)，得

$$\begin{aligned} \bar{\tau}^*d^*V = k^*\tau^*d^*V &= k^*(\tau^*1 \cup \bar{\pi}^*d^*V) = k^*(\tau^*1 \cup \bar{k}^*V) \\ &= k^*(\tau^*1 \cup V) = k^*\tau^*1 \cup V \end{aligned}$$

或 ①

$$\bar{\tau}^*d^*V = \tau^*1 \cup V. \tag{13}$$

最后，设 $U \in H^*(M)$ 任意而 $U \neq 0$. 此时可取 $U_1 \in H^*(M)$ 使 $U \cup U_1 = M_0$. 因 d^* 有遮盖性，故可选 V，$V_1 \in H^*(M^n)$ 使 $d^*V = U$，$d^*V_1 = U_1$. 由 (2)，(4)，(5)，(13)，诸式得

$$\begin{aligned} \bar{\tau}^*U \cup V_1 = \bar{\tau}^*d^*V \cup V_1 &= (\bar{\tau}^*1 \cup V) \cup V_1 = \bar{\tau}^*1 \cup (V \cup V_1) \\ &= \bar{\tau}^*d^*(V \cup V_1) = \bar{\tau}^*(U \cup U_1) = \bar{\tau}^*M_0 = M_0^n \neq 0. \end{aligned}$$

故 $\bar{\tau}^*U \neq 0$ 而 $\bar{\tau}^*: H^*(M) \subset H^*(M^n)$. 至此定理已完全证明.

---

① 直至此刻，我们实际上并没有用到系数群是 I_p 的事实，也没有用到 $H^*(M)$ 中的对偶定理和 $H^*(M)$，$H^*(M^n)$ 间的关系如 d^* 是遮盖性之类. 事实上，到这里为止的结果对于任一微分地镶嵌于一微分闭流形 N 的微分闭流形 M 而言，都依然成立：有 $\bar{\tau}^*: H^*(M) \to H^*(N)$ 存在且为 (N, M) 偶的拓扑不变量 (系数群任意).

今取 $n=3$，则按预备定理 5，$\mathfrak{R}(\mathfrak{D})\sim\mathfrak{T}(\mathfrak{D})\cup\mathfrak{T}(\mathfrak{D})$，故据§ 4 的 (16) 有

$$St_p^{2t(p-1)}\tau^*1=\tau^*Q_p^{2t(p-1)}(\mathfrak{T}(\mathfrak{D})). \tag{14}$$

两边各应用 k^* 后，依§ 1 的 $S9^\circ$ 得

$$St_p^{2t(p-1)}\bar{\tau}^*1=\bar{\tau}^*Q_p^{2t(p-1)}(\mathfrak{T}(\mathfrak{D})), \tag{15}$$

其中 $\bar{\tau}^{**}$ 定义如 (2) 式.

在 (15) 中，我们曾假定 $\mathfrak{D}$ 的级 $\geqslant 3$. 今设 $\mathfrak{D}$ 的级任意，则据 [17]，M 有一微分构造 $\mathfrak{D}'$，其级 $\geqslant 3$，且 $\mathfrak{D}'\sim\mathfrak{D}$. 由此得 $\mathfrak{T}(\mathfrak{D})\sim\mathfrak{T}(\mathfrak{D}')$，因而 $Q_p^{2t(p-1)}(\mathfrak{T}(\mathfrak{D}))=Q_p^{2t(p-1)}(\mathfrak{T}(\mathfrak{D}'))$. 今对 $\mathfrak{D}$ 而言，定义 $\bar{\tau}^*=\bar{\tau}'^*$ 如 (12)，则对此任意级的 $\mathfrak{D}$，(15) 式仍成立. 因依定理 1，如此定义之 $\bar{\tau}^*$ 与 $\mathfrak{D}$ 的选择无关，且 $\bar{\tau}^*:H^*(M)\subset H^*(M^n)$，故 (15) 亦可书为

$$Q_p^{2t(p-1)}(\mathfrak{T}(\mathfrak{D}))=\bar{\tau}^{*-1}St_p^{2t(p-1)}\bar{\tau}^*1. \tag{15$'_p$}$$

综合言之，得下述定理:

定理 2 若 M 是一个可微分的可定向闭流形，则对于 M 的任意微分构造 $\mathfrak{D}$，由 $\mathfrak{D}$ 所定切丛 $\mathfrak{T}(\mathfrak{D})$ 的法 pQ 示性类 $Q_p^{2t(p-1)}(\mathfrak{T}(\mathfrak{D}))$($p$ 为任意奇质数) 与 $\mathfrak{D}$ 的选择无关，因而是 M 的拓扑不变量. 且此等示性类可由 $(15)'_p$ 以定之.

依§ 4 的 $(19)_3$，$P_3^{4k}(\mathfrak{T}(\mathfrak{D}))$ 可由 $Q_3^{4i}(\mathfrak{T}(\mathfrak{D}))$，$i\leqslant k$，所决定，因之从定理 2 可得下述推论:

定理 3 若 M 是一个可微分的可定向闭流形，则对于 M 的任意微分构造 $\mathfrak{D}$，由 $\mathfrak{D}$ 所定切丛 $\mathfrak{T}(\mathfrak{D})$ 的法 3 Понтрягин 示性类与 $\mathfrak{D}$ 的选择无关，因而是 M 的拓扑不变量. 且此等示性类可由 $(15)'_p$ 与§ 4 的 $(19)_3$ 以定之.

参考文献

[1] Wu Wen-Tsün. Sur les puissances de Steenrod. *Colloque de Topologie de Strassbourg*, 1951.

[2] Thom R. Classes caractéristiques et *i*-carées. *C.R.Paris*, 1950, **230**: 427-429.

[3] ——. Variétés plongés et *i*-carrés. *Ibid.*, 1950, **230**: 508-511.

[4] ——. Espaces fibrés en sphéres et carrés de Steenrod. *Ann.Ec.Norm.Sup.*, 1952, **69**: 110-182.

[5] Wu Wen-Tsün. Classes caractéristiques et *i*-carrés d'une variété. *C.R.Paris*, 1950, **230**: 508-511.

[6] 吴文俊. 论 ПОНТРЯГИН 示性类 I. 数学学报, 1953, **3**: 291-314.

[7] Cartan H. Méthodes modernes en topologie algébroque. *Comm.Math.Helv.*, 1945, **18**: 1-15.

[8] Edenberg S & Steenrod N. *Foundations of Algebraic Topology*, 1952.

[9] Lefachetz S. *Algebraic Topology*, 1941.

[10] Steenrod N. Reduced powers of cohomology classés. *Cours Collége de France*, 1951.

[11] Borel A & Serre J P. Groupes de Lie et puissances réduites de Steenrod. *Amer.I.Math.*, 1953, **75**: 409-448.

[12] 吴文俊. 格拉斯曼流形中的平方运算. 数学学报，1953, **2**: 203-230. Wu Wen-Tsün. On squares in grassmanndan manifolds. *Acta Scientia Sinica*, 1953, **2**: 91-115.

[13] Chern S. Characteristic classes of Hermitian manifolds. *Annals of Math.*, 1946, **47**: 85-121.

[14] Wu Wen-Tsün. Sur les classes caractéristiques des espaces fibrés sphériques. *Actualies Scientifiques et industrialles* no.1183, Hermann, Paris, 1952.

[15] Gysin W. Zur Homologietheorie der Abbildungen und Faserungen von Mannigfaltigkeiten. *Comm.Math.Helv.*, 1941, **75**: 61-121.

[16] Chern-Spanier. The homology structure of sphere bundles. *Proc.Nat.Acad.Sci.*, 1950, **36**: 248-255.

[17] Whitney H. Differentiable manifolds. *Annals of Mathematics*, 1936, **37**: 645-680.

On ПОНТРЯГИН Classes, II

Abstract The present paper as well as the preceding one with the same title [6], contain the detailed proofs of results sketched in [1], § § 5-6.It is proved that for a closed differentiable manifold, to any odd prime p, certain combinations by means of cup products of the Понтрягин classes, reduced mod p, with respect to any given differential structure of the manifold are in reality independent of that structure and therefore are topological invariants of the manifold. For $p=3$, it turns out that the mod 3 Понтрягин classes are themselves topological invariants of the manifold.The proofs are based on the so-called "diagonal method", obtaining thus a connection between the Понтрягин classes and the Steenrod powers, of which the later one, in the case of a differentiable manifold, are of a character non-diffential, but purely topological.

17. 论ПОНТРЯГИН示性类 III*

本文继续以前二文 [1, 2] 研究微分流形上Понтрягин示性类的拓扑不变性.

本文应用了在[3]一文中首次倡用的方法，完全决定了格拉斯曼流形 $R_{n,m}$ 中的Понтрягин平方. 由此可知，在一个可微分闭流形上，Понтрягин示性类在法 4 约化后乃是这个闭流形的拓扑不变量.

在前一文中，作者曾证明若一个可微分闭流形是可定向的，那么它的法 3 Понтрягин示性类是拓扑不变量. 本文指出这定理在流形不可定向时仍然成立，因之结合本文的结果，可知一个可微分闭流形的法 12 Понтрягин示性类是拓扑不变量.

据作者所知，除了极个别的情形如Рохлин-Thom [4, 5] 对于四维可定向闭流形之外，这是关于可微分流形上Понтрягин示性类的拓扑不变性方面，直到现在所得到的最一般的结果.

§1 ПОНТРЯГИН平方 [6–10]

设 R 是一个固定的可交换环，$R_k = R/kR$ 为将 R 法 k(k 为正整数) 约化后所得的环.

设 K 是一复合形，其中每一胞腔 σ 的闭包 $Cl(\sigma)$ 都有与单纯形相同的同调群，假定对每一整数 j 有一组准同构变换

$$D^j : C_q(K) \to C_{q+j}(K \times K)$$

满足下面的性质 S1° ~S4°:

S1° 对于任一 $\sigma \in K$, $D^j\sigma$ 是 $\sigma \times \sigma$ 中的链.

S2° 若 $j < 0$ 则 $D^j = 0$.

S3° $In(D^0c) = In(c)$, $c \in C_0(K)$.

S4° $\partial D^{2i} - D^{2i}\partial = (1+T)D^{2i-1}$, $\partial D^{2i+1} + D^{2i+1}\partial = (T-1)D^{2i}$,

此处 T 为映像 $(x_1, x_2) \to (x_2, x_1)$，$x_i \in K$，所引起的链变换，1 是恒同链变换.

命 $D_j : C^q(K \times K, R) \to C^{q-j}(K, R)$ 是 D^j 的对偶变换. 若 $u \in C^m(K,R)$，$v \in C^n(K, R)$，定义 $\cup_i(u, v) \in C^{m+n-i}(K, R)$ 为 $\cup_{2i}(u, v) = (-1)^i D_{2i}(u \otimes v)$，$\cup_{2i+1}(u, v)$

---

* *Acta Math. Sinica*, 1954, 4: 323-346; *American Mathematical Society Translations, Ser.* 2, Vol. II, 155-172; *American Mathematical Society, Providence R I*, 1959.

$= (-1)^{m+n+i} D_{2i+1}(u \otimes v)$，则

$$\delta \cup_i (u, v) = \cup_i (\delta u, v) + (-1)^m \cup_i (u, \delta v) + (-1)^{m+n+i} \cup_{i-1} (u, v) + (-1)^{m+n+mn} \cup_{i-1} (v, u),$$

今称 $\cup_i$ 为 K 中的一组上 i 积. 我们知道对于每一组上 i 积，$u \to \cup_i(u, u)$，$u \in C^m(K, R)$ 引出了一组 Steenrod 平方运算

$$Sq_i = Sq^{m-i} : H^m(K, R_2) \to H^{2m-i}(K, R_2),$$

此处 Sq_i 为准同构变换，具有同伦不变性，且与 D^j 的选择无关. 又 $\cup_0$ 引出上积.

若命

$$\beta : H^m(K, R_2) \to H^{m+1}(K, R)$$

为 Bockstein 变换 (此处假定 R 无 $2a = 0$ 但 $a \neq 0$ 的元素 a)，

$$\rho_k : \quad H^*(K, R) \to H^*(K, R_k)$$

为法 k 约化，$\beta_k = \rho_k \beta$，则在 β_2 与 Sq^i 之间有下述关系:

$$\beta_2 = Sq^1, \quad \beta Sq^{2i+1} = 0, \quad \beta_2 Sq^{2i} = Sq^{2i+1}. \tag{1}$$

对任意 $u \in C^m(K, R_2)$，取 $u_0 \in C^m(K, R)$ 使 $\rho_2 u_0 = u$，并命

$$p(u_0) = \cup_0(u_0, u_0) + \cup_1(u_0, \delta u_0), \tag{2}$$

则 $u \to \rho_4 p(u_0)$ 引出了一个对应

$$\boldsymbol{p} : H^m(K, R_2) \to H^{2m}(K, R_4). \tag{2$'$}$$

$\boldsymbol{p}$ 即所谓Понтрягин平方，具有以下诸性质:

P1° $\boldsymbol{p}$ 有同伦不变性且与 D^j 的选择无关.

P2° 若 $U_i \in H^m(K, R_2)$，$a_i \in R$，则

$$m \text{ 为偶数时，} \boldsymbol{p}\left(\sum_i a_i U_i\right) = \sum_i a_i^2 \boldsymbol{p}(U_i) + \sum_{i<j} a_i a_j \theta_2(U_i U_j),$$

$$m \text{ 为奇数时，} \boldsymbol{p}\left(\sum_i a_i U_i\right) = \sum_i a_i^2 \boldsymbol{p}(U_i),$$

此处 $U_i U_j$ 指 U_i，U_j 的上积，而

$$\theta_2 : H^*(K, R_2) \to H^*(K, R_4)$$

指由自然变换 $\theta_2 : R_2 \to R_4$ 所引起的准同构变换.

P3° 若 $U \in H^m(K, R_2)$，则

$$m \text{ 为偶数时}, \quad 2\boldsymbol{p}(U) = \theta_2(U^2),$$
$$m \text{ 为奇数时}, \quad 2\boldsymbol{p}(U) = 0,$$

此处 U^2 为 U 与自身的上积.

今设两个复合形 K_α，$\alpha = 1, 2$，各有一组 $D^j_{(\alpha)}\alpha = 1,2$，其所定之上 i 积为 $\cup_i^\alpha$，依 Steenrod(［7］，§ 16)，在 $K = K_1 \times K_2$ 中可定义一组 D^j，由此在 K 中可得一组上 i 积 $\cup_i$ 满足下述 Cartan 公式 (［9］):

若 $u_\alpha \in C^{m_\alpha}(K_\alpha, R)$, $v_\alpha \in C^{n_\alpha}(K_\alpha, R)$, $u = u_1 \otimes u_2 \in C^m(K, R)$, $v = v_1 \otimes v_2 \in C^n(K, R)$, $m = m_1 + m_2$, $n = n_1 + n_2$, 则

$$\begin{aligned}\cup_i(u, v) =&(-1)^{m_2 n_1} \sum_j \cup^1_{2j}(u_1, v_1) \otimes \cup^2_{i-2j}(u_2, v_2) \\ &+(-1)^{m_2n_1+m_2n_2+m_2+n_2} \sum_j \cup^1_{2j+1}(u_1, v_1) \otimes \cup^2_{i-2j-1}(v_2, u_2).\end{aligned} \tag{3}$$

由 (2) 与 (3)，对于任意 $u_\alpha \in C^{m_\alpha}(K, R)$，$\alpha = 1, 2$，$u = u_1 \otimes u_2$，可得

$$p(u) = A_0 + A_1 + A_2 + A_3 + A_4, \tag{4}$$

此处

$$\begin{aligned}
A_0 &= (-1)^{m_1m_2}[\cup^1_0(u_1, u_1) + \cup^1_1(u_1, \delta u_1)] \otimes [\cup^2_0(u_2, u_2) + \cup^2_1(u_2, \delta u_2)], \\
A_1 &= (-1)^{m_1m_2+1} \cup^1_1 (u_1, \delta u_1) \otimes \cup^2_1(u_2, \delta u_2), \\
A_2 &= (-1)^{m_1m_2}[-1 + (-1)^{m_1}] \cup^1_0 (u_1, u_1) \otimes \cup^2_1(u_2, \delta u_2), \\
A_3 &= (-1)^{m_2(m_1+1)} \cup^1_0 (u_1, \delta u_1) \otimes \cup^2_1(u_2, u_2), \\
A_4 &= (-1)^{m_1m_2+m_1} \cup^1_1 (u_1, u_1) \otimes \cup^2_0(\delta u_2, u_2).
\end{aligned}$$

设 $\rho_2 u_\alpha \in U_\alpha \in H^{m_\alpha}(K_\alpha, R_2)$，则必有 $v_\alpha \in C^{m_\alpha+1}(K_\alpha, R)$ 使 $\delta u_\alpha = 2v_\alpha$，且 $v_\alpha \in V_\alpha = \beta U_\alpha \in H^{m_\alpha+1}(K_\alpha, R)$ 此时易见 (见 P3°)

$$\begin{aligned}
&\rho_4 A_0 \in (-1)^{m_1m_2}\boldsymbol{p}(U_1) \otimes \boldsymbol{p}(U_2) = \boldsymbol{p}(U_1) \otimes \boldsymbol{p}(U_2), \\
&\rho_4 A_1 = \rho_4 A_2 = 0, \\
&\rho_4 A_3 \in \theta_2(U_1 \rho_2 V_1 \otimes Sq_1 U_2) = \theta_2(U_1 \beta_2 U_1 \otimes Sq_1 U_2), \\
&\rho_4 A_4 \in \theta_2(Sq_1 U_1 \otimes U_2 \rho_2 V_2) = \theta_2(Sq_1 U_1 \otimes U_2 \beta_2 U_2).
\end{aligned}$$

因之，将 (4) 两边各取 ρ_4，即得下述

定理 1 若 $U_\alpha \in H^{m_\alpha}(K_\alpha, R_2)$，$\alpha = 1, 2$，则

$$\boldsymbol{p}(U_1 \otimes U_2) = \boldsymbol{p}(U_1) \otimes \boldsymbol{p}(U_2) + \theta_2(U_1\beta_2 U_1 \otimes Sq_1 U_2) + \theta_2(Sq_1 U_1 \otimes U_2 \beta_2 U_2).$$

从此定理可得下面的推论：

定理 2 若 $U_\alpha \in H^{m_\alpha}(K, R_2)$，$\alpha = 1,2$，则

$$\boldsymbol{p}(U_1 U_2) = \boldsymbol{p}(U_1)\boldsymbol{p}(U_2) + \theta_2(U_1\beta_2 U_1 Sq_1 U_2) + \theta_2(U_2\beta_2 U_2 Sq_1 U_1).$$

§1 (实) 格拉斯曼流形 $R_{n,m}$ 中的一些同调性质

我们将采用 [1] 里面的符号，并以 I 表整数加法群，I_k(k 为 >1 的整数) 表法 k 约化整数加法群，I_0 表实数域.

根据 [1] 中的定理 1，在 $r < n$ 时，或 $r = n$ 但 n 是偶数时，$R_{n,m}$ 的 r 维整数下链群有一组加法基

$$x_i^r(i = 1, \cdots, \lambda_r), \quad y_i^r(i = 1, \cdots, \mu_r), \quad z_i^r(i = 1, \cdots, \nu_r) \tag{1}$$

满足以下的边界公式 $(r \leqslant n)$：

$$\partial x_i^r = 0, \quad \partial y_i^r = 0, \quad \partial z_i^r = 2y_i^{r-1}, \tag{2}$$

此处 x_i^r，z_i^r 各取一切 Ω_0^r 和 Ω_2^r 中的初等链，而

$$\left.\begin{array}{l}\lambda_r = \Omega_0^r \text{ 中初等链的个数,} \\ \nu_r = \mu_{r-1} = \Omega_2^r \text{ 中初等链的个数.}\end{array}\right\} \tag{3}$$

在 $r = n$ 且 n 是奇数时，则 $C_n(R_{n,m})$ 有一组加法基除包括 (1) 的各链外，尚须加入 $x_0^n = [0, \cdots, 0, n]$，且此时除 (2) 外尚须满足

$$\partial x_0^n = 0. \tag{2$'$}$$

对于 $r \leqslant n$ 今命 u_i^r，v_i^r，w_i^r 为与 x_i^r，y_i^r，z_i^r 相对偶的上链，换言之 u_i^r 为由 $u_i^r(x_j^r) = \delta_{ij}$，$u_i^r(y_j^r) = 0$，$u_i^r(z_j^r) = 0$ 所定的上链，余类推. 特别有 $(i > 0)$

$$u_i^r = \{\mathrm{a}_i\}, \quad \mathrm{a}_i \in \Omega_0^r. \tag{4}$$

由上可知在 $r \leqslant n$ 时，u_i^r，v_i^r，w_i^r 作成 $R_{n,m}$ 中 r 维整数上链群的一个加法基，且其边界公式为 $(r < n)$

$$\delta u_i^r = 0, \quad \delta v_i^r = 2w_i^r, \quad \delta w_i^r = 0. \tag{5}$$

对于任一系数环 R，在 $\sum\limits_{r<s} H^r(R_{n,m},R)$ 中，定义一乘法为 $UV=0$，若 $\dim U+\dim V\geqslant s$，否则与 $H^*(R_{n,m},R)$ 中的上积相同. 如此所得之环记之为 ${}^sH^*(R_{n,m},R)$. 本节的目的，在决定 $R=I_0$ 和 I 时 ${}^nH^*(R_{n,m},R)$ 的构造.

定理 3　${}^nH^*(R_{n,m},I_0)$ 是由 $R_{n,m}$ 的实系数 P 类 P_0^{4k}, $0<k\leqslant\left[\frac{m}{2}\right]=m'$ 所产生，且以之为乘法基. 换言之，$r<n$ 时，任一 $U\in H^r(R_{n,m},I_0)$ 可表为 P_0^{4k} 的一个多项式，且此表法为唯一的.

在由定向面所成的格拉斯曼流形 $R^{\wedge}_{n,m}$(即 $R_{n,m}$ 的二叶覆叠空间) 的情形，Понтрягин 早就决定了 ${}^nH^*(R^{\wedge}_{n,m},I_0)$ 的构造 [11]. 上述定理与Понтрягин 的结果类似，而且在 $R_{n,m}$ 是可定向 (即 $m+n$ 是偶数) 时，应该也可以用同样的方法来证明它. 但Понтгрягин 的方法须用到 E. Cartan 和 de Rham 所建立的微分几何的工具，超出拓扑学的范围. 因之我们将用另法直接来证明定理 3，为此我们将先证明一个纯粹代数的结果.

假设 x_{2k}，$k=1,\ 2,\ \cdots,m'$，是一组重量是 $2k$ 的不定量，y_i，$i=1,\cdots,\ m'$，是另一组重量是 i 的不定量，而在二者之间有关系

$$(-1)^k x_{2k}=\sum_{k_1+k_2=2k}(-1)^{k_1}y_{k_1}y_{k_2},\quad (y_0=1).\tag{6}$$

设 $\varphi(x_{2k})$ 是各项重量都同是 s 的一个 x_{2k} 的多项式，以实或复数为系数，若 $\varphi(x_{2k})$ 对不定量 x_{2k} 而言不恒等于 0，则必有实或复数 $a_{2k}(k=1,\cdots,m')$ 存在使 $\varphi(a_{2k})\neq 0$. 考察方程

$$\xi^{2m'}-a_2\xi^{2m'-2}+a_4\xi^{2m'-4}-\cdots+(-1)^{m'}a_{2m'}=0,$$

命其根为 $\pm\beta_1,\cdots,\pm\beta_{m'}$，或 $\alpha_1,\cdots,\alpha_{2m'}$，此处 $\alpha_i=\beta_i$，$\alpha_{m'+i}=-\beta_i$，$i=1$，$\cdots,m'$，则 $(-1)^k a_{2k}=\int(\alpha_1\cdots\alpha_{2k})$ 为 $\alpha_i(i=1,\cdots,2m')$ 的初等对称函数之值. 命 $\beta_i(i=1,\cdots,m)$ 的初等对称函数之值为 $b_i=\int(\beta_1\cdots\beta_i)$，则显然有 $(-1)^k a_{2k}=\sum\limits_{k_1+k_2=2k}(-1)^{k_1}b_{k_1}b_{k_2}$. 因 $\varphi(a_{2k})=\varphi((-1)^k\sum(-1)^{k_1}b_{k_1}b_{k_2})\neq 0$，故 $\varphi((-1)^k\Sigma(-1)^{k_1}y_{k_1}y_{k_2})$ 对 y_k 而言不恒等于 0. 换言之，若多项式 $\varphi(x_{2k})$ 对 x_{2k} 而言不恒等于 0，则依 (6) 式代入后所得的多项式对 y_i 而言也不恒等于 0. 这就是我们要首先证明的代数结果.

定理 3 的证明　对于固定的 $r(r\equiv 0 \bmod 4)$，试考察一切作

$$(P_0^4)^{k_1}(P_0^8)^{k_2}\cdots(P_0^{4m'})k_{m'}\tag{7}$$

形状的类，此处

$$k_i\geqslant 0,\quad 4\sum_{i=1}^{m'}ik_i=r,\quad m'=\left[\frac{m}{2}\right].\tag{7$'$}$$

我们将证明: $(A_{n,m})$ 在 $r \leqslant n$ 时，这些类在 $H^r(R_{n,m}, I_0)$ 中线性无关.

为此，先设 n，m 都是偶数：$n = 2n'$，$m = 2m'$. 在 [1] 的 § 4 中我们曾定义过一个自然映像

$$k : C_{n',m'} \to R_{n,m}.$$

若命 C_0^{2i}，$0 < i \leqslant m'$，为 $C_{n',m'}$ 中的实系数 C 类，则据 [1] 的定理 7(§ 4，(6) 式) 有 ($C_0^0 =$ 实系数单位类)

$$k^* P_0^{4k} = (-1)^k \sum_{k_1+k_2=2k} (-1)^{k_1} C_0^{2k_1} C_0^{2k_2}\ . \tag{8}$$

今设 $\varphi(P_0^{4k}) \in H^r(R_{n,m}, I_0)$，此处 φ 为以上积为乘法，实数为系数的一个 P_0^{4k} 的多项式. 设 $\varphi(x_{2k})$ 不恒等于 0. 由前面代数的结果，若 y_i 如 (6) 式所示，则 $\varphi((-1)^k \sum (-1)^{k_1} y_{k_1} y_{k_2})$ 对 y_i 而言也不恒等于 0. 由假设 $r \leqslant n$，故由 [2] 的预备定理 1，$\varphi((-1)^k \sum (-1)^{k_1} C_0^{2k_1} C_0^{2k_2}) = \varphi(k^* P_0^{4k}) = k^* \varphi(P_0^{4k})$ 在 $H^r(C_{n,m} I_0)$ 中不能为 0. 于是 $\varphi(P_0^{4k})$ 在 $H^r(R_{n,m}, I_0)$ 中更不能为 0，而 $(A_{n,m})$ 得证.

其次，假定 n 是奇数而 m 是偶数. 此时可任取一整数 $n' > n$ 并考察自然映像

$$f : R_{n,m} \to R_{n',m}.$$

依照 [3] 的 § 1 中预备定理 1 的证明，我们可视 $R_{n,m}$ 为 $R_{n',m}$ 的子流形，且可取 $R_{n,m}$，$R_{n',m}$，的标准胞腔剖分 K，K' 使 K 成为 K' 的子复合形. 易见 K 与 K' 具有同样的 n 维骨架，因之对于任意系数群 G，有

$$f^* : \begin{cases} H^n(R_{n',m}, G) \subset H^n(R_{n,m}, G), \\ H^r(R_{n',m}, G) \approx H^r(R_{n,m}, G), \quad r < n. \end{cases} \tag{9}$$

今取 n' 为偶数 $> n$. 由前面已证明的 $(A_{n',m})$，命 $P_0'^{4k}$ 为 $R_{n',m}$ 中的实系数 P 类时，一切作 $(P_0'^4)^{k_1} \cdots (P_0'^{4m'})^{k_{m'}}$ 形状而 k_i 满足 (7)$'$ 的类在 $H^r(R_{n',m}, I_0)$ 中线性无关只须 $r \leqslant n'$. 故由 (9)，一切作 $f^*((P_0'^4)^{k_1} \cdots (P_0'^{4m'})^{k_{m'}})$ 形状的类在 $H^r(R_{n,m}, I_0)$ 中也线性无关只须 $r \leqslant n$. 但据 [1] 的 § 4 定理 4，$f^* P_0'^{4k} = P_0^{4k}$，故由此即得 $(A_{n,m})$.

最后，设 m 是奇数，此时我们将定义一个自然映像

$$i : R_{n,m-1} \to R_{n,m}$$

如次. 假设 $R_{n,m}$ 与 $R_{n,m-1}$ 各定义于欧氏空间 R^{n+m} 与 R^{n+m-1} 中且 R^{n+m-1} 是 R^{n+m} 的一个 $(n+m-1)$ 面. 命 $\overline{R}^1$ 为 R^{n+m} 中与 R^{n+m-1} 完全垂直的 1 面.

对于任一 $(m-1)$ 面 $X\in R_{n,m-1}$，我们就定义 $i(X)$ 为 m 面 $X\oplus \bar{R}^1 \in R_{n,m}$. 在 R^{n+m-1} 中任取一 i 面的序列:

$$R^1\subset R^2\subset\cdots\subset R^{n+m-2}. \tag{S}$$

命 $R'^i=R^{i-1}\oplus\bar{R}^1$，$i>1$，$R'^1=\bar{R}^1$. 对于 (S) 及序列 (S)′

$$R'^1\subset\cdots\subset R'^{n+m-1}(\subset R'^{n+m}=R^{n+m}) \tag{S$'$}$$

各取 $R_{n,m-1}$ 和 $R_{n,m}$ 的标准剖分 K，K'. 若 $\mathrm{a}=(a_1,\cdots,a_{m-1})\in\Omega_{n,m-1}$，命 $\mathrm{i(a)}=(0,\ a_1,\cdots,\ a_{m-1})\in\Omega_{n,m}$. 因为对于任一 $\mathrm{a}\in\Omega_{n,m-1}$，$X$ 属于假流形 $[\mathrm{a}]^*$ 的充要条件是

$$\dim(X\cap R^{a_i+i})\geqslant i,\quad i=1,\cdots,m-1, \tag{10}$$

而 (10) 显然与下组条件

$$\dim(i(X)\cap R'^{a_i+i+1})\geqslant i+1,\quad i=1,\cdots,m-1, iX\supset R'^1 \tag{10$'$}$$

同值，也就是与 $\mathrm{i}(X)\in[\mathrm{ia}]'^*$ 同值，故 i 把假流形 $[\mathrm{a}]^*$ 拓扑地映像到假流形 $[\mathrm{ia}]'^*$ 之上. 同样也易见 i 同时把 $[\mathrm{a}]^*$ 的边 $\Sigma[\mathrm{a}_{(i)}]^*$ 拓扑地映像到 $[\mathrm{ia}]'^*$ 的边 $\Sigma[(\mathrm{ia})_{(i)}]'^*$ 之上. 由此易得

$$i_\#[\mathrm{a}]=\pm[i\,\mathrm{a}]',\quad \mathrm{a}\in\Omega_{n,m-1}. \tag{11}$$

或，对于任一 $\mathrm{a}=(a_1,\cdots,a_m)\in\Omega_{n,m}$，

$$i^\#\{\mathrm{a}\}'=\begin{cases}\pm\{i^{-1}\mathrm{a}\}, & a_1=0;\\ 0, & a_1\neq 0.\end{cases} \tag{11$'$}$$

因为我们假定 m 为奇数而 $m-1$ 为偶数，故 $\mathrm{a}=(a_1,\cdots,a_m)$ 在 $R_{n,m}$ 的 Ω_0^r 中时，必有 $a_1=0$ 而 (11)′ 的上一式成立. 特别若 $\mathrm{a}=(\underbrace{0,\cdots,0,}_{m-2k}\underbrace{2,2,\cdots,2,2}_{2k})$ 时，从 (11)′ 就得

$$i^*P_0'^{4k}=\pm P_0^{4k},\quad 0<k\leqslant\left[\frac{m}{2}\right],\text{亦即}0<k\leqslant\left[\frac{m-1}{2}\right]. \tag{12}$$

因前面已证明过 $(A_{n,m-1})$ 真确，故从 (12) 知 $(A_{n,m})$ 也真确.

至此 $(A_{n,m})$ 已完全证明.

从 (4) 与 (5)，我们知道在 $r<n$ 时，$R_{n,m}$ 的 r 维上 Betti 数等于 Ω_0^r 中的个数 λ_r，也就是满足下述条件的整数组 $(r_1,\cdots,r_s)$ 的个数:

$$0<r_1\leqslant r_2\leqslant\cdots\leqslant r_s,\quad 4\sum_{i=1}^s r_i=r,\quad s\leqslant\left[\frac{m}{2}\right]=m'. \tag{13}$$

另一面，从 $(A_{n,m})$ 可知 λ_r 必须 $\geqslant \lambda'_r$ 此处 λ'_r 为作 (7) 形状的类的个数，也就是满足 $(7)'$ 的整数组 $(k_1, \cdots, k_{m'})$ 的个数.

今对任一满足 (13) 的整数组 $(r_1, \cdots, r_s)$，命

$$k_1 = r_s - r_{s-1},\ k_2 = r_{s-1} - r_{s-2}, \cdots,\ k_s = r_1, k_{s+1} = \cdots = k_{m'} = 0, \tag{14}$$

则 $(k_1, \cdots, k_{m'})$ 满足 $(7)'$. 易见 (14) 在满足 (13) 的整数组 $(r_1, \cdots, r_s)$ 与满足 $(7)'$ 的整数组 $(k_1, \cdots, k_{m'})$ 之间建立了一个一一对应，因之 $\lambda_r = \lambda'_r$ 而由 $(A_{n,m})$ 即得本定理.

今在 $H^r(R_{n,m}, I)$ 中，命由一切形状作 $(P^4)^{k_1}(P^8)^{k_2}\cdots(P^{4m'})^{k_{m'}}$ 而 k_i 满足 $(7)'$ 的元素所产生的子群为 H_1^r，又命 $H^r(R_{n,m}, I_2) = H_2^r$，则有

定理 4 在 $r < n$ 时，

$$H^r(R_{n,m}, I) \approx H_1^r \oplus \beta H_2^{r-1}. \tag{15}$$

证 设 $U \in H^r(R_{n,m}, I)$ 而 $r < n$. 由定理 3，$U_0 = \rho_0 U$ 必可表为 P_0^{4k} 的一个多项式，设为 $\varphi(P_0^{4k})$. 此时 $\rho_0(U - \varphi(P^{4k})) = 0$. 换言之，$U - \varphi(P^{4k})$ 为 $R_{n,m}$ 中的挠元素. 因之从 (5) 与 (4)，知 $U - \varphi(P^{4k})$ 必作形状 βZ，此处 $Z \in H_2$. 故任一 $U \in H^r(R_{n,m}, I)$ 必可表为 H_1^r 中一元素与 βH_2^{r-1} 中一元素之和. 因据定理 3，若 $X = \varphi(P^{4k}) \in H_1^r$ 而 $X \neq 0$，则 $\rho_0 X = \varphi(P_0^{4k})$ 亦 $\neq 0$，但 $\rho_0(\beta H_2^{r-1}) = 0$，故此表示法是唯一的，而定理 4 证毕.

因在 ${}^nH^*(R_{n,m},\ I)$ 中，$\sum\limits_{r<n} H_1^r$ 成一子环，而对于 $X \in \sum\limits_{r<n} H_1^r, Z, Z' \in \sum\limits_{r<n} H_2^{r-1}$，显然有

$$\beta Z \cup \beta Z' = \beta(Z \cup \beta_2 Z'), \quad X \cup \beta Z = \beta(\rho_2 X \cup Z),$$

因 $\beta_2 = Sq^1$，又

$$\rho_2 P^{4i} = (W^{2i})^2 + W^{2i-1} W^{2i+1}, \tag{16}$$

故 β_2，ρ_2 都是已知. 因之定理 4 实际上已完全决定了 ${}^nH^*(R_{n,m},\ I)$ 的环构造.

在以后须用到下面的

预备定理 1 在 $r \leqslant n$ 时，在 $R_{n,m}$ 的标准剖分中可取一组 r 维整系数上链群的加法基 u_i^r，v_i^r，w_i^r 使在 $r < n$ 时

$$\delta u_i^r = 0, \quad \delta v_i^r = 2w_i^{r+1},\ \delta w_i^r = 0. \tag{17}$$

而 $u_i^r (r < n)$ 所定整系数类的全体和作 $(P^4)^{k_1}(P^8)^{k_2}\cdots(P^{4m'})^{k_{m'}}$ 形状而 k_i 满足 $(7)'$ 的类的全体一致.

证 命 u_i^r, v_i^r, w_i^r 为如本节开首时所说的 r 维 $(r \leqslant n)$ 整系数上链加法基. 在 $R_{n,m}$ 中任意定义一组上链间的上积, 以 $\cup$ 表之. 命 $p_k = (0, \cdots, 0, \underbrace{2, \cdots, 2}_{2k}) \in \Omega_0^{4k}$, $0 < k \leqslant \left[\frac{m}{2}\right] = m'$, 又命在 $r < n$ 时, $\bar{u}_i^r$ 为一切作 $\{p_{k_1}\} \cup \{p_{k_2}\} \cup \cdots \cup \{p_{k_s}\}$ 形状而 $0 < k_1 \leqslant k_2 \leqslant \cdots \leqslant k_s \leqslant \left[\frac{m}{2}\right]$, $4\sum\limits_{i=1}^{s} k_i = r$, 的上链的全体. 取 $\bar{v}_i^r = v_i^r$, $\bar{w}_i^r = w_i^r$, $r \leqslant n$, $\bar{u}_i^n = u_i^n$ 则显然 $\bar{u}_i^r$, $\bar{v}_i^r$, $\bar{w}_i^r$ 可用 u_i^r, v_i^r, w_i^r 的整系数线性和来表示, 且有

$$\delta \bar{u}_i^r = 0, \quad \delta \bar{v}_i^r = 2\bar{w}_i^{r+1}, \quad \delta \bar{w}_i^r = 0 \qquad (r < n).$$

反之, 由定理 4 知, 在 $r < n$ 时, 必有整数 a_{ij}, c_{ij} 存在使 $u_i^r \sim \sum\limits_j a_{ij}^r \bar{u}_j^r + \sum\limits_j c_{ij}^r \bar{w}_j^r$ 因之有整数 d_{ij}^{r-1} 使 $u_i^r = \sum a_{ij}^r\, \bar{u}_j^r + \sum\limits_j c_{ij}^r \bar{w}_j^r + \delta \sum\limits_j d_{ij}^{r-1} v_j^{r-1} = \sum\limits_j a_{ij}^r \bar{u}_j^r + \sum\limits_j (c_{ij}^r + 2d_{ij}^{r-1})\, \bar{w}_j^r$, 故 u_i^r, v_i^r, w_i^r 也可以用 $\bar{u}_i^r$, $\bar{v}_i^r$, $\bar{w}_i^r$ 的整系数线性和来表示. 这证明了 $\bar{u}_i^r$, $\bar{v}_i^r$, $\bar{w}_i^r$ 也可以作为一组 r 维 $(r < n)$ 整系数上链群的加法基. 易 $\bar{u}_i^r$, v_i^r, $\bar{w}_i^r$ 的记号为 u_i^r, v_i^r, $w_i^r (r \leqslant n)$, 即得定理.

§2 自然映像 $h: R_{n_1,m_1} \times R_{n_2,m_2} \to R_{n,m}$

在本节中, 我们将讨论关于自然映像 $(n = n_1 + n_2, m = m_1 + m_2)$

$$h: R_{n_1,m_1} \times R_{n_2,m_2} \to R_{n,m} \tag{1}$$

的一些性质, 以为下节决定 $R_{n,m}$ 中 Понтрягин 平方的准备.

预备定理 2 若 m 为偶数时, m_1, m_2 都是偶数, m 为奇数时任意, 则在 $r \leqslant n_1$ 和 n_2 时,

$$h^*: H^r(R_{n,m}, I_0) \subset H^r(R_{n_1,m_1} \times R_{n_2,m_2}, I_0).$$

证 命 $\left[\dfrac{m}{2}\right] = m'$, $\left[\dfrac{m_i}{2}\right] = m_i'$, $i = 1, 2$, 根据假设恒有 $m' = m_1' + m_2'$. 设 $U \in H^r(R_{n,m}, I_0)$ 而 $U \neq 0$. 可设 $r < n$, 由 § 2 定理 3, $U = \varphi(P_0^{4k})$, 此处 $P_0^{4k} = \rho_0 \cdot P^{4k}$, $0 < k \leqslant m'$, 而 P^{4k} 是 $R_{n,m}$ 的整系数 P 类, φ 是以实数为系数, 上积为乘法的一个多项式, 其形式由 U 所完全确定. 由 [1] 中定理 6 的 (3) 式, 有

$$h^* P_0^{4k} = \sum_{k_1+k_2=k} P_{0(1)}^{4k_1} \otimes P_{0(2)}^{4k_2}, \tag{2}$$

此处 $P_{0(i)}^{4k}$, $0 < k \leqslant m_i'$, $i = 1, 2$, 各是 R_{n_i,m_i} 中的 P 类, 而 $P_{0(i)}^0 = 1$. 命 x_k, $0 < k \leqslant m'$, $x_k^{(i)}$, $0 < k \leqslant m_i'$, $i = 1, 2$, 各是以 k 为重量的不定量, 且其间有关系

$$x_k = \sum_{k_1+k_2=k} x_{k_1}^{(1)} x_{k_2}^{(2)}, \quad (x_0^{(1)} = 1, \quad x_0^{(2)} = 1). \tag{3}$$

由假定 $U=\varphi(P_0^{4k})\neq 0$, 故 $\varphi(x_k)$ 为每项重量都是 $\frac{r}{4}$ 的一个对 x_k 而言不恒等于 0 的实系数多项式. 照我们常用到的方法可证 $\varphi\left(\sum x_{k_1}^{(1)}x_{k_2}^{(2)}\right)$ 是一个对 $x_k^{(1)}, x_k^{(2)}$ 而言不恒等于 0 的实系数多项式. 因 $r\leqslant n_1,n_2$，故由 § $2(A_{n_i,m})$ 一切作 $P_{0(1)}^{4i_1}P_{0(1)}^{4i_2}\cdots P_{0(1)}^{4i_s}\otimes P_{0(2)}^{4j_1}P_{0(2)}^{4j_2}\cdots P_{0(2)}^{4j_t}$ 形状而 $0<i_1\leqslant\cdots\leqslant i_s\leqslant m_1'$, $0<j_1\leqslant\cdots\leqslant j_t\leqslant m_2'$, $4\sum\limits_{\lambda=1}^{s}i_\lambda+4\sum\limits_{\mu=1}^{t}j_\mu=r$ 的类在 $H^r(R_{n_1,m_1}\times R_{n_2,m_2},I_0)$ 中线性无关. 因之由 $\varphi(\sum x_{k_1}^{(1)}x_{k_2}^{(2)})$ 的不恒等于 0 得 $h^*U=h^*\varphi(P_0^{4k})=\varphi(h^*P_0^{4k})=\varphi(\sum P_{0(1)}^{4k_1}\otimes P_{0(2)}^{4k_2})\neq 0$, 即所欲证.

预备定理 2′ 设 m 为偶数时 m_1，m_2 都是偶数，m 为奇数时任意. 又设 $U=\varphi(P_0^{4k})\in H^r(R_{n,m},I_0)$, 其中 $r\leqslant n_1,n_2$，而多项式 φ 的系数都是整数. 若 φ(同类项已合并) 的系数不能都被 2^s 除尽，则 h^*U 依 (2) 式化为 $P_{0(i)}^{4k}$，$0<k\leqslant m_i'$, $i=1$, 2 的多项式时其系数也不能都被 2^s 除尽.

证 先设 $s=1$. 将 $\varphi(x_k)$ 中的系数法 2 约化后所得的多项式记为 $\varphi_2(x_k)$，则 $\varphi_2(x_k)$ 对 x_k，$0<k\leqslant m'$ 而言不恒等于 0. 由 (3) 式和 [3] 中的预备定理 7，可知 $\varphi_2(\sum x_{k_1}^{(1)}x_{k_2}^{(2)})$ 对 $x_k^{(i)}$, $0<k\leqslant m_i'$, $i=1,2$, 而言也不恒等于 0. 换言之，$\varphi(\sum x_{k_1}^{(1)}x_{k_2}^{(2)})$ 对 $x_k^{(i)}$ 而言其系数不能全被 2 除尽. 由于 (2)，这证明了 $s=1$ 时定理成立.

一般说来，若 φ 的系数不能都被 2^s 除尽，可设 $U=\varphi(P_0^{4k})=2^t\varphi'(P_0^{4k})$, 此处 $t<s$，φ' 是一个整系数多项式，其系数不能都被 2 除尽. 由上述，$\varphi'(\sum P_{0(1)}^{4k_1}\otimes P_{0(2)}^{4k_2})$ 视为 $P_{0(i)}^{4k}$, $0<i\leqslant m_i'$, $i=1$, 2 的多项式时，其系数也不能都被 2 除尽. 因之 $h^*U=2^t\varphi'(\sum P_{0(1)}^{4k_1}\otimes P_{0(2)}^{4k_2})$ 视为 $P_{0(i)}^{4k}$ 的多项式时其系数不能都被 2^{t+1}，更不能都被 2^s 所除尽，即所欲证.

预备定理 3 设 m 为偶数时 m_1, m_2 都是偶数，m 为奇数时任意，则在 $r<n_1$ 和 n_2 时，

$$h^*:\quad H^r(R_{n,m},I_4)\subset H^r(R_{n_1,m_1}\times R_{n_2,m_2},I_4).$$

证 取 $R_{n,m}$，R_{n_k,m_k}，$k=1,2$ 的标准剖分 K，K_1，K_2 并命 K' 为 K_1，K_2 的积复合形 $K'=K_1\times K_2$. 据 § 2 的预备定理 1，在 $\leqslant n$ 的维数 r 中，我们可在 K 中取一组上链群的加法基 u_i^r, v_i^r, w_i^r 使

$$\delta u_i^r=0,\quad \delta v_i^r=2w_i^{r+1}\quad \delta w_i^r=0\quad (r<n) \tag{4}$$

且 $r<n$ 时，一切 u_i^r 所定上类的全体与由一切作 $(P^4)^{k_1}\cdots(P^{4m'})^{k_{m'}}$ 形状而 $k_i\geqslant 0$, $4\sum\limits_{i=1}^{m'}ik_i=r$ 的类的全体一致. 同样在 $\leqslant n_k$ 的维数 r_k 中 $(k=1,2)$，我们也可在 K_k 中各取一组类似的上链群的加法基：$u_{i(k)}^r$, $v_{i(k)}^r$, $w_{i(k)}^r$. 于是在 $\leqslant n_1$ 和 n_2 的维数 r 中，我们可在 K' 中取一组上链群的加法基 $u_j'^r$, $v_j'^r$, w_j' 使 $u_j'^r$ 的全体即一切作

$u^{r_1}_{i_1,(1)} \otimes u^{r_2}_{i_2,(2)}$ 形状而 $r_1+r_2=r$ 的上链全体，w'^r_j 的全体即一切作 $(-1)^{r_1} u^{r_1}_{i_1(1)} \otimes w^{r_2}_{i_2,(2)}$，$w^{r_1}_{i_1,(1)} \otimes u^{r_2}_{i_2,(2)}$，$w^{r_1}_{i_1,(1)} \otimes w^{r_2}_{i_2,(2)}$，或 $w^{r_1}_{i_1,(1)} \otimes v^{r_2}_{i_2,(2)} + (-1)^{r_1-1} v^{r_1-1}_{i_1,(1)} \otimes w^{r_2+1}_{i_2,(2)}$ 形状而 $r_1+r_2=r$ 的上链全体，又 v'^r_j 的全体则是一切作 $u^{r_1}_{i_1,(1)} \otimes v^{r_2}_{i_2,(2)}$，$v^{r_1}_{i_1,(1)} \otimes u^{r_2}_{i_2,(2)}$，$v^{r_1}_{i_1,(1)} \otimes w^{r_2}_{i_2,(2)}$，或 $v^{r_1}_{i_1(1)} \otimes v^{r_2}_{i_2,(2)}$ 形状而$r_1+r_2=r$ 的上链全体. 此时 v'^r_j 的个数与 w'^{r+1}_j 的个数相同，且有下述关系：

$$\delta u'^r_j = 0, \quad \delta v'^r_j = 2w'^{r+1}, \quad \delta w'^r_i = 0 \quad (r < n_1 \text{ 和 } n_2). \tag{5}$$

又从 $u^r_{i(k)}$ 的选择知 u'^r_j 的类的全体就是一切作 $(P^4_{(1)})^{i_1} \cdots (P^{4m'_1}_{(1)})^{i_{m'_1}} \otimes (P^4_{(2)})^{j_1} \cdots (P^{4m'_2}_{(2)})^{j_{m'_2}}$ 形状而 $i_\alpha, j_\beta \geqslant 0$, $4\sum\limits_{\alpha=1}^{m'_1} \alpha i_\alpha + 4\sum\limits_{\beta=1}^{m_2} \beta j_\beta = r$ 的类的全体.

命 $h^\#: C^*(K,I) \to C^*(K',I)$ 是从属于 h 的一个链变换，也就是引起上同调群变换 h^* 的一个链变换 (这种链变换是一定存在的). 于是在 $r < n' = \min(n_1, n_2)$ 时，$h^\# u^r_i$ 必作下面的形状 (α，γ 都是整数)：

$$h^\# u^r_i = \sum_j a^r_{ij} u'^r_j + \sum_j \gamma^r_{ij} w'^r_j. \tag{6}$$

设 u^r_i 和 u'^r_j 所定的实系数类各为 U^r_i 和 U'^r_j，则由 (6) 式得 $h^* U^r_i = \sum\limits_j \alpha^r_{ij} U'^r_j$, 故对于任意整数 a^r_i，有 $h^* \sum\limits_i a^r_i U^r_i = \sum\limits_{i,j} a^r_i \alpha^r_{ij} U'^r_j$. 由于 u^r_i, u'^r_j 的选择，根据预备定理 2′ 就得到下面关于 α^r_{ij} 的性质：

在 $r < n'$ 时，若 a^r_i 不能都被 2^s 除尽，$s \geqslant 1$，则

$$\sum_i a^r_i \alpha^r_{ij} \text{ 也不能都被 } 2^s \text{ 除尽.} \tag{7}$$

其次，$h^\# w^r_i (r \leqslant n')$ 显然必作下面的形状 (c^r_{ij} 都是整数)：

$$h^\# w^r_i = \sum_j c^r_{ij} w'^r_j. \tag{8}$$

对于任意整数 c^r_i，就有 $h^\# \sum\limits_i c^r_i w^r_i = \sum\limits_{i,j} c^r_i c^r_{ij} w'^r_j$. 由 [3] 中的预备定理 7，

$$h^*: H^r(R_{n,m}, I_2) \subset H^r(R_{n_1,m_1} \times R_{n_2,m_2}, I_2), \quad r \leqslant n'.$$

故可得到下面关于 c^r_{ij} 的性质：

$$\text{在 } r \leqslant n' \text{ 时, 若 } c^r_i \text{不全为偶数, 则} \sum c_i c^r_{ij} \text{ 也不能全为偶数.} \tag{9}$$

最后我们易知应有 (d，e 都是整数)

$$h^\# v^r_i = \sum_j d^r_{ij} u'^r_j + \sum_j c^{r+1}_{ij} v'^r_j + \sum_j e^r_{ij} w'^r_j \quad (r < n'). \tag{10}$$

今设在 K 中有一 r 维的法 4 上闭链 z_4^r，$r<n$，则 z_4^r 可书为 $z_4^r=\rho_4 z^r$ 此处 z^r 为下列形状的一个整系数上链 (a_i^r，b_i^r，c_i^r 都是整数)：

$$z^r=\sum_i a_i^r u_i^r+2\sum_i b_i^r v_i^r+\sum_i c_i^r w_i^r. \tag{11}$$

若更有 $r<n'$ 则由 (6)，(8)，(10)，有

$$h^{\#}z^r=\sum_j a_j'^r u_j'^r+\sum_j b_j'^r v_j'^r+\sum_j c_j'^r w_j'^r, \tag{12}$$

此处

$$\left.\begin{aligned} a_j'^r&=\sum_i a_i^r\alpha_{ij}^r+2\sum_i b_i^r d_{ij}^r,\\ b_j'^r&=2\sum_i b_i^r c_{ij}^{r+1},\\ c_j'^r&=\sum_i a_i^r\gamma_{ij}^r+2\sum_i b_i^r e_{ij}^r+\sum_i c_i^r c_{ij}^r.\end{aligned}\right\} \tag{13}$$

今设 $z_4^r=\rho_4\quad z^r\not\sim 0$ 而 $r<n'$ 如前. 我们的目的，在证明 $h^{\#}z_4^r\not\sim 0$. 为此，试就以下三种情形分论之：

第一，设 b_i 不全为偶数. 此时，因 $r<n'$，由 (9)，$\sum b_i^r c_{ij}^{r+1}$ 也不全为偶数. 由 (13) 的第二式，$b_j'^r$ 不全为 4 的倍数，因之由 (5) 和 (12)，$\rho_4\, h^{\#}z^r=h^{\#}z_4^r\not\sim 0$.

第二，设 b_i^r 全为偶数，但 a_i^r 不全为 4 的倍数. 此时因 $r<n'$，由 (7)，知 $\sum\limits_i a_i^r\alpha_{ij}^r$ 也不能全为 4 的倍数. 由 (13) 的第一式，知 $a_j'^r$ 不能全为 4 的倍数，因之由 (5) 和 (12)，$\rho_4\, h^{\#}z^r=h^{\#}z_4^r\not\sim 0$.

第三，设 b_i^r 全为偶数，且 a_i^r 全为 4 的倍数. 此时 $z_4^r=\rho_4\ z^r=\rho_4\ \sum\limits_j c_i^r w_i^r$. 因由假设 $z_4^r\not\sim 0$ 而由 (5) $2w_i^r\sim 0$，故 c_i^r 不能全为偶数. 由 (9)，$\sum\limits_i c_i^r c_{ij}^r$ 不能全为偶数，而由 (13) 的第三式，$c_j'^r$ 也不能全为偶数. 于是由 (5) 和 (12)，知 $\rho_4 h^{\#}z^r=h^{\#}z_4^r\not\sim 0$.

至此，定理已完全证毕.

§4 $R_{n,m}$ 中的ПОНТРЯГИН平方

本节中将决定 $R_{n,m}$ 中的Понтрягин平方

$$\boldsymbol{p}:H^r(R_{n,m},I_2)\to H^{2r}(R_{n,m},I_4),\quad r\geqslant 0.$$

因为 $H^*(R_{n,m},I_2)$ 系由 $R_{n,m}$ 中的 W 类所产生，故由§ 1 的 2° 及定理 2，我们只须决定 $\boldsymbol{p}(W^i)$，$0<i\leqslant m$ 即可. 对此我们有下述

定理 5　若 W^i 是 $R_{n,m}$ 中的法 $2W$ 类，而 P^{4k} 是 $R_{n,m}$ 中整系数 P 类，则

$$\boldsymbol{p}(W^{2i+1})=\beta_4 Sq^{2i}W^{2i+1}+\theta_2(W^1 Sq^{2i}W^{2i+1}),\tag{I$_{m,i}$}$$

$$\boldsymbol{p}(W^{2i})=\rho_4 P^{4i}+\beta_4(W^{2i-1}\,W^{2i})+\theta_2\left(W^1 Sq^{2i-1}W^{2i}+\sum_{j=0}^{i-1}W^{2j}W^{4i-2j}\right),\tag{II$_{m,i}$}$$

其中 θ_2，β_4，ρ_4 的意义见 § 1.

若命 $R^{\wedge}_{n,m}$ 为 $R_{n,m}$ 的两叶覆叠空间，π 为其投影，应用 π^* 于 $(\mathrm{II})_{m,1}$，得 $\boldsymbol{p}(\pi^*W^2)=\rho_4\pi^*P^4+\theta_2\pi^*W^4$，此即最早Понгрягин所得 [12] 关于格拉斯曼流形中Понтрягин平方的公式.

我们将先列举一些证明中所要用到的公式：

$$Sq^r W^{r+1}=\sum_{i=0}^{r}W^iW^{2r+1-i},\tag{1}$$

$$\beta_2 W^{2i}=Sq^1W^{2i}=W^1W^{2i}+W^{2i+1}\tag{2}$$

$$\beta_2 W^{2i+1}=Sq^1W^{2i+1}=W^1W^{2i+1},\tag{3}$$

$$(W^{2i})^2=\rho_2P^{4i}+\beta_2(W^{2i-1}W^{2i}),\tag{4}$$

$$(W^{2i+1})^2=\beta_2(Sq^{2i}W^{2i+1})\,.\tag{5}$$

对于自然映像 $(n=n_1+n_2, m=m_1+m_2)$

$$h:R_{n_1,m_1}\times R_{n_2,m_2}\to R_{n,m}$$

有

$$h^*W^i=\sum_{i_1+i_2=i}W^{i_1}_{(1)}\otimes W^{i_2}_{(2)},\tag{6}$$

$$\begin{aligned}&h^*[P^{4k}+\beta(W^{2k-1}W^{2k})]\\&=\sum_{k_1+k_2=k}[P^{4k_1}_{(1)}+\beta(W^{2k_1-1}_{(1)}W^{2k_1}_{(1)})]\otimes[P^{4k_2}_{(2)}+\beta(W^{2k_2-1}_{(2)}W^{2k_2}_{(2)})]\\&\quad+\sum_{k_1+k_2=k-1}\beta(Sq^{2k_1}W^{2k_1+1}_{(1)})\otimes\beta(Sq^{2k_2}W^{2k_2+1}_{(2)}).\end{aligned}\tag{6$'$}$$

以上各式中，(1)，(2)，(3) 是 Sq^iW^j 公式的特例 (见 [3]). 在 [1] 中 (18)，(18)′ 所引入的 U^{4i-1}，V^{4i+1}，由上 (1) 式即为 $U^{4i-1}=Sq^{2i-1}W^{2i}+W^{2i-1}W^{2i}$，$V^{4i+1}=Sq^{2i}W^{2i+1}$，故 $\beta U^{4i-1}=\beta(W^{2i-1}W^{2i})$，因而该文的 (19)，(19)′，(27) 三式可改写为上面的 (4)，(5)，(6)′ 三式，其中 (4)，(5) 亦不难直接证明之.

此外，若 U，V 是任两空间中的法 2 上类则易见应有

$$\theta_2 U \otimes \beta_4 V = \beta_4 U \otimes \theta_2 V = \theta_2(\beta_2 U \otimes V) = \theta_2(U \otimes \beta_2 V), \tag{7}$$

$$\beta_4 U \otimes \beta_4 V = \beta_4(U \otimes \beta_2 V) = \beta_4(\beta_2 U \otimes V), \tag{8}$$

$$\theta_2 U \otimes \theta_2 V = 0. \tag{9}$$

由 (4) 及 (5) 式，可知

$$\rho_2[\boldsymbol{p}(W^{2i+1}) - \beta_4(Sq^{2i}W^{2i+1})] = (W^{2i+1})^2 - \beta_2(Sq^{2i}W^{2i+1}) = 0, \tag{10}$$

$$\rho_2[\boldsymbol{p}(W^{2i}) - \rho_4 P^{4i} - \beta_4(W^{2i-1}W^{2i})] = (W^{2i})^2 - \rho_2 P^{4i} - \beta_2(W^{2i-1}W^{2i}) = 0. \tag{10$'$}$$

因对于任意空间 E，有一正确序列

$$\cdots \to H^r(E, I_2) \xrightarrow{\theta_2} H^r(E, I_4) \xrightarrow{\rho_2} H^r(E, I_2) \xrightarrow{\beta_2} H^{r+1}(E, I_2) \to \cdots \tag{11}$$

故从 (10)，(10)$'$ 可知必有 $X^{2k} \in H^{2k}(R_{n,m}, I_2)$，$0 < k \leqslant m$，存在使

$$\boldsymbol{p}(W^{2i+1}) - \beta_4 Sq^{2i}W^{2i+1} = \theta_2 X^{4i+2}, \tag{12}$$

$$\boldsymbol{p}(W^{2i}) - \rho_4 P^{4i} - \beta_4(W^{2i-1}W^{2i}) = \theta_2 X^{4i}. \tag{12$'$}$$

定理 5 的意义，即在于给予了 X^{2k} 以一个明确的形式：

$$X^{4i+2} = W^1 Sq^{2i} W^{2i+1}, \tag{13}$$

$$X^{4i} = W^1 Sq^{2i-1} W^{2i} + \sum_{j=0}^{i-1} W^{2j} W^{4i-2j}. \tag{13$'$}$$

定理 5 的证明. 我们将对 m 实行归纳法以证之. 我们并将在 n，n_1，n_2 等都充分大的假定之下进行推理，此种限制可用以前常用之法以除去之，无关紧要.

首先，在 $m=1$ 时，(I)，(II) 只有一个公式是有意义的，即 $(\mathrm{I})_{1,0}$. 因此 $i=0$ 时，而 $\theta_2(W^1)^2 = \theta_2\,\beta_2\,W^1 = 0$，故此式可改写为

$$\boldsymbol{p}(W^1) = \beta_4 W^1. \tag{I$_{1,0}$}$$

在 $m=1$ 时，$R_{n,1}$ 是一个 n 维的投影空间，$\theta_2 H^{2k}(R_{n,1}, I_2) = 0$，故 (12) 的右端应为 0. 但此时与 $i=0$ 相当的 (12) 式即约化为 $(\mathrm{I})_{1,0}$，即所欲证.

其次，假定 $m=2$，此时只有两个公式须要证明，即 $(\mathrm{I})_{2,0}$，$(\mathrm{II})_{2,1}$. 此两式易见可简化为以下形状：

$$\boldsymbol{p}(W^1) = \beta_4 W^1, \tag{I$_{2,0}$}$$

$$\boldsymbol{p}(W^2) = \rho_4 P^4 + \theta_2[(W^1)^2 W^2]. \tag{(II)$_{2,1}$}$$

今证此两式如次. 因 $H^2(R_{n,2}, I_2)$ 以 $(W^1)^2$ 与 W^2 为加法基，$H^4(R_{n,2},I_2)$ 以 $(W^1)^4$, $(W^1)^2W^2$，$(W^2)^2$ 为加法基，而 $\theta_2(W^1)^2 = \theta_2\beta_2 W^1 = 0$，$\theta_2(W^1)^4 = \theta_2\beta_2(W^1)^3 = 0$，$\beta_4\ (W^1W^2) = \beta_4\ \beta_2 W^2 = 0$，故由 (12) 与 (12)′，必有法 2 整数 a，c_1，c_2 存在使

$$\boldsymbol{p}(W^1) = \beta_4 W^1 + a\theta_2 W^2, \tag{14}$$

$$\boldsymbol{p}(W^2) = \rho_4 P^4 + \theta_2[c_1(W^1)^2 W^2 + c_2(W^2)^2]. \tag{15}$$

今考察自然映像 $(n = n_1 + n_2)$

$$h: \quad R_{n_1,1} \times R_{n_2,1} \to R_{n,2}.$$

命 $R_{n_j,1}(j=1,2)$ 中的 W 类为 $W^1_{(j)}$，则

$$h^*W^1 = W^1_{(1)} \otimes 1 + 1 \otimes W^1_{(2)}, \quad h^*W^2 = W^1_{(1)} \otimes W^1_{(2)}, \tag{16}$$

$$\begin{aligned} h^*\boldsymbol{p}(W^1) &= \boldsymbol{p}(h^*W^1) = \boldsymbol{p}(W^1_{(1)} \otimes 1 + 1 \otimes W^1_{(2)}) = \boldsymbol{p}(W^1_{(1)} \otimes 1) + \boldsymbol{p}(1 \otimes W^1_{(2)}) \\ &= \boldsymbol{p}(W^1_{(1)}) \otimes 1 + 1 \otimes \boldsymbol{p}(W^1_{(2)}). \end{aligned}$$

由 $(\mathrm{I})_{1,0}$，即得

$$h^*\boldsymbol{p}(W^1) = \beta_4 W^1_{(1)} \otimes 1 + 1 \otimes \beta_4 W^1_{(2)}.$$

又

$$\begin{aligned} &h^*\beta_4 W^1 = \beta_4 h^* W^1 = \beta_4(W^1_{(1)} \otimes 1 + 1 \otimes W^1_{(2)}) = \beta_4 W^1_{(1)} \otimes 1 + 1 \otimes \beta_4 W^1_{(2)}, \\ &h^*\theta_2 W^2 = \theta_2 h^* W^2 = \theta_2(W^1_{(1)} \otimes W^1_{(2)}). \end{aligned}$$

因从 (14) 须有 $h^*\boldsymbol{p}(W^1) = h^*\beta_4 W^1 + a h^*\theta_2\ W^2$，故比较上三式得

$$\theta_2(a\, W^1_{(1)} \otimes W^1_{(2)}) = 0.$$

由正确序列 (11)，知必有一 $Y \in H^1(R_{n_1,1} \times R_{n_2,1},\ I_2)$, 使 $\beta_2 Y = aW^1_{(1)} \otimes W^1_{(2)}$. 但 $H^1(R_{n_1,1} \times R_{n_2,1}, I_2)$ 有两个母素 $W^1_{(1)} \otimes 1$, $1 \otimes W^1_{(2)}$ 而 $\beta_2(W^1_{(1)} \otimes 1) = (W^1_{(1)})^2 \otimes 1$, $\beta_2(1 \otimes W^1_{(2)}) = 1 \otimes (W^1_{(2)})^2$. 故欲上式成立必须 $a = 0$，$Y = 0$. 此时 (14) 式即成为 $(\mathrm{I})_{2,0}$.

同样由 $(\mathrm{I})_{1,0}$，(6)′ 等式得

$$h^*\boldsymbol{p}(W^2) = \beta_4 W^1_{(1)} \otimes \beta_4 W^1_{(2)} + \theta_2[(W^1_{(1)})^3 \otimes W^1_{(2)}] + \theta_2[W^1_{(1)} \otimes (W^1_{(2)})^3],$$

$$h^*\rho_4P^4=\beta_4W^1_{(1)}\otimes\beta_4W^1_{(2)},$$
$$h^*\theta_2[(W^1)^2W^2]=\theta_2[(W^1_{(1)})^3\otimes W^1_{(2)}+W^1_{(1)}\otimes(W^1_{(2)})^3],$$
$$h^*\theta_2(W^2)^2=0.$$

故以 h^* 应用于 (15) 的两边得

$$c_1\theta_2[(W^1_{(1)})^3\otimes W^1_{(2)}+W^1_{(1)}\otimes(W^1_{(2)})^3]=\theta_2\ [(W^1_{(1)})^3\otimes W^1_{(2)}+W^1_{(1)}\otimes(W^1_{(2)})^3].$$

由正确序列 (11) 知必有一 $Y\in H^3(R_{n_1,1}\times R_{n_2,1},I_2)$ 使

$$\beta_2Y=(c_1-1)\ [(W^1_{(1)})^3\otimes W^1_{(2)}+W^1_{(1)}\otimes(W^1_{(2)})^3]. \tag{17}$$

但 $H^3(R_{n_1,1}\times R_{n_2,1},I_2)$ 以 $(W^1_{(1)})^3\otimes 1,(W^1_{(1)})^2\otimes W^1_{(2)}W^1_{(1)}\otimes(W^1_{(2)})^2,1\otimes(W^1_{(2)})^3$ 为母素，而

$$\beta_2[(W^1_{(1)})^3\otimes 1]=(W^1_{(1)})^4\otimes 1,\beta_2[1\otimes(W^2_{(2)})^3]=1\otimes(W^1_{(2)})^4,$$
$$\beta_2[(W^1_{(1)})^2\otimes W^1_{(2)}]=\beta_2[W^1_{(1)}\otimes(W^1_{(2)})^2]=(W^1_{(1)})^2\otimes(W^1_{(2)})^2.$$

故欲 (17) 成立之唯一可能性为 $Y=0$，且 (法 2)

$$c_1=1. \tag{18}$$

今暂设 n 是一个偶数：$n=2n'$. 对于自然映像

$$k:C_{n',1}\to R_{n,2}.$$

据［1］定理 7 有

$$k^*P^4=(C^2)^2,\quad k^*W^1=0,\quad k^*W^2=\rho_2C^2,$$

其中 C^2 是 $C_{n',1}$ 的整系数 C 类，故应用 k^* 于 (15) 两边，得

$$c_2\theta_2(\rho_2C^2)^2=2c_2\rho_4(C^2)^2=0.$$

因之有 (法 2)

$$c_2=0. \tag{18$'$}$$

由 (15)，(18)，(18)$'$ 即得 $(\mathrm{II})_{2,1}$，其中 n 为偶数之限制易于除去之.

至此，在 $m=1,2$ 时，公式 (I)，(II) 已完全证明. 今设在 $l<m$ 时 $(\mathrm{I})_{l,i}$，$(\mathrm{II})_{l,i}$ 已成立而来证明 $(\mathrm{I})_{m,i}$ 与 $(\mathrm{II})_{m,i}$. 为此，试考察自然映像

$$h:\quad R_{n_1,m-2}\ \times\ R_{n_2,2}\ \to R_{n,m}.$$

在 $R_{n_1,m-2}$，$R_{n_2,2}$ 中的 W 类，P 类等将分别以 $W^i_{(1)}$，$P^{4i}_{(1)}$，与 $W^i_{(2)}$，$P^{4i}_{(2)}$ 等记之. 应用 (1)—(9) 各式，H. Cartan 关于 Sq^i 公式，§ 1 中诸公式，以及已证明的 $(\mathrm{I})_{2,0}$，$(\mathrm{II})_{2,1}$ 诸式，可得

$$h^*\boldsymbol{p}(W^{2i})=\boldsymbol{p}(W^{2i}_{(1)})\otimes 1+\boldsymbol{p}(W^{2i-2}_{(1)})\otimes\rho_4P^4_{(2)}+\beta_4Sq^{2i-2}W^{2i-1}_{(1)}\otimes\beta_4W^1_{(2)}$$

$$
\begin{aligned}
&+\theta_2[\{(W^1_{(1)})^2Sq^{2i-2}W^{2i-1}_{(1)}+W^{2i-1}_{(1)}W^{2i}_{(1)}\}\otimes W^1_{(2)}]\\
&+\theta_2[Sq^{2i-2}W^{2i-1}_{(1)}\otimes(W^1_{(2)})^3]+\theta_2[\{(W^1_{(1)}W^{2i-2}_{(1)})^2+W^{2i-2}_{(1)}W^{2i}_{(1)}\}\otimes W^2_{(2)}]\\
&+\theta_2[Sq^{2i-3}W^{2i-2}_{(1)}\otimes W^1_{(2)}(W^2_{(2)})^2]+\theta_2[(W^{2i-2})^2\otimes(W^1_{(2)})^2W_{(2)}{}^2],
\end{aligned}
$$

$$
\begin{aligned}
h^*[\rho_4P^{4i}+\beta_4(W^{2i-1}W^{2i})]=&[\rho_4P^{4i}_{(1)}+\beta_4(W^{2i-1}_{(1)}W^{2i}_{(1)})]\otimes 1+[\rho_4P^{4i-4}_{(1)}\\
&+\beta_4(W^{2i-3}_{(1)}W^{2i-2}_{(1)})]\otimes\rho_4P^4_{(2)}+\beta_4Sq^{2i-2}W^{2i-1}_{(1)}\otimes\beta_4W^1_{(2)},
\end{aligned}
$$

$$
\begin{aligned}
h^*\theta_2(W^1Sq^{2i-1}W^{2i})=&\theta_2W^1_{(1)}Sq^{2i-1}W^{2i}_{(1)}\otimes 1+\theta_2(W^1_{(1)}Sq^{2i-3}W^{2i-2}_{(1)})\otimes\rho_4P^4_{(2)}\\
&+\theta_2[\{Sq^{2i-1}W^{2i}_{(1)}+(W^1_{(1)})^2Sq^{2i-2}W^{2i-1}_{(1)}\}\otimes W^1_{(2)}]\\
&+\theta_2[Sq^{2i-2}W^{2i-1}_{(1)}\otimes(W^1_{(2)})^3]+\theta_2[(W^1_{(1)}W^{2i-2}_{(1)})^2\otimes W^2_{(2)}]\\
&+\theta_2[Sq^{2i-3}W^{2i-2}_{(1)}\otimes W^1_{(2)}(W^2_{(2)})^2]\\
&+\theta_2[(W^{2i-2}_{(1)})^2\otimes(W^1_{(2)})^2W^2_{(2)}],
\end{aligned}
$$

$$
\begin{aligned}
h^*\theta_2\sum_{j=0}^{i-1}W^{2j}W^{4i-2j}=&\theta_2\sum_{j=0}^{i-1}W^{2j}_{(1)}W^{4i-2j}_{(1)}\otimes 1+\theta_2\sum_{j=0}^{i-2}W^{2j}_{(1)}W^{4i-2j-4}_{(1)}\otimes\rho_4P^4_{(2)}\\
&+\theta_2[(Sq^{2i-1}W^{2i}_{(1)}+W^{2i-1}_{(1)}W^{2i}_{(1)})\otimes W^1_{(2)}]\\
&+\theta_2[W^{2i-2}_{(1)}W^{2i}_{(1)}\otimes W^2_{(2)}].
\end{aligned}
$$

由归纳假设 $(\mathrm{I})_{m-2,i}$ 与 $(\mathrm{II})_{m-2,i}$，即得

$$
h^*\boldsymbol{p}(W^{2i})=h^*[\rho_4P^{4i}+\beta_4(W^{2i-1}W^{2i})]+h^*\theta_2(W^1Sq^{2i-1}W^{2i}+\sum_{j=0}^{i-1}W^{2j}W^{4i-2j}).
$$

但据 § 3 预备定理 3，在 n_1，n_2 充分大时，h^* 为一对一的变换，故从上式立得 $(\mathrm{II})_{m,i}$.

$(\mathrm{I})_{m,i}$ 亦可以同法证之.

§5 微分流形上的ПОНТРЯГИН示性类

设 M 是一个 m 维可微分闭流形，$\mathfrak{D}$ 是 M 的任意一个微分构造. 并命 $\mathfrak{T}(\mathfrak{D})$ 为由 $\mathfrak{D}$ 所定的切丛，假定由映像 $\lambda: M\to R_{n,m}$ 所引起 (n 充分大). 对 $\mathfrak{T}(\mathfrak{D})$ 而言的 W 和 P 示性类将简记为 $W^i(\mathfrak{D})$ 和 $P^{4i}(\mathfrak{D})$，余类推. 我们知道 $W^i(\mathfrak{D})$ 实际上与所选择的 $\mathfrak{D}$ 无关，而是 M 的拓扑不变量 [13, 14]. 由 § 4 的 (12)′，易得

$$
\rho_4P^{4i}(\mathfrak{D})=\boldsymbol{P}(W^{2i}(\mathfrak{D}))+\beta_4(W^{2i-1}(\mathfrak{D})W^{2i}(\mathfrak{D}))+\theta_2X^{4i}(\mathfrak{D}),\tag{1}
$$

此处 $X^{4i}(\mathfrak{D})=\lambda^{*}X^{4i}$，$X^{4i}\in H^{4i}(R_{n,m},I_2)$. 因 $H^{*}(R_{n,m},I_2)$ 由 W^i 所产生，故 $X^{4i}(\mathfrak{D})$ 可表为 $W^i(\mathfrak{D})$ 的多项式 (以上积为乘法)，又因 θ_2，β_4，$\boldsymbol{p}$ 都是同伦不变的运算，因之从 (1) 式可得下定理：

定理 6 若 M 是一个可微分闭流形，$\mathfrak{D}$ 是 M 的任一微分构造，则对 $\mathfrak{D}$ 而言的法 4 Понтрягин 示性类实际上与 $\mathfrak{D}$ 的选择无关，而是 M 的拓扑不变量.

由 § 4 的 (13)′，可取

$$X^{4i}(\mathfrak{D})=W^1(\mathfrak{D})Sq^{2i-1}W^{2i}(\mathfrak{D})+\sum_{j=0}^{i-1}W^{2j}(\mathfrak{D})W^{4i-2j}(\mathfrak{D}). \tag{2}$$

因 $W^i(\mathfrak{D})$ 可从 M 的同调构造 (包括 M 的法 2 上同调环与 Steenrod 平方运算) 所明确地决定出来 [14]，因之由 (1)，(2) 两式，$P_4^{4i}(\mathfrak{D})$ 也可以用 M 的同调构造 (包括 M 的法 2 上同调环、Steenrod 平方、Понтрягин 平方与运算 β_4，θ_2) 所明确地决定.

在 [2] 中，对于任一奇质数 p，曾从 $P_p^{4i}(\mathfrak{D})=\rho_pP^{4i}(\mathfrak{D})$ 用上积作出一组法 p 类 $Q_p^{4j}(\mathfrak{D})$，并在 M 是可定向时，证明 $Q_p^{4i}(\mathfrak{D})$ 为 M 的拓扑不变量. 此定理在 M 不可定向时亦成立，换言之，有

定理 7 若 M 是一个可微分闭流形，$\mathfrak{D}$ 是 M 的任一微分构造，则对 $\mathfrak{D}$ 而言的法 p 类 $Q_p^{4j}(\mathfrak{D})$ 与 $\mathfrak{D}$ 的选择无关，而是 M 的拓扑不变量，此处 p 为任一奇质数.

证 在 M 可定向时，此即 [2] 中的主要定理. 今设 M 不可定向，则必有一个可定向流形 $\tilde{M}$ 为 M 的二叶覆叠空间. 命其投影为 π，则 π 在 $\tilde{M}$ 上从 $\mathfrak{D}$ 引起一微分构造 $\tilde{\mathfrak{D}}$ 使 $\mathfrak{T}(\tilde{\mathfrak{D}})\sim\pi^{*}\mathfrak{T}(\mathfrak{D})$. 设 $\mathfrak{D}'$ 是 M 的另一微分构造，从 $\mathfrak{D}'$ 得 $\tilde{\mathfrak{D}}'$ 等. 因 $\tilde{M}$ 可定向，故有 $Q_p^{4j}(\tilde{\mathfrak{D}}')=Q_p^{4j}(\tilde{\mathfrak{D}})$，或即

$$\pi^{*}Q_p^{4j}(\mathfrak{D}')=\pi^{*}Q_p^{4j}(\mathfrak{D}). \tag{3}$$

因 p 为奇质数，而 $\tilde{M}$ 覆叠 M 的叶数是 2，与 p 互质，故应有

$$\pi^{*}:H^{*}(M,I_p)\subset H^{*}(\tilde{M},I_p). \tag{4}$$

由 (3)，(4)，即得 $Q_p^{4j}(\mathfrak{D}')=Q_p^{4j}(\mathfrak{D})$，换言之，$Q_p^{4j}(\mathfrak{D})$ 与 $\mathfrak{D}$ 的选择无关，即所欲证.

在 $p=3$ 时，从 [2] 可知 $Q_3^{4j}(\mathfrak{D})=P_3^{4j}(\mathfrak{D})=\rho_3P^{4j}(\mathfrak{D})$，因之结合上两定理，可将定理 6 加强之如下

定理 8 若 M 是一个可微分闭流形，$\mathfrak{D}$ 是 M 的任一微分构造，则对 $\mathfrak{D}$ 而言的法 12 P 示性类 $P_{12}^{4i}(\mathfrak{D})=\rho_{12}P^{4i}(\mathfrak{D})$ 实际上与 $\mathfrak{D}$ 的选择无关，而为 M 的拓扑不变量.

参考文献

[1] 吴文俊. 论Понтрягин示性类，I. 数学学报，1953, **3**: 291-315.

[2] ——. 论Понтрягин示性类，II. 同上 1954, **4**: 171-199.

[3] ——. 格拉斯曼流形中的平方运算. 同上 1953, **2**: 205-230.

[4] Рохлин В А. Новыерезультаты теори игетырехмерных многоодразий. ДАН, 1952, (84): 221-224.

[5] Thom R. Variétés différentiables cobordantes. *C. R. Paris*, 1953, **236**: 1733-1735.

[6] Steenrod N E. Products of cocycles and extensions of mappings. *Annals of Mathematics*, 1947, **48**: 290-320.

[7] ——. Reduced powers of cohomology classes. Cours de Collège de France, 1951.

[8] ——. Reduced powers of cohomology classes. *Annals of Math.*, 1952, **56**: 47-67.

[9] Cartan H. Une théorie axiomatique des carrés de Steenrod. *C. R. Paris*, 1950, **230**: 425-427.

[10] Pontrjagin L. Mappings of the 3-dim, sphere into an n-dim. complex. ЛАН, 1942, **34**: 35-37.

[11] ——. On some topologic invariants of Riemannian manifolds. *Ibid.*, 1944, **43**: 91-94.

[12] ——. Classification of some skew products. *Ibid.*, 1945, **47**: 322-325.

[13] Thom R. Espaces fibrés en sphéres et carrés de Steenrod. *Ann. Ec. Norm. Sup.* 1952, **693**: 110-182.

[14] Wu Wen-tsün. Classes caractéristiques et *i*-carrés d'une variété. *C. R. Paris*, 1950, **230**: 508-511.

On Pontrjagin Classes III

Abstract We determine completely the Pontrjagin squares in the Grassmannian manifold. As a consequence, we prove that the Pontrjagin classes, reduced mod 4, and hence also mod 12, when combined with the preceding one in this series of papers, of a closed differentiable manifold are topological invariants of that manifold.

18. 论 Понтрягин 示性类 IV*

前　言

在前一文中 [2], 我们曾应用 Steenrod 幂以证明一个可微分闭流形上法 3 Понтрягин 示性类的拓扑不变性. 本文的目的则在提供一个不同的方法, 不假助于 Steenrod 幂以证明同一事实. 同样的方法, 亦可用于 Stiefel-Whitney 示性类, 而由此获得这些类的拓扑不变性的一个不同证明, 而在证明中避免用到 Steenrod 平方.

本文所用的方法约略可述之如次. 设 M 为一 m 维可微分闭流形, $M^p = \underbrace{M \times \cdots \times M}_{p}$, p 为一奇质数. 由 M 的一个微分构造 $\mathfrak{D}$ 出发, 在 M^p 中可定义一个"对称"的 Riemann 计量 $d\tilde{s}^2$. 对 $d\tilde{s}^2$ 而言, 令 $\tilde{N}_\varepsilon$ 为 M^p 中与对角流形相距 ε(ε 充分小) 的点所成的空间. 在 M^p 中有一周期为 p 而无不变点的拓扑变换

$$t : (a_1, \cdots, a_p) \to (a_2, \cdots, a_p, a_1),$$

t 同时亦为 $\tilde{N}_\varepsilon$ 中的拓扑变换. 令 N_ε 为 $\tilde{N}_\varepsilon$ 在 t 之下的法空间 (modular space), 则 N_ε 可视为一个在 M 上的纤维丛的丛空间, 其纤维为某种特殊的透镜空间, 投影为 π. 根据 P. A. Smith 的一般理论 [3, 4], 在 N_ε 中可定义一组上同调类 $U^i \in H^i(N_\varepsilon, I_p)$. 于是在 N_ε 中有一关系式

$$\prod_{i=1}^{q} \sum_{k=0}^{m'} (-1)^k i^{m-2k}\, U^{2m-4k} \cup \pi^* P_p^{4k}(\mathfrak{T}) = 0 \tag{1}$$

存在, 其中 $q = (p-1)/2$, m' 为 $\leqslant m/2$ 的最大整数, $\mathfrak{T}$ 为由 M 的微分构造 $\mathfrak{D}$ 所定的切丛. 上式亦可书为

$$\sum_{k=0}^{m'q} U^{2qm-4k} \cup \pi^* R_p^{4k}(\mathfrak{T}) = 0,$$

此处

$$R_p^{4k}(\mathfrak{T}) = (-1)^k \sum_{k_1+\cdots+k_q=k} 1^{m-2k_1} \cdot 2^{m-2k_2} \cdots q^{m-2k_q} P_p^{4k_1}(\mathfrak{T}) P_p^{4k_2}(\mathfrak{T}) \cdots P_p^{4k_q}(\mathfrak{T});$$

---

* *Acta Math. Sinica*, 1955, 5: 37-63; *Amer. Math. Soc. Translations, Ser.*, 2(92): 93-121.

于是我们可以证明对于 M 的任意两个微分构造所定的切丛 $\mathfrak{T}, \mathfrak{T}'$, 有

$$R_p^{4k}(\mathfrak{T}) = R_p^{4k}(\mathfrak{T}').$$

换言之, $R_p^{4k}(\mathfrak{T})$ 是 M 的拓扑不变量, 特别在 $p=3$ 时,

$$R_3^{4k}(\mathfrak{T}) = (-1)^k P_3^{4k}(\mathfrak{T}),$$

因之证明了法 3 Понтрягин 示性类的拓扑不变性, 亦即本文的主要定理.

在 §1 中, 将概述 Smith 的一般理论, 特别与 Smith 变换 μ_i 有关的部分, 其中 μ_i 的若干主要性质, 则为作者所获得 [5]. §2 将讨论某种特殊的透镜空间. §3, §4 证明与任一欧氏空间丛相配, 可作一特殊透镜空间丛, 在其丛空间中有与 (1) 类似的关系式存在. §5 中将以此为根据, 证明法 3 Понтрягин 示性类的拓扑不变性, 并约略说明 Stiefel-Whitney 示性类的情形.

以上结果系作者在 1950 年所获得但没有发表过. 在下一文中, 我们将把这个证明与 [2] 中的证明作一比较并阐明其互相沟通之处.

§1 周期变换下的空间 [3–5]

设 $\tilde{K}$ 是一抽象复合形, T 是一乘法群, 作用于 $\tilde{K}$ 上. 换言之, 与每一 $t \in T$ 相当, 有一 $\tilde{K}$ 到自身的一对一对应, 仍记之为 t, 使 $t_2(t_1\tilde{\sigma}) = (t_2t_1)\tilde{\sigma}$, $1\tilde{\sigma} = \tilde{\sigma}(\tilde{\sigma} \in \tilde{K}$, $t_1, t_2 \in T$, $1 =$ T 的么元素). 假设 $(\tilde{K}, T)$ 具有以下各项性质 ($t \in T$, $\tilde{\sigma}, \tilde{\tau} \in \tilde{K}$), 则称之为一简单组 (primitive system):

1°. $t \neq 1$ 时, $t\tilde{\sigma}$ 与 $\tilde{\sigma}$ 恒不相同;

2°. $\dim(t\tilde{\sigma}) = \dim \tilde{\sigma}$;

3°. 若 $\tilde{\tau}$ 是 $\tilde{\sigma}$ 的面, 则 $t\tilde{\tau}$ 也是 $t\tilde{\sigma}$ 的面;

4°. 若 $\dim \tilde{\sigma} = \dim \tilde{\tau} + 1$, 则 $t\tilde{\sigma}$ 与 $t\tilde{\tau}$ 的关连数 (incidence number) 等于 $\tilde{\sigma}$ 与 $\tilde{\tau}$ 的关连数.

今在 $\tilde{K}$ 中定义一同值观念, 使 $\tilde{\sigma} \sim \tilde{\tau}$ 与有 $t \in T$ 存在使 $t\tilde{\sigma} = \tilde{\tau}$ 同义, 对于任一同值组 σ 任取一确定的 $\tilde{\sigma} \in \sigma$. 此等 $\tilde{\sigma}$ 的集合称为 $\tilde{K}$ 的一个基本域 (fundamental domain), 其中维数为 r 者设为 $\{\tilde{\sigma}_i^r\}$, $i \in \Im^r$. 对于任一系数群 G, 命 $G(T)$ 为一切 $\sum g_i t_i (g_i \in G, t_i \in T)$ 形状的元素依通常加法所成的群, 并命 $n(\sum g_i t_i) = \sum g_i \in G$. 于是任一 $C_r(\tilde{K}, G)$ 中的下链可记作 $\sum\limits_{i \in \Im^r} a_i \tilde{\sigma}_i^r$, $\alpha_i \in$ G (T), 而 $\tilde{K}$ 中的边界关系可写作

$$\partial \tilde{\sigma}_i^r = \sum_{j \in \Im^{r-1}} \alpha_{ij}^r \tilde{\sigma}_j^{r-1}, \quad i \in \Im^r, \tag{1}$$

此处 $\alpha_{ij}^r \in I(T)$, I 为整数加法群. 因 $I(T)$ 的元素与 $G(T)$ 的元素可依自然定义相乘得 $G(T)$ 的元素, 故对于任一 $\alpha \in I(T)$,

$$\alpha\left(\sum_{i\in\Im^r}\alpha_i\tilde{\sigma}_i^r\right)=\sum_{i\in\Im^r}(\alpha\alpha_i)\tilde{\sigma}_i^r \quad (\alpha_i \in G(T)),$$

显然定义了 $\tilde{K}$ 中的一个系数群为 G 的下链变换. 由 α 所引起的下同调群变换将记之为 α_*.

令 K 为 $\tilde{K}$ 中一切同值组 σ 的集合. 由 2°, 任一 $\tilde{\sigma} \in \sigma$ 的维数与 $\tilde{\sigma}$ 的选择无关, 我们定义之为 σ 的维数. 若对任二 $\sigma, \tau \in K$ 有 $\tilde{\sigma} \in \sigma, \tilde{\tau} \in \tau$, 使 $\tilde{\tau}$ 是 $\tilde{\sigma}$ 的面, 则我们定义 τ 为 σ 的面. 在 K 中又定义一边界关系 ∂ 为

$$\partial\sigma_i^r=\sum_{i\in\Im^{r-1}}n(\alpha_{ij}^r)\sigma_j^{r-1}, \quad i\in\Im^r \tag{2}$$

此处 $\alpha_{ij}^r \in I(T)$ 如前述, 而 σ_i^r 为含有 $\tilde{\sigma}_i^r$ 的同值组. 于是 K 成为一复合形, 称之为 $\tilde{K}$ 在 T 下的法复合形 (modular complex), 记之为 $K=\tilde{K}_T$. 自然对应 $\pi:\ \tilde{\sigma} \to \sigma(\tilde{\sigma} \in \sigma \in K)$ 称为 $(\tilde{K},T)$ 的投影 $\pi:\ \tilde{K} \to K$. 由 π 所引起的下链变换与下同调群的变换将记之为 π 与 π_*.

今设 T 为一有限群, 则 $s=\sum\limits_{t\in T}t \in I(T)$ 有意义. 设有 $x \in C_r(K,G)$, 则必有 $\tilde{x} \in C_r(\tilde{K},G)$ 使 $\pi\tilde{x}=x$, 易见 $s\tilde{x} \in C_r(\tilde{K},G)$ 只与 x 有关, 令之为 $\bar{\pi}x$, 则 $\bar{\pi}$ 定义了一个下链变换 $\bar{\pi}:\quad C_r(K,\ G) \to C_r(\tilde{K},G)$. 显然,

$$\bar{\pi}\pi=s, \qquad \pi\bar{\pi}=n(s)\cdot 1. \tag{3}$$

我们将以 $\pi', \bar{\pi}', \alpha'(\alpha \in I(T))$ 表与 $\pi, \bar{\pi}, \alpha$ 对偶的上链变换, 而以 $\pi^*, \bar{\pi}^*, \alpha^*$ 表它们所引起的上同调群变换; 余类推.

今设 T 任意而 $(\tilde{K}_1,T)$ 是另一简单组, $K_1=(\tilde{K}_1)_T$ 为其法复合形, 投影仍记为 π. 假设集变换 $\tilde{S}:\tilde{K} \to \tilde{K}_1$ 与下链变换 $\tilde{f}:\ \tilde{K} \to \tilde{K}_1$ 能使对于任一 $t \in T$ 有 $t\tilde{S}=\tilde{S}t$ 与 $t\tilde{f}=\tilde{f}t$, 则称 $\tilde{S}$ 与 $\tilde{f}$ 为 T 变换, 记作 $\tilde{S}:(\tilde{K},T) \to (\tilde{K}_1,T)$ 与 $\tilde{f}:(\tilde{K},T) \to (\tilde{K}_1,T)$. 此时 $\tilde{S}$ 自然地引起了一个集变换 $S:K \to K_1$ 使 $\pi\tilde{S}=S\pi$. 易见对于任一 $x \in C_r(K,G)$, 若任取一 $\tilde{x} \in C_r(\tilde{K},G)$ 使 $\pi\tilde{x}=x$ 时, $\pi\tilde{f}\tilde{x} \in C_r(K_1,G)$ 将只倚赖于 x 而与 $\tilde{x}$ 的选择无关. 我们将定义 $\pi\tilde{f}\tilde{x}$ 为 fx, 于是有

$$\pi\tilde{f}=f\pi. \tag{4}$$

特别在 T 为有限群时, 可定义 $\bar{\pi}:C_r(K,G) \to C_r(\tilde{K},G)$ 与 $\bar{\pi}:C_r(K_1,G) \to C_r(\tilde{K}_1,G)$. 此时对于任一 $x \in C_r(K,G)$, 任取一 $\tilde{x} \in C_r(\tilde{K},G)$ 使 $\pi\tilde{x}=x$, 则由 (4)

式有 $\pi\tilde{f}\tilde{x}=f\pi\tilde{x}=fx$, 因之由 $\bar{\pi}$ 的定义有 $\pi fx=s\tilde{f}\tilde{x}=\tilde{f}s\tilde{x}=\tilde{f}\bar{\pi}x$, 故

$$\tilde{f}\bar{\pi}=\bar{\pi}f. \tag{4$'$}$$

若 $\tilde{K}_1$ 为 $\tilde{K}$ 的一个剖分①, 而定义此剖分的集变换 $\tilde{S}:\tilde{K}\to\tilde{K}_1$ 与下链变换 $\tilde{f}:C_r(\tilde{K})\to C_r(\tilde{K}_1)$, $\tilde{g}:C_r(\tilde{K}_1)\to C_r(\tilde{K})$ 都是 T 变换时, $\tilde{K}_1$ 将称为 $\tilde{K}$ 的一个 T 剖分. 此时易见 $\tilde{S}$, $\tilde{f}$, $\tilde{g}$ 所引起的集变换 S 与下链变换 f,g 也定义了一个 K 的剖分, 即 K_1.

以下将设 p 为一固定质数, 而 $T\approx I_p$, I_p 为法 p 整数加法群, 并设 t 为一固定的母素. 我们有时将改写 $(\tilde{K},T)$ 及 $\tilde{K}_T$ 为 $(\tilde{K},t)$ 与 $\tilde{K}_t$. 令 $(i\geqslant 0)$

$$s_i=\begin{cases}1+t+\cdots+t^{p-1}\in I(T), & i\text{ 为偶数},\\ 1-t\in I(T), & i\text{ 为奇数};\end{cases} \tag{5}$$

此处 s_{2i} 即前面的 s. 依 P. A. Smith 有

A. 设 $\tilde{x}\in C_r(\tilde{K},G)$(或 $\tilde{x}'\in C^r(\tilde{K},G)$). 若 $s_i\tilde{x}=0$(或 $s_i'\tilde{x}'=0$), 则必有 $\tilde{y}\in C_r(\tilde{K},G)$(或 $\hat{y}'\in C^r(\tilde{K},G)$) 使 $\tilde{x}=s_{i+1}\tilde{y}$ (或 $\tilde{x}'=s_{i+1}'\tilde{y}'$).

今设 $X\in H_r(K,I_p)$. 任取 $x_0\in X$, 并取 $\tilde{x}_0\in C_r(\tilde{K},I_p)$ 使 $\pi\tilde{x}_0=x_0$. 据 (3) 得 $s_0\partial\tilde{x}_0=\partial s_0\tilde{x}_0=\partial\bar{\pi}\pi\tilde{x}_0=\bar{\pi}\partial x_0=0$, 故由 (A) 有 $\tilde{x}_1\in C_{r-1}(\tilde{K},I_p)$ 使 $\partial\tilde{x}_0=s_1\tilde{x}_1$. 因 $s_1\partial\tilde{x}_1=\partial s_1\tilde{x}_1=\partial\partial\tilde{x}_0=0$, 故由 (A) 又有 $\tilde{x}_2\in C_{r-2}(\tilde{K},I_p)$ 使 $\partial\tilde{x}_1=s_2\tilde{x}_2$. 依次类推可得 $\tilde{x}_i\in C_{r-i}(\tilde{K},I_p)$ 使

$$\pi\tilde{x}_0=x_0\in X,\quad \partial\tilde{x}_i=s_{i+1}\tilde{x}_{i+1},\quad r>i\geqslant 0. \tag{6}$$

易见每一 $\pi\tilde{x}_i$ 都为一法 p 下闭链. 若另有一组 $y_0\in X$, $\tilde{y}_i\in C_{r-i}(\tilde{K},I_p)$ 使 $\pi\tilde{y}_0=y_0\in X$, $\partial\tilde{y}_i=s_{i+1}\tilde{y}_{i+1}$, $r>i\geqslant 0$, 则应用 (3), (A) 由归纳法知必有 $\tilde{z}_i\in C_{r-i-1}(\tilde{K},I_p)$, $r>i\geqslant 0$ 存在, 使 $\tilde{y}_i-\tilde{x}_i=\partial\tilde{z}_i+s_{i+1}\tilde{z}_{i+1}$, $r>i\geqslant 0$, 因之 $\pi\tilde{y}_i\sim\pi\tilde{x}_i$. 换言之, $\pi\tilde{x}_i$ 所定的法 p 下同调类与 $x_0,\tilde{x}_0,\cdots,\tilde{x}_i$ 的选择无关, 我们定义之为 $\mu_{i,t}X$, $r\geqslant i\geqslant 0$, 或在不致引起误会时, 简写为 μ_iX. 由此得一组准同态变换

$$\mu_i:\quad H_r(K,I_p)\to H_{r-i}(K,I_p),\qquad r\geqslant i\geqslant 0,$$

此处 μ_0 显为恒同变换.

同样, 设 $X^*\in H^r(K,I_p)$. 我们可取 $x_0'\in X^*$, $\tilde{x}_0'\in C^{r+i}(\tilde{K},I_p)$, $i\geqslant 0$, 使

$$\bar{\pi}'\tilde{x}_0'=x_0'\in X^*,\quad \delta\tilde{x}_i'=s_{i+1}'\,\tilde{x}_{i+1}',\quad i\geqslant 0. \tag{6$'$}$$

① 剖分的定义见 S. Lefschetz. *Algebraic Topology*, 1942: 162-163.

于是 $\bar{\pi}'\tilde{x}_i'$ 为一法 p 上闭链, 其法 p 上同调类与 x_0', $\tilde{x}_0'$, $\cdots$, $\tilde{x}_i'$ 的选择无关, 定义之为 $\mu_{i,t}^* X^*$ 或 $\mu_i^* X^*$, 即得一组准同态变换 ($\mu_0^* =$ 恒同变换):

$$\mu_i^*:\quad H^r(K,I_p)\to H^{r+i}(K,I_p),\quad i\geqslant 0.$$

设 1 为 K 的法 p 单位类, 则我们将称

$$\mu_i^* 1 = U^i(\tilde{K},t),\quad i\geqslant 0 \tag{7}$$

为 $(\tilde{K}, T)$ 对于母素 t 或 $(\tilde{K}, t)$ 的 (法 p) 示性类. 对于同一 $t\in T$ 而言, 变换 μ_i, μ_i^* 与类 U^i 具有以下各项性质:

A1°. $\mu_i\mu_j=\mu_j\mu_i$, $\mu_{2i}=(\mu_2)^i$, $\mu_{2i+1}=\mu_1(\mu_2)^i$. 在 $p=2$ 时又有 $(\mu_1)^2=\mu_2$, 而在 $p>2$ 时 $(\mu_1)^2=0$.

A2°. 若 $\tilde{f}$ 为简单组 $(\tilde{K}, T)$ 到 $(\tilde{K}_1, T)$ 的下链 T 变换, 而 f 为由 $\tilde{f}$ 所引起 K 到 K_1 的下链变换, 则 $f_*\mu_i=\mu_i f_*$, $\mu_i^* f^*=f^*\mu_i^*$.

A3°. 若 $(\tilde{K}_1, T)$ 为 $(\tilde{K}, T)$ 的 T 剖分, 由集变换 $\tilde{s}$ 与下链变换 $\tilde{f}(\tilde{K}\to\tilde{K}_1)$, $\tilde{g}(\tilde{K}_1\to\tilde{K})$ 所定义, 则对于 $\tilde{f}$, $\tilde{g}$ 所引起的下链变换 $f(K\to K_1)$, $g(K_1\to K)$ 有 $\mu_i^* f^*=f^*\mu_i^*$, $g^*\mu_i^*=\mu_i^* g^*$ 等.

A4°. 设 $K=\tilde{K}_r$, 而 K, $\tilde{K}$ 都是简单的 (simple) 复合形, 换言之, 对于任意 $\sigma\in K$ 与 $\tilde{\sigma}\in\tilde{K}$, $\mathrm{cl}\,\sigma$ 与 $\mathrm{cl}\,\tilde{\sigma}$ 都有简单的同调解. 此时在 K 与 $\tilde{K}$ 中可定义上积 $\cup$. 于是有

$$\mu_i^* X^*=\mu_i^* 1\cup X^*=U^i(\tilde{K},t)\cup X^*,\quad X^*\in H^r(K,I_p) \tag{8}$$

(参阅 [5], §1).

证 兹先设 K, $\tilde{K}$ 都是单纯复合形. 此时可将 K 与 $\tilde{K}$ 的顶点排列成序, 使投影 $\pi:\tilde{K}\to K$ 保持顶点间的次序. 今依 Whitney 方法对此等次序在 K 与 $\tilde{K}$ 的上链间定义上积 U. 于是对于任意 $\tilde{y}\in C^r(\tilde{K},I_p)$, $\tilde{z}\in C^s(\tilde{K},I_p)$ 应有

$$\left.\begin{aligned}&\bar{\pi}'(\tilde{y}\cup s'\tilde{z})=\bar{\pi}'\tilde{y}\cup\bar{\pi}'\tilde{z},\\&s_i'(\tilde{y}\cup s'\tilde{z})=s_i'\tilde{y}\cup s'\tilde{z},\quad i\geqslant 0.\end{aligned}\right\} \tag{9}$$

试证上一式如次: 设 $(a_0\cdots a_{r+s})$ 为任一 K 中的 $r+s$ 维单纯形, 而依所择顶点次序有 $a_0<\cdots<a_{r+s}$. 在 $\tilde{K}$ 中可取一 $r+s$ 维单纯形 $(\tilde{a}_0\cdots\tilde{a}_{r+s})$ 使 $\pi\tilde{a}_i=a_i$, $i=0$, 1, $\cdots$, $r+s$. 因 π 保持顶点间的次序, 故同样亦有 $\tilde{a}_0<\cdots<\tilde{a}_{r+s}$. 令 $t^j\tilde{a}_i=\tilde{a}_i^j$, 于是

$$\begin{aligned}&\bar{\pi}'(\tilde{y}\cup s'\tilde{z})(a_0\cdots a_{r+s})\\&=(\tilde{y}\cup s'\tilde{z})(\bar{\pi}(a_0\cdots a_{r+s}))=(\tilde{y}\cup s'\tilde{z})(s(\tilde{a}_0\cdots\tilde{a}_{r+s}))\end{aligned}$$

$$
\begin{aligned}
&= \sum_{j=0}^{p-1} (\tilde{y} \cup s'\tilde{z})(\tilde{a}_0^j \cdots \tilde{a}_{r+s}^j) = \sum_{j=0}^{p-1} [\tilde{y}(\tilde{a}_0^j \cdots \tilde{a}_r^j) \cdot s'\tilde{z}(\tilde{a}_r^j \cdots \tilde{a}_{r+s}^j)] \\
&= \sum_{j=0}^{p-1} [\tilde{y}(\tilde{a}_0^j \cdots \tilde{a}_r^j) \cdot \tilde{z}(s(\tilde{a}_r^j \cdots \tilde{a}_{r+s}^j))] \\
&= \sum_{j=0}^{p-1} [\tilde{y}(\tilde{a}_0^j \cdots \tilde{a}_r^j) \cdot \tilde{z}(s(\tilde{a}_r \cdots \tilde{a}_{r+s}))] \\
&= \left[\sum_{j=0}^{p-1} \tilde{y}(\tilde{a}_0^j \cdots \tilde{a}_r^j)\right] \cdot \tilde{z}(s(\tilde{a}_r \cdots \tilde{a}_{r+s})) \\
&= \tilde{y}(s(\tilde{a}_0 \cdots \tilde{a}_r)) \cdot \tilde{z}(s(\tilde{a}_r \cdots \tilde{a}_{r+s})) \\
&= \tilde{y}(\bar{\pi}(a_0 \cdots a_r)) \cdot \tilde{z}(\bar{\pi}(\tilde{a}_r \cdots \tilde{a}_{r+s})) \\
&= \bar{\pi}'\tilde{y}(a_0 \cdots a_r) \cdot \bar{\pi}'\tilde{z}(a_r \cdots a_{r+s}) \\
&= (\bar{\pi}'\tilde{y} \cup \bar{\pi}'\tilde{z})(a_0 \cdots a_{r+s}).
\end{aligned}
$$

因 $(a_0 \cdots a_{r+s})$ 为任意的, 故得 (9) 的第一式. 第二式的证明亦类似.

今取 $\tilde{u}_i \in C^i(\tilde{K}, I_p)$, $i \geqslant 0$, 使 ($1 =$ 法 p 单位上链):

$$\bar{\pi}'\tilde{u}_0 = 1, \quad \delta\tilde{u}_i = s'_{i+1}\tilde{u}_{i+1}, \quad i \geqslant 0. \tag{10}$$

又取 $x'_0 \in X^*$, $\tilde{x}'_i \in C^{r+i}(\tilde{K}, I_p)$, 使 $(6)'$ 成立. 于是由 (9) 有

$$
\begin{aligned}
\bar{\pi}'(\tilde{u}_0 \cup s'\tilde{x}'_0) &= \bar{\pi}'\tilde{u}_0 \cup \bar{\pi}'\tilde{x}'_0 = 1 \cup \bar{\pi}'\tilde{x}'_0 = \bar{\pi}'\tilde{x}'_0 = x'{}_0 \in X^*, \\
\delta(\tilde{u}_i \cup s'\tilde{x}'_0) &= \delta\tilde{u}_i \cup s'\tilde{x}'_0 + (-1)^i\tilde{u}_i \cup \delta s'\tilde{x}'_0 = s'_{i+1}\tilde{u}_{i+1} \cup s'\tilde{x}'_0 + (-1)^i\tilde{u}_i \cup s'\delta\tilde{x}'_0 \\
&= s'_{i+1}(\tilde{u}_{i+1} \cup s'\tilde{x}'_0) + (-1)^i\tilde{u}_i \cup s's'_1\tilde{x}'_1 = s'_{i+1}(\tilde{u}_{i+1} \cup s'\tilde{x}'_0), \\
\bar{\pi}'(\tilde{u}_i \cup s'\tilde{x}'_0) &= \bar{\pi}'\tilde{u}_i \cup \bar{\pi}'\tilde{x}'_0 = \bar{\pi}'\tilde{u}_i \cup x'{}_0,
\end{aligned}
$$

但由定义 $\bar{\pi}'\tilde{u}_i \in \mu_i^* 1 = U^i(\tilde{K}, t)$, $\bar{\pi}'(\tilde{u}_i \cup s'\tilde{x}'_0) \in \mu_i^* X^*$, 故从最后一式即得定理.

在一般情形时, 可取 $\tilde{K}$ 的诱导复合形 (derived complex) $\tilde{K}_1$. 此时 T 自然地作用于 $\tilde{K}_1$ 中, 使之成为一简单组 $(\tilde{K}_1, T)$, 且易见 $K_1 = (\tilde{K}_1)_r$ 即为 K 的诱导复合形. 因诱导并不改变同调群与上积, 故由前特款及 A3° 知本定理仍真确.

由 A1°, A4°, 可知在 $(\tilde{K}, t)$ 的示性类 $U^i = U^i(\tilde{K}, t)$ 之间有下述关系:

A5°. 若 $p = 2$, 则 $U^i = (U^1)^i$,

若 $p > 2$, 则 $(U^1)^2 = 0$, $U^{2i} = (U^2)^i$, $U^{2i+1} = U^1(U^2)^i$.

A6°. 若 $\tilde{L}$ 为另一复合形, T 作用于 $\tilde{K} \times \tilde{L}$ 中使 $t(\tilde{\sigma} \times \tilde{\tau}) = t\tilde{\sigma} \times \tilde{\tau}$ $(\tilde{\sigma} \in \tilde{K}, \tilde{\tau} \in \tilde{L})$, 则 $(\tilde{K} \times \tilde{L}, T)$ 亦成一简单组. 对 t 而言, 令其与 μ_i, μ_i^* 相当的变换为 $\bar{\mu}_i$, $\bar{\mu}_i^*$, 则

$$\bar{\mu}_i(X \otimes Y) = \mu_i X \otimes Y, \quad \bar{\mu}_i^*(X^* \otimes Y^*) = \mu_i^* X^* \otimes Y^*,$$

此处 $X \in H(K, I_p)$, $Y \in H(\tilde{L}, I_p)$, $X^* \in H^*(K, I_p)$, $Y^* \in H^*(\tilde{L}, I_p)$.

今设 $\tilde{K}$ 为一已有固定定向的 m 维可定向 (有限) 组合流形. 令 $\tilde{K}^*$ 为 $\tilde{K}$ 的对偶复合形. 设 $\tilde{K}$ 的 r 维胞腔为 $\tilde{\sigma}^r_\lambda$, λ 在某一与 r 有关的指数集内. 对 $\tilde{K}$ 的已定定向而言, $\tilde{\sigma}^r_\lambda \in \tilde{K}$ 的对偶将记之为 $\tilde{\tau}^{m-r}_\lambda$. 于是 $\tilde{K}$、$\tilde{K}^*$ 中的边界关系可书为

$$\left.\begin{aligned}\partial \tilde{\sigma}^r_\lambda &= \sum_\mu a^r_{\lambda\mu} \tilde{\sigma}^{r-1}_\mu, \\ \partial \tilde{\tau}^{m-r+1}_\mu &= (-1)^r \sum_\lambda a^r_{\lambda\mu} \tilde{\tau}^{m-r}_\lambda.\end{aligned}\right\} (a^r_{\lambda\mu} = \pm 1). \tag{11}$$

对于任意系数群 G 及下链 $x = \sum\limits_\lambda g_\lambda \tilde{\sigma}^r_\lambda \in C_r(\tilde{K}, G)$, 定义 $\tilde{d}x \in C^{m-r}(\tilde{K}^*, G)$ 使 $\tilde{d}x(\tilde{\tau}^{m-r}_\lambda) = g_\lambda$, 则 $\tilde{d}$ 定义了一个同构变换

$$\tilde{d} : C_r(\tilde{K}, G) \approx C^{m-r}(\tilde{K}^*, G).$$

我们知道

$$\delta \tilde{d} x = (-1)^r \tilde{d} \partial x, \quad x \in C_r(\tilde{K}, G), \tag{12}$$

因之 $\tilde{d}$ 引起了同构变换

$$\tilde{D} : H_r(\tilde{K}, G) \approx H^{m-r}(\tilde{K}^*, G).$$

设组合流形 $\tilde{K}$ 有一群 T 作用其中使 $(\tilde{K}, T)$ 成一简单组. 设 $t\tilde{\sigma}^r_\lambda = \tilde{\sigma}^r_{t(\lambda)}$, 此处 $t \in T$ 任意, 则我们将定义 $t\tilde{\tau}^{m-r}_\lambda = \tilde{\tau}^{m-r}_{t(\lambda)}$. 于是 T 亦作用于 $\tilde{K}^*$ 中. 由 (11) 的第一式有

$$\begin{aligned}\partial t \tilde{\sigma}^r_\lambda &= \partial \tilde{\sigma}^r_{t(\lambda)} = \sum_\mu a^r_{t(\lambda),t(\mu)} \tilde{\sigma}^{r-1}_{t(\mu)}, \\ t \partial \tilde{\sigma}^r_\lambda &= t \sum_\mu a^r_{\lambda\mu} \tilde{\sigma}^{r-1}_\mu = \sum_\mu a^r_{\lambda\mu} t \tilde{\sigma}^{r-1}_\mu = \sum_\mu a^r_{\lambda\mu} \tilde{\sigma}^{r-1}_{t(\mu)}.\end{aligned}$$

因 $(\tilde{K}, T)$ 为简单组故有 $\partial t = t\partial$, 因之从上二式得 $a^r_{\lambda\mu} = a^r_{t(\lambda),t(\mu)}$. 同样由 (11) 的第二式得

$$\begin{aligned}\partial t \tilde{\tau}^{m-r+1}_\mu &= \partial \tilde{\tau}^{m-r+1}_{t(\mu)} = (-1)^r \sum_\lambda a^r_{i(\lambda)t(\mu)} \tilde{\tau}^{m-r}_{t(\lambda)}, \\ t \partial \tilde{\tau}^{m-r+1}_\mu &= (-1)^r t \sum_\lambda a^r_{\lambda\mu} \tilde{\tau}^{m-r}_\lambda = (-1)^r \sum_\lambda a^r_{\lambda\mu} \tilde{\tau}^{m-r}_{t(\lambda)},\end{aligned}$$

由 $a^r_{\lambda\mu} = a^r_{t(\lambda),t(\mu)}$ 得 $\partial t \tilde{\tau}^{m-r+1}_\mu = t \partial \tilde{\tau}^{m-r+1}_\mu$. 由此知 $(\tilde{K}^*, T)$ 适合与 4° 相当的条件, 因 1°– 3° 甚为显然, 故 $(\tilde{K}^*, T)$ 亦为一简单组.

仍设 $T \approx I_p$, 且设母素 $t \in I_p$ 为保向变换, 则 $K = \tilde{K}_T$ 与 $K^* = \tilde{K}_T^*$ 亦为一可定向 m 维 (有限) 组合流形的对偶剖分, 且 $\tilde{K}$ 的已有定向引起了 K 的定向, 使投影 $\pi : \tilde{K} \to K$ 为保向的. 对此 K 的定向可如前定义对偶变换

$$
\begin{aligned}
d &: C_r(K, G) \approx C^{m-r}(K^*, G),\\
D &: H_r(K, G) \approx H^{m-r}(K^*, G).
\end{aligned}
$$

在 $G \approx I_p$ 时, 我们有以下诸关系, 其证明从略:

$$\bar{\pi}'\tilde{d} = d\pi, \tag{13}$$

$$(t')^{-1}\tilde{d} = \tilde{d}t, \tag{14}$$

$$\tilde{d}\, s_i = s_i'\bar{s}_i'\tilde{d}, \tag{15}$$

此处

$$\bar{s}_i' = \begin{cases} 1, & i \text{ 为偶数},\\ 1 + t' + \cdots + t'^{p-2}, & i \text{ 为奇数}, \end{cases} \tag{15$'$}$$

$$\bar{\pi}'\bar{s}_i' = (-1)^i\bar{\pi}'. \tag{16}$$

A7°. 对于简单组 $(\tilde{K}, t)$, $(\tilde{K}^*, t)$, 定义

$$
\begin{aligned}
\mu_i &: H_r(K, I_p) \to H_{r-i}(K, I_p),\\
\mu_i^* &: H^r(K^*, I_p) \to H^{r+i}(K^*, I_p)
\end{aligned}
$$

如前, 则

$$D\mu_i X = (-1)^{i(r+1)}\mu_i^* DX, \quad X \in H_r(\tilde{K}, I_p). \tag{17}$$

证 取 $x_0 \in X$, $\tilde{x}_i \in C_{r-i}(\tilde{K}, I_p)$ 使 (6) 成立. 由 (6) 及 (12)–(16) 各式可得

$$
\begin{aligned}
\bar{\pi}'\tilde{d}\tilde{x}_0 &= d\pi\tilde{x}_0 = dx_0 \in DX,\\
\delta(\bar{s}_1' \cdots \bar{s}_i'\tilde{d}\tilde{x}_i) &= \bar{s}_1' \cdots \bar{s}_i'\delta\tilde{d}\tilde{x}_i = (-1)^{r-i}\bar{s}_1' \cdots \bar{s}_i'\tilde{d}\partial\tilde{x}_i\\
&= (-1)^{r-i}\bar{s}_1' \cdots \bar{s}_i'\tilde{d}s_{i+1}\tilde{x}_{i+1}\\
&= (-1)^{r-i}s_{i+1}'(\bar{s}_1' \cdots \bar{s}_i'\ \bar{s}_{i+1}'\tilde{d}\tilde{x}_{i+1}),\\
\bar{\pi}'(\bar{s}_1' \cdots \bar{s}_i'\tilde{d}\tilde{x}_i) &= (-1)^{1+2+\cdots+i}\bar{\pi}'\tilde{d}\tilde{x}_i = (-1)^{1+2+\cdots+i}d\pi\tilde{x}_i.
\end{aligned}
$$

由定义 $\pi\tilde{x}_i \in \mu_i X$, 而 $\mu_i^* DX$ 含有 $(-1)^{r+(r-1)+\cdots+(r-i+1)}\bar{\pi}'(\bar{s}_1' \cdots \bar{s}_i'\ \tilde{d}\tilde{x}_i) = (-1)^{i(r+1)}d\pi\tilde{x}_i$. 由此即得定理.

今设 $\tilde{M}$ 为一紧空间, T 为 $\tilde{M}$ 中的拓扑变换群, 且每一 $t \in T$, 而 $t \neq 1$ 者皆无不变点, 此时我们称 $(\tilde{M}, T)$ 为一简单组. 若 $(\tilde{M}_1, T)$ 为另一简单组, 连续

映像 $\tilde{f}:\tilde{M}\to\tilde{M}_1$ 能使对每一 $t\in T$ 有 $\tilde{f}t=t\tilde{f}$, 则 $\tilde{f}$ 称为一 T 映像, 记作 $\tilde{f}:(\tilde{M},T)\to(\tilde{M}_1,T)$. 将 $\tilde{M}$ 中由 T 可互相变换而得的点恒同成一点得一空间 M, 称作 $(\tilde{M},T)$ 的法空间, 记作 $M=\tilde{M}_T$, 其自然投影设为 $\pi:\tilde{M}\to M$. 于是任一 T 映像 $\tilde{f}:(\tilde{M},T)\to(\tilde{M}_1,T)$ 必引起一 $f:M\to M_1$ 使 $f\pi=\pi f$. 设 $\tilde{V}_a=\{\tilde{V}_{ia}\}$ 是 $\tilde{M}$ 的一个有限开遮盖, 使 $\tilde{V}_{ia}\in\tilde{V}_a$ 时, $t\tilde{V}_{ia}$ 也 $\in\tilde{V}_a$, 且每 $\tilde{V}_{ia}$ 与任一 $t\tilde{V}_{ia}(t\neq 1)$ 不相遇, 此处 $t\in T$, 则 $\tilde{V}_a$ 将称为简单的. 此时 $V_a=\{V_{ia}=\pi\tilde{V}_{ia}\}$ 也成为 M 的一个开遮盖. 令 $\tilde{K}_a$ 为 $\tilde{V}_a$ 的神经, K_a 为 V_a 的神经, 则 t 将在 $\tilde{K}_a$ 中引起一单纯变换 t_a, 于是群 $T_a=\{t_a/t\in T\}$ 作用于 $\tilde{K}_a$ 中而使 $(\tilde{K}_a,T_a)$ 成一简单组, 且 $K_a=(\tilde{K}_a)_{T_a}$. 今设 $T\approx I_p$, 并设 $t\in T$ 为一固定母素. 此时根据 P. A. Smith, 简单的开遮盖恒存在, 且此等开遮盖的全体与一切开遮盖的全体同尾. 设 $\tilde{V}_a$, $\tilde{V}_\beta$ 都是简单的, 且 $\tilde{V}_\alpha\Subset\tilde{V}_\beta$, 我们可选择单纯映像 $\tilde{f}_{\alpha\beta}:\tilde{K}_\alpha\to\tilde{K}_\beta$ 使 $t_\beta\tilde{f}_{\alpha\beta}=\tilde{f}_{\alpha\beta}t_\alpha$, 此时 $\tilde{f}_{\alpha\beta}$ 引起 $f_{\alpha\beta}:K_\alpha\to K_\beta$. 据 A2°, 有

$$\mu_i^* f_{\alpha\beta}^*=f_{\alpha\beta}^*\mu_i^*. \tag{18}$$

今设 $X\in H^r(M,I_p)$. 取一 $\tilde{V}_\alpha$ 使 X 在 K_α 中有代表元素 X_α, 则由 (18), $\mu_i^*X_\alpha$ 在 $H^{r+1}(M,I_p)$ 中所代表的元素与 α 的选择无关, 今定义之为 μ_i^*X. 同样可定义 $\mu_i:H_r(M,I_p)\to H_{r-1}(M,I_p)$, 并可定义 $(\tilde{M},T)$ 的示性类为

$$U^i(\tilde{M},t)=\mu_i^*1\in H^i(M,I_p),\quad i\geqslant 0. \tag{19}$$

自然 μ_i, μ_i^* 与 U^i 的定义与 T 中母素 t 的选择有关.

若 $\tilde{M}$ 同时为一复合形 $\tilde{K}$ 的空间, 且 $t\in T$ 在 $\tilde{K}$ 中所引起的胞腔变换使 $(\tilde{K}, t)$ 成简单组, 则由 A2°, A3° 可知对于 $(\tilde{K}, t)$ 及 $(\tilde{M}, t)$ (即 $(\tilde{M}, T)$) 所定义的 μ_i, μ_i^* 互相重合, 而特别有 $U^i(\tilde{M},t)=U^i(\tilde{K},t)$. 关于 μ_i, μ_i^* 亦皆具有不变意义, 因之除 A3° 外, A1°– 6° 皆可推广之为 $(\tilde{M}, t)$ 中的性质, 不再赘述.

§2 透镜空间

自本节以迄本文之末, 若非另外声明, p 将指一固定的奇质数, $q=(p-1)/2$, $\varepsilon=e^{2\pi i/p}$.

设 U^n 为复 n 维的酉空间, 以 $(z_1,\cdots,z_n)$ 为一组复坐标, $\tilde{L}$ 为 U^n 中的单位球: $\sum\limits_{i=1}^n z_i\bar{z}_i=1$, 此处 $\bar{z}$ 表 z 的共轭复数. 设 l_i, m_i, $i=1,\cdots,n$, 为一组与 p 互质的整数而 $l_im_i\equiv 1 \bmod p$, 则

$$t:(z_1,\cdots,z_n)\to(\varepsilon^{m_1}z_1,\cdots,\varepsilon^{m_n}z_n) \tag{1}$$

为 $\tilde{L}$ 中的一个拓扑保向变换, 以 p 为周期且无不变点, 因之 $(\tilde{L},t)$ 为一简单组, 它的法空间 $L=\tilde{L}_t$ 将称为一个定义于 U^n 中对坐标 $(z_1,\cdots,z_n)$ 而言的 $(l_1,\cdots,l_n)$ 型

透镜空间 (Lens Spaces). $\tilde{L}$ 和 L 都是可定向流形, 具有自然定向. 定义 $\tilde{S}_k, \tilde{I}_{k,j} \subset \tilde{L}$ 及 $\tilde{P}_{k,j} \in \tilde{L}$ 如次 ($k = 1, 2, \cdots, n$; $j =$ 任意整数):

$$\tilde{S}_k : z_1 = \cdots = z_{k-1} = 0, \quad z_{k+1} = \cdots = z_n = 0, \quad |z_k| = 1;$$

$$\tilde{I}_{k,j} : z_1 = \cdots = z_{k-1} = 0, \quad z_{k+1} = \cdots = z_n = 0, \quad |z_k| = 1, \quad \frac{2\pi}{p} \cdot j \leqslant \arg z_k \leqslant \frac{2\pi}{p} \cdot (j+1);$$

$$\tilde{P}_{k,j} : z_1 = \cdots = z_{k-1} = 0, \quad z_{k+1} = \cdots = z_n = 0, \quad |z_k| = 1, \quad \arg z_k = \frac{2\pi}{p} \cdot j.$$

又定义 $[k]_j^* \subset \tilde{L}(k = 0, 1, \cdots, n-1; j =$ 任意整数) 如次 ($\circ$ 表示一种联合 (join), 其义自明):

$$[2k+1]_j^* = \tilde{S}_1 \circ \cdots \circ \tilde{S}_k \circ \ I_{k+1,j}, \quad [2k]_j^* = \tilde{S}_1 \circ \cdots \circ \tilde{S}_k \circ \tilde{P}_{k+1,j} \tag{2}$$

或即

$$\left.\begin{aligned} &[2k+1]_j^* : \frac{2\pi}{p} \cdot j \leqslant \arg z_{k+1} \leqslant \frac{2\pi}{p} \cdot (j+1), z_{k+2} = \cdots = z_n = 0, \sum_{i=1}^{n} z_i \bar{z}_i = 1, \\ &[2k]_j^* : \arg z_{k+1} = \frac{2\pi}{p} \cdot j, \ z_{k+1} = \cdots = z_n = 0, \sum_{i=1}^{n} z_i \bar{z}_i = 1. \end{aligned}\right\} \quad (2)'$$

则 $[k]_j = [k]_j^* - \sum\limits_{i=1}^{p} [k-1]_i^*$ 为 k 维的开胞腔, 而 $[k]_j (k = 0, 1, \cdots, n-1;\ j = 1, \cdots, p)$ 的集合构成 $\tilde{L}$ 的一个剖分 $\tilde{K}$. 显然 $[k]_j^*$ 为 $[k]_j$ 的闭包且为假流形, 可依 (2) 自然定向, 而 $[2n-1]_i^*$ 的自然定向彼此协合, 亦即 $\tilde{L}$ 的自然定向. 对于此等自然定向的 $[k]_j$ 而言, 有

$$\left.\begin{aligned} \partial[2k]_j &= [2k-1]_1 + \cdots + [2k-1]_p, \\ \partial[2k+1]_j &= [2k]_{j+1} - [2k]_j. \end{aligned}\right\} \tag{3}$$

变换 t 在 $\tilde{K}$ 中所引起的胞腔变换, 仍记之为 t, 则

$$\left.\begin{aligned} t[2k+1]_j &= [2k+1]_{j+m_{k+1}}, \\ t[2k]_j &= [2k]_{j+m_{k+1}}, \end{aligned}\right\} \tag{4}$$

或即

$$\left.\begin{aligned} t^{l_{k+1}}[2k+1]_j &= [2k+1]_{j+1}, \\ t^{l_{k+1}}[2k]_j &= [2k]_{j+1}. \end{aligned}\right\} \quad (4)'$$

由 (4) 或 (4)′ 知 $(\tilde{K}, t)$ 是一简单组, 它的法复合形 $K = \tilde{K}_t$ 是 L 的一个剖分. 我们将称 $\tilde{K}, K$ 各为 $\tilde{L}, L$ 对于坐标 $(z_1, \cdots, z_n)$ 的标准剖分. 令投影为 $\lambda : \tilde{K} \to K$, $\tilde{L} \to L$, 而

$$\lambda[k]_j = [k] \quad (k = 0, 1, \cdots, 2n-1), \tag{5}$$

则由 (3) 得

$$\left.\begin{aligned}\partial[2k]&=p[2k-1],\\ \partial[2k+1]&=0.\end{aligned}\right\}\tag{6}$$

由 (6), 知对于每一 k, $[k]$ 为一法 p 下闭链, 以自然定向的假流形 $[k]^*=\lambda[k]_j^*$ 为其几何代表, 而 $[k]$ 的下同调类为 $H_k(L,I_p)\approx I_p$ 的一个母素. 但这些母素与 L 的剖分 K, 因之与 U^n 中坐标 $(z_1,\cdots,z_n)$ 的选择有关.

令自然定向的 L 所定的 $2n-1$ 维基本下闭链为 x_0, 又令

$$t_j=1+t+\cdots+t^{l_j-1},\quad j=1,\cdots,n,\tag{7}$$

$$\left.\begin{aligned}\tilde{x}_{2k-1}&=(-1)^k t_n t_{n-1}\cdots t_{n-k+1}[2n-2k]_1,\\ \tilde{x}_{2k}&=(-1)^k t_n t_{n-1}\cdots t_{n-k+1}[2n-2k-1]_1;\end{aligned}\right\}(k=0,1,\cdots,n-1)\tag{8}$$

则由 (3), (4)′ 得 (符号 s_i 见 §1 (5))

$$\partial\tilde{x}_i=\ s_i\tilde{x}_{i+1},\tag{9}$$

$$\left.\begin{aligned}\lambda\tilde{x}_{2k-1}&=(-1)^k l_n l_{n-1}\cdots l_{n-k+1}[2n-2k],\\ \lambda\tilde{x}_{2k}&=(-1)^k l_n l_{n-1}\cdots l_{n-k+1}[2n-2k-1];\end{aligned}\right\}(k=0,1,\cdots,n-1).\tag{10}$$

由 (9), (10) 及 μ_i 的定义知, 若命 X 为由自然定向的 L 的 $2n-1$ 维法 p 基本下同调类, 则有下述定理:

B1°. $\mu_{2n-2k-1}X$, $\mu_{2n-2k}X$ 各含法 p 下闭链 $(-1)^{n-k}l_{k+1}\cdots l_n[2k]$ 与 $(-1)^{n-k}$ $l_{k+1}\cdots l_n[2k-1]$.

因 l_i 与 p 互质, 故得

B2°. μ_kX, $k=0,1,\cdots,2n-1$ 成为 $H(L,I_p)$ 的一组加法基.

今对 L 的自然定向而言, 令 $D:H_r(L,I_p)\approx H^{2n-1-r}(L,I_p)$ 为其对偶变换. 又令 1 为 L 的单位类, 而 $U^j=U^j(\tilde{L},t)=\mu_j^*1$ 为 $(\tilde{L},t)$ 的法 p 示性类, 则由 A4°, 5°, 7° 及上面的 B2° 可得

B3°. 透镜空间 L 中的示性类 $U^i=U^i(\tilde{L},t)$, $i=0,1,\cdots,2n-1$ 构成上同调环 $H^*(L,I_p)$ 的一组加法基. 就环构造而论, $H^*(L,I_p)$ 系由 U^1, U^2 所产生, 此处 U^1, U^2 满足

$$(U^1)^2=0,\quad (U^2)^n=0,\quad U^1U^2=U^2U^1;\tag{11}$$

$$U^{2i}=(U^2)^i,\quad U^{2i+1}=U^1(U^2)^i,\quad i=0,1,\cdots,n-1;\tag{11$'$}$$

又

$$U^k=D\mu_kX,\quad k=0,1,\cdots,2n-1.\tag{12}$$

在本文中, 一种特别的透镜空间具有重要的意义, 其定义如次: 令 R^m 为 m 维欧氏空间. 在 R^m 中任取 p 个矢量 $\mathfrak{x}_1, \cdots, \mathfrak{x}_p$, 使满足

$$\sum_{i=1}^{p}|\mathfrak{x}_i|^2=1, \tag{13}$$

$$\sum_{i=1}^{p}\mathfrak{x}_i=0. \tag{13$'$}$$

此等 (有序的) 矢量组 $(\mathfrak{x}_1, \cdots, \mathfrak{x}_p)$ 自然地成一空间 $\tilde{S}$ 与 $2qm-1$ 维球同拓. 在 $\tilde{S}$ 中定义一保向的同拓变换

$$t:(\mathfrak{x}_1,\cdots,\mathfrak{x}_p)\rightarrow(\mathfrak{x}_2,\cdots,\mathfrak{x}_p,\mathfrak{x}_1), \tag{14}$$

则 t 以 p 为周期且无不变点. 令简单组 $(\tilde{S}, t)$ 的法空间为 L_m^p, 则 L_m^p 为一透镜空间, 证之如次.

令 $(x_1, \cdots, x_m)$ 为 R^m 中的一组坐标, 又令 U^m 为 m 维酉空间如前. 对于任意 $\mathfrak{x}=(x_1,\cdots,x_m)\in R^m$, 定义 $j(\mathfrak{x})=\mathfrak{z}=(z_1,\cdots,z_m)\in U^m$ 为 $z_i=x_i$, $i=1, 2, \cdots, m$, 得映像 $\bar{f}:R^m\rightarrow U^m$. 对于任一矢量组 $(\mathfrak{x}_1,\cdots,\mathfrak{x}_p)\in\tilde{S}(\mathfrak{x}_i\in R^m)$, 令

$$\mathfrak{x}_i'=\sum_{j=1}^{p}\varepsilon^{(j-1)i}\bar{f}(\mathfrak{x}_j),\quad i=0,1,\cdots,p-1, \tag{15}$$

则 $\mathfrak{z}_i'$ 与 $\mathfrak{z}_{p-i}'$ 为 U^m 中的共轭复矢量:

$$\mathfrak{z}_{p-i}'=\bar{\mathfrak{z}}_i',\quad i=1,\cdots,q, \tag{16}$$

且有

$$p\bar{f}(\mathfrak{x}_j)=\sum_{i=0}^{p-1}\varepsilon^{-(j-1)i}\mathfrak{z}_i',\quad j=1,\cdots,p. \tag{17}$$

从 (15) 可知 (13)′ 与下式同值:

$$\mathfrak{z}_0'=0. \tag{18}$$

由 (13), (16), (18) 有

$$2\sum_{i=1}^{q}|\mathfrak{z}'_i|^2=\sum_{i=0}^{p-1}\mathfrak{z}'_i\bar{\mathfrak{z}}_i'=p\sum_{j=1}^{p}|\mathfrak{x}_j|^2=p. \tag{19}$$

由此可令

$$\mathfrak{z}_i=\sqrt{\frac{2}{p}}\mathfrak{z}_i', \tag{20}$$

于是

$$\sum_{i=1}^{q}|\mathfrak{z}_i|^2=1. \tag{21}$$

令 $\tilde{S}'$ 为一切满足 (21) 的 (有序的) 矢量组 $(\mathfrak{z}_1,\cdots,\mathfrak{z}_q)(\mathfrak{z}_i\in U^m)$ 所自然地决定的空间, 则易见 (17), (19), (20) 定义了一个 $\tilde{S}$ 到 $\tilde{S}'$ 上的拓扑变换

$$h:(\mathfrak{x}_1,\cdots,\mathfrak{x}_p)\to(\mathfrak{z}_1,\cdots,\mathfrak{z}_q). \tag{22}$$

在 h 之下, t 变为 $\tilde{S}'$ 中的一个拓扑变换

$$t'=hth^{-1}:(\mathfrak{z}_1,\cdots,\mathfrak{z}_q)\to(\varepsilon^{-1}\mathfrak{z}_1,\cdots,\varepsilon^{-i}\mathfrak{z}_i,\cdots,\varepsilon^{-q}\mathfrak{z}_q). \tag{23}$$

令 $U_q^m=\underbrace{U^m\otimes\cdots\otimes U^m}_{q}$. 在 U_q^m 中取坐标

$$(z_1^{(1)},\cdots,z_m^{(1)},z_1^{(2)},\cdots,z_m^{(2)},\cdots,z_1^{(q)},\cdots,z_m^{(q)}),$$

此处 $(z_1^{(i)},\cdots,z_m^{(i)})=\mathfrak{z}_i$, $(\mathfrak{z}_1,\cdots,\mathfrak{z}_q)\in U_q^m$, 可见 $(\tilde{S}', t')$ 亦即 $(\tilde{S}, t)$ 法空间 L_m^p 为一透镜空间, 对上述坐标而言, 其型为

$$(1',\cdots,1',\cdots,\ i',\cdots,i',\cdots,\ q',\cdots,q'),$$

其中 i' 满足

$$ii'\equiv-1 \bmod p,\quad i=1,\cdots,q,$$

且每一 i' 出现 m 次, 如所欲证.

上述空间 L_m^p, 将称为定义于 $R_p^m=\underbrace{R^m\oplus\cdots\oplus R^m}_{p}$ 中或 U_q^m 中的特殊透镜空间.

§3 复格拉斯曼流形上的一个透镜空间丛

在本节中, p, q, ε 的意义同 §2, 又所有的链、同调关系与同调群等, 都对系数群 I_p 而言, 不再明述, 亦不再明白写出.

设 U^{n+m} 为 $n+m$ 复维酉空间, 以 $(z_1,\cdots,z_{n+m})$ 为一组复坐标, U^k 为由 $z_1=\cdots=z_{n+m-k}=0$ 所定义的复 k 面. 又设 $C_{n,m}$ 为定义于 U^{n+m} 中的复格拉斯曼流形, 其中的链 $[a_1,\cdots,a_m]_c$ 及假流形 $[a_1,\cdots,a_m]_c^*(0\leqslant a_1\leqslant\cdots\leqslant a_m\leqslant n)$ 系对由序列 $U^1\subset\cdots\subset U^{n+m-1}$ 所定标准胞腔部分如 [1] §3 所定义. 令 $\tilde{C}_{n,m}$ 为在 $C_{n,m}$ 上模范球纤维丛 $\mathfrak{C}_{n,m}$ 的丛空间, 其中的点可表作 $(\mathfrak{z}, Z)$ 形状, 此处 $Z\in C_{n,m}$,

$\mathfrak{z} \in Z \subset U^{n+m}$, $|\mathfrak{z}| = 1$. 作 $\mathfrak{C}_{n,m}$ 的 q 次 Whitney 幂 $\mathfrak{C}^q_{n,m} = \underbrace{\mathfrak{C}_{n,m} \cup \cdots \cup \mathfrak{C}_{n,m}}_{q}$, 令其从空间为 $\tilde{\Delta}$, 则 $\tilde{\Delta}$ 的点可表作 $\{\mathfrak{z}_1, \cdots, \mathfrak{z}_q; Z\}$形状, 此处 $Z \in C_{n,m}$, $\mathfrak{z}_i \in Z \subset U^{n+m}$, $i = 1, \cdots, q$, 而 $\sum\limits_{i=1}^{q} |\mathfrak{z}_i|^2 = 1$.

令 $\tilde{S}$ 为 $U_q^{n+m} = (U^{n+m})^q$ 中的单位球, 其中的点可表作 $(\mathfrak{z}_1, \cdots, \mathfrak{z}_q)$ 形状, 此处 $\mathfrak{z}_i \in U^{n+m}$, $\sum\limits_{i=1}^{q} |\mathfrak{z}_i|^2 = 1$. 在 $\tilde{S}$ 中,

$$t : (\mathfrak{z}_1, \cdots, \mathfrak{z}_q) \to (\varepsilon^{-1}\mathfrak{z}_1, \cdots, \varepsilon^{-q}\mathfrak{z}_q)$$

是一个以 p 为周期而无不变点的拓扑变换. 法空间 $\tilde{S}_t$ 即 §2 末所讨论的特殊透镜空间 L^p_{n+m}. 同样, t 在 $\tilde{\Delta}$ 中也引起了一个以 p 为周期而无不变点的拓扑变换 $\{\mathfrak{z}_1, \cdots, \mathfrak{z}_q; Z\} \to \{\varepsilon^{-1}\mathfrak{z}_1, \cdots, \varepsilon^{-q}\mathfrak{z}_q; Z\}$, 仍令之为 t. 显然法空间 $\Delta = \tilde{\Delta}_t$ 是一个从 $\mathfrak{L}$ 的从空间, 其底空间为 $C_{n,m}$, $Z \in C_{n,m}$ 上的纤维, 则为定义于 $Z_q = \underbrace{Z \times \cdots \times Z}_{q}$ 中的特殊透镜空间 L^p_m. 令 $\pi : \Delta \to C_{n,m}$ 为其投影, 又令 $U^i = U^i(\tilde{\Delta}, t)$ 为 Δ 中对 t 而言的示性类, C^{2i}_p 为 $C_{n,m}$ 中的 (法 p) C 类. 本节的目的, 在证明 Δ 中有下列等式存在:

$$\prod_{i=1}^{q} \sum_{k=0}^{m} i^{m-k} U^{2m-2k} \cup \pi^* C^{2k}_p = 0. \tag{1}$$

为此, 令 $\tilde{\Lambda} = \tilde{S} \times C_{n,m}$, 其中的点可表作 $\{\mathfrak{z}_1, \cdots, \mathfrak{z}_q; Z\}$, 此处 $(\mathfrak{z}_1, \cdots, \mathfrak{z}_q) \in \tilde{S}$, $Z \in C_{n,m}$. 令 $\tilde{\Delta}_i (i = 1, \cdots, q)$ 为 $\tilde{\Lambda}$ 中由条件 $\mathfrak{z}_i \in Z$ 所定的子空间. 显然, t 在 $\tilde{\Lambda}$ 与 $\tilde{\Delta}_i$ 中各引起以 p 为周期而无不变点的拓扑变换, 仍各令之为 t, 并令

$$\Lambda = \tilde{\Lambda}_t = L^p_{n+m} \times C_{n,m}, \quad \Delta_i = (\tilde{\Delta}_i)_t, \tag{2}$$

则 Λ, Δ 都是流形, Δ_i 都是假流形, 且

$$\Delta = \Delta_1 \cap \cdots \cap \Delta_q, \tag{3}$$

又
$$\begin{cases} \text{Dim } L^p_{n+m} = 2q(n+m) - 1, \\ \text{Dim } \Lambda = 2nm + 2q(n+m) - 1, \\ \text{Dim } \Delta = 2nm + 2qm - 1, \\ \text{Dim } \Delta_i = \text{Dim } \Lambda - 2n. \end{cases}$$

令 δ_i 为 Δ_i 的自然定向在 Λ 中所定的 (法 p) 下闭链, 我们将先证明

$$\delta_i \sim \sum_{k=0}^{n} i^{n-k} \xi_{2n-2k} \times [n-k, n, \cdots, n]_c, \quad i = 1, \cdots, q; \tag{4}$$

此处 $\xi_j \in \mu_j X$, X 为自然定向的 L^p_{n+m} 所定的 $2q(n+m)-1$ 维基本下同调类. 若依 §2 末取坐标并对此作 L^p_{n+m} 的标准剖分, 则由 §2 B1° $(0 \leqslant k \leqslant n)$,

$$[2n-2k] \in (q!)^{n+m} \mu_{2q(n+m)-1-2n+2k} X, \tag{5}$$

又由 §2 B3° 及对偶定理,

$$\mu_{2n-2k} X \cdot \mu_{2q(n+m)-1-2n+2k} X = \mu_{2q(n+m)-1} X,$$

故得

$$\xi_{2n-2k} \cdot [2n-2k] \sim [0].$$

由此可知 (4) 式同值于下二在自然定向的 Λ 中的相交公式:

$$\delta_i \cdot ([2n-2k] \otimes [a_1, \cdots, a_m]_c) = \begin{cases} 0, & a_m < k, \\ i^{n-k}, & a_m = k, \end{cases} \tag{4$'$}$$

其中 $\sum\limits_{i=1}^{m} a_i = k$, $k = 0, 1, \cdots, n$. 今证 $(4)'$ 如下:

情形一: $a_m < k$.

令 $\lambda : \tilde{S} \to L^p_{n+m}$ 为 $(\tilde{S}, t)$ 的投影, 又令 $[2n-2k]^*_0$ 为由以下条件所定的一切点 $\lambda(\mathfrak{z}_1, \cdots, \mathfrak{z}_q) \in L^p_{n+m}$ 所成的假流形:

$$\left.\begin{array}{l} \mathfrak{z}_j = (z_1^{(j)}, \cdots, z_{n+m}^{(j)}), \\ \mathfrak{z}_j = 0, \quad j \neq i, \\ z_{n-k+2}^{(i)} = \cdots = z_{n+m}^{(i)} = 0, \quad \arg\ z_{n-k+1}^{(i)} = 0. \end{array}\right\} \tag{6}$$

由 §2 可知 $[2n-2k]^*_0$ 所代表的下链为

$$[2n-2k]_0 \in (-i')^{n-k} (q!)^{n+m} \mu_{2q(n+m)-1-2n-2k} X.$$

与 (5) 比较, 得

$$[2n-2k]_0 \sim (-i')^{n-k} [2n-2k]. \tag{7}$$

依 [1] 的 §3, $[a_1, \cdots, a_m]^*_c$ 中的任意元素 Z 特别须满足条件

$$Z \subset U^{a_m+m} \subset U^{m+k-1}.$$

令 $\eta^* = [2n-2k]^*_0 \times [a_0, \cdots, a_m]^*_c$. 若有 $\{\mathfrak{z}_1, \cdots, \mathfrak{z}_q : Z\} \in \eta^* \cap \Delta_i$, 则应有 $\mathfrak{z}_i \in Z \subset U^{m+k-1}$, 因之

$$z_1^{(i)} = \cdots = z_{n-k+1}^{(i)} = 0.$$

于是由 (6) 式得 $\mathfrak{z}_1=\cdots=\mathfrak{z}_q=0$ 与 $\sum\limits_{i=1}^{q}|\mathfrak{z}_i|^2=1$ 相违. 故 η^*, Δ_i 无公共交点, 而得 $(4)'$ 的上面一式.

情形二: $a_m=k$.

此时 $[a_1,\cdots,a_m]_c=[0,\cdots,0,k]_c$. 取 $[2n-2k]_0^*$ 如前, 而取 $[0,\cdots,0,k]_c^*$ 为由

$$U^{m-1}\subset Z\subset U^{m+k} \tag{8}$$

所定义的假流形. 于是由 (7), 假流形 $\eta^*=[2n-2k]_0^*\times[0,\cdots,0,k]_c^*$ 所代表的 (法 p) 下闭链 $\eta=[2n-2k]_0\otimes[0,\cdots,0,k]_c$ 与 $[2n-2k]\otimes[0,\cdots,0,k]_c$ 间有下述关系:

$$[2n-2k]\otimes[0,\cdots,0,k]_c\sim i^{n-k}\eta. \tag{7$'$}$$

今设 $\{\lambda(\mathfrak{z}_1,\cdots,\mathfrak{z}_q);Z\}\in\eta^*\cap\Delta_i$, 则由 (8) 应有 $\mathfrak{z}_i\in Z\subset U^{m+k}$, 故有 $z_1^{(i)}=\cdots=z_{n-k}^{(i)}=0$, 再由 (6), 知 $\mathfrak{z}_i$ 必须为

$$\mathfrak{z}_i^0:\ z_1^{(i)}=\cdots=z_{n-k}^{(i)}=0,\quad z_{n-k+2}^{(i)}=\cdots=z_{n+m}^{(i)}=0,\quad z_{n-k+1}^{(i)}=1.$$

又由 (8) 及 $\mathfrak{z}_i\in Z$ 知 Z 必须为下复 m 面:

$$Z_0:z_1=\cdots=z_{n-k}=z_{n-k+2}=\cdots=z_{n+1}=0.$$

因之令 $\mathfrak{z}_j^0=0,\ j\neq i$, 即知 η^*, Δ_i 有一且只有一公共交点 $\{\lambda(\mathfrak{z}_1^0,\cdots,\mathfrak{z}_q^0);Z_0\}$.

在 Λ 中, $\{\lambda(\mathfrak{z}_1^0,\cdots,\mathfrak{z}_q^0);Z_0\}$ 有一邻域 $V\times W$ 如下所定. V 为一切 $\lambda(\mathfrak{z}_1,\cdots,\mathfrak{z}_q)$ 所成的子集, 此处 $\mathfrak{z}_k=(z_1^{(k)},\cdots,z_{n+m}^{(k)})$ 如下式:

$$\begin{aligned}&\mathfrak{z}_j:z_r^{(j)}=\lambda_r^{(j)}+i\mu_r^{(j)},j\neq i,\quad r=1,\cdots,n+m,\\&\mathfrak{z}_i:\begin{cases}z_r^{(i)}=\lambda_r^{(i)}+i\mu_r^{(i)},\quad r\neq n-k+1,\\z_{n-k+1}^{(i)}=1-\lambda_{n-k+1}^{(i)}+i\mu_{n-k+1}^{(i)};\end{cases}\end{aligned}$$

其中 λ, μ 取一切满足以下诸式的实数值:

$$\begin{gathered}\sum_{(r,j)\neq(n-k+1,i)}[(\lambda_r^{(i)})^2+(\mu_r^{(i)})^2]+(\mu_{n-k+1}^{(i)})^2+(1-\lambda_{n-1+k}^{(i)})^2=1,\\1-\lambda_{n-k+1}^{(i)}>0.\end{gathered}$$

又 W 由一切下方程组所决定的复 m 面 Z 所组成:

$$Z:z_r=\alpha_{r1}z_{n-k+1}+\sum_{s=2}^{m}\alpha_{rs}z_{n+s},\quad r=1,\cdots,n-k,n-k+2,\cdots,n+1,$$

其中 α 取一切复数值.

显然 V, W 的自然定向就决定了 Λ 的自然定向. 又由 η^* 与 Δ_i 的定义, $\eta^* \cap (V \times W) = \eta'^*$ 及 $\Delta_i \cap (V \times W) = \Delta_i'$ 各由下方程组所定义:

$$\eta'^*: \begin{cases} \lambda_r^{(j)} = \mu_r^{(j)} = 0, j \neq i, \\ \lambda_r^{(i)} = \mu_r^{(i)} = 0, r = n-k+2, \cdots, n+m, \\ \mu_{n-k+1}^{(i)} = 0, \\ \alpha_{rs} = 0, 2 \leqslant s \leqslant m, r = n-k+2, \cdots, n+1, \text{或} 1 \leqslant s \leqslant m, r = 1, \cdots, n-k; \end{cases}$$

$$\Delta_i': \lambda_r^{(i)} + i\mu_r^{(i)} = \alpha_{r1}(1 - \lambda_{n-k+1}^{(i)} + i\mu_{n-k+1}^{(i)}) + \sum_{s=2}^{m} \alpha_{rs}(\lambda_{n+s}^{(i)} + i\mu_{n+s}^{(i)}),$$
$$r = 1, \cdots, n-k, n-k+2, \cdots, n+1.$$

令 $\alpha_{rs} = \beta_{rs} + i\gamma_{rs}$, 此处 β, γ 为实数, 则上式变为

$$\Delta_i': \begin{cases} \lambda_r^{(i)} = \beta_{r1}(1 - \lambda_{n-k+1}^{(i)}) - \gamma_{r1}\mu_{n-k+1}^{(i)} + \sum_{s=2}^{m} (\beta_{rs}\lambda_{n+s}^{(i)} - \gamma_{rs}\mu_{n+s}^{(i)}), \\ \mu_r^{(i)} = \beta_{r1}\mu_{n-k+1}^{(i)} + \gamma_{r1}(1 - \lambda_{n-k+1}^{(i)}) + \sum_{s=2}^{m} (\beta_{rs}\mu_{n+s}^{(i)} + \gamma_{rs}\lambda_{n+s}^{(i)}), \\ \qquad r = 1, \cdots, n-k, n-k+2, \cdots, n+1, \end{cases}$$

η'^* 与 Δ_i' 的唯一交点, 也就是 η^* 与 Δ_i 的唯一交点, 即前面所说的 $0 = \{\lambda(\mathfrak{z}_1^0, \cdots, \mathfrak{z}_q^0); Z_0\}$, 相当于所有的 $\lambda_r^{(j)}, \mu_r^{(j)}, \alpha_{rs}$ 都 $= 0$. 如欲决定二者在 0 的交点数, 可先将 η'^* 移至下列 $\eta_\varepsilon'^*$ 的位置.

$$\eta_\varepsilon'^*: \begin{cases} \lambda_r^{(i)} = \mu_r^{(i)} = 0, \quad j \neq i, \\ \lambda_r^{(i)} = \mu_r^{(i)} = 0, \quad r = n-k+2, \cdots, n+m, \\ \mu_{n-k+1}^{(i)} = \varepsilon, \\ \alpha_{rs} = 0, 2 \leqslant s \leqslant m,\ r = n-k+2 \cdots, n+1 \text{ 或 } 1 \leqslant s \leqslant m, r = 1, \cdots, n-k; \end{cases}$$

其中 $\varepsilon < 1$ 充分小. 易见 $\eta_\varepsilon'^*$ 与 Δ_i' 只交于一点 O_ε 如下:

$$O_\varepsilon: \begin{cases} \lambda_r^{(j)} = \mu_r^{(j)} = 0, (r, j) \neq (n-k+1, i), \\ \lambda_{n-k+1}^{(i)} = 1 - \sqrt{1 - \varepsilon^2}, \mu_{n-k+1}^{(i)} = \varepsilon, \\ \alpha_{rs} = 0,\ 1 \leqslant s \leqslant m,\ r = 1, \cdots, n-k, n-k+2, \cdots, n+1, \end{cases}$$

且 $\eta_\varepsilon'^*$ 与 Δ_i' 在 O_ε 的交点数, 也就是 η'^* 与 Δ_i 在 O 的交点数.

Δ_i' 在 O_ε 的切面为

$$T_i:\begin{cases}d\lambda_r^{(i)}=\sqrt{1-\varepsilon^2}d\beta_{r1}-\varepsilon d\gamma_{r1},\\ d\mu_r^{(i)}=\varepsilon d\beta_{r1}+\sqrt{1-\varepsilon^2}d\gamma_{r1},\end{cases}\quad (r=1,\cdots,n-k,n-k+2,\cdots,n+1),$$

又 $\eta_\varepsilon'^*$ 在 O_ε 的切面为

$$T_\varepsilon:\begin{cases}d\lambda_r^{(j)}=d\mu_r^{(j)}=0,j\neq i,\\ d\lambda_r^{(i)}=d\mu_r^{(i)}=0,r=n-k+2,\cdots,n+m,\\ d\mu_{n-k+1}^{(i)}=0,\\ d\beta_{rs}=d\gamma_{rs}=0,2\leqslant s\leqslant m,r=n-k+2,\cdots,n+1,\text{或 }1\leqslant s\leqslant m,r=1,\cdots,n-k.\end{cases}$$

显见 T_i 与 T_ε 线性无关, 且其相交数为 1, 亦即 $\Delta_i'\cdot\eta_\varepsilon'^*=1$, 或 $\delta_i\cdot\eta=1$. 于是由 $(7)'$ 即得 $(4)'$ 的下一式. 由此完全证明了 (4) 式.

由 (4) 式得

$$\delta_1\cdots\delta_q\sim\prod_{i=0}^{q}\sum_{k=0}^{n}i^{n-k}\cdot\xi_{2n-2k}\otimes[n-k,n,\cdots,n]_c. \tag{9}$$

由 (3) 及维数关系知应有整数 a, 使①

$$\delta_1\cdots\delta_q\sim a\delta, \tag{10}$$

此处 δ 为由自然定向的 Δ 所定的下闭链. 因由 (9),

$$\delta_1\cdots\delta_q\sim(q!)^n\xi_{2qn}\otimes[n,\cdots,n]_c+\cdots\not\sim 0,$$

故

$$a\not\equiv 0 \bmod p. \tag{10$'$}$$

今令

$$\bar\delta_i=\sum_{j=0}^{m}i^{m-j}\xi_{2m-2j}\otimes[\underbrace{n-1,\cdots,n-1}_{j},\underbrace{n,\cdots,n}_{m-j}]_c, \tag{11}$$

$$\bar\delta=\bar\delta_1\cdots\bar\delta_q. \tag{12}$$

因在 $C_{n,m}$ 中有

$$\sum_{j+k=r}[\underbrace{n-1,\cdots,n-1}_{j},\underbrace{n,\cdots,n}_{m-j}]_c\cdot[n-k,n,\cdots,n]_c\sim\begin{cases}0, & r>0,\\ [n,\cdots,n]_c, & r=0;\end{cases}$$

① 参阅 S. Lefschetz. Topology. *Amer. Math. Soc. Coll.*, 1930, 12, Ch. IV, §6.

故有

$$\delta_i\cdot\bar{\delta}_i\sim\sum_{\substack{m\geqslant j\geqslant 0\\ n\geqslant k\geqslant 0}} i^{n+m-j-k}\xi_{2m-2j}\xi_{2n-2k}$$

$$\otimes[\underbrace{n-1,\cdots,n-1}_{j},n,\cdots,n]_c\cdot[n-k,n,\cdots,n]_c$$

$$\sim\sum_{r\geqslant 0}[i^{n+m-r}\xi_{2n+2m-2r}$$

$$\otimes(\sum_{j+k=r}[\underbrace{n-1,\cdots,n-1}_{j},n,\cdots,n]_c\cdot[n-k,n,\cdots,n]_c)]$$

$$\sim i^{n+m}\xi_{2n+2m}\otimes[n,\cdots,n]_c,$$

$$a\delta\cdot\bar{\delta}\sim(q!)^{n+m}(\xi_{2n+2m})^q\otimes[n,\cdots,n]_c\sim 0.$$

由 $(10)'$, 即得

$$\delta\cdot\bar{\delta}\sim 0. \tag{13}$$

今 $\tilde{\Delta}$ 为 $C_{n,m}$ 上的球丛 $\mathfrak{C}^q_{n,m}$ 的丛空间, 令 $\tilde{\pi}$ 为其投影. 在 $\tilde{\Delta}$ 中任取一纤维 $\tilde{S}_0$. 显然 t 也是 $\tilde{S}_0$ 中的一个周期为 p 而无不变点的拓扑变换. 令 $L_0=(\tilde{S}_0)_t$, 则 L_0 为在 $C_{n,m}$ 上 L^p_m 丛的丛空间 Δ 中的一个纤维. 令 $U^i_0=U^i(\tilde{S}_0,t)$, j: $L_0\subset\Delta$, 则 U^i_0 产生 $H^*(L_0)$, 且 $j^*U^i=U^i_0$, 故由 Leray-Hirsch 的一个定理, $H^*(\Delta)$ 有一组加法基为 $\{U^i\cup\pi^*A_p\}$, 其中 $0\leqslant i\leqslant 2qm-1$, 而 $\{A_p\}$ 为 $H^*(C_{n,m})$ 的一组加法基. 由 §1, A6°, $U^i(\tilde{\Lambda},t)=V^i\times 1$, 此处 1 为 $C_{n,m}$ 中的 (法 p) 单位类, 而 $V^i=U^i(\tilde{S},t)$.

今设 $\kappa:\Delta\subset\Lambda=L^p_{n+m}\times C_{n,m}$, 而 $\omega:\Lambda\to L^p_{n+m}$, $\omega':\Lambda\to C_{n,m}$ 为其投影. 同样令 $\tilde{\kappa}:\tilde{\Delta}\subset\tilde{\Lambda}=\tilde{S}\times C_{n,m}$, 而 $\tilde{\omega}:\tilde{\Lambda}\to\tilde{S}$, $\tilde{\omega}':\tilde{\Lambda}\to C_{n,m}$ 为其投影, 则 $(\tilde{\omega}\tilde{\kappa})t=t(\tilde{\omega}\tilde{\kappa})$, 故 $(\omega\kappa)^*V^i=U^i$, 又因 $\omega'\kappa=\pi$, 故有

$$\kappa^*(V^i\otimes A_p)=\kappa^*(\omega^*V^i\cup\omega'^*A_p)=\kappa^*\omega^*V^i\cup\kappa^*\omega'^*A_p=U^i\cup\pi^*A_p. \tag{14}$$

因之 $\kappa^*H^*(\Lambda)=H^*(\Delta)$. 对偶言之, 有

$$\kappa_*:H(\Delta)\subset H(\Lambda). \tag{15}$$

今令 $\qquad D:\quad H_r(\Lambda)\approx H_{d(\Lambda)-r}(\Lambda),\qquad (d(\Lambda)=\dim\Lambda)$

$\qquad\qquad D'\quad H_r(\Delta)\approx H_{d(\Delta)-r}(\Delta),\qquad (d(\Delta)=\dim\Delta)$

各为对于自然定向而言的 Λ 与 Δ 中的对偶变换, 又令 Δ 的自然定向所表的 Δ 的

基本下闭链为 δ_0, 则由 (13) 及关于相交与上下积间关系的一般公式得

$$\begin{aligned}K_* {D'}^{-1}\kappa^* D\{\bar\delta\} &= \kappa_*(\kappa^* D\{\bar\delta\}\cap\{\delta_0\}) = D\{\bar\delta\}\cap\kappa_*(\delta_0)\\ &= \{\bar\delta\}\cdot\kappa_*\{\delta_0\} = \{\bar\delta\}\cdot\{\delta\} = 0.\end{aligned}$$

由此及 (15) 得

$$\kappa^* D\{\bar\delta\} = 0. \tag{16}$$

但由 (11) 及 §1 的 A7°,

$$\begin{aligned}D\{\bar\delta_i\} &= \sum_{k=0}^{m} i^{m-k} D\mu_{2m-2k}X\otimes D\{[\underbrace{n-1,\cdots,n-1,}_{k} n,\cdots,n]_c\}\\ &= \sum_{k=0}^{m} i^{m-k} V^{2m-2k}\otimes C_p^{2k},\end{aligned}$$

故

$$D\{\bar\delta\} = D\{\bar\delta_1\}\cdots D\{\bar\delta_q\} = \prod_{i=1}^{q}\sum_{k=0}^{m} i^{m-k} V^{2m-2k}\otimes C_p^{2k},$$

而由 (14) 得

$$\kappa^* D\{\bar\delta\} = \prod_{i=1}^{q}\sum_{k=0}^{m} i^{m-k} U^{2m-2k}\cup\pi^* C_p^{2k}.$$

于是由 (16) 即得 (1) 式, 即所欲证.

§4 与欧氏空间丛相配的一个透镜空间丛

设 $R_{n,m}$ 为定义于 $n+m$ 维欧氏空间 R^{n+m} 中的 (实) 格拉斯曼流形, $\Re_{n,m}$ 为 $R_{n,m}$ 上的模范 m 欧氏空间丛, 以 $\tilde R_{n,m}$ 为丛空间. 令一切满足

$$\mathfrak{x}_i\in X\in R_{n,m},\quad \sum_{i=1}^{p}|\mathfrak{x}_i|^2 = 1,\quad \sum_{i=1}^{p}\mathfrak{x}_i = 0 \tag{1}$$

的组 $(\mathfrak{x}_1,\cdots,\mathfrak{x}_p;X)$ 所定的空间为 $\tilde\Xi$, 则 $\tilde\Xi$ 显然是一个以 $R_{n,m}$ 为底空间, 且为与 $\Re_{n,m}$ 相配的一个 $(2qm-1)$–球纤维丛 $\tilde{\mathfrak{X}}$ 的丛空间. 在 $\tilde\Xi$ 中有一无不变点而以 p 为周期的拓扑变换

$$t':\{\mathfrak{x}_1,\cdots,\mathfrak{x}_p:X\}\to\{\mathfrak{x}_2,\cdots,\mathfrak{x}_p,\mathfrak{x}_1;X\}. \tag{2}$$

令 $\Xi=\tilde\Xi_{t'}$. 因 t' 显为保纤变换, 故 Ξ 为与 $\tilde\Xi$ 因之 $\Re_{n,m}$ 相配的一个丛 $\mathfrak{X}$ 的丛空间. 命 $\tilde S_X$ 为 $\tilde{\mathfrak{X}}$ 中 $X\in R_{n,m}$ 上的纤维球, 则由 §2, $\mathfrak{X}$ 中的纤维 $(\tilde S_X)_{t'}$ 为定义于欧氏空间 $\underbrace{X\oplus\cdots\oplus X}_{p}=(X)^p$ 中的特殊透镜空间 L_m^p.

今设 $C_{n,m}$ 为定义于酉空间 U^{n+m} 中的复格拉斯曼流形, Δ 为 §3 中定义于 $C_{n,m}$ 上的透镜空间丛 $\mathfrak{L}$ 的丛空间. 在 R^{n+m} 与 U^{n+m} 各取固定坐标后, 可定义自然映像 $\bar{l}: R^{n+m} \to U^{n+m}$ 与 $l: R_{n,m} \to C_{n,m}$(如 [1] 的 §5). 又令

$$h: (\mathfrak{x}_1, \cdots, \mathfrak{x}_p) \to (\mathfrak{z}_1, \cdots, \mathfrak{z}_q)$$

为如 §2 (22) 所定义的映像, 则

$$\{\mathfrak{x}_1, \cdots, \mathfrak{x}_p; X\} \to \{h(\mathfrak{x}_1, \cdots, \mathfrak{x}_p); l(X)\}$$

定义了映像

$$\tilde{\tilde{l}}: \tilde{\Xi} \to \tilde{\Delta}.$$

又如 §3, 令

$$t: (\mathfrak{z}_1, \cdots, \mathfrak{z}_q; Z) \to \{\varepsilon^{-1}\mathfrak{z}_1, \cdots, \varepsilon^{-q}\mathfrak{z}_q; Z\}$$

为 $\tilde{\Delta}$ 中以 p 为周期的拓扑变换, 则显然 $t\tilde{\tilde{l}} = \tilde{\tilde{l}}t'$, 故 $\tilde{\tilde{l}}$ 引起了映像

$$\tilde{l}: \Xi \to \Delta,$$

且由 §1 A2° 有

$$\tilde{l}^* U^i(\tilde{\Delta}, t) = U^i(\tilde{\Xi}, t').$$

其次令 π 与 π' 各为丛 $\mathfrak{L}$ 与 $\mathfrak{X}$ 的投影, 则因 $\tilde{l}$ 显然是保纤映像且满足 $\pi\tilde{l} = l\pi'$, 故有

$$\mathfrak{X} \sim l^* \mathfrak{L}.$$

因据 [1] §5,

$$l^* C_p^{2i} = \begin{cases} 0, & i\text{为奇数}, \\ (-1)^{i/2} P_p^{2i}, & i\text{为偶数}; \end{cases}$$

故令 $U^i = U^i(\tilde{\Xi}, t)$ 时, 从 §3 (1) 式可得

$$\prod_{i=1}^{q} \sum_{k=0}^{m} (-1)^k i^{m-2k}\, U^{2m-4k} \cup \pi'^* P_p^{4k} = 0. \tag{3}$$

今设 $\mathfrak{B} = \{V, \pi, K, R^m, O_m\}$ 为任意 m 欧氏空间丛, 并设 $\mathfrak{B} \sim g^* \mathfrak{R}_{n,m}$. 令 $\mathfrak{B}$ 中任一 $x \in K$ 上的纤维为 R_x. 取一切满足

$$x \in K, \quad \mathfrak{x}_i \in R_x, \quad \sum_{i=1}^{p} |\mathfrak{x}_i|^2 = 1, \quad \sum_{i=1}^{p} \mathfrak{x}_i = 0$$

的组 $(\mathfrak{x}_1, \cdots, \mathfrak{x}_p; R_x)$ 成一空间 $\tilde{\Xi}(\mathfrak{B})$, 并令 t 为下述 $\tilde{\Xi}(\mathfrak{B})$ 中无不变点而以 p 为周期的拓扑变换

$$t: \{\mathfrak{x}_1, \cdots, \mathfrak{x}_p; R_x\} \to \{\mathfrak{x}_2, \cdots, \mathfrak{x}_p, \mathfrak{x}_1; R_x\}.$$

又令 $\Xi(\mathfrak{B}) = (\tilde{\Xi}(\mathfrak{B}))_t$, 则 $\Xi(\mathfrak{B})$ 自然地成一与 $\mathfrak{B}$ 相配的透镜空间从 $\mathfrak{X}(\mathfrak{B})$, 且易见

$$\mathfrak{X}(\mathfrak{B}) \sim g^*\mathfrak{X}.$$

令

$$U^i = U^i(\tilde{\Xi}(\mathfrak{B}), t), \tag{4}$$

并令 φ 为 $\mathfrak{X}(\mathfrak{B})$ 的从投影, 则由 (3) 式即得

$$\prod_{i=0}^{q}\sum_{k=0}^{m'}(-1)^k i^{m-2k}U^{2m-4k} \cup \varphi^* P_p^{4k}(\mathfrak{B}) = 0, \tag{5}$$

或

$$\sum_{k=0}^{m'q} U^{2qm-4k} \cup \varphi^* R_p^{4k}(\mathfrak{B}) = 0, \tag{5$'$}$$

此处 m' 表 $\leqslant m/2$ 的最大整数, 而

$$R_p^{4k}(\mathfrak{B}) = (-1)^k \sum_{k_1+\cdots+k_q=k} 1^{m-2k_1} \cdot 2^{m-2k_2} \cdots q^{m-2k_q} P_p^{4k_1}(\mathfrak{B}) \cdot P_p^{4k_2}(\mathfrak{B}) \cdots P_p^{4k_q}(\mathfrak{B}). \tag{6}$$

特别在 $p = 3$, 因之 $q = 1$ 时,

$$R_3^{4k}(\mathfrak{B}) = (-1)^k P_3^{4k}, \quad 0 \leqslant k \leqslant m'. \tag{7}$$

§5 微分流形上的Понтрягин示性类

设 M 为一 m 维闭解析流形, 具有与其解析构造 $\mathfrak{D}$ 相容的 Riemann 计量 ds^2. 令 $\tilde{M} = \underbrace{M \times \cdots \times M}_{p}$, $d: M \to \tilde{M}$ 为对角映像, 则 $\tilde{M}$ 亦为解析流形且有自然的 "对称" Riemann 计量 $d\tilde{s}^2 = ds_1^2 + \cdots + ds_p^2$, 此处 ds_i^2 为与 ds^2 相当的 $\tilde{M}$ 中第 i 个因子 M 的 Riemann 计量, 而 p 为 > 1 的任意整数. 今对任意 $\tilde{a}$, $\tilde{b} \in \tilde{M}$(或 a, $b \in M$) 定义其距离 $\tilde{d}(\tilde{a}, \tilde{b})$(或 $d(a, b)$) 为一切在 $\tilde{M}$(或 M) 中连接 $\tilde{a}$, $\tilde{b}$(或 a, b) 两点的可距弧长各对 $d\tilde{s}^2$ 与 ds^2 而言的下限. 我们又定义 $\tilde{d}(\tilde{a})$ 为自 $\tilde{a}$ 至 $d(M)$ 任意一点 $\tilde{b}$ 的距离 $\tilde{d}(\tilde{a}, \tilde{b})$ 的下限, 称之为 $\tilde{a}$ 至 $d(M)$ 的距离. 因 M 为闭的, 故有 $\eta > 0$ 存在, 满足以下诸条件:

1°. 若 $\tilde{a}, \tilde{b} \in \tilde{M}$, 而 $\tilde{d}(\tilde{a}, \tilde{b}) \leqslant \eta$, 则有一唯一的对 $d\tilde{s}^2$ 而言的最短线 $\tilde{g}_{\tilde{a}\tilde{b}}$ 连接 $\tilde{a}$, $\tilde{b}$, 且此时 $\tilde{g}_{\tilde{a}\tilde{b}}$ 对 $d\tilde{s}^2$ 而言的长即 $\tilde{d}(\tilde{a}, \tilde{b})$.

2°. 若 $a, b \in M$, 而 $d(a,b) \leqslant \eta$, 则有一唯一的对 ds^2 而言的最短线 $g_{ab} \subset M$ 连接 a, b, 且此时 g_{ab} 对 ds^2 而言的长即为 $d(a,b)$.

3°. 若 $\tilde{a} \in \tilde{M}$ 而 $\tilde{d}(\tilde{a}) \leqslant \eta$, 则有一唯一经过 $\tilde{a}$ 的对 $d\tilde{s}^2$ 而言的最短线 $\tilde{g}_{\tilde{a}}$ 存在与 M 垂直, 且 $\tilde{g}_{\tilde{a}}$ 对 $d\tilde{s}^2$ 之长即为 $\tilde{d}(\tilde{a})$. 此最短线的垂足将记为 $d\tilde{\rho}(\tilde{a})$, $\tilde{\varphi}(\tilde{a}) \in M$.

今设 $\varepsilon > 0$ 而 $\leqslant \eta$. 定义 $\tilde{N}_\varepsilon, \tilde{V}_\varepsilon \subset \tilde{M}$ 为

$$\tilde{N}_\varepsilon = \{\tilde{a}/\tilde{a} \in \tilde{M}, \tilde{d}(\tilde{a}) = \varepsilon\}, \quad \tilde{V}_\varepsilon = \{\tilde{a}/\tilde{a} \in \tilde{M}, \tilde{d}(\tilde{a}) \leqslant \eta\}.$$

显然 $\tilde{N}_\varepsilon$ 是 M 上一个球丛 $\tilde{\mathfrak{N}}_\varepsilon$ 的丛空间. 命 $\mathfrak{T}$ 为 M 上一切单位长 (对于 ds^2 而言) 切矢所定的切球丛, 以 T 为丛空间, $a \in M$ 上的纤维球为 T_a. 今设 $\mathrm{a}_1, \cdots, \mathrm{a}_p$ 为 M 在同一点 $a \in M$ 对于 $\mathfrak{D}$ 而言的切矢, 满足

$$\sum_{i=1}^{n} \mathrm{a}_i = 0, \quad \sum_{i=1}^{p} |\mathrm{a}_i|^2 = 1,$$

则据 [2] 中 §5 预备定理 5 的证明, 可知 $\mathrm{a} = (\mathrm{a}_1, \cdots, \mathrm{a}_p)$ 可视为 $\tilde{M}$ 中与 $d(M)$ 在 $d(a)$ 垂直的一个单位法矢 (对 $d\tilde{s}^2$ 而言). 所有此等矢量 a 作成一个在 M 上的球丛 $\tilde{\mathfrak{N}}$ 的丛空间 $\tilde{N}$. 易见 $\tilde{\mathfrak{N}} \sim \tilde{\mathfrak{N}}_\varepsilon$, 丛映像 $h: \tilde{N} \to \tilde{N}_\varepsilon$ 可定义之如次. 过 $d(a)$ 作对 $d\tilde{s}^2$ 而言的最短线 $g(\mathrm{a})$ 与在 $d(a)$ 的法矢 a 相切, 在 $g(\mathrm{a})$ 上取一点使其与 $d(M)$ 的距离为 ε, 则此点即为 $h(\mathrm{a})$. 若与 §4 中 $\tilde{\mathfrak{X}}(\mathfrak{B})$ 的定义比较, 可见在 p 为奇质数时, $\tilde{\mathfrak{N}} \sim \tilde{\mathfrak{X}}(\mathfrak{T})$, 因之

$$\tilde{\mathfrak{N}}_\varepsilon \sim \tilde{\mathfrak{X}}(\mathfrak{T}) \quad (p\text{为奇质数}).$$

今在 $\tilde{M}$ 中令 t 为下述以 p 为周期的拓扑变换:

$$t: (a_1, \cdots, a_p) \to (a_2, \cdots, a_p, a_1), \quad a_i \in M.$$

因为 $d\tilde{s}^2$ 是"对称"的, 故 $t(\tilde{N}_\varepsilon) \subset \tilde{N}_\varepsilon$, 而 t 也是 $\tilde{N}_\varepsilon$ 中的一个周期为 p 的拓扑变换, 且 t 在 $\tilde{N}_\varepsilon$ 中无不变点. 因之 $(\tilde{N}_\varepsilon, t)$ 成一简单组, 其法空间将令之为 $N_\varepsilon = (\tilde{N}_\varepsilon)_t$. 显然 N_ε 是一个在 M 上的丛 $\mathfrak{N}_\varepsilon$ 的丛空间, 其投影φ定义如 $\varphi\lambda(a_1, \cdots, a_p) = \tilde{\varphi}(a_1, \cdots, a_p)$, 此处 λ 为 $(\tilde{N}_\varepsilon, t)$ 的投影, 而 $\tilde{\varphi}$ 见 3°, 是 $\tilde{\mathfrak{N}}_\varepsilon$ 的投影.

显然在 $h: \tilde{N} \to \tilde{N}_\varepsilon$ 之下, t 在 $\tilde{N}$ 中也引起了一个拓扑变换

$$hth^{-1} = \bar{t}: (\mathrm{a}_1, \cdots, \mathrm{a}_p) \to (\mathrm{a}_2, \cdots, \mathrm{a}_p, \mathrm{a}_1).$$

命 $N = \tilde{N}_{\bar{t}}$, 则依 §4, 当 p 为奇质数时, N 为透镜空间丛 $\mathfrak{X}(\mathfrak{T})$ 的丛空间. 由此易见

$$\tilde{\mathfrak{N}}_\varepsilon \sim \mathfrak{X}(\mathfrak{T}) \quad (p\text{ 为奇质数}). \tag{1}$$

今设 $\mathfrak{D}'$ 是 M 的另一解析构造, 而 $d\tilde{s}'^2$, η', ε', $N'_{\varepsilon'}=(\tilde{N}'_{\varepsilon'})_{t'}$, φ', $\mathfrak{N}'_{\varepsilon'}$ 等定义如上法, 此处 $\varepsilon, \varepsilon' \leqslant \min(\eta, \eta')$. 定义一映像 (参阅 [6] chap. IV)

$$\tilde{f}:\tilde{N}_\varepsilon \to \tilde{N}'_{\varepsilon'}$$

如次. 设 $\tilde{a}\in\tilde{N}_\varepsilon$, $\tilde{g}'_{\tilde{a}}$ 为对 $d\tilde{s}'^2$ 而言从 $\tilde{a}$ 到 $d(M)$ 的最短法线, 其足为 $d\tilde{\varphi}'(\tilde{a})\in d(M)$. 在 $\tilde{g}'_{\tilde{a}}$ 上取一点 $\tilde{a}'$ 使在 $d\tilde{s}'^2$ 下的距离 $\tilde{d}'(\tilde{a}', d\tilde{\varphi}'(\tilde{a}))=\varepsilon'$. 于是 $\tilde{a}'\in\tilde{N}'_{\varepsilon'}$. 我们将定义 $\tilde{f}$ 如 $\tilde{f}(\tilde{a})=\tilde{a}'$.

预备定理　$\tilde{\varphi}\simeq\tilde{\varphi}'\tilde{f}$.

证　对使 $\tilde{d}(\tilde{a})\leqslant\eta$, $\tilde{d}'(\tilde{a})\leqslant\eta'$ 的任意 $\tilde{a}\in\tilde{V}_\eta\cap\tilde{V}'_{\eta'}$, $\tilde{\varphi}(\tilde{a})$, $\tilde{\varphi}'(\tilde{a})\in M$ 皆有定义且 $d(\tilde{\varphi}(\tilde{a}), \tilde{\varphi}'(\tilde{a}))$ 为集合 $\tilde{V}_\eta\cap\tilde{V}'_{\eta'}$ 上的实连续函数. 特别当 $\tilde{a}\in d(M)$ 时, 有 $d(\tilde{\varphi}(\tilde{a}), \tilde{\varphi}'(\tilde{a}))=0$. 因 M, $\tilde{M}$ 都是闭的, 故 $\tilde{V}_\eta\cap\tilde{V}'_{\eta'}$ 亦然. 因之可取 $\varepsilon, \varepsilon'\leqslant\min(\eta, \eta')$ 使对任意 $\tilde{a}\in\tilde{V}_\varepsilon\cap\tilde{V}'_{\varepsilon'}$ 而言, 有 $d(\tilde{\varphi}(\tilde{a}), \tilde{\varphi}'(\tilde{a}))\leqslant\eta$. 我们假定 $\varepsilon, \varepsilon'$ 已取定使满足这一条件, 显然如此取法对结论并无影响. 此时由 3°, 对 ds^2 而言在 M 中有一唯一的最短线 $g_{\tilde{a}}$ 连接 $\tilde{\varphi}(\tilde{a})$ 与 $\tilde{\varphi}'(\tilde{a})$. 若沿 $g_{\tilde{a}}$ 作移动, 即得

$$\tilde{\varphi}\simeq\tilde{\varphi}'',$$

此处 $\tilde{\varphi}'':\tilde{N}_\varepsilon\to M$ 定义如 $\tilde{\varphi}''(\tilde{a})=\tilde{\varphi}'(\tilde{a})$, $\tilde{a}\in\tilde{N}_\varepsilon$. 但由 $\tilde{f}$ 的定义有 $\tilde{\varphi}''=\tilde{\varphi}'\tilde{f}$. 故得定理.

因 t 将 $\tilde{M}$ 中到 $d(M)$ 的对 $d\tilde{s}'^2$ 的最短法线变为到 $d(M)$ 的对 $d\tilde{s}'^2$ 的最短法线, 而其足不变, 故由 $\tilde{f}$ 的定义可知

$$\tilde{f}t=t\tilde{f}.$$

因之 $\tilde{f}$ 引起了它们法空间的映像

$$f:N_\varepsilon\to N'_{\varepsilon'}.$$

同样 $\tilde{\varphi}$, $\tilde{\varphi}'$ 都可与 t 交换, 因之也引起相当法空间的映像 $\varphi:N_\varepsilon\to M$, $\varphi':N'_{\varepsilon'}\to M$, 而 φ, φ' 显即 $\mathfrak{N}$, $\mathfrak{N}'$ 的投影. 与预备定理同法可证

$$\varphi\simeq\varphi' f,$$

因之

$$\varphi^*=f^*\varphi'^*. \tag{2}$$

令 $U^i=U^i(\tilde{N}_\varepsilon, t)$, $U'^i=U^i(\tilde{N}'_{\varepsilon'}, t)$, 则由 §1 A2°,

$$f^*U'^i=U^i. \tag{3}$$

仍设 p 为奇质数, 则由 (1), 在 N_ε 与 $N'_{\varepsilon'}$ 中依 §4 (5)′ 各有下述关系:

$$\sum_{k=0}^{m'q} U^{2qm-4k} \cup \varphi^* R_p^{4k}(\mathfrak{T}) = 0, \tag{4}$$

$$\sum_{k=0}^{m'q} U'^{2qm-4k} \cup \varphi'^* R_p^{4k}(\mathfrak{T}') = 0. \tag{4$'$}$$

将 (4)′ 两边各应用 f^*, 则由 (2), (3) 可得

$$\sum_{k=0}^{m'q} U^{2qm-4k} \cup \varphi^* R_p^{4k}(\mathfrak{T}') = 0. \tag{5}$$

因由 (1), $\mathfrak{N} \sim \mathfrak{X}(\mathfrak{T})$ 满足 Leray-Hirsch 条件, 故比较 (4), (5) 两式得

$$R_p^{4k}(\mathfrak{T}) = R_p^{4k}(\mathfrak{T}'). \tag{6}$$

特别在 $p=3$ 时, (6) 变为

$$P_3^{4k}(\mathfrak{T}) = P_3^{4k}(\mathfrak{T}'). \tag{7}$$

由此即得下述定理.

定理 一个具有解析构造的闭流形因之对于任意可微分闭流形 [7] 的 R 示性类 $R_p^{4k}(\mathfrak{T})$(p 为奇质数) 与此流形的微分构造无关, 而为此流形的拓扑不变量. 特别是法Зпонтрягин 示性类 $P_3^{4k}(\mathfrak{T})$ 为此流形的拓扑不变量.

若 $p=2$, 则 $\tilde{\mathfrak{N}}_\varepsilon \sim \mathfrak{T}$, 而 $\mathfrak{N}_\varepsilon$ 为与 $\mathfrak{T}$ 相配的投影空间丛, 此时据 [1] §7 末附注一, 在 $\mathfrak{N}_\varepsilon$ 中有关系式

$$\sum_{i=0}^{m} [W^1(\mathfrak{R})]^{m-i} \cup \pi^* W^i(\mathfrak{T}) = 0,$$

其中 $\mathfrak{R}$ 为自然地定义于 N_ε 上的 1 欧氏空间丛. 容易证明

$$W^1(\mathfrak{R}) = U^1(\tilde{N}_\varepsilon, t),$$

故上式成为

$$\sum_{i=0}^{m} (U^1)^{m-i} \cup \pi^* W^i(\mathfrak{T}) = 0,$$

此处 $U^1 = U^1(\tilde{N}_\varepsilon, t)$. 于是与前同法可得下述

定理 一个可微分闭流形的 Stiefel-Whitney 示性类与此流形的微分构造无关, 因之是该流形的拓扑不变量.

参考文献

[1] 吴文俊. 论Понтрягин示性类 (I). 数学学报, 1953, 3(4): 291-315.
[2] 吴文俊. 论Понтрягин示性类 (II). 数学学报, 1954, 4(2): 171-198.
[3] Smith P A. Fixed points of periodic transformations. *Lefschetz, Algebraic Topology*, 1942, Appendix B: 350-373.
[4] Smith P A. Periodic and nearly periodic transformations. *Lectures on Topology*, 1941: 159-190.
[5] Wu Wen-Tsün. Sur les puissances de Steenrod. *Colloque de Topologie de Strabourg*, 1947.
[6] Thom R. Espaces fibrés en sphères et carrés de steenrod. *Annales Scientifiques de L'école Normale Supérieure*, 1952, 69: 110-182.
[7] Whitney H. Differentiable manifolds. *Annals of Mathematics*, 1936, 37: 645-680.

On Pontrjagin Classes IV

Abstract The aim of this paper is to give a second proof that the Pontrjagin classes reduced mod 3 of a closed differentiable manifold are topological invariants of that manifold. The proof is different from a preceeding one in that Steenrod powers do not come explicitely in the considerations. A comparison of the two proofs which exhibits besides some general aspects about Steenrod powers will be given later.

19. 论 Понтрягин 示性类 V*

本文是这系列著作中 II [1] 的一个补充. 在 II 中 (参阅 II 的更正 [2]) 我们证明了可微分闭流形的某些示性类特别是法 3 Понтрягин 示性类的拓扑不变性. 它的证明是隐含的 (implicit). 本文目的在进一步求得这些示性类用流形同调构造来表示的显溪 (explicit) 公式, 使我们能就任意可定向的可微分闭流形的这些示性类进行具体的计算. 特别可以获得下述结果:

一个四维或五维可微分闭流形的四维法 3 Понтрягин 示性类必为 0.

以下设 M 是一 m 维闭流形, $H^*(M)$ 为以 I_p 为系数的上同调环, 此处 p 为一质数, I_p 不明写, 其他类似. 在 $p=2$ 时, 我们不作任何假定; 在 $p>2$ 时, 我们假定 M 是可定向的, 且已有固定定向. 我们将以 M_0 表 $H^m(M)$ 中 ($p>2$ 时, 与 M 固定定向相当) 的母素. 依对偶定理, 在 $H^*(M)$ 中可取一组加法基 $\{z_\alpha^i\}$, $z_\alpha^i \in H^i(M)$, $\alpha=1,\cdots,p^i$(p^i 为 M 的 i 维法 p Betti 数), 使

$$\left.\begin{aligned} &z_\alpha^i \cup z_\beta^{m-i} = \varepsilon_\alpha^i \delta_{\alpha',\beta} M_0, \\ &z_1^m = M_0, \quad z_1^0 = 1. \end{aligned}\right\} \tag{1}$$

此处 $\varepsilon_1^0 = \varepsilon_1^m = 1$, ε_α^i 与 p 互质①, 而 α' 与 α 之关系如次:

在 m 不作 $4m'+2$ 形状或 $m \neq 2i$ 时, $\alpha' = \alpha$.

在 $m=4m'+2$ 且 $i=2m'+1$ 时, $p^{2m'+1}$ 必为偶数, 此时

$$\alpha' = \begin{cases} \alpha + \dfrac{1}{2}p^{2m'+1}, & 1 \leqslant \alpha \leqslant \dfrac{1}{2}p^{2m'+1}, \\ \alpha - \dfrac{1}{2}p^{2m'+1}, & \dfrac{1}{2}p^{2m'+1} < \alpha \leqslant p^{2m'+1}. \end{cases}$$

今设

$$z_{\alpha_1}^{i_1} \cup \cdots \cup z_{\alpha_n}^{i_n} = \sum_r c_{\alpha_1,\cdots,\alpha_n;\gamma}^{i_1,\cdots,i_n} z_\gamma^{i_1+\cdots+i_n}, \quad (i_1+\cdots+i_n \leqslant m), \tag{2}$$

特别在 $i_1+\cdots+i_n=m$ 时, 应有

$$c_{\alpha_1,\cdots,\alpha_n;1}^{i_1,\cdots,i_n} = \varepsilon_{\alpha_1}^{i_1} c_{\alpha_2,\cdots,\alpha_n;\alpha'_1}^{i_2,\cdots,i_n}. \tag{3}$$

---

* *Acta Math. Sinica*, 1955, 5: 401-410; *Amer. Math. Soc. Translations, Ser.*, 1964, 2(38): 259-268.

① 在 m 不作 $4m'$ 形状或 $m \neq 2i$ 时, 并可使 $\varepsilon_\alpha^i = \pm 1$.

今对任意整数 $n>1$ 作 $\tilde{M}=\underbrace{M\times\cdots\times M}_{n}$. 在 $p>2$ 时, 由 M 的固定定向可依上式取 $\tilde{M}$ 的固定定向. 命 $\tilde{M}_0$ 为 $H^{mn}(\tilde{M})=H^{mn}(\tilde{M},I_p)(p>2$ 时与上述固定定向相当) 的母素, 则 $\tilde{M}_0=\underbrace{M_0\otimes\cdots\otimes M_0}_{n}$. 又命 $d:M\to\tilde{M}$ 为对角映像,

$$\begin{aligned}D&:H^i(M)\approx H_{m-i}(M),\\ \tilde{D}&:H^i(\tilde{M})\approx H_{mn-i}(\tilde{M}).\end{aligned}$$

各为 M 与 $\tilde{M}$(在 $p>2$ 时与 $M,\tilde{M}$ 所择固定定向相当) 的对偶变换.

设准同构变换

$$\varphi_i^*=\varphi^*:H^i(M)\to H^{i+m(n-1)}(\tilde{M}),\qquad 0\leqslant i\leqslant m,$$

具有下述两项性质:

1° $\varphi^*M_0=\tilde{M}_0$;

2° 对任意 $\tilde{V}\in H^*(\tilde{M})$, 有

$$\varphi^*d^*\tilde{V}=\tilde{V}\cup\varphi^*1.$$

则必有

(a) φ^* 为 1°, 2° 所唯一地确定,

(b) φ^*1 由以下 (4), (4)′ 两式所定,

(c) $\varphi^*:\ H^i(M)\subset H^{i+m(n-1)}(\widetilde{M})$,

(d) $\varphi^*=\tilde{D}^{-1}d_*D$.

证之如次:

$H^m(\tilde{M})$ 有一组加法基为 $\{z_{\alpha_1}^{i_1}\otimes\cdots\otimes z_{\alpha_n}^{i_n}\}$, 此处 $i_1+\cdots+i_n=m$, α_j, i_j 尽取一切可能的值. 因

$$\begin{aligned}(z_{\alpha_1}^{i_1}\otimes\cdots\otimes z_{\alpha_n}^{i_n})\cup\varphi^*1&=\varphi^*d^*(z_{\alpha_1}^{i_1}\otimes\cdots\otimes z_{\alpha_n}^{i_n})\\&=\varphi^*\left(z_{\alpha_1}^{i_1}\cup\cdots\cup z_{\alpha_n}^{i_n}\right)\\&=c_{\alpha_1,\cdots,\alpha_n;1}^{i_1,\cdots,i_n}\varphi^*M_0\\&=c_{\alpha_1,\cdots,\alpha_n;1}^{i_1,\cdots,i_n}\tilde{M}_0,\end{aligned}$$

故由对偶定理, φ^*1 由上式即被完全确定, 且知必如下式所示:

$$\varphi^*1=\sum\varepsilon_{\alpha_1,\cdots,\alpha_n}^{i_1,\cdots,i_n}c_{\alpha_1,\cdots,\alpha_n;1}^{i_1,\cdots,i_n}\qquad(z_{\alpha'_1}^{m-i_1}\otimes\cdots\otimes z_{\alpha'_n}^{m-i_n}),\tag{4}$$

$$\left.\begin{aligned}&\varepsilon^{i_1,\cdots,i_n}_{\alpha_1,\cdots,\alpha_n}=(-1)^{(m-i_1)k_2+(m-i_2)k_3+\cdots+(m-i_{n-1})k_n}\cdot(\varepsilon^{i_1}_{\alpha_1}\cdots\varepsilon^{i_n}_{\alpha_n})^{-1}\\&k_j=i_j+i_{j+1}+\cdots+i_n,\quad j=2,\cdots,n,\end{aligned}\right\}\qquad(4)'$$

其中 $\sum$ 展开于使 $i_1+\cdots+i_n=m$ 的一切可能的 α_j 与 i_j 之上. 因 d^* 有遮盖性, 故由 2° 可得任意 φ^*U, $U\in H^*(M)$, 因之得 (a).

其次, 命 $H_m(M)$, $H_{mn}(\tilde{M})$(在 $p>2$ 时与 M, $\tilde{M}$ 的所择固定定向相当) 的母素为 Z, $\tilde{Z}$, 则对于 $\tilde{V}\in H^i(M)$,

$$\begin{aligned}d_*Dd^*\tilde{V}&=d_*(d^*\tilde{V}\cap Z)=\tilde{V}\cap d_*Z=\tilde{D}\tilde{V}\cdot d_*Z\\&=\tilde{D}(\tilde{V}\cup\tilde{D}^{-1}d_*D1),\end{aligned}$$

或

$$(\tilde{D}^{-1}d_*D)d^*\tilde{V}=\tilde{V}\cup\tilde{D}^{-1}d_*D1.$$

故 $\tilde{D}^{-1}d_*D$ 满足 2°, 显然也满足 1°, 因之从 (a) 得 (d). 因 $d_*:H_i(M)\subset H_i(\tilde{M})$, 故又有 (c) .

今设 M 为可微分的, 则由 [1] §5 可定义 $\tilde{M}$ 中 $d(M)$ 的一个邻域 A, 及准同构变换

$$\begin{aligned}&\tau_i^*=\tau^*:H^i(M)\approx H^{i+m(n-1)}(A),\\&k^*:\quad H^i(A)\to H^i(\tilde{M}),\\&\bar{\tau}_i^*=\bar{\tau}^*=k^*\tau^*:\quad H^i(M)\to H^{i+m(n-1)}(\tilde{M}).\end{aligned}$$

由该文 §5 知 $(-1)^{(i+1)m(n-1)}\bar{\tau}_i^*=\varphi_i^*$ 满足 1°, 2°.

今依 [3, 4] 定义

$$\left.\begin{aligned}&\mathfrak{P}^k_p=S^{2k}_q,\qquad p=2\text{ 时},\\&\mathfrak{P}^k_pU=(-1)^{q(i-2k)(i-2k-1)/2}(q!)^{-i+2k}St^{2k(p-1)}_pU,\quad p>2\text{ 时},\end{aligned}\right\}\qquad(5)$$

此处 $U\in H^i(E,I_p)$, E 为任意空间, $q=(p-1)/2$. 则对于任意 $U\in H^i(M)$, 有

$$\begin{aligned}\mathfrak{P}^k_p\varphi^*U&=(-1)^{(i+1)m(n-1)}\mathfrak{P}^k_p\bar{\tau}^*U=(-1)^{(i+1)m(n-1)}k^*\mathfrak{P}^k_p\tau^*U\\&=(-1)^{(i+1)m(n-1)}\bar{\tau}^*(\tau^{*-1}\mathfrak{P}^k_p\tau^*U)=\varphi^*(\tau^{*-1}\mathfrak{P}^k_p\tau^*U),\end{aligned}$$

或

$$\mathfrak{P}^k_p\varphi^*H^*(M)\subset\varphi^*H^*(M).$$

同样在 $p=2$ 时, 有

$${S_q}^k\varphi^*H^*(M)\subset\varphi^*H^*(M).$$

特别应有

$$\left.\begin{aligned} p>2\text{时},\quad & \mathfrak{P}_p^k\varphi^*1=\varphi^*R_{p,k,n},\\ p=2\text{时},\quad & {S_q}^k\varphi^*1=\varphi^*R_{2,k,n},\end{aligned}\right\}\quad R_{p,k,n}\in H^*(M). \tag{6}$$

由 [1, 2] 知 $n=2$, $p=2$ 时,

$$R_{p,k,n}=W^k, \tag{7}$$

而在 $n=3$, $p>2$ 时, 则有

$$\bar{\tau}^*Q_p^{2t(p-1)}=\lambda_2^{t-m}St_p^{2t(p-1)}\bar{\tau}^*1,$$

此处 $\lambda_2=(-1)^q(q!)^2$. 因之有 $n=3$, $p>2$ 时,

$$R_{p,k,n}=Q_p^{2k(p-1)}, \tag{8}$$

其中 W^k 为 M 的 Stiefel-Whitney 示性类, 而 $Q_p^{2k(p-1)}$ 如 [1] 所示, Q_p^0 则定义为法 p 单位类.

在 [5] 中, 我们曾从 (6), (7) 算出 W^k 倚赖于 $H^*(M)$ 中环及平方构造的明确公式, 今用同样方法计算 $R_{p,k,n}$ 如次.

兹先设 $p>2$, 并设

$$R_{p,j,n}=\sum_\beta a_{\beta,j}z_\beta^{2j(p-1)}, \tag{9}$$

于是

$$\begin{aligned}\varphi^*R_{p,k,n}=&\varphi^*d^*(1\otimes\cdots\otimes 1\otimes R_{p,k,n})=(1\otimes\cdots\otimes 1\otimes R_{p,k,n})\cup\varphi^*1\\ =&(1\otimes\cdots\otimes 1\otimes\sum_\beta a_{\beta,k}z_\beta^{2k(p-1)})\\ &\cup\sum_{\substack{i_1+\cdots+i_n=m\\ \alpha_j}}\varepsilon_{\alpha_1,\cdots,\alpha_n\alpha_1,\cdots,\alpha_n;1}^{i_1,\cdots,i_n i_1,\cdots,i_n}(z_{\alpha_1'}^{m-i_1}\otimes\cdots\otimes z_{\alpha_n'}^{m-i_n})\\ =&\sum_{\substack{i_1+\cdots+i_n=m\\ \beta,\alpha_j}}\varepsilon_{\alpha_1,\cdots,\alpha_n}^{i_1,\cdots,i_n}c_{\alpha_1,\cdots,\alpha_n;1}^{i_1,\cdots,i_n}a_{\beta,k}z_{\alpha_1'}^{m-i_1}\\ &\otimes\cdots\otimes z_{\alpha_{n-1}'}^{m-i_{n-1}}\otimes z_\beta^{2k(p-1)}z_{\alpha_n'}^{m-i_n}.\end{aligned}$$

在此展开式中欲获得

$$\zeta_{\gamma,k}=z_\gamma^{2k(p-1)}\otimes z_1^m\otimes\cdots\otimes z_1^m$$

一项, 必须有

$$\begin{aligned}&i_2=\cdots=i_{n-1}=0,\quad i_n=2k(p-1),\quad i_1=m-2k(p-1),\\ &\alpha_2=\cdots=\alpha_{n-1}=1,\quad \alpha_n=\beta,\quad \alpha_1=\gamma'.\end{aligned}$$

此时与之相当的

$$\varepsilon_{\alpha_1,\cdots,\alpha_n}^{i_1,\cdots,i_n} = \left(\varepsilon_{\gamma'}^{m-2k(p-1)}\varepsilon_{\beta}^{2k(p-1)}\right)^{-1},$$
$$c_{\alpha_1,\cdots,\alpha_n;1}^{i_1,\cdots,i_n} = c_{\gamma',1,\cdots,1,\beta;1}^{i_1,0,\cdots,0,i_n} = \varepsilon_{\gamma'}^{m-2k(p-1)}\delta_{\gamma\beta}.$$

因之 $\zeta_{\gamma,k}$ 在 $\varphi^* R_{p,k,n}$ 中的系数为

$$\sum_{\substack{i_1+\cdots+i_n=m\\ \beta,\alpha_j}} \varepsilon_{\alpha_1,\cdots,\alpha_n}^{i_1,\cdots,i_n} c_{\alpha_1,\cdots,\alpha_n;1}^{i_1,\cdots,i_n} a_{\beta,k}\varepsilon_{\beta}^{2k(p-1)}$$
$$= \sum_{\beta} (\varepsilon_{\gamma'}^{m-2k(p-1)}\varepsilon_{\beta}^{2k(p-1)})^{-1} \cdot \varepsilon_{\gamma'}^{m-2k(p-1)}\delta_{\gamma\beta}\cdot a_\beta, k\cdot\varepsilon_{\beta}^{2k(p-1)} = \alpha_{\gamma,k}.$$

其次, 设

$$\mathfrak{P}_p^j z_\alpha^i = \sum_\beta s_{\alpha,\beta}^{i,j} z_\beta^{i+2j(p-1)}, \tag{10}$$

因对任意空间的任意类 U_1, U_2, 应有 [3, 4]

$$\mathfrak{P}_p^j(U_1\otimes U_2) = \sum_{j_1+j_2=j} \mathfrak{P}_p^{j_1}U_1\otimes\mathfrak{P}_p^{j_2}U_2, \tag{11}$$

$$\mathfrak{P}_p^j(U_1\cup U_2) = \sum_{j_1+j_2=j} \mathfrak{P}_p^{j_1}U_1\cup\mathfrak{P}_p^{j_2}U_2. \tag{11$'$}$$

故

$$\mathfrak{P}_p^j\varphi^*1 = \sum_{\substack{i_1+\cdots+i_n=m\\ \alpha}} \varepsilon_{\alpha_1,\cdots,\alpha_n}^{i_1,\cdots,i_n} c_{\alpha_1,\cdots,\alpha_n;1}^{i_1,\cdots,i_n}\mathfrak{P}_p^j\left(z_{\alpha_1'}^{m-i_1}\otimes\cdots\otimes z_{\alpha_n'}^{m-i_n}\right)$$
$$= \sum_{\substack{i_1+\cdots+i_n=m\\ j_1+\cdots+j_n=j\\ \alpha}} \varepsilon_{\alpha_1,\cdots,\alpha_n}^{i_1,\cdots,i_n} c_{\alpha_1,\cdots,\alpha_n;1}^{i_1,\cdots,i_n}\mathfrak{P}_p^{j_1} z_{\alpha_1'}^{m-i_1}\otimes\cdots\otimes\mathfrak{P}_p^{j_n} z_{\alpha_n'}^{m-i_n}$$
$$= \sum_{\substack{i_1+\cdots+i_n=m\\ j_1+\cdots+j_n=j\\ \alpha,\gamma}} \varepsilon_{\alpha_1,\cdots,\alpha_n}^{i_1,\cdots,i_n} c_{\alpha_1,\cdots,\alpha_n;1}^{i_1,\cdots,i_n} s_{\alpha_1',\gamma_1}^{m-i_1,j_1}\cdots$$
$$s_{\alpha_n',\gamma_n}^{m-i_n,j_n} z_{\gamma_1}^{m-i_1+2j_1(p-1)}\otimes\cdots\otimes z_{\gamma_n}^{m-i_n+2j_n(p-1)},$$

在此展开式中仍考察产生 $\zeta_{\gamma,k}$ 的项, 此时必须有

$$\left.\begin{aligned} &j_1+\cdots+j_n = j,\\ &i_2 = 2j_2(p-1),\cdots,i_n = 2j_n(p-1),\\ &i_1 = m+2j_1(p-1)-2j(p-1), \end{aligned}\right\} \tag{12}$$

$$\gamma_2 = \cdots = \gamma_n = 1, \quad \gamma_1 = \gamma.$$

由此知 $\zeta_{\gamma,k}$ 在 $\mathfrak{P}_p^j \varphi^* 1$ 的展开式中的系数为

$$\sum \left(\varepsilon_{\alpha_1}^{i_1} \cdots \varepsilon_{\alpha_n}^{i_n}\right)^{-1} c_{\alpha_1,\cdots,\alpha_n;1}^{i_1,\cdots,i_n} s_{\alpha_1',\gamma}^{m-i_1,j_1} s_{\alpha_2',1}^{m-i_2,j_2} \cdots s_{\alpha_n',1}^{m-i_n,j_n},$$

其中 i_r, j_r 须满足 (12), 而 $\sum$ 展开于一切可能的 j_r, α_r 之上. 由 (6), 与前比较知此系数即为 $a_{\gamma,k}$:

$$a_{\gamma,k} = \sum \left(\varepsilon_{\alpha_1}^{i_1} \cdots \varepsilon_{\alpha_n}^{i_n}\right)^{-1} c_{\alpha_1,\cdots,\alpha_n;1}^{i_1,\cdots,i_n} s_{\alpha_1',\gamma}^{m-i_1,j_1} s_{\alpha_2',1}^{m-i_2,j_2} \cdots s_{\alpha_n',1}^{m-i_n,j_n}. \tag{13}$$

今定义 $S_p^{2j(p-1)} \in H^{2j(p-1)}(M)$, $j \geqslant 0$ 为

$$S_p^{2j(p-1)} \cup X^{m-2j(p-1)} = \mathfrak{P}_p^j X^{m-2j(p-1)}, \tag{14}$$

此处 $X^{m-2j(p-1)} \in H^{m-2j(p-1)}(M)$ 任意. 由 (10), 可知

$$S_p^{2j(p-1)} = \sum_{\alpha} \left(\varepsilon_\alpha^{2j(p-1)}\right)^{-1} s_{\alpha_1',1}^{m-2j(p-1),j} z_\alpha^{2j(p-1)}. \tag{15}$$

又定义 $T_p^{2i(p-1)} \in H^{2i(p-1)}(M)$, $i \geqslant 0$ 为

$$T_p^{2i(p-1)} = \sum_j \mathfrak{P}_p^{i-j} S_p^{2j(p-1)}. \tag{16}$$

于是由 (3), (9), (10), (11)′, (12), (13), (15), (16) 得

$$\begin{aligned}
R_{p,j,n} &= \sum_\gamma a_{\gamma,j} z_\gamma^{2j(p-1)} \\
&= \sum \left(\varepsilon_{\alpha_1}^{i_1} \cdots \varepsilon_{\alpha_n}^{i_n}\right)^{-1} c_{\alpha_1,\cdots,\alpha_n;1}^{i_1,\cdots i_n} s_{\alpha_2',1}^{m-i_2,j_2} \cdots s_{\alpha_n',1}^{m-i_n,j_n} \left(\sum_\gamma s_{\alpha_1',\gamma}^{m-i_1,j_1} z_\gamma^{2j(p-1)}\right) \\
&= \sum \left(\varepsilon_{\alpha_1}^{i_1} \cdots \varepsilon_{\alpha_n}^{i_n}\right)^{-1} c_{\alpha_1,\cdots,\alpha_n;1}^{i_1,\cdots i_n} s_{\alpha_2',1}^{m-i_2,j_2} \cdots s_{\alpha_n',1}^{m-i_n,j_n} \mathfrak{P}_p^{j_1} z_{\alpha_1'}^{m-i_1} \\
&= \sum_{j_1} \mathfrak{P}_p^{j_1} \sum \left(\varepsilon_{\alpha_2}^{i_2} \cdots \varepsilon_{\alpha_n}^{i_n}\right)^{-1} s_{\alpha_2',1}^{m-i_2,j_2} \cdots s_{\alpha_n',1}^{m-i_n,j_n} \left(\sum_{\alpha_1} c_{\alpha_2,\cdots,\alpha_n;\alpha_1'}^{i_2,\cdots,i_n} z_{\alpha_1'}^{m-i_1}\right) \\
&= \sum \mathfrak{P}_p^{j_1} \left(\sum \left(\varepsilon_{\alpha_2}^{i_2} \cdots \varepsilon_{\alpha_n}^{i_n}\right)^{-1} s_{\alpha_2',1}^{m-i_2,j_2} \cdots s_{\alpha_n',1}^{m-i_n,j_n} z_{\alpha_2}^{i_2} \cup \cdots \cup z_{\alpha_n}^{i_n}\right) \\
&= \sum \mathfrak{P}_p^{j_1} \left(\sum \left(\varepsilon_{\alpha_2}^{i_2}\right)^{-1} s_{\alpha_2',1}^{m-i_2,j_2} z_{\alpha_2}^{i_2} \cup \cdots \cup \sum \left(\varepsilon_{\alpha_n}^{i_n}\right)^{-1} s_{\alpha_n',1}^{m-i_n,j_n} z_{\alpha_n}^{i_n}\right) \\
&= \sum \mathfrak{P}_p^{j_1} \left(S_p^{2j_2(p-1)} \cup \cdots \cup s_p^{2j_n(p-1)}\right) \\
&= \sum_{j_2'+\cdots+j_n'=j_1} \mathfrak{P}_p^{j_2'} S_p^{2j_2(p-1)} \cup \cdots \cup \mathfrak{P}_p^{j_n'} S_p^{2j_n(p-1)} \\
&= \sum_{k_1+\cdots+k_{n-1}=j} T_p^{2k_1(p-1)} \cup \cdots \cup T_p^{2k_{n-1}(p-1)}.
\end{aligned}$$

特别在 $n=3$ 时, 由 (8) 得下述 (参阅 [9]):

定理 1 对于一已有固定定向的可定向可微分闭流形, 其 Q 示性类可如下表示:

$$Q_p^{2j(p-1)}=\sum_{j_1+j_2=j}T_p^{2j_1(p-1)}\cup T_p^{2j_2(p-1)},\tag{17}$$

其中 $T_p^{2j(p-1)}$ 定义如 (14), (16).

若 $p=3$, 则因

$$Q_3^{4j}=\sum_{j_1+j_2=j}P_3^{4j_1}\cup P_3^{4j_2},$$

$$P_3^0=T_3^0=S_3^0=1.$$

故从 (17) 可得

定理 2 对于一已有固定定向的可定向可微分闭流形, 其法 3 Понтрягин示性类可表示为

$$P_3^{4j}=T_3^{4j}.\tag{18}$$

因 $T_p^{2j(p-1)}$ 由定义完全由流形的 Steenrod 幂结构与环结构所决定, 而此二者皆为流形的拓扑不变量, 因之 (17), (18) 已隐含了示性类 $Q_p^{2j(p-1)}$, 特别是法 3 Понтрягин示性类的拓扑不变性. (17), (18) 又给示了如何由 Steenrod 幂结构与环结构以具体计算 $Q_p^{2j(p-1)}$ 及 P_3^{4j} 的明确公式.

今设流形的维数 $m=2(p-1)$ 或 $2p-1$, 则由 $S_p^{2j(p-1)}$ 的定义及 $\mathfrak{P}_p^j$ 的性质 [3, 4], 得

$$S_p^{2j(p-1)}=0,\quad j>0.$$

因之 $T_p^{2j(p-1)}=0, j>0$, 而有

定理 3 设质数 $p>2$, 则在 $2(p-1)$ 或 $2p-1$ 维可微分的定向闭流形中,

$$Q_p^{2(p-1)}=0.$$

特别在 $p=3$ 时, 4 维或 5 维可微分定向闭流形的

$$P_3^4=0.$$

若 M 是不可定向的, 则可考虑 M 的二叶可定向遮盖流形 M' 及投影 π. 因 p 为奇质数而 $\pi^*: H^*(M)\subset H^*(M')$. 故从 $Q_p^{2(p-1)}(M')=\pi^*Q_p^{2(p-1)}(M)=0$ 可得 $Q_p^{2(p-1)}(M)=0$. 换言之, 上述定理不论 M 可定向与否均成立.

以上均设 $p>2$. 在 $p=2$ 时, 同样亦可计算如次. 设

$$R_{2,j,n}=\sum_{\beta} b_{\beta,j} z_{\beta}^{j},$$
$$S_q^j z_{\alpha}^i=\sum s_{\alpha,\beta}^{i,j} z_{\beta}^{i+j},$$

由计算得

$$\varphi^* R_{2,j,n}=\sum\nolimits_1 b_{\beta,j} c_{\alpha_1,\cdots,\alpha_n;1}^{i_1,\cdots,i_n} c_{\beta,\alpha_n;\gamma}^{j,m-i_n} z_{\alpha_1\cdot}^{m-i_1}\otimes\cdots\otimes z_{\alpha_{n-1}}^{m-i_{n-1}}\otimes z_{\gamma}^{j+m-i_n},$$
$$S_q^j\varphi^* 1=\sum\nolimits_2 c_{\alpha_1,\cdots,\alpha_n;1}^{i_1,\cdots,i_n} s_{\alpha_1,\beta_1}^{m-i_1,j_1}\cdots s_{\alpha_n,\beta_n}^{m-i_n,j_n} z_{\beta_1}^{m-i_1+j_1}\otimes\cdots\otimes z_{\beta_n}^{m-i_n+j_n},$$

其中 Σ_1 展开于使 $i_1+\cdots+i_n=m$ 的一切可能的 i, α 和 β 上, 又 Σ_2 展开于使 $j_1+\cdots+j_n=j$, $i_1+\cdots+i_n=m$ 的一切可能的 i, j, α, β 上.

比较两式中 $z_{\gamma}^j\otimes z_1^m\otimes\cdots\otimes z_1^m$ 的系数得

$$b_{\gamma,j}=\sum_{\substack{i_1+\cdots+i_n=m\\ \alpha}} c_{\alpha_1,\cdots,\alpha_n;1}^{i_1,\cdots,i_n} s_{\alpha_1,\gamma}^{m-i_1,i_1+j-m} s_{\alpha_2,1}^{m-i_2,i_2}\cdots s_{\alpha_n,1}^{m-i_n,i_n}.$$

定义 $U^j, V^j\in H^j(M)$ 为

$$U^j\cup X^{m-j}=S_q^j X^{m-j},\quad X^{m-j}\in H^{m-j}(M), \tag{19}$$

或即

$$U^j=\sum s_{\alpha,1}^{m-i,j} z_{\alpha}^j,$$
$$V^j=\sum S_q^{j-i} U^i, \tag{20}$$

则与前同法可得

$$R_{2,j,n}=\sum_{j_1+\cdots+j_{n-1}=j} V^{j_1}\cup\cdots\cup V^{j_{n-1}}, \tag{21}$$

特别在 $n=2$ 时, 有 [5]

定理 4　任一可微分闭流形的 W 示性类可依下式由此流形的 Steenrod 平方构造与环构造所明确决定:

$$W^j=V^j=\sum_i S_q^{j-i} U^i. \tag{22}$$

(21) 式即 [5] 中的主要公式, 凡关于流形上 W 示性类的已知性质, 特别如 Whitney 在 [6, 7] 以及 Stiefel 在 [8] 中所得的结果, 都是 (21) 的简单后果.

定理 1 与定理 4 使我们在任一闭流形, 不论其是否可微分, 都可定义 W 示性类 W^j 及 (流形可定向并对某一固定定向而言的) Q 示性类 $Q_p^{2j(p-1)}$.

参考文献

[1] 吴文俊. 论Понтрягин示性类 II. 数学学报 4 卷 2 期, 171-199.

[2] 同上更正, 数学学报 4 卷 4 期.

[3] Steenrod N E. Cyclic reduced powers of cohomology classes. *Proc.Nat.Acad.Sci.*, 1933, 39: 213-223.

[4] Borel A and Serre J P. Groupes de Lie et puissances réduites de Steenrod. *Amer.J.of Math.*, 1953, 75: 409-448.

[5] Wu Wen-tsün. Classes caractéristiques et i-carrés d'une variété. *C. R. Paris*, 1950, 230: 508-511.

[6] Whitney H. On the theory of sphere bundles. *Proc. Nat. Acad. Sci.*, 1940, 26: 148-153.

[7] Whitney H. On the topology of differentiable manifolds. Lectures in Topology. Univ, of Mich. Press, 1941.

[8] Stiefel E. Richtungsfelder und Fernparallelismus in Mannigfaltigkeiten. *Comm. Math. Helv.*, 1936, 8: 3-51.

[9] Hirzebruch F. On Steenrod's reduced powers, the index of inertia, and the Todd genus. *Proc. Nat. Acad. Sci.*, 1953, 39: 951-956.

On Pontrjagin Classes V

Abstract This paper gives the explicit formula for Q-classes introduced in II of this series of papers as follows. Let M be an oriented closed differentiable manifold of dimension m. For any odd prime p, let $S_p^{2j(p-1)}$, $T_p^{2j(p-1)} \in H^{2j(p-1)}(M, I_p)$ be defined by

$$S_p^{2j(p-1)} \cup X^{m-2j(p-1)} = \mathfrak{P}_p^i X^{m-2j(p-1)}, \quad X^{m-2j(p-1)} \in H^{m-2j(p-1)}(M, I_p),$$

and

$$T_p^{2j(p-1)} = \sum_i \mathfrak{P}_p^{j-i} S_p^{2i(p-1)},$$

in which $\mathfrak{P}_p^j$ are Steenrod powers in the new notations of J. P. Serre. Then the Q-classes are given by

$$Q_p^{2j(p-1)} = \sum_{j_1+j_2=j} T_p^{2j_1(p-1)} \cup T_p^{2j_2(p-1)}.$$

In particular, for $p = 3$, we deduce for the mod 3 Pontrjagin classes P_3^{4j} of the oriented manifold M:

$$P_3^{4j} = T_3^{4j}.$$

As a corollary, we get the following result: In an oriented closed differentiable manifold M of dimension $2(p-1)$ or $2p-1$,

$$Q_p^{2(p-1)} = 0.$$

In particular, we have $P_3^4 = 0$ in an oriented closed differentiable manifold of dimension 4 or 5.

20. On the Realization of Complexes in Euclidean Spaces I*

Abstract It was early known that any n-dimensional abstract complex may be realized in a $(2n+1)$-dimensional Euclidean space R^{2n+1}. From this theorem, whose proof is quite simple, it follows that the $(2n+1)$-dimensional Euclidean space contains in reality all imaginable n-dimensional complexes. However, the complete recognization of all n-dimensional complexes in an Euclidean space of a given dimension m where $m < 2n+1$, is a problem much more difficult which cannot, it seems, be solved completely in the near future. Among the miscellaneous results so far obtained along this line the most remarkable one is no doubt that of Van Kampen [3, 4] and Flores [5], who first proved the existence of n-dimensional complexes which, even under further subdivisions, cannot be realized in an R^{2n}.

The invariant by means of which Van Kampen was able to conclude the non-realizability of a (finite simplicial) n-dimensional complex in an R^{2n} may be described as follows. Denote the k-dimensional simplexes of the given n-dimensional complex K by S_i^k. Any two simplexes of K with no vertices in common will be said to be disjoint. Let A be the set of all unordered index pairs (i, j), corresponding to pairs of disjoint n-dimensional simplexes S_i^n and S_i^n. Construct a vector space $\mathfrak{L}$ on the ring of integers with dimension equal to the number of elements in A. Any vector of $\mathfrak{L}$ may then be represented by a system of integers (α_{ij}) where $(i, j) \in A$. To each pair of disjoint simplexes S_α^{n-1} and S_l^n in K a certain vector $V_{l\alpha} = (\alpha_{ij})$ of $\mathfrak{L}$ may be determined in the following manner. If both $i, j \neq l$ or one of them, say $j = l$, but S_a^{n-1} is not a face of S_i^n, then we put $\alpha_{ij} = 0$. Otherwise we put $\alpha_{il} = \pm 1$ (with sign conveniently chosen). Two vectors P, P' of $\mathfrak{L}$ will then be said to be equivalent, if $P - P'$ is a certain linear combination with integral coefficients of vectors of form $V_{l\alpha}$ above defined. The vectors of $\mathfrak{L}$ are thus distributed in such equivalence classes.

Take now an arbitrary simplicial subdivision K_1 of K and try to realize K_1 in R^{2n}

* First published in Chinese in *Acta Mathemalica Sinica*, 1955, V(4): 505-552. *Amer. Math. Soc. Translations, Ser.*, 1968, 2(78): 137-184.

as far as possible. We shall obtain then some "almost true" realization such that parts $'S_i^k$ and $'S_j^l$, corresponding to disjoint S_i^k and S_i^l of K will be disjoint in R^{2n} when $k+l<2n$, while they intersect only in isolated points when $k=l=n$. With respect to an orientation arbitrarily chosen of R^{2n}, $'S_i^n$ and $'S_j^n$ determine then a definite intersection number $\pm\alpha_{ij}$ (with sign conveniently chosen). These numbers determine in turn a vector $P=(\alpha_{ij})$ of $\mathfrak{L}$. Van Kampen's work shows that, whatever be the subdivision K_1 of K and the "almost true" realization of K_1 in R^{2n}, the corresponding vectors P always belong to one and the same equivalence class in $\mathfrak{L}$. It follows that this equivalence class is an invariant of the complex K. It is evident that the belonging of the zero vector to this invariant equivalence class is a necessary condition for the existence of "true" realization of K in R^{2n}. It is this invariant which has enabled Van Kampen to assure the existence of n-dimensional complexes non-realizable in R^{2n}. On the other hand, Van Kampen failed to ascertain whether the above necessary condition is also sufficient and the problem of characterizing n-dimensional complexes in R^{2n} remains unsettled up to the present. Moreover the method of Van Kampen-Flores cannot be seen to be readily generalizable to the realizability in R^m, m being arbitrary. We remark also that whether Van Kampen's invariant is a "topological" invariant of the space of K, or even whether it is a combinatorial invariant of K_1, cannot be decided from his works.

At the time of Van Kampen and Flores the cohomology theory has not yet been created. To get a deeper insight of their results we will reformulate them in the modern terminology of cohomology. Their statements will then become clear and natural as follows. From the given simplicial complex (of any dimension) let us construct a subcomplex $\tilde{K}^*$ of $K\times K$, consisting of all cells $\sigma\times\tau$ such that σ, τ are disjoint in K. Identify each pair $\sigma\times\tau$ and $\tau\times\sigma$ of $\tilde{K}^*$ to the same cell $\sigma*\tau=\tau*\sigma$, we get a cell complex K^*. Suppose that the cells $\sigma\in K$ are oriented and let us orient the cells $\sigma*\tau$ of K^* as $\sigma\times\tau$ in the product complex $K\times K$, such that

$$\sigma*\tau=(-1)^{\dim\sigma\dim\tau}\tau*\sigma. \tag{1}$$

Then for dim $K=n$, any vector P in $\mathfrak{L}$ may be regarded as an integral $2n$-dimensional cocycle of K^*, and the equivalence of vectors in $\mathfrak{L}$ is the same as the cohomologousness of their corresponding cocycles. It follows that Van Kampen's invariant is essentially an integral cohomology class in K^*.

From this reformulation we may naturally extend Van Kampen's method to the realization problem of complexes of arbitrary dimension in Euclidean space of arbitrary dimension m. For this let us take an arbitrary simplicial subdivision K_1 of K and try to realize K_1 as much as possible in R^m, such that any two simplexes of K_1 are in general position whenever possible. Let the chain in R^m thus obtained, corresponding to any $\sigma \in K$, be denoted by σ'. Then, with respect to a fixed orientation of R^m, to any two disjoint simplexes σ, τ in K with sum of dimensions just equal to m, there corresponds a definite intersection number $\phi\ (\sigma', \tau')$. Let $I_{(m)}$ be either the additive group of integers I or the group of integers mod 2 I_2, depending on m, and $\rho_{(m)}$ the corresponding identity or reduction mod 2. Then an m-dimensional cochain $\varphi \in C^m(K^*, I_{(m)})$ may be defined by

$$\varphi(\sigma * \tau) = \varepsilon_r \rho_{(m)} \phi(\sigma', \tau'), \quad \dim \sigma = r, \quad \dim \tau = m - r, \tag{2}$$

where $\varepsilon_r = +1$ or -1, depending on r. To make φ a cocycle on coefficient group $I_{(m)}$ and to make the definition of φ consistent with (1) we should take ε_r such that

$$\rho_{(m)}\varepsilon_r + \rho_{(m)}\varepsilon_{r+1} = 0,$$

$$\rho_{(m)}\varepsilon_r = \rho_{(m)}\varepsilon_{m-r}.$$

If we make the further restriction that $\varepsilon_0 = +1$, then to make the above equations consistent, we may take

$$I_{(m)} = \begin{cases} I, & \text{when } m = \text{even}, \\ I_2, & \text{when } m = \text{odd} \end{cases}$$

and to choose ε_r to be, say $(-1)^r$. We thus obtain an integral cocycle φ in the case that m be even, while only a mod 2 cocycle φ in the contrary case. Just as in the special case considered by Van Kampen, it turns out that these cocycles, whatever the subdivision K_1 and the "almost true" realization may be, always belong to one and the same cohomology class $\Phi^m \in H^m\ (K^*, I_{(m)})$. Moreover, it may be shown that so far as $m > 1$, any cocycle in Φ^m may be realized as one arisen from some subdivision K_1 of K and some almost true realization of K_1 in R^m. However, this is not true for $m = 1$, as seen from very simple examples.

The series of classes $\Phi^m \in H^m\ (K^*, I_{(m)})$ will be called in the present work the *imbedding classes* of K. The vanishing of Φ^m is evidently a necessary condition for the realizability of K in R^m. We have $2\Phi^m = 0$, when m is even; but in general Φ^m are nontrivial and thus they serve as effective tools in the study of realization problems.

We remark that we may define, just as in (2), with respect to any simplicial subdivision K_1 of K and any "almost true" realization of K_1 in R^m, a certain integral

m-dimensional cochain $\tilde{\varphi}$ in $\tilde{K}^*$ by

$$\tilde{\varphi}(\sigma \times \tau) = (-1)^{\dim \sigma} \phi(\sigma', \tau'), \quad \sigma \times \tau \in \tilde{K}^*, \quad \dim \sigma + \dim \tau = m.$$

It is true that $\tilde{\varphi}$ is always an *integral* cocycle and its class $\tilde{\Phi}^m$ is uniquely determined. However, it turns out that $\tilde{\Phi}^m$ is always 0 (what is not easily seen from the definition itself) and therefore the complex $\tilde{K}^*$ is not so useful as K^*, so far as the realization problem is concerned.

Let $R^1 \subset R^2 \subset \cdots \subset R^m$ be a sequence of linear subspaces of increasing dimensionality in R^m. By trying to realize the complex K in a certain canonical manner such that $K^0 \subset R^1$, $K^1 \subset R^3$, etc., representative cocycles in Φ^m may be explicitly constructed. This not only gives the means to compute effectively these classes in every concrete case, but also makes it possible to derive a series of properties of Φ^m which are not easy to foresee, e.g., $\frac{1}{2}\delta\Phi^{2m-1} = \Phi^{2m}$, $\Phi^i \cup \Phi^j = \Phi^{i+j} \bmod 2$. This also enables us to determine, for some particular complexes generalized from those of Van Kampen, exactly the lowest dimension of R^m in which they may be realized. It seems that this cannot be done with any other known methods.

At last we should point out that the realization problem is in reality "topological", but not "homotopic" in character. For example, a segment and a triangle have the same homotopy type, but the former may be realized in R^1 while the latter cannot. It follows that the problem cannot be completely solved without the aid of topological invariants which are in general not invariants of homotopy types. In a previous paper [4] the author has described a general method of constructing such invariants. The above-mentioned groups $H^m(K^*, G)$ (and H^m $(\tilde{K}^m, G)$) are particular cases of these invariants and we may thus legitimitely write $H^{m,2}$ (K, G) or $H^{m,2}$ (P, G) instead of H^m (K^*, G) where $P = \bar{K}$ is the space of K. Similarly we write $H^m(\tilde{K}^*, G) = \tilde{H}^{m,2}(K, G) = \tilde{H}^{m,2}(P, G)$. Based on [6] we may prove that $\Phi^m \in H^{m,2}(K, I_{(m)}) \approx H^{m,2}(P, I_{(m)})$ are not only combinatorial invariants of K but also topological invariants of P, an important point completely disregarded by Van Kampen in the special case studied by him. On the other land, Φ^m are not invariants of homotopy type of P. It seems that this is the very reason for the successfulness of methods, originated from Van Kampen and developed here.

We restrict ourselves in the present paper to give a basis of the whole theory and leave to later considerations the study of relations of the imbedding classes with Steenrod squares and also with Stiefel-Whitney classes in the case of a manifold. We leave also to a later occasion the proof of the sufficiency of our condition for the realizability in certain extreme cases.

§1 Linear realization of complexes in Euclidean spaces

In what follows, K will be a finite Euclidean simplicial complex,① and R^m a Euclidean space of dimension m.

Suppose given in R^m a Euclidean simplicial complex K', which is isomorphic to K under the correspondence $T: K \to K'$, then we shall say that $K' = TK$ is a *linear realization* of K in R^m. Denote the topological map induced by T of the spaces $\bar{K}$, $\bar{K}'$ of K, K' by $\bar{T}: \bar{K} \equiv \bar{K}'$, then $\bar{T}$ or T will be called a *linear imbedding* of K in R^m.

It is known that any abstract simplicial complex of dimension r may be realized as the associated abstract complex of a Euclidean simplicial complex in R^{2r+1} of dimension $2r+1$, but not necessarily so in Euclidean spaces of lower dimension [3-5]. From this we may draw two conclusions. First, the problem of existence of Euclidean complex in an R^m associated with a given abstract simplicial complex, is equivalent to the problem of linear realizability of Euclidean simplicial complexes in R^m. For that reason, whenever we speak of complexes, we mean Euclidean complexes in a Euclidean space of sufficiently high dimension, and a subdivision will always mean a Euclidean subdivision. Secondly, a Euclidean complex K in general has no linear realization in R^m if $m < 2 \dim K + 1$. To study this problem, we shall recapitulate and introduce some concepts as follows.

Let σ^r, τ^s be two Euclidean simplexes in R^m, of dimensions r and s respectively. If for any r'-dimensional face σ' of σ^r and any s'-dimensional face τ' of τ, the linear subspace determined by σ' and τ' has a dimension min $(r'+s'+1, m)$, or in other words, if any $r'+1$ vertices of σ^r and $s'+1$ vertices of τ^s are linearly independent so far as $r'+s'+1 \leqslant m$, then σ^r, σ^s are said to be *in general position.*

Suppose given in R^m a set of points $v'_1, \cdots, v'_n$ and a set of geometric simplexes② $\sigma'_1, \cdots, \sigma'_s$ spanned by these points of which the totality K' satisfies the following

① We consider only finite complexes in this work, so that the modifier "finite" will be omitted throughout.

② For the definition of geometric simplex cf. [2], pp. 607. The geometric simplex spanned by $a_0, \cdots, a_r$ of R^m will be denoted by $(a_0, \cdots, a_r)$.

conditions: 1°. If σ_i' is spanned by $v_{i_0}', \cdots, v_{i_k}'$ and $k \leqslant m$, then $v_{i_0}', \cdots, v_{i_k}'$ are linearly independent so that σ_i' may be considered as a Euclidean simplex. 2°. If $\sigma_i' \in K'$, then any face of σ_i' is in K' too. 3°. If $\sigma_i', \sigma_j' \in K'$ are both Euclidean simplexes and have no vertices in common, then σ_i', σ_j' are in general position. In such case we shall say that K' is an *almost Euclidean simplicial complex* in R^m, and $\bar{K}' = \sum_i \bar{\sigma}_i'$ is defined as the *space* of K'.

Suppose that to a Euclidean simplicial K we have in R^m an almost Euclidean simplicial complex K' isomorphic to K, i.e., K, K' have same number of vertices v_i, v_i', $i = 1, \cdots, n$ and the 1-1-correspondence $v_i \leftrightarrow v_i'$ between these vertices is such that $(v_{i_0}, \cdots, v_{i_r}) \in K$ is equivalent to $(v_{i_0}', \cdots, v_{i_r}') \in K'$. Let the induced correspondence be $T: K \to K'$, then we will define $K' = TK$ as an *almost linear realization* of K in R^m. The continuous map $\bar{T}: \bar{K} \to \bar{K}'$ induced by T and also T itself will then be called an *almost linear imbedding* of K in R^m.

Evidently a linear realization (or linear imbedding) of K in R^m is also an almost linear realization (or almost linear imbedding) of K in R^m, but the converse is not true. It is easy to see that K has almost linear realizations in R^m of arbitrary dimension m, though it has linear realizations only in R^m of dimension m sufficiently high. We shall introduce in what follows some invariants of K through its almost linear realizations in R^m with the aim to study the linear realizability of K in R^m.

Since a complex is equivalent to its subdivisions from the point of view of combinatorial topology, we shall introduce the following concepts.①

Suppose given a simplicial subdivision K_1 of K and a linear (or almost linear) realization $K_1' = TK_1$ of K_1 in R^m, then we shall say that K_1' is a *semi-linear* (or *almost semi-linear*) *realization* of K in R^m *through its subdivision* K_1, and T or the induced topological map $\bar{T}: \bar{K} \equiv \bar{K}_1'$ (or continuous map $\bar{T}: \bar{K} \to \bar{K}_1'$) will be defined as a *semi-linear* (or *almost semi-linear*) *imbedding* of K in R^m *through its subdivision* K_1.

Let K be an almost Euclidean simplicial complex in R^m and o be a point of R^m. The $(r+1)$-dimensional geometric simplex $o\,\sigma$ spanned by o and any r-dimensional geometric simplex σ of K will be called the central projection of σ from o. The totality of all such simplexes $o\,\sigma$ and the simplexes of K form a simplicial complex, called the *central projection* of K from o and denoted by oK. In general, oK is not

① From some examples given by Cairns and Van Kampen[7,8], it may be seen that the problem of realization would be topologically meaningless by considering only the original complex without introducing further subdivisions.

an almost Euclidean complex even if K be so. However, we have the following

(A) Lemma. If K is an almost Euclidean simplicial complex in R^m and L a subcomplex of K, then there exist points o in R^m such that $oL + K$ is an almost Euclidean complex. Moreover, such points o may be chosen in any neighbourhood of any point o' of R^m.

Proof. Consider any pair of simplexes σ^r, τ^s in K having no vertices in common, for which the sum of dimensions $r+s$ is $\leqslant m-2$. Since K is almost Euclidean, the linear subspace $P\ (\sigma, \tau)$ determined by σ, τ has a dimension $r+s+1 \leqslant m-1$. Again the linear subspace $P(\sigma)$ determined by any simplex σ of dimension $r \leqslant m-1$ in K has a dimension $r \leqslant m-1$. Hence in any neighbourhood of o' there exist points not belonging to any of such linear subspaces $P\ (\sigma, \tau)$ or $P(\sigma)$. Evidently any such point may be chosen to be a point o as required in the Lemma.

(B) Lemma. Let K be a Euclidean simplicial complex, L a subcomplex of K, and L_* the subcomplex formed of all simplexes of L which has no vertex in common with any simplex of $K-L$. Let $f: \bar{K} \to R^m$ be a continuous map such that $f/\bar{L}$ is an almost linear imbedding of L in R^m, i.e., $f(\bar{L})$ is the space of an almost Euclidean complex L' in $R^m: \bar{L}' = f(\bar{L})$ and L' is isomorphic to L under the map f. Then for any $\varepsilon > 0$, there is an ε-approximation $\bar{T}: \bar{K} \to R^m$ of f such that $\bar{T}$ is an almost semi-linear imbedding of K in R^m through a subdivision K_1, $\bar{T}/\bar{L} \equiv f$, and K_1 has a subdivision L_1 of L as a subcomplex which coincides with L_* on $\bar{L}_*$.

Proof. Since K is finite, we have $\delta > 0$ such that for any two points $x, y \in \bar{K}$, $\rho(x,y) < \delta$ would imply $\rho(f(x), f(y)) < \varepsilon/5$. Take now a simplicial subdivision K_0 of K such that L_* is a subcomplex of K_0 and any simplex of K_0 on $\overline{K_0 - L}$ has a diameter $< \delta$. The part of K_0 on $\bar{L}$ is a subdivision of L which will be denoted by L_0. Let K_1 be the subdivision of K_0 obtained by constructing central subdivisions① of simplexes of $K_0 - L_*$, the centre of $\sigma \in K_0$ being o_σ. The part of K_1 on $\bar{L}$ will be denoted by L_1. Under f, L_0 and L_1 will correspond respectively to a simplicial subdivision L_0' of L' and its central subdivision L_1'. By convenient choice of K_0 and centres o_σ, we may make L_0' and L_1' the almost Euclidean complexes which will be supposed to be so. Arrange now the simplexes in K_0 but not in L_0 in an order $\sigma_1 < \cdots < \sigma_n$, such that those of lower dimension will precede those of higher dimension, but otherwise arbitrary. By (A), we may take successively points

① By a central subdivision we mean a subdivision analogous to the construction in barycentric subdivision. The only difference is that here any interior point of corresponding simplex, not necessarily the barycenter, may be used as the centre of projection in the construction.

$o'_1, \cdots, o'_n$ to satisfy the following conditions:

1°. o'_i is in the $\varepsilon/5$ neighbourhood of $f(o_{\sigma_i})$.

2°. If we define $T(o_{\sigma_i}) = o'_i$, $i = 1, \cdots, n$ and $T(\sigma) = f(\sigma)$ for $\sigma \in L_1$, then T determines an almost semi-linear imbedding of K_0 in R^m through K_1.

Evidently $\bar{T}/\bar{L} \equiv f$. Let $x \in \bar{\tau} = (v_0 \cdots v_r) \in K_1 - L_1$, v_j being vertices of K_1, then

$$\begin{aligned}
\rho(\bar{T}(v_j), \bar{T}(v_k)) &\leqslant \rho(\bar{T}(v_j), f(v_j)) + \rho(f(v_j), f(v_k)) + \rho(f(v_k), \bar{T}(v_k)) \\
&< \frac{\varepsilon}{5} + \frac{\varepsilon}{5} + \frac{\varepsilon}{5} = 3\frac{\varepsilon}{5}, \\
\rho(\bar{T}(x), \bar{T}(v_0)) &\leqslant \text{Diam } \bar{T}\bar{\tau} = \max_{i,k} \rho(\bar{T}(v_j), \bar{T}(v_k)) < 3\frac{\varepsilon}{5}, \\
\rho(\bar{T}(x), f(x)) &\leqslant \rho(\bar{T}(x), \bar{T}(v_0)) + \rho(\bar{T}(v_0), f(v_0)) + \rho(f(v_0), f(x)) \\
&< 3\frac{\varepsilon}{5} + \frac{\varepsilon}{5} + \frac{\varepsilon}{5} = \varepsilon.
\end{aligned}$$

Hence $\bar{T}$ is an ε-approximation of f and (B) is proved.

§2 The imbedding cochain of an almost semi-linear realization

Let R^m be a Euclidean space of dimension m with fixed orientation, K, L be two Euclidean simplicial complexes in R^m and $x = \sum a_i\sigma_i$, $y = \sum b_j\tau_j$ be two chains on integer coefficients in K, L of dimension r, s respectively with $r + s = m$. The subcomplexes of K, L determined by those σ_i, τ_j for which a_i, b_j are $\neq 0$ will be denoted by $|x|$ and $|y|$ respectively. Suppose that $|x|$, $|y|$ are in general position, i.e., any simplex σ of $|x|$ and any simplex τ of $|y|$ are in general position, then with respect to the oriented R^m, the chains x, y have an intersection number ([2] Chap. 11)

$$\phi(x, y) = \sum a_i b_j \phi(\sigma_i, \tau_j),$$

which is bilinear and possesses the following three properties (dim $x = r$, dim $y = s$ and all intersection numbers are supposed to be defined):

$$\phi(x, y) = (-1)^{rs}\phi(y, x), \quad r + s = m, \tag{1}$$

$$\phi(x, \partial y) = (-1)^r \phi(\partial x, y), \quad r + s = m + 1, \tag{2}$$

and finally, change the orientation of R^m and denote the intersection number with respect to this otherwise oriented R^m by ϕ', then

$$\phi(x,y) = -\phi'(x,y), \quad r+s=m. \tag{3}$$

We may also extend the definition of intersection number $\phi(x,y)$ of chains x, y with sum of dimensions $r+s=m$ for which only the conditions $\overline{\partial x}\cap\bar{y}=\bar{x}\cap\overline{\partial y}=\varnothing$ are satisfied by considering them as singular chains. The properties (1), (2) and (3) hold still for such intersection numbers.

In what follows ϕ_m will be used to denote the intersection number or the intersection number reduced mod 2, according as m is even or odd.

For an arbitrary Euclidean simplicial complex K, define now two abstract complexes $\tilde{K}^*$ and K^* as follows. First, we define $\tilde{K}^*$ as the subcomplex of the product complex $K\times K$ consisting of all cells $\sigma\times\tau$ for which σ, $\tau\in K$ have no vertices in common. Orient each cell $\sigma\times\tau$ of $\tilde{K}^*$ in the usual manner we would have

$$\partial(\sigma\times\tau) = \partial\sigma\times\tau + (-1)^r\sigma\times\partial\tau, \quad r=\dim\sigma \tag{4}$$

(σ, τ are oriented cells of K). In $\tilde{K}^*$, $\sigma\times\tau\to\tau\times\sigma$ defines a cell map t of period 2 and having no fixed cells, which induces a chain map given by

$$t_{\#}(\sigma\times\tau) = (-1)^{rs}(\tau\times\sigma), \quad r=\dim\sigma, \quad s=\dim\tau. \tag{5}$$

With respect to t, $\tilde{K}^*$ has a modular complex $\tilde{K}^*/t = K^*$, which is obtained by identifying each pair of cells $\sigma\times\tau$, $\tau\times\sigma$ in $\tilde{K}^*$ with a single cell $\sigma*\tau$ (or what is the same, $\tau*\sigma$). Orient $\sigma*\ \tau$ now as $\sigma\times\tau$, then by (4), (5) we have ($r=$ dim σ, $s=$ dim τ):

$$\partial(\sigma*\tau) = \partial\sigma*\tau + (-1)^r\sigma*\partial\tau. \tag{6}$$

$$\sigma*\tau = (-1)^{rs}\tau*\sigma. \tag{7}$$

We may regard $\tilde{K}^*$ and K^* as Euclidean complexes. Then by [6], the homotopy type of $\tilde{K}^*$ and K^*, in particular the homology and cohomology groups of $\tilde{K}^*$ and K^*, are all topological invariants of $\bar{K}=P$, a *fortiori* combinatorial invariants of K. For this reason we shall adopt the following notations: $H^r(\tilde{K}^*,G)=\tilde{H}^{r,2}(K,G)=\tilde{H}^{r,2}(P,G)$, $H^r(K^*,G)=H^{r,2}(K,G)=H^{r,2}(P,G)$.

Consider now any almost semi-linear realization $K_1'=TK_1$ of a simplicial complex K in R^m through a simplicial subdivision K_1 of K. Let the chain map induced by the subdivision $K \to K_1$ be Sd, and the chain map from K_1 to K_1' induced by T be $T_\#$. Write for simplicity $T_\#\ Sd$ by T. Similar notations will be used throughout this work. For any oriented cell $\sigma \times \tau$ of dimension m in $\tilde{K}^*$, $T\sigma$ and $T\tau$ are in general position since σ, τ have no common vertex in K and T is almost semi-linear. It follows that we may define an integer by

$$\tilde{\varphi}_T(\sigma \times \tau) = (-1)^{\dim \sigma} \phi(T\sigma, T\tau). \tag{8}$$

Then $\tilde{\varphi}_T$ is a cochain in $\tilde{K}^*$ on integer coefficients.

Let

$$I_{(m)} = \begin{cases} I, & m \text{ even}, \\ I_2, & m \text{ odd}. \end{cases}$$

Let $\rho_{(m)} : I \to I_{(m)}$ be the identity or the reduction mod 2. Further let $\phi_{(m)} = \rho_{(m)}\phi$ so that $(-1)^r \phi_m = (-1)^s \phi_m$ for $r + s = m$. Then we have always

$$(-1)^s \phi_m(T\tau, T\sigma) = (-1)^s \cdot (-1)^{rs} \phi_m(T\sigma, T\tau) = (-1)^{rs}(-1)^r \phi_m(T\sigma, T\tau)$$

for any $\sigma * \tau \in K^*$ with $\dim \sigma = r$, $\dim \tau = s$ and $r + s = m$. Comparing with (7) we see that ($\sigma * \tau \in K^*$, $\dim \sigma +$ $\dim \tau = m$)

$$\varphi_T(\sigma * \tau) = (-1)^{\dim \sigma} \phi_m(T\sigma, T\tau) \tag{9}$$

defines unambiguously in K^* a cochain $\varphi_T \in C^m$ $(K^*, I_{(m)})$.

Let $\xi \times \eta$ be any cell of dimension $m + 1$ in $\tilde{K}^*$. Then

$$\begin{aligned} \delta\tilde{\varphi}_T(\xi \times \eta) &= \tilde{\varphi}_T\partial(\xi \times \eta) = \tilde{\varphi}_T(\partial\xi \times \eta) + (-1)^{\dim \xi} \tilde{\varphi}_T(\xi \times \partial\eta) \\ &= (-1)^{\dim \xi - 1} \phi(T\partial\xi, T\eta) + \phi(T\xi, T\partial\eta) \\ &= (-1)^{\dim \xi - 1} \cdot (-1)^{\dim \xi} \phi(T\xi, \partial T\eta) + \phi(T\xi, T\partial\eta) = 0. \end{aligned}$$

Similarly we have $\delta\varphi_T(\xi * \eta) = 0 \bmod I_{(m)}$. Hence $\tilde{\varphi}_T$, φ_T are all m-dimensional cocycles of $\tilde{K}^*$ and K^* on coefficient groups I and $I_{(m)}$ respectively. If T is a semi-linear imbedding, then $\tilde{\varphi}_T$ and φ_T are evidently 0. Hence from the definition of $\tilde{\varphi}_T$ and φ_T we see that they may serve as a measure of T to the deviation from a true semi-linear imbedding. We shall accordingly define $\tilde{\varphi}_T$ and φ_T as the *imbedding cochains* of the almost semi-linear imbedding (or realization) T. The above results may then be written as follows:

Theorem 1 With respect to R^m with a fixed orientation, the imbedding cochains $\tilde{\varphi}_T \in C^m(\tilde{K}^*)$ and $\varphi_T \in C^m(K^*, I_{(m)})$ of an almost semi-linear realization T of a Euclidean simplicial complex K in R^m are all cocycles (and may be thus called the *imbedding cocycles* of T).

The definition of imbedding cocycles depends on the orientation of R^m. By (3) we have

Theorem 2 With respect to R^m with the two opposite orientations, the two imbedding cocycles $\tilde{\varphi}_T$ and $\tilde{\varphi}'_T$ (or φ_T and φ'_T) of an almost semi-linear imbedding T of K in R^m differ at most by a sign:

$$\tilde{\varphi}_T = -\tilde{\varphi}'_T, \quad \varphi_T = -\varphi'_T. \tag{10}$$

§3 Definition of imbedding classes

Let K, R^m be the same as in the preceding section, K_1, K_2 be two simplicial subdivisions of K, and $T_1K_1 = K'_1$, $T_2K_2 = K'_2$ be two almost semi-linear realizations of K in R^m through K_1 and K_2 respectively. The aim of the present section is to prove that the imbedding cocycles of T_1 and T_2 are cohomologous to each other. We shall suppose in what follows that $\bar{T}_1\bar{K}_1$ and $\bar{T}_2\bar{K}_2$ are disjoint. As this may be achieved by at most a parallel translation of $\bar{T}_2\bar{K}_2$ and as the imbedding cocycles $\tilde{\varphi}_{T_2} = \tilde{\varphi}_2$, $\varphi_{T_2} = \varphi_2$ remain unchanged after the parallel translation, there will be no loss of generality in making this supposition.

Arrange now the simplexes of K in an order $\sigma_1 < \sigma_2 < \cdots$, such that simplexes of lower dimension will precede those of higher dimension, but otherwise arbitrary. Let $[1, 2]$ be the closed interval $1 \leqslant t \leqslant 2$, and $K\times [1, 2]$ the complex with usual cell decomposition which will be considered as a Euclidean complex. Let J_0 be the set of all indices i for which $\dim \sigma_i = 0$, $J_r = J_{r-1}+(r)$, $L_0 = K\times(1)+K\times(2)+\sum\limits_{i\in J_0} \sigma'_i\times[1,2]$ and $L_r = L_{r-1} + \sigma_r \times [1,2]$, $r = 1, 2, \cdots$. We shall construct now for each $r = 0$, $1, 2, \cdots$, a simplicial subdivision $L_{r,0}$ of L_r and an almost semi-linear realization $H_rL_{r,0} = L'_{r,0}$ of L_r in R^m through the subdivision $L_{r,0}$ such that the following conditions are satisfied:

1°. $L_{0,0} = K_1 \times (1) + K_2 \times (2) + \sum\limits_{i\in J_0} \sigma_i \times [1,2]$.

2°. $H_0(\tau_j \times (j)) = T_j(\tau_j)$, $\tau_j \in K_j$, $j = 1$, 2 and for $i \in J_0$, $\bar{H}(\bar{\sigma}_i \times [1,2])$ is a simple broken line l_i.

3°. $L_{r-1,0}$ is a subcomplex of $L_{r,0}$ and $H_r/L_{r-1,0} \equiv H_{r-1}$.

4°. If i, $j \in J_r$, dim σ_i+ dim $\sigma_j = m-2$, and σ_i, σ_j have no vertex of K in common, then $\bar{H}_r(\bar{\sigma}_i \times [1,2] \cap \bar{H}_r(\bar{\sigma}_i \times [1,2]) = \varnothing$.

For the construction let us first draw in R^m for each $i \in J_0$ a simple broken line l_i joining $T_j(\bar{\sigma}_i \times (j))$, $j=1$, 2, such that these l_i together with $T_1K_1+T_2K_2$ form an almost Euclidean complex. In the case $m \geqslant 2$, we shall choose l_i to be disjoint from each other. Define now $L_{0,0}$ and H_0 according to 1°, 2°, then 1°—4° are all satisfied for them. Suppose now $L_{i,0}$ and H_i have been constructed for $i \leqslant r-1$ which satisfy 1°—4° and let us define $L_{r,0}$ and H_r as follows. If $r \in J_0$, then $L_{r,0}=L_{0,0}$, $H_r \equiv H_0$ having been defined. Furthermore, as the case dim $\sigma_r > m-2$ is trivial, we shall suppose in what follows dim $\sigma_r = e \leqslant m-2$ and >0.

Put $d=m-2-e \leqslant m-3$. Any two simplexes ξ', η' in $L'_{r-1,0}$ not belonging to $H_{r-1}(K_1 \times (1)+K_2 \times (2))$ and having a dimension $\leqslant d+1$ have an intersection which determines a linear subspace P (ξ', η') with a dimension $\leqslant$ max $(-1, 2\ (d+1)-m)$. Let ζ' be any simplex in $L'_{r-1,0}$ lying on $\bar{H}_{r-1}(\bar{\sigma}_r \times (1)+\bar{\sigma}_r \times (2)+\bar{\sigma}'_r \times [1,2])$, then ζ' and P (ξ', η') will determine a linear subspace P (ξ', η', ζ') with a dimension $\leqslant$ max $(-1, 2\ (d+1)-m)+e+1=$ max $(d+1, e) \leqslant m-2$. Take now a point o_r in the interior of $\bar{\sigma}_r\times$ [1, 2] and form the central projection of the boundary of $\bar{\sigma}_r\times$ [1, 2] with centre o_r, thus obtaining a simplicial subdivision $L_{r,1}$ of $L_{r-1,0}+\sigma_r\times$ [1, 2] which contains $L_{r-1,0}$ as a subcomplex. By § 1 (A) we may choose a point o'_r in R^m not lying on any linear subspaces P (ξ', η', ζ') such that $H'_r/L_{r-1,0} \equiv H_{r-1}$, $H'_r(o_r)=o'_r$ will induce an almost semi-linear realization $H'_rL_{r,1}=L'_{r,1}$ of L_r in R^m through $L_{r,1}$. Then $L_{r,1}$ and H'_r satisfy 1°, 2°, 3° and for i, $j \in J_{r-1}$ also satisfy 4°. Suppose now i_1, $i_2, \cdots \in J_{r-1}$ be the totality of indices for which dim $\sigma_{i_\mu}=d$ and σ_{i_μ}, σ_r have no common vertex in K. Then for each μ, $\bar{H}'_r(\bar{\sigma}_r \times [1,2]) \cap \bar{H}'_r(\bar{\sigma}_{i_\mu} \times [1,2])$ consists of at most finite number of points $p'_{1\mu}, \cdots, p^r_{s_\mu,\mu}$. Owing to the choice of o'_r, $\bar{H}'^{-1}_r(p'_{h\mu})$ consists of only one point in $\sum\limits_\lambda \bar{\sigma}_{i_\mu} \times [1,2]$, say $p_{h\mu}$, where $h=1, \cdots, s_\mu$. Consider now any fixed index $i=i_\lambda$. In $L_{r-1,0}$, let us take any d-dimensional simplex η lying on $\bar{\sigma}_i \times (2)$. Take also point x interior to η with $\bar{H}'_r(x)=x'$. If $d=0$, we shall denote by p_1 that point among $p_{h\lambda}=p_h$ $(h=1, \cdots, s_\lambda=s$ on $\sigma_i\times$ [1, 2]) which is nearest to $x=\sigma_i\times$ (2) from $\sigma_i\times$ (1) to x, and denote by B the part of $\sigma_i\times$ [1, 2] from p_1 to x. Then $\bar{H}'_r(B)=B' \subset l_i$ is a simple broken line. If $d>0$, then as $2(d+1)-m \leqslant (d+1)-2$, we may still join p_1 to x by a simple broken line B lying wholly on $\bar{\sigma}_i \times [1,2]$ such that for any point $y \in \bar{\sigma}_{i_\mu} \times [1,2]$, we shall have $\bar{H}'_r(y) \notin B'=\bar{H}'_r(B)$ so far as $y \notin B$. It follows that, whatever the case may be, B has always a neighborhood N in $\bar{\sigma}_i \times [1,2]$ such that $\bar{H}'_r(N) = N'$ is disjoint from $\bar{H}'_r(\sum\limits_{\mu} \bar{\sigma}_{i_\mu} \times [1,2] - N)$; $\bar{H}'_r/N$ is one to one, and B does not pass through the points $p_2, \cdots, p_s$. Let the vertices of the broken line B be successively $x, x_1, \cdots, x_k = p_1$ and suppose that the segment xx_1 is in the $(d+1)$-dimensional simplex ξ_1, the segment $x_{j-1}x_j$ is in the $(d+1)$-dimensional simplex ξ_j of $L_{r-1,0}$ lying on $\sigma_i \times [1, 2]$ $(j = 2, \cdots, k)$, and x_j is an interior point of the d-dimensional simplex η_j which is the face common to ξ_j and ξ_{j+1} $(j = 1, 2, \cdots, k-1)$. Put $\bar{H}'_r(x_j) = x'_j$, $\bar{H}'_r(\xi_j) = \xi'_j$, $\bar{H}'_r(\eta_j) = \eta'_j$, $\bar{H}'_r(\eta) = \eta'$. Prolong x'_1x' to x'_0 such that $x'x'_0$ meets $\bar{H}'_r(\sum\limits_{\mu} \bar{\sigma}_{i_\mu} \times [1,2])$ only in x'. Denote by $z_1, \cdots, z_t$ the totality of points in $\bar{H}'^{-1}_r(p'_1) \cap \bar{\sigma}_r \times [1,2]$ and denote by ζ_i the $(e+1)$-dimensional simplex of $L_{r,1}$ lying on $\bar{\sigma}_r \times [1,2]$ which contains z_i in its interior. The integer t will be called the multiplicity index of p'_1. Consider any $\zeta_1 = \zeta$, and denote by P'_k the $(e+1)$-dimensional linear subspace determined by $\zeta' = \bar{H}'_r(\zeta)$. Through each x'_j $(0 \leqslant j < k)$ draw now an $(m-1)$-dimensional linear subspace P'_j such that for $0 < j < k$, P'_j contains η'_j with $\bar{\xi}'_j$, $\bar{\xi}'_{j+1}$ on opposite sides of P'_j, and P'_0 meets $x'_0x'_1$ only in x'_0. Take an $(e+1)$-dimensional simplex τ containing $z = z_1$ in its interior and contained in $\bar{\zeta}$, with diameter less than a sufficiently small $\varepsilon > 0$. Put $\bar{\tau}' = \bar{H}'_r(\bar{\tau})$. For any $y \in \bar{\tau}$ we shall draw a broken line $B'_y = y'_k \cdots y'_1y'_0x'_0$ such that $y'_j \in P'_j$, $y'_k = \bar{H}'_r(y)$ and $y'_jy'_{j-1}$ is parallel to $x'_jx'_{j-1}$ $(j = 1, \cdots, k)$. Evidently for ε sufficiently small we may make $B'_y \cap \sum\limits_{\mu} \bar{H}'_r(\bar{\sigma}_{i_\mu} \times [1,2]) = \phi$ for $y \in \bar{\tau}$. After the choice of such an ε we may define a continuous map $\bar{H}''_r : \bar{L}_{r,1} \to R^m$ by $\bar{H}''_r/\bar{L}_{r,1} - \bar{\tau} \equiv \bar{H}'_r$, while for any $y \in \bar{\tau}$, $\bar{H}''_r$ maps linearly yz_1 to the segment B'_y. By §1 (B), we may construct an arbitrarily small approximation $\bar{H}'''_r$ of $\bar{H}''_r(\bar{H}'''_r/\bar{L}_{r,1} - \bar{\tau} \equiv \bar{H}''_r)$ such that $\bar{H}'''_r\bar{L}_{r,1}$ is the space of an almost Euclidean complex, and $L_{r,1}$ has a simplicial subdivision $L_{r,2}$ having $L_{r-1,0}$, as a subcomplex, while $\bar{H}'''_r$ is the continuous map associated with the simplicial map $H'''_r(L_{r,2}) = L'_{r,2}$. By taking the approximation sufficiently small, we may make $\bar{H}'''_r(\bar{\sigma}_r \times [1,2]) \cap \bar{H}'''_r(\bar{\sigma}_i \times [1,2])$ consist of only $k-1$ points $p'_2, \cdots, p'_k$, or though of the same number of points $p'_1, \cdots, p'_k$ as before, the multiplicity index of p'_1 is decreased by 1, while the number of intersecting points and the multiplicity indices of $\bar{H}'''_r$ $(\bar{\sigma}_r \times [1, 2])$ and $\bar{H}'''_r(\bar{\sigma}_{i_\mu} \times [1,2])$ (besides at p'_1) are all unchanged.

Proceeding successively with the same process, we shall make the number of intersections of the images in R^m of $\bar{\sigma}_r \times [1, 2]$ and $\bar{\sigma}_i \times [1, 2]$ reduce to 0. Using the same procedure to each σ_{i_μ} we get finally a complex $L_{r,0}$ and a realization $H_rL_{r,0} = L'_{r,0}$ which satisfies the conditions 1°—4°.

By induction on r, we get finally a simplicial subdivision L of $K\times$ [1, 2] and an almost semi-linear realization $HL = L'$ of $K\times$ [1, 2] in R^m through L. By 4°, this H will satisfy the following condition: If dim σ_i+ dim $\sigma_j = m-2$ and σ_i, σ_j have no common vertex in K, then

$$\bar{H}(\bar{\sigma}_i \times [1,2]) \cap \bar{H}(\bar{\sigma}_j \times [1,2]) = \varnothing. \tag{1}$$

Now let the chain map induced by the subdivision of $K\times$ [1, 2] into L be denoted by Sd, and that induced by H be denoted by $H_{\#}$. Orient [1, 2] by the direction from 1 to 2 and put for simplicity

$$h\sigma_i = (-1)^{\dim \sigma_i} H_{\#} Sd(\sigma_i \times [1,2]),$$

then

$$\partial h\sigma_i = T_2\sigma_i - T_1\sigma_i - h\partial\sigma_i, \quad \sigma \in K. \tag{2}$$

By (1), we get further: If dim σ_i+ dim $\sigma_j = m-2$ and σ_i, σ_j have no common vertex in K, then

$$\phi(h\sigma_i, h\sigma_j) = 0. \tag{3}$$

Define now a cochain $\tilde{\psi} \in C^{m-1}(\tilde{K}^*)$ as follows:

$$\tilde{\psi}(\sigma_i \times \sigma_j) = \phi(\partial h\sigma_i, h\sigma_j), \quad \dim\sigma_i + \dim\sigma_j = m-1, \quad \sigma_i \times \sigma_j \in \tilde{K}^*. \tag{4}$$

Since

$$\begin{aligned}\phi_m(\partial h\sigma_i, h\sigma_j) &= (-1)^{\dim\ \sigma_i+1}\phi_m(h\sigma_i, \partial h\sigma_j)\\ &= (-1)^{\dim\ \sigma_i+1}\cdot(-1)^{(\dim\ \sigma_i+1)\ \dim\ \sigma_j}\phi_m(\partial h\sigma_j, h\sigma_i)\\ &= (-1)^{\dim\ \sigma_i\ \dim\ \sigma_j}\phi_m(\partial h\sigma_j, h\sigma_i),\end{aligned}$$

we know, by comparing with (7) of §2, that we may define unambiguously a cochain $\psi \in C^{m-1}(K^*, I_{(m)})$ by

$$\psi(\sigma_i * \sigma_j) = \phi_m(\partial h\sigma_i, h\sigma_j), \quad \dim\sigma_i + \dim\sigma_j = m-1, \quad \sigma_i * \sigma_j \in K^*. \tag{5}$$

For any $\sigma_k \times \sigma_l \in \tilde{K}^*$ with dim σ_k+ dim $\sigma_l = m$ we have now

$$\begin{aligned}\delta\psi(\sigma_k \times \sigma_l) = \tilde{\psi}\partial(\sigma_k \times \sigma_l) &= \tilde{\psi}(\partial\sigma_k \times \sigma_l) + (-1)^{\dim\ \sigma_k}\tilde{\psi}(\sigma_k \times \partial\sigma_l)\\ &= \phi(\partial h\partial\sigma_k, h\sigma_l) + (-1)^{\dim\ \sigma_k}\phi(\partial h\sigma_k, h\partial\sigma_l),\end{aligned}$$

or

$$\delta\tilde{\psi}(\sigma_k \times \sigma_l) = (-1)^{\dim \sigma_k}[\phi(h\partial\sigma_k, \partial h\sigma_l) + \phi(\partial h\sigma_k, h\partial\sigma_l)]. \tag{6}$$

On the other hand, since $\bar{T}_1\bar{K}_1$ and $\bar{T}_2\bar{K}_2$ are disjoint,

$$\phi(T_1\sigma_k, T_2\sigma_l) = \phi(T_2\sigma_k, T_1\sigma_l) = 0. \tag{7}$$

By (2), (3), (4), (7) and the definitions of $\tilde{\varphi}_{T_i} = \tilde{\varphi}_i$ ($i = 1, 2$) we get

$$\begin{aligned}
&(-1)^{\dim \sigma_k}[\tilde{\varphi}_1(\sigma_k \times \sigma_l) + \tilde{\varphi}_2(\sigma_k \times \sigma_l)] = \phi(T_1\sigma_k, T_1\sigma_l) + \phi(T_2\sigma_k, T_2\sigma_l)\\
=&\phi(T_2\sigma_k - T_1\sigma_k - h\partial\sigma_k, T_2\sigma_l - T_1\sigma_l - h\partial\sigma_l) - \phi(h\partial\sigma_k, h\partial\sigma_l)\\
&+ \phi(T_2\sigma_k - T_1\sigma_k, h\partial\sigma_l) + \phi(h\partial\sigma_k, T_2\sigma_l - T_1\sigma_l)\\
=&\phi(\partial h\sigma_k, \partial h\sigma_l) - \phi(h\partial\sigma_k, h\partial\sigma_l) + \phi(\partial h\sigma_k + h\partial\sigma_k, h\partial\sigma_l)\\
&+ \phi(h\partial\sigma_k, \partial h\sigma_l + h\partial\sigma_l) = \phi(\partial h\sigma_k, h\partial\sigma_l) + \phi(h\partial\sigma_k, \partial h\sigma_l).
\end{aligned}$$

Comparing with (6) we get

$$\tilde{\varphi}_1(\sigma_k \times \sigma_l) + \tilde{\varphi}_2(\sigma_k \times \sigma_l) = \delta\tilde{\psi}(\sigma_k \times \sigma_l).$$

Since $\sigma_k \times \sigma_l$ is an arbitrary m-cell of $\tilde{K}^*$, we have

$$\tilde{\varphi}_1 + \tilde{\varphi}_2 \sim 0. \tag{8}$$

In particular, if T_2 is obtained from T_1 by parallel translations in R'' so that $\tilde{\varphi}_1 = \tilde{\varphi}_2$, (8) becomes

$$2\tilde{\varphi}_1 \sim 0. \tag{9}$$

From (8), (9) we get therefore

$$\tilde{\varphi}_1 \sim \tilde{\varphi}_2. \tag{10}$$

Similarly, we have also

$$2\varphi_1 \sim 0, \quad \bmod \mathrm{I}_{(m)}, \tag{11}$$

$$\varphi_1 \sim \varphi_2, \quad \bmod \mathrm{I}_{(m)}. \tag{12}$$

The above imbedding cocycles $\tilde{\varphi}_T$, φ_T are defined with respect to R^m with a fixed orientation. If we reverse the orientation of R^m and denote the corresponding imbedding cocycles of T by $\tilde{\varphi}'_T$, φ'_T, then we have by Theorem 2 of § 2, $\tilde{\varphi}'_T = -\tilde{\varphi}_T$, $\varphi'_T = -\varphi_T$. Hence by (9), (11) we have

$$\tilde{\varphi}'_T \sim \tilde{\varphi}_T, \tag{13}$$

$$\varphi'_T \sim \varphi_T, \quad \text{mod I}_{(m)}. \tag{14}$$

From (9)-(14) we get therefore the following two theorems:

Theorem 3 The imbedding cocycles $\tilde{\varphi}_T$ and φ_T of an almost semi-linear realization T of a Euclidean simplicial complex K in an oriented R^m each belong to a fixed cohomology class $\tilde{\Phi}^m \in \tilde{H}^{m,2}(K)$ and $\Phi^m \in H^{m,2}\ (K,\ I_{(m)})$. Moreover, these classes are independent of the chosen orientation of R^m and the chosen realization T.

Definition The cohomology class Φ^m in the above theorem will be called the m-dimensional *imbedding class* of K, $m > 0$.

Theorem 4 All imbedding classes on integer coefficients have order 2:

$$2\Phi^m = 0 \quad (m \text{ even} > 0). \tag{15}$$

Remark. We have also $2\tilde{\Phi}^m = 0$, $m > 0$. However, we shall prove later that we have always $\tilde{\Phi}^m = 0$ (cf. Theorem 9 of §5 and Theorem 16 of §8). Hence in reality $\tilde{\Phi}^m$ are of no significance at all.

From the definition of imbedding cocycles and imbedding classes we have

Theorem 5 A necessary condition that a Euclidean simplicial complex K may have a linear (or semi-linear) realization in R^m, is that

$$\Phi^m = 0. \tag{16}$$

We shall see later that in certain cases, this condition is also sufficient.

Evidently $\tilde{K}^*$ is a two-sheeted covering complex of K^*. Denote the projection by π:

$$\pi(\sigma \times \tau) = \sigma * \tau \quad (\sigma \times \tau \in \tilde{K}^*).$$

Then by the definition of $\tilde{\varphi}_T$, φ_T we have evidently $\pi^{\#}\varphi_T = \rho_{(m)}\tilde{\varphi}_T$. Hence we have

Theorem 6 Let π be the covering projection of $\tilde{K}^*$ on K^*, then

$$\pi^*\Phi^m = \tilde{\Phi}^m \quad (m \text{ even} > 0), \tag{17}$$

$$\pi^*\Phi^m = \rho_2\tilde{\Phi}^m \quad (m \text{ odd}), \tag{18}$$

in which ρ_2 denotes reduction mod 2. Suppose for the moment $\tilde{\Phi}^m = 0$ (cf. the remark above), then (17) and (18) may be reduced simply to

$$\pi^*\Phi^m = 0, \quad m > 0. \tag{19}$$

§4 The realizability of any cocycle in the imbedding classes

We have proved that the m-dimensional imbedding cochains $\tilde{\varphi}_T$ (or φ_T) of K are all cocycles and belong to one and the same cohomology class $\tilde{\Phi}^m \in \tilde{H}^{m,2}(K)$ (or $\Phi^m \in H^{m,2}$ $(K, I_{(m)})$. Conversely, by the definition of $\tilde{\varphi}_T$ in §2, any "imbedding cocycle" $\tilde{\varphi}_T$ in $\tilde{\Phi}^m$ must satisfy the conditions ($\sigma_i \times \sigma_j \in \tilde{K}^*$, dim σ_i+ dim $\sigma_j = m$)

$$\tilde{\varphi}_T(\sigma_i \times \sigma_j) = (-1)^{m+\dim\sigma_i \ \dim\sigma_j}\tilde{\varphi}_T(\sigma_j \times \sigma_i).$$

Hence if we change $\tilde{\varphi}_T$ into an arbitrary coboundary $\delta\tilde{\psi}$, $\tilde{\psi} \in C^{m-1}(\tilde{K}^*)$, the cocycle $\tilde{\varphi}_T + \delta\tilde{\psi}$ of $\tilde{\Phi}^m$ thus obtained is in general no more an imbedding cocycle, and is not necessarily realized as one of an almost semi-linear realization of K in R^m. On the contrary, for the class Φ^m we have the following

Theorem 7 If $m > 1$, then any cocycle in Φ^m may be realized as an imbedding cocycle. In other words, there must exist an almost semi-linear realization of K in the oriented R^m, with any given cocycle in Φ^m as its imbedding cocycle.

Remark. This theorem is not true for $m = 1$. For example, let K be a one-dimensional complex consisting of three vertices a, b, c and two segments ab, ac. Since K may be realized in R^1, we have $\Phi^1 = 0$. Hence defining a mod 2 cochain $\psi \in C^0$ (K^*, I_2) by $\psi(b * c) = 1$, $\psi(a * b) = \psi(a * c) = 0$, we would have $\varphi = \delta\psi \in \Phi^1$. Suppose that there exists an almost semi-linear realization T of K in R^1 with imbedding cocycle $\varphi_T = \varphi$. Let $T(a) = a'$, $T(b) = b'$ and $T(c) = c'$. Then since $\varphi_T(b * (ac)) = \delta\psi(b * (ac)) = \psi(b * c) \neq 0$, we have $\rho_2\phi(b', T(ac)) \neq 0$ so that b' must lie between a' and c'. Similarly, since $\varphi_T(c * (ab)) \neq 0$, c' must lie between a' and b'. But this is impossible. Consequently $\varphi \in \Phi^1$ cannot bc rcalizcd as an imbedding cocycle.

Proof of Theorem 7. Consider any almost semi-linear realization $T_0K_0 = K_0'$ of K in oriented R^m through a subdivision K_0 of K with corresponding imbedding cocycle $\varphi_{T_0} = \varphi_0 \in \Phi^m$. Denote the simplexes of K by $\sigma_1, \sigma_2, \cdots$ with dim $\sigma_i = d_i$. Consider an arbitrary but fixed $(m-1)$-cell $\sigma_i * \sigma_j$ in $K^* : d_i + d_j = m - 1$, and define a cochain $\chi_{i,j} \in C^{m-1}$ $(K^*, I_{(m)})$ by

$$\chi_{i,j}(\sigma_k * \sigma_l) = \begin{cases} 0, & \sigma_k * \sigma_l \neq \sigma_i * \sigma_j \text{ or } \sigma_j * \sigma_i, \\ 1, & \sigma_k * \sigma_l = \sigma_i * \sigma_j. \end{cases} \tag{1}$$

Our object is to modify T_0 to an almost semi-linear realization of K in the oriented R^m through a subdivision K_1 of K such that the imbedding cocycle of T_1 is

$$\varphi_{T_1} = \varphi_1 = \varphi_0 + c\delta\chi_{i,j}, \tag{2}$$

where $c = \pm 1$ is arbitrarily but previously assigned. Since $\sigma_i * \sigma_j$ is an arbitrary $(m-1)$-cell of K^*, we may start from φ_0 and arrive at any cocycle in Φ^m, and our theorem would be proved in this manner.

For this purpose let us remark first that, as $m > 1$ by hypothesis, we may suppose $d_i > 0$. Consider any simplexes of K_0 of dimension d_i, d_j with $\bar{\tau}_j \subset \bar{\sigma}_i$, $\bar{\tau}_j \subset \bar{\sigma}_j$ respectively. Let $\tau'_i = T_0\tau_i$, $\tau'_j = T_0\tau_j$. In each of τ_i, τ_j take an interior point x_0, x such that $x'_0 = \bar{T}_0(x_0) \notin \bar{T}_0(\overline{K_0^{m-1}} - \bar{\tau}_i), x' = \bar{T}_0(x) \notin \bar{T}_0(\overline{K_0^{m-1}} - \bar{\tau}_j)$. Construct a simple broken line B, with successive vertices $x'_0, x'_1, \cdots, x'_n$, such that the following conditions are satisfied:

1°. $x' \in x'_{n-1}x'_n$, $x'_{n-1}x'_n$ is orthogonal to the linear subspace R determined by τ'_j and $B \cap \bar{\tau}'_j = (x')$.

2°. B is disjoint to the space of $|T_0K_0 - T_0St_0\tau_i - T_0St_0\tau_j|^{m-2}$ in which St_0 denotes the star in K_0. Finally,

3°. $x'_0x'_1$ is orthogonal to the linear subspace Q_0 determined by τ'_i and meets $\bar{T}_0\overline{K_0^{m-1}}$ only in the point x'_0.

Let Q be the d_i-dimensional linear subspace in R^m passing through x' and completely orthogonal to R and $x'_{n-1}x'_n$. Let T be an orthogonal transformation of R^m, transforming Q_0 to Q, x'_0 to x', and the $x'_0x'_1$ direction to $x'x'_n$ direction. For any point $y_0 \in \overline{St_0\tau_i}$ and any $\varepsilon \geqslant 0$ and $\leqslant 1$, let $y_0(\varepsilon)$ be the point on x_0y_0 with $x_0y_0(\varepsilon) : x_0y_0 = \varepsilon$. Let B_{y_0} be the broken line with successive vertices $y'_0 = \bar{T}_0(y_0)$, $x'_0, x'_1, \cdots, x'_{n-1}, y'_{n-1}, y' = T(y'_0), y'_n$ and x'_n, where $y'_{n-1}y'$, $y'y'_n$ are parallel and equal to $x'_{n-1}x'$, $x'x'_n$ respectively. From 1°— 3° and the above construction we see that

4°. If ε is sufficiently small, then for any $y_0 \in \overline{St_0\tau_i}$, $B_{y_0(\varepsilon)}$ is disjoint from the space of $|T_0K_0 - T_0St_0\tau_i - T_0St_0\tau_j|^{m-2}$.

5°. If $\varepsilon > 0$ is sufficiently small, then for any $y_0 \in \bar{\tau}_i$, $B_{y_0(\varepsilon)}$ is disjoint from $\bar{\tau}'_j$.

Now for any $\tau_k \in Cl\ St_0\tau_i$, let $\tau_{k,\varepsilon}$ be the contraction of τ_k with centre x_0 and ratio of contraction $\varepsilon : 1$ $(0 \leqslant \varepsilon \leqslant 1)$. In particular, $\tau_{k,1} = \tau_k$, $\bar{\tau}_{k,0} = (x_0)$. Let L_ε be the complex formed of all $\tau_{k,\varepsilon}$ for which $\tau_k \in Cl\ St_0\tau_i$. Construct a simplicial subdivision K_{00}, with both $K_0 - St_0\tau_i$ and L_ε as its subcomplexes. In $\bar{K}_{00}\times$ [0, 1] identify each segment $(z) \times [0,1]$ with a single point, where $z \in \overline{K_{00} - St_{00}\tau_{i,\varepsilon}}$ $(St_{00}$

being star in K_{00}), obtaining thus a space $\bar{M}_{\varepsilon 0}$ and a natural map $f_\varepsilon : \bar{K}_{00} \times [0,1] \to \bar{M}_{\varepsilon 0}$. Under f_ε, $K_{00} \times [0,1]$ will induce on $M_{\varepsilon 0}$ a cell decomposition such that f_ε is a cell-map. Denote this cell decomposition of $\bar{M}_{\varepsilon_0}$ by $M_{\varepsilon 0}$. Then the parts of M_{ε_0} on $f_\varepsilon(\bar{K}_{00} \times (0))$and $f_\varepsilon(\bar{K}_{00} \times (1))$ are isomorphic to $K_{00} \times (0)$ and $K_{00}\times$ (1) respectively under the map f_ε.

Define now a map $\bar{H}^0 : \bar{M}_{\varepsilon_0} \to R^m$ as follows. First, for any $y_0 \in \overline{St_0\tau_i}$, let α_{y_0} be a linear map of [0, 1] to B_{y_0} such that α_{y_0} $(i/2\ (n+2)\ )$, $i = 0,\ 1,\ \cdots,\ n+2$ are successively $y'_0,\ x'_0,\ \cdots,\ x'_{n-1},\ y'_{n-1},\ y'_n$ while $\alpha_{y_0}(1) = x'_n$ (for the symbols cf. above). If $z \in f_\varepsilon\ (\overline{K_{00} - St_{00}\tau_{i,\varepsilon}})$, then define $\bar{H}^0(z) = \bar{T}_0(z)$. Finally, let $\tau_k \in St_0\tau_i$, y_0 be any point $\in \bar{\tau}_{k,\varepsilon}$, but not interior to $\bar{\tau}_{i,\varepsilon}$, then for any $t \in$ [0, 1], $\bar{H}^0$ linearly map $f_\varepsilon(y_0 x_0,\ t)$ to the broken line

$$B_{y_{0,t}} = \alpha_{y_0}\left[0, \frac{t}{2}\right] + \alpha_{y_0}\left(\frac{t}{2}\right)\alpha_{x_0}\left(\frac{t}{2}\right),$$

such that

$$\bar{H}^0 f_\varepsilon(y_0(\eta), t) = \alpha_{y_0}(1-\eta), \quad 1 - \frac{t}{2} \leqslant \eta \leqslant 1,$$
$$\bar{H}^0 f_\varepsilon(y_0(\eta), t) = \alpha_{y_0(2\eta/(2-t))}\left(\frac{t}{2}\right), \quad 0 \leqslant \eta \leqslant 1 - \frac{t}{2}.$$

The map $\bar{H}^0$ thus defined is evidently continuous. Let $g_t : \bar{K} \to \bar{K} \times [0,1]$ be the map $g_t(z) = (z,t)$, $z \in \bar{K}$, and $\bar{H}^0_t : \bar{K} \to R^m$ be the map $\bar{H}^0_t = \bar{H}^0 f_\varepsilon g_\varepsilon$. Then $\bar{H}^0_0 \equiv \bar{T}_0$, and $\bar{H}^0_1$ maps $y_0 x_0$ to B_{y_0}, where $y_0 \in \bar{\tau}_{k,\varepsilon}-$Int $\bar{\tau}_{i,\varepsilon}$, $\tau_k \in St_0\tau_l$. By 4°, 5° we have then

4°_0. If ε is sufficiently small, then for any $\tau_k \in St_0\tau_i$ $\tau_l \in |K_0 - St_0\tau_i - St_0\tau_j|^{m-2}$, we have

$$\bar{H}^0_t(\bar{\tau}_{k,\varepsilon}) \cap \bar{T}_0\bar{\tau}_l = \varnothing, \quad t \in [0,1].$$

5°_0. If $\varepsilon > 0$ is sufficiently small, then $\bar{H}^0_1(\bar{\tau}_{i,\varepsilon})$ is disjoint from $\bar{T}_0\bar{\tau}_j$.

Choose now $\varepsilon > 0$ sufficiently small such that 4°_0 and 5°_0 are both satisfied. By §1 (B), there exist a simplicial subdivision M_ε of $M_{\varepsilon 0}$ and an arbitrarily small approximation $\bar{H} : \bar{M}_{\varepsilon 0} \to R^m$ of $\bar{H}^0$ such that $\bar{H} \equiv \bar{H}^0/f_\varepsilon(\bar{K}_{00} \times (0))$, and $\bar{H}$ be an almost semi-linear imbedding of $M_{\varepsilon 0}$ in R^m through M_ε. This approximation $\bar{H}$ of $\bar{H}^0$ may be chosen so small that the following conditions corresponding to 4°_0 and 5°_0 will be supposed to be satisfied.

6°. For any $\tau_k \in St_0\tau_i$ and $\tau_l \in |K_0 - St_0\tau_i - St_0\tau_j|^{m-2}$ we have $(\overline{H}_t = \bar{H} f_\varepsilon g_t)$

$$\bar{H}_t(\bar{\tau}_{k,\varepsilon}) \cap \bar{T}_0\tau_l = \varnothing, \quad t \in [0,1].$$

7°. $\bar{H}_1(\bar{\tau}_{i,\varepsilon}) \cap \bar{T}_0\bar{\tau}_j = \varnothing$.

Define now a map $\bar{T}_1 : \bar{K} \to R^m$ by $\bar{T}_1 = \bar{H}_1 f_\varepsilon g_1$, then $\bar{T}_1$ determines an almost semi-linear realization T_1 of K in R^m through a certain subdivision K_1. Let Sd and Sd' be the chain maps induced by the subdivisions K or $K_0 \to K_{00}$ and $M_{\varepsilon 0} \to M_\varepsilon$ respectively. For any chain c of K or K_0, $HSd'f_\varepsilon(Sdc \times [0,1])$ is a chain in R^m, which will be denoted for simplicity by $(-1)^{\dim c}$ h (c). Then we have

$$\partial h(c) = T_1(c) - T_0(c) - h\partial(c), \quad c \in C^*(K_0) \text{ or } C^*(K), \tag{3}$$

$$h(\tau) = 0, \quad \tau \in K_0 \text{ but } \notin St_0\tau_i. \tag{4}$$

By 5°, 7° and the construction we see easily $\phi_m(h\tau_i, T_0\sigma_j) = \pm 1$. By conveniently choosing the orthogonal transformation T of R^m to be orientation-preserving or orientation-reversing, we may always make① ($c = \pm 1$ as in (1))

$$\phi(h\tau_i, T_0\sigma_j) = (-1)^{d_i+1}c. \tag{5}$$

Let $\tau_k \in St_0\tau_i$, $\tau_l \in K_0 - St_0\tau_i - St_0\tau_j$, where dim τ_k+ dim $\tau_l = m-1$. Then since dim $\tau_k \geqslant$ dim $\tau_i > 0$ we have dim $\tau_l \leqslant m-2$ and $\tau_l \in |K_0 - St_0\tau_i - St_0\tau_j|^{m-2}$, Hence, by 6°,

$$\phi(h\tau_k, T_0\tau_l) = 0. \tag{6}$$

Let the imbedding cocycle of T_1 be $\varphi_{T_1} = \varphi_1$. We shall prove that for any cell $\sigma_k * \sigma_l \in K^*$ with $d_k + d_l = m$, we have

$$\varphi_1(\sigma_k * \sigma_l) - \varphi_0(\sigma_k * \sigma_l) = c\delta\chi_{i,j}(\sigma_k * \sigma_l). \tag{7}$$

Case Ⅰ. $\sigma_k, \sigma_l \notin St\sigma_i$.

In that case $\bar{T}_1/\bar{\sigma}_k \equiv \bar{T}_0$, $\bar{T}_1/\bar{\sigma}_l \equiv \bar{T}_0$. Hence $\varphi_1(\sigma_k * \sigma_l) = \varphi_0(\sigma_k * \sigma_l)$. As $\delta\chi_{i,j}(\sigma_k * \sigma_l) = 0$, we have (7).

Case Ⅱ. $\sigma_k \in St\sigma_i$, $\sigma_l = \sigma_j$.

We have $d_k = d_i + 1$, snd $\bar{T}_1/\bar{\sigma}_l \equiv \bar{T}_1/\bar{\sigma}_j \equiv \bar{T}_0/\bar{\sigma}_j$. Let

$$\partial\sigma_k = a\sigma_i + \zeta, \quad a = \pm 1 \tag{8}$$

(ζ contains no term involving σ_i), then

$$\delta\chi_{i,j}(\sigma_k * \sigma_l) = \chi_{i,j}(\partial\sigma_k * \sigma_l) = a\chi_{i,j}(\sigma_i * \sigma_j) = a. \tag{9}$$

① This is not nocssarily possible when $m = l$ (cf. the remark before the proof).

Let $\tau_k \in St_0\tau_i$ be the d_k-dimensional simplex of K_0 with $\bar{\tau}_k \subset \bar{\sigma}_k$, and let τ_i, τ_k be oriented concordantly with σ_i, σ_k respectively, then by (8) we get

$$\partial\tau_k = a\tau_i + \zeta', \quad a = \pm 1 \tag{10}$$

(ζ' contains no term involving τ_i). By (3)—(6) and (10), we get

$$\begin{aligned}
&(-1)^{d_k}[\varphi_1(\sigma_k * \sigma_l) - \varphi_0(\sigma_k * \sigma_l)] = \phi_m(T_1\sigma_k, T_1\sigma_l) - \phi_m(T_0\sigma_k, T_0\sigma_l)\\
=&\phi_m(T_1\sigma_k - T_0\sigma_k, T_0\sigma_j) = \phi_m(T_1\tau_k - T_0\tau_k, T_0\tau_l)\\
=&\phi_m(\partial h\tau_k, T_0\sigma_j) + \phi_m(h\partial\tau_k, T_0\sigma_j) = \pm\phi_m(h\tau_k, \partial T_0\sigma_j)\\
&+ a\phi_m(h\tau_i, T_0\sigma_l) + \phi_m(h\zeta', T_0\sigma_l) = (-1)^{d_i+1}ac = (-1)^{d_k}ac.
\end{aligned}$$

Comparing with (9), we get (7).

Case III. $\sigma_l \in St\sigma_i$, $\sigma_k = \sigma_j$.

According to Case II, we have

$$\begin{aligned}
\varphi_1(\sigma_k * \sigma_l) - \varphi_0(\sigma_k * \sigma_l) &= (-1)^{d_k d_l}[\varphi_1(\sigma_l * \sigma_k) - \varphi_0(\sigma_l * \sigma_k)]\\
&= (-1)^{d_k d_l}c\delta\chi_{i,j}(\sigma_l * \sigma_k) = c\delta\chi_{i,j}(\sigma_k * \sigma_l),
\end{aligned}$$

i.e., the equation (7).

Case IV. $\sigma_k = \sigma_i$, $\sigma_l \in St\sigma_j$.

We have then $d_i = d_j + 1$. Let (ζ contains no term involving σ_j)

$$\partial\sigma_l = a\sigma_j + \zeta, \quad a = \pm 1.$$

Then

$$\begin{aligned}
\delta\chi_{i,j}(\sigma_k * \sigma_l) = \delta\chi_{i,j}(\sigma_i * \sigma_l) &= (-1)^{d_i}\chi_{i,j}(\sigma_i * \partial\sigma_l)\\
&= (-1)^{d_i}a\chi_{i,j}(\sigma_i * \sigma_j) = (-1)^{d_i}a.
\end{aligned}$$

Since $\bar{T}_1/\bar{\sigma}_l \equiv \bar{T}_0/\bar{\sigma}_l$, we have as before

$$\begin{aligned}
&(-1)^{d_i}[\varphi_1(\sigma_k * \sigma_l) - \varphi_0(\sigma_k * \sigma_l)] = \phi_m(T_1\sigma_i - T_0\sigma_i, T_0\sigma_l)\\
=&\phi_m(\partial h\sigma_i, T_0\sigma_l) + \phi_m(h\partial\sigma_i, T_0\sigma_l) = (-1)^{d_i+1}\phi_m(h\sigma_i, T_0\partial\sigma_l)\\
=&(-1)^{d_i+1}[a\phi_m(h\sigma_i, T_0\sigma_l) + \phi_m(h\sigma_i, T_0\zeta)] = ac.
\end{aligned}$$

By comparing, we get again (7).

Case V. $\sigma_l = \sigma_i$, $\sigma_k \in St\sigma_j$.

By Case IV we have

$$\varphi_1(\sigma_k * \sigma_l) - \varphi_0(\sigma_k * \sigma_l) = (-1)^{d_k d_l}[\varphi_1(\sigma_l * \sigma_k) - \varphi_0(\sigma_l * \sigma_k)]$$
$$=(-1)^{d_k d_l} c\chi_{i,j}(\sigma_l * \sigma_k) = c\chi_{i,j}(\sigma_k * \sigma_l),$$

i.e., the equation (7).

Case VI. The other cases.

We have then necessarily $\sigma_k \in St\sigma_i$, $\sigma_k \neq \sigma_i$, $\sigma_l \neq \sigma_j$, or $\sigma_k = \sigma_i$, $\sigma_l \notin St\sigma_j$, or the cases with the role of k, l interchanged. For the two preceding cases we have always $\sigma_l \notin St\sigma_j$. Hence, let $\tau_k \in St_0\tau_i$ be the d_k-dimensional simplex of K_0 with $\bar{\tau}_k \subset \bar{\sigma}_k$ and let τ_i, τ_k, be oriented as σ_l, σ_k, we would have

$$\phi_m(\partial h\tau_k, T_0\sigma_l) = \phi_m(h\partial\tau_k, T_0\sigma_l) = 0.$$

As $T_1\sigma_l = T_0\sigma_l$, we have $\varphi_1(\sigma_k * \sigma_l) - \varphi_0(\sigma_k * \sigma_l) = 0$. On the other hand, that $\delta\chi_{i,j}(\sigma_k * \sigma_l) = 0$ is evident. Hence (7) is true. The same holds on interchanging the role of k and l.

From the all possible cases considered above, we see that (7) is always true for any $\sigma_k * \sigma_l \in K^*$ with $d_k + d_l = m$. Hence (2) is established and our theorem is completely proved.

§5 Relations between Φ^{2m-1} and Φ^{2m} : $\frac{1}{2}\delta\Phi^{2m-1} = \Phi^{2m}$

In this section the vertices of a Euclidean simplicial complex K will be arranged in a definite order $a_1 < a_2 < \cdots < a_N$, and any simplex $\sigma \in K$ will be written in the normal form $\sigma = (a_{i_0} \cdots a_{i_r})$ with $i_0 < \cdots < i_r$ and oriented accordingly. The dimension of σ will be denoted by $d(\sigma)$, the barycentre of σ by o_σ, and the barycentric subdivision of K by K_1. The simplexes of K will also be arranged in an order $<$ such that $\sigma = (a_{i_0}, \cdots, a_{i_r}) < \tau = (a_{j_0}, \cdots, a_{j_s})$ if and only if either $d(\sigma) = r < d(\tau) = s$, or $r = s$ and t exists with $i_0 = j_0, \cdots, i_{t-1} = j_{t-1}$ but $i_t < j_t$.

Theorem 8 Between the imbedding classes $\Phi^{2m-1} \in H^{2m-1,2}(K, I_2)$ and $\Phi^{2m} \in H^{2m,2}(K)$ of K we have the following relation:

$$\frac{1}{2}\delta\Phi^{2m-1} = \Phi^{2m}. \tag{1}$$

Proof. Let R^{2m} be a Euclidean space of dimension $2m$, with a rectangular system of coordinates $(x_1, \cdots, x_{2m})$. Let R^s be the linear subspace of dimension s defined by $x_{s+1} = \cdots = x_{2m} = 0$ which is separated by R^{s-1} into two parts $R^s_+ : x_s > 0$ and $R^s_- : x_s < 0$, and will be oriented according as the ordered sequence of coordinates $x_1, \cdots, x_s$. Let l_s $(1 \leqslant s \leqslant m-1)$ be the line $x_1 = \cdots = x_{2s} = 1$, $x_{2s+2} = \cdots = x_{2m} = 0$ so that $l_s \subset R^{2s+1}$ and $l_s \cap R^{2s-1} = \varnothing$. Define an almost semi-linear realization $TK_1 = K_1'$ of K in R^{2m} through K_1 as follows. Let us take on R^1 a set of mutually different points $A_1, A_2, \cdots, A_N$ and on l_s a set of mutually different points $A_{i_0 \cdots i_s}$ $(1 \leqslant i_0 < \cdots < i_s \leqslant N)$, $1 \leqslant s \leqslant m-1$. Then

$$T(a_i) = A_i,$$
$$T(o_\sigma) = A_{i_0 \cdots i_s} = A_\sigma (\sigma = (a_{i_0} \cdots a_{i_s}) \in K)$$

defines uniquely a semi-linear realization T of K^{m-1} in $R^{2m-1} \subset R^{2m}$ through its barycentric subdivision. For any $\sigma = (a_{i_0} \cdots a_{i_s}) \in K$ with $d(\sigma) = s \geqslant m$ we may choose by §1 (A) a point $A_\sigma = A_{i_0 \cdots i_s}$, in R^{2m}_+ so that on defining

$$T(o_\sigma) = A_\sigma, \quad \sigma \in K, \quad d(\sigma) \geqslant m,$$

we may extend the above defined semi-linear realization to an *almost* semi-linear realization $TK_1 = K_1'$ of K in R^{2m} through K_1. Similarly, for any $\sigma \in K$ with $d(\sigma) \geqslant m$, we may also choose a point A'_σ in R^{2m-1}_+ such that

$$\begin{cases} T'(a_i) = T(a_i) = A_i, & & \\ T'(o_\sigma) = T(o_\sigma) = A_\sigma, & d(\sigma) \leqslant m-1, & \sigma \in K, \\ T'(o_\sigma) = A'_\sigma, & d(\sigma) \geqslant m, & \sigma \in K, \end{cases}$$

will define an *almost* semi-linear realization $T'K_1 = K_1''$ of K in R^{2m-1} through K_1.

With respect to R^{2m} and R^{2m-1} already oriented, T defines an imbedding cocycle $\varphi_T = \varphi^{2m} \in \Phi^{2m}$ and T' defines an imbedding cocycle $\varphi_{T'} = \varphi^{2m-1} \in \Phi^{2m-1}$. Let ϕ^s and Lk^s denote the intersection number and the linking number in the oriented R^s respectively. If $\sigma * \tau \in K^*$, $d(\sigma) + d(\tau) = 2m-1$ and $d(\sigma) < m-1$, $d(\tau) > m$, then by construction $\bar{T}'\bar{\sigma} = \bar{T}\bar{\sigma} \subset R^{2m-3}$, $T'(Int\bar{\tau}) \cap R^{2m-3} = \varnothing$, so that $\bar{T}'\bar{\sigma} \cap \bar{T}'\bar{\tau} = \varnothing$ and we have

$$\phi^{2m-1}(T'\sigma, T'\tau) = 0, \quad d(\sigma) < m-1, \quad d(\sigma) + d(\tau) = 2m-1. \tag{2}$$

Define now $\varphi_0^{2m-1} \in C^{2m-1}(K^*)$ by $(\sigma * \tau \in K^*, d(\sigma) + d(\tau) = 2m-1, d(\sigma) < d(\tau))$:

$$\varphi_0^{2m-1}(\sigma * \tau) = (-1)^m \phi^{2m-1}(T'\sigma, T'\tau). \tag{3}$$

Then it is easy to see that (3) remains true for $d(\sigma) > d(\tau)$. By (2) we have then ($\sigma * \tau \in K^*, d(\sigma) + d(\tau) = 2m - 1$):

$$\varphi_0^{2m-1}(\sigma * \tau) = \varphi_0^{2m-1}(\tau * \sigma) = 0, \quad d(\sigma) < m-1, \quad d(\tau) > m, \tag{3$'$}$$

and

$$\rho_2 \varphi_0^{2m-1} = \varphi^{2m-1}, \tag{4}$$

where ρ_2 denotes reduction mod 2. Our object is to prove that

$$\delta \varphi_0^{2m-1} = 2\varphi^{2m}. \tag{4$'$}$$

For this purpose let $\xi * \eta \in K^*$, $d(\xi) + d(\eta) = 2m$, and consider the various possible cases as follows:

Case I. $d(\xi) < d(\eta)$ so that $d(\xi) < m$, $d(\eta) > m$.

We have the

$$\begin{aligned}
\delta \varphi_0^{2m-1}(\xi * \eta) &= \varphi_0^{2m-1} \partial(\xi * \eta) \\
&= \varphi_0^{2m-1}(\partial \xi * \eta) + (-1)^{d(\xi)} \varphi_0^{2m-1}(\xi * \partial \eta) \\
&= (-1)^m \phi^{2m-1}(T'\partial\xi, T'\eta) + (-1)^m \cdot (-1)^{d(\xi)} \phi^{2m-1}(T'\xi, T'\partial\eta) \\
&= (-1)^m \phi^{2m-1}(T'\partial\xi, T'\eta) + (-1)^m \phi^{2m-1}(\partial T'\xi, T'\eta) \\
&= 2(-1)^m \phi^{2m-1}(T'\partial\xi, T'\eta).
\end{aligned}$$

As $d(\xi) < m$, any face in $\partial\xi$ has a dimension $< m-1$, so that by (2) we have

$$\delta \varphi_0^{2m-1}(\xi * \eta) = 0. \tag{5}$$

On the other hand, as $d(\xi) < m$, we have by construction $\bar{T}\bar{\xi} \subset R^{2m-1}$, $\bar{T}(Int\bar{\eta}) \cap R^{2m-1} = \varnothing$, so that $\bar{T}\bar{\xi} \cap \bar{T}\bar{\eta} = \varnothing$ and we have

$$\varphi^{2m}(\xi * \eta) = (-1)^{d(\xi)} \phi^{2m}(T\xi, T\eta) = 0. \tag{6}$$

Comparing (5) and (6), we get

$$\delta \varphi_0^{2m-1}(\xi * \eta) = 2\varphi^{2m}(\xi * \eta) = 0. \tag{7}$$

Case II. $d(\xi) > d(\eta)$ so that $d(\xi) > m$, $d(\eta) < m$.

We have then

$$\begin{aligned}
\delta \varphi_0^{2m-1}(\xi * \eta) &= (-1)^{d(\xi)d(\eta)} \delta \varphi_0^{2m-1}(\eta * \xi), \\
\varphi^{2m}(\xi * \eta) &= (-1)^{d(\xi)d(\eta)} \varphi^{2m}(\eta * \xi).
\end{aligned}$$

Hence by (5) and (6), we again get (7).

Case III. $d(\xi) = d(\eta) = m$.

By (3) we have then

$$\begin{aligned}\delta\varphi_0^{2m-1}(\xi * \eta) &= \varphi_0^{2m-1}(\partial\xi * \eta) + (-1)^m \varphi_0^{2m-1}(\xi * \partial\eta)\\ &= (-1)^m \phi^{2m-1}(T'\partial\xi, T'\eta) + \phi^{2m-1}(T'\xi, T'\partial\eta).\end{aligned}$$

or

$$\delta\varphi_0^{2m-1}(\xi * \eta) = 2\phi^{2m-1}(T'\xi, T\partial\eta). \tag{8}$$

On the other hand, let B_η be the reflection of A_η with respect to R^{2m-1}, then $A_\eta T\partial\eta - B_\eta T\partial\eta$ is a cycle on integer coefficients, $B_\eta \bar{T}\bar{\eta}'$ is disjoint from $\bar{T}\bar{\xi}$, and $T\partial\xi = T'\partial\xi$. Hence

$$\begin{aligned}\varphi^{2m}(\xi * \eta) &= (-1)^m \phi^{2m}(T\xi, T\eta) = (-1)^m \phi^{2m}(T\xi, A_\eta T\partial\eta)\\ &= (-1)^m \phi^{2m}(T\xi, A_\eta T\partial\eta - B_\eta T\partial\eta)\\ &= (-1)^m Lk^{2m}(\partial T\xi, A_\eta T\partial\eta - B_\eta T\partial\eta)\\ &= (-1)^m Lk^{2m}(\partial T'\xi, A_\eta T\partial\eta - B_\eta T\partial\eta)\\ &= (-1)^m \phi^{2m}(T'\xi, A_\eta T\partial\eta - B_\eta T\partial\eta)\\ &= (-1)^m \phi^{2m}(T'\xi, A_\eta T\partial\eta).\end{aligned}$$

It may be seen that the last expression is the same as ϕ^{2m-1} $(T'\xi,\ T\partial\eta)$, so that we have

$$\varphi^{2m}(\xi * \eta) = \phi^{2m-1}(T'\xi, T\partial\eta). \tag{9}$$

Comparing (8) and (9) we get

$$\delta\varphi_0^{2m-1}(\xi * \eta) = 2\varphi^{2m}(\xi * \eta). \tag{10}$$

From (7) and (10), we see that for any $\xi * \eta \in K^*$ with $d(\xi)+d(\eta) = 2m$, we have always (10). Hence we get (4)′. From (4) and (4)′ we get (1) and the theorem is proved.

Theorem 9 $\tilde{\Phi}^{2m} = 0$.

Proof. Let $\pi : \tilde{K}^* \to K^*$ be the covering projection. Then by Theorem 6,

$$\pi^* \Phi^{2m-1} = \rho_2 \tilde{\Phi}^{2m-1}, \quad \pi^* \Phi^{2m} = \tilde{\Phi}^{2m}.$$

As $\pi^* \cdot \dfrac{1}{2}\delta = \dfrac{1}{2}\delta \cdot \pi^*$, we get by Theorem 8,

$$\frac{1}{2}\delta(\rho_2 \tilde{\Phi}^{2m-1}) = \tilde{\Phi}^{2m}.$$

As $\tilde{\Phi}^{2m-1}$ is a cohomology class on integer coefficients, we have $\tilde{\Phi}^{2m} = 0$.

§6 Explicit expressions of certain representative cocycles in Φ^{2m-1}

We will make the same assumptions and use the same notations about K as in the preceding section. For any $\sigma * \tau \in K^*$, let $\{\sigma * \tau\}$ denote the cochain on integral coefficients of K^* which takes the value 1 on the cell $\sigma * \tau$ and the value 0 on all other cells of K^*. The purpose of this section is to prove the following

Theorem 10 The $(2m-1)$-dimensional imbedding class Φ^{2m-1} of K has a representative cocycle

$$\varphi^{2m-1} = \rho_2 \sum \{(a_{i_0} \cdots a_{i_{m-1}}) * (a_{j_0} \cdots a_{j_m})\}, \tag{1}$$

in which $\sum$ is extended over all possible sets of indices (i, j) for which $j_0 < i_0 < j_1 < \cdots < i_{m-1} < j_m$. Similarly, $\tilde{\Phi}^{2m-1}$ has also a representative cocycle φ^{2m-1} given by

$$\begin{aligned}&\tilde{\varphi}^{2m-1}\\ =&\Sigma[\{(a_{i_0} \cdots a_{i_{m-1}}) \times (a_{j_0} \cdots a_{j_m})\} - \{(a_{j_0} \cdots a_{j_m}) \times (a_{i_0} \cdots a_{i_{m-1}})\}],\end{aligned} \tag{1$'$}$$

in which $\sum$ has the same meaning as in (1).

The proof of this theorem will be divided into several steps as follows.

1°. **Lemma.** Let $R^{2h-3} \subset R^{2h-1}$ and σ, τ be Euclidean simplexes of dimension r, s respectively, with $r + s = 2h - 2$. Suppose $\bar{\sigma}^{\cdot}$ and $\bar{\tau}^{\cdot}$ to be disjoint, and T a semi-linear imbedding of $\sigma^{\cdot} + \tau^{\cdot}$ in R^{2h-3}. Let l be a line in R^{2h-1} not meeting and also not parallel to R^{2h-3}, and A_0. A_1, A_2 be three mutually different points on l. Orient R^{2h-3} as an oriented simplex ξ in R^{2h-3} and orient R^{2h-1} as the oriented simplex $A_1A_2\xi$. Then

$$\begin{aligned}&\phi^{2h-1}(A_1A_2T\partial\sigma, A_0T\partial\tau)\\ &= \begin{cases} (-1)^r Lk^{2h-3}(T\partial\sigma, T\partial\tau), & \text{if } A_0 \text{ lies between } A_1, A_2,\\ 0, & \text{if otherwise,}\end{cases}\end{aligned} \tag{2}$$

or what is the same,

$$\begin{aligned}&\phi^{2h-1}(A_0T\partial\tau, A_1A_2T\partial\sigma)\\ &= \begin{cases} Lk^{2h-3}(T\partial\tau, T\partial\sigma), & \text{if } A_0 \text{ lies between } A_1, A_2,\\ 0, & \text{if otherwise,}\end{cases}\end{aligned} \tag{2$'$}$$

in which ϕ^k and Lk^k denote respectively the intersection number and the linking number in the oriented R^k $(k = 2h-1, 2h-3)$.

Proof. Suppose first A_0 does not lie between A_1 and A_2. If $A_1A_2\bar{T}\bar{\sigma}^{\cdot}$ and $A_0\bar{T}\bar{\tau}^{\cdot}$ have an intersecting point, then there must exist points x, y in $\bar{T}\bar{\sigma}^{\cdot}$, $\bar{T}\bar{\tau}^{\cdot}$ respectively and a point z on l between A_1, A_2, such that the segments zx and A_0y will meet in the above point. As $A_0 \neq z$ and $x \neq y$ by hypothesis, the two lines $A_0z = l$ and $xy \subset R^{2h-3}$ would have intersecting points, contrary to supposition. Hence $A_1A_2\bar{T}^{\cdot}\bar{\sigma}^{\cdot}$ and $A_0\bar{T}\bar{\tau}$ are disjoint and we get the lower half of (2) or (2)$'$.

Next suppose A_0 lies between A_1, A_2. We may then take in R^{2h-3} a point $A \notin \bar{T}(\bar{\sigma}\cdot+\bar{\tau}\cdot)$. Prolong A_0A to A_0' and join A_1A_0', $A_0'A_2$. Let $C = A_0'A_1 + A_1A_2 + A_2A_0'$. Take also in R^{2h-3} a point O' such that O' $(T\sigma\cdot+T\tau\cdot)$ is an almost Euclidean complex. Define a semi-linear imbedding T' of σ in R^{2h-3} through its barycentric subdivision by $T'/\sigma\cdot \equiv T$ and $T'(O_\sigma) = O'$. Then

$$
\begin{aligned}
Lk^{2h-3}(T\partial\sigma, T\partial\tau) &= Lk^{2h-3}(T'\partial\sigma, T\partial\tau)\\
&= Lk^{2h-3}(\partial T'\sigma, T\partial\tau) = \phi^{2h-3}(T'\sigma, T\partial\tau).
\end{aligned}
$$

Since the 2-dimensional simplex $A_1A_2A_0'$ and R^{2h-3} meet in the single point A, we see according to the chosen orientations of R^{2h-3}, R^{2h-1} that

$$
\begin{aligned}
\phi^{2h-3}(T'\sigma, T\partial\tau) &= \phi^{2h-1}(CT'\sigma, T\partial\tau) = \phi^{2h-1}(CT'\sigma, \partial(A_0T\partial\tau))\\
&= (-1)^r\phi^{2h-1}(\partial(CT'\sigma), A_0T\partial\tau) = (-1)^r\phi^{2h-1}(C\partial T'\sigma, A_0T\partial\tau)\\
&= (-1)^r\phi^{2h-1}(CT\partial\sigma, A_0T\partial\tau).
\end{aligned}
$$

Hence

$$
\begin{aligned}
Lk^{2h-3}(T\partial\sigma, T\partial\tau) =&(-1)^r\phi^{2h-1}(CT\partial\sigma, A_0T\partial\tau)\\
=&(-1)^r[\phi^{2h-1}(A_1A_2T\partial\sigma, A_0T\partial\tau) + \phi^{2h-1}(A_0'A_1T\partial\sigma, A_0T\partial\tau)\\
&+ \phi^{2h-1}(A_2A_0'T\partial\sigma, A_0T\partial\tau)]. \qquad (3)
\end{aligned}
$$

Denote by R^{2h-2} the linear subspace determined by A_0A_0' and R^{2h-3}. Consider any point z of $A_0'A_1 + A_0'A_2$, any point x of $\bar{T}\bar{\sigma}\cdot$, and any point y of $\bar{T}\bar{\tau}\cdot$. In case $z \neq A_0'$, zx will meet R^{2h-2} only in $x \in R^{2h-3}$ while $A_0y \subset R^{2h-2}$, $A_0y \cap R^{2h-3} = (y) \neq (x)$. In case $z = A_0'$, $A_0'x$ and A_0y will lie wholly in R^{2h-2} but on opposite sides of R^{2h-3}. Hence whatever $z \in A_0'A_1 + A_1'A_2$, $x \in \bar{T}\bar{\sigma}\cdot$ and $y \in \bar{T}\bar{\tau}\cdot$ may be, zx and A_0y are always non-intersecting. Consequently $A_0'A_1\bar{T}\bar{\sigma}^{\cdot} + A_0'A_2\bar{T}\bar{\sigma}\cdot$ and $A_0\bar{T}\bar{\tau}'$ are disjoint and we have

$$\phi^{2h-1}(A_0'A_1T\partial\sigma, A_0T\partial\tau)=\phi^{2h-1}(A_2A_0'T\partial\sigma, A_0T\partial\tau)=0.$$

It follows that (3) becomes the upper half of (2) and our Lemma is proved.

2°. Let R^{2m-1} be a Euclidean space of dimension $2m-1$, having a rectangular system of coordinates $(x_1,\cdots,x_{2m-1})$. Let R^s $(1\leqslant s\leqslant 2m-1)$ be the s-dimensional linear subspace of R^{2m-1} defined by $x_{s+1}=\cdots=x_{2m-1}=0$, l_s $(1\leqslant s\leqslant m-1)$ the line $x_1=\cdots=x_{2s}=1$, $x_{2s+2}=\cdots=x_{2m-1}=0$, such that $l_s\subset R^{2s+1}$ and l_s meet R^{2s} in the single point $O_s=(\underbrace{1,\cdots,1}_{2s},\underbrace{0,\cdots,0}_{2m-2s-1})$. Corresponding to each $s\geqslant 1$ and $\leqslant m-1$ and each $\sigma=(a_{i_0},\cdots,a_{i_s})\in K$, we shall take a point $A_{i_0\cdots i_s}=A_\sigma$ with $x_{2s+1}(A_\sigma)>0$ such that $\sigma<\tau$ (τ is another s-dimensional simplex of K) would imply $x_{2s+1}(A_\sigma)<x_{2s+1}(A_\tau)$, or in simpler form, $A_\sigma<A_\tau$. Take also points $A_i=(i,\underbrace{0,\cdots,0}_{2m-2})$ on R^1. Define now an almost semi-linear imbedding T of K in R^{2m-1} through its barycentric subdivision K_1 by

$$\begin{aligned}&T(a_i)=A_i,\\&T(o_\sigma)=A_\sigma,\quad \sigma\in K,\quad 0<\dim\sigma\leqslant m-1,\end{aligned}$$

and for $\sigma\in K$, let $\dim\sigma\geqslant m$, $T(o_\sigma)=A_\sigma$ be a point conveniently chosen in $R_+^{2m-1}(x_{2m-1}>0)$. We have then

$$T(a_{i_0}\cdots a_{i_h})=A_{i_0\cdots i_h}T\partial(a_{i_0}\cdots a_{i_h}),\tag{4}$$

and

$$T\partial(a_{i_0}\cdots a_{i_h})=\partial\sum_{r,s}(-1)^{r+s}A_{i_0\cdots\hat{i}_r\cdots i_h}A_{i_0\cdots\hat{i}_s\cdots i_h}T(a_{i_0}\cdots\hat{a}_{i_r}\cdots\hat{a}_{i_s}\cdots a_{i_h}),\tag{5}$$

in which $h\leqslant m$ and $\hat{a}_i(\hat{i}_t)$ means that $a_i(i_t)$ does not appear in the corresponding sequence.

Proof. (4) is quite evident. The right hand side of (5) is

$$\begin{aligned}&=-\sum_{r,s}(-1)^{r+s}A_{i_0\cdots\hat{i}_r\cdots i_h}T(a_{i_0}\cdots\hat{a}_{i_r}\cdots\hat{a}_{i_s}\cdots a_{i_h})\\&\quad+\sum_{r,s}(-1)^{r+s}A_{i_0\cdots\hat{i}_s\cdots i_h}T(a_{i_0}\cdots\hat{a}_{i_r}\cdots\hat{a}_{i_s}\cdots a_{i_h})\\&\quad+\sum_{r,s}(-1)^{r+s}A_{i_0\cdots\hat{i}_r\cdots i_h}A_{i_0\cdots\hat{i}_s\cdots i_h}\partial T(a_{i_0}\cdots\hat{a}_{i_r}\cdots\hat{a}_{i_s}\cdots a_{i_h}).\end{aligned}$$

Denote the three terms successively by $\sum_1$, $\sum_2$ and $\sum_3$, then

$$\Sigma_3 = \sum_{r,s,t} (-1)^{r+s+t} (A_{i_0\cdots\hat{i}_r\cdots i_h} A_{i_0\cdots\hat{i}_s\cdots i_h} - A_{i_0\cdots\hat{i}_r\cdots i_h} A_{i_0\cdots\hat{i}_t\cdots i_h}$$
$$+ A_{i_0\cdots\hat{i}_s\cdots i_h} A_{i_0\cdots\hat{i}_t\cdots i_h}) T(a_{i_0} \cdots \hat{a}_{i_r} \cdots \hat{a}_{i_s} \cdots \hat{a}_{i_t} \cdots a_{i_h}).$$

As $A_{i_0\cdots\hat{i}_\lambda\cdots i_h} (\lambda = r, s, t)$ are all on the line l_{h-1}, we have $(r < s < t)$

$$A_{i_0\cdots\hat{i}_r\cdots i_h} A_{i_0\cdots\hat{i}_s\cdots i_h} - A_{i_0\cdots\hat{i}_r\cdots i_h} A_{i_0\cdots\hat{i}_t\cdots i_h} + A_{i_0\cdots\hat{i}_s\cdots i_h} A_{i_0\cdots\hat{i}_t\cdots i_h} = 0.$$

Hence

$$\Sigma_3 = 0. \tag{6}$$

Next we have

$$\sum\nolimits_1 + \sum\nolimits_2 = \sum_{r,s} (-1)^{r+s} A_{i_0\cdots\hat{i}_r\cdots i_h} [-T(a_{i_0} \cdots \hat{a}_{i_r} \cdots \hat{a}_{i_s} \cdots a_{i_h})$$
$$+ T(a_{i_0} \cdots \hat{a}_{i_s} \cdots \hat{a}_{i_r} \cdots a_{i_h})]$$
$$= \sum (-1)^r A_{i_0\cdots\hat{i}_r\cdots i_h} T\partial(a_{i_0} \cdots \hat{a}_{i_r} \cdots a_{i_h})$$
$$= T \sum_r (-1)^r (a_{i_0} \cdots \hat{a}_{i_r} \cdots a_{i_h}),$$

or

$$\sum\nolimits_1 + \sum\nolimits_2 = T\partial(a_{i_0} \cdots a_{i_h}). \tag{6$'$}$$

From (6) and (6)′ we get (5).

3°. Let us orient R^s, $s = 1, \cdots, 2m-1$, as its coordinate sequence $x_1, \cdots, x_s$ and orient l_s by the increasing values of x_s. For any $0 \leqslant h \leqslant m-1$ and any two simplexes $\sigma = (a_{i_0} \cdots a_{i_{m-1}})$, $\tau = (a_{i_0} \cdots a_{j_m})$ of K having no vertices in common, let

$$\phi^{2m-2h-1}(T(a_{i_h} \cdots a_{i_{m-1}}),$$
$$\sum_{r,s} (-1)^{r+s} A_{j_h\cdots\hat{j}_r\cdots j_m} A_{j_h\cdots\hat{j}_s\cdots j_m} T(a_{j_h} \cdots \hat{a}_{j_r} \cdots \hat{a}_{j_s} \cdots a_{j_m})) = I_h. \tag{7$_h$}$$

Then

$$I_h = \begin{cases} (-1)^{(m-h)(m-h-1)/2}, & j_h < i_h < j_{h+1} < \cdots < i_{m-1} < j_m, \\ 0, & \text{otherwise}. \end{cases} \tag{8$_h$}$$

Proof. For $h = m-1$, we have

$$I_{m-1} = \phi^1(T(a_{i_{m-1}}), -A_{j_m} A_{j_{m-1}}) = \phi^1(A_{i_{m-1}}, A_{i_{m-1}} A_{j_m}).$$

Since $A_{i_{m-1}}$, $A_{j_{m-1}}$, A_{j_m} are all on the line R^1 and $A_{i_{m-1}}$ lies between $A_{j_{m-1}}$ and A_{j_m} if and only if $j_{m-1} < i_{m-1} < j_m$, we have

$$I_{m-1} = \begin{cases} +1, & j_{m-1} < i_{m-1} < j_m, \\ 0, & i_{m-1} < j_{m-1} \text{ or } i_{m-1} > j_m. \end{cases}$$

Hence $(8)_{m-1}$ is true. Suppose now $(8)_{h+1}, \cdots, (8)_{m-1}$have been proved and let use prove $(8)_h$ as follows.

Case I. $i_h < j_h$ or $i_h > j_{h+1}$.

By construction $A_{i_h \cdots i_{m-1}}$, $A_{j_h \cdots \hat{j}_r \cdots j_m}$ and $A_{j_h \cdots \hat{j}_s \cdots j_m}$ are all on the line $l_t(t = m-h-1)$ and for $i_h < j_h$, we have always $A_{i_h \cdots i_{m-1}} < A_{j_h \cdots \hat{j}_r \cdots j_m}$, $r = h, \cdots, m$, while for $i_h > j_{h+1}$, we have always $A_{i_h \cdots i_{m-1}} > A_{j_h \cdots \hat{j}_r \cdots j_m}$, $r = h, \cdots, m$. Hence whatever be k and $r \neq s$ $(h \leqslant k \leqslant m-1,\ h \leqslant r < s \leqslant m)$, we have

$$A_{i_h \cdots i_{m-1}} \bar{T}(a_{i_h} \cdots \hat{a}_{i_k} \cdots a_{i_{m-1}}) \cap A_{j_h \cdots \hat{j}_r \cdots j_m} A_{j_h \cdots \hat{j}_s \cdots j_m} \bar{T}(a_{j_h} \cdots \hat{a}_{j_r} \cdots \hat{a}_{j_s} \cdots a_{j_m}) = \varnothing.$$

Hence by (4) we get

$$\begin{aligned} I_h = & \phi^{2m-2h-1}\Big(A_{i_h \cdots i_{m-1}} T \sum_k (-1)^{k-h} (a_{i_h} \cdots \hat{a}_{i_k} \cdots a_{i_{m-1}}), \\ & \sum_{r\ s} (-1)^{r+s} A_{j_h \cdots \hat{j}_r \cdots j_m} A_{j_h \cdots \hat{j}_s \cdots j_m} T(a_{j_h} \cdots \hat{a}_{j_r} \cdots \hat{a}_{j_s} \cdots a_{j_m})\Big) = 0. \end{aligned}$$

This is the lower half of $(8)_h$.

Case II. $j_h < i_h < j_{h+1}$.

Again by (4) we get

$$\begin{aligned} I_h = \phi^{2m-2h-1}\Big[& A_{i_h \cdots i_{m-1}} \partial T(a_{i_h} \cdots a_{i_{m-1}}), \\ & \sum_{r,s} (-1)^{r+s} A_{j_h \cdots \hat{j}_r \cdots j_m} A_{j_h \cdots \hat{j}_r \cdots j_m} T(a_{j_h} \cdots \hat{a}_{j_r} \cdots \hat{a}_{j_s} \cdots a_{j_m})\Big]. \end{aligned}$$

Since $A_{j_h \cdots j_{m-1}} < A_{j_h \cdots \hat{j}_r \cdots j_m}$ for $r > h$, we have for r, $s > h$,

$$\begin{aligned} & \phi^{2m-2h-1}[A_{i_h \cdots i_{m-1}} \partial T(a_{i_h} \cdots a_{i_{m-1}}), \\ & A_{j_h \cdots \hat{i}_r \cdots j_m} A_{j_h \cdots \hat{j}_r \cdots j_m} T(a_{j_h} \cdots \hat{a}_{j_r} \cdots \hat{a}_{j_s} \cdots a_{j_m})] = 0 \end{aligned}$$

As

$$A_{j_{h+1} \cdots j_m} A_{j_h \cdots \hat{j}_s \cdots j_m} = A_{j_{h+1} \cdots j_m} A_{j_h j_{h+2} \cdots j_m} + A_{j_h j_{h+2} \cdots j_m} A_{j_h \cdots \hat{j}_s \cdots j_m},$$

the expression for I_h may be further simplified as

$$\begin{aligned} I_h =& \phi^{2m-2h-1}\Big[A_{i_h\cdots i_{m-1}}\partial T(a_{i_h}\cdots a_{i_{m-1}}),\\ &(-1)^h\sum_s(-1)^s A_{j_{h+1}\cdots j_m}A_{j_h\cdots\hat{j}_s\cdots j_m}T(a_{j_{h+1}}\cdots\hat{a}_{j_s}\cdots a_{j_m})\Big]\\ =&(-1)^h\phi^{2m-2h-1}\Big[A_{i_h\cdots i_{m-1}}\partial T(a_{i_h}\cdots a_{i_{m-1}}),\\ &\sum_s(-1)^s A_{j_{h+1}\cdots j_m}A_{j_h i_{h+2}\cdots j_m}T(a_{j_{h+1}}\cdots\hat{a}_{j_s}\cdots a_{j_m})\Big]\\ =&-\phi^{2m-2h-1}[A_{i_h\cdots i_{m-1}}\partial T((a_{i_h}\cdots a_{i_{m-1}}),\\ &A_{j_{h+1}\cdots j_m}A_{j_h j_{h+1}\cdots j_m}\partial T((a_{j_{h+1}}\cdots a_{j_m})]. \end{aligned}$$

Applying the Lemma of 1°, and (4), (5) we get

$$\begin{aligned} I_h =& Lk^{2m-2h-3}[\partial T(a_{i_h}\cdots a_{i_{m-1}}),\partial T(a_{j_{h+1}}\cdots a_{j_m})]\\ =& Lk^{2m-2h-3}\Big[\sum_k(-1)^{k-h}T(a_{i_h}\cdots\hat{a}_{i_k}\cdots a_{i_{m-1}}),\\ &\partial\sum_{r,s}(-1)^{r+s}A_{j_{h+1}\cdots\hat{j}_r\cdots j_m}A_{j_{h+1}\cdots\hat{j}_s\cdots j_m}T(a_{j_{h+1}}\cdots\hat{a}_{j_r}\cdots\hat{a}_{j_s}\cdots a_{j_m})\Big]\\ =&(-1)^{m-1}\phi^{2m-2h-3}\Big[\sum_k(-1)^k A_{i_h\cdots\hat{i}_k\cdots i_{m-1}}T\partial(a_{i_h}\cdots\hat{a}_{i_k}\cdots a_{i_{m-1}}),\\ &\sum_{r,s}(-1)^{r+s}A_{j_{h+1}\cdots\hat{j}_s\cdots j_m}A_{j_{h+1}\cdots\hat{j}_s\cdots j_m}T(a_{j_{h+1}}\cdots\hat{a}_{j_r}\cdots\hat{a}_{j_s}\cdots a_{j_m})\Big]. \end{aligned}$$

For $k>h$, we have $i_h<j_{h+1}$ and $A_{i_h\cdots\hat{i}_k\cdots i_{m-1}}<A_{j_{h+1}\cdots\hat{j}_r\cdots j_m}r=h+1,\cdots,m$, hence $(k>h)$

$$\begin{aligned} &\phi^{2m-2h-3}[A_{i_h\cdots\hat{i}_k\cdots i_{m-1}}T\partial(a_{i_h}\cdots\hat{a}_{i_k}\cdots a_{i_{m-1}}),\\ &\qquad A_{j_{h+1}\cdots\hat{j}_r\cdots j_m}A_{j_{h+1}\cdots\hat{j}_s\cdots j_m}T(a_{j_{h+1}}\cdots\hat{a}_{j_r}\cdots\hat{a}_{j_s}\cdots a_{j_m})]=0, \end{aligned}$$

and I_h may be simplified as

$$\begin{aligned} I_h =&(-1)^{m+h+1}\phi^{2m-2h-3}\Big[A_{i_{h+1}\cdots i_{m-1}}T\partial(a_{i_{h+1}}\cdots a_{i_{m-1}}),\\ &\sum_{r,s}(-1)^{r+s}A_{j_{h+1}\cdots\hat{j}_r\cdots j_m}A_{j_{h+1}\cdots\hat{j}_s\cdots j_m}T(a_{j_{h+1}}\cdots\hat{a}_{j_r}\cdots\hat{a}_{j_s}\cdots a_{j_m})\Big]\\ =&(-1)^{m+h+1}\phi^{2m-2h-3}\Big[T(a_{i_{h+1}}\cdots a_{i_{m-1}}),\\ &\sum_{r,s}(-1)^{r+s}A_{j_{h+1}\cdots\hat{j}_r\cdots j_m}A_{j_{h+1}\cdots\hat{j}_s\cdots j_m}T(a_{i_{h+1}}\cdots\hat{a}_{j_r}\cdots\hat{a}_{j_s}\cdots a_{j_m})\Big]\\ =&(-1)^{m+h+1}I_{h+1}. \end{aligned}$$

By induction hypothesis $(8)_{h+1}$, we get $(8)_h$.

4°. We now prove Theorem 1° as follows.

Take the point $A'_{i_0\cdots i_{m-1}}: x_{2m-1}(A'_{i_0\cdots i_{m-1}})<0$ symmetric to $A_{i_0\cdots i_{m-1}}$ with respect to $o_{m-1}=l_{m-1}\cap R^{2m-2}$ and define a semi-linear realization T'_σ of $\sigma=(a_{i_0}\cdots a_{i_{m-1}})\in K$ in R^{2m-1} through the barycentric subdivision of σ by

$$T'_\sigma(a_i)=A_i;\quad i=i_0,\cdots,i_{m-1};\quad T'_\sigma|\sigma\equiv T;\quad T'_\sigma(0_\sigma)=A'_\sigma.$$

Then $T_\sigma-T'_\sigma\sigma$ is a cycle. For any $\tau\in K$ having no vertex in common with σ where $d(\tau)+d(\sigma)\leqslant 2m-1$, we have $\bar{T}'_\sigma\bar{\sigma}\cap\bar{T}\bar{\tau}=\varnothing$, hence $\phi^{2m-1}(T'_\sigma\sigma,T\tau)=0$.

Let $\sigma=(a_{i_0}\cdots a_{i_{m-1}})$, $\tau=(a_{j_0}\cdots a_{j_m})$, $\sigma\times\tau\in\tilde{K}^*$, then

$$\begin{aligned}&(-1)^{m-1}\tilde{\varphi}_T^{2m-1}(\sigma\times\tau)=\phi^{2m-1}(T\sigma,T\tau)=\phi^{2m-1}(T\sigma-T'{}_\sigma\sigma,T\tau)\\&=\phi^{2m-1}\left[T\sigma-T'{}_\sigma\sigma,T\tau-\sum_{r,s}(-1)^{r+s}A_{j_0\cdots\hat{j}_r\cdots j_m}A_{j_0\cdots\hat{j}_s\cdots j_m}T(a_{j_0}\cdots\hat{a}_{i_r}\cdots\hat{a}_{j_s}\cdots a_{i_m})\right]\\&\quad+\phi^{2m-1}\left[T\sigma-T'{}_\sigma\sigma,\sum_{r,s}(-1)^{r+s}A_{j_0\cdots\hat{j}_r\cdots j_m}A_{j_0\cdots\hat{j}_s\cdots j_m}T(a_{j_0}\cdots\hat{a}_{j_r}\cdots\hat{a}_{j_s}\cdots a_{j_m})\right].\end{aligned}$$

By (5), $T\sigma-T'_\sigma\sigma$ and $T\tau-\sum\limits_{r,s}(-1)^{r+s}A_{j_0\cdots\hat{j}_r\cdots j_m}A_{j_0\cdots\hat{j}_s\cdots j_m}T(a_{j_0}\cdots\hat{a}_{i_r}\cdots\hat{a}_{j_s}\cdots a_{j_m})$ are both cycles, so that the first term of the right hand side in the above expression vanishes. As $\bar{T}'_\sigma\bar{\sigma}$ is disjoint from $A_{j_0\cdots\hat{j}_r\cdots j_m}A_{j_0\cdots\hat{j}_s\cdots j_m}\bar{T}(a_{j_0}\cdots\hat{a}_{j_r}\cdots\hat{a}_{j_s}\cdots a_{j_m})$, the above expression may be simplified as

$$\begin{aligned}(-1)^{m-1}\tilde{\varphi}_T^{2m-1}(\sigma\times\tau)=\phi^{2m-1}\Big[&T\sigma,\\&\sum_{r,s}(-1)^{r+s}A_{j_0\cdots\hat{j}_r\cdots j_m}A_{j_0\cdots\hat{j}_s\cdots j_m}T(a_{j_0}\cdots\hat{a}_{j_r}\cdots\hat{a}_{j_s}\cdots a_{j_m})\Big].\end{aligned}$$

By (7), (8) of 3°, we get

$$\tilde{\varphi}_T^{2m-1}(\sigma\times\tau)=\begin{cases}(-1)^{(m-1)(m-2)/2}, & j_0<i_0<j_1<\cdots<i_{m-1}<j_m,\\0, & \text{otherwise}.\end{cases}\tag{9}$$

From this we get further

$$\begin{aligned}&\tilde{\varphi}_T^{2m-1}(\tau\times\sigma)=-\tilde{\varphi}_T^{2m-1}(\sigma\times\tau)\\&=\begin{cases}-(-1)^{(m-1)(m-2)/2}, & j_0<i_0<j_1<\cdots<i_{m-1}<j_m,\\0, & \text{otherwise}.\end{cases}\end{aligned}\tag{9'}$$

Next let $\sigma\times\tau\in\tilde{K}^*$, $d(\sigma)+d(\tau)=2m-1$, while $d(\sigma)<m-1$, $d(\tau)>m$. Then $\bar{T}\bar{\sigma}\subset R^{2m-3}$ and $\bar{T}(\text{Int}\bar{\tau})\cap\bar{T}\bar{\sigma}=\varnothing$. Hence

$$\tilde{\varphi}_T^{2m-1}(\sigma\times\tau)=0,\quad d(\sigma)<m-1,\quad d(\tau)>m.\tag{10}$$

Similarly,

$$\tilde{\varphi}_T^{2m-1}(\tau \times \sigma) = 0, \quad d(\sigma) < m-1, \quad d(\tau) > m. \tag{10$'$}$$

From (9), (9)′, (10) and (10)′ we get

$$\tilde{\varphi}_T^{2m-1} = \sum (-1)^{(m-1)(m-2)/2}[\{(a_{i_0} \cdots a_{i_{m-1}}) \times (a_{j_0} \cdots a_{j_m})\} - \{(a_{j_0} \cdots a_{j_m}) \times (a_{i_0} \cdots a_{i_{m-1}})\}],$$

in which $\sum$ is extended over all possible sets of indices (i, j) for which $j_0 < i_0 < j_1 < \cdots < i_{m-1} < j_m$. Now by Theorem 4 and the remark below $\tilde{\varphi}_T$ and $-\tilde{\varphi}_T$ are both cocycles in $\tilde{\Phi}^{2m-1}$. Hence $\tilde{\Phi}^{2m-1}$ has a representative cocycle as given by (1)′.

Similarly we have a representative cocycle in Φ^{2m-1} as given in (1). Our theorem is thus completely proved.

§7 Explicit expressions for certain representative cocycles in Φ^{2m}

The notations will be the same as in the preceding section.

Theorem 11 In Φ^{2m} there is a representative cocycle

$$\varphi^{2m} = \sum \{(a_{i_0} \cdots a_{i_m}) * (a_{j_0} \cdots a_{j_m})\}, \tag{1}$$

in which $\sum$ is extended over all possible sets of indices (i, j) such that $i_0 < j_0 < \cdots < i_m < j_m$.

Proof. By Theorem 10 of §6, Φ^{2m-1} has a representative cocycle

$$\varphi^{2m-1} = \rho_2 \sum \{(a_{i_0} \cdots a_{i_{m-1}}) * (a_{j_0} \cdots a_{j_m})\},$$

in which $\sum$ is extended over all possible sets of indices (i, j) such that $j_0 < i_0 < j_1 < \cdots < i_{m-1} < j_m$. Define now $\varphi_0^{2m-1} \in C^{2m-1}(K^*)$ by

$$\varphi_0^{2m-1} = \sum \{(a_{i_0} \cdots a_{i_{m-1}}) * (a_{j_0} \cdots a_{j_m})\}, \tag{2}$$

in which $\sum$ is as before, then

$$\rho_2 \varphi_0^{2m-1} = \varphi^{2m-1}. \tag{3}$$

By Theorem 7 of §5, $\Phi^{2m} = \dfrac{1}{2}\delta\Phi^{2m-1}$, hence Φ^{2m} has a representative cocycle φ_0^{2m} such that

$$\delta\varphi_0^{2m-1} = 2\varphi_0^{2m}. \tag{4}$$

We prove now φ_0^{2m} is the same as φ^{2m} in (1) as follows.

Let $\sigma * \tau = (a_{i_0} \cdots a_{i_p}) * (a_{j_0} \cdots a_{j_q}) \in K^*$, $i_0 < j_0$, $p + q = 2m$, then

$$\begin{aligned}\delta\varphi_0^{2m-1}(\sigma * \tau) =&\varphi_0^{2m-1}(\partial(a_{i_0} \cdots a_{i_p}) * (a_{j_0} \cdots a_{j_q})) \\ &+ (-1)^p \varphi_0^{2m-1}((a_{i_0} \cdots a_{i_p}) * \partial(a_{j_0} \cdots a_{j_q})).\end{aligned} \tag{5}$$

or

$$\delta\varphi_0^{2m-1}(\sigma * \tau) = \sum\nolimits_1 + (-1)^p \sum\nolimits_2,$$

where

$$\left.\begin{aligned}\sum\nolimits_1 &= \sum_r (-1)^r \varphi_0^{2m-1}((a_{i_0} \cdots \hat{a}_{i_r} \cdots a_{i_p}) * (a_{j_0} \cdots a_{j_q})), \\ \sum\nolimits_2 &= \sum_r (-1)^r \varphi_0^{2m-1}((a_{i_0} \cdots \hat{a}_{i_r} \cdots a_{i_q}) * (a_{j_0} \cdots a_{j_p})).\end{aligned}\right\} \tag{6}$$

Consider now various possible cases as follows (i_0 always $< j_0$) :

Case I. $p = q = m$, $i_0 < j_0 < \cdots < i_m < j_m$.

By (2) we have

$$\varphi_0^{2m-1}((a_{i_0} \cdots \hat{a}_{i_r} \cdots a_{i_m}) * (a_{j_0} \cdots a_{j_m})) = \begin{cases} 1, & r = 0, \\ 0, & r > 0; \end{cases}$$

$$\varphi_0^{2m-1}((a_{j_0} \cdots \hat{a}_{j_r} \cdots a_{j_m}) * (a_{i_0} \cdots a_{i_m})) = \begin{cases} 1, & r = m, \\ 0, & r < m. \end{cases}$$

Hence $\sum_1 = 1$, $\sum_2 = (-1)^m$, and (5) becomes

$$\delta\varphi_0^{2m-1}(\sigma * \tau) = 2.$$

Comparing with (1) we get

$$\delta\varphi_0^{2m-1}(\sigma * \tau) = 2\varphi^{2m}(\sigma * \tau). \tag{7}$$

Case II. $p = q = m$, and $i_0 < j_0 < \cdots < i_m < j_m$ does not hold.

In that case there exists either an index s with no j_k satisfying $i_s < j_k < i_{s+1}$ so that

$$\varphi_0^{2m-1}(a_{j_0} \cdots \hat{a}_{j_r} \cdots a_{j_m}) * (a_{i_0} \cdots a_{i_m})) = 0, \tag{8}$$

or an index s with no i_k satisfying $j_s < i_k < j_{s+1}$. In the second alternative we have still (8) for $r \neq s$, $s+1$, while for $r = s$, $s+1$, $\varphi_0^{2m-1}((a_{j_0} \cdots \hat{a}_{j_s} \cdots a_{j_m}) * (a_{i_0} \cdots a_{i_m}))$ and $\varphi_0^{2m-1}((a_{j_0} \cdots \hat{a}_{j_{s+1}} \cdots a_{j_m}) * (a_{i_0} \cdots a_{i_m}))$ have the same value 0 or 1 so that $\sum_2 = (-1)^s + (-1)^{s+1} = 0$. Hence we have always $\sum_2 = 0$.

As $i_0 < j_0$, we have $\varphi_0^{2m-1}((a_{i_0}\cdots\hat{a}_{i_r}\cdots a_{i_m}) * (a_{j_0}\cdots a_{j_m})) = 0$ for $r > 0$. It is also $= 0$ for $r = 0$ since $j_0 < i_1 < j_1 < \cdots < i_m < j_m$ is not true. Hence we have always $\sum_1 = 0$.

It follows that $\delta\varphi_o^{2m-1}(\sigma * \tau) = 0$. As $\varphi^{2m}(\sigma * \tau) = 0$ by (1), we get (7).

Case III. $p < m - 1$ or $q < m - 1$.

In that case each term of $\sum_1$, $\sum_2$ is 0 and we get still (7) by comparing with (1).

Case IV. $p = m - 1$, $q = m + 1$.

As $i_0 < j_0 < j_1$, we have $\sum_2 = 0$. Comparing with (1), we get (7).

Case V. $p = m + 1$, $q = m - 1$.

In that case $\sum_2 = 0$, while

$$\sum\nolimits_1 = \sum_r (-1)^r \varphi_0^{2m-1}((a_{j_0}\cdots a_{j_{m-1}}) * (a_{i_0}\cdots\hat{a}_{i_r}\cdots a_{i_{m+1}})).$$

Now we have necessarily an index s with no j_k satisfying $i_s < j_k < i_{s+1}$ $(0 \leqslant s \leqslant m)$. Then $\varphi_0^{2m-1}((a_{j_0}\cdots a_{j_{m-1}}) * (a_{i_0}\cdots\hat{a}_{i_r}\cdots a_{i_{m+1}}))$ are all 0 for $r \neq s$, $s+1$ while for $r = s$, $s + 1$ they are both of same value 0 or 1. We have thus always $\sum_1 = 0$. Comparing with (1) we get again (7).

Thus whatever the case may be, we get always (7) for any $\sigma*\tau \in K^*(d(\sigma)+d(\tau) = 2m)$. Hence φ_0^{2m} in (4) coincides with φ^{2m} of (1) and the theorem is proved.

§8 Relations between Φ^i and their topological invariance

As before let the vertices of K be arranged in a fixed order $a_1 < a_2 < \cdots$, and all simplexes of K be written in normal form $(a_{i_0}\cdots a_{i_k})$, with $i_0 < i_1 < \cdots < i_k$. Since $\tilde{K}^*$ is a two-sheeted covering complex of K^*, we may define as in [9] § 1 chain transformations

$$t : C(\tilde{K}^*, G) \to C(\tilde{K}^*, G),$$
$$\pi : C(\tilde{K}^*, G) \to C(\tilde{K}^*, G),$$

and

$$\bar{\pi} : C(K^*, G) \to C(\tilde{K}^*, G),$$

such that for $(a_{i_0}\cdots a_{i_p})$, $(a_{j_0}\cdots a_{j_q}) \in K$ with no vertices in common, we have

$$t((a_{i_0}\cdots a_{i_p}) \times (a_{j_0}\cdots a_{j_q})) = (-1)^{pq}((a_{j_0}\cdots a_{j_q}) \times (a_{i_0}\cdots a_{i_p})), \tag{1}$$

$$\pi((a_{i_0}\cdots a_{i_p})\times(a_{j_0}\cdots a_{j_q}))=(a_{i_0}\cdots a_{i_p})*(a_{j_0}\cdots a_{j_q}), \tag{2}$$

$$\bar{\pi}((a_{i_0}\cdots a_{i_p})*(a_{j_0}\cdots a_{j_q}))=(a_{i_0}\cdots a_{i_p})\times(a_{j_0}\cdots a_{j_q}) + (-1)^{pq}(a_{j_0}\cdots a_{j_q})\times(a_{i_0}\cdots a_{i_p}). \tag{3}$$

Put $s=1+t$, $d=1-t$, and let π', $\bar{\pi}'$, t', s', d' be the dual cochain transformations of π, $\cdots$. For any classes $X\in H^r(K^*, I_2)$ and $Y\in H^r(K^*)$, let us take $\tilde{x}_0$, $\tilde{x}_1\in C^*(\tilde{K}^*, I_2)$ and $\tilde{y}_0$, $\tilde{y}_1$, $\tilde{y}_2\in C^*(\tilde{K}^*)$ such that

$$\bar{\pi}'\tilde{x}_0\in X,\quad \delta\tilde{x}_0=d'\tilde{x}_1, \tag{4}$$

$$\bar{\pi}'\tilde{y}_0\in Y,\quad \delta\tilde{y}_0=d'\tilde{y}_1,\quad \delta\tilde{y}_1=s'\tilde{y}_2. \tag{4$'$}$$

Then $\bar{\pi}'\tilde{x}_1$, and $\bar{\pi}'\tilde{y}_2$ are cocycles mod 2 and integral respectively whose classes are independent of the choice of x, y and may be denoted by $\mu^*X\in H^{r+1}(K^*, I_2)$ and $\nu^*Y\in H^{r+2}(K^*)$. By [9] §1 (or [10] §1), we have (1 denotes the unit class mod 2 or integral in K^*)

$$\mu^*X=\mu^*1\cup X. \tag{5}$$

Similarly we have (cf. [10] §1) :

$$\nu^*Y=\nu^*1\cup Y, \tag{6}$$

and

$$\rho_2\nu^*=(\mu^*)^2\rho_2. \tag{7}$$

We shall now first establish the following

Theorem 12 Let Φ^0 be the integral unit class of K^*, then

$$\Phi^{2i+1}=\mu^*\rho_2\Phi^{2i},\quad i\geqslant 0, \tag{8}$$

$$\Phi^{2i+2}=\nu^*\Phi^{2i},\quad i\geqslant 0, \tag{9}$$

$$\rho_2\Phi^{2i+2}=\mu^*\Phi^{2i+1},\quad i\geqslant 0. \tag{10}$$

Proof. Define $\tilde{\varphi}^i\in C^i(\tilde{K}^*)$, $i\geqslant 0$, as follows:

$$\tilde{\varphi}^{2m-1}((a_{i_0}\cdots a_{i_{m-1}})\times(a_{j_0}\cdots a_{j_m}))=\begin{cases}1, & j_0<i_0<\cdots<j_m,\\ 0, & \text{otherwise;}\end{cases} \tag{11}$$

$$\tilde{\varphi}^{2m-1}((a_{i_0}\cdots a_{i_p})\times(a_{j_0}\cdots a_{j_q}))=0,\quad p+q=2m-1,\quad (p,q)\neq(m-1,m), \tag{11$'$}$$

$$\tilde{\varphi}^{2m}((a_{i_0}\cdots a_{i_m})\times(a_{j_0}\cdots a_{j_m}))=\begin{cases}1, & i_0<j_0<\cdots<j_m,\\ 0, & \text{otherwise};\end{cases} \tag{12}$$

$$\tilde{\varphi}^{2m}((a_{j_0}\cdots a_{j_p})\times(a_{i_0}\cdots a_{i_q}))=0,\quad p+q=2m,\quad (p,q)\neq(m,m). \tag{12$'$}$$

Then we shall show that

$$\delta\tilde{\varphi}^{2m-1}=s'\tilde{\varphi}^{2m}, \tag{I}$$

and

$$\delta\tilde{\varphi}^{2m}=(-1)^m d'\tilde{\varphi}^{2m+1}. \tag{II}$$

Let us first prove (I). Consider any $(a_{i_0}\cdots a_{i_p})\times(a_{j_0}\cdots a_{j_q})\in\tilde{K}^*p+q=2m$, then we have

$$\delta\tilde{\varphi}^{2m-1}((a_{i_0}\cdots a_{i_p})\times(a_{j_0}\cdots a_{j_q}))=\sum\nolimits_1+(-1)^p\sum\nolimits_2, \tag{13}$$

where

$$\sum\nolimits_1=\sum_r(-1)^r\tilde{\varphi}^{2m-1}((a_{i_0}\cdots\hat{a}_{i_r}\cdots a_{i_p})\times(a_{j_0}\cdots a_{j_q})), \tag{13$_1$}$$

$$\sum\nolimits_2=\sum_r(-1)^r\tilde{\varphi}^{2m-1}((a_{i_0}\cdots a_{i_p})\times(a_{j_0}\cdots\hat{a}_{j_r}\cdots a_{j_q})). \tag{13$_2$}$$

For the calculation of (13) let us consider various possible cases as follows.

Case I. $(p,q)\neq(m,m)$.

In that case $(p-1,q)\neq(m-1,m)$, hence $\sum_1=0$ by (11)$'$. If $(p,q)\neq(m-1,m+1)$, then by (11)$'$ $\sum_2=0$ too. Suppose now $p=m-1$, $q=m+1$, then there exists an index s with no i_k satisfying $j_s<i_k<j_{s+1}$. For $r\neq s$, $s+1$ we have then $\tilde{\varphi}^{2m-1}((a_{i_0}\cdots a_{i_{m+1}})\times(a_{j_0}\cdots\hat{a}_{j_r}\cdots a_{j_{m+1}}))=0$, while for $r=s$, $s+1$, $\tilde{\varphi}^{2m-1}(a_{i_0}\cdots a_{i_{m-1}})\times(a_{j_0}\cdots\hat{a}_{j_s}\cdots a_{j_m}))$ and $\tilde{\varphi}^{2m-1}((a_{i_0}\cdots a_{i_{m-1}})\times(a_{j_0}\cdots\hat{a}_{j_{s+1}}\cdots a_{j_m}))$ are either both 0 or both 1. Hence $\sum_2=0$ always and we have

$$\delta\tilde{\varphi}^{2m-1}((a_{i_0}\cdots a_{i_p})\times(a_{j_0}\cdots a_{j_q}))=0,\quad (p,q)\neq(m,m),\quad p+q=2m.$$

Case II. $(p,q)=(m,m)$.

By (11)$'$, $\sum_2$ is evidently 0.

If there is an s with no i_k satisfying $j_s<i_k<j_{s+1}$, then $\sum_1=0$ by (11).

If there is an s with no j_k satisfying $i_s<j_k<i_{s+1}$, then by (11), $\tilde{\varphi}^{2m-1}((a_{i_0}\cdots\hat{a}_{i_r}\cdots a_{i_m})\times(a_{j_0}\cdots a_{j_m}))=0$ for $r\neq s$, $s+1$, while for $r=s$, $s+1$, they are either both 0 or both 1. Hence $\sum_1=0$ always.

If there is no such s, then either $i_0 < j_0 < \cdots < i_m < j_m$ or $j_0 < i_0 < \cdots < j_m < i_m$. By (10), we have then

$$\begin{aligned}&\tilde{\varphi}^{2m-1}((a_{i_0}\cdots\hat{a}_{i_r}\cdots a_{i_m})\times(a_{j_0}\cdots a_{j_m}))\\ &=\begin{cases}1, & r=0\\ 0, & r>0\end{cases}\quad(i_0<j_0<\cdots<i_m<j_m),\\ &=\begin{cases}1, & r=m\\ 0, & r<m\end{cases}\quad(j_0<i_0<\cdots<j_m<i_m).\end{aligned}$$

Hence

$$\sum\nolimits_1=\begin{cases}1, & i_0<j_0<\cdots<i_m<j_m,\\ (-1)^m, & j_0<i_0<\cdots<j_m<i_m,\\ 0, & \text{otherwise}.\end{cases}$$

Combining the above cases together, we have

$$\begin{aligned}&\delta\tilde{\varphi}^{2m-1}((a_{i_0}\cdots a_{i_p})\times(a_{j_0}\cdots a_{j_q}))\\ &=\left\{\begin{array}{ll}1, & p=q=m, i_0<j_0<\cdots<i_m<j_m,\\ (-1)^m, & p=q=m, j_0<i_0<\cdots<j_m<i_m,\\ 0, & \text{otherwise}.\end{array}\right\}\end{aligned}\tag{14}$$

Next, by (1) and (12), (12)′, we have

$$\begin{aligned}&s'\tilde{\varphi}^{2m}((a_{i_0}\cdots a_{i_m})\times(a_{j_0}\cdots a_{j_m}))\\ &=\tilde{\varphi}^{2m}((a_{i_0}\cdots a_{i_m})\times(a_{j_0}\cdots a_{j_m}))+(-1)^m\tilde{\varphi}^{2m}((a_{j_0}\cdots a_{j_m})\times(a_{i_0}\cdots a_{i_m}))\\ &=\begin{cases}1, & i_0<j_0<\cdots<i_m<j_m,\\ (-1)^m, & j_0<i_0<\cdots<j_m<i_m,\\ 0 & \text{otherwise}.\end{cases}\end{aligned}$$

$$s'\tilde{\varphi}^{2m}((a_{i_0}\cdots a_{i_p})\times(a_{j_0}\cdots a_{j_q}))=0\quad (pq)\neq(m,m).$$

Comparing with (14), we get (I).

Next let us prove (II) as follows. Consider any cell $(a_{i_0}\cdots a_{i_p})\times(a_{j_0}\cdots a_{j_q})\in\tilde{K}^*$, $p+q=2m+1$, we have

$$\delta\tilde{\varphi}^{2m}((a_{i_0}\cdots a_{i_p})\times(a_{j_0}\cdots a_{j_q}))=\sum\nolimits_1+(-1)^p\sum\nolimits_2,\tag{15}$$

where

$$\sum\nolimits_1=\sum(-1)^r\tilde{\varphi}^{2m}((a_{i_0}\cdots\hat{a}_{i_r}\cdots a_{i_p})\times(a_{j_0}\cdots a_{j_q})),\tag{16$_1$}$$

$$\sum\nolimits_2 = \sum (-1)^r \tilde{\varphi}^{2m}((a_{i_0} \cdots a_{i_p}) \times (a_{j_0} \cdots \hat{a}_{j_r} \cdots a_{j_q})). \tag{16$_2$}$$

Case I. $(p, q) \neq (m, m+1)$ or $(m+1, m)$.

We have evidently $\sum_1 = \sum_2 = 0$ by $(12)'$.

Case II. $(p, q) = (m, m+1)$.

By $(12)'$ we have $\sum_1 = 0$, while by (12) we have by the same method as above

$$\sum\nolimits_2 = \begin{cases} 1, & j_0 < i_0 < \cdots < i_m < j_{m+1}, \\ 0, & \text{otherwise}. \end{cases}$$

Case III. $(p, q) = (m+1, m)$.

In that case we have $\sum_2 = 0$, while

$$\sum\nolimits_1 = \begin{cases} (-1)^{m+1}, & i_0 < j_0 < \cdots < j_m < i_{m+1}, \\ 0, & \text{otherwise}. \end{cases}$$

Combining together all the preceding cases, we get

$$\delta\tilde{\varphi}^{2m}((a_{i_0} \cdots a_{i_p}) \times (a_{j_0} \cdots a_{j_q}))$$
$$= \left\{ \begin{array}{lll} (-1)^m, & (p, q) = (m, m+1), & j_0 < i_0 < \cdots < i_m < j_{m+1}, \\ (-1)^{m+1}, & (p, q) = (m+1, m), & i_0 < j_0 < \cdots < j_m < i_{m+1}, \\ 0, & \text{otherwise}. & \end{array} \right\} \tag{17}$$

Next by (1), (11) and $(11)'$, we have

$$d'\tilde{\varphi}^{2m+1}((a_{i_0} \cdots a_{i_p}) \times (a_{j_0} \cdots a_{j_q}))$$
$$= \tilde{\varphi}^{2m+1}((a_{i_0} \cdots a_{i_p}) \times (a_{j_0} \cdots a_{j_q})) - (-1)^{pq}\tilde{\varphi}^{2m+1}((a_{j_0} \cdots a_{j_q}) \times (a_{i_0} \cdots a_{i_p}))$$
$$= \begin{cases} 1, & (p, q) = (m, m+1), \quad j_0 < i_0 < \cdots < i_m < j_{m+1}, \\ -1, & (p, q) = (m+1, m), \quad i_0 < j_0 < \cdots < j_m < i_{m+1}, \\ 0, & \text{otherwise}. \end{cases}$$

Combining with (17), we get (II).

Define now $\varphi^i \in C^i(K^*)$ by

$$\varphi^i = \bar{\pi}'\tilde{\varphi}^i, \quad i \geqslant 0. \tag{18}$$

Then by (3), (11), $(11)'$, (12) and $(12)'$, we have

$$\varphi^{2m-1}((a_{i_0} \cdots a_{i_p}) * (a_{j_0} \cdots a_{j_q}))$$
$$= \begin{cases} 1, & p = m-1, q = m, j_0 < i_0 < j_1 < \cdots < i_{m-1} < j_m, \\ 0, & \text{otherwise}, \end{cases} \quad (i_0 > j_0); \tag{19}$$

$$\varphi^{2m}((a_{i_0}\cdots a_{i_p})*(a_{j_0}\cdots a_{j_q}))$$
$$=\begin{cases}1, & p=q=m, i_0<j_0<\cdots<i_m<j_m,\\ 0, & \text{otherwise},\end{cases}\qquad (i_0<j_0). \tag{20}$$

By Theorem 10 of §6 and Theorem 11 of §7,

$$\rho_2\varphi^{2m-1}\in\Phi^{2m-1},\quad m>0, \tag{21}$$

and

$$\varphi^{2m}\in\Phi^{2m},\quad m>0. \tag{22}$$

Moreover φ^0 is the integral unit cocycle on K^*. Hence from (I), (II), (18), (21), (22) and the definitions of μ^* and ν^*, we get (8), (10) and also $\nu^*\Phi^{2m}=(-1)^*\Phi^{2m+2}$. By Theorem 4 of §3, $2\Phi^{2m+2}=0$. Hence the last equation is the same as (9) and our theorem is proved.

From the above theorem and (5), (6), (7) we get also the following theorems:

Theorem 13

$$\Phi^{2i}\cup\Phi^{2j}=\Phi^{2i+2j}, \tag{23}$$

$$\rho_2\Phi^{2i}\cup\Phi^{2j+1}=\Phi^{2i+2j+1}, \tag{24}$$

$$\Phi^{2i+1}\cup\Phi^{2j+1}=\rho_2\Phi^{2i+2j+2}. \tag{25}$$

Theorem 14 Denote by $()^i$ the i-fold powers by cup products, then

$$\Phi^{2i}=(\Phi^2)^i,\quad i>0, \tag{26}$$

$$\Phi^{2i+1}=(\Phi^1)^{2i+1},\quad i\geqslant 0, \tag{27}$$

$$\varrho_2\Phi^{2r}=(\Phi^1)^{2i},\quad i>0. \tag{28}$$

Theorem 15 If $\Phi^m=0$, then for any $i>0$, we have $\Phi^{m+i}=0$.

Theorem 16 $\tilde{\Phi}^{2m-1}=0$, for $m>0$.

Proof. Define $\tilde{\varphi}^{2m-1}$ by (11) and (11)′, we have by Theorem 10 of §6 ($\sum$ being extended over all possible sets of indices (i,j) with $j_0<i_0<j_1<\cdots<i_{m-1}<j_m$)

$$d'\tilde{\varphi}^{2m-1}$$
$$=\sum\{(a_{i_0}\cdots a_{i_{m-1}})\times(a_{j_0}\cdots a_{j_m})\}-\sum\{(a_{j_0}\cdots a_{j_m})\times(a_{i_0}\cdots a_{i_{m-1}})\}\in\tilde{\Phi}^{2m-1}.$$

But by (II) we have $d'\tilde{\varphi}^{2m-1}\sim 0$. Hence $\tilde{\Phi}^{2m-1}=0$.

Combining this theorem with Theorem 9 of §5, we see that $\tilde{\Phi}^m$ are always $= 0$ $(m > 0)$. Hence $\tilde{\Phi}^m$ are practically useless. On the other hand, as we shall see in the two following sections, Φ^m are generally $\neq 0$ and play an important role in the study of realization of complexes.

Theorem 17 All the imbedding classes $\Phi^m \in H^{m,2}(K, I_{(m)}) = H^{m,2}(P, I_{(m)})$, $m > 0$, of a complex K are topological invariants of the polyhedron $\bar{K} = P$.

Proof. Let L be another simplicial subdivision of P. As the construction of K^* and $\tilde{K}^*$ from K, let L^*, $\tilde{L}^*$ be the corresponding complexes constructed from L and let $\omega : \tilde{L}^* \to L^*$ be the covering projection. By [6], we know that the spaces $\bar{\tilde{L}}^*$, $\bar{\tilde{K}}^*$ of the complexes $\tilde{L}^*$, $\tilde{K}^*$ have same homotopy type, and the same is true for the spaces $\bar{L}^*$, $\bar{K}^*$ of L^*, K^*. Moreover, the identity of these homotopy types may be realized by continuous maps $\tilde{f} : \bar{\tilde{L}}^* \to \bar{\tilde{K}}^*$ and $f : \bar{L}^* \to \bar{K}^*$ such that $\pi\tilde{f} = f\omega$. It follows that μ^*, ν^* are commutative with f^* (cf. [9] §1 or [10] §1) and by Theorem 12 we have (Φ° denotes the integral unit class) :

$$\begin{aligned} f^*\Phi^{2i}(K) &= f^*(\nu^*)^i\Phi^\circ(K) = (\nu^*)^i f^*\Phi^\circ(K) = (\nu^*)^i\Phi^\circ(L) = \Phi^{2i}(L), \\ f^*\Phi^{2i+1}(K) &= f^*(\mu^*)^{2i+1}\rho_2\Phi^\circ(K) = (\mu^*)^{2i+1}f^*\rho_2\Phi^\circ(K) \\ &= (\mu^*)^{2i+1}\rho_2\Phi^\circ(L) = \Phi^{2i+1}(L). \end{aligned}$$

Since $f^* : H^{m,2}(K, I_{(m)}) \approx H^{m,2}(L, I_{(m)})$ may be considered as the identity homomorphism of $H^{m,2}(P, I_{(m)})$, the last two equations show that $\Phi^m(K)$ and $\Phi^m(L)$, $m > 0$, are identical elements in $H^{m,2}(P, I_{(m)})$. In other words, $\Phi^m(K) \in H^{m,2}(P, I_{(m)})$, $m > 0$, are independent of the subdivision K of P and are therefore topological invariants of P.

The above theorem may also be slightly extended as follows.

Let $P \subset Q$ be a regular pair of finite polyhedrons so that P, Q have simplicial subdivisions L, K respectively for which L is a regular subcomplex of K (cf. [6]). Construct complexes L^* and K^* as before, then L^* is a subcomplex of K^* and the inclusion map i will induce homomorphisms

$$i^* : H^*(K^*, G) \to H^m(L^*, G),$$

or

$$i^* : H^{m,2}(K, G) \to H^{m,2}(L, G). \tag{29}$$

As in [6], these homomorphisms are really independent of the choice of the subdivisions

K, L and may thus be written as

$$i^* : H^{m,2}(Q, G) \to H^{m,2}(P, C). \tag{30}$$

As in the preceding theorem we may then prove the following

Theorem 18 Let $P \subset Q$ (or $L \subset K$) be a regular pair of finite polyhedrons (or a regular pair of finite simplicial complexes). Define i^* as the homomorphisms in (29), (30) induced by the inclusion map $i : P \subset Q$ (or $i : L \subset K$), then

$$i^* \Phi^m(Q) = \Phi^m(P),$$

or

$$i^* \Phi^m(K) = \Phi^*(L).$$

§9 Complexes realizable in $\boldsymbol{R}^{m+1}$ but not in $\boldsymbol{R}^m$

Given integers $n > 0$ and $N \geqslant n$. Let us take $N+1$ linearly independent points $a_0, \cdots, a_N$ in R^N which span an N-dimensional simplex Δ_N. The n-dimensional skeleton of Δ_N is an n-dimenional complex $K_{N,n}$. Using the notations $K \subset R^m$ and $K \not\subset R^m$ to denote that K can or cannot be semi-linearly realized in R^m, we have the following

Theorem 19

$$K_{N,n} \subset R^{2n+1}, \quad N \geqslant 2n+2, \tag{1}$$

$$K_{m+2,n} \subset R^{m+1}, \quad 2n \geqslant m \geqslant n, \tag{2}$$

$$K_{n+1,n} \subset R^{n+1}, \tag{3}$$

$$K_{n,n} \subset R^n; \tag{4}$$

$$K_{N,n} \not\subset R^{2n}, \quad N \geqslant 2n+2, \tag*{$(1)'_n$}$$

$$K_{m+2,n} \not\subset R^m, \quad 2n \geqslant m \geqslant n, \tag*{$(2)'_{m,n}$}$$

$$K_{n+1,n} \not\subset R^n, \tag*{$(3)'$}$$

$$K_{n,n} \not\subset R^{n-1}. \tag*{$(4)'$}$$

In the proof below, a_i will be arranged in the order $a_0 < \cdots < a_N$ and φ^m will denote the representative cocycle in Φ^m as asserted in Theorem 10 of §6 and Theorem 11 of §7. All simplexes of $K_{N,n}$ will also be supposed to be written in normal forms $(a_{i_0} \cdots a_{i_k})$: $0 \leqslant i_0 < \cdots < i_k \leqslant N$.

Proof of (1).

This is a classical result, of which the proof is quite simple (cf. [1] Chap. 1 and [2] Chap. 3 §2).

Proof of (2).

In R^{m+1} let us take $m+2$ linearly independent points $a'_0, \cdots, a'_{m+1}$ which span a simplex Δ'_{m+1}, and a point a'_{m+2} in the interior of Δ'_{m+1}. Let $K'_{m+2,n}$ be the complex formed of all k-dimensional simplexes $(0 \leqslant k \leqslant n)$ with vertices taken from a'_i $(i = 0, 1, \cdots, m+2)$. Then $T(a_i) = a'_i$, $i = 0, 1, \cdots, m+2$, define a linear realization $K'_{m+2,n} = TK_{m+2,n}$ of $K_{m+2,n}$ in R^{m+1}.

(3) and (4) are evident.

Before proceeding to the proof of (1)′—(4)′, let us first remark that (1)′ is the well-known result of Van Kampen and Flores [3-5] (3)′ states that an n-sphere is not imbeddable in R^n and (4)′ states that an n-simplex is not imbeddable in R^{n-1}. Both (3)′ and (4)′ are classical results, of which the proof of the former depends on Alexander's duality theorem, and that of the latter is a consequence of a theorem of Brouwer connected with theory of dimension. In what follows we shall give (1) ′—(4)′a unified proof which makes it plausible that in all these cases the non-imbeddability is owing to the same fact, namely, the corresponding imbedding class Φ^m is $\neq 0$ (cf. Theorem 5 of §3).

Proof of (1)′.

Evidently it is sufficient to prove that $K_{2n+2,n} \not\subset R^n$.

For this let us consider in $K^*_{2\pi+2,n}$ the following integral chain

$$z = \sum_{i_0<j_0} (a_{i_0} \cdots a_{i_n}) * (a_{j_0} \cdots a_{j_n}), \tag{5}$$

or what is the same,

$$z = \frac{1}{2} \sum \varepsilon_{i_0 j_0} (a_{i_0} \cdots a_{i_n}) * (a_{j_0} \cdots a_{j_n}), \tag{5$'$}$$

in which the preceding $\sum$ is extended over all possible cells with $i_0 < j_0$, the second $\sum$ is extended over all possible cells, and $\varepsilon_{i_0 j_0} = +1$ or $(-1)^n$ according as $i_0 < j_0$ or $i_0 > j_0$.

Let us now calculate ∂z. Consider any cell $(a_{k_0} \cdots a_{k_{n-1}}) * (a_{l_0} \cdots a_{l_n}) \in K^*_{2n+2,n}$. Let r, s be the two remaining numbers after removing $k_0, \cdots, k_{n-1}$, $l_0, \cdots, l_n$ from $0, 1, \cdots, 2n+2$. Suppose $r < s$ and $k_0 < \cdots < k_{\alpha-1} < r < k_\alpha < \cdots < k_{\beta-1} < s < k_\beta < \cdots < k_{n-1}$. Then the term $(a_{k_0} \cdots a_{k_{n-1}}) * (a_{l_0} \cdots a_{l_n})$ in ∂z is produced from the following terms in $(5)'$:

$$\partial(a_{k_0}\cdots a_{k_{\alpha-1}}a_r a_{k_\alpha}\cdots a_{k_{n-1}})*(a_{l_0}\cdots a_{l_n}),$$
$$\partial(a_{k_0}\cdots a_{k_{\beta-1}}a_s a_{k_\beta}\cdots a_{k_{n-1}})*(a_{l_0}\cdots a_{l_n}),$$
$$(a_{l_0}\cdots a_{l_n})*\partial(a_{k_0}\cdots a_{k_{\alpha-1}}a_r a_{k_\alpha}\cdots a_{k_{n-1}}),$$
$$(a_{l_0}\cdots a_{l_n})*\partial(a_{k_0}\cdots a_{k_{\beta-1}}a_s a_{k_\beta}\cdots a_{k_{n-1}}).$$

Hence the coefficient λ of the term $(a_{k_0}\cdots a_{k_{n-1}})*(a_{l_0}\cdots a_{l_n})$ in ∂z is given by the following:

for $k_0 < r$,

$$\begin{aligned}\lambda =&\frac{1}{2}[(-1)^{\alpha}\varepsilon_{k_0l_0}+(-1)^{\beta}\varepsilon_{k_0l_0}+(-1)^{n}\cdot(-1)^{\alpha}\cdot(-1)^{n(n-1)}\varepsilon_{l_0k_0}\\ &+(-1)^{n}\cdot(-1)^{\beta}\cdot(-1)^{n(n-1)}\varepsilon_{l_0k_0}]=[(-1)^{\alpha}+(-1)^{\beta}]\varepsilon_{k_0l_0},\end{aligned}$$

for $r < k_0 < s$,

$$\begin{aligned}\lambda =&\frac{1}{2}[\varepsilon_{rl_0}+(-1)^{\beta}\varepsilon_{k_0l_0}+(-1)^{n}\cdot(-1)^{n(n-1)}\varepsilon_{l_0r}\\ &+(-1)^{n}\cdot(-1)^{\beta}\cdot(-1)^{n(n-1)}\varepsilon_{l_0k_0}]=\varepsilon_{rl_0}+(-1)^{\beta}\varepsilon_{k_0l_0},\end{aligned}$$

and for $r < s < k_0$,

$$\begin{aligned}\lambda =&\frac{1}{2}[\varepsilon_{rl_0}+\varepsilon_{sl_0}+(-1)^{n}\cdot(-1)^{n(n-1)}\varepsilon_{l_0r}\\ &+(-1)^{n}\cdot(-1)^{n(n-1)}\varepsilon_{l_0s}]=\varepsilon_{rl_0}+\varepsilon_{sl_0}.\end{aligned}$$

Consequently we have always $\lambda\equiv 0 \bmod 2$, and $\rho_2 z$ is a mod 2 cycle.

For any $r\geqslant 0$ and $\leqslant 2n+2$, define now

$$\alpha_r(s)=\begin{cases} s, & 0\leqslant s\leqslant r-1,\\ s+1, & r\leqslant s\leqslant 2n+1.\end{cases}$$

By Theorem 11 of §7, the $2n$-dimensional imbedding class $\varPhi^{2n}$ of $K_{2n+2,n}$ has a representative cocycle

$$\varphi^{2n}=\sum_{r=0}^{2n+2}\{(a_{\alpha_r(0)}a_{\alpha_r(2)}\cdots a_{\alpha_r(2n)})*(a_{\alpha_r(1)}a_{\alpha_r(3)}\cdots a_{\alpha_r(2n+1)})\}.$$

Hence $\varphi^{2n}(z)=2n+3$ and $\rho_2\varphi^{2n}(\rho_2 z)\neq 0$. It follows that $\rho_2\varPhi^{2n}\neq 0$ and we have $K_{2n+2,n}\not\subset R^{2n}$. i.e., $(1)'$, by Theorem 5 of §3.

Proof of $(2)'_{2n-1,n}$, i.e., $K_{2n+1,n} \not\subset R^{2n-1}$.

This may be derived from $(1)'$. Suppose we have a realization $T : K_{2n+1,n} \subset R^{2n-1}$. Consider R^{2n-1} as a linear subspace of R^{2n} and $K_{2n+1,n}$ as a subcomplex of $K_{2n+2,n}$. Take a point $a'_{2n+2} \in R^{2n}$ and $\notin R^{2n-1}$. Then by setting $T(a_{2n+2}) = a'_{2n+2}$, we may extend T to a realization $T : K_{2n+2,n} \subset R^{2n}$, contrary to $(1)'$. Hence $K_{2n+1,n} \not\subset R^{2n-1}$.

We may also give a direct proof as follows:

By Theorem 10 of §6, $\varPhi^{2n-1}$ has a representative cocycle

$$\begin{aligned}\varphi^{2n-1} &= \rho_2 \sum \{(a_{i_0} \cdots a_{i_{n-1}}) * (a_{j_0} \cdots a_{j_n})\} \\ &= \rho_2 \sum_{r=0}^{2n+1} \{(a_{\alpha_r(1)} a_{\alpha_r(3)} \cdots a_{\alpha_r(2n-1)}) * (a_{\alpha_r(0)} a_{\alpha_r(2)} \cdots a_{\alpha_r(2n)})\},\end{aligned}$$

in which the first $\sum$ is extended over all possible sets of indices (i, j) with $0 \leqslant j_0 < i_0 < j_1 < \cdots < i_{n-1} < j_n \leqslant 2n+1$, and α_r is defined by

$$\alpha_r(s) = \begin{cases} s, & 0 \leqslant s \leqslant r+1, \\ s+1, & r \leqslant s \leqslant 2n. \end{cases}$$

Consider now in K^* a $(2n-1)$-dimensional integral chain

$$z = \sum_{j_0>0} (a_{i_0} \cdots a_{i_{n-1}}) * (a_{j_0} \cdots a_{j_n}),$$

in which $\sum$ is extended over all possible sets of indices (i, j) with $j_0 > 0$. It is easy to see that $\rho_2 z$ is a mod 2 cycle and $\varphi^{2n-1}(\rho_2 z) = 1$ mod 2. It follows that $\varphi^{2n-1} \not\sim 0$ or $\varPhi^{2n-1} \neq 0$ and $K_{2n+1,n} \not\subset R^{2n-1}$, as we require to prove.

Proof of $(1)'_{m,n}$.

When $n = 1$, we have $m = 1, 2$ and $(2)'$ becomes $K_{3,1} \not\subset R^1$ and $K_{4,1} \not\subset R^2$ which is $(2)'_{2n-1,n}$ and $(1)'_n$ in the case $n = 1$ and hence is already proved.

Suppose now $(2)'_{m,n-1}$ for $2(n-1) \geqslant m \geqslant n-1$ has already been proved. Prove now $(2)_{m,n}$ as follows.

When $m = 2n$, $(2)'_{m,n}$ is the same as $(1)'_n$ and has been proved. The case $m = 2n-1$ has also been proved. We may suppose therefore $m \leqslant 2n-2$. In that case $K_{m+2,n}$ has a subcomplex $K_{m+2,n-1}$. By induction hypothesis $K_{m+2,n-1} \not\subset R^m$, hence we have *àfortiori* $K_{m+2,n} \not\subset R^m$, what we require to prove.

We may also reason as follows. Consider the following assertion:

$$\varPhi^m(K_{m+2,n}) \neq 0, \quad 2n \geqslant m \geqslant n. \qquad (2)''_{m,n}.$$

When $m = 2n$ or $m = 2n - 1$ and in particular for $n = 1$, we know already that $(2)''_{m,n}$ is true. Suppose now $(2)''_{m,n-1}$ is true and $m \leqslant 2(n-1)$. Then by induction hypothesis $\Phi^m(K_{m+2,n-1}) \neq 0$. As $K_{m+2,n-1}$ is a subcomplex of $K_{m+2,n}$, we have by Theorem 18 of §8, $\Phi^m(K_{m+2,n}) \neq 0$ and $(2)''_{m,n}$ is also true.

From $(2)''_{m,n}$ we get $(2)'_{m,n}$. From this reasoning it is seen that the truth of $(2)'_{m,n}$ is again due to $\Phi^m \neq 0$.

Proof of (3) ′.

Suppose first $n = 2n'$. Then

$$\varphi^n = \varphi^{2n'} = \{(a_0a_2\cdots a_{2n'}) * (a_1a_3\cdots a_{2n'+1})\} \in \Phi^n,$$

and

$$z = \rho_2 \sum_{i_0<j_0} (a_{i_0}\cdots a_{i_{n'}}) * (a_{j_0}\cdots a_{j_{n'}}) + \rho_2 \sum_{r=1}^{n'} \sum (a_{k_0}\cdots a_{k_{n'-r}}) * (a_{l_0}\cdots a_{l_{n'+r}}),$$

may be easily seen to be a mod 2 cycle, in which the first $\sum$ is extended over all possible sets of indices (i, j) with $i_0 < j_0$, and the second $\sum$ in the second term is extended over all possible sets of indices (k, l). As $\rho_2\varphi^{2n}(z) = 1$ mod 2, we have $\varphi^n \not\sim 0$ or $\Phi^n \neq 0$ and hence $K_{n+1,n} \not\subset R^n$.

Suppose next $n = 2n' - 1$, then

$$\varphi^n = \varphi^{2n'-1} = \rho_2\{(a_1a_3\cdots a_{2n'-1}) * (a_0a_2\cdots a_{2n'})\} \in \Phi^n.$$

Moreover

$$z = \rho_2 \sum_{r=0}^{n'-1} \sum (a_{i_0}\cdots a_{i_{n'-r-1}}) * (a_{j_0}\cdots a_{j_{n'+r}})$$

is a mod 2 cycle, in which $\sum$ is extended over all possible sets of indices (i, j). Since $\varphi^n(z) = 1$ mod 2 we have $\varphi^n \not\sim 0$ or $\Phi^n \neq 0$ and again $K_{n+1,n} \not\subset R^n$.

Proof of $(4)'$.

First, (4) may be derived from $(3)'$. Consider R^{n-1} as a linear subspace of R^n and $K_{n,n}$ as a subcomplex of $K_{n+1,n}$. If there exists a realization $T : K_{n,n} \subset R^{n-1}$, then on taking a point $a'_{n+1} \in R^n$ but $\notin R^{n-1}$ and setting $T(a_{n+1}) = a'_{n+1}$, we get an extension of T to a realization $T : K_{n+1,n} \subset R^n$, contrary to $(3)'$. Hence $K_{n,n} \not\subset R^{n-1}$.

Given now a direct proof by the unified method as follows.

Suppose first n be even: $n = 2n'$. By Theorem 10 of §6 we have

$$\varphi^{n-1} = \varphi^{2n'-1} = \rho_2\{(a_1a_3\cdots a_{2n'-1}) * (a_0a_1\cdots a_{2n'})\} \in \Phi^{n-1}.$$

Put

$$z = \sum_{r=1}^{n'} \sum (a_{i0} \cdots a_{i_{n'-r}}) * (a_{j_0} \cdots a_{j_{n'+r+1}}),$$

in which $\sum$ is extended over all possible sets of indices (i, j). Then $\rho_2 z$ is a mod 2 cycle and $\varphi^{n-1}(\rho_2 z) = 1$ mod 2. Hence $\Phi^{n-1} \neq 0$ and we have $K_{n,n} \not\subset R^{n-1}$.

Next suppose n be odd: $n = 2n' + 1$. By Theorem 11 of §7 we have

$$\varphi^{n-1} = \varphi^{2n'} = \{(a_0 a_2 \cdots a_{2n'}) * (a_1 a_3 \cdots a_{2n'+1})\} \in \Phi^{2n}.$$

Put

$$z = \sum_{r=1}^{n'} \sum (a_{i_0} \cdots a_{i_{n'-r}}) * (a_{j_0} \cdots a_{j_{n'+r}}) + \sum_{i_0 < j_0} (a_{i_0} \cdots a_{i_{n'}}) * (a_{j_0} \cdots a_{j_{n'}}),$$

in which the second $\sum$ is extended over all possible sets of indices (i, j) and the last $\sum$ is extended over all possible sets of indices (i, j) with $i_0 < j_0$. Then $\rho_2 z$ is a mod 2 cycle and $\rho_2 \varphi^{n-1}(\rho_2 z) = 1$ mod 2. Hence $\Phi^{n-1} \neq 0$ and we have again $K_{n,n} \not\subset R^{n-1}$.

Our theorem is now completely proved. From the proof we have furthermore the following theorems:

Theorem 20 The necessary and sufficient condition for $K_{N,n} \subset R^m$ is $\Phi^m(K_{N,n}) = 0$.

Theorem 21 To any $n > 0$ and $m \leqslant 2n$ and $\geqslant n - 1$, there exist complexes $K(m, n) \subset R^{m+1}$. But $\not\subset R^m$. In other words, for $n - 1 \leqslant m \leqslant 2n$, R^{m+1} contains always more n-dimensional conplexes than does R^m.

The complexes $K(m, n)$ in this theorem may be taken to be $K_{m+2,n}$ for $2n \geqslant m \geqslant n$ and $K_{n,n}$ for $m = n - 1$.

§10 Another example of Van Kampen [3] and its generalization

In this section we shall apply the theory developed in the preceding sections to give an alternative proof of the non-imbeddability of another n-dimensional complex in R^{2n} also due to Van Kampen. We give also its generalizations.

The n-dimensional complex K_n in this second example of Van Kampen is constructed in the following manner. Consider $n+1$ sets of triple of points $a_i^{(0)}$, $a_i^{(1)}$, $a_i^{(2)}$ $(i = 0, 1, \cdots, n)$. Take one point from each of these $n+1$ sets to form an n-simplex, say $(a_0^{(i_0)}a_1^{(i_1)}\cdots a_n^{(i_n)})$ $(i = 0,\ 1$ or $2)$. Then K_n is formed by all these simplexes as well as all their faces. We have then

Theorem 22 (Van Kampen) [3] The Van Kampen complex K_n defined above is non-imbeddable in R^{2n}.

Proof. Arrange the vertices of K_n in an order such that $a_i^{(k)} < a_j^{(l)}$ if and only if either $i < j$ or $i = j$ and $k < l$ $(i, j = 0, 1, \cdots, n;\ k, l = 0, 1$ or $2)$. By Theorem 11 of §7, the $2n$-dimensional imbedding class $\varPhi^{2n} \in H^{2n,2}(K_n)$ of K_n has with respect to this ordering of vertices a representative cocycle

$$\varphi^{2n} = \sum\{(a_0^{(i_0)}a_1^{(i_1)}\cdots a_n^{(i_n)}) * (a_0^{(j_0)}a_1^{(j_1)}\cdots a_n^{(j_n)})\},$$

in which $\sum$ is extended over all possible sets of indices $(i,\ j)$ with $i_0 < j_0,\ i_1 < j_1,\ \cdots,\ i_n < j_n$. Next construct a $2n$-dimensional chain

$$z = \sum(a_0^{(i_0)}a_1^{(i_1)}\cdots a_n^{(i_n)}) * (a_0^{(j_0)}a_1^{(j_1)}\cdots a_n^{j_n}),$$

in which $\sum$ is extended over all possible sets of indices $(i,\ j)$ with $i_0 < j_0$. It is easy to see that $\rho_2 z$ is a mod 2 cycle and $\varphi^{2n}(z) = 3^{n+1} = 1 \bmod 2$. Hence $\varPhi^{2n} \neq 0$ and $K_n \not\subset R^{2n}$, what we require to prove.

Let us now extend Van Kampen's example in the following manner. Suppose we are given $p+1$ sets $(p \geqslant -1)$ of triple of points $a_i^{(0)},\ a_i^{(1)},\ a_i^{(2)},\ i = 0,\ 1,\ \cdots,\ p$; and q sets $(q \geqslant 0)$ of pairs of points $a_j^{(0)},\ a_j^{(1)},\ j = p+1,\ \cdots,\ p+q$. Take one point from each of these $p+q+1$ sets and form a $(p+q)$ -dimensional simplex. The complex formed by all these simplexes as well as their faces will be denoted by $A_{p,q}$. In particular, $A_{p,0}$ is the above-defined Van Kampen's complex K_p, and $A_{-1,q}$ is a subdivision of the $(q-1)$ -dimensional sphere.

Theorem 23 For $p \geqslant 0$, we have $A_{p,q} \subset R^{2p+q+1}$, but $\varPhi^{2p+q}(A_{p,q}) \neq 0$, so that $A_{p,q} \not\subset R^{2p+q}$.

Theorem 24 If $n \leqslant m \leqslant 2n$, then the n-dimensional complex $\subset R^{m+1}$ but $\not\subset R^m$ as stated in Theorem 21 may also be taken as $A_{m-n,2n-m}$.

Theorem 25 The necessary and sufficient condition for $A_{p,q} \subset R^m$ is $\varPhi^m = 0$.

The last two theorems are both simple consequences of Theorem 23. In order to prove Theorem 23, we shall prove first a general result as follows. Given a simplicial complex L and two points $b_0,\ b_1$, the set of all simplexes $b_0\sigma,\ b_1\sigma$ and τ $(\sigma,\ \tau \in L)$ forms a simplicial complex K, written as $K = b_0L + b_1L$. In fact, K is the join complex of L and the 0-sphere $\{b_0b_1\}$. We have then

Theorem 26 Let K be the join complex of the simplicial complex L and the 0-sphere $S^0 = \{b_0, b_1\}$. If $\rho_2\Phi^{2m}(L) \neq 0$, then $\Phi^{2m+1}(K) \neq 0$. If $\Phi^{2m-1}(L) \neq 0$, then $\rho_2\Phi^{2m}(K) \neq 0$.

Proof. Let us arrange the vertices of L in an order $a_1 < a_2 < \cdots$ and the vertices of K in the order $b_0 < b_1 < a_1 < a_2 < \cdots$. With respect to such ordering of vertices, the imbedding classes $\Phi^m(L)$, $\Phi^m(K)$ of L, K would have respectively representative cocycles $\varphi^m(L)$ and $\varphi^m(K)$ as follows.

$$\varphi^{2m}(L) = \sum\nolimits_1 \{(a_{i_0} \cdots a_{i_m}) * (a_{j_0} \cdots a_{j_m})\}, \tag{1}$$

$$\varphi^{2m-1}(L) = \rho_2 \sum\nolimits_2 \{(a_{i_0} \cdots a_{i_{m-1}}) * (a_{j_0} \cdots a_{j_m})\}, \tag{2}$$

$$\begin{aligned}\varphi^{2m+1}(K) =& \rho_2 \sum_k \sum_1 \{(a_{i_0} \cdots a_{i_m}) * (b_k a_{j_0} \cdots a_{j_m})\} \\ &+ \rho_2 \sum\nolimits_2 \{(b_1 a_{i_0} \cdots a_{i_{m-1}}) * (b_0 a_{j_0} \cdots a_{j_m})\} \\ &+ \rho_2 \sum\nolimits_3 \{(a_{i_0} \cdots a_{i_m}) * (a_{j_0} \cdots a_{j_{m+1}})\},\end{aligned} \tag{3}$$

$$\begin{aligned}\varphi^{2m}(K) =& \sum_k \sum\nolimits_2 \{(b_k a_{i_0} \cdots a_{i_{m-1}}) * (a_{j_0} \cdots a_{j_m})\} \\ &+ \sum\nolimits_4 \{(b_0 a_{i_0} \cdots a_{i_{m-1}}) * (b_1 a_{j_0} \cdots a_{j_{m-1}})\} \\ &+ \sum\nolimits_1 \{(a_{i_0} \cdots a_{i_m}) * (a_{j_0} \cdots a_{j_m})\}.\end{aligned} \tag{4}$$

In these equations $\sum_1, \cdots, \sum_4$ are to be extended over all possible sets of indices (i, j) satisfying respectively:

$$i_0 < j_0 < i_1 < \cdots < i_m < j_m, \qquad (\textstyle\sum_1)$$

$$j_0 < i_0 < j_1 < \cdots < i_{m-1} < j_m, \qquad (\textstyle\sum_2)$$

$$j_0 < i_0 < j_1 < \cdots < i_m < j_{m+1}, \qquad (\textstyle\sum_3)$$

$$i_0 < j_0 < i_1 < \cdots < i_{m-1} < j_{m-1}. \qquad (\textstyle\sum_4)$$

To any chain $z = \sum_i c_i(\sigma_i * \tau_i) \in C_r(L^*)$, define now a chain $B_0 z \in C_{r+1}(K^*)$ by

$$B_0 z = \sum_i c_i(\sigma_i * b_0\tau_i + b_0\sigma_i * \tau_i).$$

Then $B_0 : C_r(L^*) \to C_{r+1}(K^*)$ is a homomorphism satisfying

$$\partial B_0 z = B_0 \partial z \mod 2. \tag{5}$$

For any $\sigma * \tau \in L^*$, dim σ+ dim $\tau = r$, $c \in C_r(L^*)$, we have also

$$\{\sigma * b_0\tau\}(B_0c) = \{b_0\sigma * \tau\}(B_0c) = \{\sigma * \tau\}(c) \mod 2, \tag{6}$$

and

$$\{\sigma * b_1\tau\}(B_0c) = \{b_1\sigma * \tau\}(B_0c) = 0. \tag{7}$$

Suppose now $\rho_2\Phi^{2m}(L) \neq 0$. Then in L^* there must exist a mod 2 cycle $\rho_2 z$, $z \in C_{2m}(L^*)$, such that $\rho_2\varphi^{2m}(L)(\rho_2 z) \neq 0$. By (1), (3), (6) and (7) we get

$$\varphi^{2m+1}(K)(\rho_2 B_0 z) = \rho_2\varphi^{2m}(L)\ (\rho_2 z) \neq 0.$$

By (5), $\rho_2 B_0 z$ is a mod 2 cycle in K^*. Hence the last equation shows that $\varphi^{2m+1}(K) \not\sim 0$ or $\Phi^{2m+1}(K) \neq 0$. In the same manner we may prove that $\Phi^{2m-1}(L) \neq 0$ implies $\rho_2\Phi^{2m}(K) \neq 0$.

Proof of Theorem 23.

When $q = 0$, $A_{p,0}$ is the same as the Van Kampen complex K_p in Theorem 22. By that theorem we have already $\rho_2\Phi^{2p}(A_{p,0}) \neq 0$. As

$$A_{p,q} = a_q^{(0)} A_{p,q-1} + a_q^{(1)} A_{p,q-1},$$

by induction on applying successively Theorem 26 we obtain that $\Phi^{2p+q}(A_{p,q}) \neq 0$. Consequently $A_{p,q} \not\subset R^{2p+q}$.

On the other hand, let us take a rectangular system of coordinates $(x_1, \cdots, x_{2p+q+1})$ in R^{2p+q+1}. Let R^{2p+1} be the linear subspace of R^{2p+q+1} defined by $x_1 = \cdots = x_q = 0$, and R_i^1 $(i = 1, 2, \cdots, q)$ the line defined by $x_j = 0\ j \neq i$. Let us take on each line R_i^1 two points $b_i^{(0)} = (\underbrace{0, \cdots, 0}_{i-1}, 1, 0, \cdots, 0)$ and $b_i^{(1)} = (\underbrace{0, \cdots, 0}_{i-1}, -1, 0, \cdots, 0)$. As $A_{p,0}$ is a complex of dimension p, there must exist a realization $T : A_{p,0} \subset R^{2p+1}$. On setting $T(a_{p+i}^{(0)}) = b_i^{(0)}$, $T(a_{p+i}^{(1)}) = b_i^{(1)}$, $i = 1, 2, \cdots, q$, we get then an extension of $T : A_{p,q} \subset R^{2p+q+1}$. The theorem is thus completely proved.

Remark. The theorem 19 of §9 concerning the realization problem of complex $K_{N,n}$ in R^m is in reality settled by known method (principally the method of Van Kampen-Flores). However, the realization problem of the complex $A_{p,q}$ in R^m, so far as the author knows, seems impossible to be settled by any known methods (Alexander, Van Kampen, Flores, Thom, etc.).

References

[1] Pontrjagin L. *Combinatorial topology,* 1947.
[2] Alexandroff P and Hopf H. *Topologie* I, 1935.
[3] Van Kampen E R. *Abh. Math. Sem. Hamburg*, 1932, 9: 72-78.
[4] Van Kampen. Berichtigung dazu, *Loc. Cit.* 152-153.
[5] Flores A I. *Erg. Math. Kolloqu*, 1935, 6: 4-7.
[6] Wu Wen-tsün. *Acta Math. Sinica.* 1953, 3: 261-290.
[7] Cairns S S. *Annals of Math.*, 1940, 41: 792-795.
[8] Van Kampen E R. *Lectures in Topology*, 1941: 311-314.
[9] Wu Wen-tsün. *Acta Math. Sinica.*, 1955, 5: 37-63.
[10] Wu Wen-tsün. *Sur les puissances de Steenrod. Colloque de Topologie* (Strasbourg), 1951.

21. On the Imbedding of Polyhedrons in Euclidean Spaces*

1. For the imbedding or realisation of complexes or more general spaces in Euclidean spaces we may cite the following known results:

(a) In 1932, van Kampen [1] (also Flores [2]) proved the existence of n-dimensional finite complexes not (rectilinearly) realisable in Euclidean space of dimension $2n$.

(b) In the case of differentiable manifolds: M. Whitney [3] has introduced a system of invariants $\bar{W}^m \in H^m$ $(M$, mod 2) and proved that

$$(1) \qquad \bar{W}^k = 0, \quad k \geqslant m - n$$

are necessary conditions for an n-dimensional closed-differentiable mauifold to be (differentiably) realisable in R^m.

(c) For compact Hausdorff spaces X, Thom ([4], Theorem III. 25) has proved the following theorem: Let

$$Q^i : H^r(X, \text{mod } 2) \to H^{r+1}(X, \text{mod } 2)$$

be certain operations deduced from Steenrod squares, then

$$(2) \qquad Q^i H^r(X, \text{mod } 2) = 0, \quad 2i + r \geqslant m$$

are necessary conditions for X to be (topologically) realisable in R^m.

In the various theories listed above, not only the methods used are quite different from each other, but also the domain of spaces considered and the realisation concepts involved are not the same. In what follows we shall show that these theories, so different in appearance, are merely the particular phases of a more general theory.

2. Let us recall some facts about the theory of P. A. Smith [5] concerning spaces① under periodic transformations. Let $\tilde{Y}$ be a space and t a homeomorphism of $\tilde{Y}$ into itself without fixed points and of period 2. Let $Y = \tilde{Y}/t$ be the space of orbits of $\tilde{Y}$ under t and $\pi : \tilde{Y} \to Y$ the natural projection, so that $\tilde{Y}$ is a two-sheeted covering

* *Bull. Acad. Polon. Sci.* C1. III., 1956, 4: 573-577.

① By spaces we shall always mean Hausdorff spaces.

space of Y with projection π. With respect to the pair $(\tilde{Y}, t)$ or the triple $(\tilde{Y}, Y, \pi)$ we may define according to P. A. Smith a system of cohomology classes

$$A^m(\tilde{Y}, Y, \pi) = A^m(\tilde{Y}, t) \in H^m(Y, I_{(m)}), \quad m \geqslant 0,$$

where $I_{(m)}$ = group of integers I for m even, $I_{(m)}$ = group I_2 of integers mod 2 for m odd, and H represents a singular homology system. If $(\tilde{Y}', t')$ is another such pair with corresponding projection $\pi' : \tilde{Y}' \to Y' = \tilde{Y}'/t'$, and $\tilde{f} : \tilde{Y} \to \tilde{Y}'$, $f : Y \to Y'$ are maps such that $\pi'\tilde{f} = f\pi$, then $(\tilde{f}, f) : (\tilde{Y}, Y, \pi) \to\to (\tilde{Y}', Y', \pi')$ will be called simply a map. For such a map, we have then

$$f^* A^m(\tilde{Y}', Y', \pi') = A^m(\tilde{Y}, Y, \pi), \quad m \geqslant 0. \tag{3}$$

Now, for any space X let $\tilde{X}^*$ be the space of all ordered pairs (x_1, x_2) and X^* the space of all non-ordered pairs $\{x_1, x_2\}$, where $x_1, x_2 \in X$ and $x_1 \neq x_2$. If $t : \tilde{X}^* \to \tilde{X}^*$ denotes the transformation $t(x_1, x_2) = (x_2, x_1)$ and $\pi : \tilde{X}^* \to X^*$ the projection $\pi(x_1, x_2) = \{x_1, x_2\}$, then we have a pair $(\tilde{X}^*, t)$ or a triple $(\tilde{X}^*, X^*, \pi)$, which (and hence also the classes $A^m(\tilde{X}^*, X^*, \pi) = A^m(\tilde{X}^*, t)$) depend evidently only on the topological type of the space X. We may therefore denote these classes by

$$\Phi^m(X) \in H'(X^*, I_{(m)}).$$

Our theory may then be formulated in terms of the following

Main Theorem *For a space X to be topologically realisable in a Euclidean space R^m of dimension m it is necessary that*

$$\Phi^m(X) = 0. \tag{4}$$

Proof. Suppose that X is realisable in R^m and g is such a realisation. Let S^{m-1} be the unit sphere in R^m, and t' the antipodal map of S^{m-1} so that S^{m-1}/t' is an $(m-1)$-dimensional projective space P^{m-1}, for which the projection $S^{m-1} \to P^{m-1}$ will be denoted by π'. Now for any point $(x_1 x_2) \in \tilde{X}^*$ let us draw a half-line through the origin of R^m, parallel to the line joining $g(x_1)$ and $g(x_2)$ and pointing in the same direction from $g(x_1)$ to $g(x_2)$. Denote the point of intersection of this half-line with S^{m-1} by $\tilde{f}\ (x_1, x_2)$ and the point $\pi'\tilde{f}\ (x_1, x_2)$ by $f\{x_1, x_2\}$, then we get a map

$$(\tilde{f}, f) : (\tilde{X}^*, X^*, \pi) \to (S^{m-1}, P^{m-1}, \pi\prime).$$

As $A^m\ (S^{m-1}, P^{m-1}, \pi')$ is evidently 0, we get (4) from (3).

That the theorems of Whitney and Thom (the latter in the case of finite polyhedrons only) are both consequences of the above main theorem may be seen from the following two propositions:

Theorem 1 *Let X be a closed differentiable manifold of dimension n, then* (1) *is a consequence of* (4).

Theorem 2 *For any finite polyhedron X,* (2) *is a consequence of* (4).

The proofs of these two theorems are too complicated to be given here. We shall restrict ourselves to the following remarks:

1° Examples may be constructed which show, in the case of closed differentiable manifolds, that neither (1) is a consequence of (2), nor (2) is a consequence of (1). It follows that even in the case of differentiable realisation of differentiable manifolds, the methods of Whitney and Thom are quite independent of one another, while both of them may be deduced from our main theorem.

2° Our main theorem shows that Whitney's condition (1) is necessary not merely for the *differentiable* realisation of the differentiable manifold in the corresponding Euclidean space, but also for the *topological* realisation.

3. Consider now the connection between our theory and that of van Kampen [6].

For this purpose let us consider a finite polyhedron X with a simplicial subdivision K, its reduced product $\tilde{K}^*$ formed by cells of the form $\sigma \times \tau$ with σ, $\tau \in K$ having no common vertices and its reduced symmetric product K^* obtained from $\tilde{K}^*$ by the identification of each pair $\sigma \times \tau$ and $\tau \times \sigma$ into the same cell $\sigma * \tau$ (or $\tau * \sigma$). If we orient $\sigma * \tau$ as the usual product $\sigma \times \tau$, then algebraically we would have

$$\sigma * \tau = (-1)^{\dim\sigma\dim\tau} \tau * \sigma. \tag{5}$$

As the complex $\tilde{K}^*$ is a two-sheeted covering complex over K^* with natural projection, say ω, the triple① $(|\tilde{K}^*|, |K^*|, \omega)$ gives rise to a system of classes $\Phi^m(K) = A^m(|\tilde{K}^*|, |K^*|, \omega) \in H^m(|K^*|, I_{(m)})$. It is easy to see that $(|\tilde{K}^*|, |K^*|, \omega)$ is a "deformation retract" of $(\tilde{X}^*, X^*, \pi)$ so that we have canonically

$$H^m(|K^*|, I_{(m)}) \approx H^m(X^*, I_{(m)}), \tag{6}$$

and $\Phi^m(K)$, $\Phi^m(X)$ correspond under this isomorphism.

Let us now try to imbed rectilinearly K in R^m as far as possible, so that under such a "pseudo-imbedding", say f, the images $f\sigma$, $f\tau$ of every pair of simplexes σ, τ

① We denote the space of a complex K by $|K|$.

of K, having no common vertices and with dim $\sigma+$ dim $\tau \leqslant m$, are always Euclidean simplexes in general position. For any cell $\sigma * \tau \in K^*$ of dimension m, $f\sigma$ and $f\tau$ then have a well-defined index of intersection, $I(f\sigma, f\tau)$, with respect to a fixed orientation of R^m.

In view of (5), the equations $\varphi_f(\sigma * \tau) = (-1)^{\dim \sigma}\rho(m)I(f\sigma, f\tau)$, in which $\rho(m)$ is the identity for $m =$ even and reduction mod 2 for $m =$ odd, then define consistently an m-dimensional co-chain φ_f of K^* on the coefficient group $I_{(m)}$. This co-chain may serve as a primary measure of the deviation of the pseudo-imbedding f from a true one. We then have the following

Theorem 3 *The co-chain φ_f associated to a pseudo-imbedding $f : K \to R^m$ is always a co-cycle with its co-homology class $\Phi^m(K) \in H^m(|K^*|, I_{(m)})$, independent of the pseudo-imbedding f chosen, and coincides with $\Phi^m(X)$ under the natural isomorphism* (6).

For dim $K = n$ and $m = 2n$ the class $\Phi^{2n}(X)$ is in reality the same as the original van Kampen invariant introduced by him to establish his theorem. It follows that from our main theorem, as well as from Theorem 3, we get van Kampen's theorem in this extreme case.

Remark. Even in this extreme case our theorem contains a little more than that of van Kampen in the sense that $\Phi^{2n} = 0$ is necessary not only for *rectilinear* realisability of a certain subdivision of X in R^{2n}, but also for the *topological* realisation of X in R^{2n}.

4. In certain extreme cases the condition (4) in our main theorem is not only necessary, but also sufficient for X to be realisable in R^m. In fact, we have

Theorem 4 *If X is a finite polyhedron of dimension $n > 2$, then*

$$\Phi^{2n}(X) = 0$$

is a necessary and sufficient condition for X to be realisable in R^{2n}.

This theorerh furnishes us an intrinsic characterisation of n-dimensional finite polyhedrons in $2n$-dimensional Euclidean spaces ($n > 2$). Our proof depends on the realisability ([6], Theorem 7) of every co-cycle in $\Phi^{2n}(K)$. (X is a finite polyhedron with a simplicial subdivision K) as a co-cycle φ_f defined in par. 3 and also on some special device used by Whitney [7] to prove his theorem stating that every n-dimensional differentiable manifold is differentiably realisable in R^{2n}.

Theorem 4 remains true in case $n = 1$, as follows easily from a wellknown theorem of Kuratowski [8] about planar Peano continuum. In fact, this theorem may be reformulated in the following form:

Theorem 4′ *For a Peano continuum X to be planar it is necessary and sufficient that* $\Phi^2(X) = 0$.

In particular, when X is a finite linear graph, this gives us a "quantitative" criterion of its planarity and also effective means to determine it in a finite number of steps, in contrast to the original "qualitative" criterion of Kuratowski (also Whitney, MacLane and perhaps others).

Remark. The case when $n = 2$ has not yet been settled.

References

[1] van Kampen. *Komplexe in euklidischen Räumen.* Abh. Math. Sem. Hamburg, 1932, 9: 72; Berichtigung dazu, ibid, 152.

[2] Flores A I. *Über die Existenz n-dimensionaler Komplexe, die nicht in den R^{2n} topologisch einbettbar sind.* Erg. Math. Kolloq, 1935, 6: 4.

[3] Whitney H. *On the topology of differentiable manifolds.* Michigan Lectures in Topology, 1941: 101.

[4] Thom R. *Espaces fibrés en sphères et carrés de Steenrod.* Ann. Ec. Norm. Sup., 1952, 69(3): 110.

[5] Smith P A. *Periodic and nearly periodic transformations.* Michigan Lectures in Topology, 1941: 159.

[6] Wu Wen-tsün. *On the realisation of complexes in Euclidean spaces* I. Acta Mathematica Sinica, 1955, 5: 505.

[7] Whitney H. *The self-intersections of a smooth n-manifold in 2n-space.* Annals of Math., 1944, 45: 220.

[8] Kuratowski K. *Sur le problème des courbes gauches en topologie.* Fund. Math., 1930, 15: 271.

22. On the Realization of Complexes in Euclidean Spaces II*

Abstract For the realization problem of complexes or more general spaces in Euclidean spaces we may cite the following results:

1° In 1932 van Kampen [1] proved the existence of n-dimensional complexes K not realizable in Euclidean space of dimension $2n$.

His proof depends on certain invariants deduced from the reduced 2-fold symmetric product K^* of K. The author [2] has pointed out that van Kampen's invariant is merely the extreme one of a system of invariants $\Phi^m \in H^m(K^*, I_{(m)})$, ($I_{(m)} = I$ or I_2, depending on m being even or odd), namely Φ^{2n}, and $\Phi^m = 0$ is a necessary condition for K to be realizable in R^m. We proved also in [2] the topological invariance of Φ^m, i.e., the independence of Φ^m of the chosen subdivision K of the space of K, while van Kampen, in the extreme case $m = 2n$ considered by him, has not even proved its combinatorial invariance.

2° In the case of differentiable manifold, Whitney (cf. e.g., [3]) has introduced a system of invariants which the present author has called the dual Whitney classes $\bar{W}^m$, and proved that

$$\bar{W}^k = 0, \quad k \geqslant m - n \tag{1}$$

are necessary conditions for an n-dimensional closed differentiable manifold to be realizable in R^m.

3° For compact Hausdorff spaces X, Thom has proved ([4] Th. III. 25) the following theorem: Let Q^i be certain operations deduced from Steenrod squares, then

$$Q^i H^r(X, I_2) = 0, \quad 2i + r \geqslant m \tag{2}$$

are necessary conditions for X to he realizable in R^m. The operations Q^i were previously introduced by the author by applying Smith's theory of

* First published in Chinese in *Acta Mathematica Sinica, 1957, VII(1): 79-101*; *Amer. Math. Soc. Translations, Ser.*, 1968, 2(78): 185-208.

periodic transformations and will henceforth be denoted by Sm^i (cf. [5] and [6]).

Besides, Flores [7] got also the same results as van Kampen's concerning the existence of n-dimensional complexes K not realizable in R^{2n}. However, what he used to prove his results is got by imbedding the first K in R^{2m+1}, and is actually not an invariant, but depends in general on the way of realization of K in R^{2n+1}. This will be explained further in the sequel.

In the various theories listed above, not only the methods used are quite different from one another, but also the objects and the realization concept involved are not the same. For example, the theory of van Kampen studies about the semi-linear realization of finite complexes, the theory of Whitney is applicable only for the differentiable realization of differentiable manifolds, and the theory of Thom is concerned with topological realization of more general spaces.

The present paper gives a general theory including the various theories cited above as its particular cases. This theory may be formulated in terms of the following easily proved fundamental theorem:

For any Hausdorff space X let $\tilde{X}^*$ be the space of all ordered pairs (x_1, x_2), where x_1, $x_2 \in X$ and $x_1 \neq x_2$. Let $t : \tilde{X}^* \equiv \tilde{X}^*$ be the transformation $t(x_1, x_2) = (x_2, x_1)$, X^* the modular space $\tilde{X}^*/t$. With respect to the pair $(\tilde{X}^*, t)$, we may define according to P. A. Smith[8] a system of cohomology classes

$$\Phi^m(X) \in H^m(X^*, I_{(m)}),$$

where H represents singular homology system. Then $\Phi^m(X) = 0$ is a necessary condition for X to be topologically realizable in R^m.

If X be a finite polyhedron, then according to results of [2] , the classes Φ^m defined above are the same as the classes Φ^m defined in that paper in a quite different manner. It follows that $\Phi^m = 0$ is not only a necessary condition for X to possess a subdivision K such that K is semilinearly realizable in R^m, as shown in [2] , but also one for X itself to be topologically realizable in R^m.

If X is a closed differentiable manifold of dimension n, then we prove that from $\rho_2\, \Phi^m(X) = 0$ we get (1), but not vice versa. Our fundamental theorem has therefore Whitney's theorem as its consequence. It follows also that Whitney's condition (1) is not only necessary for the differentiable realization, but also necessary for topological realization of X in R^m.

We prove also that if X be a finite polyhedron, then from $\rho_2\Phi^m(X) = 0$ we get (2). Hence our fundamental theorem has also, at least in the case of finite polyhedron, Thom's theorem as its consequence.

In the case of differentiable manifold, Whitney [9] has introduced the concept of "immersion" in a Euclidean space. He proved that any differentiable manifold of dimension n may be immersed in R^{2n-1}, but said nothing about the criterion of immersibility in R^m of arbitrary dimension m. We extend the concept of immersion to arbitrary spaces and our method for the study of realization may also be applied to the immersion problem of finite polyhedrons. For such a space X, let us take a subdivision K, and construct a certain "tube" $K^{(0)}$ of the diagonal in the two-fold symmetric product $K * K$. Then there exists in $K^{(0)}$, again by the theory of P. A. Smith, a system of cohomology classes $\Psi^m(K^{(0)}) \in H^m(K^{(0)}, I_{(m)})$. According to [10], the homotopy type of $|K^{(0)}|$ as well as $\Psi^m(K^{(0)})$ is topological invariants of X, and therefore $\Psi^m(K^{(0)})$ may also be denoted by $\Psi^m(X)$. Our fundamental theorem about immersion may then be stated as follows:

If a finite polyhedron X may be immersed in R^m, then $\Psi^m(X) = 0$.

As before, we may deduce from this theorem results analogous to theorems of Thom and Whitney in the case of realization.

Our method may also be used to study the problem of isotopy or isoposition. For a topological realization f of a Hausdorff space X in R^m, we may define a cohomology class $\Theta_f^{m-1} \in H^{m-1}(\tilde{X}^*)$The definition of Θ_f^{m-1} was essentially due to Flores[7] . But Flores discussed mainly about the case where Θ_f^{m-1} is independent of the realization f and is thus topological invariant of X. We prove on the other hand that Θ_f^{m-1} depends generally on the realization f and is an invariant of isotopy as well as isoposition. More precisely, $\theta_f^{m-1} = \theta_g^{m-1}$ (or $\Theta_f^{m-1} = \pm\ \Theta_g^{m-1}$) is a necessary condition for the two realizations $f(X)$ and $g(X)$ in R^{m-1} to be isotopic (or in the same position). A consequence of this is that some known results as theorem of Adkisson and McLane about planar or spherical figures (of. [11] and literatures listed there) may be formulated in a "quantitative" instead of "qualitative" form.

We remark finally that the realization theorem of Thom is proved as consequence of our fundamental realization theorem only in the case of finite polyhedrons. This is due to the fact that we use the singular homology system. If another homology system is used, it would be possible to get Thom's theorem for more general spaces. As this concerns only the choice of homology systems, and as the corresponding theorem for immersion seems to be difficult to extend to more general spaces other than finite polyhedrons, our discussions will be restricted mainly to the latter class of spaces.

§1 Smith homomorphisms

Let $\tilde{E}$ be a Hausdorff space, and t a topological transformation of $\tilde{E}$ with period n (n an integer > 1) such that any power t^i of t has no fixed points for $i \not\equiv 0 \bmod n$. We say in such case that $(\tilde{E}, t)$ is a simple system. Let $E = \tilde{E}/t$ be the modular space of $\tilde{E}$ with respect to t and $\pi : \tilde{E} \to E$ the natural projection. Let $S(\tilde{E})$, $S(E)$ be the singular complexes of corresponding spaces defined by means of ordered singular simplexes. Then t operates also in $S(\tilde{E})$, having period n, with no fixed simplexes under any power t^i of t, for $i \not\equiv 0 \bmod n$, and the modular complex of $S(\tilde{E})$ with respect to t is no other than $S(E)$. We shall denote by $C^*(\tilde{E}, G)$ or $C^r(\tilde{E}, G)$ (G =the coefficient group) the cochain groups in $S(\tilde{E})$, and denote by $\bar{C}^*(\tilde{E}, G)$ or $\bar{C}^r(\tilde{E}, G)$ the groups of cochains which take values 0 on all singular simplexes disjoint from certain compact sets (varied with the cochains) in $\tilde{E}$. The cohomology groups derived from these cochain groups will be denoted by $H^*(\tilde{E}, G)$, $\bar{H}^*(\tilde{E}, G)$, etc. Similarly, we have also $H^*(E, G)$, $\bar{H}^*(E, G)$, etc. $\bar{H}^*$ is the so-called compact singular cohomology groups (cf. [12]).

For any γ-dimensional ordered singular simplex σ of $S(E)$, there are just n singular simplexes $\tilde{\sigma}_1, \cdots, \tilde{\sigma}_n$ in $S(\tilde{E})$ such that $\pi\tilde{\sigma}_i = \sigma$. Then for any cochain $\tilde{\varphi} \in C^r(\tilde{E}, G)$, $\bar{\pi}^{\#}\tilde{\varphi}(\sigma) = \sum\limits_i \tilde{\varphi}(\tilde{\sigma}_i)$ defines a cochain map $\bar{\pi}^{\#} : C^r(\tilde{E}, G) \to C^r(E, G)$. As usual put $s = 1 + t + \cdots + t^{n-1}$, $d = 1 - t$. According to P. A. Smith, for any cocycle x (on arbitrary coefficient group G) we may take a system of cochains $\tilde{y}$, $\tilde{y}_1$, $\cdots$ in $S(\tilde{E})$ such that (cf. [8])

$$\pi^{\#}x = s\tilde{y}, \delta\tilde{y} = d\tilde{y}_1, \delta\tilde{y}_1 = s\tilde{y}_2, \cdots .$$

Put $x_i = \bar{\pi}^{\#}\, \tilde{y}_i$, then $x \to x_i$ will induce a system of homomorphisms

$$\mu_i = \mu_i(\tilde{E}, t) : \quad H^r(E, G) \to H^{r+i}(E,\ G_{(i)}) \ ,$$

where

$$G_{(i)} = \begin{cases} G, & i \text{ even}, \\ G/nG, & i \text{ odd}. \end{cases}$$

When $x \in \bar{C}^r(E, G)$, we may take $\tilde{y}$, $\tilde{y}_1$, $\cdots$ all in $\bar{C}^*(\tilde{E}, G)$ so that $x \to x_i$ induces also a system of homomorphisms

$$\bar{\mu}_i = \bar{\mu}_i(\tilde{E}, t) : \bar{H}^r(E,\ G) \to \bar{H}^{r+i}(E, G_{(i)}).$$

Between these homomorphisms the following relations exist:

$$\left.\begin{array}{c} \mu_{2i} = (\mu_2)^i, \quad \mu_{2i+1} = (\mu_2)^i \mu_1 = \mu_1(\mu_2)^i, \\ \bar{\mu}_{2i} = (\bar{\mu}_2)^i, \quad \bar{\mu}_{2i+1} = (\bar{\mu}_2)^i \mu_1 = \bar{\mu}_1(\mu_2)^i, \\ \mu_i\, j^* = j^*\, \bar{\mu}_i, \end{array}\right\} \tag{1}$$

in which $j^* : \bar{H}^*(E, G) \to H^*(E, G)$ is the natural homomorphism. Denote by ρ_n the homomorphism of cohomology groups induced by the reduction mod n, $G \to G/nG$ or the identity $G \to G$, we have also

$$a\rho_n \mu_2 = (\mu_1)^2, \quad a\rho_n \bar{\mu}_2 = (\bar{\mu}_1)^2,$$

where

$$a = \begin{cases} 0, & n \text{ odd}, \\ -\dfrac{n}{2}, & n \text{ even}. \end{cases}$$

Let I be the additive group of integers, and 1 the integral unit class in $C^\circ(E)$, then the classes $\mu_i 1 \in H^i(E, I_{(i)})$ will be denoted by

$$A^i = A^i(\tilde{E}, t) = \mu_i 1.$$

For any class $U \in \bar{H}^*(E,\ G)$ we have then

$$\bar{\mu}_i U = A^i \smile U.$$

Remark that as A^i, U belong to H^* and $\bar{H}^*$ respectively, so that their $\smile$-product is defined and belongs to $\bar{H}^* : H^* \smile \bar{H}^* \subset \bar{H}^*$, the coefficient groups being paired in the usual manner: $I_{(i)} \smile G \subset G_{(i)}$. This theorem has been proved for the special case of a complex and $G = I$ (cf. [5] or [13]). The proof in the general case is essentially the same.

Let $(\tilde{E}', t')$ be another simple system, $E' = \tilde{E}'/t'$, and t' has also the same period n. If the continuous map $\tilde{f} : \tilde{E} \to \tilde{E}'$ satisfies $t'\tilde{f} = \tilde{f}t$, then $\tilde{f}$ will be called a P-map.

In such case $\tilde{f}$ will induce a continuous map $f : E \to E'$ with $f\pi = \pi'\tilde{f}$, where π, π' are corresponding projections. Moreover, we have

$$f^*\mu_i(\tilde{E}', t') = \mu_i(\tilde{E}, t)f^*,$$

in particular $f^*A^i(\tilde{E}', t') = A^i(\tilde{E}, t)$. If f is admissible, i.e., the inverse image of any compact set in E' is also a compact set in E, then we have also

$$f^*\bar{\mu}_i(\tilde{E}', t') = \tilde{\mu}_i(\tilde{E}, t)f^*.$$

Similar to a P-map, we may also naturally define P-deformation and P-homotopy type into which we do not enter.

§2 The fundamental realization theorem

Let X be a Hausdorff space and n be a prime. Let $\underbrace{X \times \cdots \times X}_{n} = \tilde{X}_n$ and $d_n : X \to \tilde{X}_n$ be the diagonal map. Define $t_n : \tilde{X}_n \to \tilde{X}_n$ by $t_n(x_1, \cdots, x_n) = (x_n, x_1, \cdots, x_{n-1})$. Put $\tilde{X}_n^* = \tilde{X}_n - d_nX$, then $(\tilde{X}_n^*, t_n)$ is a simple system. Let $X_n = \tilde{X}_n/t_n$, $X_n^* = \tilde{X}_n^*/t_n$ be the corresponding modular spaces, then the classes $A^i(\tilde{X}_n^*, t_n) \in H^i(X_n^*, I_{(i)})$ in X_n^* are topological invariants of X and may be denoted by $\Phi_n^i(X)$. In particular, they will be simply denoted by $\Phi^i(X)$ when $n = 2$.

Let K be a simplicial complex, and $\tilde{K}_n^*$ be the subcomplex of $\tilde{K}_n = \underbrace{K \times \cdots \times K}_{n}$ formed of all cells $\sigma_1 \times \cdots \times \sigma_n$ for which $\sigma_i \in K$ and no vertex of K is common to all these σ_i. Define $t_n : |\tilde{K}_n| \to |\tilde{K}_n|$ as above① and let $K_n^* = \tilde{K}_n^*/t_n$ be the modular complex of the simple system $(\tilde{K}_n^*, t_n)$ (n a prime). In [2] we have proved that from the consideration of almost semi-linear realizations of K in R^m we get a system of cohomology classes $\Phi^m(K) \in H^m\ (K_2^*, I_{(m)})$, called the imbedding classes of K, whose vanishing is a necessary condition for K to have a semi-linear realization in R^m. In the same paper we have also proved that $\Phi^m(K) = A^m(\tilde{K}_2^*, t_2)$. Since $|\tilde{K}_2^*|$ is a P-deformation retract of $|\tilde{K}|_2^*$, we may identify $H^*(|K|_2^*, G)$ and $H^*(K_2^*, G)$, and then $A^m(\tilde{K}_2^*, t_2) = A^m(|\tilde{K}|_2^*, t_2)$, i.e., $\Phi^m(K)$ coincides with the above-defined $\Phi^m(|K|)$. For this reason, even in the case of an arbitrary Hausdorff space X, $\Phi^m(X) \in H^m(X_2^*, I_{(m)})$ will also be called the *imbedding classes* of X.

The above extended definition of imbedding classes is justified by the following theorem:

① The space of a complex K will be denoted by $|K|$.

Theorem 1 If a Hausdorff space X is (topologically) realizable in a Euclidean space R^m of dimension m, then $\Phi^m(X) = 0$.

Proof. Suppose that X is realizable in R^m so that a topological map $f : X \to R^m$ exists. Let 0 be the origin of R^m, S^{m-1} the unit sphere in R^m with centre 0, and t the antipodal map of S^{m-1} onto itself. The modular space $P^{m-1} = S^{m-1}/t$ of the simple system (S^{m-1}, t) is then a projective space of dimension $m-1$. Consider any point $(x_1, x_2) \in \tilde{X}_2^*$. As $f(x_1) \neq f(x_2)$, we may draw a half-line starting from 0 and parallel to the directed line from $f(x_1)$ to $f(x_2)$. Denote the intersecting point of this half-line with S^{m-1} by $\tilde{g}(x_1, x_2)$, then $\tilde{g}(x_2, x_1) = t\tilde{g}(x_1, x_2)$ so that $\tilde{g}$ is a P-map and induces thus a map $g : X_2^* \to P^{m-1}$. It follows that

$$g^* A^i(S^{m-1}, t) = A^i(\tilde{X}_2^*, t_2) = \Phi^i(X), \quad i \geqslant 0.$$

Since $A^i(S^{m-1}, t) = 0$ for $i \geqslant m$, the theorem is proved.

By the Menger-Nöbeling imbedding theorem①, we get the following

Corollary If X is a normal space with countable basis having dimension m, then $\Phi^{2m+1}(X) = 0$.

Remark 1. By [2], for any integer m there exist complexes K of dimension m with $\Phi^{2m}(|K|) \neq 0$.

Remark 2. If X is a finite polyhedron, then by [2] , $\Phi^m(X) = 0$ is a necessary condition for X to possess a simplicial subdivision K semi-linearly imbeddable in R^m. The theorem here shows that the condition is necessary not only for semi-linear but also for topological imbedding.

§3 Immersion and immersion classes

Let X be a Hausdorff space and $f : X \to R^m$ a continuous map possessing the following property: there exists an open covering $\mathfrak{A} = \{U_i\}_{i \in I}$ of X such that for any two points x, $y \in$ the same $U_i \in \mathfrak{A}$, we have $f(x) \neq f(y)$. In such case we shall say that f is an *immersion* written $f : X \subset\!\!\!\subset R^m$. The aim of this section is to introduce certain invariants to measure the possibility of immersing X in a Euclidean space, at least in the case that X is a finite polyhedron.

For this purpose let us consider a simplicial subdivision K of X and construct $\tilde{K}_2 = K \times K$ as well as $\tilde{K}_2^*$ as in §2. Subdivide the diagonal d_2X as K such that d_2X

① Cf. e.g. Hurewicz and Wallman's *Dimension Theory*, Theorem V3, 1941.

is decomposed into simplexes of the form $d_2\sigma$, where $\sigma \in K$. Denote this complex by d_2K, then, according to [10], $\tilde{K}_2$ has a cell decomposition with subcomplexes d_2K and $\tilde{K}_2^*$ as follows. Let σ, σ_1, $\sigma_2 \in K$ have the following property: σ_1, σ_2 have no common vertex, and σ and σ_1 (resp. σ and σ_2) span a simplex σ_1' (resp. σ_2') of K. In the cell $\sigma_1' \times \sigma_2'$ let us take a point $x_0 \in |\sigma|$ and a point $x_1 \in |\sigma_1 \times \sigma_2|$, and denote by x_r the point on segment x_0x_1 dividing it in the ratio $r : 1-r (0 \leqslant r \leqslant 1)$. When x_0, x_1 run over $|\sigma|$ and $|\sigma_1 \times \sigma_2|$ respectively, the points $x_{1/2}$ will form a $(d+d_1+d_2)$-cell $[\sigma, \sigma_1 \times \sigma_2]$, where $d = \dim \sigma$, $d_i = \dim \sigma_i$, $i = 1, 2$. Similarly, the points x_r, $r \geqslant 1/2$ (resp. $r \leqslant 1/2$) form a $(d+d_1+d_2+1)$-cell $[\sigma, \sigma_1 \times \sigma_2]^+$ (resp. $[\sigma, \sigma_1 \times \sigma]^-$). The set of all cells in $\tilde{K}_2^*$, d_2K and those of the forms $[\sigma, \sigma_1 \times \sigma_2]^+$, $[\sigma, \sigma_1 \times \sigma_2]^-$ and $[\sigma, \sigma_1 \times \sigma_2]$ forms then a cell subdivision of $\tilde{K}_2$ which will be denoted by $\omega\tilde{K}_2$. The subcomplex of $\omega\tilde{K}_2$ formed by all cells $[\sigma, \sigma_1 \times \sigma_2]$ will be denoted by $\tilde{K}_2^{(0)}$ and the subcomplex formed of all cells of $\tilde{K}_2^*$ as well as all $[\sigma, \sigma_1 \times \sigma_2]^+$ and $[\sigma, \sigma_1 \times \sigma_2]$ will be denoted by $\tilde{K}_2^{(+)}$. Evidently t_2 will operate in $\tilde{K}_2^{(+)}$ and $\tilde{K}_2^{(0)}$ with modular complexes $K_2^{(+)} = \tilde{K}_2^{(+)}/t_2$ and $K_2^{(0)} = \tilde{K}_2^{(0)}/t_2$ respectively. It may be seen that the systems $(|\tilde{K}_2^{(+)}|, t_2)$, $(|\tilde{K}_2^*|, t_2)$ and $(\tilde{X}_2^*, t_2)$ all have the same P-homotopy type. By [10] , we know also that the P-homotopy type of $(|\tilde{K}_2^{(0)}|, t_2)$ is always the same whatever the chosen subdivision K may be, and is thus a topological invariant of X. In particular, the classes $A^i(|\tilde{K}_2^{(0)}|, t_2) \in H^i(|K_2^{(0)}|, I_{(i)})$ are all topological invariants of X and will be denoted by $\Psi^i(X)$.

Let the cell in $K_2^{(0)}$ which is obtained by identifying the cells $[\sigma, \sigma_1 \times \sigma_2]$ and $[\sigma, \sigma_2 \times \sigma_1]$ in $\tilde{K}_2^{(0)}$ be denoted by $[\sigma, \sigma_1 * \sigma_2]$. Then the mod 2 boundary relations in $K_2^{(0)}$ may be written symbolically as

$$\partial[\sigma, \sigma_1 * \sigma_2] = [\partial\sigma, \sigma_1 * \sigma_2] + [\sigma, \partial(\sigma_1 * \sigma_2)] \bmod 2. \tag{1}$$

Now each point x of a cell $[\sigma, \sigma_1 \times \sigma_2] \in \tilde{K}_2^{(0)}$ is the mid-point of a certain segment x_0x_1, where $x_0 \in |\sigma|$ and $x_1 \in |\sigma_1 \times \sigma_2|$. Then $x \to x_1$ defines a cell-map

$$\tilde{f} : \tilde{K}_2^{(0)} \to \tilde{K}_2^*.$$

As $\tilde{f}$ is evidently a P-map, it induces also a cell-map

$$f : K_2^{(0)} \to K_2^*$$

such that

$$f^*A^i(\tilde{K}_2^*, t_2) = A^i(\tilde{K}_2^{(0)}, t_2),$$

or

$$f^*\Phi^i = \Psi^i. \tag{2}$$

By definition and from dimensionality considerations we see that

$$f_\#(\sigma, \sigma_1 * \sigma_2) = \begin{cases} 0, & \dim\ \sigma > 0 \\ \sigma_1 * \sigma_2, & \dim\ \sigma = 0 \end{cases} \pmod 2. \tag{3}$$

Theorem 2 If a finite polyhedron X is immersible in R^m, then

$$\Psi^m(X) = 0.$$

Proof. Let $f : X \subset\!\!\!\!\!\!\propto R^m$, and $\mathfrak{A} = \{U_i\}$ be the open covering of X as required in the definition of an immersion, of which the number of U_i may be supposed to be finite. Take a sufficiently small simplicial subdivision K such that for any point $(x, y) \in |\tilde{K}_2^{(0)}|$, we should have x, $y \in$ the same $U_i \in \mathfrak{A}$. In such case we have $f(x) \neq f(y)$. Hence, let $\tilde{g}(x, y)$ be the intersecting point with the unit sphere S^{m-1} in R^m by the half-line through origin O of R^m and parallel to the directed line from $f(x)$ to $f(y)$, we get a continuous map $\tilde{g} : |\tilde{K}_2^{(0)}| \to S^{m-1}$. Let t be the antipodal map of S^{m-1}, then $\tilde{g}$ is a P-map of the system $(|\tilde{K}_2^{(0)}|, t_2)$ in the system (S^{m-1}, t). Similar to the proof of Theorem 1 we get then $\Psi^m(X) = 0$.

On account of this theorem, we shall call $\Psi^m(X)$ the *immersion classes* of the finite polyhedron X.

Let the finite polyhedron X be of dimension m. Take a simplicial subdivision K of X and let $f : K \subset R^{2m+1}$ be a linear realization of K in R^{2m+1}. Since for any two simplexes σ, $\tau \in K$ with vertices in common the linear subspace $P(\sigma, \tau)$ determined by them has a dimension $\leqslant 2m$, there must exist lines l in R^{2m+1} not parallel to any one of such linear subspaces $P(\sigma, \tau)$. Let R^{2m} be a $2m$-dimensional linear subspace of R^{2m+1} orthogonal to such a line l and φ be the orthogonal projection of R^{2m+1} onto R^{2m}. Then $\varphi f : X \to R^{2m}$ is evidently an immersion. We have therefore the following

Corollary For any finite polyhedron X of dimension m we have always $\Psi^{2m}(X) = 0$.

On the other hand, we shall prove that for any $n \geqslant 1$, there exist n-dimensional complexes not immersible in R^{2n-1} as follows.

Let $K_{N,n}$ be the complex formed of all n-dimensional simplexes as well as their faces with vertices taken from a_0, a_1, $\cdots$, a_N. In [2] we have proved that $K_{2n+1,n}$

is realizable in R^{2n}, but not in R^{2n-1}. In reality, $K_{2n+1,n}$ is not only non-realizable in R^{2n-1}, but also non-immersible in R^{2n-1}. To prove it, let us consider in $K_2^{(0)}$ the chain

$$z=\sum[(a_0),(a_{i_0}\cdots a_{i_{n-1}})*(a_0a_{j_0}\cdots a_{j_{n-1}})],$$

in which $\sum$ is extended over all possible sets of indices $(i,\ j)$ such that $0<i_0<\cdots<i_{n-1}\leqslant 2n+1$, $0<j_0<\cdots<j_{n-1}\leqslant 2n+1$ and no i is the same as any j. By (1), we know that z is a mod 2 cycle. By Theorem 10 of [2] , $\Phi^{2n-1}(K)\in H^{2n-1}(K_2^*,I_2)$ has a representative cocycle

$$\varphi^{2n-1}=\sum\rho_2\{(a_{i_0}\cdots a_{i_{n-1}})*(a_{j_0}\cdots a_{j_n})\},$$

in which $\sum$ is extended over all possible sets of indices $(i,\ j)$ with $0\leqslant j_0<i_0<j_1<\cdots<i_{n-1}<j_n\leqslant 2n+1$. Define $f:\ |K_2^{(0)}|\to|K_2^*|$ as above, we see that $\varphi^{2n-1}(f_\# z)=2n+1 \bmod 2\neq 0$. Hence, $f^*\Phi^{2n-1}=\Psi^{2n-1}\neq 0$ and by Theorem 2 we have $K_{2n+1,n}\not\subset\!\!\!\!\propto R^{2n-1}$.

Similarly, in Theorem 19 of [2] we have proved that for $n>1$, $K=K_{2n,n}$ is $\subset R^{2n-1}$, but $\not\subset R^{2n-2}$. Let us prove now $K_{2n,n}\not\subset\!\!\!\!\propto R^{2n-2}$ as follows. In that case

$$z=\sum[(a_0),(a_{i_0}\cdots a_{i_{n-1}})*(a_{j_0}\cdots a_{j_{n-1}})]$$

is a mod 2 cycle in $K_2^{(0)}$, in which $\sum$ is extended over all possible sets of indices $(i,\ j)$ such that $i_0<j_0, 0\leqslant i_0<\cdots<i_{n-1}\leqslant 2n$, $0\leqslant j_0<\cdots<j_{n-1}\leqslant 2n$, and no i is equal to any j. On the other hand, by Theorem 11 of [2] , $\Phi^{2n-2}(K)\in H^{2n-2}(K_2^*)$ has a representative cocycle

$$\varphi^{2n-2}=\sum\{(a_{i_0}\cdots a_{i_{n-1}})*(a_{j_0}\cdots a_{j_{n-1}})\},$$

in which $\sum$ is extended over all possible sets of indices $(i,\ j)$ with $0\leqslant i_0<j_0<i_1<\cdots<j_{n-1}\leqslant 2n$. Define f as before, then we have $\rho_2\varphi^{2n-2}(f_\# z)=2n+1 \bmod 2\neq 0$. Therefore $f^*\Phi^{2n-2}=\Psi^{2n-2}\neq 0$ and we have $K_{2n,n}\not\subset\!\!\!\!\propto R^{2n-2}$.

In general, we have the following

Theorem 3 For $n\leqslant m\leqslant 2n-1$ we have

$$K_{m+2,n}\subset R^{m+1}\quad\text{but}\quad\not\subset\!\!\!\!\propto R^m. \qquad (3)_{m,n}$$

Proof. The case where $n=1$, is evident. For $m=2n-1$ or $2n-2$, $(3)_{m,n}$ has already been proved. Suppose now $(3)_{m,n-1}$ has been proved with corresponding $\Psi\neq$

0. Then for $n-1 \leqslant m \leqslant 2n-3$, we have by induction hypothesis $\Psi^m\ (K_{m+2,n-1}) \neq 0$, a fortiori we have $\Psi^m(K_{m+2,n}) \neq 0$ and therefore $(3)_{m,n}$ is also true.

Remark In Theorem 19 of [2] we proved only

$$K_{m+2,n} \not\subset R^n (n \leqslant m \leqslant 2n-1).$$

§4 Relations with steenrod squares

Let H still denote singular homology and $\bar{H}$ compact singular homology as in §1. Then for any finite complex and its subcomplex with spaces X and Y, we have (R being the coefficient ring)

$$H^*(X, Y; R) \approx \bar{H}^*(X - Y; R). \tag{1}$$

In what follows the coefficient group will always be I_2 and will therefore be omitted throughout.

For any finite polyhedron X let us consider now the following commutative diagram:

$$\begin{array}{ccccc}
\bar{H}^r(X_2^*) & \xrightarrow{\bar{\mu}_1} & \bar{H}^{r+1}(X_2^*) & \xrightarrow{\pi^*} & \bar{H}^{r+1}(\tilde{X}_2^*) \\
 & & \| & & \\
 & & H^{r+1}(X_2, d_2X) & & \\
 & & \uparrow \bar{\delta}^* & & \\
H^r(\tilde{X}_2) & \xrightarrow{i^*} & H^r(X) & \xrightarrow{\tilde{\delta}^*} & H^{r+1}(\tilde{X}_2, d_2X)
\end{array}$$

In the above diagram $\bar{\delta}^*$, $\tilde{\delta}^*$ are coboundary homomorphisms, $\bar{\mu}_1 = \bar{\mu}_1(\tilde{X}_2^*, t_2)$ and $\pi : \tilde{X}_2^* \to X_2^*$ is natural projection. From the theory of Smith we know that the first row $\bar{\mu}_1, \pi^*$ is exact, and that the second row i^*, $\tilde{\delta}^*$ is also exact, being sequence of the pair $(X \times X,\ X)$. Since $i^* H^r(\tilde{X}_2) = H^r(X)$, we have $\tilde{\delta}^* = 0$ and hence $\pi^* \bar{\delta}^* = 0$. It follows that

$$\bar{\delta}^* H^r(X) \subset \bar{\mu}_1 \bar{H}^r(X_2^*). \tag{2}$$

Suppose now $\rho_2 \Phi^m(X) = 0$. Put $\bar{\mu}_i = \bar{\mu}_i(\tilde{X}_2^*, t_2)$, then we have (see §1 (1), (2))

$$\bar{\mu}_{m-1} \bar{\delta}^* H^*(X) \subset \bar{\mu}_{m-1} \bar{\mu}_1 \bar{H}^*(X_2^*) = \bar{\mu}_m \bar{H}^*(X_2^*) = \rho_2 \Phi^m(X) \smile \bar{H}^*(X_2^*),$$

consequently

$$\bar{\mu}_{m-1} \bar{\delta}^* = 0. \tag{3}$$

For any class $U \in H^r(X)$, there exists a relation (cf. [14] or [15])

$$\sum_{j=0}^{r} \bar{\mu}_{r-j}\bar{\delta}^* Sq^j U = 0. \tag{4}$$

Let $Sm^i : H^r(X) \to H^{r+i}(X)$ be the Smith operations determined by the following relations (cf. Theorem 2 of [6]) :

$$\sum_{i_1+i_2=i} Sq^{i_1} Sm^{i_2} = \begin{cases} 0, & i > 0, \\ 1, & i = 0, \end{cases} \tag{5}$$

in which 1 denotes identity homomorphism. For the classes $Sm^i U \in H^{r+i}(X)$, we have similarly

$$\sum_{j=0}^{r+i} \bar{\mu}_{r+i-j}\bar{\delta}^* Sq^j Sm^i U = 0, \tag*{(4)$_i$}$$

Let k, r be integers satisfying $m+1 \geqslant 2k+r \geqslant m \geqslant r+1$. Applying $\bar{\mu}_{m-r-2i-1}$ to $(4)_i$, $i = 0, 1, \cdots, k-1$, adding all these relations, and noting that $\bar{\mu}_r\bar{\mu}_s = \bar{\mu}_{r+s}$, we have

$$\sum_{s=0}^{l} \bar{\mu}_{m-s-1}\bar{\delta}^* \sum_{i=0}^{k-1} Sq^{s-i} Sm^i U = 0, \quad l = r+2k-2.$$

By (5) this equation may also be written as

$$\bar{\mu}_{m-1}\bar{\delta}^* U = \sum_{s=k}^{l} \bar{\mu}_{m-s-1}\delta^* V_s, \tag{6}$$

where

$$V_s = \sum_{i=k}^{s} Sq^{s-i} Sm^i U, \quad s = k, \cdots, l.$$

By (3), the right-hand side of (6) is 0. Since by hypothesis $2k + r \geqslant m$, we have for the term $\bar{\mu}_{m-k-1}\ \bar{\delta}^* V_k$ of the right-hand side of (6) with the greatest lower index of $\bar{\mu}$, $m-k-1 < k+r = \dim\ V_k$. It follows that (cf. [14] or [15])

$$V_s = 0, \quad l \geqslant s \geqslant k.$$

This system of equations is evidently equivalent to the following system:

$$Sm^s U = 0, \quad l \geqslant s \geqslant k.$$

From the above we get

Theorem 4 If $\rho_2\Phi^m(X) = 0$ for a finite polyhedron X, then for any integers k, r satisfying $2k + r \geqslant m$, we have

$$Sm^k H^r(X) = 0, \quad 2k + r \geqslant m. \tag{7}$$

From the main Theorem 1 we have then (cf. [4] Th. III. 25) the following

Corollary (Thom) If a finite polyhedron X is realizable in R^m, then (7) is true.

Theorem 5 If $\rho_2\Psi^{m-1}(X) = 0$ for a finite polyhedron X, then (7) is true.

Proof. Take a simplicial subdivision K of X and construct as in §3 the complexes $\tilde{K}_2^*$, $\tilde{K}_2^{(+)}$, $\tilde{K}_2^{(0)}$ and $K_2^* = \tilde{K}_2^*/t_2$, etc. Let $\tilde{j} : (\tilde{X}_2, d_2X) \subset (\tilde{X}_2, d_2X \cup |\tilde{K}_2^*|)$ be the inclusion map, which induces the homomorphisms $\tilde{j}^* : H^*(\tilde{X}_2, d_2X \cup |\tilde{K}_2^*|) \to H^*(\tilde{X}_2, d_2X)$, or what is the same,

$$\tilde{j}^* : \bar{H}^*(\tilde{Y}) \to \bar{H}^*(\tilde{X}_2^*),$$

where $\tilde{Y} = \tilde{X}_2 - d_2X \cup |\tilde{K}_2^*|$. Similarly we have the inclusion map $j : (X_2, d_2X) \subset (X_2, d_2X \cup |K_2^*|)$ and homomorphisms

$$j^* : \bar{H}^*(Y) \to \bar{H}^*(X_2^*),$$

where $Y = X_2 - d_2X \cup |K_2^*| = \tilde{Y}/t_2$. Now there are topological maps $\tilde{i} : \tilde{Y} \equiv |\tilde{K}_2^{(0)}| \times I$, $i : Y \equiv |K_2^{(0)}| \times I$, where I is an open segment, and $t_2\tilde{i} \equiv t_2/|\tilde{K}_2^{(0)}| \otimes 1$ (1 being the identity map), we have on setting $\bar{\mu}_i(\tilde{Y}, t_2) = \bar{\mu}_{i,Y}, \bar{\mu}_i(|\tilde{K}_2^{(0)}|, t_2) = \mu_i(|\tilde{K}_2^{(0)}|, t_2) = \mu_i^{(0)}$,

$$i^* : H^*(|K_2^{(0)}|) \otimes \bar{H}^*(I) \approx \bar{H}^*(Y). \tag{8}$$

Moreover, we have

$$\bar{\mu}_{i,Y} i^* = i^*(\mu_i^{(0)} \otimes 1), \tag{8$'$}$$

in which 1 is the identity homomorphism. Consider now the following commutative diagram:

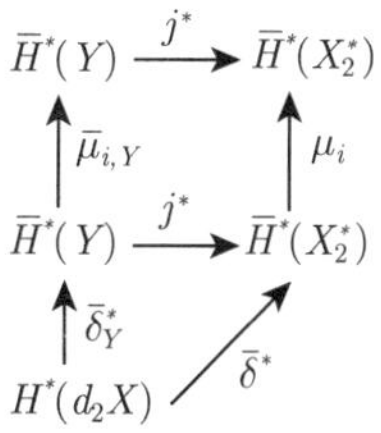
$\bar{H}^*(Y)$
j^*
$\bar{H}^*(X_2^*)$
$\bar{\mu}_{i,Y}$
μ_i
$\bar{H}^*(Y)$
j^*
$\bar{H}^*(X_2^*)$
δ_Y^*
δ^*
$H^*(d_2X)$

In the above diagram $\bar{\mu}_i = \bar{\mu}_i(\tilde{X}_2^*, t_2)$, $\bar{\delta}_Y^*$, $\tilde{\delta}^*$ are all coboundary homomorphisms

Now by hypothesis $\rho_2 \Psi^{m-1}(X) = 0$, or $\mu_{m-1}^{(0)} H^*(|K_2^{(0)}|) = 0$. Hence by (8) and (8)′ we have $\bar{\mu}_{m-1,Y} = 0$. In the above diagram setting $i = m-1$, we get

$$\bar{\mu}_{m-1} \bar{\delta}^* H^*(d_2 X) = 0,$$

i.e., (3). As in the proof of the above Theorem 4 we have then (7) This proves our theorem.

Corollary If a finite polyhedron X may he immersed in R^{m-1}, then (7) is true.

§5 Relations with Whitney classes in case of a differentiable manifold

In this section, the coefficient group will still he supposed to be I_2 and will be omitted throughout.

Let X be a closed differentiable manifold of dimension n, with class $\geqslant 3$. Take a Riemannian metric ds^2 of X and let $d\tilde{s}^2$ be the corresponding symmetric Riemannian metric of $X \times X$. Take $\varepsilon > 0$ sufficiently small such that through any point x in $X \times X$ having a distance $d(x) \leqslant \varepsilon$ from $d_2 X$, there is one and only one geodesic perpendicular to $d_2 X$ with length $= d(x)$. Let $\tilde{Y}$ be the space of all points x in $X \times X$ for which $0 < d(x) < \varepsilon/2$, and $\tilde{T}$ be the space of all points x with $d(x) = \varepsilon/2$. Put $\tilde{M} = \tilde{X}_2^* - \tilde{Y}$, then t_2 will operate in $\tilde{M}$, $\tilde{T}$ to form simple systems. Let $M = \tilde{M}/t_2$, $T = \tilde{T}/t_2$, then we see easily that, by taking sufficiently small simplicial subdivision K of X compatible with its given differential structure, $|\tilde{K}_2^{(0)}|$ will have the same P-homotopy type as $\tilde{T}$, and $|K_2^{(0)}|$ and T will have the same homotopy type. It follows that

$$A^i(\tilde{T}, t_2) = A^i(|\tilde{K}_2^{(0)}|, t_2) = \Psi^i(X).$$

On the other hand, we know that T is the bundle space of a bundle of (n-1)-dimensional projective spaces over X, satisfying the Leray-Hirsch condition. Moreover, denoting by W^i the mod 2 reduced Stiefel-Whitney classes of X, W^i would be completely determined by the following equation (cf. e.g. [16]):

$$\sum_{i=0}^{n} \rho_2 \Psi^i(X) \smile g_0^* W^{n-i} = 0, \tag{1}$$

in which $g_0\colon T \to X$ is the natural projection. From (1) we get

$$\rho_2\Psi^{n+i}(X) = \sum_{j=1}^{n} \rho_2\Psi^{n-j}(X) \smile g_0^* U_{j,i}, \tag{2}$$

where $U_{j,i} \in H^{i+j}(X)$ and satisfies

$$\begin{cases} U_{j,0} = W^j, \quad j \geqslant 1; \\ U_{j,i+1} = U_{j+1,i} + U_{1i} \smile W^j, \quad j \geqslant 1. \end{cases} \tag{3}$$

Therefrom we get

$$\sum_{\substack{i+j=k \\ i,j\geqslant 0}} U_{1,i} \smile W^j = W^{k+1}.$$

Comparing with the definition of dual Whitney classes $\bar{W}^i$, we have therefore

$$U_{1,i} = \bar{W}^{i+1}. \tag{4}$$

From (2) we know that if $\rho_2\Psi^{m-1} = 0$, then $U_{j,i} = 0$, $i \geqslant m-n-1$, $n \geqslant j \geqslant 1$ and hence by (4), $\bar{W}^k = 0$, $k \geqslant m-n$. Conversely, if $\bar{W}^k = 0$ $k \geqslant m-n$, then by (2), (3) and (4) we have also $\rho_2\Psi^{m-1} = 0$. Consequently we have

Theorem 6 If $\rho_2\Psi^{m-1} = 0$ for an n-dimensional closed differentiable manifold X, then

$$\bar{W}^k = 0, \quad k \geqslant m-n.$$

The converse is also true.

By the principal Theorem 2 we have the following

Corollary If an n-dimensional closed differentiable manifold X is immersible in R^{m-1}, then

$$\bar{W}^k = 0, \quad k \geqslant m-n. \tag{5}$$

Whitney has proved that any n-dimensional closed differentiable manifold may be imbedded in R^{2n} ([17]) and immersed in R^{2n-1} ([9]). The above theorem shows that the immersion theorem of Whitney cannot be further improved in general. For example, the example given in [3] §25 is a 4-dimensional manifold with $\bar{W}^3 \neq 0$, and therefore it is non-immersible in R^6 by theorem 6. This answers also a problem of Whitney (cf. the introduction of [9]).

Theorem 7 If $\rho_2\Phi^m = 0$ for a closed differentiable manifold X. then $\rho_2\Psi^{m-1} = 0$.

Proof. $\tilde{M}=\tilde{X}_2^*-\tilde{Y}$ is a bounded $2n$-dimensional manifold with $\tilde{T}$ as boundary, similarly M is a bounded manifold with T as boundary. Take now two models $\tilde{M}_1$ and $\tilde{M}_2$ of $\tilde{M}$, each homomorphic to $\tilde{M}$ under h_j, and with boundaries $\tilde{T}_1$, $\tilde{T}_2$ respectively. Identify now $\tilde{T}_1$ and $\tilde{T}_2$ under $h_2^{-1}h_1$ with $\tilde{T}'$, then from $\tilde{M}_i$ we get a $2n$-dimensional closed differentiable manifold $\tilde{M}'$. In $\tilde{M}'$ we define a periodic transformation t' by

$$t'(\tilde{x}_1)=h_2^{-1}t_2h_1(\tilde{x}_1),\quad t'(\tilde{x}_2)=h_1^{-1}t_2h_2(\tilde{x}_2),\quad \tilde{x}_i\in\tilde{M}_i. \tag{6}$$

Let $M'=\tilde{M}'/t'$, then M' may be regarded as the $2n$-dimensional closed manifold obtained from $\tilde{M}$ by identifying $\tilde{T}$ under t_2 with $T=\tilde{T}/t=\tilde{T}'/t'$. Define $\tilde{h}:\tilde{M}'\to\tilde{M}$ by $\tilde{h}(\tilde{x}_i)=h_i(\tilde{x}_i)$, $\tilde{x}_i\in\tilde{M}_i$, then $\tilde{h}t'=t_2\tilde{h}$. Hence, $\tilde{h}$ induces a map $h:M'\to M$ and we have

$$\rho_2A^m(\tilde{M}',t')=\rho_2h^*A^m(\tilde{M},t_2)=h^*\rho_2\Phi^m=0,\quad \mu'_m=\mu_m(\tilde{M}',t'). \tag{7}$$

With respect to the simple systems ($\tilde{M}'$, t') and ($\tilde{T}'$, t'), we may also define the Smith homomorphisms between homology groups (coefficient groups being I_2)①:

$$\nu_j'=\nu_j(\tilde{M}',t'):H_r(M')\to H_{r-i}(M'),$$

and

$$\nu_i''=\nu_i(\tilde{T}',t'):H_r(T)\to H_{r-i}(T).$$

In [13] we have proved that (as the coefficient groups are I_2, it is not necessary to suppose the manifolds to be orientable)

$$\mathscr{D}'\nu_i'=\mu_i'\mathscr{D}',\quad \mathscr{D}''\nu_i''=\mu_i''\mathscr{D}'', \tag{8}$$

where

$$\mu_i''=\mu_i(\tilde{T}',t')$$

and

$$\mathscr{D}':H_r(M')\approx H^{2n-r}(M'),$$
$$\mathscr{D}'':H_r(T)\approx H^{2n-1-r}(T)$$

are the dual isomorphisms in the closed manifolds M' and T respectively.

Construct now simplicial subdivisions of T and M', denoted still by T and M', such that T is a subcomplex of M'. Construct also simplicial subdivisions of $\tilde{T}'$

① In [13], μ_i, ν_i are denoted by μ_i^* and μ_i respectively.

and $\tilde{M}'$, still denoted by $\tilde{T}'$ and $\tilde{M}'$, which are covering complexes of T and M' respectively.

Let the fundamental cycles of $\tilde{M}'$, etc., be respectively $\tilde{z}' \in \tilde{Z}' \in H_{2n}(\tilde{M}'), \tilde{z}'' \in \tilde{Z}'' \in H_{2n-1}(\tilde{T}')$, $z' \in Z' \in H_{2n}(M')$, and $z'' \in Z'' \in H_{2n-1}(T)$. We may write $\tilde{z}' = \tilde{z}'_1 + \tilde{z}'_2$, where $\tilde{z}'_i$ are chains in $\tilde{M}_i$. Then $t'\ \tilde{z}'_1 = \tilde{z}'_2$, $\partial \tilde{z}'_1 = \partial \tilde{z}'_2 = \tilde{z}''$, $\pi \tilde{z}'_1 = \pi \tilde{z}'_2 = z'$, where $\pi : \tilde{M}' \to M'$ is the natural projection. We may also write $\tilde{z}'' = (1+t')\tilde{c}''$, where $\tilde{c}''$ is a chain in $\tilde{T}'$, and $\pi \tilde{c}'' = z''$. Whence we know that if $\tilde{j} : \tilde{T}' \subset \tilde{M}'$ and $j : T \subset M'$ are inclusion maps, then

$$j_* \nu''{}_i = \nu'{}_i j_*,$$
$$j_* Z'' = \nu'{}_1 Z'.$$

Consequently by (7), (8) we have (1 denotes the mod 2 unit class of M')

$$j_* \nu''_{m-1} Z'' = \nu'_m Z' = \mathscr{D}'^{-1} \mu'_m 1 = 0,$$

or

$$\rho_2 j_* \mathscr{D}''^{-1} \Psi^{m-1} = 0.$$

It follows that in order to prove our theorem, it is sufficient to prove that

$$j_* : H(T) \subset H(M'). \tag{9}$$

For this purpose let us first define a continuous map

$$\tilde{g} : \tilde{M} \to X \times X$$

as follows. For any point $\tilde{x} \in \tilde{M}$ having a distance $d(\tilde{x}) = \varepsilon$ to d_2X, let $l(\tilde{x})$ be the geodesic through $\tilde{x}$ and perpendicular to d_2X, and let $\tilde{x}' \in \tilde{T}$ be the mid-point of this geodesic. Let us map now any point $\tilde{y}$ on the geodesic arc between $\tilde{x}$, $\tilde{x}'$ to that point $\tilde{g}(\tilde{y})$ on $l(\tilde{x})$ such that the arc length from $\tilde{x}$ to $\tilde{g}\,(\tilde{y})$ is two-fold as that from $\tilde{x}$ to $\tilde{y}$. On the other hand, if $\tilde{x}$ has a distance $> \varepsilon$ to d_2X, then we put $\tilde{g}\,(\tilde{x}) = \tilde{x}$. In this manner we get the above-mentioned map $\tilde{g}$. As the chosen Riemannian metric in $X \times X$ is symmetric, we have for any $\tilde{x} \in \tilde{T}$, $\tilde{g}(\tilde{x}) = \tilde{g}\, t_2(\tilde{x})$, It follows that $\tilde{g}$ will induce a continuous map

$$g : M' \to X \times X.$$

If we regard T as the bundle space of a bundle of $(n-1)$-dimensional projective spaces over d_2X or X, $g/T = g_0 : T \to d_2X$ is no other than the bundle projection.

It is known that $H^*\ (T)$ is generated by all classes Ψ^i and g_0^*U, where $U \in H^*(X)$. As

$$j^*A^i(\tilde{M}', t') = A^i(\tilde{T}', t') = \Psi^i,$$

and

$$j^*g^*(1 \otimes U) = g_0^*d_2^*(1 \otimes U) = g_0^*U,$$

it follows that

$$j^*H^*(M') = H^*(T).$$

Dually, the last equation is equivalent to (9). Our theorem is thus completely proved.

By Theorem 6 and the main Theorem 2, we get the following corollaries:

Corollary 1 If $\rho_2\Phi^m = 0$ for an n-dimensional closed differentiable manifold X, then (5) is true.

Corollary 2 If an n-dimensional closed differentiable manifold is realizable in R^m, then (5) is true.

Remark 1. Whitney has proved that (5) is a necessary condition for an n-dimensional closed differentiable manifold to be "differentiably" realizable in R^m. The above Corollary 2 shows that this condition is not only necessary for "differentiable" realizability, but also for "topological" realizability.

Remark 2. By Theorem 3 of §3, $K = K_{m+1,n}$ has the following property. $\rho_2\Psi^{m-1} \neq 0$ and $\Phi^m = 0$. It follows that Theorem 7 is not true for general finite polyhedrons. In reality, the property of X being a manifold is essentially used in the proof of Theorem 7.

§6 The classes Φ^m of projective spaces

In general, the calculation of the classes Φ^m is quite difficult. As an example, we shall now determine $\rho_2\Phi^m(P^n) = \rho_2\Phi^m$ of an n-dimensional projective space P^n as follow s, n being supposed to be $\geqslant 2$.

Consider P^n as the space of all lines in R^{n+1} through its origin, then $\tilde{P}_2^{n*}$ is the space of all ordered pairs $(l_1,\ l_2)$, in which $l_1,\ l_2$ are any two *different* lines in R^{n+1} through its origin, and $t_2 = t : \tilde{P}_2^{n*} \to \tilde{P}_2^{n*}$is defined by $t(l_1, l_2) = (l_2, l_1)$. For any point $(l_1, l_2) \in \tilde{P}_2^{n*}$, let $[l_1, l_2]$ be the plane determined by $l_1,\ l_2$. In this plane we may take two lines $l_1',\ l_2'$ through the origin, perpendicular to each other, and such that the acute angle θ_1 between l_1' and l_1 is equal to the acute angle θ_2 between l_2' and l_2. Between l_i and $l_i'(i = 1, 2)$ we may construct a line $l_i^{(\tau)}$ which divides θ_i into two angles with ratio $\tau : 1-\tau$, where $0 \leqslant \tau \leqslant 1$. Then $h_\tau : (l_1,\ l_2) \to (l_1^{(\tau)}, l_2^{(\tau)})$ is a P-deformation of $\tilde{P}_2^{n*}$. Denote $h_1(\tilde{P}_n^{n*})$ by $\tilde{P}_{n,2}$, then $\tilde{P}_{n,2}$ is the space of all ordered pairs $(l_1,\ l_2)$, where $l_1,\ l_2$ are any two mutually perpendicular lines in R^{n+1} through its origin and t will still operate in $\tilde{P}_{n,2}$. Denote the modular space of $(\tilde{P}_{n,2},\ t)$ by $P_{n,2} = \tilde{P}_{n,2}/t$, and the projection $\tilde{P}_{n,2} \to P_{n,2}$ by λ. Then we have

$$A^i(\tilde{P}_{n,2}, t) = A^i(\tilde{P}_2^{n*}, t) = \varPhi^i, \tag{1}$$

and

$$\lambda^*\varPhi^i = 0, \quad i > 0. \tag{2}$$

Let $R_{n-1,2}$ be the Grassmann manifold defined in R^{n+1}, and $\tilde{\pi} : \tilde{P}_{n,2} \to R_{n-1,2}$ be the natural map defined by $\tilde{\pi}(l_1, l_2) = [l_1, l_2]$. We may also define $\pi : P_{n,2} \to R_{n-1,2}$ by $\pi\lambda = \tilde{\pi}$. Evidently $P_{n,2}$ is the bundle space of an S^1-bundle over $R_{n-1,2}$ with π as the projection map of the bundle. Let S_0^1 be a fixed fibre of this bundle, and $j : S_0^1 \to P_{n,2}$ the inclusion map, then $j^*\varPhi^1$ is a generator of $H^1(S_0^1)$①. It follows that this bundle satisfies the condition of Leray-Hirsch and therefore $H^*(P_{n,2})$ is additively isomorphic to $H^*(R_{n-1,2}) \otimes H^*(S^1)$, and also

$$\pi^* : H^*(R_{n-1,2}) \subset H^*(P_{n,2}), \tag{3}$$

$$(\varPhi^1)^2 + \pi^*A^1 \smile \varPhi^1 + \pi^*A^2 = 0, \quad A^i \in H^i(R_{n-1,2}), \quad i = 1, 2. \tag{4}$$

By (2) and (3) we get further

$$\tilde{\pi}^*A^2 = \lambda^*\pi^*A^2 = 0. \tag{5}$$

Since $\tilde{P}_{n,2}$ is the bundle space of the projective-line bundle associated with the canonical S^1-bundle over $R_{n-1,2}$, this bundle has also the Leray-Hirsch property and the bundle projection $\tilde{\pi}$ satisfies

$$\tilde{\pi}^* : H^*(R_{n-1,2}) \subset H^*(\tilde{P}_{n,2}).$$

By (5) we get $A^2 = 0$, hence (4) becomes

$$(\varPhi^1)^2 + \pi^*A^1 \smile \varPhi^1 = 0.$$

Since $n \geqslant 2$ by hypothesis, we have $(\varPhi^1)^2 = \rho_2\varPhi^2 \neq 0$ and from the above equation and (3) we know that $A^1 \neq 0$. As $H^1(R_{n-1,2})$ is generated by W-class W^1 of $R_{n-1,2}$, we have necessarily $A^1 = W^1$, and (4) becomes

$$(\varPhi^1)^2 + \pi^*W^1 \smile \varPhi^1 = 0,$$

① All coefficient groups will be supposed to be I_2, so that it will be omitted throughout.

whence

$$(\Phi^1)^r = \pi^*(W^1)^{r-1} \smile \Phi^1.$$

From the known cohomology ring structure of $R_{n-1,2}$ (cf. [18] or [19]), we know that for $n = 2^a + s$, $s < 2^a$, we have

$$(W^1)^{r-1} \begin{cases} \neq 0, & r-1 \leqslant 2n-2s-2, \\ = 0, & r-1 > 2n-2s-2. \end{cases}$$

It follows that

$$\rho_2 \Phi^r = (\Phi^1)^r \begin{cases} \neq 0, & r \leqslant 2n-2s-1, \\ = 0, & r > 2n-2s-1. \end{cases}$$

Remark 1. From the above equation we know that $\rho_2 \Phi^{2n-2s} = 0$, but we do not know whether Φ^{2n-2s} is also 0 or not.

Remark 2. By the principal Theorem 1 we know that for $n = 2^a + s, s < 2^a$,

$$P^n \not\subset R^{2n-1-2s}.$$

This may also be deduced from Whitney's condition. For, let C^i be the generator of $H^i(P^n)$, then the Whitney polynomial of P^n is given by

$$W(P^n) = (1 + C^1 t)^{n+1},$$
$$\bar{W}(P^n) = (1 + C^1 t)^{-(n+1)} = \sum \binom{n+i}{n} C^i t^i.$$

By elementary theory of numbers we know that for

$$n = 2^{a_1} + \cdots + 2^{a_\alpha} \quad (a_1 > \cdots > a_\alpha),$$
$$m = 2^{b_1} + \cdots + 2^{b_\beta} \quad (b_1 > \cdots > b_\beta),$$

the necessary and sufficient condition for $\binom{n}{m} \not\equiv 0 \bmod 2$ is that each b_i should be equal to a certain a_i. Hence, for $n = 2^a + s$, $s < 2^a$, we have ($i \leqslant n$)

$$\binom{n+i}{n} \begin{cases} \neq 0 \mod 2, & i = n-1-2s, \\ = 0 \mod 2, & i > n-1-2s. \end{cases}$$

Hence, $\bar{W}^{n-1-2s}(P^n) \neq 0$ and we have $P^n \not\subset R^{2n-1-2s}$.

§7 Isotopy and iso-position

Let f_0 and f_1 be two realizations (resp. immersions) of a Hausdorff space X in R^m. Let I be the segment $[0, 1]$. If there exists a continuous map $F : X \times I \to R^m$ such that $F|X \times (0) \equiv f_0$, $F|X \times (1) \equiv f_1$, and for each $t \in I$, $F|X \times (t) \equiv f_t$ is a realization (resp. immersion) of X in R^m, then we will say that the realizations (resp. immersions) f_0 and f_1 are *isotopic.*

Let f_0 and f_1 be two realizations of X in R^m. If there exists a topological transformation h of R^m onto itself such that $hf_0 \equiv f_1$, then the two realizations will be said to be in *iso-position.* We will say also that f_0, f_1 are in *orientation-preserving* or *orientation-reversing iso-position* according as h is orientation-preserving or orientation reversing.

Suppose now given $f : X \subset R^m$ or $f : X \propto\!\!\subset R^m$. Let S^{m-1} be the unit sphere of R^m. Choose a fixed orientation of R^m and let S be the generator of $H^{m-1}(S^{m-1})$ corresponding to this orientation. In the case of $f : X \subset R^m$, let us draw for each point $(x_1, x_2) \in \tilde{X}_2^*$ a half-line through the origin O and parallel to the oriented line from $f(x_1)$ to $f(x_2)$, which meets S^{m-1} in a point $\tilde{f}(x_1, x_2)$. The map $\tilde{f} : \tilde{X}_2^* \to S^{m-1}$ will be said to be associated with f, and $\tilde{f}^*S$ will be denoted by $\theta_f \in H^{m-1}(\tilde{X}_2^*)$. If the orientation of R^m is changed, then θ_f also changes by a sign. Similarly, let $f : X \propto\!\!\subset R^m$ and X be a finite polyhedron with a sufficiently small simplicial subdivision K, then we may define in a similar manner a continuous map $\tilde{f} : |\tilde{K}_2^{(0)}| \to S^{m-1}$ and the class $\tilde{f}^*S$ will be denoted by $\theta'_f \in H^{m-1}(|\tilde{K}_2^{(0)}|)$. In [10] we have proved that $H^{m-1}(|\tilde{K}_2^{(0)}|)$ is independent of the chosen subdivision K of X and by the method of [10] we may see that θ'_f is also independent of the subdivision K and depends only on X, on the immersion f and on the chosen orientation of R^m.

Theorem 8 With respect to a fixed orientation of R^m, the class θ_f(resp. θ'_f) is an isotopy invariant of the realization $f : X \to R^m$ (resp. the immersion $f : X \propto\!\!\subset R^m$, in that case X should be a finite polyhedron).

Proof. Let $f_0, f_1 : X \subset R^m$ be isotopic so that there exists a map $F : X \times I \to R^m$ as in the beginning of this section. For any $t \in I$, let $f_t|X \equiv F \quad X \times (t)$, then map $\tilde{f}_t : \tilde{X}_2^* \to S^{n-1}$ associated with f_t gives rise to a homotopy between $\tilde{f}_0$, $\tilde{f}_1$, so that $\tilde{f}_0^*\, S = \tilde{f}_1^*\, S$, i.e., $\theta_{f_0} = \theta_{f'_1}$, or θ_f is an isotopy invariant of f. In the case that X is a finite polyhedron and $f : X \propto\!\!\subset R^m$ is an immersion, that θ'_f is an isotopy invariant may be proved similarly.

Theorem 9 With respect to a fixed orientation of R^m, $\pm\ \theta_f$ (resp. $+\theta_f$) is an iso-position invariant (resp. orientation-preserving iso-position invariant) of the realization $f : X \subset R^m$.

Proof. Let $\tilde{R}^*$ be the space of all ordered pairs $(x_1,\ x_2)$, where $x_i \in R^m$ and $x_1 \neq x_2$. Let $\tilde{R}'^*$ be the subspace of $\tilde{R}^*$ consisting of all points $(x_1,\ x_2)$ with length of $x_1x_2 = 1$. For any point $(x_1, x_2) \in \tilde{R}^*$, let y be the mid-point of x_1x_2, and $y_1,\ y_2$ be points on half-lines yx_1 and yx_2 respectively such that yy_1 and yy_2 are both of length $1/2$. Denote by $y_i(t)$ that point on x_iy_i such that $x_iy_i(t) : y_i(t)y_i = t : 1-t$, then $d_t : (x_1,\ x_2) \to (y_1(t),\ y_2(t))$ is a deformation-retraction of $\tilde{R}^*$ onto $\tilde{R}'^*$. Since $\tilde{R}'^*$ is the bundle space of an $(m-1)$-sphere bundle over R^m as base space, we have by a theorem of Feldban $\tilde{R}'^* \equiv R^m \times S^{m-1}$, and we have $H^*(\tilde{R}^*) \approx H^*(\tilde{R}'^*) \approx H^*(S^{m-1})$, where H denotes a singular homology system. For any point $(x_1, x_2) \in \tilde{R}^*$, let us draw through O a half-line parallel to the directed line from x_1 to x_2, which intersects the unit sphere S^{m-1} in a point $g(x_1, x_2)$. Then the continuous map $g : \tilde{R}^* \to S^{m-1}$ induces $g^* : H^*(S^{m-1}) \approx H^*(\tilde{R}^*)$ which is the same as the isomorphism described above. Fixing now an orientation of S^{m-1} and letting S be the generator of $H^{m-1}(S^{m-1})$ corresponding to this orientation, then $g^*S \in H^{m-1}(\tilde{R}^*)$ and will be denoted by $\varGamma$.

If h is a topological transformation of R^m onto itself, h will induce a topological map $\tilde{h}$ of $\tilde{R}^*$ onto $\tilde{R}^*$. It is easy to see that for fixed orientation of R^m, we have $\tilde{h}^*\varGamma = +\varGamma$ or $-\ \varGamma$, accordjng as h is orientation-preserving or not.

Suppose now $f,\ f'$ be two iso-positional realizations of the Hausdorff space X into R^m, so that $h : R^m \equiv R^m$ exists with $hf = f'$. Let $\tilde{f},\ \tilde{f}'$ be maps of $\tilde{X}_2^*$ into $\tilde{R}^*$ induced by f and f' respectively: $\tilde{f}(x_1,x_2) = (f(x_1),\ f(x_2))$, $\tilde{f}'(x_1,x_2) = (f'(x_1),\ f'(x_2))$, $(x_1, x_2) \in \tilde{X}_2^*$, then we have evidently $\theta_f = \tilde{f}^*\varGamma$, $\theta_{f'} = \tilde{f}'^*\varGamma$. We have also $\tilde{h}\tilde{f} = \tilde{f}'$. Hence, $\theta_f = +\theta_{f'}$ or $-\theta_{f'}$, according as h is orientation-preserving or not. From this we get the theorem.

Example 1 Given a circle C. Take three points $a_1,\ a_2,\ a_3$ on C, then $H_1(\tilde{C}_2^*)$ is generated by a single generator Z which has the following singular chain as a representative; namely,

$$\begin{aligned} z =& a_1 \times (a_2a_3) + a_2 \times (a_3a_1) + a_3 \times (a_1a_2) \\ &+ (a_2a_3) \times a_1 + (a_3a_1) \times a_2 + (a_1a_2) \times a_3. \end{aligned}$$

Consider now a circle D in the plane R^2. Orient R^2 arbitrarily and let D be oriented accordingly. This oriented D will be denoted by $+D$. Denote the oriented circle C with $a_1,\ a_2,\ a_3$ in positive order by $+C$. Define now two realizations $f,\ f' : X \subset R^2$

by $f(+C) = +D$, $f'(+C) = -D$. Then with respect to the above orientation of R^2, we have $\theta_f(Z) = +1$, $\theta_{f'}(Z) = -1$. Hence, by Theorem 9, f, f' cannot be in orientation-preserving iso-position.

Example 2 Let X be the space consisting of a circle C and an isolated point P. Orient C and define Z as in Ex. 1. Then $H_1(\tilde{X}_2^*)$ has three generators of which one is Z and a second one is Z' with a representative cycle $z' = P \times (+C)$. In the oriented plane R^2 take an oriented circle $+D$ as in Ex. 1. Take also a point Q inside D and a point Q' outside D. Define realizations f_1, f_1', f_2, f_2': $X \subset R^2$ by

$$f_1(+C) = f_1'(+C) = +D, \quad f_2(+C) = f_2'(+C) = -D,$$
$$f_1(P) = f_2(P) = Q, \quad f_1'(P) = f_2'(P) = Q'.$$

Then with respect to the above orientation of R^2, we have

$$\theta_{f_1}(Z') = +1, \quad \theta_{f_2}(Z') = -1, \quad \theta_{f_1'}(Z') = \theta_{f_2'}(Z') = 0,$$
$$\theta_{f_1}(Z) = \theta_{f_1'}(Z) = +1, \quad \theta_{f_2}(Z) = \theta_{f_2'}(Z) = -1.$$

From these equations we know that f_1 and f_2 (also f_1' and f_2'), though in iso-position, are not in orientation-preserving iso-position, while f_1 or f_2 is not in iso-position with either f_1' or f_2'.

Example 3 Let K be a complex consisting of three segments (a_0a_1), (a_0a_2), (a_0a_3) joined together at a_0, and $X = |K|$. In $\tilde{X}_2^*$ there is a 1-dimensional cycle non-homologous to 0, namely,

$$\begin{aligned} z =& a_1 \times (a_2a_0) + a_1 \times (a_0a_3) + a_2 \times (a_3a_0) + a_2 \times (a_0a_1) \\ &+ a_3 \times (a_1a_0) + a_3 \times (a_0a_2) + (a_2a_0) \times a_1 + (a_0a_3) \times a_1 \\ &+ (a_3a_0) \times a_2 + (a_0a_1) \times a_2 + (a_1a_0) \times a_3 + (a_0a_2) \times a_3. \end{aligned} \tag{1}$$

The cycle z represents a generator Z of $H_1(\tilde{X}_2^*)$.

In the plane R^2 take four points b_0, b_1, b_2, b_3 and three simple arcs $\beta_1 = \widehat{b_0b_1}$, $\beta_2 = \widehat{b_0b_2}$, $\beta_3 = \widehat{b_0b_3}$ such that β_1, β_2, β_3 do not meet one another except at b_0. Such a figure is called a triod and will be denoted by $[\beta_1, \beta_2, \beta_3]$, ($X$ is also a triod). Define now a realization f: $X \subset R^2$ by $f(a_0) = b_0$, $f(a_0\,a_i) = \beta_i$, $i = 1, 2, 3$. Fix an orientation of R^2, and define θ_f with respect to this orientation. Let us rotate β_1 about b_0 first to β_2, and then to β_3. Then, if this rotation is concordant with the chosen orientation of R^2, we should have $\theta_f(Z) = +1$. Otherwise we have $\theta_f(Z) = -1$.

By the three examples given above we may prove the following

Theorem 10 Let the Peano continuum M be realizable in the plane. Then the necessary and sufficient condition for any two realizations f, f' of M in R^2 to be in iso-position is that $\theta_f = \pm\theta_{f'}$ (with respect to the same fixed orientation of R^2).

Proof. The necessity has already been given in Theorem 9. Let us prove the sufficiency part as follows.

Let $A = [\alpha_1, \alpha_2, \alpha_3]$ and $B = [\beta_1, \beta_2, \beta_3]$ be any two triods contained in M. In $H_1(\tilde{A}_2^*)$ and $H_1(\tilde{B}_2^*)$ and consequently in $H_1(\tilde{M}_2^*)$, there are elements α, β containing cycles corresponding to (1) as representative cycles. Consider the values a, b of θ_f on α, β and the values a', b' of $\theta_{f'}$ on α, β, where a, b, a', b' are all $= \pm 1$. By Ex. 3 we see that if $f'f^{-1}$ preserves the relative sense (cf. [20]) of $f(\alpha)$, $f(\beta)$, then either $a' = a$, $b' = b$ or $a' = -a$, $b' = -b$. In the contrary case we should have either $a' = +a$, $b' = -b$ or $a' = -a$, $b' = +b$. It follows that when $\theta_f = \pm\theta_{f'}$, $f'f^{-1}$ preserves the relative sense of any pair of triods in $f(M)$.

By Theorems 10 and 11 of [20], we see that if k is the boundary of a complementary domain of $f(M)$, then $f'f^{-1}(k)$ is also the boundary of a complementary domain of $f'(M)$, and vise versa. Suppose that k is the boundary of the infinite complementary domain of $f(M)$, while $f'f^{-1}(k)$ is the boundary of a finite complementary domain of $f'(M)$. Then in M there must exist a simple closed curve C and also a point P not on C, such that $f(C) \subset k$, $f(P)$ lies outside of $f(C)$ and $f'(P)$ lies inside $f'(C)$. By Ex. 2, for $X = C + P$, we should have $\theta_f(X) \neq \pm\theta_{f'}(X)$, and *a fortiori* $\theta_f \neq \pm\theta_{f'}$, with respect to the same fixed orientation of R^2. As this is in contradiction with the hypothesis, we see that $f'f^{-1}$ preserves also the boundaries of finite complementary domains of $f(M)$. By Theorem 2 of [20] , we may then extend the topological map $f'f^{-1}$ from $f(M)$ to $f'(M)$ to the whole plane. Denote the extended topological map by h, then $hf = f'$ and f, f' are in iso-position, what we require to prove.

Moreover, by Ex. 1, 2, we see also that with respect to the same fixed orientation of the plane, $\theta_f = +\theta_{f'}$ is the necessary and sufficient condition for f, f' to be in orientation-preserving iso-position and $\theta_f = -\theta_{f'}$ is the necessary and sufficient condition for f, f' to be in orientation-reversing iso-position.

The above theorem may be regarded as a quantitative reformulation of a known theorem of Adkisson-McLane (Theorem 2 of [20]). In the same manner, another theorem of Adkisson-McLane about realizations on a sphere (Theorem 1 of [20]) and a well-known theorem of Kuratowski about realization in the plane ([21]) the original statements of which are qualitative in character, may be given quantitative reformulations as follows:

Theorem 10′ (Adkisson-McLane) Suppose a Peano-continuum M is realizable in the sphere S^2 and let f, f', be two such realizations. For any $p \notin f(M)$, let us consider f as a realization f_p of M in $R_p^2 = S^2 - (p)$. Take a fixed orientation of S^2 and orient R_p^2 concordantly with corresponding class θ_{f_p}. Similarly we define classes $\{\theta_{f'_{p'}}\}$ for $p' \notin f'(M)$ (S^2 to be oriented as before). Then a necessary and sufficient condition for f and f' to be in iso-position is that there exists a 1-1-correspondence between the sets $\{\theta_{f_p}\}$ and $\{\theta_{f'_{p'}}\}$, or more simply, there exist a point $p \notin f(M)$ and a point $p' \notin f'(M)$ such that $\theta_{f_p} = \theta_{f'_{p'}}$.

Theorem 11 (Kuratowski) For a Peano continuum M containing only a finite number of simple closed curves to be realizable in the plane, it is necessary and sufficient that $\Phi^2 = 0$.

References

[1] van Kampen. *Abh. Math. Sem.* Hamburg, 1932, 9: 72-78, 152-153.

[2] Wu Wen-tsün. *Acta Math. Sinica.*, 1955, 5: 505-552.

[3] Whitney H. *Michigan Lectures in Topology,* 1941: 101-141.

[4] Thom R. *Ann. Ec. Norm. Sup.*, 1952, 69: 110-182.

[5] Wu Wen-tsün. Sur les puissances de Steenrod. *Colloque de Topologie de Strasbourg,* 1951.

[6] Wu Wen-tsün. Relations between Smith operations and Steenrod powers. *Acta Math. Sinica.,* 1957, 7.

[7] Flores A I. *Erg. math. Kolloqu.,* 1935, 6: 4-7.

[8] Smith P A. Fixed points of periodic transformations. *Algebraic Topology* by S. Lefschetz, 1941: 350-393.

[9] Whitney H. *Annals of Math.*, 1944, 45: 247-293.

[10] Wu Wen-tsün. *Acta Math. Sinica.,* 1953, 3: 261-290.

[11] McLane S. and Adkisson V W. Extensions of homomorphisms of the sphere. *Michigan Lectures in Topology,* 1941: 223-236.

[12] Cartan H. *Algebraic Topology.* Harvard University, 1949.

[13] Wu Wen-tsün. *Acta Math. Sinica.,* 1955, 5: 37-63.

[14] Thom R. Une thérie intrinsêque de puissance de Steenrod. *Colloque de Topologie de Strasbourg,* 1951.

[15] Bott R. *Annals of Math.,* 1953, 59: 579-590.

[16] Wu Wen-tsün. *Acta Scientia Sinica.*, 1954, 3: 353-367.

[17] Whitney H. *Annals of Math.*, 1944, 45: 220-246.

[18] Wu Wen-tsün. Sur les classes caractéristiques des espaces fibrés. *Act. Sci. et Ind.* (1183) Paris: Hermann, 1952.

[19] Chern S. *Annals of Math.*, 1948, 49: 362-372.

[20] Adkisson V W & McLane S. *Duke Math. J.*, 1940, 6: 216-228.

[21] Kuratowski C. *Fund. Math.*, 1930, 15: 271-283.

23. On the $\Phi_{(p)}$-Classes of a Topological Space*

§ 1. Let $\bar{K}$ be a simplicial complex and t a simplicial map of $\bar{K}$ onto itself with prime period p. The system ($\bar{K}$, t) will be said to be *simple of period* p if for any $\tilde{\sigma} \in \bar{K}$, $t^i\tilde{\sigma} \neq \tilde{\sigma}$ for $0 < i < p$. For such a simple system ($\bar{K}$, t) of prime period p, let $K = \bar{K}/t$ be the *modular complex* obtained from $\bar{K}$ by identifying all simplexes $\tilde{\sigma}$, $t\tilde{\sigma}$, $\cdots$, $t^{p-1}\tilde{\sigma} \in \bar{K}$ to a single one and $\pi^{\#} : C^r(K, G) \to C^r(\bar{K}, G)$ the cochain map induced by the natural projection $\pi : \bar{K} \to K$. Define homomorphisms $\bar{\pi}^{\#} : C^r(\bar{K}, G) \to C^r(K, G)$ by $\bar{\pi}^{\#}\tilde{u}(\sigma) = \sum_{i=0}^{p-1} \tilde{u}(t^i\tilde{\sigma})$, where $\tilde{u} \in C^r(\bar{K}, G)$, σ is an r-cell of K, and $\tilde{\sigma}$ any simplex of $\bar{K}$ with $\pi(\tilde{\sigma}) = \sigma$. Then $\bar{\pi}^{\#}$ are also cochain maps and are all onto. For the 0-cocycle $u \in C^0(K)$ taking value 1 on all vertices of K, let us take a cochain $\tilde{u}_0 \in C^0(\bar{K})$ with $\bar{\pi}^{\#}\tilde{u}_0 = u$. Then we may find successively a system of cochains $\tilde{u}_i \in C^i(\bar{K})$ such that $\delta\, \tilde{u}_i = {s_{i+1}}^{\#}\, \tilde{u}_{i+1}$, $i \geqslant 0$, where $s_i^{\#}$ are the cochain maps induced by $s_i = 1 - t$ or $1 + t + \cdots + t^{p-1}$ according as i is odd or even. We may see that $u_i = \rho_{(i)}\bar{\pi}^{\#}\, \tilde{u}_i$ is a cocycle in $C^i(K, I_{(i)})$ whose class $A^i(\bar{K}, t) \in H^i(K, I_{(i)})$ is independent of the choice of $\tilde{u}_j$, in which $I_{(i)}$ is the group of integers I for i even, and is the group I_p of integers mod p for i odd; and $\rho_{(i)}$ are the cochain maps induced by the natural homomorphisms $I \to I_{(i)}$. These classes $A^i(\bar{K}, t)$ will be called the *Smith classes* of the system ($\bar{K}$, t).

Given two simple systems ($\bar{K}$, t) and ($\bar{K}'$, t') of same prime period p, a simplicial map $\tilde{f} : \bar{K} \to \bar{K}'$ will be called a *P-map* of the systems if $\tilde{f}t = t'\tilde{f}$. Such a P-map will induce then a cell map $f : K \to K'$ of their modular complexes such that $\pi'\tilde{f} = f\pi$, where π, π' are the corresponding projections. We have then

$$f^*A^i(\bar{K}', t') = A^i(\bar{K}, t). \tag{1}$$

§ 2. In what follows a space will always mean a normal space with countable basis so that it is metrizable. A system ($\tilde{X}$, t) consisting of a space $\tilde{X}$ and a transformation t of $\tilde{X}$ onto itself with prime period p will be called a *simple system of period* p if t has no fixed points. For such a simple system the space $X = \tilde{X}/t$ obtained from $\tilde{X}$ by identifying all sets of points $\tilde{x}$, $t\tilde{x}$, $\cdots$, $t^{p-1}\tilde{x}(\tilde{x} \in \tilde{X})$ to a single point will be called

* *Sci. Record (N. S.)*, 1957, I(6): 377-380.

the *modular space* of the system and the natural projection $\pi : \tilde{X} \to X$ the *projection* of the system. An open covering (finite or infinite) $\bar{\mathfrak{U}} = \{\tilde{U}_\alpha\}$ of $\tilde{X}$, consisting of mutually distinct non-empty open sets $\tilde{U}_\alpha$ will be called a *P-covering* of $\tilde{X}$ if $\tilde{U}_\alpha \in \mathfrak{U}$ implies $t\tilde{U}_\alpha \in \mathfrak{U}$, and no set $\tilde{U}_\beta \in \mathfrak{U}$ can meet both $\tilde{U}_\alpha$ and $t^i\tilde{U}_\alpha$ for any $\tilde{U}_\alpha \in \mathfrak{U}$ and $0 < i < p$. Denote by $\mathfrak{U}$ the covering of X consisting of mutually distinct sets $U_\alpha = \pi(\tilde{U}_\alpha) = \cdots = \pi(t^{p-1}\tilde{U}_\alpha)$, which will be called the *projection* of $\mathfrak{U}$. Denote by $\bar{K}_\mathfrak{U}$, $K_\mathfrak{U}$ the nerve complexes of the coverings $\bar{\mathfrak{U}}$ and $\mathfrak{U}$. Then t induces. in a natural manner a simplicial map $t_\mathfrak{U}$ of $\bar{K}_\mathfrak{U}$ on itself such that $(\bar{K}_\mathfrak{U}, t_\mathfrak{U})$ is a simple system of period p in the sense of § 1 and with modular complex canonically isomorphic to $K_\mathfrak{U}$ so that they may be identified. The projection $\pi_\mathfrak{U} : \bar{K}_\mathfrak{U} \to K_\mathfrak{U}$ of the system $(\bar{K}_\mathfrak{U}, t)$ is then given by the simplicial map $\pi_\mathfrak{U}(\tilde{U}_\alpha) = U_\alpha$. Let $\overline{\mathfrak{B}} = \{\bar{V}_\beta\}$ be another P-covering of $\tilde{X}$ which is a refinement of $\bar{\mathfrak{U}}$ and $\mathfrak{B} = \{V_\beta\}$ be the projection of $\overline{\mathfrak{B}}$. Construct $\bar{K}_\mathfrak{B}$, $K_\mathfrak{B}$, $t_\mathfrak{B}$ and $\pi_\mathfrak{B}$ as before. Let $\tilde{\tau}_{\mathfrak{U}\mathfrak{B}} : \bar{K}_\mathfrak{B} \to \bar{K}_\mathfrak{U}$ and $\tau_{\mathfrak{U}\mathfrak{B}} : K_\mathfrak{B} \to K_\mathfrak{U}$ be the simplicial maps chosen so that $\tilde{V}_\beta \subset \tilde{\tau}_{\mathfrak{U}\mathfrak{B}}(\tilde{V}_\beta)$, $\tilde{\tau}_{\mathfrak{U}\mathfrak{B}}(\tilde{V}_\beta) = \tilde{U}_\alpha$ implies $\tilde{\tau}_{\mathfrak{U}\mathfrak{B}}(t\tilde{V}_\beta) = t\tilde{U}_\alpha$, and $\tau_{\mathfrak{U}\mathfrak{B}}(\pi\tilde{V}_\beta) = \pi\tilde{\tau}_{\mathfrak{U}\mathfrak{B}}(\bar{V}_\beta)$. Then $\tilde{\tau}_{\mathfrak{U}\mathfrak{B}}$ is a P-map of the system $(\bar{K}_\mathfrak{B}, t_\mathfrak{B})$ in $(\bar{K}_\mathfrak{U}, t_\mathfrak{U})$ with induced map $\tau_{\mathfrak{U}\mathfrak{B}}$. By §1, we have then

$$\tau^*_{\mathfrak{U}\mathfrak{B}} = A^i(\bar{K}_\mathfrak{U}, t_\mathfrak{U}) = A^i(\bar{K}_\mathfrak{B}, t_\mathfrak{B}). \tag{2}$$

Now the sets of all P-coverings of $\tilde{X}$ and their projections are cofinal in the sets of all open coverings of $\tilde{X}$ and X respectively and may serve to define their respective Cêch cohomology groups $H^i(\tilde{X}, G)$ and $H^i(X, G)$. By (2), the classes $A^i(\bar{K}_\mathfrak{U}, t_\mathfrak{U})$ will give rise to same cohomology classes in the limit which will then be denoted by $A^i(\tilde{X}, t) \in H^i(X, I_{(i)})$ and will be called the *Smith classes* of the simple system $(\tilde{X}, t)$.

Given two simple systems $(\tilde{X}, t)$ and $(\tilde{X}', t')$ of same prime period p, a continuous map $\tilde{f} : \tilde{X} \to \tilde{X}'$, will be called a *P-map* of the systems if $\tilde{f}t = t'\tilde{f}$. Such a P-map will induce then a continuous map $f : X \to X'$ of their modular spaces such that $\pi'\tilde{f} = f\pi$, where π, π' are the corresponding projections. We have then

$$f^* A^i(\tilde{X}', t') = A^i(\tilde{X}, t) \tag{1$'$}$$

§ 3. For a space X, let us form the p-fold product $\tilde{X}_p = X_p \times \cdots \times X_p$ of X by itself and let $\tilde{\Delta}_p$ be the diagonal consisting of all points $(x_1, \cdots x_p)$ with $x_1 = \cdots = x_p \in X$. The space $\tilde{X}^*_p = \tilde{X}_p - \tilde{\Delta}_p$ will be called the *p-fold reduced product* of X. Suppose p is a prime. Then the transformation t of $\tilde{X}_p$ onto itself defined by $t(x_1, \cdots, x_p) = (x_2, \cdots, x_p, x_1)$, when restricted to $\tilde{X}^*_p$ will make $(\tilde{X}^*_p, t)$ a simple system of prime period p. The modular space $X_p^* = \tilde{X}_p^*/t$ of the system $(\tilde{X}_p^*, t)$ will be called the *p-fold reduced cyclic product* of X. The Cêch cohomology groups $H^r(X_p^*, G)$ are evidently topological invariants of X itself and will be denoted simply by $H_{(p)}^r(X, G)$. The Smith classes $A^i(\tilde{X}_{p,}^* t) \in H^i(X_p^*, I_{(i)}) = H^i{}_{(p)}(X, I_{(i)})$ will be called the $\Phi_{(p)}$-*classes* of X and will be denoted by $\Phi_{(p)}^i(X)$. As $\Phi_{(p)}^0(X) \neq 0$, and $\Phi_{(p)}^i(X) = 0$ implies $\Phi_{(p)}^j(X) = 0$, for all $j \geqslant i$, we shall call the least integer n, if it exists, for which $\Phi_{(p)}^n(X) = 0$, the $\Phi_{(p)}$-*index* of X, and shall denote it by $I_p(X)$. If there does not exist such an integer n, we shall set $I_p(X) = \infty$. The integer p is always a prime.

Theorem 1 For an Euclidean space R^N of dimension N or a simplex Δ^N of dimension N, we have $(N > 0)$

$$I_p(R^N) = I_p(\Delta^N) = N(p-1).$$

Proof. Let us consider $X = R^N$ as a vector space so that $\tilde{X}_p^*$ is the space of all ordered sets of p vectors $(\mathfrak{B}_1, \cdots, \mathfrak{B}_1)$ of R^N with $\mathfrak{B}_i$ not all equal. The p-fold reduced cyclic product X_p^* of X is then the modular space $\tilde{X}_P^*/t$ where t is given by $t(\mathfrak{B}_1, \cdots, \mathfrak{B}_p) = (\mathfrak{B}_2, \cdots, \mathfrak{B}_p, \mathfrak{B}_1)$ In $\tilde{X}_p^*$ we have a subspace $\tilde{S}_p$ defined by the relations $\sum_{i=1}^{p} \mathfrak{B}_i = 0$ and $\sum_{i=1}^{p} |\mathfrak{B}_i|^2 = 1$, where $|\mathfrak{B}|$ denotes the length of a vector $\mathfrak{B}$. It may be seen that $\tilde{S}_p$ is a sphere of dimension $N(p-1)-1$ and the modular space $S_p = \tilde{S}_p/t$ is a lens space, for which

$$A^i(\tilde{S}_p, t) \neq 0 \text{ for } i \leqslant N(p-1) - 1.$$

A fortiori, we have $\Phi_{(p)}^i(R^N) = A^i(\tilde{X}_p^*, t) \neq 0$ for $i \leqslant N(p-1) - 1$, i.e.,

$$I_P(R^N) \geqslant N(p-1). \tag{3}$$

Now consider $Y = \Delta^N$ as a complex K with the usual subdivision and let $\bar{K}_p^*$ be the subomplex of the usual p-fold product complex of K which consists of all cells $\sigma_1 \times \cdots \times \sigma_p (\sigma_i \in K)$ with no vertex of K common to all these σ_i. It may be shown that the space $|\bar{K}_p^*|$ of $\bar{K}_p^*$ is a deformation retract of $\tilde{Y}_p^*$ for which the deformation commutes with the cyclic transformation t in $\tilde{Y}_p^*$. It follows that $A^i(\tilde{Y}_p^*, t)$ may be identified with $A^i(|\bar{K}_p^*|, t)$. As $\bar{K}_p^*$ is of a dimension $N(p-1)-1$, we have $A^i(\tilde{Y}_p^*, t) = A^i(|\bar{K}_p^*|, t) = 0$ for $i \geqslant N(p-1)$, i.e.,

$$I_p(\Delta^N) \leqslant N(p-1). \tag{4}$$

Now each of the spaces R^N and Δ^N may be considered as a subspace of the other so that from definition we have necessarily $I_p(R^N) = I_p(\Delta^N)$. This combined with (3) and (4) proves our theorem.

§ 4. A space X will be said to be (*topologically*) *realizable* in a space Y if there exists a continuous map f of X in Y such that X and $f(X)$ are homeomorphic under the map f. The map f, if it exists, will then be called a (*topological*) *realization* of X in Y. Suppose that X is realizable in Y with a realization f, then for any prime p, f will induce in a natural manner a continuous map $\bar{F}$ of $\tilde{X}_p^*$ in $\tilde{Y}_P^*$ which is a P-map of the system $(\tilde{X}_p^*, t_X)$ in the system $(\tilde{Y}_p^*, t_Y)$ where t_X and t_Y are the corresponding cyclic transformations in $\tilde{X}_p^*$ and $\tilde{Y}_p^*$ as in §2. Let $F : X_p^* \to Y_p^*$ be the map induced by $\bar{F}$, then by $(1)'$, $F^*\Phi^i_{(p)}(Y) = \Phi^i_{(p)}(X)$ so that $I_p(X) \leqslant I_p(Y)$. From Theorem 1 we get then

Theorem 2 If a space X is realizable in a Euclidean space R^N of dimension N, then

$$I_p(X) \leqslant N(p-1), \quad \text{i.e.,} \, \Phi^{N(P-1)}_{(P)}(X) = 0.$$

§ 5. For any space X let

$$Sm^i_{(p)} : H^r(X, I_p) \to H^{r+i}(X, I_p)$$

be the set of homomorphisms as defined in [3], then as a consequence of Theorem 2 we have (Cf. [2] Th. III §25 and [4] II §4 for the case $p = 2$ under the condition that X be locally contractible),

Theorem 3 If a compact separable Hausdorff space X is realizable in R^N, then

$$Sm^k_{(2)} H^r(X, I_2) = 0, \quad \text{for } 2k + r \geqslant N$$

and

$$Sm^{2k}_{(p)} H^r(X, I_p) = 0, \quad \text{for } 2pk + (p-1)r \geqslant N(p-1), \quad p > 2.$$

In particular, we should have $H^r(X, I_p) = 0$ for $r \geqslant N$, $p \geqslant 2$.

Similarly, we may deduce from Theorem 2 conditions for a closed differentiable manifold to be realizable in R^N in terms of its Whitney classes and Pontrjagin classes which we shall give later.

§ 6. As a further application of the $\Phi_{(p)}$-classes, we have

Theorem 4 The dimension of a finite polytope X is the least integer $n \geqslant 0$ such that $\Phi^{(n+1)(p-1)-1}_{(p)}(X) = 0$ for any prime p sufficiently large.

Proof. Let $\dim X = n$, then by dimensionality considerations we have $\Phi^i_{(p)}(X) = 0$ for $i \geqslant np + 1$, so that $\Phi^{(n+1)(p-1)-1}_{(p)}(X) = 0$ for $p > n + 2$. On the other hand, by theorem 1, $\Phi^{n(p-1)-1}_{(p)}(X) \neq 0$ for any prime p if $n > 0$. This gives the theorem.

Problem Does the above theorem remain true for arbitrary normal spaces with countable basis?

References

[1] Smith P A. Fixed Points of Periodic Transformations. In Lefschetz, *Algebraic Topology*, 1941: 350-373.

[2] Thom R. Espaces fibrés en sphères et carrés de. Steenrod. *Ann. Ec. Norm. Sup.*, 1952, 69(3): 110-182.

[3] Wu Wen-tsün. Sur les puissances de Steenrod. Colloque de Topologie de Strasbourg, 1951.

[4] ____. On the Realization of Complexes in Euclidean Spaces, I, II. *Acta Math. Sinica,* 1955, 5: 505-552; 1957, 7: 79-101.

[5] ____. On the Imbedding of Polyhedrons in Euclidean Spaces. *Bull. Acad. Polonaise des Sc.*, Cl. III, 1956, 4: 573-577.

24. On the Relations between Smith Operations and Steenrod Powers*

Let K be a complex, I_p the field of integers mod p, p being a prime. The so-called Steenrod powers [4]

$$\mathrm{St}^k_{(p)} = \mathrm{St}^k : H^r(K, I_p) \to H^{r+k}(K, I_p)$$

are deduced from the consideration of the pth power $K = \underbrace{K \times \cdots \times K}_{p}$ under the cyclic transformation① $t(a_1, \ldots, a_p) = (a_p, a_1, \ldots, a_{p-1})$, $a_i \in \bar{K}$. On the other hand, from K^p under the transformation t, we may introduce in a natural manner according to the theory of P. A. Smith [3], [2] a system of homomorphisms

$$\mathrm{Sm}^{(p)}_k = \mathrm{Sm}_k : H_r(K, I_p) \to H_{r-k}(K, I_p).$$

The question what relations exist between the Smith operations Sm_k and the Steenrod powers St^k naturally arises. The author discovered formerly [6] that these two systems of operations are actually equivalent, in the sense that one is determined by the other, and found the mode of their mutual determination. This furnishes a more natural and simpler definition of Steenrod powers and makes it directly connected with the theory of Smith. However, the original proof of the author depends on the intrinsic axiomatic theory of Steenrod powers of Thom [5], which is quite complicated. We need therefore a direct proof without the use of Thom's theory, which is the object of the present paper.

§1 The definition of Sm^r

Let K be a finite simplicial complex, $K^p = \underbrace{K \times \ldots \times K}_{p}$ its pth power subdivided as a product complex, t: $\bar{K}^p \to \bar{K}^p$ the transformation defined by $t(a_1, \ldots, a_p) =$

* *Acta Math. Sinica*, 1957, 7: 235-241; *Fund. Math.*, 1957, 44: 262-269; *Amer. Math. Soc. Translations, Ser.*, 1964, 2(38): 269-276.

①For a complex K, $\overline{K}$ means the space of K.

$(a_p, a_1, \ldots, a_{p-1})$, $a_i \in \bar{K}$, where p is a fixed prime. Let $\Delta : \bar{K} \to \bar{K}^p$ be the diagonal map, then $\Delta(\bar{K})$ may be subdivided as K, and the complex isomorphic to K thus obtained will be denoted by $\Delta(K)$. Take a subdivision $\tilde{K}^p$ of K^p such that $\Delta(K)$ is a subcomplex of $\tilde{K}^p$ and that t is a cell map of $\tilde{K}^p$. Let ω be the corresponding chain mapping. Then for any $x_i \in C_{r_i}(K)$ we have

$$t\omega(x_1 \otimes \cdots \otimes x_p) = (-1)^{r_p(r_1+\cdots+r_{p-1})}\omega(x_p \otimes x_1 \otimes \cdots \otimes x_{p-1}),$$

where t denotes also the chain mapping induced by t in $\tilde{K}^p$.Put as usual $d = 1-t, s = 1+t+\cdots+t^{p-1}$: then for any $x \in Z_q(K, I_p), \omega x^p = \omega\underbrace{(x \otimes \cdots \otimes x)}_{p}$ is always a d-cycle (i.e., a cycle z with $dz = 0$) and according to P. A. Smith[3], [2] we have a sequence (we shall call it the *sequence associated with* x):

$$\omega x^p = s x_0 + x'_0,$$

$$\partial x_{2i-1} = s x_{2i} + x'_{2i}, \quad i > 0, \tag{1}$$

$$\partial x_{2i} = d x_{2i+1} + x'_{2i-1}, \quad i \geqslant 0,$$

in which $x'_j \subset \Delta(K)$, $x_j \subset \tilde{K}^p - \Delta(K)$, and

$$\dim x_j = \dim x'_j = pq - j. \tag{2}$$

Put $\bar{d} = 1 + 2t + 3t^2 + \cdots + (p-1)t^{p-2}$, then for coefficients mod p we have for the operations $\bar{d}$, $d\bar{d} = s$. If we set

$$s_i = \begin{cases} s, & \text{for } i \text{ even,} \\ d, & \text{for } i \text{ odd,} \end{cases}$$

then (1) may also be written as

$$\omega x^p = s x_0 + x'_0, \quad \partial x_i = s_{i+1} x_{i+1} + x'_{i+1}, \quad i \geqslant 0. \tag{1$'$}$$

Definiftion $S_k x = \Delta^{-1} x'_{k+(p-1)q} \in Z_{q-k}(K, I_p)$.

Remark 1. The cycle $S_k x$ is completely determined by x (as well as the subdivision $\tilde{K}^p$) and lies in the smallest complex $|x|$ determined by x. It follows that $x'_j = 0$ for $j < (p-1)q$, and $S_k x$ have a meaning only for $0 \leqslant k \leqslant q$. In particular, we have $S_0 x = x$ for $q = 0$.

Lemma $S_k : Z_q(K, I_p) \to Z_{q-k}(K, I_p)$ *is a homomorphism such that*

$$S_k[B_q(K, I_p)] \subset B_{q-k}(K, I_p), \quad k < q, \tag{3}$$

$$S_q[Z_q(K, I_p)] \subset B_0(K, I_p), \quad q > 0. \tag{4}$$

Proof. Consider any $x, y \in Z_q(K, I_p)$. We have

$$\omega(x+y)^p - \omega x^p - \omega y^p = sz,$$

where $z = \omega(x \otimes y \otimes \cdots \otimes y + \cdots)$ is a cycle $\subset \tilde{K}^p - \Delta(K)$. Hence the sequence associated with $x + y$ will be obtained by adding the corresponding equations of the sequences associated with x and y. We have therefore $S_k(x+y) = S_k(x) + S_k(y)$. Similarly we have $S_k(-x) = (-1)^p S_k x = -S_k x$ (coefficient group I_p). Hence S_k is a homomorphism.

Let σ be a $(q+1)$-dimensional simplex of K: then $S_k(\partial\sigma)$ is a cyele in the complex determined by σ and hence for $k < q$, $S_k(\partial\sigma) \sim 0$ in this complex, *a fortiori* ~ 0 in K. As S_k is already known to be a homomorphism, we get (3).

Let $x \in Z_q(K, I_p), q > 0$. By (1′) we get

$$S_q x = \Delta^{-1} x'_{pq} \in C_0(K, I_p) \quad \text{and} \quad \partial x_{pq-1} = s_{pq} x_{pq} + x'_{pq}.$$

Hence $KI(x'_{pq}) = KI(-s_{pq}x_{pq}) \equiv 0 \bmod p$ and we have $\Delta^{-1}x'_{pq} \sim 0 \bmod p$ in K if K is connected. The general case follows then from the fact that S_k is a homomorphism. This proves (4), q. e. d.

From the lemma it follows that S_k induces a system of homomorphisms

$$\mathrm{Sm}_k : H_q(K, I_p) \to H_{q-k}(K, I_p), \quad q \geqslant k \geqslant 0.$$

Furthermore we have

$$\mathrm{Sm}_q / H_q(K, I_p) = \begin{cases} 0, & q > 0, \\ 1, & q = 0, \end{cases}$$

in which 1 means the identity.

Since I_p is a field, we have, dual to Sm_k, a system of homoniorphisms

$$\mathrm{Sm}^k : H^{q-k}(K, I_p) \to H^q(K, I_p)$$

such that

$$\mathrm{Sm}^q / H^0(K, I_p) = \begin{cases} 0, & q > 0, \\ 1, & q = 0, \end{cases}$$

where 1 is again the identity.

Remark 2. Let $\tilde{K}$ be a complex, t a cell map of $\tilde{K}$ with prime period p such that any face of a cell fixed under t is also a fixed cell of t. The set of all fixed cells under t then forms a (closed) subeomplex L. Let ϱ and $\bar{\varrho}$ denote either $d = 1 - t$ and $s = 1 + t + \cdots + t^{p-1}$ or s and d. Also, let H_k^{ϱ} be the special homology group of Smith determined from the k-dimensional cycles x satisfying $\varrho x = 0$. Then by Smith we have some homomorphisms

$$\mu_{\varrho} : H_k^{\varrho}(\tilde{K}, I_p) \to H_{k-1}^{\bar{\varrho}}(\tilde{K}, I_p),$$

and

$$\nu_{\varrho} : H_k^{\varrho}(\tilde{K}, I_p) \to H_k(L, I_p).$$

In particular if $\tilde{K}$ is a subdivision $\tilde{K}^p$ of $K^p = K \times \cdots \times K$, and t: $\tilde{K}^p \to \tilde{K}^p$, as given at the beginning of this section, such that $L = \Delta(K)$, then for any cycle x mod p of K, ωx^p may be considered as a d-cycle mod p and the d-homology class of ωx^p depends only on the homology class of x, whence it may be denoted by αX(however, α is not necessarily a homomorphism). Then Sm_k may be defined as

$$\mathrm{Sm}_k = \nu_{\varrho}\mu_{\varrho} \ldots \mu_d \mu_s \mu_d \alpha,$$

in which μ occur $(p-1)q - k$ times ($q = \dim X$) alternatively as μ_s and μ_d, and ϱ is d or s according as k is odd or even.

§2 Relations beetwen Sm^i and St^j

The Steenrod pth powers (p being a prime)

$$\mathrm{St}^j : H^r(K, I_p) \to H^{r+j}(K, I_p)$$

may be defined as follows: Form K^p and its subdivision $\tilde{K}^p$ with $\Delta(K)$ as a subcomplex as in §1. As in the original proof of Steenrod we may show that there exists a system of homomorphisms① (q, j are arbitrary integers)

$$D^j : C_q(\tilde{K}^p) \to C_{q+j}(K^p)$$

satisfying the following conditions②:

1° $D^j = 0,\ j < 0$.

2° $\mathrm{In}\ D^0 c = \mathrm{In}\ c,\ c \in C_0(\tilde{K}^p)$.

① In the case of $p = 2$, D^j have been introduced by R.Bott [1].

② In what follows we write for simplicity $\sigma_1 \cdots \sigma_p$ instead of $\sigma_1 \otimes \cdots \otimes \sigma_p$, and thus for the others. Fundamenta Mathematicae T XLIV.

3° If $\tau \in |\omega(\sigma_1 \cdots \sigma_p)|$, $\sigma_i \in K$, then $D^j\tau \subset |\sigma_1 \cdots \sigma_p|$.

4° $\partial D^{2j} = D^{2j}\partial + \sum_{a=0}^{p-1} t^{-a}D^{2j-1}t^a, \partial D^{2j+1} = -D^{2j+1}\partial + (t^{-1}D^{2j}t - D^{2j})$.

From 1°—4° we may get the following property of D^0:

$$D^0\omega(\sigma_1 \cdots \sigma_p) = \sigma_1 \cdots \sigma_p, \quad \sigma_i \in K. \tag{1}$$

Proof. If $\sum \dim\ \sigma_i = 0$, (1) is true by 2° and 3°. Suppose that (1) has been proved for $\sum \dim\ \sigma_i < k$. Let $\sigma_i \in K$, with $\dim\ \sigma_i = d_i$, $d_1 + \cdots + d_i = r_i$, $\sum \dim\ \sigma_i = k$: then by 3° we must have $D^0\omega(\sigma_1 \cdots \sigma_p) = \lambda\sigma_1 \cdots \sigma_p$, where λ is a certain integer. From (1) and the first formula of 4°, we get

$$\begin{aligned}\lambda\partial(\sigma_1 \cdots \sigma_p) &= \partial D^0\omega(\sigma_1 \cdots \sigma_p) = D^0\partial\omega(\sigma_1 \cdots \sigma_p)\\ &= D^0\omega\partial(\sigma_1 \cdots \sigma_p) = \sum (-1)^{r_{i-1}} D^0\omega(\sigma_1 \cdots \partial\sigma_i \cdots \sigma_p),\end{aligned}$$

which, by the induction hypothesis, is

$$= \sum (-1)^{r_{i-1}}\sigma_1 \cdots \partial\sigma_i \cdots \sigma_p = \partial(\sigma_1 \cdots \sigma_p).$$

As $\partial(\sigma_1 \cdots \sigma_p) \neq 0$, we get $\lambda = 1$ and (1) is proved.

Let

$$D_j : C^{q+j}(K^p, R) \to C^q(\tilde{K}^p, R)$$

be the homomorphisms dual to D^j(R is a commutative ring with unit element). D_j must satisfy the following conditions corresponding to 1°—4°:

$\bar{1}$° $D_j = 0$, $j < 0$.

$\bar{2}$° If $u \in C^0(K^p, R)$ takes the same constant value $\alpha \in R$ on all vertices of K^p, then $D_0u \in C^0(\tilde{K}^p, R)$ takes the same value α on all vertices of $\tilde{K}^p$.

$\bar{3}$° $D_jc \subset |\omega'c|$, where $c \in C^q(K^p, R)$, and ω' is the dual of ω.

$\bar{4}$° $D_{2j}\delta = \delta D_{2j} + \sum_{a=0}^{p-1} t^aD_{2j-1}t^{-a}, D_{2j+1}\delta = -\delta D_{2j+1} + (tD_{2j}t^{-1} - D_{2j})$, here t stands also for the cochain mapping induced by the cell mapping t in the complex K^p or $\tilde{K}^p$.

In particular for c with $tc = c$ and $\delta c = 0$, we have

$$\delta D_ic = -s_iD_{i-1}c. \tag{2}$$

Let $U \in H^q(K, I_p)$: then by definition $\mathrm{St}^j_{(p)}U = \mathrm{St}^jU \in H^{q+j}(K, I_p)$ is the class uniquely determined by the cocycle $\Delta^{-1}D_{p-1,q-j}u^p$($u^p$ stands for $u \otimes \cdots \otimes u$), which is independent of the chosen cocycle $u \in U$. It is easy to see that this definition of St^j coincides with the original one of Steenrod.

In the above discussion the subdivision $\tilde{K}^p$ of K^p is rather arbitrary, subject only to conditions already stated. Now take a *canonical subdivision* $\tilde{K}^p$ of K^p as follows. The complex $\tilde{K}^p$ will be formed by the following three sets of cells: (a) $\Delta(\sigma)$, $\sigma \in K$; (b) $\sigma_1 \times \cdots \times \sigma_p$, where $\sigma_i \in K$ and $\sigma_1, \cdots, \sigma_p$ have no vertices common to all of them; (c) $\Delta(\sigma) \circ (\sigma_1 \times \cdots \times \sigma_p)$, where $\sigma, \sigma_i \in K$, $\sigma_1 \times \cdots \times \sigma_p$ a cell of type (b), and for each i, σ and σ_i span a simplex τ_i of $K(\sigma, \sigma_i$ may have common vertices). The symbol $\circ$ means the join operation.

In the case (c)we have $\Delta(\sigma) \circ (\sigma_1 \times \cdots \times \sigma_p) \subset |\omega(\tau_1 \times \cdots \times \tau_p)|$. Let dim $\sigma = d$, dim $\sigma_i = d_i$, dim $\Delta(\sigma) \circ (\sigma_1 \times \cdots \times \sigma_p) = q$, dim $(\tau_1 \times \cdots \times \tau_p) = r$: then $q = d + d_1 + \cdots + d_p + 1$, $r \leqslant pd + d_1 + \cdots + d_p + p$, and consequently $r \leqslant pq$. In view of 2°, 3°, it follows that for any cell $\xi \in \tilde{K}^p$ of type (c) we have $D^j \xi = 0$, $j > (p-1)q$. As the same is true for cells of type (a) or (b), with $j > (p-1)q$, we have for any $c \in C_q(\tilde{K}^p)$,

$$D^j c = 0, \quad j > (p-1)q,$$
$$D^0 c = c, \quad q = 0.$$

Dually, we have for any $c \in C^q(K^p, R)$,

$$D_j c = 0, \quad pj > (p-1)q, \tag{3}$$

$$D_0 c = c, \quad q = 0. \tag{3'}$$

From (3) and $\bar{4}°$ it follows that for any $u \in Z^q(K, I_p)$ we have for either $p > 2$ or $p = 2$,

$$-dD_{(p-1)q}u^p = (tD_{(p-1)q}t^{-1} - D_{(p-1)q})u^p = \delta D_{(p-1)q+1}u^p = 0,$$

or

$$dD_{(p-1)q}u^p = 0, \quad u \in Z^q(K, I_p). \tag{4}$$

Theorem 1 *The two sets of operations* $\{\mathrm{Sm}^i\}$ *and* $\{\mathrm{St}^j\}$, *are equivalent to each other in the sense that either one may be determined by the other. The mode of mutual determination is given by the following system of equations:*

$$A^k = \sum_{j=0}^{k} (-1)^j \mathrm{Sm}^{k-j} \mathrm{St}^j = \begin{cases} 0, & k > 0 \\ 1, & k = 0 \end{cases} \quad (p = 2 \quad \text{or} \quad pk \text{ odd}), \tag{5}$$

$$B^k = \sum_{j=0}^{k} \mathrm{Sm}^{2k-2j}\mathrm{St}^{2j} = \begin{cases} 0, & k > 0 \\ 1, & k = 0 \end{cases} \quad (p > 2), \tag{6}$$

in which 1 *means identity.*

For example let us prove (6) as follows:

Let $U\in H^q(K, I_p)$, $X \in H_r(K, I_p)$, $r = q + 2k$. Take $u\in$ U, $x \in X$ and form the sequence associated with x as given in §1 (1). By $\bar{4}°$ and (2), we then have①

$$\begin{aligned} &\mathrm{Sm}^{2k-2j}\mathrm{St}^{2j}U \cdot X = \mathrm{St}^{2j}U \cdot \mathrm{Sm}_{2k-2j}X \\ =& D_{(p-1)q-2j}u^p \cdot x'_{(p-1)r+2k-2j} \\ =& \delta D_{(p-1)q-2j}u^p \cdot x_{(p-1)r+2k-2j-1} - sD_{(p-1)q-2j}u^p \cdot x_{(p-1)r+2k-2j} \\ =& -sD_{(p-1)q-2j-1}u^p \cdot x_{(p-1)r+2k-2j-1} - sD_{(p-1)q-2j}u^p \cdot x_{(p-1)r+2k-2j} \\ =& -\bar{d}\delta D_{(p-1)q-2j-1}u^p \cdot x_{(p-1)r+2k-2j-2} - sD_{(p-1)q-2j}u^p \cdot x_{(p-1)r+2k-2j} \\ =& +sD_{(p-1)q-2j-2}u^p \cdot x_{(p-1)r+2k-2j-2} - sD_{(p-1)q-2j}u^p \cdot x_{(p-1)r+2k-2j}. \end{aligned}$$

Adding together the above equations, we get

$$B^kU \cdot X = sD_{(p-1)q-2k-2}u^p \cdot x_{(p-1)r-2} - sD_{(p-1)q}u^p \cdot x_{(p-1)r+2k}.$$

Now take $\tilde{K}^p$ as the canonical subdivision of K^p so that, if we apply (4), the last term in the above equation vanishes. Applying again $\bar{4}°$ and § 1 (1) and noting that $x'_j = 0$, $j < (p-1)r$, we may successively reduce the resulting equations as follows:

$$\begin{aligned} B^kU \cdot X &= D_{(p-1)q-2k-2}u^p \cdot sx_{(p-1)r-2} = D_{(p-1)q-2k-4}u^p \cdot sx_{(p-1)r-4} \\ &= \cdots = D_{-2kp}u^p \cdot sx_0. \end{aligned}$$

Hence for $k > 0$, $B^kU \cdot X = 0$. Since both X and U are arbitrary, we get the first equation of (6). If $k = 0$, then by (1) we get

$$B^0U \cdot X = D_0u^p \cdot sx_0 = D_0u^p \cdot \omega x^p = u^p \cdot D^0(\omega x^p) = u^p \cdot x^p = (u \cdot x)^p = u \cdot x,$$

since the coefficients considered are in the group I_p. It follows that $B^0U \cdot X = U \cdot X$ for any U, X, and we get the second formula of (6).

The proof of (5) is similar and will thus be omitted.

① For a cohomology class U (or a cocycle u) and a homology class of the same dimension X or (a cycle x) on coefficient group I_p, $U \cdot X$ (or $u \cdot x$) will mean the value of U on X (or of u on x).

Theorem 2 *The sets of operations* $\{\mathrm{Sm}^i\}$ *and* $\{\mathrm{St}^j\}$ *may also be mutually determined by the following relations:*

$$\bar{A}^k = \sum_{j=0}^{k} (-1)^j \mathrm{St}^{k-j}\mathrm{Sm}^j = \begin{cases} 0, & k>0 \\ 1, & k=0 \end{cases} \quad (p=2 \quad \text{or} \quad pk \text{ odd}), \tag{$\bar{5}$}$$

$$\bar{B}^k = \sum_{j=0}^{k} \mathrm{St}^{2k-2j}\mathrm{Sm}^{2j} = \begin{cases} 0, & k>0 \\ 1, & k=0 \end{cases} \quad (p>2). \tag{$\bar{6}$}$$

Proof. For example let us prove $(\bar{5})$ in the case pk = odd. The proof of other cases is similar. Suppose that $(\bar{5})$ and $(\bar{6})$ have been proved in the case $\leqslant k-1$. Let δ_i denote the homomorphism 0 or 1 according as $i>0$ or $=0$: then by (5) and the induction hypothesis we have

$$\begin{aligned}
&\sum_{j>0} (-1)^j \mathrm{St}^{k-j}\mathrm{Sm}^j\mathrm{St}^0 \\
&= -\sum_{j>0}\left(\mathrm{St}^{k-2j}\sum_{i>0}\mathrm{Sm}^{2j-2i}\mathrm{St}^{2i}\right) + \\
&\quad + \sum_{j>0}\left(\mathrm{St}^{k-2j+1}\sum_{i>0}(-1)^i\mathrm{Sm}^{2j-i-1}\mathrm{St}^i\right) \\
&= -\sum_{i>0}\left(\sum \mathrm{St}^{k-2j}\mathrm{Sm}^{2j-2i}\right)\mathrm{St}^{2i} + \sum_{i>0}\left(\sum (-1)^i \mathrm{St}^{k-2j+1}\mathrm{Sm}^{2j-i-1}\right)\mathrm{St}^i \\
&= -\sum_{i>0}\left(\sum (-1)^j \mathrm{St}^{k-j}\mathrm{Sm}^{j-2i}\right)\mathrm{St}^{2i} - \sum_{i>0}\left(\sum \mathrm{St}^{k-2j+1}\mathrm{Sm}^{2j-2i}\right)\mathrm{St}^{2i-1} \\
&= -\sum_{i>0}\delta_{k-2i}\mathrm{St}^{2i} - \sum_{i>0}\delta_{k+1-2i}\mathrm{St}^{2i-1} = -\mathrm{St}^k.
\end{aligned}$$

Hence $\overline{A}^k = 0$. As $\bar{A}^0 = \bar{B}^0 = 1$, $(\bar{5})$ is proved by induction on k.

Remark. Let $\beta = ((1/p)\delta)_p$ be the homomorphism of Bockstein: then $\beta\mathrm{St}^{2k} = -\mathrm{St}^{2k+1}$. Hence by comparing $\beta\bar{B}^k = 0$ with $(\bar{5})$ we know that in case of pk being odd $(\bar{5})$ may also be written as

$$\bar{A}^k = \sum_{j=0}^{k} \mathrm{St}^{k-j}\mathrm{Sm}^j = 0.$$

Similarly for pk odd (5) may also be written as

$$A^k = \sum_{j=0}^{k} \mathrm{Sm}^{k-j}\mathrm{St}^j = 0.$$

References

[1] Bott R. *On symmetric products and the Steenrod squares.* Ann. of Math, 1953, 57: 579-590.

[2] Richardson M and Smith P A. *Periodic transformation of complexes.* Ann. of Math, 1938, 39: 611-633.

[3] Smith P A. *Fixed points of periodic transformations.* Appendix B, 351-373 to S. Lefschetz. Algebraic Topology, New York, 1942.

[4] Steenrod N E. *Reduced powers of cohomology classes.* Paris, 1951.

[5] Thom R. *Une théorie intrinsèque de puissances de Steenrod.* Colloque de Topologie de Strasbourg, 1951.

[6] Wu Wen-tsün. *Sur les puissances de Steenrod.* Colloque de Topologie de Strasbourg, 1951.

25. On the Realization of Complexes in Euclidean Spaces III*

Let K be a finite simplicial complex. We can always view K as a Euclidean complex in a Euclidean space of sufficiently high dimension N. Let its underlying space be denoted by $\bar{K}$. In studying whether K can be imbedded in the Euclidean space R^m of some fixed dimension m, we have introduced the following definitions (in [1] the notation is slightly different).

Let $T : \bar{K} \to R^m$ be a topological mapping such that for every $\sigma \in K$, T/σ is a linear mapping, then T is called a linear imbedding of K. If $T : \bar{K} \to R^m$ is a linear imbedding of some simplicial subdivision K' of K, then T is called a semilinear imbedding of K in R^m through the subdivision K'. Again, if $T : \bar{K} \to R^m$ is a continuous mapping such that for any $\sigma \in K$, T/σ is a linear mapping (perhaps degenerate), T/σ is nondegenerate if dim $\sigma \leqslant m$, and for any simplices σ and r with no common vertices, and with dim $\sigma +$ dim $r \leqslant m$, $T(\sigma)$ and $T(r)$ are in general position, then T is called a linear pseudoimbedding. If $T : \bar{K} \to R^m$ is a linear pseudoimbedding of some subdivision K', then call T a semilinear pseudoimbedding of K through K'.

In [1], for a semilinear pseudoimbedding of K in R^m, we have introduced a system of invariants $\Phi^m(K) \in H^m(K^*, l_{(m)})$, $m > 0$, called the imbedding index of K, where K^* is the reduced two-fold symmetric product of K, $l_{(m)}$ is the additive group of integers if m is even, integers mod 2 if m is odd. We have also proved that $\Phi^m(K) = 0$ is a necessary condition for K to be semilinearly imbeddable in R^m. The purpose of this paper is to prove that these conditions are also sufficient in certain extreme cases.

More explicitly, let K^1 be a 1-dimensional complex, then $\Phi^1(K^1) = 0$ is the necessary and sufficient condition for K^1 to be semilinearly imbeddable in R^1. This is obvious. Similarly $\Phi^2(K^1) = 0$ is a necessary and sufficient conditiou for K^1 to be semilinearly imbeddable in R^2. As was pointed out in [2], this is another way of stating Kuratowski's Theorem. Otherwise, if the dimension n of K is greater than

* *Acta Mathematica Sinica*, 1958, 8(1): 79-94; *Sci. Sinica*, 1959, 8: 133-150.

2, then $\Phi^{2n}(K) = 0$ is a necessary and sufficient condition for K to be semilinearly imbeddable in R^{2n}. This is the main theorem of this paper (see §2, Theorem 1). This theorem was first studied in the work of Van Kampen (see [3]). The invariant introduced by Van Kampen in [3] is another way of expressing $\rho_2\Phi^{2n}(K)$, where ρ_2 is reduction mod 2. There are errors in both his statement and proof of the theorem (see [4]). But the method of our proof of the theorem still follows mainly the original proof of Van Kampen; where the mistake occurs, we use a construction of Whitney's to correct it.

In the following, a complex will always mean a finite Euclidean simplicial complex in Euclidean space.

§1 Several constructions

In order to prove the main theorem of this paper, namely Theorem 1 of §2, we need the following simple constructions:

A. Tube consttuction. Let C be an infinitely differentiable simple arc in R^m (i.e. the image in R^m of a line segment under an infinitely differentiable topological mapping), with endpoints a_0 and a_1. Let L_0^n and L_1^n be two n-dimensional linear subspaces orthogonal to C at a_0 and a_1, respectively, let S_0^{n-1} and S_1^{n-1} be $(n-1)$-spheres in L_0^n and L_1^n, with centers a_0 and a_1, and each with sufficiently small radius $\varepsilon > 0$. Assume further that each S_i^{n-1} is oriented. We will prove that if $m \geqslant n+2$, there always exists an n-dimensional differentiable tube T in the ε-neighborhood of C, i.e. T is the image under a differentiable topological mapping of the topological product of an $(n-1)$-spheres with a line segment, such that the two ends of T are S_0^{n-1} and S_1^{n-1} respectively, and after suitably orienting T, $\partial T = S_1^{n-1} - S_0^{n-1}$.

To prove this, write the points of C as a_t, $0 \leqslant t \leqslant 1$. Let L_t^{m-1} be the $(m-1)$-dimensional linear subspace orthogonal to C at a_t. Let the Grassmann manifold of all oriented n-planes of L_t^{m-1} through a_t be denoted by $\tilde{R}^{(t)}_{m-n-1,n}$. Then there exists a fiber space L on C in which the fibers are the image of the Grassmann manifold $\tilde{R}^{(t)}_{m-n-1,n}$, whose projection $\pi : L \to C$ is defined by $\pi^{-1}(a_t) = \tilde{R}^{(t)}_{m-n-1,n}$. Since C is a simple arc, this fiber bundle has a product structure. Also from the assumption that $m-n-1 \geqslant 1$, $\tilde{R}^{(t)}_{m-n-1,n}$ is an arcwise connected space, so any cross section on a_0, a_1 may be extended to a cross section on C. In particular, take the cross section $f(a_0) = L_0^n$, $f(a_1) = L_1^n$ on a_0, a_1, where the orientation of L_0^n, L_1^n corresponds to that of ${S_0}^{n-1}$, S_1^{n-1}, it may be extended to a cross section f on C, $f(a_t) = L_t^n$, such that L_t^n is an oriented n-plane of L_t^{m-1} through a_t. From a theorem of Steenrod, we can make f into an infinitely differentiable cross section. Let S_t^{n-1} be an $(n-1)$ -sphere of radius ε, center a_t in L_t^n, and T be the space formed by the S_t^{n-1}, for $0 \leqslant t \leqslant 1$. Then for sufficiently small ε, T is a differentiable tube with ends S_0^{n-1} and S_1^{n-1} in the ε-neighborhood of C. Furthermore, with a suitable orientation, we have $\partial T = S_1^{n-1} = S_0^{n-1}$ as required.

B. Whitney construction. Let σ_1 and σ_2 be two n-simplices in the complex K with no common vertex and $n > 2$, and let T be a continuous mapping of $\bar{K}$ into R^{2n} satisfying the following conditions:

1°. The restriction of T to the $(n-1)$-skeleton of K is a linear imbedding.

2°. The restriction of T to the interior of any n-simplex of K is a differentiable topological mapping.

3°. T has only double points, but no triple points.

4°. $T(\sigma_1)$ and $T(\sigma_2)$ intersect in an even number of points $q_1, \cdots, q_r, q_1', \cdots, q_r'$. At each point q_i or q_i', the tangent planes of $T(\sigma_1)$ and $T(\sigma_2)$ intersect only in the point q_i or q_i'. Furthermore, with respect to a fixed orientation of R^{2n}, the intersection numbers of $T(\sigma_1)$ and $T(\sigma_2)$ (σ_1 and σ_2 are already oriented) are $+1$ at each q_i, and -1 at each q_i'.

Whitney's construction (see [5], §10-12) permits us to alter T into another continuous mapping T', preserving properties 1°– 3° and changing 4° to

$\bar{4}$°. $T'(\sigma_1)$ and $T'(\sigma_2)$ do not meet. Furthermore, the alteration only occurs on σ_1, and σ_2, and does not change the other double points. In other words, we have

$\bar{5}$°. T and T' agree on $\bar{K}$-Int σ_1 − Int σ_2. Furthermore, apart from eliminating the double points q_i and q_i', T and T' have the same remaining double points.

We briefly state the construction of T'.

We may construct a simple differentiable arc B_i ($i = 1, 2$), connecting $q_1 = q$ and $q_1' = q'$ in $T(\sigma_i)$ such that B_i does not go through any other double points in $T(\sigma_i)$, and B_1 and B_2 have no common points other tnan the two endpoints q and q'. Let E^{2n} be a $2n$-dimensional Euclidean space with Cartesian coordinates $(x_1, \cdots, x_{2n})$, E^2 will be the (x_1, x_2) plane. Let A_1 be the line segment $0 \leqslant x_1 \leqslant 1$, $x_2 = x_3 = \cdots = x_{2n} = 0$, whose endpoints are $r = (0, \cdots, 0)$ and $r' = (1, 0, \cdots, 0)$, A_2 be a smootn curve: $x_2 = \lambda(x_1)$, $0 \leqslant x_1 \leqslant 1$, connecting r and r' in $E^2(x_2 \geqslant 0)$. Let $A = A_1 + A_2$, and τ be a 2-cell formed by a sufficiently small neighborhood of A and the interior of A. It follows from [5], §10, that for $n > 2$, we may construct a differentiable mapping ψ of τ, such that $\psi(r) = q$, $\psi(r') = q'$, $\psi(A_1) = B_1$, $\psi(A_2) = B_2$; the differentiable 2-cell $\psi(\tau) = \sigma$ intersects $T(\sigma_i)$ only at B_i for $i = 1, 2$, it does not intersect $T(\bar{K} - \text{Int } \sigma_1 = \text{Int } \sigma_2)$; and at any point of B_i, the tangent planes of $T(\sigma_i)$ and of σ only intersect at the tangent line of B_i at that point.

Since the intersection numbers of the two oriented cells $T(\sigma_1)$ and $T(\sigma_2)$ at q and q' are $+1$ and -1 respectively, and $n > 2$, it follows from [5], §11, that we may define a system of vector fields $W_1(q^*), \cdots, W_{2n}(q^*)$ on $\psi(\tau) = \sigma$, where all the $W_i(q^*)$ are linearly independent at each $q^* \in \sigma$, depending continuously and smoothly on q^*, and satisfying the following conditions:

(a) Let $e_i(r^*)$, $i = 1, \cdots, 2n$ be the unit vectors parallel to x_i at $r^* \in \tau$. Then $W_1(q^*)$ and $W_2(q^*)$ are the images of $e_1(r^*)$ and $e_2(r^*)$ respectively, under the vector mapping induced by ψ, where $q^* = \psi(r^*) \in \sigma$.

(b) When $q^* \in B_1$, $W_3(q^*)$, $\cdots$, $W_{n+1}(q^*)$ are the tangent vectors of $T(\sigma_1)$ at q^*.

(c) When $q^* \in B_2$, $W_{n+2}(q^*)$, $\cdots$,$W_{2n}(q^*)$ are the tangent vectors of $T(\sigma_2)$ at q^*.

Now for $r^* \in \tau$, define

$$\psi(r^* + \sum_{i=3}^{2n} a_i e_i(r^*)) = \psi(r^*) + \sum_{i=3}^{2n} a_i \omega_i(\psi(r^*)).$$

Then in a sufficiently small neighborhood U of $\tau \subset E^{2n}$, ψ is one-to-one. Now in each σ_i, take a sufficiently small neignbornood M_i of $C_i = T^{-1}(B_i)$; such that $T(M_i) \subset V = \psi(U)$, $i = 1, 2$. Let π be the projection: $(x_1, x_2, \cdots, x_{2n}) \to (x_1, 0, x_3, \cdots, x_{2n})$. We may assume M_i to be sufficiently small so that $\pi(N_1)$ and $\pi(N_2)$ only intersect on the x_1-axis, where $N_i = \psi^{-1}(T(M_i))$, $i = 1, 2$. This is always possible in view of conditions (a), (b) and (c).

Now take $\varepsilon > 0$ sufficiently small, and construct a continuously differentiable real function $\nu(x)$, such that

$$|\nu(x)| \leqslant 1, \nu(0) = 1, \text{and when } |x| \geqslant \varepsilon^2, \nu(x) = 0$$

Construct another continuously differentiable function $x_2 = \mu(x_1)$ such that

$$\varepsilon > \mu(x_1) - \lambda(x_1) > 0, \quad 0 \leqslant x_1 \leqslant 1,$$
$$\mu(x_1) = 0, x_1 < -\varepsilon \text{ or } x_1 > 1 + \varepsilon.$$

For any point $(x_1, \cdots, x_{2n}) = r^*$ in N_2, let

$$\theta(r^*) = r^* - \nu(x_3^2 + \cdots + x_{2^n}^2)\mu(x_1)e_2.$$

Then $\theta(N_2)$ does not meet N_1. Now alter $T : \bar{K} \to R^{2n}$ to $T_1 : \bar{K} \to R^{2n}$, so that T_1 agrees with T on $\bar{K} - M_2$, and on M_2,

$$T_1/M_2 = \theta\psi^{-1}T.$$

Then, if ε is sufficiently small, T_1 is obviously continuous, and is a differentiable topological mapping on σ_1 and σ_2, does not have $q = q_1$ and $q' = q_1'$ as double points, but otherwise has the same double points as T.

Now use the same technique on $(q, q') = (q_2, q_2'), \cdots, (q_r, q_r')$ to finally obtain a mapping T', satisfying 1°– 3°, $\bar{4}^\circ$ and $\bar{5}^\circ$ as required.

C. Van Kampen construction. Let σ_1 and σ_2 be two n-simplices of a $2n$-dimensional complex K, $n > 2$, having common vertices. Let T be a continuous mapping of $\bar{K}$ to R^{2n} satisfying the followin conditions:

1°. T is a semilinear pseudoimbedding of K through some subdivision K'.

2°. T is a semilinear imbedding of the subcomplex K_1 (or K_2) defined by σ_1 (or σ_2) and its faces through the subdivision K_1' (or K_2') of the above K' restricted to σ_1 (or σ_2).

3°. T has only double points, but no triple points.

In this situation, Van Kampen's construction (see proof of Lemma 2 of [4]) permits us to change T to another continuous mapping T' satisfying the following conditions:

$\overline{1^\circ}$. T' is a semilinear pseudoimbedding of K through some subdivision.

$\overline{2^\circ}$. T' is a semilinear imbedding of the subcomplex L formed by σ_1, σ_2 and their faces through the subcomplex induced by the above subdivision of K restricted to L.

$\overline{3^\circ}$. T' coincides with T on $\bar{K} - \text{Int}\,\sigma_1 - \text{Int}\,\sigma_2$. Moreover, T' has no triple points and has the same double points as T except that the original double points common to $T(\sigma_1)$ and $T(\sigma_2)$ have been removed.

Following the original construction of Van Kampen, we reconstruct the above T' as follows:

Let $x_i = T(x_{1i}) = T(x_{2i})$, $i = 1, \cdots, r$, where $x_{1i} \in \text{Int}\,\sigma_1$ and $x_{2i} \in \text{Int}\,\sigma_2$, be the double points of T formed by the intersection of $T(\sigma_1)$ and $T(\sigma_2)$. We will change T step by step to diminish the number of common double points in $T(\sigma_1)$ and $T(\sigma_2)$, and make the last mapping so obtained satisfy $\overline{1^\circ} - \overline{3^\circ}$.

For this let O be a common vertex of σ_1 and σ_2, and let τ_1 be an n simplex of σ_1 belonging to K_1' and having O as a vertex. In $T(\sigma_1)$ construct a broken line l_1

from x_1 to an interior point x_1' of $T(r_1)$ such that the broken line does not go through $x_2, \cdots, x_r$ or any other double points in $T(\sigma_1)$. Also construct a sufficiently small linear tube C_1 of l_1 with one end the boundary of a sufficiently small neighborhood V_1 of x_1 in $T(\sigma_2)$ and the other end the boundary of a sufficiently small n-dimensional convex neighborhood V_1' which intersects $T(\tau_1)$ only at x_1', and such that C_1 does not contain any double points of T. Now alter T to T_1 so that T_1 maps $\sigma_2 \bigcap T^{-1}(V_1)$ to $C_1 + V_1'$ and on the remaining $\bar{K} - \sigma_2 \cap$ Int $T^{-1}(V_1)$, T_1 is the same as T. Then this new mapping T_1 still has properties 1°-3°. Only its double points $x_1, \cdots, x_r$ have been changed to $x_1', x_2, \cdots, x_r$. The same technique can be applied to $x_2, \cdots, x_r$. Hence we may assume from the start that the double points $x_1, \cdots, x_r$ all lie in $T(\tau_1)$ and on any line segment $O'x_i$ has no double points except for x_i. Here $O' = T(O)$. Similarly, by slightly moving $T(\sigma_2)$ if necessary, we may assume that T has the following additional properties:

4°. For any $\tau' \in K_2'$ which does not have O as a vertex the linear subspace spanned by $T(\tau')$ does not pass through O'.

5°. For any $\tau \in K'$ and any $\tau' \in K_2'$ such that τ' does not have O as a vertex, and τ, τ' have no common vertex, then O' and the subspace spannedby $T(\tau)$ and that spanned by $T(\tau')$ are in general position.

Now let ξ_1 be the n-dimensional simplex in K_2' with x_{21} as an interion point. Obviously, ξ_1 does not have O as a vertex. Hence there exists a chain of n-simplices, $\xi_1, \cdots, \xi_s$ in K_2', such that every two consecutive simplices ξ_i, ξ_{i+1} $(i = 1, \cdots, s-1)$ have an $(n-1)$-simplex η_i in common, ξ_s has O as a vertex, but none of the other ξ_i's $(i < s)$ has O as a vertex. It follows from 5° that for any simplex $\tau \in K'$, $\tau \neq \xi_1$, when τ and ξ_1 have no vertex in common or only one, the linear subspace $L(O, r)$ spanned by O' and $T(\tau)$ intersect the linear subspace $L(\xi_1)$ spanned by $T(\xi_1)$ in a line segment $s(\tau)$ at most. When the common face of τ and ξ_1 has dim $\geqslant 1$, L (O, τ) and $L(\xi_1)$ intersect in this common face only. Hence, because of the assumption that $n > 2$, we can construct a line segment in $T(\xi_1)$ through x_1, which intersects the interior of $T(\eta_1)$, but intersects all the line segments $s(\tau)$ only at x_1, and does not pass through any other double points in $T(\xi_1)$. Denote by y_1 the intersection of this line segment with $T(\eta_1)$, and by y_0 the intersection with another $(n-1)$-face $T(\eta_0)$ of $T(\xi_1)$. Again from 4° and 5°, for any simplex $\tau \in K'$, $\tau \neq \xi_2$, the linear subspace spanned by O' and $T(\tau)$ intersect $L(\xi_2)$ at most in a line segment or a face of $T(\xi_2)$. Since $n > 2$, in the interior of $T(\xi_2)$ we may construct a line through y_1 which does not go through any double points, which does not intersect any other such line segments, and which intersects $T(\eta_2)$ at y_2. Continuing this construction, we obtain a broken line $l = y_0y_1 \cdots y_{s-1}$, where y_i is an interior point of $T(\eta_i)$. Now in $T(\eta_0)$, construct a sufficiently small $(n-1)$ -simplex η_0' with y_0 as an interior point. Through each point y_0' on the boundary of η_0', construct a straight line parallel to y_0y_1, which intersect $T(\eta_1)$ at y_1'; through y_1' construct a straight line parallel to y_1y_2, which intersect $T(\eta_2)$ at y_2'; follow the same method, construct y_3', $\cdots$, y_{s-1}'. Then for each $i = 1, 2, \cdots, s-1$, all the y_i' form the boundary of an $(n-1)$-simplex η_i' in $T(\eta_i)$, all the line segments $y_{i-1}'y_i'$ form a tube ξ_1' in $T(\xi_i)$ with $y_{i-1}y_i$ as axis. Let $\xi' = \xi_1' + \cdots + \xi_{s-1}$, and $C = \xi' + \eta_0'$. Then, because of the construction, 4°, 5°and the assumption that $O'x_1$ contains no double point other than x_1, we know that if we choose η_0' sufficiently small, C does not contain any double points. If we project C with O' as center, the resultant cone $\tilde{C}$ is an n-cell, which only intersects $T(\bar{K})$ on $C + O'\dot{\eta}_{s-1}'$, and only intersects $T(\bar{K}_1)$ at O'.

Let $I(\xi')$ be the portion of $T(\xi_1)+\cdots+T(\xi_{s-1})$ enclosed by $\xi'+\eta_0'+\eta_{s-1}'$. Now alter T to T_1 so that T_1 coincides with the original T on $\bar{K} - \sigma_2 \cap \text{Int } T^{-1}(l(\xi') + O'\eta_{s-1}')$, and T_1 maps $\sigma_2 \bigcap T^{-1}(I(\xi') + O'\eta_{s-1}')$ to $\tilde{C}$. Then the new mapping so obtained still satisfies 1°– 3°, but x_1 is no longer a double point of T_1. T_1 has no new double points.

Now use this method on each of x_2, $\cdots$, x_r; we obtain a mapping T', which satisfies $\overline{1°} - \overline{3°}$ as required.

§2 Main theorem—the necessary and sufficient condition for $K^n \subset R^{2n}$ when $n > 2$

In [1], we have introduced a system of invariants, $\Phi^m(K) \in H^m(K^*, I_{(m)})$, $m > 0$, of a finite simplicial complex K, called the m-dimensional imbedding index of K. Here K^* is the reduced two-fold symmetric product of K, $I_{(m)}$ is the additive group of integers or the mod 2 integer group, according as m is even or odd. We have also proved that $\Phi^m(K) = 0$ is a necessary condition that K be semilinearly imbeddable in R^m. In [2], we indicated that a famous theorem of Kuratowski about space curves, in the case of a complex, can be restated in the following manner: $\Phi^2(K) = 0$ is a necessary and sufficient condition for a 1-dimensional complex to be imbeddable in the plane. The purpose of this section is to prove the following theorem:

Theorem 1 *For a finite simplicial complex K of dimension $n > 2$ to be semilinearly imbeddable in R^{2n}, it is necessary and sufficient that $\Phi^{2n}(K) = 0$.*

The necessity of the condition $\Phi^{2n}(K) = 0$ has already been stated. We will prove the sufficiency as follows. We will assume that K is a Euclidean complex in a Euclidean space R^N of sufficiently high dimension N.

Let R^{2n-1} be an $(n-1)$-dimensional linear subspace of R^{2n} dividing R^{2n} into two halfspaces, R^{2n}_+ and R^{2n}_-. Let T be a linear imbedding of the $(n-1)$-skeleton K^{n-1} of K into R^{2n-1}. For any n-simplex σ of K, let $\bar{O}_\sigma$ be the barycenter of σ, let O_σ be the barycenter of the n-simplex in R^{2n-1} determined by $T(\dot{\sigma})$, let P_σ be the n-dimensional linear subspace spanned by $T(\dot{\sigma})$, and let S_σ be the unit $(n-1)$-sphere in P_σ with center O_σ. For each $x \in S_\sigma$, construct the halfline $O_\sigma x$ from O_σ to x, intersecting $T(\dot{\sigma})$ at x'. Let the distance between O_σ and x' be $\rho_\sigma(x)$. Then ρ_σ is a continuous function on the differentiable manifold S_σ. Choose arbitrary $\varepsilon > 0$, then, by an approximation theorem of Whitney, we may easily construct a continuously differentiable function f on S_σ, such that for any $x \in S_\sigma$, $\varepsilon > \rho_\sigma(x) - f(x) > 0$. For any $\alpha > 0$, let $f_{\sigma,\alpha}(x)$ be the point on the halfline $O_\sigma x$ whose distance from O_σ is $\alpha f(x)$. Then for $\alpha > 1$, $f_{\sigma,\alpha}$ is a differentiable topological mapping of S_σ to P_σ.

Now, through O_σ construct a halfline L in R^{2n}_+, orthogonal to R^{2n-1}, and choose an arbitrary point O'_σ on it. Through O'_σ construct the n-dimensional linear space P'_σ parallel to P_σ. Assume the previously choosen ε is sufficiently small. For any $x \in S_\sigma$, let the straight line through $f_{\sigma,\varepsilon}(x)$ and orthogonal to R^{2n-1} intersect P'_σ at $f'_{\sigma,\varepsilon}(x)$, let the line segment connecting $f_{\sigma,\ 1-\varepsilon}(x)$ and $f'_{\sigma,\varepsilon}(x)$ be l_x, and let π_x be the orthogonal projection onto $x'O_\sigma$ of the broken line with consecutive vertices x', $f_{\sigma,1-\varepsilon}(x)$, $f'_{\sigma,\varepsilon}(x)$ and O'_σ. Define $T(\bar{K}^{n-1})$ as before. Extend T to the interior of each n-simplex $\sigma \in K$ as follows: for any $\bar{x}' \in \dot{\sigma}$, if $x' = T(\bar{x}')$, x is the intersection of the halfline $O_\sigma x'$ with S_σ, and $T_{\bar{x}'}$ is the linear mapping of the line segment $\bar{x}'\bar{O}_\sigma$ to the line segment $x'O_\sigma$, then $T/\bar{x}'\bar{S}_\sigma \equiv \pi_x^{-1}T_{\bar{x}}$. Next, on the plane determined by $O_\sigma x$, l_x and $O'_\sigma f'_{\sigma,\varepsilon}(x)$, construct a circular arc C_x tangent to the straight line $O_\sigma x$ at $f_{\sigma,1-\varepsilon/2}(x)$ and tangent to l_x at p_x. Also construct a circular arc C'_x tangent to $O'_\sigma f'_{\sigma,\varepsilon}(x)$ at O'_σ and tangent to l_x at p'_x. Let l'_x be the segment of l_x between p_x and p'_x, and let π_x be the orthogonal projection of the union of the line segment $x'f_{\sigma,1-\varepsilon/2}(x)$, the arc C_x, the segment l'_x and the arc C'_x onto $x'O_\sigma$. Define $T'(\bar{K}^{n-1})$ as $T(\bar{K}^{n-1})$, and extend T' to the interior of each n-simplex $\sigma \in K$, so that for any $\bar{x}' \in \dot{\sigma}$, if x', x and $T_{\bar{x}'}$, are as previously defined, then $T'/\bar{x}'\bar{O} = \pi'^{-1}_x T_{\bar{x}}$. It is obvious that by choosing ε sufficiently small and suitably choosing O'_σ, we may make T into a semilinear pseudoimbedding through some subdivision of K, with no triple points, and any double points must be in the interior of the segments l'_x. Similarly, T' has no triple points, its double points agree with those of T, its estriction to the interior of every n-simplex $\sigma \in K$ is a differentiable topological mapping, T' is sufficiently close to T, and in a sufficiently small neighborhood of $\bar{K}^{n-1}$, T' coincides with T.

In the following, $\bar{x} \in \bar{K}$ will be called a singularity of T or T' if its image, $T(\bar{x}) = T'(\bar{x})$ is a double point of both T and T'.

Choose an arbitrary orientation of R^{2n} and let ϕ be the integer intersection number with respect to this orientation. Let σ_i, $i = 1, \cdots, r$ be all the n-simplices of K, and τ_α, and let $\alpha = 1, \cdots, s$ be all the $(n-1)$-simplices of K, each arbitrarily oriented, so that for any $\sigma_i * \sigma_j$ or $\sigma_i * \tau_\alpha \in K^*$, we have

$$\left.\begin{aligned} \sigma_i * \sigma_j &= (-1)^{\dim \sigma_i \dim \sigma_j} \sigma_j * \sigma_i = (-1)^n \sigma_j * \sigma_i, \\ \sigma_i * \tau_\alpha &= (-1)^{\dim \sigma_i \dim \tau_\alpha} \tau_\alpha * \sigma_i = \tau_\alpha * \sigma_i. \end{aligned}\right\} \tag{1}$$

According to the definition in [1], for any arbitrary $2n$-cell $\sigma_i * \sigma_j \in K^*$,

$$\varphi_T(\sigma_i * \sigma_j) = (-1)^n \phi(T\sigma_i, T\sigma_j) \tag{2}$$

determines the imbedding cochain $\varphi_T \in \Phi^{2n}(K) \in H^{2n}(K^*)$ of a semilinear pseudoimbedding T. But by assumption, $\Phi^{2n}(K) = 0$, so there exists a $\chi \in C^{2n-1}$, such that

$$\delta\chi = \varphi_T, \tag{3}$$

where χ is defined as

$$\chi(\sigma_i * \tau_\alpha) = C_{i\alpha}, \sigma_i * \tau_\alpha \in K^*. \tag{4}$$

Let J denote the set of all pairs of indices (i, α), such that $\sigma_i * \tau_\alpha \in K^*$. Let $a_i = \sum_\alpha |C_{i\alpha}|$, $b_\alpha = \sum_i |C_{i\alpha}|$, each summed over all α or all i, such that $(i, \alpha) \in J$. In each σ_i, choose a_i distinct interior points, all distinct from any singularities: $\bar{x}_{i_{\alpha,1}}, \cdots, \bar{x}_{i\alpha,|C_{i\alpha}|}$, $(i, \alpha) \in J$. In each τ_α, also choose b_α distinct interior points: $\bar{y}_{i_{\alpha,1}}, \cdots, \bar{y}_{i_\alpha}$, $|C_{i\alpha}|$, $(i, \alpha) \in J$. Let $T(\bar{x}_{i_{\alpha,k}}) = x_{i\alpha,k}$ and $T(\bar{y}_{i_{\alpha,k}}) = y_{i\alpha,k}$. For each $y_{i\alpha,\ k}$ construct an $(n+1)$-dimensional linear space $P_{i_{\alpha,k}}$ orthogonal to $T(\tau_\alpha)$ and passing through $y_{i\alpha,k}$. Construct an n-sphere $S_{i\alpha,k}$ in $P_{i\alpha,K}$ of sufficiently small radius $\varepsilon_{i\alpha,k} > 0$ and center at $y_{i\alpha,k}$, which intersects R^{2n-1} in the $(n-1)$-sphere $S^{(0)}_{i\alpha,k}$, dividing $S_{i\alpha,k}$ into two hemispheres: $S^+_{i\alpha,k} \subset R^{2n}_+$ and $S^-_{i\alpha,k} \subset R^{2n}_-$. Let $z_{i\alpha,k}$, be the intersection of $S^+_{i\alpha,k}$ with the halfline through $y_{i\alpha,k}$, orthogonal to R^{2n-1} and in R^{2n}_+. Let $z^+_{i\alpha,k}$ be a sufficiently small spherical neighborhood of $z_{i\alpha,k}$ in $S^+_{i\alpha,k}$. Let $z^-_{i\alpha,k}$ be what remains of $S_{i\alpha,k}$ after taking away the interior of $z^+_{i\alpha,k}$. Let the common boundary of $z^+_{i\alpha,k}$ be $z^{(0)}_{i\alpha,k}$. We orient $P_{i\alpha,k}$ so that

$$\phi(P_{i\alpha,k}, T\tau_\alpha) = (-1)^n \text{sgn } C_{i\alpha}, \tag{5}$$

where sgn $C_{i\alpha} = +1$, -1 or 0 according as $C_{i\alpha} > 0$, < 0 or $= 0$. Orient $S_{i\alpha,k}$ with respect to the orientation of $P_{i\alpha,k}$. Then, by taking sufficiently small $\varepsilon_{i\alpha,k} > 0$, we may construct, in $T(\sigma_i)$, $(n-1)$-spheres $S'_{i\alpha,k}$ with center $x_{i\alpha,k}$ with radius $\varepsilon_{i\alpha,k}$ such that the $(S'_{i\alpha,k})'s$ are mutually disjoint on $T(\sigma_i)$ and they do not contain any double points in any of their interiors. Let $T^i_{i\alpha,k}$ be the interior of $S'_{i\alpha,k}$, with the orientation induced from that of σ_i. Let $T^e_{i\alpha,k}$ be the intersection of $S'_{i\alpha,k}$ and $\dot{\sigma}_i$, also with the orientation of σ. In R^{2n}_+, construct an (infinitely) differentiable simple arc $C_{i\alpha,k}$, connecting $x_{i\alpha,k}$ and $z_{i\alpha,k}$ so that the arcs mutually do not meet, and meet $T(\bar{K})$, $T'(\bar{K})$ and all the $S_{i\alpha,k}$ only at $x_{i\alpha,k}$ and $z_{i\alpha,k}$. Following construction A of §1, we may construct a differentiable tube $C_{i\alpha,k}$ in a sufficiently small neighborhood of $C_{i\alpha,k}$, connecting $S'_{i\alpha,k}$ and $z^{(0)}_{i\alpha,k}$, such that after suitably orienting $C_{i\alpha,k}$, the orientation on $S'_{i\alpha,k}$ induced by $C_{i\alpha,k}$ and $T^i_{i\alpha,k}$ are the same, while the orientation on $z^{(0)}_{i\alpha,k}$ induced by $C_{i\alpha,k}$ and $z^-_{i\alpha,k}$ are opposite. In other words, $T^e_{i\alpha,k} + C_{i\alpha,k} + z^-_{i\alpha,k}$ is a relative cycle mod $\dot{\sigma}$. Now alter T'/σ_i to T'', mapping $T'^{-1}(T^i_{i\alpha,k}) \subset \sigma_i$ to $C_{i\alpha,k} + z^-_{i\alpha,k}$, $(i, \alpha) \in J$; T'' coincides with T' on the remainder of σ_i. The mapping may be smoothed out near $S'_{i\alpha,k}$ and $z^{(0)}_{i\alpha,k}$ by at most infinitesimal changes, so we may assume that T'' is a differentiable topological mapping in the interior of σ_i. By choosing all the $\varepsilon_{i\alpha,k}$'s sufficiently small, we may also assume that all the $C_{i\alpha,k}$ are mutually disjoint, and only meet $T'(\bar{K})$ at $S'_{i\alpha,k}$ Then, $T'' : \bar{K} \to R^{2n}$ is continuous and possesses the following properties: T'' is a differentiable topological mapping in the interior of each σ_i; T'' has no tripie points, and each of its double points is in the interior of both $T''\sigma_i$ and $T''\sigma_j$; the tangent planes of $T''\sigma_i$ and $T''\sigma_j$ at this double point intersect only at this point. Also, since for each $\sigma_i \in K$, T'' coincides with T' or T in a sufficiently small neighorhoo of $\dot{\sigma}_i$, T'' is a linear imbedding on a sufficiently small neighborhood of $\bar{K}^{n-1}$.

Let $\sigma_i * \sigma_j \in K^*$, and

$$\partial\sigma_i = \sum_\alpha \eta_{i\alpha}\tau_\alpha, \partial\sigma_j = \sum_\beta \eta_{j\beta}\tau_\beta \quad (\eta_{i\alpha}, \eta_{j\beta} = \pm 1 \text{ or } 0),$$

Let $T_{i\beta,k}$ denote the union of the interior of $S_{i\beta,k}$ in $P_{i\beta,k}$ and $S_{i\beta,k}$, so that $T_{i\beta,k}$ is compatible with $S_{i\beta,k}$, and we have $(k = 1, 2, \cdots, |C_{i\beta}|)$

$$\begin{aligned}\phi(S_{i\beta,k}, T'\sigma_j) &= \phi(\partial T_{i\beta,k}, T'\sigma_j) = (-1)^{n+1}\phi(T_{i\beta,k}, \partial T'\sigma_j)\\ &= (-1)^{n+1}\eta_{j\beta}\phi(P_{i\beta,k}, T'\tau_\beta) = (-1)^{n+1}\eta_{j\beta}\phi(P_{j\beta,k}, T\tau_\beta)\\ &= -\eta_{j\beta}\text{sgn } C_{i\beta},\end{aligned}$$

Similarly, we have ($l = 1, 2, \cdots, |C_{j\alpha}|$),

$$\begin{aligned}\phi(T''\sigma_i, S_{j\alpha,l}) &= \phi(T''\sigma_i, \partial T_{j\alpha,l}) = (-1)^n\phi(\partial T''\sigma_i, T_{j\alpha,l})\\ &= (-1)^n\eta_{i\alpha}\phi(T''\tau_\alpha, P_{j\alpha,l}) = (-1)^n\eta_{i\alpha}\phi(T\tau_\alpha, P_{j\alpha,l})\\ &= (-1)^n\eta_{i\alpha}\cdot(-1)^{(n-1)(n+1)}\phi(P_{j\alpha,1}, T\tau_\alpha)\\ &= (-1)^{n+1}\eta_{i\alpha}\text{sgn } C_{j\alpha}.\end{aligned}$$

Hence

$$\begin{aligned}&\phi(T''\sigma_i, T''\sigma_j) - \phi(T'\sigma_i, T'\sigma_j)\\ =&\phi(T''\sigma_i - T'\sigma_i, T'\sigma_j) + \phi(T''\sigma_i + T''\sigma_j - T'\sigma_j)\\ =&\sum_{\beta,k}\phi(C_{i\beta,k} - T^i_{i\beta,k} + S_{i\beta,k} - z^+_{i\beta,k}, T'\sigma_j)\\ &+\sum_{\alpha,l}\phi(T''\sigma_i, C_{j\alpha,l} - T^i_{j\alpha,l} + S_{j\alpha,l} - z^+_{j\alpha,l}),\end{aligned}$$

Since $C_{i\beta,k}$, $T^i_{i\beta,k}$ and $z^+_{i\beta,k}$ do not meet $T'\sigma_j$, and $C_{j\alpha,l}$, $T^i_{j\alpha,l}$ and $z^+_{j\alpha,l}$ also do not meet $T''\sigma_i$, we may simplify the above formula to

$$\begin{aligned}&=\sum_{\beta,k}\phi(S_{i\beta,k}, T'\sigma_j) + \sum_{\alpha,l}\phi(T''\sigma_i, S_{j\alpha,l})\\ &=\sum_\beta -|C_{i\beta}|\,\eta_{j\beta}\text{sgn } C_{i\beta} + \sum_\alpha(-1)^{n+1}|C_{j\alpha}|\,\eta_{i\alpha}\text{sgn } C_{j\alpha}\\ &=-\sum_\beta C_{i\beta}\eta_{j\beta} + (-1)^{n+1}\sum_\alpha C_{j\alpha}\eta_{i\alpha}.\end{aligned} \tag{6}$$

On the other hand, we have

$$\begin{aligned}(-1)^n\phi(T'\sigma_i, T'\sigma_j) &= (-1)^n\phi(T\sigma_i, T\sigma_j) = \varphi_T(\sigma_i * \sigma_j)\\ &= \delta\chi(\sigma_i * \sigma_j) = \chi(\partial\sigma_i * \sigma_j) + (-1)^n\chi(\sigma_i * \partial\sigma_j)\\ &= \sum_\alpha \eta_{i\alpha}\chi(\tau_\alpha * \sigma_j) + (-1)^n\sum_\beta \eta_{j\beta}\chi(\sigma_i * \tau_\beta)\\ &= \sum_\alpha \eta_{i\alpha}\chi(\sigma_j * \tau_\alpha) + (-1)^n\sum_\beta \eta_{j\beta}\chi(\sigma_i * \tau_\beta)\\ &= \sum_\alpha C_{j\alpha}\eta_{i\alpha} + (-1)^n\sum_\beta C_{i\beta}\eta_{j\beta}.\end{aligned} \tag{7}$$

By comparing (6) and (7), we get

$$\phi(T''\sigma_i, T''\sigma_j) = 0, \quad \sigma_i * \sigma_j \in K^*.$$

Since the intersection number of $T''\sigma_i$ and $T''\sigma_j$ at each double point is ± 1, we know that $T''\sigma_i$ and $T''\sigma_j$ must intersect at an even number of points. Let this number be $2n_{ij}$. Then the intersection number at each of n_{ij} double points is $+1$, and the intersection number at each of the remainine n_{ij} double points is -1. Since T'' is a differentiable topological mapping in the interiors of σ_i and σ_j, and at every double point, the tangent planes of $T''\sigma_i$ and $T''\sigma_j$ intersect only at this point, by construction B of §1, we may alter the restriction of T'' on σ_i and σ_j, to obtain a mapping which is still a differentiable topological mapping on σ_i and σ_j, but whose images of σ_i and σ_j will be disjoint; and there are no new double points resulting from the intersection of the images of σ_i and σ_j with the image of any other σ_k. Now apply this construction to every $\sigma_i * \sigma_j \in K^*$ to obtain a mapping $T_0 : \bar{K} \to R^{2n}$, with the following properties:

1°. T_0 is a differentiable topological mapping in the interior of every σ_i.

2°. T_0 is a linear imbedding on a sufficiently small neighborhood of $\bar{K}^{n-1}$.

3°. T_0 has no triple points. For every double point, $T_0(x) = T_0(y) = p$, x and y must be in the interior of two separate simplices σ_i and σ_j, and σ_j have a ccmmon vertex.

Because of 1°, 2° and a theorem of Cairns (e.g., see [6], Theorem 2), we may construct a sufficiently close approximation T_0' of T_0 such that for any σ_i, T_0' coincides with T_0 in some sufficiently small neighborhood of $\dot{\sigma}_i$. By choosing this approximation sufficiently close, we may obviously still preserve property 3°. So following the Van Kampen construction (§1, construction C), we may alter T_0' to remove all the double points. The resulting mapping $h : \bar{K} \to R^{2n}$ is a semilinear imbedding of K through some division of it.

Thus, we have obtained a semilinear imbedding of K to R^{2n} and proved the sufficiency of the theorem.

For an arbitrary Hausdorff space X, we have introduced a system of topological invariants $\Phi^m(X) \in H^m(X^*, I_{(m)})$, $m > 0$ of X, called the m-dimensional imbedding index of X. Here, X^* is the twofold symmetric product of X. We have also proved that when X is a finite polyhedron and K is a simplicial subdivision of it, under a fixed isomorphism $H^m(X^*, I_{(m)}) \approx H^m(K^*, I_{(m)})$, $\Phi^m(X)$ and $\Phi^m(K)$ are cohomologous. Hence the above theorem has the following corollary:

Theorem 1′ *For a finite polyhedron X of dimension $n > 2$ to be semilinearly imbeddable in R^{2n}. it is necessary and sufficient that $\Phi^m(X) = 0$.*

§3 Some suficient conditions for $K^n \subset R^{2n}$

The purpose of this section is to derive, from the main Theorem 1 of last section, some sufficient conditions for an n-dimensional finite complex K to be semilinearly imbeddable in R^{2n}. These conditions are either determined by the homology of K, or easily derived from the complex structure of K, see Theorems 2-6 below.

Theorem 2 *A finite simplicial complexof dimension $n \neq 0$ is semilinearly imbeddable in R^{2n} if H_n $(K, \bmod 2) = 0$.*

Proof. The theorem is obvious if $n = 1$. So assume $n > 2$. Let $\tilde{K}^*$ be the subcomplex of $K \times K$ spanned by all cells $\sigma \times \tau$, where σ and $\tau \in K$ have no common vertex. By assumption,

$$H_{2n}(K \times K, \bmod 2) \approx H_n(K, \bmod 2) \otimes H_n(K, \bmod 2) = 0,$$
$$H_{2n+1}(K \times K, \tilde{K}^*; \bmod 2) = 0$$

and from the exact sequence of $(K \times K, \tilde{K}^*)$,

$$\cdots \to H_{2n+1}(K \times K, \tilde{K}^*; \bmod 2) \to H_{2n}(\tilde{K}^*, \bmod 2)$$
$$\to H_{2n}(K \times K, \bmod 2) \to \cdots$$

it follows that $H_{2n}(\tilde{K}^*, \bmod 2) = 0$. Since the highest dimension of $\tilde{K}^*$ is $2n$, this implies that $Z_{2n}(\tilde{K}^*, \bmod 2) = 0$, where Z_{2n} denotes the group of $2n$-cycles. In other words, there exist no mod 2 $2n$-cycles in $\tilde{K}^*$.

Obviously, $\tilde{K}^*$ is a twofold covering complex of K^*, where the covers for $\sigma * \tau \in K^*$ are the cell $\sigma \times \tau$ and $\tau \times \sigma$. If $\rho_2 \Sigma_{i=1}^s \sigma_i * \tau_i$ is a mod 2 $2n$-cycle of K^*, where the $\sigma_i * \tau_i$ are mutually disjoint, and ρ_r denotes reduction mod r, then $\rho_2 \Sigma \sigma_i \times \tau_i + \rho_2 \Sigma \tau_i \times \sigma_i$ will be a nonzero mod 2 $2n$-cycle of K^*, contradicting the above. So we must have $Z_{2n}(K^*, \bmod 2) = 0$, and hence $Z_{2n}(K^*, \bmod 2m) = 0$ for any integer $m \geqslant 0$. So, assuming ϕ to be a cocycle of Φ^{2^n}, ϕ mod 2 is orthogonal to all mod $2m$ $2n$-cycles of K^*, i.e. $\phi \cdot z = 0 \bmod 2m$ for all $z \in Z_{2n}(K^*, \bmod 2m) = 0$. Next, from Theorem 8 of [1], $\Phi^{2n} = \frac{1}{2}\delta\Phi^{2n-1}$, so $\phi = \frac{1}{2}\delta\phi'$, where $\rho_2\phi' \in \Phi^{2n-1} \in H^{2n-1}(K^*, \bmod 2)$. Thus, for any arbittary mod $(2m+1)$ $2n$-cycle z of K^*, if we let z' be an integral chain such that $\rho_{2m+1} z' = z$, we have $\phi \cdot z' = \frac{1}{2}\delta\phi' \cdot z' = \frac{1}{2}\phi' = 0$, mod $2m+1$.

Hence, whether m is even or odd, ϕ is orthogonal to any mod m $2n$-cycle of K^*. From a theorem of Whitney [Whitney. *On matrices of integers and combinational topology.* Duke Math., 1937, J.3: 35-45], ϕ is a coboundary, or $\Phi^{2^n} = 0$. Since $n > 2$, the theorem follows from Theorem 1.

Theorem 2 has the following corollary.

Theorem 2′ *A finite polyhedron X of dimension $n \neq 2$ can be topologically imbedded in R^{2n} if H_n $(X, \text{mod } 2) = 0$.*

Theorem 3 *A finite simplicial complex K of dimension $n \neq 2$ can be semilinearly imbedded in R^{2n} if $H^n(K) = 0$.*

This theorem obviously follows from Theorem 2 and the lemma:

Lemma *If, for an n-dimensional finite complex K, $H^n(K) = 0$,* then $H_n(K,$ mod 2$) = 0$.

Proof. From the universal coefficient theorem, we have

$$H_n(K, \text{mod } 2) \approx H_n(K) \otimes I_2 + \text{Tor}(H_{n-1}(K), I_2),$$
$$H^n(K) \approx \text{Hom}(H_n(K), I) + \text{Ext}(H_{n-1}(K), I),$$

where I is the additive group of integers and I_2 is the mod 2 integer group. By formula (2) and the assumption that $H^n(K) = 0$, it follows that Hom $(H_n(K), I) = 0$ and Ext $(H_{n-1}(K), I) = 0$. From the former, it follows that $H_n(K)$ must be a finite group. Since K is of dimension n, there is no torsion in the nth dimension, so $H_n(K) = 0$. From the latter, it follows that $H_{n-1}(K)$ has no element of finite order, hence Tor $(H_{n-1}(K), I_2) = 0$. From formula (1), we have $H_n(K,$ mod $2) = 0$.

Theorem 3′ *A finite polyhedron X of dimension $n \neq 2$ can be topologically imbedded in R^{2n} if $H^n(X) = 0$.*

In *Topologie* I of Alexandroff and Hopf (AH for brevity), Chapter 7, §l, there was defined the socalled closed complex and irreducible closed complex (irreduzible geschlossene Komplexe). We will prove.

Theorem 4 *If a finite simplicial complex K of dimension $n \neq 2$ is an irreducible closed complex, then K can be sernilinearly imbedded in R^{2n}.*

Proof. If $n = 1$, then from p.284 of AH, Theorem 12, K is a simple closed polygon, so the theorem is obvious. So we may assume $n > 2$. It follows from AH, Chapter 7, §1 No. 4 Theorem 5 and Chapter 7, §1 No. 5 that K has a natural modulus m, $m = 0$ or $m \geqslant 2$. When $m \geqslant 2$, there exist a mod m n-cycle ${\rho_m}^z$ in K, where $z = \Sigma a_i \sigma_i$, the $a_i' s$ are nonzero integers, a_i and m are relatively prime and $\sum$ is summed over all n-simplices σ_i of K; so that for any coefficient group G, $H_n(K, G) = Z_n$ $(K,\ G)$ consists of all the cycles gz, where $g \in G$ and $mg = 0$. So when m is odd, $H_n(K,\ \mathrm{mod}\ 2) = 0$, and it follows from Theorem 3 that K can be semilinearly imbedded in R^{2n}. If $m \geqslant 2$ is even, then $Z_n(K,\ \mathrm{mod}\ 2)$ has only one nonzero mod 2 cycle, $z_2 = \rho_2\Sigma a_i\alpha_i = \rho_2\Sigma\sigma_i$. From the Künneth Theorem, we know that $Z_{2n}(K \times K,\ \mathrm{mod}\ 2) = H_{2n}(K \times K,\ \mathrm{mod}\ 2)$ also has only one nonzero mod 2 cycle, $z_2 \otimes z_2 = \rho_2\Sigma\sigma_i \times \sigma_j$. Now let $\tilde{z}^* = \rho_2 \sum' \alpha_{ij}\sigma_i \times \sigma_j$ be a mod 2 $2n$-cycle or $\tilde{K}^*$, where $\alpha_{ij} = 0$ or 1 and $\sum'$ is summed over all pairs of indices $(i,\ j)$ such that σ_i and σ_j have no common vertices. Then, viewing $\tilde{z}^*$ as a mod 2 cycle of $K \times K$, we should have $\tilde{z}^* = a(z_2 \otimes z_2)$, where $a = 0$ or 1. But $z_2 \otimes z_2$ has terms of the form $\sigma_i \times \sigma_i$, while $\tilde{z}^*$ cannot have such terms, so a must be 0, and $H_{2n}(\tilde{K}^*,\ \mathrm{mod}\ 2) = Z_{2n}$ $(\tilde{K}^*,\ \mathrm{mod}\ 2) = 0$. From reasoning similar to that in the proof of Theorem 2, we have $\Phi^{2^n}(K) = 0$. Hence, from Theorem 1, K may be semilinearly imbedded in R^{2n}.

Next, assume $m = 0$. Then from Theorem 4 of AH, Chapter 7, §1 No.4, K has an n-cycle with integral coefficient, $z = \sum \sigma_i$, where the $\sum$ is summed over all n-simplices σ_i of K, each with suitable orientation, such that for any coefficient group G, $H_n(K, G) = Z_n(K, G)$ consists of all cycles gz, where $g \in G$ is arbitrary. In particular, $Z_n(K,\ \mathrm{mod}\ 2)$ has only one nonzero mod 2 cycle, $\rho_2 z = \rho_2 \sum \sigma_i$. By the same reasoning, we have $H_{2n}(\tilde{K}^*,\ \mathrm{mod}\ 2) = 0$, bence $\Phi^{2n}(K) = 0$, and so K can be semilinearly imbedded in R^{2n}. The theorem is now completely proved.

From AH, Chapter 10, §3 No. 5 and Theorem 4 of AH, Chapter 13, §4 No. 4, whether a finite complex K is a closed complex or not is a topological invariant of $\bar{K}$. Similarly, from AH, Chapter 8, §4 No. 7, whether a finite complex K is an irreducible closed complex or not, is also a topological invariant $\bar{K}$. Let us call the space underlying a closed complex or an irreducible closed complex, a closed polyhedron or an irreducible closed polyhedron. The Theorem 4 has the following corollary:

Theorem 4′ *If X is an irreducible closed polyhedron of dimension $n \neq 2$, then X can be topologically imbedded in R^{2n}.*

Theorem 5 (Van Kampen) *If any $(n-1)$-simplex of the n-dimensional complex K is at most the face of two n-simplices, then K can be semilinearly imbedded in R^{2n}.*

Proof. The theorem is obvious if $n < 2$.

Now assume $n = 2$. Also assume first that K is a 2-dimensional homogeneous complex. Then K must be constructed as follows: let K_i, $i = 1, \cdots, r$ be some complexes, obtained by suitably subdividing some connected surfaces, with or without boundary; then K is obtained by identifying some vertices of the (K_i)'s. From the well-known result of the classification of surfaces, we know that there exist semilinear imbeddings $T_i : \bar{K}_i \to R^4$. We may assume that the $T_i(\bar{K}_i)$'s are mutually disjoint. Then the $T_i's$ together determine a semilinear imbedding $T : \Sigma\bar{K}_i \to R^4$, where $T/\bar{K}_i \equiv T_i$. Now let $\overline{a}_j$, $j = 1, 2, \cdots, N$, be all the vertices of the $(K_i)'s$ that will become the Vertices of K only after identification. Assume also that all the $\overline{a}_j$ are divided into several families $(\overline{a}_{j_1}, \cdots, \overline{a}_{j_t})$, $j_1 < j_2 < \cdots < j_t$, where the vertices in the same family will be identified, while those of different families will not be identified. Let $T(\bar{a}_j) = a_j$. For each family $(\overline{a}_{j_1}, \cdots, \overline{a}_{j_t})$, we may construct a family of simple broken lines $l_{j_2}, \cdots, l_{j_t}$, connecting $a_{j_1}, a_{j_2}; \cdots ; a_{j_1}, a_{j_t}$ respectively, such that these broken lines and $T(\bar{K}_i)$ are mutually disjoint, except for the common endpoints. Let $j = j_2, \cdots$, or j_t, $\bar{a}_j$ be a vertex of K_{ij} and V_j be a sufficiently small linear neighborhood of a_j in $T(\bar{K}_{ij})$. Then, in a sufficiently small neighborhood of each broken line l_{j_r}, we may construct a "linear conical surface" C_{j_r} with "vertex" a_{j_1}, "base" the boundary of V_{j_r}, "axis" the broken line l_{j_r}, and we can assume that these conical surfaces are mutually disjoint except for the vertices, and only meet $T(\sum \bar{K}_i)$ at the vertices and the bases. Alter T to $T' : \sum \bar{K}_i \to R^4$ so that T' coincides with T on $\sum K_i - \sum$ Int V_j, and T' maps V_j to C_j. Then T' can be viewed as a semilinear imbedding of K into R^4 as required. Next, still assume $n = 2$, but let K be arbitrary. Let K' be the subcomplex of K, consisting of all 2-simplices of K and their faces. From the preceding, we may construct a semilinear imbedding $T : \bar{K}' \to R^4$. Obviously, this imbedding can be extended to a semilinear imbedding of K. Hence the theorem is proved for $n = 2$.

Now we prove the theorem, assuming $n > 2$.

First, assume K is a regular connected (regulärer zusammenhängender, see AH, Chapter 4) n-dimensional homogeneous complex. Then, for any two n-simplices σ and $\sigma' \in K$ with no common vertices, there exists a chain of n-simplices in K, $\sigma_1 = \sigma$, $\sigma_2, \cdots, \sigma_r = \sigma'$ such that σ_i and σ_{i+1} $(i = 1, \cdots, r-1)$ have only an $(n-1)$-simplex τ_i as their common face. Among $\sigma_1, \cdots, \sigma_{r-1}$, let σ_s be the last simplex having no common vertex with σ_r, $1 \leqslant s < r-1$. Then after suitably orienting the σ's and τ's in K^*, we have

$$\{\sigma_1 * \sigma_r\} - \{\sigma_2 * \sigma_r\} = \delta\{\tau_1 * \sigma_r\},$$

$$\{\sigma_2 * \sigma_r\} - \{\sigma_3 * \sigma_r\} = \delta\{\tau_2 * \sigma_r\},$$

$$\{\sigma_{s-1} * \sigma_r\} - \{\sigma_s * \sigma_r\} = \delta\{\tau_{s-1} * \sigma_r\},$$

$$\{\sigma_s * \sigma_r\} = \delta\{\tau_s * \sigma_r),$$

there $\{\xi * \eta\}$ $(\xi, \eta \in K, \xi * \eta \in K^*)$ represents the integral cochain with value 1 on $\xi * \eta$, but value 0 on any other cell of K^*. Adding these formulas, we see that the cochain $\{\sigma_1 * \sigma_r\} = \{\sigma * \sigma'\}$ is a coboundary. Hence $H^{2n}(K^*) = 0$, and in particular, $\Phi^{2n}(K) = 0$. By the assumption that $n > 2$ and Theorem 1, K can be semilinearly imbedded in R^{2n}.

Next, consider the general case for $n > 2$. Let K' be the homogeneous complex consisting of all the n-simplices of K and their faces. Let L be the subcomplex consisting of all the r-simplices, $r \leqslant n-1$, which are not faces of any n-simplices. According to AH, Chapter 4, §5 No. 8, K' can be decomposed into regulärer connected regular components (regulärer Komponenten) $K_1, \cdots, K_s$, such that

$$K = K_1 + \cdots + K_s + L,$$

where the common portion of any two subcomplexes on the right-hand side is a subcomplex of dimension at most $n-2$. Thus

$$K^* = \sum_{i \leqslant j} K_{ij}^* + \sum_i L_i^* + L^*,$$

where K_{ij}^*, L_i^* and L^* are formed by cells of the form $\sigma * \tau$, σ and τ have no common vertices, $\sigma \in K_i$, $\tau \in K_j$, or $\sigma \in K_i$ and $\tau \in L$, or $\sigma, \tau \in L$. It is easy to see that L^* is a subcomplex of dimension at most $2n-2$, each L_i^* is a subcomplex of dimension at most $2n-1$ and the common portion of any two subcomplexes among the (K_{ij}^*)'s and (L_i^*)'s is a subcomplex of dimension at most $2n-2$. Hence, any mod $2m$ $(m \geqslant 0)$, $2n$-cycle z of K^* can be written as

$$z = \sum_{i \leqslant j} z_{ij},$$

where each z_{ij} is a mod $2m$ chain of K_i^*, indeed it is a mod $2m$ cycle.

Now order the vertices of K in a sequence $a_0 < a_1 < \cdots < a_n$, such that the vertices of K_i $(i = 1, \cdots, s-1)$ are all in front of those of K_{1+1}, but not in front of those of $K_1, \cdots$, or K_{i-1}. At the end of the sequence are the vertices that are in L, but not in any of the K_i's. Otherwise, the ordering is arbitrary. It follows from Theorem 11 of [1], that, with respect to this ordering, there exists a cocycle in $\Phi^{2n}(K)$,

$$\varphi^{2n} = \sum \{(a_{i_0} \cdots a_{i_n}) * (a_{j_0} \cdots a_{j_n})\},$$

where $\sum$ is summed over all possible groups of indices (i, j) such that $i_0 < j_0 < i_1 < \cdots < i_n < j_n$. Let $\sigma_i = \{a_{i_0}, \cdots, a_{i_n}) \in K_i$, $\sigma_j = (a_{j0}, \cdots, a_{j_n}) \in K_j$, $i_0 < \cdots < i_n$, $j_0 < \cdots < j_n$ and $\sigma_i * \sigma_j \in K_{ij}$. If $i < j$, then, since K_i and K_j has at most an $(n-2)$-simplex in common, by choice of the order of the vertices, we cannot have either $i_0 < j_0 < i_1 < \cdots < i_n < j_n$ or $j_0 < i_0 < j_1 < \cdots < j_n < i_n$. So $\phi^{2n}(\sigma_i * \sigma_j) = 0$. Therefore, by formula (3), we have $\rho_{2m}\phi^{2n}(z) = \sum \rho_{2m}\phi^{2n}(z_{ii})$. Now let λ_i be the inclusion mapping of K_{ii}^* into K^*. Then obviously, $\lambda_i^{\#}\phi^{2n} \in \Phi^{2n}(K) \in H^{2n}(K_{ii}^*)$. But from the last part of this proof, $\Phi^{2n}(K_i) = 0$. Hence $\lambda_i^{\#}\phi^{2n} \sim 0$, and $\rho_{2m}\phi^{2n}(z_{ii}) = \lambda_i^{\#}\rho_{2m}\phi^{2n}(z_{ii}) = 0 \bmod 2m$. Thus $\rho_{2m}\phi^{2n}(z) = 0$, i.e. $\phi^{2n} \bmod 2m$ is orthogohal to any mod $2m$ $2n$-cycles of K^*. Analugous to the last part of the proof of Theorem 2, we have, for any integer $m = 0$ or $\geqslant 2$, that $\phi^{2n} \bmod m$ is orthogonal to any mod m $2n$-cycles of K^* Thus, $\phi^{2n} \sim 0$, or $\Phi^{2n}(K) = 0$. By the assumption that $n > 2$ and Theorem 1, K can be semilinearly imbedded in R^{2n}. Thus Theorem 5 is completely proved.

Remark. Theorem 5 above is Theorem 4 of Van Kampen's original paper [3]. His corrected prove (see Lemma 6 of [4]), uses a deformation theorem of [3] ([3], Theorem 2), so is different from the proof given here.

Theorem 5 has the following corollary:

Theorem 6 *Every combinatorial manifold (with or without boundary) of dimension n can be semilinearly imbedded in R^{2n}.*

Bibliography

[1] Wu Wen-jun (Wu Wen-chün). *On the realization of complexes in Euclidean spaces.* I. Acta Math. Sinica, 1955, 5: 505-552. MR 17, 883.

[2] ——. *On the realization of complexes in Euclidean spaces.* II. Acta. Math. Sinica, 1957, 7: 79-101. MR 20 # 3536.

[3] Van Kampen E R. *Komplexe in euklidischen Räumen.* Abh. Math. Sem. Hamburg, 1932, 9: 72-78.

[4] ——. Supplement to the preceding, ibid, 152-153.

[5] Whitney H. *The self-intersections of a smooth n-manifold in 2n-space.* Ann. of Math., 1944, 45: 220-246. MR 5, 273.

[6] Whitehead J H C. *On C^1-complexes.* Ann. of Math., 1940, 41: 809-824. MR 2, 73.

26. On the Reduced Products and the Reduced Cyclic Products of a Space*

Given a topological space or a system of topological spaces, we may associate with it in an invariant manner various spaces, e.g., the universal covering space of a space in case it is arcwise connected, the space of closed curves with fixed initial and end points of a space (the so-called "espace de lacets"), the topological product of two or more spaces, etc. It is usually of importance to study the topological properties of such associated spaces in terms of those of the original spaces. A typical example is furnished by the so-called Künneth formulas of product spaces which express the homology of the topological product $X_1 \times X_2$ of two spaces in terms of the homology of the two spaces X_1 and X_2 themselves. In particular, the integral cohomology group $H^*(X_1 \times X_2)$ contains as subgroup the tensor product

$$H^*(X_1) \otimes H^*(X_2)$$

of those of X_1 and X_2. However, it seems to be more interesting to take another point of view. We may study the properties of these associated spaces not as our "final objective" but only as "intermediary procedures" with the aim to get more knowledge of the original spaces. A typical example is again furnished by the Künneth formulas of product spaces. Let $\tilde{X}_2$ be the topological product of the space X by itself, $d : X \to \tilde{X}_2$ the diagonal map defined by $d(x) = (x, x)$, where $x \in X$. For any two integral cohomology classes $U_i \in H^{r_i}(X)$, we may determine by the Künneth formulas an integral cohomology class $U_1 \otimes U_2 \in H^{r_1+r_2}(\tilde{X}_2)$. It follows that $d^*(U_1 \otimes U_2) \in H^{r_1+r_2}(X)$ is a well-defined class of X which depends on U_1 and U_2 and the product $\smile$ defined by $U_1 \smile U_2 = d^*(U_1 \otimes U_2)$ induces then a ring structure in the cohomology group $H^*(X)$. Similarly, the consideration of the p-fold topological product $\tilde{X}_p = \underbrace{X \times \cdots \times X}_{p}$ of a space X with respect to the cyclic transformation $t : (x_1, \cdots, x_p) \to (x_2, \cdots, x_p, x_1), x_i \in X$, has led to the discovery of the so-called

---

* *Jber Deutsch. Math. Verein*, 1958, 61, Abt. 1: 65-75. Nach einem Vortrag auf der Tagung der Deutschen Mathematiker-Vereinigung in Dresden im September 1957.

Steenrod powers which are homomorphisms $St^i_{(p)} : H^r(X, I_p) \to H^{r+i}(X, I_p)$①. We may also remark that the homotopy groups $\pi_i(X)$ of an arcwise connected space X were originally defined by Hurewicz as fundamental groups of the $(i-1)$-fold espace de lacets of X. Thus, almost all the most powerful invariants in algebraic topology are more or less connected with the general principle given above.

Let X be an arbitrary space, $\tilde{X}_p$ the p-fold topological product $\underbrace{X \times \cdots \times X}_{p}$ of X with itself, $\tilde{\Delta}_p$ the diagonal in $\tilde{X}_p$ consisting of all points $(x_1, \cdots, x_p)$ with $x_1 = \cdots = x_p \in X$, and

$$t : (x_1, \cdots, x_p) \to (x_2, \cdots, x_p, x_1)$$

the cyclic transformation in $\tilde{X}_p$. The space obtained from $\tilde{X}_p$ by identifying points transformable into each other by t or its powers is called the p-fold *cyclic product* of X and will be denoted by $X_p (= \tilde{X}_p/t)$. Denote the natural projection of $\tilde{X}_p$ onto X_p by π. The image $\pi(\tilde{\Delta}_p)$ of $\tilde{\Delta}_p$ will be called the *diagonal* in X_p and will be denoted by Δ_p. The spaces obtained from $\tilde{X}_p$ and X_p by removing the diagonals $\tilde{\Delta}_p$ and Δ_p will be called the *p-fold reduced product* and the *p-fold reduced cyclic product* of X and will be denoted by $\tilde{X}^*_p$ and X^*_p, respectively. The study of $\tilde{X}_p$ and X_p, as already shown above, has led us to the notions of cohomology rings and cohomology operations like Steenrod powers. We shall show in what follows that the study of the reduced products and reduced cyclic products may furnish new topological invariants of the given space and show their applications to certain kinds of problems②.

First of all let us remark that, while the homotopy types of the spaces $\tilde{X}_p$ and X_p are completely determined by the homotopy type of X itself, this is not the case for the reduced products $\tilde{X}^*_p$ and X^*_p. For example, a segment and a triangle are both contractible to a point, but their 2-fold reduced products are of different homotopy types, one being connected while the other is not. It follows that, while the cohomology ring structure and the Steenrod powers arisen from the study of p-fold products are not only topological invariants, but also invariants of homotopy type, we would expect that the study of the reduced products will lead us to topological invariants which are not invariants of the homotopy type. Such invariants, of which we know very little except isolated ones like dimension, would be useful in problems

① I_p denotes the group of integors mod Pand I denotes the additive groups of integers.

② The products $\tilde{X}^*_p$ and X^*_p for a finite polyhedron, X, were first introduced and studied from the above point of view in [1].

of a topological, but not of a homotopic character, e.g., the problem of topological classification of spaces, the problem of imbedding one space in another, the problem of isotopy, etc.

A simple invariant [1] arisen in this manner is the Euler-Poincaré characteristic $\chi(\tilde{X}_p^*) = \tilde{\chi}_p^*(X)$, if it is defined, of the p-fold reduced product $\tilde{X}_p^*$ of X. If X is a finite polyhedron with a simplicial decomposition K, then $\chi_p^*(X)$ may be determined in terms of the simplicial structure of K. For this let us state first the following①.

Lemma[1] For a finite simplicial complex K let $\tilde{K}_p^*$ be the subcomplex of the p-fold product complex $\tilde{K}_p = \underbrace{K \times \cdots \times K}_{p}$ formed of all cells $\sigma_1 \times \cdots \times \sigma_p$ for which no vertex of K is common to all the simplexes σ_i. Then $|\tilde{K}_p^*|$ is a deformation retract of $|\tilde{K}|_p^*$ for which the deformation may be chosen to be commutable with the cyclic transformation $t: (x_1, \cdots, x_p) \to (x_2, \cdots, x_p, x_1)$ in $|\tilde{K}|_p^*$.

Let $a_{d_1\cdots d_p}(K)$ be the number of ordered p-tuples of simplexes $(\sigma_1, \cdots, \sigma_p)$ of K such that σ_i is of dimension d_i and no vertex of K is common to all of them. Then by the lemma above we have

$$\tilde{\chi}_p^*(X) = \sum (-1)^{d_1+\cdots+d_p} a_{d_1\cdots d_p}(K).$$

Let us remark that the usual definition of homology and cohomology groups of a complex requires only the knowledge of the face relations of two simplexes in the complex for which the dimensions differ only by I. W. Mayer [2] has indroduced a kind of homology which involves the face relations between simplexes of arbitrary dimensions. It turns out, however, that the Mayer homology groups are determined by the usual homology groups and give us nothing new [3]. On the other hand, the invariants $\tilde{\chi}_p^*(X)$ which also involve general face relations are topological invariants but not invariants of homotopy type so that they are independent of all known homology and homotopy invariants. More generally, we may set forth the question as follows: What invariants can we obtian from the general face relations in the combinatorial structure of a complex? In particular, let $a_{ij,k}(K)$ be the number of ordered pairs of an i-simplex and a j-simplex of K which have precisely a k-simplex $(-1 \leqslant k)$ as their common face. We may then ask about all possible linear combinations $\sum \alpha_{ij,k} a_{ij,k}(K)$ which will be a combinatorial (or a topological) invariant of K(or $|K|$). We have known already three such invariants, namely, the Euler-Poincaré characteristic of $K: \chi(K) = \sum (-1)^i a_{ii,i}(K)$; the square of $\chi(K): \chi^2(K) = \sum (-1)^{i+j} a_{ij,k}(K)$ and

① The space of a complex K will be denoted by $|K|$.

the expression

$$\tilde{\chi}_p^*(K) = \sum (-1)^{i+j} a_{ij,-1}(K)$$

as introduced above. It has been proved by K. C. Lee [4] that any combinatorial invariant of the form $\sum \alpha_{ij,k} a_{ij,k}(K)$ of K must be a linear combination of $\chi(K), \chi^2(K)$ and $\tilde{\chi}_2^*(K)$ and thus be a topological invariant of $|K|$. This may serve as an illustrative example of a systematic procedure to derive invariants from the given combinatorial structure of a complex.

Next we may consider the homology and cohomology groups of $\tilde{X}_p^*$ and X_p^*. To fix the ideas, let H denote the singular cohomology groups. We shall put $H^r(\tilde{X}_p^*, G) = \tilde{H}_{(p)}^r(X, G)$ and

$$H^r(X_p^*, G) = H_{(p)}^r(X, G).$$

In the case that X is a closed differentiable manifold and p a prime, it may be shown that $\tilde{H}_{(p)}^r(X, I_p)$ and $H_{(p)}^r(X, I_p)$ may be determined in terms of the cohomology ring on I_p and the Steenrod powers of the manifold X such that they furnish us nothing new. I do not know whether $\tilde{H}_{(p)}^r$ and $H_{(p)}^r$, over an arbitrary coefficient group G, for a closed differentiable manifold, are also completely determined by the known cohomology properties of X. It is sure anyhow that for an arbitrary space, even for a finite polyhedron X, it is impossible to express $\tilde{H}_{(p)}^r(X, G)$ and $H_{(p)}^r(X, G)$ in terms of the homology and homotopy properties of X alone. For example, let X be the space consisting of three arcs oa, ob and oc joined at the same point o. Then X is contractible to a point and hence all its homology and homotopy characters are trivial. However, the singular homology group $H_1(\tilde{X}_2^*)$ contains a non-zero class with a representative cycle given by

$$\begin{aligned}(a) \otimes (bo) + (a) \otimes (oc) + (b) \otimes (co) + (b) \otimes (oa) + (c) \otimes (ao)\\ + (c) \otimes (ob) + (bo) \otimes (a) + (oc) \otimes (a) + (co) \otimes (b)\\ + (oa) \otimes (b) + (ao) \otimes (c) + (ob) \otimes (c).\end{aligned}$$

Similarly, it may be shown that $\tilde{H}_{(2)}^1(X)$ is not trivial. In the case that X is a finite polyhedron we may determine $\tilde{H}_{(p)}^r(X, G)$ by means of the combinatorial structure of a simplicial subdivision of X, and the same is true for $H_{(p)}^r(X, G)$ at least if p is a prime (cf. the lemma above). In any case it is difficult to get general information about these groups. We shall thus, instead of studying the whole groups, turn our attention to some particular elements of these groups which have proved to be useful in certain kinds of problems.

Henceforth let us make the assumption that p is a prime and X is a Hausdorff space. In this case the cyclic transformation

$$t : (x_1, \cdots, x_p) \to (x_2, \cdots, x_p, x_1)$$

is a homeomorphic transformation without fixed points in $\tilde{X}_2^*$ and with prime period p. The reduced cyclic product X_p^* is the modular space of $\tilde{X}_p^*$ with respect to t and is also a Hausdorff space. By the theory of P. A. Smith [5] concerning periodic transformations, we may then attach to the triple $(\tilde{X}_p^*, X_p^*, t)$ a system of classes

$$\Phi_{(p)}^i(X) = A^i(\tilde{X}_p^*, X_p^*, t) \in H^i(X_p^*, I_{(i)}) = H_{(p)}^i(X, I_{(i)}),$$

in which $I_{(i)} = I$ for i even and $= I_p$ for i odd, in the following manner. Let $\tilde{\mathfrak{U}} = \{\tilde{U}_\alpha\}$ be an open covering of the reduced product $\tilde{X}_p^*$ such that if $\tilde{U}_\alpha \in \tilde{\mathfrak{U}}$, it follows that $t\tilde{U}_\alpha \in \tilde{\mathfrak{U}}$ and that $\tilde{U}_\alpha, t\tilde{U}_\alpha, \cdots, t^{p-1}\tilde{U}_\alpha$ are mutually disjoint. The covering $\mathfrak{U}$ of the reduced cyclic product X_p^* consisting of all open sets $U_\alpha = \pi\tilde{U}_\alpha, \tilde{U}_\alpha \in \tilde{\mathfrak{U}}$, where π is the natural projection $\tilde{X}_p^* \to X_p^*$, will be called the *projection* of the covering $\tilde{\mathfrak{U}}$. Let $S(\tilde{X}_p^*)$ be the singular complex of $\tilde{X}_p^*$ consisting of all singular simplexes $(\sigma, \tilde{f})$ where σ is a Euclidean simplex with ordered vertices and $\tilde{f}$ a continuous map of σ into $\tilde{X}_p^*$. Two singular simplexes $(\sigma, \tilde{f})$ and $(\sigma', \tilde{f}')$ will be identified as usual if a barycentric map g of σ onto σ', keeping the ordering of vertices, exists with $\tilde{f}'g = \tilde{f}$. Let $S_{\tilde{\mathfrak{U}}}(\tilde{X}_p^*)$ be the subcomplex of $S(\tilde{X}_p^*)$ consisting of only those singular simplexes $(\sigma, \tilde{f})$ for which $\tilde{f}(\sigma)$ is contained in some open set of the covering $\tilde{\mathfrak{U}}$. Similarly, let $S(X_p^*)$ be the singular complex of X_p^* and let $S_{\mathfrak{U}}(X_p^*)$ be the subcomplex of $S(X_p^*)$ consisting of only those singular simplexes (σ, f) of X_p^* for which $f(\sigma)$ is contained in some open set of the covering $\mathfrak{U}$. Now the cyclic transformation t in $\tilde{X}_p^*$ induces evidently a "simplicial" map of the complex $S_{\tilde{\mathfrak{U}}}(\tilde{X}_p^*)$ into itself having no fixed simplexes and with prime period p, which will still be denoted by t. Identifying the simplexes of $S_{\tilde{\mathfrak{U}}}(\tilde{X}_p^*)$ which are transformable into each other by t, we get a complex $S_{\tilde{\mathfrak{U}}}(\tilde{X}_p^*)/t$ (the modular complex of $S_{\tilde{\mathfrak{U}}}(\tilde{X}_p^*)$ with respect to t) which is canonically isomorphic to $S_{\mathfrak{U}}(X_p^*)$ and hence may be identified with it. Put $s = 1+t+\cdots+t^{p-1}$ and $d = 1-t$ and denote by $s^\#$, $d^\#$ the cochain maps induced by them, then the following sequence is exact:

$$\cdots \xrightarrow{d^\#} C^r(S_{\tilde{\mathfrak{U}}}(\tilde{X}_p^*)) \xrightarrow{s^\#} C^r(S_{\tilde{\mathfrak{U}}}(\tilde{X}_p^*)) \xrightarrow{d^\#} C^r(S_{\tilde{\mathfrak{U}}}(\tilde{X}_p^*)) \xrightarrow{s^\#} \cdots$$

Let the natural projection of $S_{\tilde{\mathfrak{U}}}(\tilde{X}_p^*)$ onto $S_{\mathfrak{U}}(X_p^*)$ be π. It induces a cochain map

$$\pi^\# : C^r(S_{\mathfrak{U}}(X_p^*)) \to C^r(S_{\tilde{\mathfrak{U}}}(\tilde{X}_p^*)).$$

Let us also define a system of homomorphisms

$$\bar{\pi}^{\#}: C^r(S_{\tilde{\mathfrak{U}}}(\tilde{X}_p^*)) \to C^r(S_{\mathfrak{U}}(X_p^*))$$

by setting

$$\bar{\pi}^{\#}\tilde{u}(\sigma, f) = \sum_{i=0}^{p-1} \tilde{u}(\sigma, t^i\tilde{f})$$

where

$$\tilde{u} \in C^r(S_{\tilde{\mathfrak{U}}}(\tilde{X}_p^*)) \quad \text{and} \quad \pi(\sigma, \tilde{f}) = (\sigma, f), \text{ i. e.}, \pi\tilde{f} = f.$$

Then $\bar{\pi}^{\#}$ is also a cochain map, is onto, and for the unit cocycle φ^0 of $S_{\mathfrak{U}}(X_p^*)$, which assumes the value 1 on all 0-simplexes of $S_{\mathfrak{U}}(X^*)$, we may find a system of cochains $\tilde{\varphi}^i \in C^i(S_{\tilde{\mathfrak{U}}}(\tilde{X}_p^*))$ such that

$$\bar{\pi}^{\#}\tilde{\varphi}^0 = \varphi^0, \quad \delta\tilde{\varphi}^i = s_{i+1}^{\#}\tilde{\varphi}^{i+1}, \quad i \geqslant 0,$$

where $s_i = s$ for i even and $= d$ for i odd. It may be shown that, denoting by $\varrho_{(i)}$ the reduction mod p for i odd and the identity for i even, $\varrho_{(i)}\bar{\pi}^{\#}\varphi^i$ is a cocycle of $C^r(S_{\mathfrak{U}}(X_p^*), I_{(i)})$ and whose class is independent of the choice of the cochains $\tilde{\varphi}^i$. Now

$$H^r(\tilde{X}_p^*, G) = H^r(S(\tilde{X}_p^*), G)$$

is known to be canonically isomorphic to the groups $H^r(S_{\tilde{\mathfrak{U}}}(\tilde{X}_p^*), G)$. Similarly $H^r(X_p^*, G) \approx H^r(S_{\mathfrak{U}}(X_p^*), G)$ canonically. Thus the classes of $\varrho_{(i)}\pi^{\#}\tilde{\varphi}^i$ may be considered as classes of $H^i(X_p^*, I_{(i)}) = H^i_{(p)}(X, I_{(i)})$, which, by definition, will be our classes $\Phi^i_{(p)}(X)$. We have for these classes the following properties:

$$\Phi^{2i}_{(p)}(X) \smile \Phi^{2j}_{(p)}(X) = \Phi^{2i+2j}_{(p)}(X),$$

$$\varrho_p\Phi^i_{(p)}(X) \smile \varrho_p\Phi^j_{(p)}(X) = \varrho_p\Phi^{i+j}_{(p)}(X),$$

$$\Phi^{2i}_{(p)}(X) = \beta_p\Phi^{2i-1}_{(p)}(X),$$

$$\Phi^0_{(p)}(X) = \text{integral unit class of } X_p^*,$$

in which ϱ_p, denotes reduction mod p if necessary, and

$$\beta_p: H^r_{(p)}(X, I_p) \to H^{r+1}_{(p)}(X)$$

denotes the Bockstein homomorphism. It follows that $\Phi^i_{(p)}(X) = 0$ implies $\Phi^j_{(p)}(X) = 0$ for all $j \geqslant i$. As $\Phi^0_{(p)}(X) \neq 0$ we can introduce the following

Definition The least integer n, if it exists, for which $\Phi^n_{(p)}(X) = 0$ will be called the *p-index* of the space X, and will be denoted by $I_p(X)$. We set $I_p(X) = \infty$ if no such integer exists.

We can only determine $I_p(X)$ for few elementary spaces. In particular we have [6, 10]

Theorem 1 For the N-dimensional Euclidean space R^N or the N-dimensional simplex Δ^N,

$$I_p(R^N) = I_p(\Delta^N) = N(p-1).$$

In other words,

$$\Phi^i_{(p)}(R^N) = 0 \quad \text{for} \quad i \geqslant N(p-1),$$

while

$$\Phi^{N(p-1)}_{(p)}(R^N) \neq 0.$$

Similarly for Δ^N.

For the proof let us remark first that each of the spaces R^N and Δ^N may be considered as a subspace of the other so that, by definition, we have necessarily

$$\text{(a)} \qquad I_p(R^N) = I_p(\Delta^N).$$

Now consider $X = \Delta^N$ as a complex K with the usual subdivison and let $\tilde{K}^*_p$ be the subcomplex of the usual p-fold product complex of K which consists of all cells $\sigma_1 \times \cdots \times \sigma_p$ with no vertex of K common to all these $\sigma_i \in K$. By the lemma mentioned above, the space $|\tilde{K}^*_p|$ of $\tilde{K}^*_p$ is a deformation retract of $\tilde{X}^*_p$ for which the deformation commutes with the cyclic transformation t in $\tilde{X}^*_p$. As $\tilde{K}^*_p$ is of a dimension $N(p-1)-1$ we may deduce that

$$\Phi^i_{(p)}(X) = 0 \quad \text{for} \quad i \geqslant N(p-1),$$

i.e.,

$$\text{(b)} \qquad I_p(\Delta^N) \leqslant N(p-1).$$

On the other hand, let us consider the p-fold product $\tilde{Y}_p$ of $Y = R^N$ as a Euclidean space of dimension Np and the diagonal $\tilde{\Delta}_p$ of $\tilde{Y}_p$ as a linear subspace of $\tilde{Y}_p$. The end-points of unit vectors orthogonal to $\tilde{\Delta}_p$ is then a sphere $\tilde{S}$ of dimension $N(p-1)-1$ on which the cyclic transformation t in $\tilde{Y}^*_p$ is a transformation of prime period p without fixed points, such that the modular space $\tilde{S}/t$ is a lens space. From this we may deduce that

$$\text{(c)} \qquad I_p(R^N) \geqslant N(p-1).$$

Theorem 1 follows then from (a), (b), and (c).

As a first application of our theorem we may deduce the following [6].

Theorem 2 The dimension of a finite polyhedron X is the least integer $n \geqslant 0$ such that for each prime number p, sufficiently large ($> n+2$), we have $\Phi_{(p)}^{(n+1)(p-1)-1}(X) = 0$[or $H_{(p)}^{(n+1)(p-1)-1}(X, I_p) = 0$].

Proof. Let dim $X < n$. Then X_p^* is of dimension

$$\leqslant (n-1)p < n(p-1) - 1$$

for $p > n+1$. Hence $H_{(p)}^{n(p-1)-1}(X, I_p) = 0$ if dim $X < n$ and $p > n+1$. On the other hand we have always $\Phi_{(p)}^{n(p-1)-1}(X) \neq 0$ so that $H_{(p)}^{n(p-1)-1}(X, I_p) \neq 0$ if dim $X \geqslant n > 0$ for any prime $p (> 2)$. This proves our theorem.

Whether the theorem above is also true for more general spaces is unknown to the author. It seems very likely to be true and it would show that the topological invariant "dimension" may be expressed in terms of the homology invariants of the reduced cyclic product of a space.①

A second application of our theorem concerns the problem of imbedding. A space X will be said to be (topologically) *imbeddable* in a space Y if there exists a continuous map of X into Y such that X and $f(X)$ are homeomorphic under f. The map f, if it exists, will then be called an *imbedding* of X in Y. Suppose that X is imbeddable in Y with an imbedding f, then for any prime p, f will induce in a natural manner a continuous map $\tilde{F}_p$ of $\tilde{X}_p^*$ into $\tilde{Y}_p^*$ which commutes with the cyclic transformations t_X in $\tilde{X}_p^*$ and t_Y in $\tilde{Y}_p^*$: $t_Y \tilde{F}_p = \tilde{F}_p t_X$. Let F_p $X_p^* \to Y_p^*$ be the map induced by $\tilde{F}_p$, then $F_p^* \Phi_{(p)}^i(Y) = \Phi_{(p)}^i(X)$ so that $\Phi_{(p)}^i(Y) = 0$ would imply $\Phi_{(p)}^i(X) = 0$, or in other words, $I_p(X) \leqslant I_p(Y)$. From Theorem 1 we obtain therefore [6, 10]

Theorem 3 If a space X is imbeddable in a Euclidean space R^N of dimension N, then

$$I_p(X) \leqslant N(p-1), \quad \text{i.e.,} \quad \Phi_{(p)}^{N(p-1)}(X) = 0.$$

---

① Added in proof. The theorem has been proved for normal spaces with countable basis with the definition of classes $\Phi_{(p)}^i(X)$ slightly modified. Cf. On the dimension of a normal space with countable basis. Science Record, New Series, 1958, 2: 65-69.

This theorem, simple as it appears, contains many known results about imbedding as its consequences. For example, for any space X let [11]

$$Sm^i_{(p)} : H^r(X, I_p) \to H^{r+i}(X, I_p)$$

be the set of homomorphisms defined by

$$\sum_{j=0}^{k} (-1)^j St^{k-j}_{(p)} Sm^j_{(p)} = \begin{cases} 0, & k > 0 \\ \text{identity}, & k = 0 \end{cases} \quad (p = 2 \text{ or } pk \text{ odd}).$$

$$\sum_{j=0}^{k} St^{2k-2j}_{(p)} Sm^{2j}_{(p)} = \begin{cases} 0, & k > 0 \\ \text{identity}, & k = 0 \end{cases} \quad (p > 2).$$

Then we have the following theorem of which the case $p = 2$ (with the further supposition that X be locally contractible) was originally proved by R. Thom based on a method entirely different from ours (Cf. [7] Th. III. 25):

Theorem 4[10] If a compact space X is imbeddable in R^N, then (H^r denotes the Čech cohomology group)

$$Sm^k_{(p)} H^r(X, I_2) = 0 \quad \text{for} \quad 2k + r \geqslant N,$$

and

$$Sm^{2k}_{(p)} H^r(X, I_p) = 0 \quad \text{for} \quad 2kp + (p-1)r \geqslant N(p-1), \quad p > 2.$$

In particular, on putting $k = 0$, we should have $H^r(X, I_p) = 0$ for $r \geqslant N, p \geqslant 2$ if X is imbeddable in R_N. This classical theorem was originally proved as a consequence of the Alexander duality theorem.

For a closed differentiable manifold M^n of dimension n we have a system of invariants $W^i \in H^i(M^n, I_2)$ called the (reduced mod 2) Stiefel-Whitney classes of M^n. Define $\bar{W}^i \in H^i(M^n, I_2)$ by

$$\sum_{i+j=k} W^i \smile \bar{W}^j = \begin{cases} 0, & k > 0 \\ \text{unit class} \mod 2, & k = 0. \end{cases}$$

Then Whitney has proved that

$$\bar{W}^k = 0, \quad k \geqslant N - n,$$

is a necessary condition for M^n to be *differentiably* imbeddable in R^N. This may be shown to be a consequence of our Theorem 3 for $p = 2$ and moreover, this condition is also necessary for *topological* imbedding [9]. For $p \geqslant 3$, our theorem leads also to conditions in terms of the so-called Pontrjagin classes which we do not discuss here.

The conditions $\Phi_{(p)}^{N(p-1)}(X) = 0$ for the imbedding of a space X in R^N as given in Theorem 3 are not only necessary but also sufficient in certain extreme cases. For example, we have [8, 9, 10, 12]

Theorem 5 For a finite polytope X of dimension $n(> 2)$ to be imbeddable in R^{2n} it is necessary and sufficient that

$$\Phi_{(2)}^{2n}(X) = 0.$$

Theorem 6 For a Peano continuum X with only a finite number of simple closed curves to be imbeddable in the plane it is necessary and sufficient that

$$\Phi_{(2)}^{2}(X) = 0.$$

Let us also indicate some applications of the reduced products of a space to the problem of *isotopy* or *iso-position* [9]. Let f_0 f_1 be two imbeddings of a space X in a space Y. Then f_0, f_1 will be said to be *isotopic* if there exists a continuous map F of the topological product of X by the interval $I = [0,1]$ into Y such that for any

$$x \in X, \quad f_0(x) = F(x,0), \quad f_1(x) = F(x,1),$$

and for any $t \in I$, the map $f_t : X \to Y$ defined by $f_t(x) = F(x_1 t)$, $x \in X$, is an imbedding of X in Y. The imbeddings f_0, f_1 of X in Y will be said to be in *iso-position* if there exists a topological map h of Y onto itself such that $f_1 \equiv hf_0$. Now consider any imbedding f of a space in a Euclidean space R^N of dimension N. Let S^{N-1} be the unit sphere in R^N. For any point $(x_1, x_2) \in \tilde{X}_2^*$ such that $x_1, x_2 \in X$ but $x_1 \neq x_2$, the half line through the origin of R^N and parallel to the directed line from $f(x_1)$ to $f(x_2)$ will intersect the sphere S^{N-1} in a point $\tilde{f}(x_1, x_2)$. This gives rise to a continuous map $\tilde{f} : \tilde{X}_2^* \to S^{N-1}$. With respect to a definite orientation of S^{N-1} let the generator of $H^{N-1}(S^{N-1})$ be denoted by s. Then the class $\Theta_f = \tilde{f}^*(s) \in H^{N-1}(\tilde{X}_2^*)$ up to a sign may be shown to be an isotopy invariant as well as an iso-position invariant of the imbedding f. In certain extreme cases, this invariant is already sufficient to characterise the imbedding with respect to isotopy or iso-position for example, a certain theorem of MacLane-Adkisson [13] may be reformulated in the following form:

Theorem 7 If a Peano continuum X is imbeddable in a plane, then any two imbeddings f, g of X in the plane will be in iso-position if and only if $\Theta_f = \pm\Theta_g$.

We have depicted above some applications that may be derived from the consideration of the reduced products and the reduced cyclic products of a space. Though our results obtained are fragmentary as well as meager, I hope that what we have said above may give some indications of the fruitfulness of the concepts of these reduced products and reduced cyclic products which are worthy of further investigations.

Literature

[1] Wu Wen-tsün. Topological invariants of a new type of finite polyhedrons. Acta Math. Sinica, 1953, 3: 261-290.

[2] Mayer W. A new homology theory, I, II. Ann. of Math., 1942, 43: 370-380, 594-605.

[3] Spanier E H. The Mayer homology theory. Bull. Amer. Math. Soc., 1949, 55: 102-112.

[4] Lee K C. Über die Eindeutigkeit von einigen kombinatorischen Invarianten endlicher Komplexe. Science Record, New Series 1, 1957.

[5] Smith P A. Fixed points of periodic transformations, in Lefschetz. Algebraic Topology, 1942: 350-373.

[6] Wu Wen-tsün. On the $\Phi_{(p)}$-classes of a topological space. Science Record, New Series 1, 1957.

[7] Thom R. Espaces fibrés en sphéres et carrés de Steenrod. Ann. Ec. Norm. Sup., 1952, 69(3): 110-182.

[8] Van Kampen. Komplexe in Euclidischen Räumen. Abh. Math. Sem. Hamburg, 1932, 9: 72-78; Berichtigung dazu, ibid., 152-153.

[9] Wu Wen-tsün. On the realization of complexes in Euclidean spaces, I-III. Acta Math. Sinica, 1955, 5: 505-552; 1957, 7: 79-101; 1958, 8. (An English translation of I, II will appear soon in Acts, Sci. Sinica.)

[10] —. A theory of imbedding and immersion in Euclidean spaces. Peking, 1957 (mimeographed).

[11] —. Relations between Smith operations and Steenrod powers. Fund. Math., 1957, 44: 262-269.

[12] Kuratowski C. Sur les courbes gauches en topologie. Fund. Math., 1930, 15: 271-283.

[13] MacLane S, Adkisson V W. Extensions of homeomorphisms of the sphere. Michigan Lectures in Topology, 1941: 223-236.

27. On the Dimension of a Normal Space with Countable Basis*

In a preceding paper[5] we have introduced the notions of $\Phi_{(p)}$-classes and $\Phi_{(p)}$-indices for a normal space with countable basis. We have also shown that these classes serve to characterize the dimension of a finite polyhedron P as the smallest integer $n \geqslant 0$ such that $\Phi_p^{(n+1)(p-1)-1}(P) = 0$ for any prime p sufficiently large. In other words, the dimension of a finite polyhedron P is equal to $[I_p(P)/(p-1)]$ for p sufficiently large, where $[x]$ denotes the greatest integer $\leqslant x$. The purpose of this note is to prove that the same is true for the dimension of any normal space with countable basis (cf. [1] or [2]) if the notions of $\Phi_{(p)}$-classes and $\Phi_{(p)}$-indices are slightly modified. Throughout the present note, p will denote a prime integer and the same notations as in [5] will be used. Furthermore, a "space" will always mean a normal space with countable basis.

§ 1. An open covering $\mathfrak{U} = \{U_i\}$ of a space X is called *point-finite* if each point $x \in X$ is contained in, at most, a finite number of the open sets of the covering $\mathfrak{U}$. It is called *locally finite* if every open set U_i of $\mathfrak{U}$ meets at most a finite number of open sets U_i of $\mathfrak{U}$. Let $\mathfrak{U} =\{U_i\}$ be a point-finite open covering of X, and $K_{\mathfrak{U}}$, the nerve complex of $\mathfrak{U}$ with U_i as vertices. Let $K'_{\mathfrak{U}}$ be the *first-derived*[3] of $K_{\mathfrak{U}}$ such that the vertices σ' of $K'_{\mathfrak{U}}$ are in 1-1-correspondence with simplexes σ of $K_{\mathfrak{U}}$ and the simplexes of $K'_{\mathfrak{U}}$ are of the form $(\sigma'_1 \ \cdots \ \sigma'_r)$ with the simplexes $\sigma_i \in K_{\mathfrak{U}}{}'$ corresponding to its vertices σ'_i in a relation $\sigma_1 \prec \cdots \prec \sigma_r$. We shall call the simplex $(\sigma'_1 \cdots \sigma'_r)$ as well as its faces the *subsimplexes* of the simplex σ_r of $K_{\mathfrak{U}}$. An open covering $\mathfrak{U}'$ of X will then be said to be a *first strong-derived* of $\mathfrak{U}$ if it is possible to label its sets as U'_σ in 1-1-correspondence with the simplexes σ of $K_{\mathfrak{U}}$ such that the nerve complex $K_{\mathfrak{U}'}$ of $\mathfrak{U}'$ becomes isomorphic to $K'_{\mathfrak{U}}$ under this correspondence and moreover $U'_\sigma \subset U_{i_0} \cap \cdots \cap U_{i_r}$, for $\sigma = (U_{i_0} \cdots U_{i_r}) \in K_{\mathfrak{U}}$. Any first strong-derived of a point-finite open covering $\mathfrak{U}$ is itself point-finite and a first strong-derived $\mathscr{B}'$ of an r-th strong-derived $\mathscr{B}$ of $\mathfrak{U}$ will be called an $(r+1)$ – *th strong-derived of* $\mathfrak{U}$. If $\tau \in K_{\mathfrak{B}}$ is a subsimplex (suppose already defined) of a simplex $\sigma \in K_{\mathfrak{U}}$, then any

* *Science Record*, 1958, II (2): 65-69.

subsimplex $\in K_{\mathcal{B}'} \approx K'_{\mathcal{B}}$ of τ will also be called a *subsimplex* of σ and its vertices will be called *subvertices* of σ. As in [4] Chap. III we have then the following

Lemma Any locally-finite open covering of a space has a first strong-derived and hence also r-th strong-deriveds for all $r \geqslant 1$.

§ 2. For a space X, let $\tilde{X}_p^*$ be the p-fold reduced product and X_p^* the p-fold reduced cyclic product of X as in [5]. The pair of a locally-finite open covering $\mathfrak{U} = \{U_i\}$ of X and an r-th strong-derived $\mathfrak{B} = \{V_\tau\}$ of $\mathfrak{U}$ will be said to be *subordinate* to an open covering $\tilde{\mathfrak{U}} = \{\tilde{U}_a\}$ of $\tilde{X}_p^*$ if the following condition (C) is satisfied:

(C) For any p-uple $(\sigma_1, \cdots, \sigma_p)$ of simplexes of $K_{\mathfrak{U}}$ with no vertex of $K_{\mathfrak{U}}$ common to all these simplexes σ_i, and any p-uple of vertices $(V_{\tau_1}, \cdots, V_{\tau_p})$ of $K_{\mathcal{B}}$ with V_{τ_j} as a subvertex of σ_i, $i = 1, \cdots, p$, the open set $V_{\tau_1} \times \cdots \times V_{\tau_p}$ is contained in a certain open set $\tilde{U}_a$ of the covering $\tilde{\mathfrak{U}}$.

A P-covering $\tilde{\mathfrak{U}}$ of $\tilde{X}_p^*$ will be said to be *admissible* if there exists a locally-finite open covering $\mathfrak{U}$ of X and an r-th strong-derived $\mathfrak{B}$ of $\mathfrak{U}$ with $r \geqslant p$ such that the pair $(\mathfrak{U}, \mathcal{B})$ is subordinate to $\tilde{\mathfrak{U}}$. The projection of an *admissible P-covering* of $\tilde{X}_p^*$ will then be called an *admissible P-covering* of X_p^*. It may be seen that admissible P-coverings of $\tilde{X}_p^*$ and X_p^* do exist always. The limit group of the system of all cohomology groups $H^r(K_{\tilde{\mathfrak{U}}}, G)$ (resp. $H^r(K_{\mathfrak{U}}, G)$), over the directed set of all admissible P-covering $\tilde{\mathfrak{U}}$ of $\tilde{X}_p^*$ (resp. $\mathfrak{U}$ of X_p^*) ordered by inclusion, will be called the *modified* $\tilde{H}^r_{(p)}$ -groups (resp. *modified* $H^r_{(p)}$-groups) of X over the coefficient group G and will be denoted by $\tilde{\mathscr{H}}^r_{(p)}(X, G)$ (resp. $\mathscr{H}^r_{(p)}(X, G)$). The limit of the Smith classes $A^i(K_{\tilde{\mathfrak{U}}}, t) \in H^i_{(p)}(K_{\pi(\tilde{\mathfrak{U}})}, I_{(i)})$ of the simple systems $(K_{\tilde{\mathfrak{U}}}, t)$, where $\tilde{\mathfrak{U}}$ are admissible P-coverings of $\tilde{X}_p^*$ and $\pi(\tilde{\mathfrak{U}})$ are. the projections of $\tilde{\mathfrak{U}}$, will be called the *modified Smith classes* of X and will be denoted by $\varPhi^i_{(p)}(X) \in \mathscr{H}^i_{(p)}(X, I_{(i)})$. The least integer $n \geqslant 0$, if it exists, for which $\check{\varPhi}^n_{(p)}(X) = 0$, will then be called the *modified* $\varPhi_{(p)}$-index of X and will be denoted by $\vartheta_p(X)$. In the case that X is a finite polytope, it may be seen that the system of all admissible P-coverings of $\tilde{X}_p^*$(or X_p^*) is cofinal in the system of all P-coverings of $\tilde{X}_p^*$(or X_p^*) so that the modified groups $\tilde{\mathscr{H}}^r_{(p)}(X, G)$, $\mathscr{H}^r_{(p)}(X, G)$, the modified Smith classes $\check{\varPhi}^i_{(p)}(X)$, and the modified indices $\vartheta_e(X)$ coincide with the groups $\tilde{H}^r_{(p)}(X, G)$, $H^r_{(p)}(X, G)$, the classes $\varPhi^i_{(p)}(X)$ and the indices $I_p(X)$ respectively as already introduced in [5]. With these modified notions the theorem to be proved may then be stated in the following manner:

Theorem 1 The dimension of a space X is the least integer $n \geqslant 0$ such that $\check{\varPhi}^{(n+1)(p-1)-1}_{(p)}(X, I_p) = 0$ for p sufficiently large. In other words,

$$\operatorname{Dim} X = [\vartheta_p(X)/(p-1)], \quad p^{\text{sufficiently large}}.$$

We may also characterize the dimension by means of the modified $H_{(p)}$-groups, namely

Theorem 2 The dimension of a space X is the least integer $n \geqslant 0$ such that for the modified $H_{(p)}$-groups of X we have

$$\mathscr{H}_{(p)}^{(n+1)(p-1)-1}(X, I_p) = 0$$

for p sufficiently large.

These theorems will furnish us with intrinsic "homological" definitions of the dimension of a space.

§ 3. The Theorems 1 and 2 are easy consequences of the following

Theorem 3 If the dimension of a space X is $\geqslant n > 0$, then

$$\Phi_{(p)}^{n(p-1)-1}(X) \neq 0 \tag{1}$$

for any prime p.

To prove this, let $\tilde{\mathfrak{U}}$ be an arbitrary admissible P-covering of $\tilde{X}_p^*$, $\mathfrak{U} = \{U_i\}$ a locally finite covering① of X and $\mathfrak{B} = \{V_\tau\}$ an r-th strong-derived of $\mathfrak{U}$ with $r \geqslant p$ such that the pair $(\mathfrak{U}, \mathfrak{B})$ is subordinate to $\tilde{\mathfrak{U}}$. Let $K = K_{\mathfrak{U}}$ be the nerve complex of $\mathfrak{U}$ which may be considered as a Euclidean simplicial subdivision of a Euclidean polyhedron P. The product space $\tilde{P}_p = \underbrace{P \times \cdots \times P}_{p}$ has then a cell subdivision $\tilde{K}_p = \underbrace{K \times \cdots \times K}_{p}$ consisting of all cells of the form $\sigma_1 \times \cdots \times \sigma_p$, where $\sigma_i \in K$. Let $\tilde{K}_p^*$ be the subcomplex of $\tilde{K}_p$ consisting of these cells $\sigma_1 \times \cdots \times \sigma_p$ for which no vertex of K is common to all these σ_i. Let the vertices of each simplex of K be arranged in a definite order $\prec$ such that the order of vertices of the face of a simplex coincides with the one induced by that of the simplex. Such an order will be called provisionally a *local order* of vertices in K. With respect to definite local order of vertices in K we may then define a simplicial subdivision $\tilde{K}_p^\Delta$ of $\tilde{K}_p$ as follows. The vertices of $\tilde{K}_p^\Delta$ are the vertices $U_{i_1} \times \cdots \times U_{i_p} = U_{i_1 \cdots i_p}$ of $\tilde{K}_p$, where U_i are vertices of K and the simplexes of $\tilde{K}_p^\Delta$ are these spanned by sets of vertices $U_{i'_1 \cdots i'_p}, U_{i''_1 \cdots i''_p}, \cdots, U_{i_1^{(r)} \cdots i_p^{(r)}}$ such that for each $k \leqslant p$ and $\geqslant 1$, $U_{i'_k}, U_{i'_k}, \cdots, U_{i_k^{(r)}}$ span a simplex of K with $U_{i'_k} \preceq U_{i'_k} \preceq \cdots \preceq U_{i_k^{(r)}}$ in the local order of vertices of K already chosen. Let $t' : \tilde{P}_p \to \tilde{P}_p$ be the cyclic

① Without loss of generality, we may suppose that $\mathfrak{U}$ consists of countably infinite number of open sets.

transformation in $\tilde{P}_p$ defined by $t'(y_1, \cdots, y_p) = (y_2, \cdots, y_p, y_1)$, $y_i \in P$, and $\tilde{P}_p^* = \tilde{P}_p$-Diag. be the p-fold reduced product of P. Then t' will induce a cyclic transformation in $\tilde{P}_p^*$ as well as in $\tilde{K}_p^*$ which will still be denoted by t'. As $\tilde{K}_p^*$ may be seen to be a deformation retract of $\tilde{P}_p^*$ for which the deformation is commutative with t', the Smith classes $A^i(\tilde{K}_p^*, t')$ and $A^i(\tilde{P}_p^*, t')$ of corresponding simple systems may be identified. Consider now the nerve complex $L = K_{\mathfrak{B}}$ of $\mathfrak{B}$ in which local ordering of vertices may be introduced in a natural manner. Let complex $\tilde{L}_p^{\Delta}$ be constructed in the similar manner for L as the complex $\tilde{K}_p^{\Delta}$ for K, then $\tilde{L}_p^{\Delta}$ is also a simplicial subdivision of $\tilde{K}_p$. Denote now by $\tilde{L}_p^{*\Delta}$ the subcomplex of $\tilde{L}_p^{\Delta}$ which covers the complex $\tilde{K}_p^*$. As $\tilde{L}_p^{*\Delta}$ is a simplicial subdivision of $\tilde{K}_p^*$, the Smith classes $A^i(\tilde{L}_p^{*\Delta}, t)$, where the transformation t' in $\tilde{L}_p^{*\Delta}$ is again induced by t' of $\tilde{P}_p^*$, may be identified with $A^i(\tilde{K}_p^*, t')$. The vertices of $\tilde{L}_p^{*\Delta}$ are just those $V_{\tau_1} \times \cdots \times V_{\tau_p} = V_{\tau_1 \cdots \tau_p}$ for which V_{τ_j} is a subvertex of $\sigma_i \in K$ $(i = 1, 2, \cdots, p)$ where $\sigma_1 \times \cdots \times \sigma_p \in \hat{K}_p^*$, i.e., no vertex of K is common to all these σ_i. Since the pair $(\mathfrak{U}, \mathfrak{B})$ is subordinate to $\tilde{\mathfrak{U}}$, we see by condition (C) that to each vertex $V_{\tau_1 \cdots \tau_p}$ of $\tilde{L}_p^{*\Delta}$ may be taken an open set $\tilde{U}_{\tau_1 \cdots \tau_p}$ of the covering $\tilde{\mathfrak{U}}$ such that $V_{\tau_1} \times \cdots \times V_{\tau_p} \subset \tilde{U}_{\tau_1 \cdots \tau_p}$. The correspondence $j : V_{\tau_1 \cdots \tau_p} \to \tilde{U}_{\tau_1 \cdots \tau_p}$ induces then a simplicial map of $\tilde{L}_p^{*\Delta}$ in $K_{\tilde{\mathfrak{U}}}$. Now the cyclic transformation t of $\tilde{X}_p = \underbrace{X \times \cdots \times X}_{p}$ defined by $t(x_1, \cdots, x_p) = (x_2, \cdots, x_p, x_1)$, $x_i \in X$, will induce cyclic transformations in $K_{\tilde{\mathfrak{U}}}$ which will still be denoted by t. Then $j : \tilde{L}_p^{*\Delta} \to K_{\tilde{\mathfrak{U}}}$ may be so defined to commute with t, t' that we have for the Smith classes of the corresponding simple systems

$$j^* A^i(K_{\tilde{\mathfrak{U}}}, t) = A^i(\tilde{L}_p^{*\Delta}, t'). \tag{2}$$

Now by hypothesis the dimension of X is $\geqslant n > 0$. It follows that Dim $P \geqslant n$ and hence by theorems already proved in [5] we have

$$A^i(\tilde{K}_p^*, t') \neq 0, \quad i \leqslant n(p-1) - 1.$$

From what is described above, we would have also

$$A^i(\tilde{L}_p^{*\Delta}, t') \neq 0, \quad i \leqslant n(p-1) - 1.$$

By (2) we have therefore

$$A^i(K_{\tilde{\mathfrak{U}}}, t) \neq 0, \quad i \leqslant n(p-1) - 1. \tag{3}$$

As $\tilde{\mathfrak{U}}$ is an arbitrary admissible P-covering of $\tilde{X}_p^*$, we get from (2) by passing to the limit the relation (1). This proves Theorem 3.

References

[1] Alexandroff P S. *Proc. Royal Soc. London*, 1947, 189: 11-39.
[2] Hurewicz W and Wallman H. *Dimension Theory*. Princeton, 1942.
[3] Lefschetz S. *Algebraic Topology*. Princeton, 1942.
[4] ——. *Topics in Topology*. Princeton, 1942.
[5] Wu W T. *Science Record*. New Ser., 1957, 1(6): 15-18.

28. On the Isotopy of C^r-Manifolds of Dimension n in Euclidean $(2n+1)$-Space*

Let M be a differentiable manifold of class $r(1 \leqslant r \leqslant \infty)$. A C^r-map of M in a Euclidean space R^N of dimension N is called a *regular C^r-imbedding* if it is topological and the rank of the map at each point is equal to the dimension of M. Any two such regular C^r-imbeddings f_0, f_1 of M in R^N are said to be *regularly C^r-isotopic* if there exists a C^r-map F of $M\times[0, 1]$ in R^N such that $F(x,i) == f_i(x)$, $x \in M$, $i = 0, 1$ and that for each $0 \leqslant t \leqslant 1$, the map f_t: $M \to R^N$ defined by $f_t(x) = F(x,t)$ is a regular C^r-imbedding. It is well known that any two regular C^r-imbeddings of an M in R^N are regularly C^r-isotopic if $N \geqslant 2 \dim M + 2$[3]. On the other hand, if $N = 2 \dim M + 1$, then in the case $\dim M = 1$, there are an infinity of such imbeddings of an $M = S^1$ (a circle) in R^3 not (regularly) isotopic to each other, the study of which forms the subject of the theory of knots. Artint[1] has raised the question about the existence of high-dimensional knots, in particular, the existence of (regular) imbeddings of an n-sphere $(n > 1)$ in R^{2n+1} which are not (regularly) isotopic to an ordinary n-sphere in R^{2n+1}. We shall prove in what follows that such high-dimensional knots of S^n in R^{2n+1} $(n > 1)$ can not exist. In fact, we have the following more general

Theorem *Any two regular C^r-imbeddings of a compact C^r-manifold M of dimension $n > 1$ in R^{2n+1} are regularly C^r-isotopic $(1 \leqslant r \leqslant\leqslant \infty)$.*

§ 1. Let L be thc linc $-\infty < t < +\infty$. We shall put $\tilde{M} = M \times L$, $M_t = M \times (t)$, $\tilde{M}^- = \tilde{M} - M_0 - M_1$. Let $Q = Q^n$ (resp. $Q\prime = Q'^n$) be the cube $-1 < x_i < +1$, $\left(\text{resp. } -\frac{1}{2} < x_i < +\frac{1}{2}\right)$, $i = 1, \cdots, n$, in a Euclidean space R^n with coordinates $x_1, \cdots, x_n$. Then the differential structure on $\tilde{M}^-$ deduced from that of M may be defined by a count-able system of topological maps $\theta_a : Q \times L_a \equiv U_a \times L_a \subset \tilde{M}^-$ with $U_a \subset M$, $\theta_a(Q \times (t)) \equiv U_a \times (t)$, $L_a = [t_a < t < t'_a] \subset L - (0) - (1)$ such that the following conditions are verified: 1°. The system of sets $U'_a \times L'_a = \theta_a(Q' \times L'_a)$

---

* *Science Record*, 1958, II(9): 271-275.

forms a covering of $\tilde{M}^-$, where $L'_\alpha = \left[t_a - -\frac{t'_\alpha - t_\alpha}{4} < t < t'_\alpha + \frac{t'_\alpha - t_\alpha}{4}\right]$; 2°. Each θ_a may be extended to a C^r-map of $\overline{Q} \times \overline{L}_a$ in $\tilde{M}^-$. 3°. Each $p \in \tilde{M}^-$ is only in a finite number of $U_a \times L_\alpha$.

Let R^N be a Euclidean space with coordinates $y_1, \cdots, y_N$. We shall put $R^{N+1} = R^N \times L$, $R^N_t = R^N \times (t)$. A map f of $\tilde{M}$ (or $\tilde{M}^-$) in R^{N+1} will then be said to be *admissible* if $f(M_t) \subset R^N_t$. For any admissible C^r-map $f : \tilde{M}^- \to R^{N+1}$ and a system of positive continuous functions① $\eta = \{\eta^{(1)}(\tilde{p}) | 0 \leqslant l \leqslant r\}$ on $\tilde{M}^-$, we shall denote by $W\ (f,\ \eta)$ the set of all admissible C^r-maps g of $\tilde{M}^-$ in R^{N+1} such that for any α, we have

$$(*) ||D_k g(\theta_a(x,t)) - D_k f(\theta_a(x,t))|| < \eta^{(\sigma_k)}(\theta_a(x,t)),$$

for $0 \leqslant \sigma_k \leqslant r$, $(x,t) \in Q \times L_a$, where $k = (k_1, \cdots, k_n, k_0)$, $D_k == \frac{\partial^{k_1+\cdots+k_n+k_0}}{\partial x_1^{k_1} \cdots \partial x_n^{k_n} \partial t^{k_0}}$, $\sigma_k = k_1 + \cdots + k_n + k_0$, and $||v||$ denotes the length of a vector v in R^{N+1}. The set $A_r{}^- = A_r(\tilde{M}^-, R^{N+1})$ of all admissible C^r-maps of $\tilde{M}^-$ in R^{N+1} will then become a topological space if we take for any $f \in A_r^-$ the system of all $W(f,\eta)$ as a complete system of neighbourhoods of f. We remark that the set A_r^- as well as the topology in A_r^- is independent of the choice of maps θ_a which define the differential structure of $\tilde{M}^-$.

Proposition 1　$A_r^- = A_r(\tilde{M}^-, R^{N+1})$ is a Baire space.

Proof. Let O_1, O_2, $\cdots$ be a countable number of everywhere dense open sets in A_r^- and let $W(f,\eta)$ be given, where $f \in A_r^-$, and $\eta = \{\eta^{(l)}(\tilde{p}) | 0 \leqslant l \leqslant r\}$ is a system of positive continuous functions on $\tilde{M}^-$. We may then find successively systems of positive continuous functions $\eta_i = \{\eta_i^{(l)}(\tilde{p}) | 0 \leqslant l \leqslant r\}$, $i \geqslant 0$, on $\tilde{M}^-$ and maps $g_i \in A_r^-$, $i \geqslant 0$, such that

$$\eta_t^{(t)}(\tilde{p}) < \frac{1}{2}\eta_{i-1}^{(l)}(\tilde{p}), \quad i > 0; \quad \eta_0^{(l)}(\tilde{p}) < \frac{1}{2}\eta^{(l)}(\tilde{p});$$

and

$$g_0 = f, \quad W(g_i, \eta_i) \subset O_i \cap W(g_{i-1}, \eta_{i-1}), \quad i > 0.$$

For any $(x,t) \in Q \times L_a$, $i > 0$, we have then

$$||D_k g_i(\theta_\alpha(x,t)) - D_k g_{i-1}(\theta_a(x,t))|| < \eta_{i-1}^{(\sigma_k)}(\theta_a(x,t)), \quad 0 \leqslant \sigma_k \leqslant r.$$

This shows that for any $\tilde{p} \in \tilde{M}^-$, $g_1(\tilde{p})$, $g_2(\tilde{p})$, $\cdots$ converge to a limit $g(\tilde{p})$ with $g : \tilde{M}^- \to R^{N+1}$ evidently admissible. The same relations show also that $g \in W(f,\eta)$

① If $r = \infty$, $0 \leqslant l \leqslant r$ means $0 \leqslant l < \infty$. Similarly for analogous expressions below.

and $\in O_i$ for all i. Hence the intersection of all O_i is. everywhere dense in A_r^-, and A_r^- is a Baire space.

For each α let R_a be the subset of A_r^- consisting of all maps $g \in A_r^-$ which are regular on $\bar{U}'_a \times \bar{L}'_a$. Then we have

Proposition 2 Each subset R_a is everywhere dense and open in A_r^- if $N \geqslant 2n+1$.

Proof. That R_a is open is evident. To prove that R_a is everywhere dense, let us consider an arbitrary neighbourhood $W(f,\eta)$ in A_r^-, where $\eta = \{\eta^{(l)}(\tilde{p}) | 0 \leqslant l \leqslant r\}$. Put $a_\alpha^{(l)} = \min \eta^{(l)}(\tilde{p}) > 0$ and take a C^r-function $\lambda_a(\tilde{p})$ on $\tilde{M}^-$ with $0 \leqslant \lambda_a \leqslant 1$, $\lambda_a = 1$ on $\bar{U}'_a \times \bar{L}'_a$, and $\lambda_a = 0$ on $\tilde{M}^- - U_a \times L_a$. We shall now define successively admissible C^s-maps $g_i : Q \times L_a \to R^{N+1} = R^N \times L$, $0 \leqslant i \leqslant n$, $s = \max\,(r, 2)$, such that the following conditions are verified ($(x,t) \in Q \times L_a$):

1°. $||D_k[\lambda_a(\theta_a(x,t))\{g_0(x,t) - f(\theta_\alpha(x,t))\}]|| < \dfrac{1}{n+1} a_\alpha^{(\sigma_k)}, \quad 0 \leqslant \sigma_k \leqslant r.$

2_i°. $||D_k[\lambda_a(\theta_a(x,t))\{g_i(x,t) - g_{i-1}(x,t)\}]|| < \dfrac{1}{n+1} a_a^{(\sigma_k)}, \quad 0 \leqslant \sigma_k \leqslant r.$

3_i°. The vectors $\dfrac{\partial g_i(x,t)}{\partial x_1}, \cdots, \dfrac{\partial g_i(x,t)}{\partial x_i}$ are linearly independent for any $(x,t) \in \bar{Q}' \times \bar{L}'_\alpha$.

That an admissible g_0 exists verifying 1° results from the usual approximation theory of functions, if $r = 1$. For $r \geqslant 2$, we may set simply $g_0 = f(\theta_a)$. Suppose that $g_1, \cdots, g_{i-1}$ $(i \leqslant n)$ have been already defined which satisfy 2_j° and 3_j° for $j \leqslant i-1$. In particular, by 3_{i-1}°, the vectors $\dfrac{\partial g_{i-1}(x,t)}{\partial x_1}, \cdots, \dfrac{\partial g_{i-1}(x,t)}{\partial x_{i-1}}$ will determine an $(i-1)$ -dimensional linear space in R_0^N for any $(x,\, t) \in \bar{Q}' \times \bar{L}'_\alpha$. Since $n + i < N$ for $i \leqslant n$, there exists in R_0^N, as may be shown by the method in [3] §21, a vector $v \in R_0^N$ with $||v|| > 0$ arbitrarily small such that $g_i(x,t) = g_{i-1}(x,t) + x_i v$, where $(x,t) \in Q \times L_a$, $x = (x_1, \cdots, x_n)$, satisfies 3_i°. With $||v||$ sufficiently small 2_i° will also be satisfied. We remark that the maps $g_1, \cdots, g_n$ successively defined in this manner are all admissible by the very procedure of construction.

Define now g: $\tilde{M}^- \to R^{N+1}$ by

$$g(\tilde{p}) = \begin{cases} f(\tilde{p}) + \lambda_a(\tilde{p})[g_n(\theta_a^{-1}(\tilde{p})) - f(\tilde{p})], & \tilde{p} \in U_a \times L_\alpha, \\ f(\tilde{p}), & \tilde{p} \in \tilde{M}^- - U_a \times L_a. \end{cases}$$

We have then $g \in W(f,\eta)$ and g is regular on $\bar{U}'_a \times \bar{L}'_\alpha$, i.e. R_a is everywhere dense in A_r^-. q.e.d.

§ 2. We are now in a position to prove our theorem as follows.

Let f_0, $f_1 : M \to R^N$ ($N = 2n+1$, dim $M = n$) be two regular C^{r-} imbeddings. Define $f' : \tilde{M} \to R^{N+1} = R^N \times L$ as the admissible map such that for each $p \in M$, $f'((p) \times L)$ is the line joining $f_0(p) \times (0)$ and $f_1(p) \times (1)$. Take a positive continuous function $\eta(\tilde{p})$ on $\tilde{M}^-$, say $\eta(p,t) = |t(1-t)|$, $t \neq 0$, 1, $p \in M$, which approaches 0 when $\tilde{p}$ approaches points of M_0 or M_1. By Propositions 1 and 2, there exists then an $f \in W(f', \eta) \cap \prod\limits_a R_a$, i.e. an admissible map $f \in W(f', \eta)$ regular on $\tilde{M}^-$, where η denotes the system of positive continuous functions $\eta^{(l)}(\tilde{p})$ on $\tilde{M}^-$, each equal to $\eta(\tilde{p})$. Extend f to $\tilde{M}$ by setting $f(p,0) = f_0(p)$, $f(p,1) = f_1(p)$, $p \in M$, then f is a C^r-map of $\tilde{M}$ in R^{N+1} which is admissible as well as regular on $\tilde{M}$ (cf. [3], Lemma 10).

Let $\eta(\tilde{p}) > 0$ be as before. Then we see as in [3] §25 that there exists a positive continuous function $\eta\prime(\tilde{p}) < \eta(\tilde{p})$ on $\tilde{M}^-$ and a field of $(N-n)$ -elements $P(\tilde{p})$ in R^{N+1} such that: (a) the function $\tilde{p} \to P(\tilde{p})$ is of class C^r; (b) $P(\tilde{p})$ passes through $f(\tilde{p})$, lying in R^N_t if $\tilde{p} = (p,t)$, and is transversal to $f(\tilde{M}^-)$ at $f(\tilde{p})$; (c) every $\tilde{p} \in \tilde{M}^-$ has a neighbourhood $V(\tilde{p})$ such that the parts $R(\tilde{p}')$ of points $\tilde{y} \in P(\tilde{p}')$ with a distance $\rho(\tilde{y}, f(\tilde{p}')) < \eta'(\tilde{p}')$ from $f(\tilde{p}') \in f(V(< \tilde{p}))$ are mutually disjoint and fill out a neighbourhood of $f(V(\tilde{p}))$ in R^{N+1}. Let T be the subspace of $\tilde{M}^- \times R^{N+1}$ consisting of all points $(\tilde{p}, \tilde{y})$ with $\tilde{y} \in R(\tilde{p})$. Let H be the set of all regular C^r-homeomorphisms τ of T on itself satisfying the conditions $\tau(R(\tilde{p})) = R(\tilde{p})$ and $\tau(\tilde{p}, \tilde{y}) = (\tilde{p}, \tilde{y})$ for each $\tilde{p} \in \tilde{M}^-$, $\tilde{y} \in R(\tilde{p})$ with $\rho(f(\tilde{p}), \tilde{y}) \geqslant \frac{1}{2}\eta'(\tilde{p})$. This set H will become a Baire space by introducing certain topology. As in [4], Chap.I, there exists then a $\tau \in H$ such that, if a map $g : \tilde{M}^- \to R^{N+1}$ is defined by $(\tilde{p}, g(\tilde{p})) = \tau(\tilde{p}, f(\tilde{p}))$, then g is an admissible regular C^r-map verifying the following condition: If $g(\tilde{p}_1) = g(\tilde{p}_2)$ is a self-intersection of $\tilde{M}^-$ under g, then the images under g of the two pieces of $\tilde{M}^-$ about $\tilde{p}_1$ and $\tilde{p}_2$ are transversal to each other. Extend now g to the whole $\tilde{M}$ by setting $g(p,0) = f(p,0) = f_0(p)$, and $g(p,1) = f(p,1) = f_1(p)$, $p \in M$, then g is an admissible regular C^r-map of $\tilde{M}$ in R^{N+1}. Moreover, only a finite number of self-intersections $g(p_1,t) = g(p_2,t)(p_1 \neq p_2 \in M)$ can exist for $0 \leqslant t \leqslant 1$, and we may suppose that no three points of $\tilde{M}^-$ can have their images coincide, by a further deformation if necessary.

Suppose provisionally n odd and M orientable. Then to each self-intersection may be associated a unique index $\varepsilon = +1$ or -1, namely, the index of intersection of the two pieces nearby of $\tilde{M}$ coherently oriented (R^{N+1} definitely oriented). Consider any such self-intersection $(y,t) = g(p_1,t) = g(p_2,t) \in R^N_t$, with $p_1 \neq p_2 \in M$. Let us take an arbitrary point $\tilde{y}' = (y', t)$ on $g(M_t)$ different from all self- intersections and a small neighbourhood V about this point containing no self-intersections of g. As in §2 of [4], we may then slightly modify g to a regular C^r-map g' such that $g'/\tilde{M} - g^{-1}(V) \equiv g$, and $(y', t) = g'(p_1', t) == g\prime(p_2', t)$, $(p_1' \neq p_2' \in M)$, is a new self-intersection created with associated index $= -\varepsilon$. This map g' may be taken still to be admissible in view of the equations defining the new self-intersection. (§2 of [4]). In M_t take now two simple arcs B_i joining p_i and p_i' $(i = 1, 2)$ extended slightly beyond them so that $g'(B_i) = C_i$ are mutually disjoint and contain no other self-intersections except the points $(y,\ t)$ and $(y',\ t)$. As $n \geqslant 2$, $N = 2n+1 \geqslant 5$, there exists a 2-cell σ in R_t^N intersecting $g'(\tilde{M})$ transversally only in C_i (cf. [4] §10). As $\dim \tilde{M}^- = n+1 > 2$ and the self-intersections $(y,\ t)$ and $(y',\ t)$ have their associated indices with opposite signs, we may get rid of them by deforming g' to g'', with other self-intersections unchanged, as in [4] §12. Moreover, this deformation may be so taken that the resulting map g'' is still admissible. The cases where n is even or M nonorientable $(n > 1)$ may be treated by a device as in [4], and the same conclusion still holds. Proceeding in the same manner for any other remaining self-intersections in $R^N \times$ [0, 1], we get finally an admissible regular C^r-map $h : M \to R^{N+1}$ which is topological on $M \times$ [0, 1].

By construction we have $h(p, t) = f_t(p)$ for $t = 0$ or 1, $p \in M$. Defining now $f_t : M \to R^N$ by $f_t(p) = h(p, t)$, $0 \leqslant t \leqslant 1$, $p \in M$, we get a regular C^r-isotopy between f_0 and f_1 as required. This proves our theorem.

Remark. The above proof shows also the essential difference between $n = 1$ and $n > 1$ for imbeddings of an n-manifold in R^{2n+1}.

References

[1] Artin E. *Abh. Math. Semin.*. Hamburg Univ., 1925, 4: 174-177.

[2] Thom R. *Comm. Math. Helv.*, 1954, 28: 17-86.

[3] Whitney H. *Annals of Math.*, 1936, 37: 645-680.

[4] Whitney H. *Annals of Math.*, 1944, 45: 220-246.

29. On the Realization of Complexes in Euclidean Spaces II*

Abstract For the realization problem of complexes or more general spaces in Euclidean spaces we may cite the following results: 1° In 1932 van Kampen [1] proved the existence of n-dimensional complexes K not realizable in Euclidean space of dimension $2n$.

His proof depends on certain invariants deduced from the reduced 2-fold symmetric product K^* of K. The author [2] has pointed out that van Kampen's invariant is merely the extreme one of a system of invariants $\Phi^m \in H^m(K^*, I_{(m)})$, ($I_{(m)} = I$ or I_2, depending on m being even or odd), namely Φ^{2n}, and $\Phi^m = 0$ is a necessary condition for K to be realizable in R^m. We proved also in [2] the topological invariance of Φ^m, i.e., the independence of Φ^m of the chosen subdivision K of the space of K, while van Kampen, in the extreme case $m = 2n$ considered by him, has not even proved its combinatorial invariance.

2° In the case of differentiable manifold, Whitney (cf. e.g., [3]) has introduced a system of invariants which the present author has called the dual Whitney classes $\bar{W}^m$, and proved that

$$\bar{W}^k = 0, \quad k \geqslant m - n \tag{1}$$

are necessary conditions for an n-dimensional closed differentiable manifold to be realizable in R^m,

3° For compact Hausdorff spaces X, Thom has proved ([4] Th. III. 25) the following theorem: Let Q^i be certain operations deduced from Steenrod squares, then

$$Q^i H^r(X, I_2) = 0, \quad 2i + r \geqslant m \tag{2}$$

are necessary conditions for X to be realizable in R^m. The operations Q^i were previously introduced by the author by applying Smith's theory of

* First published in Chinese in *Acta Math. Sinica*, 1957, VII(1): 79-101; *Sci. Sinica*, 1958, 7: 365-387.

periodic transformations and will henceforth be denoted by Sm^i (cf. [5] and [6]).

Besides, Flores [7] got also the same results as van Kampen's concerning the existence of n-dimensional complexes K not realizable in R^{2n}. However, what he used to prove his results is got by imbedding the first K in R^{2m+1}, and is actually not an invariant, but depends in general on the way of realization of K in R^{2n+1}. This will be explained further in the sequel.

In the various theories listed above, not only the methods used are quite different from one another, but also the objects and the realization concept involved are not the same. For example, the theory of van Kampen studies about the semi-linear realization of finite complexes, the theory of Whitney is applicable only for the differentiable realization of differentiable manifolds, and the theory of Thom is concerned with topological realization of more general spaces.

The present paper gives a general theory including the various theories cited above as its particular cases. This theory may he formulated in terms of the following easily proved fundamental theorem:

For any Hausdorff space X let $\tilde{X}^*$ be the space of all ordered pairs (x_1, x_2), where x_1, $x_2 \in X$ and $x_1 \neq x_2$. Let $t : \tilde{X}^* \equiv \tilde{X}^*$ be the transformation $t(x_1, x_2) = (x_2, x_1)$, X^* the modular space$\tilde{X}^*, t$. With respect to the pair $(\tilde{X}^*, t)$, we may define according to P. A. Smith [8] a system of cohomology classes

$$\Phi^m(X) \in H^m(X, I_{(m)}),$$

where H represents singular homology system. Then $\Phi^m(X) = 0$ is *a* necessary condition for X to be topologically realizable in R^m.

If X be a finite polyhedron, then according to results of [2] , the classes Φ^m defined above are the same as the classes Φ^m defined in that paper in a quite different manner. It follows that $\Phi^m = 0$ is not only a necessaiy condition for X to possess a subdivision K such that K is semilinearly realizable in R^m, as shown in [2] , but also one for X itself to be topologically realizable in R^m.

If X is a closed differentiable manifold of dimension n, then we prove that from $\rho_2\Phi^m(X) = 0$ we get (1), but not vice versa. Our fundamental theorem has therefore Whitney's theorem, as its consequence. It follows also that Whithey's condition (1) is not only necessary for the differentiable realization, but also necessary for topological realization of X in R^m.

We prove also that if X be a finite polyhedron, then from $\rho_2\Phi^m(X) = 0$ we get (2). Hence our fundamental theorem has also, at least in the case of finite polyhedron, Thom's theorem as its consequence.

In the case of differentiable manifold, Whitney [9] has introduced the concept of "immersion" in a Euclidean space. He proved that any differentiable manifold of dimension n may be immersed in R^{2n-1}, but said nothing about the criterion of immersibility in R^m of arbitrary dimension m. We extend the concept of immersion to arbitrary spaces and our method for the study of realization may also be applied to the immersion problem of finite polyhedrons. For such a space X, let us take a subdivision K, and construct a certain "tube" $K^{(0)}$ of the diagonal in the two-fold symmetric product $K*K$. Then there exist; in $K^{(0)}$, again by the theory of P. A. Smith, a system of cohomology classes $\Psi^m(K^{(0)}) \in H^m(K^{(0)}, I_{(m)})$. According to [10], the homotopy type or $|K^{(0)}|$ as well as $\Psi^m(K^{(0)})$ is topological invariants of X, and therefore $\Psi^m(K^{(0)})$ may also be denoted by $\Psi^m(X)$. Our fundamental theorem about immersion may then be stated as follows:

If a finite polyhedron X may be immersed in R^m, then $\Psi^m(X) = 0$.

As before, we may deduce from this theorem results analogous to theorems of Thom and Whitney in the case of realization.

Our method may also be used to study the problem of isotopy or isoposition. For a topological realization f of a Hausdorff space X in R^m, we may define a cohomology class $\Theta_f^{m-1} \in H^{m-1}(\tilde{X}^*)$. The definition of Θ_f^{m-1} was essentially due to Flores [7]. But Flores discussed mainly about the case where Θ_f^{m-1} is independent of the realization f and is thus topological invariant of X. We prove on the other hand that Θ_f^{m-1} depends generally on the realization f, and is an invariant of isotopy as well as isoposition. More precisely, $\theta_f^{m-1} = \theta_g^{m-1}$(or $\Theta_f^{m-1} = \pm\Theta_g^{m-1}$) is a necessary condition for the two realizations $f(X)$ and $g(X)$ in R^{m-1} to be isotopic (or in the same position). A consequence of this is that some known results as theorem of Adkisson and McLane about planar or spherical figures (of. [11] and literatures listed there) may de formulated in a "quantitative" instead of "qualitative" form.

We remark finally that the realization theorem of Thom is proved as consequence of our fundamental realization theorem only in the case of finite polyhedrons. This is due to the fact that we use the singular homology system. If another homology system is used, it would be possible to get Thom's theorem for more general spaces. As this concerns only the choice of homology systems, and as the corresponding theorem for Immersion seems to be difficult to extend to more general spaces other than finite polyhedrons, our discussions will be restricted mainly to the latter class of spaces.

§1 Smith homomorphisms

Let $\tilde{E}$ be a Hausdorff space, and t a topological transformation of $\tilde{E}$ with period n (n an integer > 1) such that any power t^i of t has no fixed points for $i \not\equiv 0 \bmod n$. We say in such case that $(\tilde{E}, t)$ is a simple system. Let $E = \tilde{E}/t$ be the modular space of $\tilde{E}$ with respect to t and $\pi : \tilde{E} \to E$ the natural projection. Let $S(\tilde{E})$, $S(E)$ be the singular complexes of corresponding spaces defined by means of ordered singular simplexes. Then t operates also in $S(\tilde{E})$, having period n, with no fixed simplexes under any power t^i of t, for $i \not\equiv 0 \bmod n$, and the modular complex of $S(\tilde{E})$ with respect to t is no other than $S(E)$. We shall denote by $C^*(\tilde{E}, G)$ or $C^r(\tilde{E}, G)$ ($G ==$ the coefficient group) the cochain groups in $S(\tilde{E})$, and denote by $\bar{C}^*(\tilde{E}, G)$ or $\bar{C}^r(\tilde{E}, G)$ the groups of cochains which take values 0 on all singular simplexes disjoint from certain compact sets (varied with the cochains) in $\tilde{E}$. The cohomology groups derived from these cochain groups will be denoted by $H^*(\tilde{E}, G)$, $\bar{H}^*(\tilde{E}, G)$, etc. Similarly, we have also $H^*(E, G)$, $\bar{H}^*(E, G)$, etc. $\bar{H}^*$ is the so-called compact singular cohomology groups (cf. [12]).

For any γ-dimensional ordered singular simplex σ of $S(E)$, there are just n singular simplexes $\tilde{\sigma}_1, \cdots, \tilde{\sigma}_n$ in $S(\tilde{E})$ such that $\pi\tilde{\sigma}_i = \sigma$. Then for any cochain $\tilde{\varphi} \in C^r(\tilde{E}, G)$, $\bar{\pi}^{\#}\tilde{\varphi}(\sigma) = \sum\limits_i \tilde{\varphi}(\tilde{\sigma}_i)$ defines a cochain map $\bar{\pi}^{\#} : C^r(\tilde{E}, G) \to C^r(E, G)$. As usual put $s = 1 + t + \cdots + t^{n-1}$, $d = 1 - t$. According to P. A. Smith, for any cocycle x (on arbitrary coefficient group G) we may take a system of cochains $\tilde{y}$, $\tilde{y}_1, \cdots$ in $S(\tilde{E})$ such that (cf. [8])

$$\pi^{\#}x = s\tilde{y}, \quad \delta\tilde{y} = d\tilde{y}_1, \quad \delta\tilde{y}_1 = s\tilde{y}_2, \cdots .$$

Put $x_i = \bar{\pi}^{\#}\tilde{y}_i$, then $x \to x_i$ will induce a system of homomorphisms

$$\mu_i = \mu_i(\tilde{E}, t) : H^r(E, G) \to H^{r+i}(E, G_{(i)}),$$

where

$$G_{(i)} = \begin{cases} G, & i \text{ even}, \\ G/nG, & i \text{ odd}. \end{cases}$$

When $x \in \bar{C}^r(E, G)$, we may take $\tilde{y}$, $\tilde{y}_1$, $\cdots$ all in $\bar{C}^*(\tilde{E}, G)$ so that $x \to x_i$ induces also a system of homomorphisms

$$\bar{\mu}_i = \bar{\mu}_i(\tilde{E}, t) : \bar{H}^r(E, G) \to \bar{H}^{r+i}(E, G_{(i)}).$$

Between these homomorphisms the following relations exist:

$$\left.\begin{aligned} \mu_{2i} = (\mu_2)^i, \quad \mu_{2i+1} = (\mu_2)^i \mu_1 = \mu_1(\mu_2)^i, \\ \bar{\mu}_{2i} = (\bar{\mu}_2)^i, \quad \bar{\mu}_{2i+1} = (\bar{\mu}_2)^i \mu_1 = \bar{\mu}_1(\mu_2)^i, \\ \mu_i j^* = j^* \bar{\mu}_i, \end{aligned}\right\} \tag{1}$$

in which $j^* : \bar{H}^*(E, G) \to H^*(F, G)$ is the natural homomorphism. Denote by ρ_n the homomorphism of cohomology groups induced by the reduction mod n, $G \to G/nG$ or the identity $G \to G$, we have also

$$a\rho_n\mu_2 = (\mu_1)^2, \quad a\rho_n\bar{\mu}_2 = (\bar{\mu}_1)^2,$$

where

$$a = \begin{cases} 0, & n \quad \text{odd}, \\ -\dfrac{n}{2}, & n \quad \text{even}. \end{cases}$$

Let I be the additive group of integers, and 1 the integral unit class in $C^\circ(E)$, then the classes $\mu_i 1 \in H^i(E, I_{(i)})$ will be denoted by

$$A^i = A^i(\tilde{E}, t) = \mu_i 1.$$

For any class $U \in \bar{H}^*(E,\ G)$ we have then

$$\bar{\mu}_i U = A^i \smile U.$$

Remark that as A^i, U belong to H^* and $\bar{H}^*$ respectively, so that their $\smile$-product is defined and belongs to $\bar{H}^*$: $H^* \smile \bar{H}^* \subset \bar{H}^*$, the coefficient groups being paired in the usual manner: $I_{(i)} \smile G \subset G_{(i)}$. This theorem has been proved for the special case of a complex and $G = I$ (cf. [5] or [13]). The proof in the general case is essentially the same.

Let $(\tilde{E}'t')$ be another simple system, $E' = \tilde{E}'/t'$, and t' has also the same period n. If the continuous map $\tilde{f} : \tilde{E} \to \tilde{E}'$ satisfies $t'\tilde{f} = \tilde{f}t$, then $\tilde{\tilde{f}}$ will be called a P-map. In such case $\tilde{f}$ will induce a continuous map $f : E \to E'$ with $f\pi = \pi'\tilde{f}$, where π, π' are corresponding projections. Moreover, we have

$$f^*\mu_i(\tilde{E}', t') = \mu_i(\tilde{E}, t)f^*,$$

in particular $f^*A^i(\tilde{E}', t') = A^i(\tilde{E}, t)$. If f is admissible, i.e., the inverse image of any compact set in E' is also a compact set in E, then we have also

$$f^*\bar{\mu}_i(\tilde{E}', t') = \tilde{\mu}_i(\tilde{E}, t)f^*.$$

Similar to a P-map, we may also naturally define P-deformation and P-homotopy type into which we do not enter.

§2 The fundamental realization theorem

Let X be a Hausdorff space and n be a prime. Let $\underbrace{X \times \cdots \times X}_{n} = \tilde{X}_n$ and $d_n : X \to \tilde{X}_n$ be the diagonal map. Define $t_n : \tilde{X}_n \to \tilde{X}_n$ *by* $t_n(x_1, \cdots, x_n) = (x_n, x_1, \cdots, x_{n-1})$. Put $\tilde{X}_n^* = \tilde{X}_n - d_nX$, then $(\tilde{X}_n^*, t_n)$ is a simple system. Let $X_n = \tilde{X}_n/t_n$, $X_n^* = \tilde{X}_n^*/t_n$ be the corresponding modular spaces, then the classes $A^i(\tilde{X}_n^*, t_n) \in H^i(X_n^*, I_{(i)})$ in X_n^* are topological invariants of X and may be denoted by $\Phi_n^i(X)$. In particular, they will be simply denoted by $\Phi^i(X)$ when $n = 2$.

Let K be a simplicial complex, and $\tilde{K}_n^*$ be the subcomplex of $\tilde{K}_n == \underbrace{K \times \cdots \times K}_{n}$ formed of all cells $\sigma_1 \times \cdots \times \sigma_n$ for which $\sigma_i \in K$ and no vertex of K is common to all these σ_i. Define $t_n : |\tilde{K}_n| \to |\tilde{K}_n|$ as above① and let $K_n^* = \tilde{K}_n^*/t_n$ be the modular complex of the simple system $(\tilde{K}_n^*, t_n)$ (n a prime). In [2] we have proved that from the consideration of almost semi-linear realizations of K in R^m we get a system of cohomology classes $\Phi^m(K) \in H^m(K_2^*, I_{(m)})$, called the imbedding classes of K, whose vanishing is a necessary condition for K to have a semi-linear realization in R^m. In the same paper we have also proved that $\Phi^m(K) = A^m(\tilde{K}_2^*, t_2)$. Since $|\tilde{K}_2^*|$ is a P-deformation retract of $|\tilde{K}|_2^*$, we may identify $H^*(|K|_2^*, G)$ and $H^*(K_2^*, G)$, and then $A^m(\tilde{K}_2^*, t_2) = A^m(|\tilde{K}|_2^*, t_2)$, i.e., $\Phi^m(K)$ coincides with the

① The space of a complex K will be denoted by $|K|$.

above-defined $\Phi^m(|K|)$. For this reason, even in the case of an arbitrary Hausdorff space X, $\Phi^m(X) \in H^m(X_2^*, I_{(m)})$ will also be called the *imbedding classes* of X.

The above extended definition of imbedding classes is justified by the following theorem:

Theorem 1 If a Hausdorff space X is (topologically) realizable in a Euclidean space R^m of dimension m, then $\Phi^m(X) = 0$.

Proof. Suppose that X is realizable in R^m so that a topological map $f : X \to R^m$ exists. Let 0 be the origin of R^m, S^{m-1} the unit sphere in R^m with centre 0, and t the antipodal map of S^{m-1} onto itself. The modular space $P^{m-1} = S^{m-1}/t$ of the simple system (S^{m-1}, t) is then a proiective space of dimension $m-1$. Consider any point $(x_1, x_2) \in \tilde{X}_2^*$. As $f(x_1) \neq f(x_2)$, we may draw a half-line starting from 0 and parallel to the directed line from $f(x_1)$ to $f(x_2)$. Denote the intersecting point of this half-line with S^{m-1} by $\tilde{g}(x_1, x_2)$, then $\tilde{g}(x_2, x_1) = t\tilde{g}(x_1, x_2)$ so that $\tilde{g}$ is a P-map and induces thus a map $g : X_2^* \to P^{m-1}$. It follows that

$$g^* A^i(S^{m-1}, t) = A^i(\tilde{X}_2^*, t_2) = \Phi^i(X), \quad i \geqslant 0.$$

Since $A^i(S^{m-1}, t) = 0$ for $i \geqslant m$, the theorem is proved.

By the Menger-Nöbeling imbedding theorem①, we get the following

Corollary If X is a normal space with countable basis having dimension m, then $\Phi^{2m+1} \cdot (X) = 0$.

Remark 1. By [2] , for any integer m there exist complexes K of dimension m with $\Phi^{2m}(|K|) \neq 0$.

Remark 2. If X is a finite polyhedron, then by [2] , $\Phi^m(X) = 0$ is a necessary condition for X to possess a simplicial subdivision K semi-linearly imbeddable in R^m. The theorem here shows that the condition is necessary not only for semi-linear but also for topological imbedding.

§3 Immersion and immersion classes

Let X be a Hausdorff space and $f : X \to R^m$ a continuous map possessing the following property: there exists an open covering $\mathfrak{A} = \{U_i\}_{i \in I}$ of X such that for any two points x, $y \in$ the same $U_i \in \mathfrak{A}$, we have $f(x) \neq f(y)$. In such case we shall say

① Cf. e.g. Hurewicz and Wallman's *Dimension Theory*, Theorem V3, 1941.

that f is an *immersion* written $f : X \subset\!\!\!\!\propto R^m$. The aim of this section is to introduce certain invariants to measure the possibility of immersing X in a Euclidean space, at least in the case that X is a finite polyhedron.

For this purpose let us consider a simplicial subdivision K of X and construct $\tilde{K}_2 = K \times K$ as well as $\tilde{K}_2^*$ as in §2. Subdivide the diagonal d_2X as K such that d_2X is decomposed into simplexes of the form $d_2\sigma$, where $\sigma \in K$. Denote this complex by d_2K, then, according to [10] , $\tilde{K}_2$ has a cell decomposition with subcomplexes d_2K and $\tilde{K}_2^*$ as follows. Let σ, σ_1, $\sigma_2 \in K$ have the following property: σ_1, σ_2 have no common vertex, and σ and σ_1 (resp. σ and σ_2) span a simplex σ_1' (resp. σ_2') of K. In the cell $\sigma_1' \times \sigma_2'$ let us take a point $x_0 \in |\sigma|$ and a point $x_1 \in |\sigma_1 \times \sigma_2|$, and denote by x_r the point on segment x_0x_1 dividing it in the ratio $r : 1 - r(0 \leqslant r \leqslant 1)$. When x_0, x_1 run over $|\sigma|$ and $|\sigma_1 \times \sigma_2|$ respectively, the points $x_{1/2}$ will form a $(d + d_1 + d_2)$-cell $[\sigma, \sigma_1 \times \sigma_2]$, where $d = \dim \sigma$, $d_i = \dim \sigma_i$, $i = 1, 2$. Similarly, the points $x_r, r \geqslant 1/2$ (resp. $r \leqslant 1/2$) form a $(d + d_1 + d_2 + 1)$-cell $[\sigma, \sigma_1 \times \sigma_2]^+$ (resp. $[\sigma, \sigma_1 \times \sigma]^-$). The set of all cells in $\tilde{K}_2^*$, d_2K and those of the forms $[\sigma, \sigma_1 \times \sigma_2]^+$, $[\sigma, \sigma_1 \times \sigma_2]^-$ and $[\sigma, \sigma_1 \times \sigma_2]$ forms then a cell subdivision of $\tilde{K}_2$ which will be denoted by $\omega\tilde{K}_2$. The subcomplex of $\omega\tilde{K}_2$ formed by all cells $[\sigma, \sigma_1 \times \sigma_2]$ will be denoted by $\tilde{K}_2^{(0)}$ and the subcomplex formed of all cells of $\tilde{K}_2^*$ as well as all $[\sigma, \sigma_1 \times \sigma_2]^+$ and $[\sigma, \sigma_1 \times \sigma_2]$ will be denoted by $\tilde{K}_2^{(+)}$. Evidently t_2 will operate in $\tilde{K}_2^{(+)}$ and $\tilde{K}_2^{(0)}$ with modular complexes $K_2^{(+)} = \tilde{K}_2^{(+)}/t_2$ and $K_2^{(0)} = \tilde{K}_2^{(0)}/t_2$ respectively. It may be seen that the systems $(|\tilde{K}_2^{(+)}|, t_2)$, $(|\tilde{K}_2^*|,\ t_2)$ and $(\tilde{X}_2^*, t_2)$ all have the same P-homotopy type. By [10] , we know also that the P-homotopy type of $(|\tilde{K}_2^{(0)}|, t_2)$ is always the same whatever the chosen subdivision K may be, and is thus a topological invariant of X. In particular, the classes $A^i(\ |\tilde{K}_2^{(0)}|, t_2) \in H^i(\ |K_2^{(0)}|, I_{(i)})$ are all topological invariants of X and will be denoted by $\Psi^i(X)$.

Let the cell in $K_2^{(0)}$ which is obtained by identifying the cells $[\sigma, \sigma_1 \times \sigma_2]$ and $[\sigma, \sigma_2 \times \sigma_1]$ in $\tilde{K}_2^{(0)}$ be denoted by $[\sigma, \sigma_1 * \sigma_2]$. Then the mod 2 boundary relations in $K_2^{(0)}$ may be written symbolically as

$$\partial[\sigma, \sigma_1 * \sigma_2] = [\partial\sigma, \sigma_1 * \sigma_2] + [\sigma, \partial(\sigma_1 * \sigma_2)] \bmod 2. \tag{1}$$

Now each point x of a cell $[\sigma, \sigma_1 \times \sigma_2] \in \tilde{K}_2^{(0)}$ is the mid-point of a certain segment x_0x_1, where $x_0 \in |\sigma|$ and $x_1 \in |\sigma_1 \times \sigma_2|$. Then $x \to x_1$ defines a cell-map

$$\tilde{f} : \tilde{K}_2^{(0)} \to \tilde{K}_2^*.$$

As $\tilde{f}$ is evidently a P-map, it induces also a cell-map

$$f : K_2^{(0)} \to K_2^*$$

such that

$$f^* A^i(\tilde{K}_2^*, t_2) = A^i(\tilde{K}_2^{(0)}, t_2),$$

or

$$f^* \Phi^i = \Psi^i. \tag{2}$$

By definition and from dimensionality considerations we see that

$$f_\#(\sigma, \sigma_1 * \sigma_2) = \begin{cases} 0, & \dim \sigma > 0 \\ \sigma_1 * \sigma_2, & \dim \sigma = 0 \end{cases} \quad (\text{mod} 2). \tag{3}$$

Theorem 2 If a finite polyhedron X is immersible in R^m, then

$$\Psi^m(X) = 0.$$

Proof. Let $f : X \subset\!\!\!\!\propto R^m$, and $\mathfrak{A} = \{U_i\}$ be the open covering of X as required in the definition of an immersion, of which the number of U_i may be supposed to be finite. Take a sufficiently small simplicial subdivision K such that for any point $(x, y) \in |\tilde{K}_2^{(0)}|$, we should have x, $y \in$ the same $U_i \in \mathfrak{A}$. In such case we have $f(x) \neq f(y)$. Hence, let $\tilde{g}\,(x,\ y)$ be the intersecting point with the unit sphere S^{m-1} in R^m by the half-line through origin O of R^m and parallel to the directed line from $f(x)$ to $f(y)$, we get a continuous map $\tilde{g} :\ |\tilde{K}_2^{(0)}| \to S^{m-1}$. Let t be the antipodal map of S^{m-1}, then $\tilde{g}$ is a P-map of the system $(|K_2^{(0)}|, t_2)$ in the system (S^{m-1}, t). Similar to the proof of Theorem 1 we get then $\Psi^m(X) = 0$.

On account of this theorem, we shall call $\Psi^m(X)$ the *immersion classes* of the finite polyhedron X.

Let the finite polyhedron X be of dimension m. Take a simplicial subdivision K of X and let $f : K \subset R^{2m+1}$ be a linear realization of K in R^{2m+1}. Since for any two simplexes σ, $\tau \in K$ with vertices in common the linear subspace $P(\sigma, \tau)$ determined by them has a dimension $\leqslant 2m$, there must exist lines l in R^{2m+1} not parallel to any one of such linear subspaces $P(\sigma, \tau)$. Let R^{2m} be a $2m$-dimensional linear subspace of R^{2m+1} orthogonal to such a line l and φ be the orthogonal projection of R^{2m+1} onto R^{2m}. Then $\varphi f : x \to R^{2m}$ is evidently an immersion. We have therefore the following

Corollary For any finite polyhedron X of dimension m we have always $\Psi^{2m}(X)=0$.

On the other hand, we shall prove that for any $n \geqslant 1$, there exist n-dimensional complexes not immersible in R^{2n-1} as follows.

Let $K_{N,n}$ be the complex formed of all n-dimensional simplexes as well as their faces with vertices taken from $a_0, a_1, \cdots, a_N$. In [2] we have proved that $K_{2n+1,n}$ is realizable in R^{2n}, but not in R^{2n-1}. In reality, $K_{2n+1,n}$ is not only non-realizable in R^{2n-1}, but also non-immersible in R^{2n-1}. To prove it, let us consider in $K_2^{(0)}$ the chain

$$z=\sum[(a_0),(a_{i_0}\cdots a_{i_{n-1}})*(a_0a_{j_0}\cdots a_{j_{n-1}})],$$

in which $\sum$ is extended over all possible sets of indices (i,j) such that $0<i_0<\cdots<i_{n-1}\leqslant 2n+1$, $0<j_0<\cdots<j_{n-1}\leqslant 2n+1$ and no i is the same as any j. By (1), we know that z is a mod 2 cycle. By Theorem 10 of [2] , $\Phi^{2n-1}(K)\in H^{2n-1}(K_2^*,I_2)$ has a representative cocycle

$$\varphi^{2n-1}=\sum\rho_2\{(a_{i_0}\cdots a_{i_{n-1}})*(a_{j_0}\cdots a_{j_n})\},$$

in which $\sum$ is extended over all possible sets of indices (i,j) with $0\leqslant j_0<i_0<j_1<\cdots<i_{n-1}<j_n\leqslant 2n+1$. Define $f:\ |K_2^{(0)}|\ \to\ |K_2^*|$ as above, we see that $\varphi^{2n-1}(f_\# z)=2n+1 \bmod 2\neq 0$. Hence, $f^*\Phi^{2n-1}=\Psi^{2n-1}\neq 0$ and by Theorem 2 we have $K_{2n+1,n}\not\subset\!\!\!\!\!\subset R^{2n-1}$.

Similarly, in Theorem 19 of [2] we have proved that for $n>1$, $K=K_{2n,n}$ is $\subset R^{2n-1}$, but $\not\subset R^{2n-2}$. Let us prove now $K_{2n,n}\not\subset\!\!\!\!\!\subset R^{2n-2}$ as follows. In that case

$$z=\sum[(a_0),(a_{i_0}\cdots a_{i_{n-1}})*(a_{j_0}\cdots a_{j_{n-1}})]$$

is a mod 2 cycle in $K_2^{(0)}$, in which $\sum$ is extended over all possible sets of indices (i,j) such that $i_0<j_0$, $0\leqslant i_0<\cdots<i_{n-1}\leqslant 2n$, $0\leqslant j_0<\cdots<j_{n-1}\leqslant 2n$, and no i is equal to any j. On the other hand, by Theorem 11 of [2] , $\Phi^{2n-2}(K)\in H^{2n-2}(K_2^*)$ has a representative cocycle

$$\varphi^{2n-2}=\sum\{(a_{i_0}\cdots a_{i_{n-1}})*(a_{j_0}\cdots a_{j_{n-1}})\},$$

in which $\sum$ is extended over all possible sets of indices (i,j) with $0\leqslant i_0<j_0<i_1<\cdots<j_{n-1}\leqslant 2n$. Define f as before, then we have $\rho_2\varphi^{2n-2}(f_\# z)=2n+1 \bmod 2\neq 0$. Therefore $f^*\Phi^{2n-2}=\Psi^{2n-2}\neq 0$ and we have $K_{2n,n}\not\subset\!\!\!\!\!\subset R^{2n-2}$.

In general, we have the following

Theorem 3 For $n \leqslant m \leqslant 2n-1$ we have

$$K_{m+2,n} \subset R^{m+1} \quad \text{but} \quad \not\subset R^m. \tag{$(3)_{m,n}$}$$

Proof. The case where $n=1$, is evident. For $m=2n-1$ or $2n-2$, $(3)_{m,n}$ has already been proved. Suppose now $(3)_{m,n-1}$ has been proved with corresponding $\Psi \neq 0$. Then for $n-1 \leqslant m \leqslant 2n-3$, we have by induction hypothesis $\Psi^m(K_{m+2,n-1}) \neq 0$, a fortiori we have $\Psi^m(K_{m+2,n}) \neq 0$ and therefore $(3)_{m,n}$ is also true.

Remark. In Theorem 19 of [2] we proved only

$$K_{m+2,n} \not\subset R^m (n \leqslant m \leqslant 2n-1).$$

§4 Relations with steenrod squares

Let H still denote singular homology and $\overline{H}$ compact singular homology as in §1. Then for any finite complex and its subcomplex with spaces X and Y, we have (R being the coefficient ring)

$$H^*(X,Y;R) \approx \bar{H}^*(X-Y;R). \tag{1}$$

In what follows the coefficient group will always be I_2 and will therefore be omitted throughout.

For any finite polyhedron X let us consider now the following commutative diagram:

$$\begin{array}{ccccc}
\bar{H}^r(X_2^*) & \xrightarrow{\bar{\mu}_1} & \bar{H}^{r+1}(X_2^*) & \xrightarrow{\pi^*} & \bar{H}^{r+1}(\tilde{X}_2^*) \\
 & & \Vert & & \\
 & & H^{r+1}(X_2, d_2X) & & \\
 & & \uparrow & & \\
H^r(\tilde{X}_2) & \xrightarrow{i^*} & H^r(X) & \xrightarrow{\tilde{\delta}^*} & H^{r+1}(\tilde{X}_2, d_2X)
\end{array}$$

In the above diagram $\bar{\delta}^*$, δ^* are coboundary homomorphisms, $\overline{\mu}_1 == \bar{\mu}_1(\tilde{X}_2^*, t_2)$ and $\pi: \tilde{X}_2^* \to X_2^*$ is natural projection. From the theory of Smith we know that the first row $\overline{\mu}_1$, π^* is exact, and that the second row i^*, $\tilde{\delta}^*$ is also exact, being sequence of the pair $(X \times X, X)$. Since $i^*H^r(\tilde{X}_2) = H^r(X)$, we have $\tilde{\delta}^* = 0$ and hence $\pi^*\bar{\delta}^* = 0$. It follows that

$$\bar{\delta}^* H^r(X) \subset \bar{\mu}_1 \bar{H}^r(X_2^*). \tag{2}$$

Suppose now $\rho_2\Phi^m(X)=0$. Put $\bar{\mu}_i=\bar{\mu}_i(\tilde{X}_2^*,t_2)$, then we have (see §1(1), (2))

$$\bar{\mu}_{m-1}\bar{\delta}^*H^*(X)\subset\bar{\mu}_{m-1}\bar{\mu}_1\bar{H}^*(X_2^*)=\bar{\mu}_m\bar{H}^*(X_2^*)=\rho_2\Phi^m(X)\smile\bar{H}^*(X_2^*),$$

consequently

$$\bar{\mu}_{m-1}\bar{\delta}^*=0. \tag{3}$$

For any class $U\in H^r(X)$, there exists a relation (cf. [14] or [15])

$$\sum_{j=0}^{r}\bar{\mu}_{r-j}\bar{\delta}^*Sq^jU=0. \tag{4}$$

Let $Sm^i: H^r(X)\to H^{r+i}(X)$ be the Smith operations determined by the following relations (cf. Theorem 2 of [6]):

$$\sum_{i_1+i_2=i}Sq^{i_1}Sm^{i_2}=\begin{cases}0, & i>0,\\ 1, & i=0,\end{cases} \tag{5}$$

in which 1 denotes identity homomorphism. For the classes $Sm^iU\in H^{r+l}(X)$, we have similarly

$$\sum_{j=0}^{r+i}\bar{\mu}_{r+i-j}\bar{\delta}^*Sq^jSm^iU=0. \tag*{(4)$_i$}$$

Let k, r be integers satisfying $m+1\geqslant 2k+r\geqslant m\geqslant r+1$. Applying $\bar{\mu}_{m-r-2i-1}$ to $(4)_i$, $i=0,1,\cdots,k-1$, adding all these relations, and noting that $\bar{\mu}_r\bar{\mu}_s=\bar{\mu}_{r+s}$, we have

$$\sum_{s=0}^{l}\bar{\mu}_{m-s-1}\bar{\delta}^*\sum_{i=0}^{k-1}Sq^{s-i}Sm^iU=0,\quad l=r+2k-2.$$

By (5) this equation may also be written as

$$\bar{\mu}_{m-1}\bar{\delta}^*U=\sum_{s=k}^{l}\bar{\mu}_{m-s-1}\delta^*V_s, \tag{6}$$

where

$$V_s=\sum_{l=k}^{s}Sq^{s-i}Sm^iU,\quad s=k,\cdots,l.$$

By (3), the right-hand side of (6) is 0. Since by hypothesis $2k+r\geqslant m$, we have for the term $\bar{\mu}_{m-k-1}\tilde{\delta}^*V_k$ of the right-hand side of (6) with the greatest lower index of $\bar{\mu}$, $m-k-1<k+r=\dim V_k$. It follows that (cf. [14] or [15])

$$V_s=0,\quad l\geqslant s\geqslant k.$$

This system of equations is evidently equivalent to the following system:

$$Sm^sU = 0, \quad l \geqslant s \geqslant k.$$

From the above we get

Theorem 4 If $\rho_2\Phi^m(X) = 0$ for a finite polyhedron X, then for any integers k, r satisfying $2k + r \geqslant m$, we have

$$Sm^kH^r(X) = 0, \quad 2k + r \geqslant m. \tag{7}$$

From the main Theorem 1 we have then (cf. [4] Th. III. 25) the following

Corollary (Thom) If a finite polyhedron X is realizable in R^m, then (7) is true.

Theorem 5 If $\rho_2\Psi^{m-1}(X) = 0$ for a finite polyhedron X, then (7) is true.

Proof. Take a simplicial subdivision K of X and construct as in §3 the complexes $\tilde{K}_2^*$, $\tilde{K}_2^{(+)}$, $\tilde{K}_2^{(0)}$ and $K_2^* = \tilde{K}_2^*/t_2$, etc. Let $\tilde{j} : (\tilde{X}_2, d_2X) \subset\subset (\tilde{X}_2, d_2X \cup |\tilde{K}_2^*|)$ be the inclusion map, which induces the homomorphisms $\tilde{j}^* : H^*(\tilde{X}_2, d_2X \cup |\tilde{K}_2^*|) \to H^*(\tilde{X}_2, d_2X)$, or what is the same.

$$\tilde{j}^* : \bar{H}^*(\tilde{Y}) \to \bar{H}^*(\tilde{X}_2^*),$$

where $\tilde{Y} = \tilde{X}_2 - \ d_2X \cup |\tilde{K}_2^*|$. Similarly we have the inclusion map $j : (X_2, d_2X) \subset (X_2, d_2X \cup |K_2^*|)$ and homomorphisms

$$j^* : \bar{H}^*(Y) \to \bar{H}^*(X_2^*),$$

where $Y = X_2 - d_2X \cup |K_2^*| = \tilde{Y}/t_2$. Now there are topological maps $\tilde{i} : \tilde{Y} \equiv |\tilde{K}_2^{(0)}| \times I$, $i : Y \equiv |K_2^{(0)}| \times I$, where I is an open segment, and $t_2\tilde{i} \equiv t_2/|\tilde{K}_2^{(0)}| \otimes 1$ (1 being the identity map), we have on setting $\bar{\mu}_i(\tilde{Y}, t_2) == \bar{\mu}_{i,Y}, \bar{\mu}_i(|\tilde{K}_2^{(0)}|, t_2) = \mu_i(|\tilde{K}_2^{(0)}|, t_2) = \mu_i^{(0)}$,

$$i^* : H^*(|K_2^{(0)}|) \otimes \bar{H}^*(I) \approx \bar{H}^*(Y). \tag{8}$$

Moreover, we have

$$\bar{\mu}_{i,Y}i^* = i^*(\mu_i^{(0)} \otimes 1), \tag*{(8)$'$}$$

in which 1 is the identity homomorphism. Consider now the following commutative diagram:

$$\begin{array}{ccc} \bar{H}^*(Y) & \xrightarrow{1^*} & \bar{H}^*(X_2^*) \\ \uparrow \bar{\mu}_{i,Y} & & \uparrow \mu_i \\ \bar{H}^*(Y) & \xrightarrow{1^*} & \bar{H}^*(X_2^*) \\ \uparrow \tilde{\delta}_Y^* & \nearrow \bar{\delta}^* & \\ H^*(d_2X) & & \end{array}$$

In the above diagram $\bar{\mu}_i = \bar{\mu}_i(\tilde{X}_2^*, t_2)$, $\tilde{\delta}_Y^*$, $\bar{\delta}^*$ are all coboundary homo-morphisms.

Now by hypothesis $\rho_2 \Psi^{m-1}(X) = 0$, or $\mu_{m-1}^{(0)} H^*(|K_2^{(0)}|) = 0$. Hence by (8) and (8)$'$ we have $\bar{\mu}_{m-1,Y} = 0$. In the above diagram setting $i = m-1$, we get

$$\bar{\mu}_{m-1} \bar{\delta}^* H^*(d_2X) = 0,$$

i.e., (3). As in the proof of the above Theorem 4 we have then (7) This proves our theorem.

Corollary If a finite polyhedron X may he immersed in R^{m-1}, then (7) is true.

§5 Relations with whitney classes in case of a differentiable manifold

In this section, the coefficient group will still he supposed to be I_2 and will be omitted throughout.

Let X be a closed differentiable manifold of dimension n, with class $\geqslant 3$. Take a Riemannian metric ds^2 of X and let $d\tilde{s}^2$ be the corresponding symmetric Riemannian metric of $X \times X$. Take $\varepsilon > 0$ sufficiently small such that through any point x in $X \times X$ having a distance $d(x) \leqslant \varepsilon$ from d_2X, there is one and only one geodesic perpendicular to d_2X with length $= d(x)$. Let $\tilde{Y}$ be the space of all points x in $X \times X$ for which $0 < d(x) < \varepsilon/2$, and $\tilde{T}$ be the space of all points x with $d(x) = \varepsilon/2$. Put $\tilde{M} = \tilde{X}_2^* - \tilde{Y}$, then t_2 will operate in $\tilde{M}$, $\tilde{T}$ to form simple systems. Let $M = \tilde{M}/t_2$, $T = \tilde{T}/t_2$, then we see easily that, by taking sufficiently small simplicial subdivision K of X compatible with its given ditferential structure, $|\tilde{K}_2^{(0)}|$ will have the same P-homotopy type as $\tilde{T}$, and $|K_2^{(0)}|$ and T will have the same homotopy type. It follows that

$$A^i(\tilde{T}, t_2) = A^i(|\tilde{K}_2^{(0)}|, t_2) = \Psi^i(X).$$

On the other hand, we know that T is the bundle space of a bundle of $(n-1)$-dimensional projective spaces over X, satisfying the Leray-Hirsch condition. Moreover, denoting by W^i the mod 2 reduced Stiefel-Whitney classes of X, W^i would be completely determined by the following equation (cf. e.g. [16]) :

$$\sum_{i=0}^{n} \rho_2 \Psi^i(X) \smile g_0^* W^{n-i} = 0, \tag{1}$$

in which $g_0 : T \to X$ is the natural projection. From (1) we get

$$\rho_2 \Psi^{n+i}(X) = \sum_{j=1}^{n} \rho_2 \Psi^{n-j}(X) \smile g_0^* U_{j,i}, \tag{2}$$

where $U_{j,i} \in H^{i+j}(X)$ and satisfies

$$\begin{cases} U_{j,0} = W^i, & j \geqslant 1; \\ U_{j,i+1} = U_{j+1,i} + U_{1i} \smile W^j, & j \geqslant 1. \end{cases} \tag{3}$$

Therefrom we get

$$\sum_{\substack{i+j=k \\ i,j \geqslant 0}} U_{1,i} \smile W^j = W^{k+1}.$$

Comparing with the definition of dual Whitney classes $\bar{W}^i$, we have therefore

$$U_{1,i} = \bar{W}^{i+1}. \tag{4}$$

From (2) we know that if $\rho_2 \Psi^{m-1} = 0$, then $U_{j,i} = 0$, $i \geqslant m-n-1$, $n \geqslant j \geqslant 1$ and hence by (4), $\bar{W}^k = 0$, $k \geqslant m-n$. Conversely, if $\overline{W}^k = 0$ $k \geqslant m-n$, then by (2), (3) and (4) we have also $\rho_2 \Psi^{m-1} = 0$. Consequently we have

Theorem 6　If $\rho_2 \Psi^{m-1} = 0$ for an n-dimensional closed differentiable manifold X, then

$$\bar{W}^k = 0, \quad k \geqslant m-n.$$

The converse is also true.

By the principal Theorem 2 we have the following

Corollary　If an n-dimensional closed differentiable manifold X is immersible in R^{m-1}, then

$$\bar{W}^k = 0, \quad k \geqslant m-n. \tag{5}$$

Whitney has proved that any n-dimensional closed differentiable manifold may be imbedded in R^{2n} ([17]) and immersed in R^{2n-1} ([9]). The above theorem shows that the immersion theorem of Whitney can not be further improved in general. For example, the example given in [3] §25 is a 4-dimensional manifold with $\bar{W}^3 \neq 0$, and therefore it is non-immersible in R^6 by Theorem 6. This answers also a problem of Whitney (cf. the introduction of [9]).

Theorem 7 If $\rho_2\Phi^m$=0 for a closed differentiable manifold X, then $\rho_2\Psi^{m-1}$=0.

Proof. $\tilde{M} = \tilde{X}_2^* - \tilde{Y}$ is a bounded $2n$-dimensional manifold with $\tilde{T}$ as boundary, similarly M is a bounded manifold with T as boundary. Take now two models $\tilde{M}_1$ and $\tilde{M}_2$ of $\tilde{M}$, each homomorphic to $\tilde{M}$ under h_j, and with boundaries $\tilde{T}_1$, $\tilde{T}_2$ respectively. Identify now $\tilde{T}_1$ and $\tilde{T}_2$ under $h_2^{-1}h_1$ with $\tilde{T}'$, then from $\tilde{M}_i$ we get a $2n$-dimensional closed differentiable manifold $\tilde{M}'$. In $\tilde{M}'$ we define a periodic transformation t' by

$$t'(\tilde{x}_1) = h_2^{-1}t_2h_1(\tilde{x}_1), \quad t'(\tilde{x}_2) = h_1^{-1}t_2h_2(\tilde{x}_2), \quad \tilde{x}_i \in \tilde{M}_i. \tag{6}$$

Let $M' = \tilde{M}'/t'$, then M' may be regarded as the $2n$-dimensional closed manifold obtained from $\tilde{M}$ by identifying $\tilde{T}$ under t_2 with $T = \tilde{T}/t = \tilde{T}'/t'$. Define $\tilde{h} : \tilde{M}' \to \tilde{M}$ by $\tilde{h}(\tilde{x}_i) = h_i(\tilde{x}_i)$, $\tilde{x}_i \in \tilde{M}_i$, then $\tilde{h}t' = t_2\tilde{h}$. Hence, $\tilde{h}$ induces a map $h : M' \to M$ and we have

$$\rho_2 A^m(\tilde{M}', t') = \rho_2 h^* A^m(\tilde{M}, t_2) = h^*\rho_2\Phi^m = 0, \quad \mu'_m = \mu_m(\tilde{M}', t'). \tag{7}$$

With respect to the simple systems $(\tilde{M}', t')$ and $(\tilde{T}', t')$, we may also define the Smith homomorphisms between homology groups (coerricient groups deing I_2)①:

$$v'_j = v_j(\tilde{M}', t') : H_r(M') \to H_{r-i}(M'),$$

and

$$v''_i = v_i(\tilde{T}', t') : H_r(T) \to H_{r-i}(T).$$

In [13] we have proved that (as the coefficient groups are I_2, it is not necessary to suppose the manifolds to be orientable)

$$\mathscr{D}'v'_i = \mu'_i\mathscr{D}', \quad \mathscr{D}''v''_i = \mu''_i\mathscr{D}'', \tag{8}$$

where

$$\mu''_i = \mu_i(\tilde{T}', t')$$

and

$$\mathscr{D}' : \quad H_r(M') \approx H^{2n-r}(M'),$$
$$\mathscr{D}'' : \quad H_r(T) \approx H^{2n-1-r}(T)$$

① In [13] , μ_i, v_i are denoted by μ_1^* and μ_i respectively.

are the dual isomorphisms in the closed manifolds M' and T respectively.

Construct now simplicial subdivisions of T and M', denoted still by T and M', such that T is a subcomplex of M'. Construct also simplicial subdivisions of $\tilde{T}'$ and $\tilde{M}'$, still denoted by $\tilde{T}'$ and $\tilde{M}'$, which are covering complexes of T and M' respectively.

Let the fundamental cycles of $\tilde{M}'$, etc. , be respectively $\tilde{z}' \in \tilde{Z}' \in H_{2n}(\tilde{M}')$, $\tilde{z}'' \in \tilde{Z}'' \in H_{2n-1}(\tilde{T}')$, $z' \in Z' \in H_{2n}(M')$, and $z'' \in Z'' \in H_{2n-1}(T)$. We may write $\tilde{z}' = \tilde{z}_1' + \tilde{z}_2'$, where $\tilde{z}_i'$ are chains in $\tilde{M}_i$. Then $t'\tilde{z}_1' = \tilde{z}_2'$, $\partial\tilde{z}_1' = \partial\tilde{z}_2' = \tilde{z}''$, $\pi\tilde{z}_1' = \pi\tilde{z}_2' = z'$, where $\pi : \tilde{M}' \to M'$ is the natural projection. We may also write $\tilde{z}'' = (1+t')\tilde{c}''$, where $\tilde{c}''$ is a chain in $\tilde{T}'$, and $\pi\tilde{c}'' = z''$. Whence we know that if $\tilde{j} : \tilde{T}' \subset \tilde{M}'$ and $j : T \subset M'$ are inclusion maps, then

$$j_* {v''}_i = {v'}_i j_*,$$
$$j_* Z'' = {v'}_1 Z'.$$

Consequently by (7), (8) we have (1 denotes the mod 2 unit class of M')

$$j_* v''_{m-1} Z'' = v'_m Z' = \mathscr{D}'^{-1} \mu'_m 1 = 0,$$

or

$$\rho_2 j_* \mathscr{D}''^{-1} \Psi^{m-1} = 0.$$

It follows that in order to prove our theorem, it is sufficient to prove that

$$j_* : H(T) \subset H(M'). \tag{9}$$

For this purpose let us first define a continuous map

$$\tilde{g} : \tilde{M} \to X \times X$$

as follows. For any point $\hat{x} \in \tilde{M}$ having a distance $d(\tilde{x}) = \varepsilon$ to $d_2 X$, let $l(\tilde{x})$ be the geodesic through $\tilde{x}$ and perpendicular to $d_2 X$, and let $\hat{x}' \in \tilde{T}$ be the mid-point of this geodesic. Let us map now any point $\tilde{y}$ on the geodesic arc between $\tilde{x}$, $\tilde{x}'$ to that point $\tilde{g}(\tilde{y})$ on $l(\tilde{x})$ such that the arc length from $\tilde{x}$ to $\tilde{g}(\tilde{y})$ is two-fold as that from $\tilde{x}$ to $\tilde{y}$. On the other hand, if $\tilde{x}$ has a distance $> \varepsilon$ to $d_2 X$, then we put $\tilde{g}(\tilde{x}) = \tilde{x}$. In this manner we get the above-mentioned map $\tilde{g}$. As the chosen Riemannian metric in $X \times X$ is symmetric, we have for any $\tilde{x} \in \tilde{T}$, $\tilde{g}(\tilde{x}) = \tilde{g}t_2(\tilde{x})$. It follows that $\tilde{g}$ will induce a continuous map

$$g : M' \to X \times X.$$

If we regard T as the bundle space of a bundle of $(n-1)$-dimensional projective spaces over d_2X or X, $g/T = g_0 : T \to d_2X$ is no other than the bundle projection.

It is known that $H^*(T)$ is generated by all classes Ψ^i and g_0^*U, where $U \in H^*(X)$. As

$$j^*A^i(\tilde{M}', t') = A^i(\tilde{T}', t') = \Psi^i,$$

and

$$j^*g^*(1 \otimes U) = g_0^*d_2^*(1 \otimes U) = g_0^*U,$$

it follows that

$$j^*H^*(M') = H^*(T).$$

Dually, the last equation is equivalent to (9). Our theorem is thus completely proved.

By Theorem 6 and the main Theorem 2, we get the following corollaries:

Corollary 1 If $\rho_2\Phi^m = 0$ for an n-dimensional closed differentiable manifold X, then (5) is true.

Corollary 2 If an n-dimensional closed differentiable manifold is realizable in R^m, then (5) is true.

Remark 1. Whitney has proved that (5) is a necessary condition for an n-dimensional closed differentiable manifold to be "differentiably" realizable in R^m. The above Corollary 2 shows that this condition is not only necessary for "differentiable" realizability, but also for "topological" realizability.

Remark 2. By Theorem 3 of §3, $K = K_{m+1,n}$ has the following property. $\rho_2\Psi^{m-1} \neq 0$ and $\Phi^m = 0$. It follows that Theorem 7 is not true for general finite polyhedrons. In reality, the property of X being a manifold is essentially used in the proof of Theorem 7.

§6 The classes Φ^m of projective spaces

In general, the calculation of the classes Φ^m is quite difficult. As an example, we shall now determine $\rho_2\Phi^m(P^n) = \rho_2\Phi^m$ of an n-dimensional projective space P^n as follows, n being supposed to be $\geqslant 2$.

Consider P^n as the space of all lines in R^{n+1} through its origin, then $\tilde{P}_2^{n*}$ is the space of all ordered pairs (l_1, l_2), in which l_1, l_2 are any two *different* lines in R^{n+1} through its origin, and $t_2 = t : \tilde{P}_2^{n*} \to \tilde{P}_2^{n*}$is defined by $t(l_1, l_2) = (l_2, l_1)$. For any point $(l_1, l_2) \in \tilde{P}_2^{n*}$, let $[l_1, l_2]$ be the plane determined by l_1, l_2. In this plane we may take two lines l_1', l_2' through the origin, perpendicular to each other, and such that the acute angle θ_1 between l_1' and l_1 is equal to the acute angle θ_2 between l_2' and l_2. Between l_i and $l_i'(i=1,2)$ we may construct a line $l_i^{(\tau)}$ which divides θ_i into two angles with ratio $\tau : 1-\tau$, where $0 \leqslant \tau \leqslant 1$. Then $h_\tau : (l_1, l_2) \to (l_1^{(\tau)}, l_2^{(\tau)})$ is a P-deformation of $\tilde{P}_2^{n*}$. Denote $h_1(\tilde{P}_n^{n*})$ by $\tilde{P}_{n,2}$, then $\tilde{P}_{n,2}$ is the space of all ordered pairs (l_1, l_2), where l_1, l_2 are any two mutually perpendicular lines in R^{n+1} through its origin and t will still operate in $\tilde{P}_{n,2}$. Denote the modular space of $(\tilde{P}_{n,2}, t)$ by $P_{n,2} = \bar{P}_{n,2}/t$, and the projection $\tilde{P}_{n,2} \to P_{n,2}$ by λ. Then we have

$$A^i(\tilde{P}_{n,2}, t) = A^i(\tilde{P}_2^{n*}, t) = \varPhi^i, \tag{1}$$

and

$$\lambda^* \varPhi^i = 0, \quad i > 0. \tag{2}$$

Let $R_{n-1,2}$ be the Grassmann manifold defined in R^{n+1}, and $\tilde{\pi} : \tilde{P}_{n,2} \to R_{n-1,2}$ be the natural map derined by $\tilde{\pi}(l_1, l_2) = [l_1, l_2]$. we may also define $\pi : P_{n,2} \to R_{n-1,2}$ by $\pi\lambda = \tilde{\pi}$. Evidently $P_{n,2}$ is the bundle space of an S^1-bundle over $R_{n-1,2}$ with π as the projection map of the bundle. Let S_0^1 be a fixed fibre of this bundle, and $j : S_0^1 \to P_{n,2}$ the inclusion map, then $j^*\varPhi^1$ is a generator of $H^1(S_0^1)$①. It follows that this bundle satisfies the condition of Leray-Hirsch and therefore $H^*(P_{n,2})$ is additively isomorphic to $H^*(R_{n-1,2}) \otimes H^*(S^1)$, and also

$$\pi^* : H^*(R_{n-1,2}) \subset H^*(P_{n,2}), \tag{3}$$

$$(\varPhi^1)^2 + \pi^* A^1 \smile \varPhi^1 + \pi^* A^2 = 0, \quad A^i \in H^i(R_{n-1,2}), \quad i = 1, 2. \tag{4}$$

By (2) and (3) we get further

$$\tilde{\pi}^* A^2 = \lambda^* \pi^* A^2 = 0. \tag{5}$$

Since $\tilde{P}_{n,2}$, is the bundle space of the projective-line bundle associated with the canonical S^1-bundle over $R_{n-1,2}$, this bundle has also the Leray-Hirsch property and the bundle projection $\tilde{\pi}$ satisfies

$$\tilde{\pi}^* : H^*(R_{n-1,2}) \subset H^*(\tilde{P}_{n,2}).$$

By (5) we get $A^2 = 0$, hence (4) becomes

$$(\varPhi^1)^2 + \pi^* A^1 \smile \varPhi^1 = 0.$$

① All coefficient groups will be supposed to be I_2, so that it will be omitted throughout.

Since $n \geqslant 2$ by hypothesis, we have $(\Phi^1)^2 = \rho_2\Phi^2 \neq 0$ and from the above equation and (3) we know that $A^1 \neq 0$. As $H^1(R_{n-1,2})$ is generated by W-class W^1 of $R_{n-1,2}$, we have necessarily $A^1 = W^1$, and (4) becomes

$$(\Phi^1)^2 + \pi^* W^1 \smile \Phi^1 = 0,$$

whence

$$(\Phi^1)^r = \pi^*(W^1)^{r-1} \smile \Phi^1.$$

From the known cohomology ring structure of $R_{n-1,2}$ (cf. [18] or [19]), we know that for $n = 2^a + s$, $s < 2^a$, we have

$$(W^1)^{r-1} \begin{cases} \neq 0, & r-1 \leqslant 2n-2s-2, \\ = 0, & r-1 > 2n-2s-2. \end{cases}$$

It follows that

$$\rho_2\Phi^r = (\Phi^1)^r \begin{cases} \neq 0, & r \leqslant 2n-2s-1, \\ = 0, & r > 2n-2s-1. \end{cases}$$

Remark 1. From the above equation we know that $\rho_2\Phi^{2n-2s} = 0$, but we do not know whether Φ^{2n-2s} is also 0 or not.

Remark 2. By the principal Theorem 1 we know that for $n = 2^a + s, s < 2^a$,

$$P^n \not\subset R^{2n-1-2s}.$$

This may also be deduced from Whitney' s condition. For, let C^i be the generator of $H^i(P^n)$, then the Whitney polynomial of P^n is given by

$$W(P^n) = (1 + C^1 t)^{n+1},$$

$$\bar{W}(P^n) = (1 + C^1 t)^{-(n+1)} = \sum \binom{n+i}{n} C^i t^i.$$

By elementary theory of numbers we know that for

$$n = 2^{a_1} + \cdots + 2^{a_\alpha} \quad (a_1 > \cdots > a_\alpha),$$

$$m = 2^{b_1} + \cdots + 2^{b_\beta} \quad (b_1 > \cdots > b_\beta),$$

the necessary and sufficient condition for $\binom{n}{m} \not\equiv 0 \bmod 2$ is that each b_i should be equal to a certain a_i. Hence, for $n = 2^a + s$, $s < 2^a$, we have $(i \leqslant n)$

$$\binom{n+i}{n} \begin{cases} \neq 0 \mod 2, & i = n-1-2s, \\ = 0 \mod 2, & i > n-1-2s. \end{cases}$$

Hence, $\bar{W}^{n-1-2s}(P^n) \neq 0$ and we have $P^n \not\subset R^{2n-1-2s}$.

§7 Isotopy and iso-position

Let f_0 and f_1 be two realizations (resp. immersions) of a Hausdorff space X in R^m. Let I be the segment $[0, 1]$. If there exists a continuous map $F : X \times I \to R^m$ such that $F|X \times (0) \equiv f_0$, $F|X \times\times(1) \equiv f_1$, and for each $t \in I$, $F|X \times (t) \equiv f_t$ is a realization (resp. immersion) of X in R^m, then we will say that the realizations (resp. immersions) f_0 and f_1 are *isotopic.*

Let f_0 and f_1 be two realizations of X in R^m. If there exists a topological transformation h of R^m onto itself such that $hf_0 \equiv f_1$, then the two realizations will be said to be in *iso-position.* We will say also that f_0, f_1 are in *orientation-preserving or orientation-reversing iso-position* according as h is orientation-preserving or orientation reversing.

Suppose now given $f : X \subset R^m$ or $f : X \subset\!\!\!\!\propto R^m$. Let S^{m-1} be the unit sphere of R^m. Choose a fixed orientation of R^m and let S be the generator of $H^{m-1}(S^{m-1})$ corresponding to this orientation. In the case of $f : X \subset R^m$, let us draw for each point $(x_1, x_2) \in \tilde{X}_2^*$ a half-line through the origin O and parallel to the oriented line from $f(x_1)$ to $f(x_2)$, which meets S^{m-1} in a point $\tilde{f}(x_1, x_2)$. The map $\tilde{\tilde{f}} :\ \tilde{X}_2^* \to S^{m-1}$ will be said to be associated with f, and $\tilde{f}^*S$ will be denoted by $\theta_f \in H^{m-1}(\tilde{X}_2^*)$. If the orientation of R^m is changed, then θ_f also changes by a sign. Similarly, let $f : X \subset\!\!\!\!\propto R^m$ and X be a finite polyhedron with a sufficiently small simplicial subdivision K, then we may define in a similar manner a continuous map $\tilde{f} : |\tilde{K}_2^{(0)}| \to S^{m-1}$ and the class $\tilde{f}^*S$ will be denoted by $\theta'_f \in H^{m-1}(|\tilde{K}_2^{(0)}|)$. In [10] we have proved that $H^{m-1}(|\tilde{K}_2^{(0)}|)$ is independent of the chosen subdivision K of X and by the method of [10] we may see that θ'_f is also independent of the subdivision K and depends only on X, on the immersion f and on the chosen orientation of R^m.

Theorem 8 With respect to a fixed orientation of R^m, the class θ_f(resp.θ'_f) is an isotopy invariant of the realization $f : X \to R^m$ (resp. the immersion f : $X \subset\!\!\!\!\propto R^m$, in that case X should be a finite polyhedron).

Proof. Let $f_0, f_1 : X \subset R^m$ be isotopic so that there exists a map $F :\ X \times I \to R^m$ as in the beginning of this section. For any $t \in I$, let $f_t|X \equiv F \quad X \times (t)$, then map $\tilde{f}_t : \tilde{X}_2^* \to S^{n-1}$ associated with f_t gives rise to a homotopy between $\tilde{f}_0$, $\tilde{f}_1$, so that $\tilde{f}_0^*S = \tilde{f}_1^*S$, i.e., $\theta_{f_0} = \theta_{f'_1}$, or θ_f is an isotopy invariant of f. In the case that X is a finite polyhedron and $f : X \subset\!\!\!\!\propto R^m$ is an immersion, that θ'_f is an isotopy invariant may be proved similarly.

Theorem 9 With respect to a fixed orientation of R^m, $\pm\theta_f$ (resp. $+\theta_f$) is an iso-position invariant (resp. orientation-preserving iso-position invariant) of the realization $f : X \subset R^m$.

Proof. Let $\tilde{R}^*$ be the space of all ordered pairs (x_1, x_2), where $x_i \in R^m$ and $x_1 \neq x_2$. Let $\tilde{R}'^*$ be the subspace of $\tilde{R}^*$ consisting of all points (x_1, x_2) with length of $x_1x_2 = 1$. For any point $(x_1, x_2) \in \tilde{R}^*$, let y be the mid-point of x_1x_2, and y_1, y_2 be points on half-lines yx_1 and yx_2 respectively such that yy_1 and yy_2 are both of length 1/2. Denote by $y_i(t)$ that point on x_iy_i such that $x_iy_i(t) : y_i(t)y_i = t : 1-t$, then $d_t : (x_1, x_2) \to (y_1(t), y_2(t))$ is a deformation-retraction of $\tilde{R}^*$ onto $\tilde{R}'^*$. Since $\tilde{R}'^*$ is the bundle space of an $(m-1)$-sphere bundle over R^m as base space, we have by a theorem of Feldban $\bar{R}'^* \equiv R^m \times S^{m-1}$, and we have $H^*(\tilde{R}^*) \approx H^*(\tilde{R}'^*) \approx H^*(S^{m-1})$, where H denotes a singular homology system. For any point $(x_1, x_2) \in \tilde{R}^*$, let us draw through O a half-line parallel to the directed line from x_1 to x_2, which intersects the unit sphere S^{m-1} in a point $g(x_1, x_2)$. Then the continuous map $g : \tilde{R}^* \to S^{m-1}$ induces $g^* : H^*(S^{m-1}) \approx H^*(\tilde{R}^*)$ which is the same as the isomorphism described above. Fixing now an orientation of S^{m-1} and letting S be the generator of $H^{m-1}(S^{m-1})$ corresponding to this orientation, then $g^*S \in H^{m-1}(\tilde{R}^*)$ and will be denoted by $\varGamma$.

If h is a topological transformation of R^m onto itself, h will induce a topological map $\tilde{h}$ of $\tilde{R}^*$ onto $\tilde{R}^*$. It is easy to see that for fixed orientation of R^m, we have $\tilde{h}^*\varGamma = +\varGamma$ or $-\varGamma$, according as h is orientation-preserving or not.

Suppose now f, f' be two iso-positional realizations of the Hausdorff space X into R^m, so that $h : R^m \equiv R^m$ exists with $hf = f'$. Let $\tilde{f}$, $\tilde{f}'$ be maps of $\tilde{X}_2^*$ into $\tilde{R}^*$ induced by f and f' respectively: $\tilde{f}(x_1,x_2) = (f(x_1), f(x_2))$, $\tilde{f}'(x_1, x_2) = (f'(x_1), f'(x_2))$, $(x_1, x_2) \in \tilde{X}_2^*$, then we have evidently $\theta_f = \tilde{f}^*\varGamma$, $\theta_{f'} = \tilde{f}'^*\ \varGamma$. We have also $\tilde{h}\tilde{f} = \tilde{f}'$. Hence, $\theta_f = +\ \theta_{f'}$ or $-\theta_{f'}$, according as h is orientation-preserving or not. From this we get the theorem.

Example 1 Given a circle C. Take three points a_1, a_2, a_3 on C, then $H_1(\tilde{C}_2^*)$ is generated by a single generator Z which has the following singular chain as a representative; namely,

$$z = a_1 \times (a_2a_3) + a_2 \times (a_3a_1) + a_3 \times (a_1a_2)$$
$$+ (a_2a_3) \times a_1 + (a_3a_1) \times a_2 + (a_1a_2) \times a_3.$$

Consider now a circle D in the plane R^2. Orient R^2 arbitrarily and let D be oriented accordingly. This oriented D will be denoted by$+D$. Denote the oriented circle C with a_1, a_2, a_3 in positive order by $+C$. Define now two realizations f, $f' : X \subset R^2$ by $f(+C) = +D$, $f'(+C) = -D$. Then with respect to the above orientation of

R^2, we have $\theta_f(Z) = +1$, $\theta_{f'}(Z) = -1$. Hence, by Theorem 9, f, f' cannot be in orientation-preserving iso-position.

Example 2 Let X be the space consisting of a circle C and an isolated point P. Orient C and define Z as in Ex.1. Then $H_1(\tilde{X}_2^*)$ has three generators of which one is Z and a second one is Z' with a representative cycle $z' = P \times (+C)$. In the oriented plane R^2 take an oriented circle $+D$ as in Ex.1. Take also a point Q inside D and a point Q' outside D. Define realizations f_1, f_1', f_2, f_2': $X \subset R^2$ by

$$f_1(+C) = f'_1(+C) = +D, \quad f_2(+C) = f'_2(+C) = -D,$$
$$f_1(P) = f_2(P) = Q, \quad f'_1(P) = f'_2(P) = Q'.$$

Then with respect to the above orientation of R^2, we have

$$\theta_{f_1}(Z') = +1, \quad \theta_{f_2}(Z') = -1, \quad \theta_{f'_1}(Z') = \theta_{f'_2}(Z') = 0,$$
$$\theta_{f_1}(Z) = \theta_{f'_1}(Z) = +1, \quad \theta_{f_2}(Z) = \theta_{f'_2}(Z) = -1.$$

From these equations we know that f_1 and f_2 (also f_1' and f_2'), though in iso-position, are not in orientation-preserving iso-position, while f_1 or f_2 is not in iso-position with either f_1' or f_2'.

Example 3 Let K be a complex consisting of three segments (a_0a_1), (a_0a_2), (a_0a_3) joined together at a_0, and $X = |K|$. In $\tilde{X}_2^*$ there is a 1-dimensional cycle non-homologous to 0, namely,

$$\begin{aligned} z = {} & a_1 \times (a_2a_0) + a_1 \times (a_0a_3) + a_2 \times (a_3a_0) + a_2 \times (a_0a_1) \\ & + a_3 \times (a_1a_0) + a_3 \times (a_0a_2) + (a_2a_0) \times a_1 + (a_0a_3) \times a_1 \\ & + (a_3a_0) \times a_2 + (a_0a_1) \times a_2 + (a_1a_0) \times a_3 + (a_0a_2) \times a_3. \end{aligned} \tag{1}$$

The cycle z represents a generator Z of $H_1(\tilde{X}_2^*)$.

In the plane R^2 take four points b_0, b_1, b_2, b_3 and three simple arcs $\beta_1 = \widehat{b_0b_1}$, $\beta_2 = \widehat{b_0b_2}$, $\beta_3 = \widehat{b_0b_3}$ such that β_1, β_2, β_3 do not meet one another except at b_0. Such a figure is called a triod and will be denoted by $[\beta_1, \beta_2, \beta_3]$, ($X$ is also a triod). Define now a realization $f : X \subset R^2$ by $f(a_0) = b_0$, $f(a_0a_i) = \beta_i$, $i = 1, 2, 3$. Fix an orientation of R^2, and define θ_f with respect to this orientation. Let us rotate β_1 about b_0 first to β_2, and then to β_3. Then, if this rotation is concordant with the chosen orientation of R^2, we should have $\theta_l(Z) = +1$. Otherwise we have $\theta_f(Z) = -1$.

By the three examples given above we may prove the following

Theorem 10 Let the Peano continuum M be realizable in the plane. Then the necessary and sufficient condition for any two realizations f, f' of M in R^2 to be in iso-position is that $\theta_f = \pm\theta_{f'}$ (with respect to the same fixed orientation of R^2).

Proof. The necessity has already been given in Theorem 9. Let us prove the sufficiency part as follows.

Let $A = [\alpha_1, \alpha_2, \alpha_3]$ and $B = [\beta_1, \beta_2, \beta_3]$ be any two triods contained in M. In $H_1(\tilde{A}_2^*)$ and $H_1(\tilde{B}_2^*)$ and consequently in $H_1(\tilde{M}_2^*)$, there are elements α, β containing cycles corresponding to (1) as representative cycles. Consider the values a, b of θ_f on α, β and the values a', b' of $\theta_{f'}$ on α, β, where a, b, a', b' are all $=\pm$ 1. By Ex.3 we see that if $f'f^{-1}$ preserves the relative sense (cf. [20]) of $f(\alpha)$, $f(\beta)$, then either $a' = a$, $b' = b$ or $a' = -a$, $b' = -b$. In the contrary case we should have either $a' = +a$, $b' = -b$ or $a' = -a$, $b' = +b$. It follows that when $\theta_f = \pm\theta_{f'}$, $f'f^{-1}$ preserves the relative sense of any pair of triods in $f(M)$.

By Theorems 10 and 11 of [20] we see that if k is the boundary of a complementary domain of $f(M)$, then $f'f^{-1}(k)$ is also the boundary of a complementary domain of $f'(M)$, and vise versa. Suppose that k is the boundary of the infinite complementary domain of $f(M)$, while $f'f^{-1}(k)$ is the boundary of a finite complementary domain of $f'(M)$. Then in M there must exist a simple closed curve C and also a point P not on C, such that $f(C) \subset k$, $f(P)$ lies outside of $f(C)$ and $f'(P)$ lies inside $f'(C)$. By Ex.2, for $X = C + P$, we should have $\theta_f(X) \neq \pm\theta_{f'}(X)$, and a *fortiori* $\theta_f \neq \pm\theta_{f'}$, with respect to the same fixed orientation of R^2. As this is in contradiction with the hypothesis, we see that $f'f^{-1}$ preserves also the boundaries of finite complementary domains of $f(M)$. By Theorem 2 of [20] , we may then extend the topological map $f'f^{-1}$ from $f(M)$ to $f'(M)$ to the whole plane. Denote the extended topological map by h, then $hf = f'$ and f, f' are in iso-position, what we require to prove.

Moreover, by Ex.1, 2, we see also that with respect to the same fixed orientation of the plane, $\theta_f = +\theta_{f'}$ is the necessary and sufficient condition for f, f' to be in orientation-preserving iso-position and $\theta_f = -\theta_{f'}$ is the necessary and sufficient condition for f, f' to be in orientation-reversing iso-position.

The above theorem may be regarded as a quantitative reformulation of a known theorem of Adkisson-McLane (Theorem 2 of [20]). In the same manner, another theorem of Adkisson-McLane about realizations on a sphere (Theorem 1 of [20]) and a well-known theorem of Kuratowski about realization in the plane ([21]) the original statements of which are qualitative in character, may be given quantitative reformulations as follows:

Theorem 10′(Adkisson-McLane) Suppose a Peano-continuum M is realizable in the sphere S^2 and let f, f', be two such realizations. For any $p \notin f(M)$, let us consider f as a realization f_p of M in $R_p^2 = S^2 - (p)$. Take a fixed orientation of S^2 and orient R_p^2 concordantly with corresponding class θ_{f_p}. Similarly we define classes $\{\theta_{t'_{p'}}\}$ for $p' \notin f'(M)$ (S^2 to be oriented as before). Then a necessary and sufficient condition for f and f' to be in iso-position is that there exists a 1-1-correspondence between the sets $\{\theta_{f_2}\}$ and $\{\theta_{f'_{p'}}\}$, or more simply, there exist a point $p \notin f(M)$ and a point $p' \notin f'(M)$ such that $\theta_{f_p} = \theta_{f'_{p'}}$.

Theorem 11(Kuratowski) For a Peano continuum M containing only a finite number of simple closed curves to be realizable in the plane, it is necessary and sufficient that $\Phi^2 = 0$.

References

[1] van Karapen. *Abh. Math. Sem.* Hamburg., 1932, 9: 72-78, 152-153.

[2] Wu Wen-tsün. *Acta Math. Sinica.*, 1955, 5: 505-552.

[3] Whitney H. *Michigan Lectures in Topology*, 1941: 101-141.

[4] Thom R. *Ann. Ec. Norm. Sup.*, 1952, 69: 110-182.

[5] Wu Wen-tsün. Sur les puissances de Steenrod. *Colloque de Topologie de Strasbourg*, 1951.

[6] Wu Wen-tsün. Relations between Smith operations and Steenrod powers. *Acta Math. Sinica.*, 1957, 7.

[7] Flores A I. *Erg. math. Kolloqu.*, 1935, 6: 4-7.

[8] Smith P A. Fixed points of periodic transformations. *Algebraic Topology* by S. Lefschetz, 1941: 350-393.

[9] Whitney H. *Annals of Math.*, 1944, 45: 247-293.

[10] Wu Wen-tsün. *Acta Math. Sinica.*, 1953, 3: 261-290.

[11] McLane S and Adkisson V W. Extensions of homomorphisms of the sphere. *Michigan Lectures in Topology*, 1941: 223-236.

[12] Cartan H. *Algebraic Topology*. Harvard University, 1949.

[13] Wu Wen-tsün. *Acta Math. Sinica.*, 1955, 5: 37-63.

[14] Thom R. Une thérie intrinsêque de puissance de Steenrod. *Colloque de Topologie de Strasbourg*, 1951.

[15] Bott R. *Annals of Math.*, 1953, 59: 579-590.

[16] Wu Wen-tsün. *Acta Scientia Sinica.*, 1954, 3: 353-367.

[17] Whitney H. *Annals of Math.*, 1944, 45: 220-246.

[18] Wu Wen-tsün. Sur les classes caractéristiques des espaces fibrés. *Act-Sci. et Ind.* (1183) Paris: Hermann, 1952.

[19] Chern S. *Annals of Math.*, 1948, 49: 362-372.
[20] Adkisson V W & McLane S. *Duke Math. J.*, 1940, 6: 216-228.
[21] Kuratowski C. *Fund. Math.*, 1930, 15: 271-283.

30. On the Isotopy of Complexes in a Euclidean Space I*

The author has shown that the notion of reduced products of a space may be quite useful for topological problems of a non-homotopic character, and has applied this notion to the study of imbedding, immersion and isotopy of a space in a Euclidean space and also some other related questions (cf. e.g.[1], and the references given there). Recently the author has shown further [2] that given any two linear imbeddings of a finite complex in a Euclidean space, invariants may be deduced from the reduced product of the complex whose vanishing gives the necessary conditions for the two imbeddings to be linearly isotopic, which turn out to be also sufficient in the critical case: dimension of the Euclidean space = 2× dimension of the complex+1, the dimension of the complex being supposed to be > 1. The details of this study are given in two papers of which the present part I introduces the notion of linear isotopy and the above-mentioned invariants of pairs of linear imbeddings, and part II will give the sufficiency theorem in the critical case.

§1 Preliminaries about spaces with periodic transformations

Let $\tilde{K}$ be an abstract complex consisting of cells σ_α^j with dimension i, face relations $<$ and incidence numbers $[\sigma_\alpha^i, \sigma_\beta^{i-1}]$, verifying usual conditions, similarly for $\tilde{K}'$. We shall say that T is a *celloperator* of $\tilde{K}$ in $\tilde{K}'$ if T is a correspondence of cells of $\tilde{K}$ in $\tilde{K}'$ which preserves dimensions, face relations and also incidence numbers up to a sign. The last statement means that numbers $\varepsilon_\alpha^i = \pm 1 (\varepsilon_\alpha^0 = +1)$ exist such that

$$[T\{\sigma_\alpha^i\},\ T\{\sigma_\beta^{i-1}\}] = \varepsilon_\alpha^i \cdot \varepsilon_\beta^{i-1} \cdot [\sigma_\alpha^i, \sigma_\beta^{i-1}].$$

---

* First published in Chinese in *Acta Math. Sinica*, 1959, 9(4): 475-493; *Sci. Sinica*, 1960, 9: 21-46.

The cell-operator T will induce homomorphisms in the chain groups and cochain groups of $\tilde{K}$ which will again be denoted by T. For any set of cell-operators $T_1, \cdots, T_r$ of $\tilde{K}$ in $\tilde{K}'$ and any integers $a_1, \cdots, a_r$ the sum $\sum a_i T_i$ will be called an *operator* of $\tilde{K}$ in $\tilde{K}'$ which will induce in a natural manner homomorphisms in the chain and cochain groups of $\tilde{K}$ in $\tilde{K}'$ again denoted by the same symbol.

Let $\tilde{K}$ be a complex with a cell-operator T of $\tilde{K}$ in itself without fixed cells and having a period p, p being a fixed prime. We shall say that $(\tilde{K}, T)$ is a *simple system of period p*. As usual, we shall denote by ρ one of the operators

$$d = 1 - T,$$

and

$$s = 1 + T + \cdots + T^{p-1},$$

and then the other by $\bar{\rho}$. Let $C^r(\tilde{K})$ and $Z^r(\tilde{K})$ be the integral cochain groups and groups of integral cocycles in the dimension $r \geqslant 0$ respectively. The elements in ${}^\rho C^r(\tilde{K}, T) = \mathrm{Ker}\rho \cap C^r(\tilde{K}) = \bar{\rho} C^r(\tilde{K})$ and ${}^\rho Z^r(\tilde{K}, T) = \mathrm{Ker}\rho \cap Z^r(\tilde{K})$ are then called ρ-cochains and *ρ-cocycles* in dimension r respectively. As $\delta \bar{\rho} C^{r-1}(\bar{K}) = \delta^\rho C^{r-1}(\tilde{K}) \subset {}^\rho Z^r(\tilde{K}, T)$ we may define *special cohomology groups* ${}^\rho H^r(\tilde{K}, T)$ as

$$ {}^\rho H^r(\tilde{K}, T) = \begin{cases} {}^\rho Z^r(\tilde{K}, T)/\delta \bar{\sigma} C^{r-1}(\tilde{K}), & r > 0, \\ {}^\rho Z^r(\tilde{K}, T), & r = 0, \end{cases} $$

of which the elements are called the *ρ-cohomologyclasses*. We shall sometimes use the notations $z \tilde{\rho} 0$(or $z_1 \tilde{\rho} z_2$) if $z \in \delta^\rho C^r(\tilde{K}, T)$ (or $z_1 - z_2 \in \delta^\rho C^r(\tilde{K}, T)$). We say then that z is *ρ-cohomologous to* 0 (or z_1, z_2 are *ρ-cohomologous*).

For the special cohomology groups we have the following exact sequence of Smith-Richardson, as reformulated by Thom:

$$ {}^{\bar{\rho}} H^0(\tilde{K}, T) \to H^0(\tilde{K}) \to {}^\rho H^0(\tilde{K}, T) \to \cdots $$
$$ \to H^r(\tilde{K}) \xrightarrow{\alpha_\rho} {}^\rho H^r(\tilde{K}, T) \xrightarrow{\beta_\rho} {}^{\bar{\rho}} H^{r+1}(\tilde{K}, T) \xrightarrow{\gamma_\rho} H^{r+1}(\tilde{K}) \to \cdots \quad (1)$$

The homomorphisms occurring in the sequence are defined as follows: α_ρ is induced by the correspondence $u \to \bar{\rho} u$ for $u \in Z^r(\tilde{K})$, β_ρ is induced by $\bar{\rho} v \to \delta v$ mod $\delta^{\bar{\rho}} C^r(\tilde{K}, T)$ for $\bar{\rho} v \in {}^\rho Z^r(\tilde{K}, T)$, and γ_ρ is induced by $w \to w$ for $w \in {}^{\bar{\rho}} Z^{r+1}(\tilde{K}, T)$.

Let $(\tilde{K}, T)$ and $(\tilde{K}', T')$ be two simple systems of the same prime period p and $\tilde{f} : \tilde{K} \to \tilde{K}'$ a cell operator which commutes with T and T'. Then $\tilde{f}$ will be called a *system map* and will be denoted as $\tilde{f} :(\tilde{K}, T) \to (\tilde{K}', T')$. Any such map will induce in a natural manner homomorphisms

$$^\rho\tilde{f}^* : {}^\rho H^r(\tilde{K}', T') \to {}^\rho H^r(\tilde{K}, T).$$

It is easy to see that $^\rho\tilde{f}^*$ commutes with the homomorphisms α_ρ, β_ρ and γ_ρ as defined above.

Let $\tilde{X}$ be a space having a homoeomorphism T without fixed points and with a prime period p. We shall say that $(\tilde{X}, T)$ is a *simple system of period* p. The homoeomorphism T will induce a cell-operator in the singular complex $S(\tilde{X})$ without fixed cells and having the same period p which will be denoted by the same letter T. The groups $^\rho H^r(S(\tilde{X}), T)$, etc., will then be denoted simply by $^\rho H^r(\tilde{X}, T)$, etc. In particular, if $\tilde{X}$ is itself the space of a complex $\tilde{K}$, i.e., $\tilde{X} = |\tilde{K}|$, and if the operator T without fixed cells and having prime period p in $\tilde{K}$ is induced by a homoeomorphism T without fixed points and having period p in the space $\tilde{X}$, then there exist canonical isomorphisms

$$\tilde{\lambda}: \quad {}^\rho H^r(\tilde{K},\ T) \approx {}^\rho H^r(\tilde{X},\ T)$$

so that they will often be identified. In that case we shall say that the system $(\tilde{K}, T)$ is *associated* with the system $(\tilde{X}, T)$.

Let $(\tilde{X}, T)$ and $(\tilde{X}', T')$ be two simple systems of the same prime period p and $\tilde{f}$ a continuous map of $\tilde{X}$ into $\tilde{X}'$ which commutes with T and T'. In that case we shall say that $\tilde{f}$ is a *system map* of the corresponding systems and we shall use the notation $\tilde{f} : (\tilde{X}, T) \to (\tilde{X}', T')$. For any such system map $\tilde{f}$ it will induce homomorphisms

$$^\rho\tilde{f}^* : {}^\rho H^r(\tilde{X}', T') \to {}^\rho H^r(\tilde{X}, T).$$

In particular, if $\tilde{X} = |\tilde{K}|$, $\tilde{X}' = |\tilde{K}'|$, $(\tilde{K}, T)$ and $(\tilde{K}', T')$ associated with $(\tilde{X}, T)$ and $(\tilde{X}', T')$ respectively, and if $\tilde{f} : (\tilde{X}, T) \to (\tilde{X}', T')$ induces also the cell operator $\tilde{g} : (\tilde{K}, T) \to (\tilde{K}', T')$, then we shall say that the system map $\tilde{g}$ is *associated* with the system map $\tilde{f}$. In that asec we may see that $^\rho\tilde{f}^*$ and $^\rho\tilde{g}^*$ commute with the canonical homomorphisms $\tilde{\lambda}$ defined above.

Two system maps $\tilde{f}_i : (\tilde{X}, T) \to (\tilde{X}', T')$, $i = 0, 1$, will be said to be *system-homotopic* if the following holds: Denote by $(\tilde{X} \times I, T)$ the simple system such that $T(x, t) = (T(x), t)$, $x \in \tilde{X}$, $t \in I$, where I is the unit interval [0, 1]. Then there exists a system map $\tilde{F} : (\tilde{X} \times I, T) \to (\tilde{X}', T')$ such that $\tilde{F}(x, 0) = f_0(x)$, $\tilde{F}(x, 1) = f_1(x)$ for all $x \in \tilde{X}$. The map $\tilde{F}$ will then be called a *system-homotopy* between the system maps $\tilde{f}_0$ and $\tilde{f}_1$.

A simple system $(\tilde{X}, T)$ will be said to be a *subsystem* of a simple system $(\tilde{X}', T')$ if $\tilde{X}$ is a subspace of $\tilde{X}'$, and $T = T'/\tilde{X}$. The subsystem $(\tilde{X}, T)$ of $(\tilde{X}', T')$ will be said to be a *system-deformation-retract* of $(\tilde{X}', T')$ if there exists a system-map $\tilde{F}$: $(\tilde{X}' \times I, T') \rightarrow (\tilde{X}', T')((\tilde{X}' \times I, T')$ defined as before) such that, defining $\tilde{f}_t : \tilde{X}' \rightarrow \tilde{X}'$ by $\tilde{f}_t(x') = \tilde{F}(x', t)$, $t \in I$, $x' \in \tilde{X}'$, then $\tilde{f}_0$ is the identical map while $F_1(\tilde{X}') \subset \tilde{X}$, $\tilde{f}_i(\tilde{X}) \subset \tilde{X}$, $t \in I$. The map $\tilde{F}$ will then be called a *system-deformation-retraction* of $(\tilde{X}', T')$ in $(\tilde{X}, T)$.

The following two propositions are simple consequences of the definitions:

Proposition 1 If the system-maps $\tilde{f}_0$, $\tilde{f}_1 : (\tilde{X}, T) \rightarrow (\tilde{X}', T')$ of the two simple systems are system-homotopic, then ${}^\rho\tilde{f}_0^* = {}^\rho\tilde{f}_1^*$.

Proposition 2 If the simple system $(\tilde{X}, T)$ is a system-deformation-retract of the simple system $(\tilde{X}', T')$, then we have

$$^\rho\tilde{i}^* : \quad {}^\rho H^r(\tilde{X}', T') \approx {}^\rho H^r(\tilde{X}, T),$$

where $\tilde{i} : (\tilde{X}, T) \rightarrow (\tilde{X}', T')$ is induced by the inclusion map $\tilde{i} : \tilde{X} \subset \tilde{X}'$.

Proposition 3 For the $(N-1)$-sphere S^{N-1} where $N > 1$ and a homoeomorphism T with prime period p and without fixed points of S^{N-1}①, we have

$$\left.\begin{array}{l}
{}^sH^1(S^{N-1},T)\approx{}^dH^2(S^{N-1},T)\approx{}^sH^3(S^{N-1},T)\approx\cdots\approx{}^{\rho_N}H^{N-2}(S^{N-1},T)\approx Z_p,\\
{}^dH^0(S^{N-1},T)\approx Z,\\
{}^sH^0(S^{N-1},T) = {}^dH^1(S^{N-1},T) = {}^sH^2(S^{N-1},T) = \cdots = {}^{\bar{\rho}_N}H^{N-2}(S^{N-1},T) = 0,\\
{}^{\rho_N}H^{N-1}(S^{N-1},T) \approx Z,\\
{}^{\bar{\rho}_N}H^{N-1}(S^{N-1},T) \approx Z_p,\\
{}^\rho H^r(S^{N-1},T) = 0,\ r \geqslant N,
\end{array}\right\} \tag{2}$$

in which z is the group of integers, z_k the group of integers mod k, and

$$\rho_N = \begin{cases} d, & N \quad \text{even}, \\ s, & N \quad \text{odd}. \end{cases}$$

$$\bar{\rho}_N = \begin{cases} s, & N \quad \text{even}, \\ d, & N \quad \text{odd}. \end{cases}$$

In particular, we have for $p = 2$,

$$\rho_N = 1 + (-1)^{N-1}T, \quad \bar{\rho}_N = 1 + (-1)^N T.$$

Proof. Let z^0 be the unit cocycle in S^{N-1} which takes value 1 on all of its points. Then z^0 is a d-cocycle and it generates ${}^dH^0(S^{N-1}, T) = {}^dZ^0(S^{N-1}, T) \approx Z$,

① For N odd it is known that p is necessarily $= 2$ which does not concern us.

while ${}^S H^0(S^{N-1},T)=0$. The homomorphism $\alpha_d : H^0(S^{N-1}) \to {}^d H^0(S^{N-1},T)$ sends the ordinary cohomology class of z^0 to $p\{z^0\}_d$, where $\{z^0\}_d$ is the d-cohomology class containing the d-cocycle z^0.

Suppose for the moment $N>2$. Then from the exactness of the sequence (all ${}^\rho H^r$, H^r, are abbreviations for ${}^\rho H^r(S^{N-1},T)$, and $H^r(S^{N-1})$)

$$0={}^s H^0 \to H^0 \overset{\alpha_d}{\to} {}^d H^0 \to {}^s H^1 \to H^1 = 0$$

and the known homomorphism α_d it follows that ${}^S H^1 \approx Z_p$.

From the exactness of the sequence

$$0={}^s H^0 \to {}^s H^1 \to H^1=0,$$

it follows that ${}^d H^1=0$. Finally, from the exactness of the sequence

$$H^r \to {}^\rho H^r \to {}^{\bar\rho} H^{r+1} \to H^{r+1},$$

it follows that ${}^\rho H^r \approx {}^{\bar\rho} H^{r+1}$, $0<r<N-2$. Combining together, we get $(2)_1$—$(2)_3$, where $N \geqslant 2$ (the case $N=2$ is trivial).

From the exactness of $H^{N-1}(\approx Z) \to {}^\rho H^{N-1} \to {}^{\bar\rho} H^N(=0)$ we see that ${}^\rho H^{N-1}$ can only be 0, or Z, or Z_k for some $k>1$. Suppose again $N>2$, and consider the exact sequence

$$\begin{aligned} H^{N-2}(=0) &\to {}^{\rho_N} H^{N-2}(\approx Z_p) \overset{\beta}{\to} {}^{\bar\rho_N} H^{N-1} \overset{\gamma}{\to} H^{N-1}(\approx Z) \overset{\alpha}{\to} {}^{\rho_N} H^{N-1} \\ &\to {}^{\bar\rho_N} H^N(=0). \end{aligned}$$

As β is isomorphism into we see that ${}^{\bar\rho_N} H^{N-1}$ can only be a Z_k for some $k>1$. We have then necessarily $\gamma=0$. It follows then that $\beta : {}^{\rho_N} H^{N-2} \approx {}^{\bar\rho_N} H^{N-1} \approx Z_p$ and $\alpha : H^{N-1} \approx {}^{\rho_N} H^{N-1} \approx Z$. This proves $(2)_4$ and $(2)_5$ for $N>2$. Suppose now $N=2$ and consider the exact sequence

$$H^0(\approx z) \overset{\alpha}{\to} {}^d H^0(\approx z) \overset{\beta}{\to} {}^s H^1 \overset{\gamma}{\to} H^1(\approx z) \overset{\alpha}{\to} {}^d H^1 \to {}^s H^2(=0).$$

As ${}^s H^1$ contains the subgroup ${}^d H^0/\alpha H^0 \approx Z_p$, it can only be a Z_k for some $k>1$. As before we have then $\gamma=0$ and $\alpha : H^1 \approx {}^d H^1 \approx Z$, ${}^s H^1 \approx {}^d H^0/\alpha H^0 \approx Z_p$. This again proves $(2)_4$ and $(2)_5$ for $N=2$.

As $(2)_6$ is evident, the proposition is completely proved.

Remark. From the proof we see that

$$\alpha=\alpha_{\rho_N} : H^{N-1}(S^{N-1}) \approx {}^{\rho_N} H^{N-1}(S^{N-1},T) \approx Z.$$

Therefore, if S^{N-1} is oriented with $\sum^{N-1}$ as the generator of $H^{N-1}(S^{N-1})$ corresponding to that orientation, $\alpha(\sum^{N-1})$ is a definite generator of ${}^{\rho N}H^{N-1}(S^{N-1},T)$ which will be denoted by ${}^{\rho_N}\Theta^{N-1}(\overset{+}{S}{}^{N-1},T)$, $\overset{+}{S}{}^{N-1}$ being the sphere S^{N-1} definitely oriented. If S^{N-1} changes its orientation, then ${}^{\rho_N}\Theta^{N-1}(\overset{+}{S}{}^{N-1},T)$ changes sign.

§2 Reduced product of a space

For any space X and any prime p let $\tilde{X}_p^*$ be the *p-fold reduced product* of X which is the p-fold topological product $\underbrace{X\times\cdots\times X}_{p}$ with the diagonal Δ_X removed. Let T_X be the cyclic transformation in $\tilde{X}_p^*$ defined by

$$T_X(x_1,\cdots,x_p)=(x_2,\cdots,x_p,x_1),\quad (x_1,\cdots,x_p)\in\tilde{X}_p^*.$$

Then $(\tilde{X}_p^*,T_X)$ is a simple system and we shall set

$${}^{\rho}H^r(\tilde{X}_p^*,T_X)={}^{\rho}\tilde{H}_{(p)}^r(X).$$

Let R^N be a Euclidean space of dimension N. In $\underbrace{R^N\times\cdots\times R^N}_{p}$ let us consider the subspace $\tilde{L}_{p,N}$ of all p-tuples $(x_1,\cdots,x_p)$ of vectors $x_i\in R^N$ such that $\sum\limits_{i=1}^{p}x_i=0$, $\sum\limits_{i=1}^{p}|x_i|^2=1$, in which $|x|$ denotes the length of a vector $x\in R^N$. Then it is evident that $\tilde{L}_{p,N}\subset(\tilde{R}^N)_p^*$ and $\tilde{L}_{p,N}$ is a sphere of dimension $(p-1)N-1$ and $T=T_{R^N}$ is also a homoeomorphism of $\tilde{L}_{p,N}$ onto itself so that $(\tilde{L}_{p,N},T)$ is a subsystem of $((\tilde{R}^N)_p^*,T)$. The following theorem is fundamental in the imbedding theory developed by the author of which a simple proof is given below.

Theorem 1 $(\tilde{L}_{p,N},T)$ is a system-deformation-retract of the system $((\tilde{R}^N)_p^*,T)$.

Proof. Let 0 be the origin of the product space $\underbrace{R^N\times\cdots\times R^N}_{p}$ considered as a pN-dimensional Euclidean space R^{pN} and Δ^N the diagonal of this product space which is defined by a system of linear equations,

$$x_1=x_2=\cdots=x_p,\quad (x_1,\cdots,x_p)\in R^{pN}.$$

For any point $x\in\Delta^N$, let P_x be the $(p-1)N$-dimensional linear subspace of R^{pN} through the point x and completely orthogonal to the linear subspace Δ^N. Let $S_x^{(p-1)N-1}$ be the unit sphere in P_x with x as centre and U the union of all such spheres

$S_x^{(p-1)N-1}$, $x \in \Delta^N$. It is clear that $T(S_x^{(p-1)N-1}) \subset S_x^{(p-1)N-1}$ for each $x \in \Delta^N$ and so both $(S_x^{(p-1)N-1}, T)$ and (U,T) are subsystems of the system $((\tilde{R}^N)_p^*, T)$. It may be seen that the system $(S_0^{(p-1)N-1}, T)$ coincides with $(\tilde{L}_{p,N}, T)$. For, given any point $x = (x_1, \cdots, x_p)$ in $S_0^{(p-1)N-1}$, where $x_i \in R^N$, we have $\sum_{i=1}^{p} |x_i|^2 = 1$ and the scalar product $x \cdot y = 0$ for any point $y = (e, \cdots, e) \in \Delta^N$, where $e \in R^N$. This gives $x_1 + \cdots + x_p = 0$ and hence $x \in \tilde{L}_{p,N}$. The converse is also true, so that $\tilde{L}_{p,N} = S_0^{(p-1)N-1}$ and the systems $(\tilde{L}_{p,N}, T)$, $(S_e^{(p-1)N-1}, T)$ coincide.

For any point $x \in (\tilde{R}^N)_p^*$ let $\pi(x)$ be the orthogonal projection of x on the linear subspace Δ^N. Let $g(x)$ be the point of intersection with $S_{\pi(x)}^{(p-1)N-1}$ of the ray from $\pi(x)$ to x. By deforming each point x linearly on the segment joining x to $g(x)$, we see that the system (U,T) is a system-deformation-retract of the system $((\tilde{R}^N)_p^*, T)$. Now all spheres $S_x^{(p-1)N-1}$ are homoeomorphic to $S_0^{(p-1)N-1}$ under the correspondence $y \to y - x$ where $y \in S_x^{(p-1)N-1}$ so that $x = \pi(y)$.

For all $x \in \Delta^N$ let us deform now any point $y \in S_x^{(p-1)N-1}$ linearly on the segment joining y to $y - x \in S_0^{(p-1)N-1}$. It is evident that this gives a system-retraction of (U,T) into $(S_0^{(p-1)N-1}, T)$ so that the latter is a system-deformation-retract of the former. Combining these two deformations together, we see that $(S_0^{(p-1)N-1}, T)$, i.e., $(\tilde{L}_{p,N}, T)$, is a system-deformation-retract of the system $((\tilde{R}^N)_p^*, T)$. This proves our theorem.

The deformation $F : (\tilde{R}^N)_p^* \times I \to (\tilde{R}^N)_p^*$ which gives the above system-deformation-retraction of $((\tilde{R}^N)_p^*, T)$ into $(\tilde{L}_{p,N}, T)$ may be given explicitly as follows. For any point $x = (x_1, \cdots, x_p) \in (\tilde{R}^N)_p^*$ let $O_x = \frac{1}{p}(x_1 + \cdots + x_p)$. Then $\pi(x) = (O_x, \cdots, O_x)$ and $g(x) = \pi(x) + \frac{1}{d_x} \cdot (x - \pi(x))$, where $d_x = \sqrt{\sum_{i=1}^{p} |x_i - O_x|^2}$ is the distance between x and $\pi(x)$ in R^{pN}. The required system-deformation-retraction F is then given by

$$F(x,t) = \begin{cases} \pi(x) + \left[2t\left(\dfrac{1}{d_x} - 1\right) + 1\right] \cdot (x - \pi(x)), & 0 \leqslant t \leqslant \dfrac{1}{2}, \\ \pi(x) + \dfrac{1}{d_x}(x - \pi(x)) - (2t-1) \cdot \pi(x), & \dfrac{1}{2} \leqslant t \leqslant 1, \end{cases}$$

in which $x \in (\tilde{R}^N)_p^*$.

As $\tilde{L}_{p,N}$ has a dimension $= (p-1)N - 1$, we get the following

Corollary 1 ${}^{\rho}\tilde{H}_{(p)}^{r}(R^N) = 0$, $r \geqslant (p-1)N$.

As already described in [3] and [4], many known theorems by Van Kampen,

Whitney and Thom about imbeddings are consequences of the above corollary in the case $p = 2$ of which an alternative proof has also been given in the above papers.

When $p = 2$, the map $(x_1, x_2) \to \sqrt{2}x_1$ gives a homoeomorphism of $\tilde{L}_{2,N}$ onto the unit sphere S^{N-1} in R^N and turns the cyclic transformation $T:(x_1, x_2) \to (x_2, x_1)$ in $\tilde{L}_{2,N}$ into the antipodal map T_a of S^{N-1}. Hence we get easily

Corollary 2 ${}^{\rho}\tilde{H}^r_{(2)}(R^N) \approx {}^{\rho}H^r(S^{N-1}, T_a)$, the isomorphism is induced by the map $j : (\tilde{R}^N)^*_2 \to S^{N-1}$ where $j(x_1, x_2)$ is the intersecting point with S^{N-1} of the ray through the origin of R^N and parallel to the ray from x_1 to x_2, S^{N-1} being the unit sphere in R^N.

Let $\overset{+}{R}{}^N$ be the space R^N with a certain definite orientation, then the $((p-1)N-1)$-sphere $\tilde{L}_{p,N}$ has also a definite orientation which determines in turn, by §1, a definite generator ${}^{\rho_{(p-1)N}}\Theta^{(p-1)N-1}(\tilde{L}^+_{p,N}, T)$ of ${}^{\rho_{(p-1)N}}H^{(p-1)N-1}(\tilde{L}^+_{p,N}, T)$, $\tilde{L}^+_{p,N}$ being the space $\tilde{L}_{p,N}$ with the orientation determined as follows: the diagonal Δ^N oriented as $\overset{+}{R}{}^N$ followed by the oriented $\tilde{L}^+_{p,N}$, should give the same orientation of the product space $\underbrace{R^N \times \cdots \times R^N}_{p}$ when each factor R^N is oriented as $\overset{+}{R}{}^N$. Denote by $\tilde{i} : (\tilde{L}_{p,N}, T)^p \to ((\tilde{R}^N)^*_p, T)$ the inclusion map, then the class $(\tilde{i}^*)^{-1}{}^{\rho_{(p-1)N}}\Theta^{(p-1)N-1}\cdot(\hat{L}^+_{2,N}, T) \in {}^{\rho_{(p-1)N}}\tilde{H}^{(p-1)N-1}_{(p)}(R^N)$ will be denoted henceforth by

$$ {}^{\rho_{(p-1)N}}\Theta^{(p-1)N-1}_{(p)}(\overset{+}{R}{}^N) \varepsilon {}^{\rho_{(p-1)N}}\tilde{H}^{(p-1)N-1}_{(p)}(R^N). $$

Remark that if the orientation of R^N is changed, ${}^{\rho_{(p-1)N}}\Theta^{(p-1)N-1}_{(p)}(\overset{+}{R}{}^N)$ changes sign for $p = 2$, but remains the same for $p > 2$.

Let us orient the unit sphere S^{N-1} in R^N coherently with that of R^N, then it is evident that

$$ {}^{\rho_N}j^{*\rho_N}\Theta^{N-1}_{(2)}(\overset{+}{S}{}^{N-1}, T_a) = {}^{\rho_N}\Theta^{N-1}_{(2)}(\overset{+}{R}{}^N), $$

in which $j : ((\tilde{R}^N)^*_2, T) \to (S^{N-1}, T_a)$ is the map as in Corollary 2, while $\overset{+}{S}{}^{N-1}$ is the sphere S^{N-1} with the orientation coherent with that of $\overset{+}{R}{}^N$.

§3 Imbedding of a space in the other

A space X will be said to be *imbeddable* in a space Y if there exists a continuous map f of X in Y such that the spaces X and $f(X)$ are homoeomorphic under the map f. Such a map f will then be called an *imbedding* of X in Y. Two *imbeddings* f, g of X in Y are said to be *isotopic* if there exists a continuous mapping $F: X \times I \to Y$ such that, defining $F_t: X \to Y$ by $F_t(x) = F(x,t)$, $t \in I = [0,1]$, $x \in X$, each F_t will be an *imbedding* of X in Y and furthermore, $F_0 \equiv f$, $F_1 = g$. The map F will then be called an *isotopy* between f, g (or from f to g).

If $f: X \to Y$ is an imbedding, then f will induce for each prime p a system map $\hat{f}_p: (\tilde{X}_p^*, T_X) \to (\tilde{Y}_p^*, T_Y)$ defined by $\tilde{f}_p(x_1, \cdots, x_p) = (f(x_1), \cdots, f(x_p))$, $(x_1, \cdots, x_p) \in \tilde{X}_p^*$. We have therefore systems of homomorphisms:

$$^\rho\tilde{f}_p^*: \quad {}^\rho\tilde{H}_{(p)}^r(Y) \to {}^\rho\tilde{H}_{(p)}^r(X).$$

Theorem 2 The homomorphisms $^\rho\tilde{f}_p^*$ are isotopy invariants of imbeddings of X in Y, i.e., $^\rho\tilde{f}_p^*$ is identical with $^\rho\tilde{g}_p^*$ for any two isotopic imbeddings f and g of X in Y.

Proof. Let $F: X \times I \to Y$ be an isotopy between the imbeddings f and g. Then $\tilde{F}_p((x_1, \cdots, x_p), t) = (F(x_1, t), \cdots, F(x_p, t))$, where $(x_1, \cdots, x_p) \in \tilde{X}_p^*, t \in I$, will define a map $\tilde{F}_p: \tilde{X}_p^* \times I \to \tilde{Y}_p^*$ which is a system-homotopy between the system-maps $\tilde{f}_p$ and $\tilde{g}_p$. The theorem follows therefore from Proposition 1 of §1.

Take in particular $Y = R^N$, then as consequences of Theorem 1 in §2, we get the following corollaries of the above theorem 2.

Theorem 3 For a space X to be imbeddable in a Euclidean space R^N of dimension N, it is necessary that

$$^\rho\tilde{H}_{(p)}^r(X) = 0, \quad \text{for} \quad r \geqslant (p-1)N.$$

Theorem 4 The classes

$$^{\rho_{(p-1)N}}\Theta_{(p)}^{(p-1)N-1}(X, f) = {}^{\rho_{(p-1)N}}\tilde{f}_p^{*\rho_{(p-1)N}}\Theta_{(p)}^{(p-1)N-1}(\overset{+}{R}{}^N)$$

are isotopy invariants of the imbedding f of the space X in R^N, i.e.,

$$^{\rho_{(p-1)N}}\Theta_{(p)}^{(p-1)N-1}(X, f) = {}^{\rho_{(p-1)N}}\Theta_{(p)}^{(p-1)N-1}(X, g)$$

for any two isotopic imbeddings f and g of X in R^N, in which $\overset{+}{R}{}^N$ denotes the Euclidean space R^N with a definite orientation.

Definition If f is an imbedding of a space X in R^N, then the classes

$$^{\rho_{(p-1)N}}\Theta_{(p)}^{(p-1)N-1}(X, f) = {}^{\rho_{(p-1)N}}\tilde{f}_p^{*\rho_{(p-1)N}}\Theta_{(p)}^{(p-1)N-1}(\overset{+}{R}{}^N)$$

will be called *isotopy classes* of the imbedding f of X in R^N with respect to the given orientation of R^N.

Remark. In the case $p = 2$, we shall also use the notation

$$^{\rho_N}\Theta_{(2)}^{N-1}(X, f) = {}^{\rho_N}\Theta_f^{N-1}(X).$$

§4 Linear imbeddings of complexes and their linear isotopies

Let K be a finite simplicial complex which will be considered as a Euclidean complex in a certain Euclidean space of sufficiently high dimension. The space of K will be denoted by $|K|$. A p-tuple of simplexes $(\sigma_1, \cdots, \sigma_p)$ of K will be said to be *non-diagonic* if no vertex of K is common to all of these p simplexes, though fewer than p of these simplexes may have vertices in common. To each p we may then associate with K the *p-fold reduced product* $\tilde{K}_p^*$ of K consisting of all cells $\sigma_1 \times \cdots \times \sigma_p$ of the p-fold product complex $\underbrace{K \times \cdots \times K}_{p}$ for which the p-tuple $(\sigma_1, \cdots, \sigma_p)$ is non-diagonic. The cyclic transformation $T_{|K|}$ in $|\tilde{K}|_p^*$ will induce a cell map T_K in $\tilde{K}_p^*$ given by

$$T_K(\sigma_1 \times \cdots \times \sigma_p) = (-1)^{\dim \sigma_1\ (\dim \sigma_2 + \cdots + \dim \sigma_p)} \cdot \sigma_2 \times \cdots \times \sigma_p \times \sigma_1.$$

For a prime p we shall set $^{\rho}H^r(\tilde{K}_p^*, T_K) = {}^{\rho}\tilde{H}_{(p)}^r(K)$.

The following theorem has been given in [3] :

Theorem 5 Let p be a prime. Then the simple system $(|\tilde{K}_p^*|, T_{|K|})$ is a system-deformation-retract of the simple system $(|\tilde{K}|_p^*, T_{|K|})$ so that there exist canonical isomorphisms

$$^{\rho}\gamma_p^* : \quad {}^{\rho}\tilde{H}_{(p)}^r(|K|) \approx {}^{\rho}\tilde{H}_{(p)}^r(K).$$

For an imbedding f of $|K|$ in an oriented R^N we shall set

$$^{\rho}\gamma_p^{*\rho(p-1)N}\Theta_{(p)}^{(p-1)N-1}(|K|, f) = {}^{\rho(p-1)N}\ \Theta_{(p)}^{(p-1)N-1}(K, f),$$

and in the case of $p = 2$, also $^{\rho_N}\Theta_{(2)}^{N-1}(K, f) = {}^{\rho_N}\Theta_f^{N-1}(K)$.

Definition The classes $^{\rho(p-1)N}\Theta_{(p)}^{(p-1)N-1}(K, f) \in {}^{\rho_{(p-1)N}}\tilde{H}_{(p)}^{(p-1)N-1}(K)$ will be called *isotopy classes* of the imbedding f of the complex K in the oriented R^N.

A continuous map f of $|K|$ in R^N will be called *a linear map* of the complex K in R^N if $f/|\sigma|$ is linear for each simplex σ of K. An imbedding f of $|K|$ in R^N will be called a *linear imbedding* of the complex K in R^N if f is also a linear map of K in R^N. The set $C_N(|K|)$ of all continuous maps of $|K|$ in R^N will become a metric space under the usual metric

$$d(f,g) = \max_{x\in|K|} d(f(x),g(x)), \quad f,g\in C_N(|K|),$$

where $d(y,z)$ means the distance between the points $y,z\in R^N$. Denote the subspace of $C_N(|K|)$ consisting of all linear maps (resp. linear imbeddings) of K in R^N by $L_N(K)$ (resp. $I_N(K)$). It is easy to see that for $f,\ g\in L_N(K)$, $d(f,g)$ is equal to the maximum of the distances $d(f(a),\ g(a))$ for all vertices a of K.

Let $f\in I_N(K)$ be given. For any non-diagonic pair of simplexes σ, τ of K let $\delta_{\sigma,\tau}>0$ be the distance between the sets $f(|\sigma|)$ and $f(|\tau|)$, i.e., the minimum of all distances $d(x,y)$, where $x\in f(|\sigma|)$, $y\in f(|\tau|)$. The minimum of all these numbers $\delta_{\sigma,\tau}$ is then >0 and will be denoted henceforth by δ_f.

Theorem 6 The set $I_N(K)$ is open in the space $L_N(K)$.

Proof. Nothing needs to be proved if $I_N(K)$ is void. Suppose on the contrary that $f\in I_N(K)$ is a linear imbedding of K in R^N. Consider now any linear map $g\in L_N(K)$ of K in R^N for which $d(f,g)<\frac{1}{2}\delta_f$. We shall prove that $g\in I_N(K)$.

To see this let us remark first that for any non-diagonic pair of simplexes σ, τ of K, the sets $g(|\sigma|)$ and $g(|\tau|)$ must be disjoint. For, for any points $x\in|\sigma|$, $y\in|\tau|$, we have $d(g(x),\ g(y))\geqslant d(f(x),\ f(y))-d(f(x),\ g(x))-d(f(y),\ g(y))\geqslant\delta_f-2d(f,g)>0$. It follows that g/σ is a non-degenerate linear map for each $\sigma\in K$. Hence, if $x\neq y\in |K|$ be such that $g(x)=g(y)$, then x, y should lie in the interior of simplexes $\sigma\neq\tau$ respectively which have necessarily vertices in common. Let 0 be a common vertex of σ, τ with σ', τ' as its opposite faces. The ray l from $g(0)$ to $g(x)=g(y)$ will meet both $g(|\sigma'|)$ and $g(|\tau'|)$. Suppose that l meets first $g(|\sigma'|)$ at the point z (or meets $g(|\sigma'|)$ and $g(|\tau'|)$ at the same point z), then $x_1=g^{-1}(z)\cap|\sigma|$ and $y_1=g^{-1}(z)\cap|\tau|$ are such that $x_1\neq y_1$, $x_1\in|\sigma'|$, $y_1\in|\tau|$ and $g(x_1)=g(y_1)$. The pair (x,y) has therefore been replaced by the pair (x_1,y_1) which lie in the interior of simplexes of smaller sum of dimensions as before. Proceeding in this manner we get finally a pair of points $x_r\neq y_r$ of $|K|$ lying respectively in the interior of a non-diagonic pair of simplexes of K. As already remarked, this is impossible, however. Hence $x\neq y\in|K|$ implies $g(x)\neq g(y)$ and $g\in I_N(K)$, q. e. d.

Two linear imbeddings f, g of K in R^N will be said to be *linearly isotopic* if there exists an isotopy F of f, g such that F is a linear mapping of a certain simplicial subdivision M' of the product complex $M = K \times I$ in R^N. It is evident that linear isotopy is an equivalence relation. It is also evident that F and the subdivision M' of $K \times I$ may be so chosen as to contain $K \times (0)$ and $K \times (1)$ as subcomplexes.

Theorem 7 If f is a linear imbedding of K in R^N, then any linear imbedding g sufficiently near to f $\left(\text{more precisely with } d(f,g) < \frac{1}{2}\delta_f\right)$ is linearly isotopic to f.

Proof. As in the proof of Theorem 6 we see that for any $g \in L_N(K)$ with $d(f,g) < \frac{1}{2}\delta_f$ we should have $g \in I_N(K)$. Consider now any such linear imbedding $g \in I_N(K)$ with $d(f,g) < \frac{1}{2}\delta_f$. We shall prove that g is linearly isotopic to f. For this let the vertices of K be $a_1, \cdots, a_r$. Define $f_i : |K| \to R^N$ as the linear map with $f_i(a_j) = g(a_j)$, for $j \leqslant i$, and $f_i(a_j) = f(a_j)$, for $j > i$. Then clearly $d(f, f_i) < \frac{1}{2}\delta_f$ so that all f_i are linear imbeddings of K in R^N. Define now $F_i : |K| \times I \to R^N$ as follows. Take a simplicial subdivision K_i' of $K \times I$ with $(a_j) \times (0)$, $(a_j) \times (1)$, for all j and $(a_k) \times \left(\frac{1}{2}\right)$ for all $k \neq i$ but only those as its vertices. Then F_i is the linear map of K' in R^N with

$$F_i\left((a_j) \times \left(\frac{1}{2}\right)\right) = F_i((a_j) \times (0)) = F_i((a_j) \times (1)) = f_i(a_j),$$

$$j \neq i \quad \text{and} \quad (f_0 = f),$$

$$F_i((a_i) \times (0)) = f_{i-1}(a_i), \quad F_i((a_i) \times (1)) = f_i(a_i).$$

It is easy to see that F_i is a linear isotopy between f_{i-1} and f_i and the set of $F_1, \cdots, F_r$ together gives a linear isotopy between f and g, q.e.d.

Theorem 8 Let F be a linear isotopy between two linear imbeddings f and g of K in R^N so that F is a linear map of a certain simplicial subdivision M' of $M = K \times I$ in R^N. Then any linear map F' of M' in R^N sufficiently near to F in the topology of $L_N(M')$ is also a linear isotopy between the linear imbeddings f' and g', where $f'(x) = F'(x, 0)$, $g'(x) = F'(x, 1)$, for $x \in |K|$.

Proof. Let π be the projection of $|M|$ on I. For any point x of M' we shall set $t_x = \pi(x)$. Define now a map $\overline{F}$ of $|M|$ in $R^{N+1} = R^N \times L$ where L is the line $-\infty < t < +\infty$ by $\bar{F}(x) = (F(x), t_x)$, $x \in |M|$. Then $\overline{F}$ is evidently a *linear* map of M' in R^{N+1}, since F is one of M' in R^N. Moreover $\overline{F}$ is a linear *imbedding* since F is an isotopy of f, g. Let $\delta = \delta_{\overline{F}} > 0$ be determined so that any linear map $\tilde{F}'$ of M' in R^{N+1} with $d(\bar{F}, \bar{F}') < \frac{1}{2}\delta$ will be a linear imbedding of M' in R^{N+1}. Consider now any linear map F' of M' in R^N with $d(F,F') < \frac{1}{2}\delta$. Define $\bar{F}' : |M'| \to R^{N+1}$ by $\bar{F}'(x) = (F'(x), t_x)$, $x \in |M'|$. Then

$$d(\bar{F},\bar{F}') = \max_{x\in|M'|} d(\bar{F}(x),\bar{F}'(x)) = \max_{x\in|M'|} d((F(x),t_x),(F'(x),t_x))$$
$$= \max_{x\in|M'|} d(F(x),F'(x)) = d(F,F') < \frac{1}{2}\delta.$$

Therefore $\bar{F}'$ is a linear imbedding of M' in R^{N+1}. From this it follows that F' is a linear isotopy between f', g' and the theorem is proved.

§5 Normal pairs of linear imbeddings

Let K be as before a finite simplicial complex in a Euclidean space of sufficiently high dimension. Two linear imbeddings (resp. linear mappings) f, g of K in R^N will be said to be *normal* with respect to a non-diagonic pair of simplexes σ, τ of K if (i) $f(|\sigma|)$ and $f(|\tau|)$ are disjoint, so are $g(|\sigma|)$ and $g(|\tau|)$, and (ii) for any points $x \in |\sigma|$, $y \in |\tau|$, the line joining $g(x)$ and $g(y)$ is not parallel to the line joining $f(x)$ and $f(y)$. The two linear imbeddings (resp. linear mappings) f, g of K in R^N are said to be *normal* with respect to K if they are normal with respect to any non-diagonic pair of simplexes σ, τ of K for which $\dim\sigma + \dim\tau \leqslant N-2$.

Lemma 1　If (f,g) is a normal pair of linear imbeddings (resp. linear mappings) of a finite simplicial complex K in R^N, then any pair of linear imbeddings (resp. linear mappings) f', g' of K in R^N sufficiently near to f and g in the topology of $L_N(K)$ is also normal.

Proof. This follows directly from the definition.

Lemma 2　Let K be the complex consisting of two disjoint simplexes $\sigma = (a_0\cdots a_p)$ and $\tau = (b_0\cdots b_q)$ as well as their faces. Let f, g be two linear imbeddings (resp. linear mappings) of K in R^N of a dimension $N \geqslant p+q+2$. If f, g are normal with respect to the pair of disjoint simplexes σ and $\xi = (b_1\cdots b_q)$, then arbitrarily near to $f(b_0)$ there exist points b_0' such that, defining a linear imbedding (resp. linear mapping) f' of K in R^N by setting $f'(a_i) = f(a_i)$, $f'(b_j) = f(b_j)$, $j \neq 0$ and $f'(b_0) = b_0'$, the pair f', g will be normal with respect to the pair (σ,τ).

Proof. By hypothesis $g(|\sigma|)$ and $g(|\xi|)$ are disjoint. As $N \geqslant p+q+2$, we may suppose without loss of generality, by a slight deformation of $g(b_0)$ if necessary, that $g(|\sigma|)$ and $g(|\tau|)$ are disjoint. For $x \in |\sigma|$, $y \in |\tau|$, let $L_{x,y}$ be the line through $f(x)$ parallel to the line joining $g(x)$ and $g(y)$. Given $x \in |\sigma|$, $z \in |\xi|$, $1 \geqslant \mu > 0$, let $M_{x,z,\mu}$

be the set of all points $u \in R^N$ with $\mu u+(1-\mu)f(z) \in L_{x,y}$, where $y=\mu b_0+(1-\mu)z$. Then $M_{x,z,\mu}$ is the set of all points u with

$$\mu u+(1-\mu)f(z)-f(x)=c[g(y)-g(x)],$$

or

$$u=\frac{c}{\mu}[\mu g(b_0)+(1-\mu)g(z)-g(x)]-\frac{1}{\mu}[(1-\mu)f(z)-f(x)].$$

for certain real numbers c. The union M of all such $M_{x,z,\mu}$ is easily seen to be a point set of dimension $\leqslant p+q+1<N$. Hence arbitrarily near to $g(b_0)$ we may take $b_0' \notin M$ such that the linear map f' defined by $f'(a_i)=f(a_i)$, $f'(b_j)=f(b_j)$, $j \neq 0$, and $f'(b_0)=b_0'$ is a linear imbedding if f is itself so. If f', g are not normal with respect to the pair σ, τ then $x \in |\sigma|$, $y \in |\tau|$ exist with the line joining $f'(x)=f(x)$ and $f'(y)$ parallel to the line joining $g(x)$ and $g(y)$ or $f'(y) \in L_{x,y}$. By hypothesis f, g and hence f', g are normal with respect to the pair σ, ξ, so we should have $y \notin |\xi|$. Whence $y=\mu b_0+(1-\mu)z$ for certain $z \in |\xi|$ and μ, $1 \geqslant \mu>0$. We should have then $f'(y)=\mu b_0'+(1-\mu)f(z) \in L_{x,y}$ or $b_0' \in M_{x,z,\mu} \subset M$ which is impossible. Therefore f', g are normal with respect to σ, τ, q. e. d.

Theorem 9 Arbitrarily near to two linear imbeddings f, g of K in R^N (in the topology of $L_N(K)$) there exist normal pairs of linear imbeddings f' and g' of K in R^N which are linearly isotopic to f and g respectively.

Proof. Let $\varepsilon>0$ be previously given. Let $\delta>0$ be a number such that for any $f', g' \in L_N(K)$ with $d(f,f')<\delta$, $d(g,g')<\delta$ we should have $f', g' \in I_N(K)$ and f', g' linearly isotopic to f, g respectively, e.g., $\delta=\min\left(\frac{1}{2}\delta_f, \frac{1}{2}\delta_g\right)$ (see Theorems 6, 7 of §4). Arrange now the totality of simplexes of K in an order

$$\sigma_0<\sigma_1<\cdots<\sigma_r,$$

such that simplexes of lower dimensions precede those of higher dimensions but otherwise arbitrary. Let K_i be the subcomplex of K consisting of the simplexes $\sigma_0, \sigma_1, \cdots, \sigma_i (i \leqslant r)$. Define now successively linear maps f_i of K in R^N with $d(f_{i-1}, f_i)<\min\left(\frac{1}{\gamma}\delta, \frac{1}{\gamma}\varepsilon\right)$ and f_i, g normal with respect to K_i as follows. Take $f_0=f$ tor which f_0 and g are trivially normal with respect to $K_0=(\sigma_0)$. Suppose that $f_0, f_1, \cdots, f_{j-1}$ have been defined verifying the required conditions. Let a_j be any vertex of σ_j. By the lemmas above we may then take a point a_j' in

R^N with $d(a_j, a'_j) < \min\left(\frac{1}{\gamma}\delta, \frac{1}{\gamma}\varepsilon\right)$ such that, defining the linear mapping f_j of K in R^N by $f_j(a_j) = a'_j$ but $f_j(a) = f_{j-1}(a)$ for any other vertex a of K, the pair f_j, g will be normal with respect to K_j. Proceeding in this manner we get finally a mapping $f' = f_r$ such that f', g are normal with respect to $K_r = K$ and $d(f, f') \leqslant \sum_{i=1}^{r} d(f_{i-1}, f_i) < \min(\delta, \varepsilon)$. The pair f' and $g' = g$ meet then the requirement of the theorem.

Remark. From the proof we see that the theorem remains true if we replace the pair of linear imbeddings by a pair of linear mappings for which the totality of image points of the vertices of K under the pair of mappings are in general position.

§6 Isotopy class of a pair of linear imbeddings

Let R^{N+1} be the product space $R^N \times L$, where$L$ is a line $-\infty < t< +\infty$. The subspace $R^N \times (t)$ will be denoted by R_t^N. For any space X and any two continuous maps f and g of X in R^N let us define the map $F:\ X \times L \to R^{N+1}$ in the manner that $F(X \times (t)) \subset R_t^N$, $F(x, 0)= (f(x), 0)$, $F(x, 1) = (g(x), 1)$, and that F maps $(x) \times L$ to the line joining $F(x, 0)$ and $F(x, 1)$ where $x \in X$. The map F thus defined will be said to be *associated* with f and g.

Let R^N be now oriented previously and R^{N+1} be oriented as the product space $R^N \times L$, where L is oriented by increasing t. Let K be a finite simplicial complex as before and f, g be two linear imbeddings of K in R^N which are *normal* with respect to K. For the associated map $F:\ |K| \times L \to R^{N+1}$ the point sets $F(|\xi| \times L)$ and $F(|\eta| \times L)$ are then disjoint for any $(N-2)$-cell $\xi \times \eta$ of the two-fold reduced complex $\tilde{K}^*_{(2)} = \tilde{K}^*$. For if not, $x \in |\xi|$ and $y \in |\eta|$ would exist with $F((x) \times L)$ and $F((y) \times L)$ intersecting each other. The line joining $F((x) \times (0))$ and $F((y) \times (0))$ is then parallel to the line joining $F((x) \times (1))$ and $F((y) \times (1))$, and the line joining $f(x)$ and $f(y)$ would be parallel to the line joining $g(x)$ and $g(y)$, contrary to the hypothesis of normality of the pair f, g. It follows that for any $(N-1)$-cell $\sigma \times \tau$ of $\tilde{K}^*$, the point-sets $|\partial F(\sigma \times I)|$ and $|F(\tau \times I)|$ are disjoint, so are the sets $|F(\sigma \times I)|$ and $|\partial F(\tau \times I)|$. Consequently with respect to the above orientation of R^{N+1}, the index of intersection of $F(\sigma \times I)$ and $F(\tau \times I)$, considered as singular chains in R^{N+1}, are well defined. We may thus define an $(N-1)$-dimensional cochain $\theta_{f,g}^{N-1}$ in the complex $\tilde{K}^*$ by setting

$$\theta_{f,g}^{N-1}(\sigma \times \tau) = \phi((F(\sigma \times I), F(\tau \times I)), \tag{1}$$

for any $(N-1)$-cell $\sigma \times \tau$ of $\tilde{K}^*$. The symbol ϕ means index of intersection in the oriented R^{N+1}. Let T_K be the cell map in $\tilde{K}^*$ given as before by

$$T_K(\sigma \times \tau) = (-1)^{\dim \sigma \dim \tau} \tau \times \sigma, \quad \sigma \times \tau \in \tilde{K}^*,$$

and ρ_N the operator

$$\rho_N = 1 + (-1)^{N-1} T_K.$$

Then for any $(N-1)$-cell $\sigma \times \tau \in \tilde{K}^*$, we have

$$\begin{aligned}
\rho_N \theta_{f,g}^{N-1}(\sigma \times \tau) &= \theta_{f,g}^{N-1} \rho_N(\sigma \times \tau) \\
&= \theta_{f,g}^{N-1}(\sigma \times \tau) + (-1)^{N-1+\dim \sigma \dim \tau} \theta_{f,g}^{N-1}(\tau \times \sigma) \\
&= \phi(F(\sigma \times I), F(\tau \times I)) \\
&\quad + (-1)^{N-1+\dim \sigma \dim \tau} \phi(F(\tau \times I), F(\sigma \times I)) \\
&= (-1)^{(\dim \sigma+1)(\dim \tau+1)} \cdot \phi(F(\tau \times I), F(\sigma \times I)) \\
&\quad + (-1)^{N-1+\dim \sigma \dim \tau} \cdot \phi(F(\tau \times I), F(\sigma \times I)) \\
&= 0,
\end{aligned}$$

since $\dim \sigma + \dim \tau = N-1$. Furthermore, for any N-cell $\xi \times \eta \in \tilde{K}^*$, we have

$$\begin{aligned}
\delta \theta_{f,g}^{N-1}(\xi \times \eta) &= \theta_{f,g}^{N-1}(\partial \xi \times \eta) + (-1)^{\dim \xi} \cdot \theta_{f,g}^{N-1}(\xi \times \partial \eta) \\
&= \phi(F(\partial \xi \times I), F(\eta \times I)) + (-1)^{\dim \xi} \cdot \phi(F(\xi \times I), F(\sigma \eta \times I)).
\end{aligned}$$

As f, g are both linear imbeddings so that $|F(\xi \times \partial I)|$ and $|F(\eta \times I)|$ are disjoint, so are $|F(\xi \times I)|$ and $|F(\eta \times \partial I)|$. Hence the last equation reduces to

$$\begin{aligned}
\delta \theta_{f,g}^{N-1}(\xi \times \eta) = \phi(\partial F(\xi \times I), F(\eta \times I)) + (-1)^{\dim \xi} \\
\cdot \phi(F(\xi \times I), \partial F(\eta \times I)) = 0.
\end{aligned} \tag{2}$$

It follows that $\theta_{f,g}^{N-1}$ is a ρ_N-cocycle of $\tilde{K}^*$ which will be called the *isotopy cocycle* of the normal pair of linear imbeddings f, g. The ρ_N-cohomology-class of $\theta_{f,g}^{N-1}$ will then be called the *isotopy class* of the normal pair f, g and will be denoted by ${}^{\rho_N}\Theta_{f,g}^{N-1}(K) \in {}^{\rho_N}\tilde{H}_{(2)}^{N-1}(K)$.

Remark. The ρ_N-cocycle $\theta_{f,g}^{N-1}$ is still well defined by (1) even if (f,g) is only a normal pair of linear mappings, but with $f(\sigma)$, $f(\tau)$ (and also $g(\sigma)$, $g(\tau)$) disjoint for any $(N-1)$-cell $\sigma \times \tau$ of $\tilde{K}^*$. The only change necessary in the proof is concerning $\delta\theta_{f,g}^{N-1} = 0$. In fact, for any N-cell $\xi \times \eta \in \tilde{K}^*$, it is easy to verify that

$$\phi(F(\xi \times \partial I), F(\eta \times I)) = \phi'(g(\xi), g(\eta)) - \phi'(f(\xi), f(\eta)),$$

and

$$\phi(F(\xi \times I), F(\eta \times \partial I)) = (-1)^{\dim \eta + 1} \cdot [\phi'(g(\xi), g(\eta)) - \phi'(f(\xi), f(\eta))],$$

where ϕ' denotes the index of intersection in the oriented R^N. It follows that (2) remains true.

By Theorem 2 of §5, arbitrarily near to twolinear imbeddings f, g of K in R^N there exist normal pairs of linear imbeddings f' and g' of K in R^N which are linearly isotopic to f and g respectively. Concerning the classes ${}^{\rho_N}\Theta_{f',g'}^{N-1}(K)$ we have

Theorem 10 Let f, g be any pair of linear imbeddings of K in R^N. Then for any normal pair of linear imbeddings f', g' of K in R^N linearly isotopic to f and g respectively, the isotopy class ${}^{\rho_N}\Theta_{f',g'}^{N-1}(K)$ is independent of the normal pair f', g' chosen.

Proof. Let us remark first that if (f', g') and (f'', g'') are two normal pairs of linear imbeddings of K in R^N such that f', f'' and similarly g', g'' are sufficiently near in the topology of $L_N(K)$, then ${}^{\rho_N}\Theta_{f',g'}^{N-1}(K) = {}^{\rho_N}\Theta_{f'',g''}^{N-1}(K)$. To see this, let $F' : |K| \times I \to R^{N+1}$ and $F'' : |K| \times I \to R^{N+1}$ be the map associated with these two normal pairs. Define as before

$$\theta_{f',g'}^{N-1}(\sigma \times \tau) = \phi(F'(\sigma \times I), F'(\tau \times I)),$$
$$\theta_{f'',g''}^{N-1}(\sigma \times \tau) = \phi(F''(\sigma \times I), F''(\tau \times I)),$$

for any $(N-1)$-cell $\sigma \times \tau$ of $\tilde{K}^*$. Then for (f'', g'') sufficiently near to (f', g') so that F'' is sufficiently near to F' (in the topology of $C_{N+1}(|K| \times I)$), we should have

$$\phi(F'(\sigma \times I), F'(\tau \times I)) = \phi(F''(\sigma \times I), F''(\tau \times I))$$

so that $\theta_{f',g'}^{N-1} = \theta_{f'',g''}^{N-1}$, or ${}^{\rho_N}\Theta_{f',g'}^{N-1}(K) = {}^{\rho_N}\Theta_{f'',g''}^{N-1}(K)$.

Now let $J = [0, 1]$ and

$$M = K \times J.$$

Let (f', g') and (f'', g'') be two normal pairs of linear imbeddings of K in R^N each linearly isotopic to f and g respectively. By at most small deformations, we may suppose without loss of generality that the totality of image points of all vertices of K under f', g', f'', g'' are in general position (cf. the remark just made). As f' and g' are linearly isotopic to f'' and g'' respectively, there exist simplicial subdivisions M'_f and M'_g of M with $K \times (0)$ and $K \times (1)$ as subcomplexes and linear maps h_f, h_g of M'_f, M'_g in R^N such that

$$h_f(x,0) = f'(x), \quad h_f(x,1) = f''(x),$$
$$h_g(x,0) = g'(x), \quad h_g(x,1) = g''(x),$$

for any $x \in |K|$. By taking common subdivision if necessary we may assume that M'_f coincide with M'_g, which will be denoted by M'. Deform now slightly the image points under h_f and h_g of the vertices of M' not in $K \times (0)$ or $K \times (1)$ so that they are in general position. By successively applying lemma 2 of §5 h_f and h_g may then be deformed into a normal pair of linear maps which will still be denoted by h_f and h_g (the vertices of $K \times (0)$ and $K \times (1)$ will be kept undeformed), cf. the remark at the end of §5. By Theorem 8 of §4, the final linear map h_f and h_g will still be linear isotopies between the pairs (f', f'') and (g', g'') respectively, so long as the deformations are sufficiently small. Let $j_i : K \to M'$ be the simplicial map defined by $j_i(a) = (a, i)$, $i = 0$ or 1, and a an arbitrary vertex of K. Let $\tilde{j}_i : \tilde{K}^* \to \tilde{M}'^* = \tilde{M}'^*_{(2)}$ be the cell-map induced by j_i. Let $|J| : |K| \times 1 \to |M| = |K| \times J$ be the map $|J|(x,t) = (x,t)$, $x \in |K|$, $t \in I$, and $|\tilde{J}| : |\tilde{K}|^* \times I \to |\tilde{M}|^*$ the map induced by $|J|$, $|\tilde{j}_i| : |\tilde{K}|^* \to |\tilde{M}|^*$ the map given by $|\tilde{j}_i|(x) = |\tilde{J}|(x,i)$, $i = 0$ or 1. Then $|\tilde{J}|$ gives a system-homotopy between $|\tilde{j}_0|$ and $|\tilde{j}_1|$ of the systems $(|\tilde{K}|^*, T_{|K|})$ in $(|\tilde{M}|^*, T_{|M|})$ so that by Proposition 1 of §1, they induce identical homomorphisms of special cohomology groups: ${}^{\rho}|\tilde{j}_0|^* = {}^{\rho}|\tilde{j}_1|^*$. As the special cohomology groups of $(\tilde{K}^*, T_K)$ and $(|\tilde{K}|^*, T_{|K|})$ are canonically isomorphic, similarly for $(\tilde{M}'^*, T_{M'})$ and $(|\tilde{M}'|^*, T_{|M'|})$, we have also ${}^{\rho}\tilde{j}_0^* = {}^{\rho}\tilde{j}_1^*$. Now the ρ_N-cocycle $\theta^{N-1}_{h_f,h_g} \in {}^{\rho_N}Z^{N-1}(\tilde{M}'^*)$ is well defined (cf. the remark before). By the very definition we have

$$\hat{j}_0^{\#}\theta^{N-1}_{h_f,h_g} = \theta^{N-1}_{f',g'}, \quad \hat{j}_1^{\#}\theta^{N-1}_{h_f,h_g} = \theta^{N-1}_{f'',g''}.$$

Since ${}^{\rho}\tilde{j}_0^* = {}^{\rho}\tilde{j}_1^*$ we get $\theta^{N-1}_{f',g'}\tilde{\rho}_N\theta^{N-1}_{f'',g''}$ or ${}^{\rho_N}\Theta^{N-1}_{f',g'}(K) = {}^{\rho_N}\Theta^{N-1}_{f'',g''}(K)$. This proves our theorem.

From the above theorem we see that the following definition is justified:

Definition Let f, g be any pair of linear imbeddings of a finite simplicial complex K in an oriented R^N. Then the isotopy class ${}^{\rho_N}\Theta^{N-1}_{f',g'}(K) \in {}^{\rho_N}\tilde{H}^{N-1}_2(K)$

of any normal pair of linear imbeddings f', g' linearly isotopic to f and g respectively will simply be called the *isotopy class* of the pair f, g and will be denoted by ${}^{\rho_N}\Theta_{f,g}^{N-1}(K) \in {}^{\rho_N}\tilde{H}_2^{N-1}(K)$.

With the above definition the Theorem 10 may then somewhat be strcngthened to the following

Theorem 10′ Let (f, g) and (f', g') be two pairs of linear imbeddings of K in R^N where f' is linearly isotopic to f and g' is linearly isotopic to g. Then ${}^{\rho_N}\Theta_{f,g}^{N-1}(K) = {}^{\rho_N}\Theta_{f',g'}^{N-1}(K)$.

The following Theorem 11 will show that the isotopy class ${}^{\rho_N}\Theta_{f,g}^{N-1}(K)$ will furnish necessary conditions for two linear imbeddings f, g of K in R^N to be linearly isotopic. As we shall see later, (Theorem 14 of §8), these necessary conditions are somewhat stronger than those furnished by the isotopy classes ${}^{\rho_N}\Theta_f^{N-1}(K)$ (see Theorem 4 of §3), so long as linear isotopics of linear imbeddings are concerned.

Theorem 11 If two linear imbeddings f, g of a finite simplicial complex K in a Euclidean space R^N are linearly isotopic, then ${}^{\rho_N}\Theta_{f,g}^{N-1}(K) = 0$.

Proof. As the linear imbeddings (g, g) are respectively linearly isotopic to the linear imbeddings (f, g), we have by Theorem 10′, ${}^{\rho_N}\Theta_{f,g}^{N-1}(K) = {}^{\rho_N}\Theta_{g,g}^{N-1}(K)$. Hence it is sufficient to prove that ${}^{\rho_N}\Theta_{g,g}^{N-1}(K) = 0$. Now by Theorem 9 of §5, there exists arbitrarily near to g normal pair of linear imbeddings g', g'' of K in R^N. Let $G : |K| \times I \to R^{N+1}$ and $G' : |K| \times I \to R^{N+1}$ be the maps associated with the pairs (g, g) and $(g'g'')$ respectively. It is clear that, if we take g', g'' sufficiently near to g, then for any $(N-1)$-cell $\sigma \times \tau$ of $\tilde{K}^*$, we should have

$$\phi(G'(\sigma \times I), G'(\tau \times I)) = \phi(G(\sigma \times I), G(\tau \times I)).$$

As the right-hand side is evidently 0, it follows that

$$\theta_{g',g''}^{N-1}(\sigma \times \tau) = \phi(G'(\sigma \times I), G'(\tau \times I)) = 0$$

for any $(N-1)$-cell $\sigma \times \tau$ of $\tilde{K}^*$. Hence $\theta_{g',g''}^{N-1} = 0$ or ${}^{\rho_N}\Theta_{g',g''}^{N-1}(K) = 0$. By definition we have ${}^{\rho_N}\Theta_{g,g}^{N-1}(K) = 0$ and the theorem is proved.

§7 An alternative definition of ${}^{\rho_N}\Theta_{f,g}^{N-1}(K)$ and some of their properties

Let f, g be any two linear imbeddings of a finite simplicial complex K in an oriented R^N. A linear map F of a certain simplicial subdivision M' of $M = K \times I$ in

$R^{N+1}=R^N\times L$ will be said to be a *joining map* of the pair f, g if $F(x,0)=(f(x),0)$, $F(x,1)=(g(x),1)$, $F(x,t)\subset R_t^N$, where $x\in|K|$, $t\in I$. It will be said to be *fine* if for each $(N-2)$-cell $\xi\times\eta$ of $\tilde{K}^*$, the singular chains $F(\xi\times I)$ and $F(\eta\times 1)$ have their point-sets disjoint from each other, where $F(\xi\times I)$ (similarly for $F(\eta\times I)$) denotes the singular chain $FSd(\xi\times I)$, Sd being the subdivision from M to M'. It is easy to see that fine joining maps of the pair f, g exist always. For any $(N-1)$ -cell $\sigma\times\tau$ of $\tilde{K}^*$, the index of intersection with respect to the orientation of R^{N+1}, as described before of the singular chains $F(\sigma\times I)$ and $F(\tau\times I)$ are then well defined and we have a cochain $\theta_F^{N-1}\in C^{N-1}(\tilde{K}^*)$ defined by

$$\theta_F^{N-1}(\sigma\times\tau)=\phi(F(\sigma\times I),F(\tau\times I)),\tag{1}$$

As before, we see that θ_F^{N-1} is a ρ_N-cocycle of $\tilde{K}^*$.

Theorem 12 Given two linear imbeddings f, g of a finite simplicial complex K in an oriented R^N, the ρ_N-cohomology-class of the ρ_N-cocycle θ_F^{N-1} determined by a fine joining map F of the pair f, g is independent of the fine joining map F chosen and coincides with the isotopy class ${}^{\rho_N}\Theta_{f,g}^{N-1}(K)$ of the pair f, g.

Proof. Let F_0 and F_1 be any two fine joining maps of the pair f, g which may be considered as linear maps of the same simplicial subdivision M' of $M=K\times I$ in R^{N+1}. Let $J=[0,1]$, M'' a certain simplicial subdivision of the product complex $K\times J$ containing $K\times(0)$ and $K\times(1)$ as subcomplexes, and $\tilde{M}$ a certain common simplicial subdivision of the product complex $M''\times I$ as well as (where $K\times I\times J$ is canonically identified to $K\times J\times I$) $M'\times J$. Let H be any linear map of $\tilde{M}$ in R^{N+1} such that H coincides with F_0 and F_1 on $|M'|\times(0)$ and $|M'|\times(1)$ respectively and H $(|M''|\ \times\ (t))\subset R_t^N$, $t\in I$. Deform the image under H of all vertices of $\tilde{M}$ slightly so that the resulting linear map H' of $\tilde{M}$ in R^{N+1} thus obtained will have the totality of images of all vertices of $\tilde{M}$ in general position. Denote the restriction of H' on $M'\times(0)$ and $M'\times(1)$ by F_0' and F_1' respectively. The deformation will be chosen sufficiently small so that for any $(N-1)$-cell $\sigma\times\tau$ of $\tilde{K}^*$, the singular chains $F_0'(\sigma\times(0)\times I)$ and $F_0'(\tau\times(0)\times I)$ have a well-defined index of intersection equal to that of $F_0(\sigma\times I)$ and $F_0(\tau\times I)$. Similarly the singular chains $F_1'(\sigma\times(1)\times I)$ and $F_1'(\ \tau\times(1)\times I)$ will have a well-defined index of intersection equal to that of $F_1(\sigma\times I)$ and $F_1(\tau\times I)$. We have then for any $(N-1)$-cell $\sigma\times\tau$ of $\tilde{K}^*$,

$$\theta_{F_0}^{N-1}(\sigma\times\tau)=\phi(F'_0(\sigma\times(0)\times I),F'_0(\tau\times(0)\times I)),$$
$$\theta_{F_1}^{N-1}(\sigma\times\tau)=\phi(F'_1(\sigma\times(1)\times I),F'_1(\tau\times(1)\times I)).$$

Define now a cochain θ^{N-1} in $\tilde{M}''^*$ such that for any $(N-1)$-cell $\tilde{\sigma}\times\tilde{\tau}$ of $\tilde{M}''^*$, we have

$$\theta^{N-1}(\tilde{\sigma}\times\tilde{\tau})=\phi(H'(\tilde{\sigma}\times I),H'(\tilde{\tau}\times I)).$$

As before, θ^{N-1} is a ρ_N-cocycle in $\tilde{M}''^*$ (cf. the remark in §6). Let j_0, j_1 be the simplicial maps of K in M'' defined by $j_0(a)=(a,0)$, $j_1(a)=(a,1)$ for any vertex a of K and denote the cell-maps of $\tilde{K}^*$ in $\tilde{M}''^*$ induced by them by $\tilde{j}_0$ and $\tilde{j}_1$ respectively. It is clear that

$$\hat{j}_0^{\#}\theta^{N-1}=\theta_{F_0}^{N-1}\quad\text{and}\quad\hat{j}_1^{\#}\theta^{N-1}=\theta_{F_1}^{N-1}.$$

It follows as in the proof of Theorem 10 of §6 that $\theta_{F_0}^{N-1}\tilde{\rho}_N\theta_{F_1}^{N-1}$, and therefore the ρ_N-cohomology-class of a fine joining map F of the pair f, g is independent of the map F.

Now by Theorem 9 of §5, arbitrarily near to f, g there exist linear imbeddings f', g' of K in R^N which form a normal pair. We may take e.g. f', g' such that $d(f,f')<\frac{1}{2}\delta_f$, $d(g,g')<\frac{1}{2}\delta_g$. The mappings f', g' are linearly isotopic to f and g respectively. There exist therefore simplicial subdivisions M_0 and M_1 of $K\times\left[0,\frac{1}{3}\right]$ and $K\times\left[\frac{2}{3},1\right]$ respectively and also linear maps h_0 of M_0 in R^N and h_1 of M_1 in R^N which realize these linear isotopies with $h_0\left(|\sigma|\times\left[0,\frac{1}{3}\right]\right)$ disjoint from $h_0\left(|\tau|\times\left[0,\frac{1}{3}\right]\right)$ and $h_1\left(|\tau|\times\left[\frac{2}{3},1\right]\right)$ disjoint from $h_1\left(|\tau|\times\left[\frac{2}{3},1\right]\right)$ for any non-diagonic pair of cells σ, τ of K. Let $F':|K|\times I\to R^{N+1}$ be the map associated with the normal pair f', g' so that ${}^{\rho_N}\Theta_{f,g}^{N-1}(K)$ contains a ρ_N-cocycle $\theta_{f',g'}^{N-1}$ defined by

$$\theta_{f',g'}^{N-1}(\sigma\times\tau)=\phi(F'(\sigma\times I),F'(\tau\times I))$$

for any $(N-1)$-cell $\sigma\times\tau$ of $\tilde{K}^*$. For F' we may take a sufficiently small simplicial subdivision M' of $K\times I$ and a linear map F'' of M' in R^{N+1} sufficiently near to F' and coinciding with F' on $|K|\times(0)+|K|\times(1)$ which gives a fine joining map of f', g'. The map F'' will be chosen so near to F' that $\phi(F''(\sigma\times I),\ F''(\tau\times I))$ will be well defined and equal to $\phi(F'(\sigma\times I),\ F'(\tau\times I))$ for any $(N-1)$-cell $\sigma\times\tau$ of $\tilde{K}^*$ so that we have also

$$\theta_{f',g'}^{N-1}(\sigma\times\tau)=\phi(F''(\sigma\times I),F''(\tau\times I)).$$

Define now a map F of $|K| \times I$ in R^{N+1} by setting ($x \in |K|$)

$$F(x,t) = \begin{cases} (h_1(x,t),t), & \frac{2}{3} \leqslant t \leqslant 1, \\ (h_0(x,t),t), & 0 \leqslant t \leqslant \frac{1}{3}, \\ (\pi F''(x,3t-1),t), & \frac{1}{3} \leqslant t \leqslant \frac{2}{3}, \end{cases}$$

in which π is the natural projection of $R^{N+1} = R^N \times L$ on R^N. Let $\tilde{M}$ be a simplicial subdivision of $K \times I$ which is simultaneously a subdivision of M_0 on $K \times \left[0, \frac{1}{3}\right]$, of M_1 on $K \times \left[\frac{2}{3}, 1\right]$, and also M'' on $K \times$ [1/3, 2/3] which is isomorphic to M' on $K \times 1$ under the map $h: (x,t) \to (x, 3t-1)$, $x \in |K|$, $\frac{1}{3} \leqslant t \leqslant \frac{2}{3}$. Then F is clearly a linear map of the subdivision $\tilde{M}$ of $K \times I$ in R^{N+1} which gives a fine joining map of the pair f, g owing to our choice of f', g', h_0, h_1. It follows that

$$\theta_F^{N-1}(\sigma \times \tau) = \phi(F(\sigma \times I), F(\tau \times I))$$

for any $(N-1)$-cell $\sigma \times \tau$ of $\tilde{K}^*$. However, it is evident that

$$\phi(F''(\sigma \times I), F''(\tau \times I)) = \phi(F(\sigma \times I), F(\tau \times I))$$

for any such cell $\sigma \times \tau$. Hence we get

$$\theta_F^{N-1} = \theta_{f',g'}^{N-1} \in {}^{\rho_N}\Theta_{f,g}^{N-1}(K).$$

Combining with the first part of the present proof, we see that the ρ_N-cohomology-class of the ρ_N-cocycle θ_F^{N-1} determined by *any* fine joining map F of the pair f, g is equal to the isotopy class ${}^{\rho_N}\Theta_{f,g}^{N-1}(K)$. Our theorem is thus completely proved.

Remark. From the above theorem we may also define the isotopy class ${}^{\rho_N}\Theta_{f,g}^{N-1}$. (K) of a pair of linear imbeddings f, g of a finite simplicial complex K in an oriented R^N as the ρ_N-cohomology-class of the ρ_N -cocycle θ_F^{N-1} in $\tilde{K}^*$ defined by (1) for any fine joining map F of the pair f, g.

Theorem 13 For any linear imbeddings f, g and h of a finite simplicial complex in an oriented R^N we have

$$ {}^{\rho_N}\Theta_{f,f}^{N-1}(K) = 0, \tag{2}$$

$$ {}^{\rho_N}\Theta_{f,g}^{N-1}(K) = -{}^{\rho_N}\Theta_{g,f}^{N-1}(K), \tag{3}$$

$$ {}^{\rho_N}\Theta_{f,g}^{N'-1}(K) + {}^{\rho_N}\Theta_{g,h}^{N-1}(K) = {}^{\rho_N}\Theta_{f,h}^{N-1}(K). \tag{4}$$

Proof. We may assume without loss of generality that each pair of the imbeddings f, g and h is normal. Let F_1 and F_2 be the maps of $|K| \times I$ in R^{N+1} associated with the pairs (f, g) and (g, h) respectively, so that

$$\theta_{f,g}^{N-1} \in {}^{\rho_N}\Theta_{f,g}^{N-1}(K), \quad \theta_{g,h}^{N-1} \in {}^{\rho_N}\Theta_{g,h}^{N-1}(K),$$

where

$$\theta_{f,g}^{N-1}(\sigma \times \tau) = \phi(F_1(\sigma \times I), F_1(\tau \times I)),$$
$$\theta_{g,h}^{N-1}(\sigma \times \tau) = \phi(F_2(\sigma \times I), F_2(\tau \times I))$$

for any $(N-1)$-cell $\sigma \times \tau$ of $\tilde{K}^*$. Arbitrarily near to F_1 and F_2 (in the topology of $C_{N+1}(|K| \times I)$) we may find linear maps F_1' and F_2' of certain simplicial subdivisions M_1' and M_2' of $K \times I$ in R^{N+1} which give fine joining maps of the pairs (f, g) and (g, h) respectively and coincide with F_1 and F_2 on $|K| \times (0) + |K| \times (1)$. For F_1', F_2' sufficiently near to F_1, F_2 we should have

$$\phi(F_1'(\sigma \times I), F_1'(\tau \times I)) = \phi(F_1(\sigma \times I), F_1(\tau \times I))$$

so that

$$\theta_{f,g}^{N-1}(\sigma \times \tau) = \phi(F_1'(\sigma \times I), F_1'(\tau \times I)),$$

for any $(N-1)$-cell $\sigma \times \tau$ of $\tilde{K}^*$ and similarly

$$\theta_{g,h}^{N-1}(\sigma \times \tau) = \phi(F_2'(\sigma \times I), F_2'(\tau \times I)).$$

Now let M' be a simplicial subdivision of $K \times I$ which is simultaneously a subdivision of the subdivision M_1'' on $K \times \left[0, \frac{1}{2}\right]$ isomorphic to M_1' on $K \times [0,1]$ under the map $h_1 : (x, t) \to (x, 2t)$, and also of the subdivision M_2'' on $K \times \left[\frac{1}{2}, 1\right]$ isomorphic to M_2' on $K \times [0,1]$ under the map $h_2 : (x, t) \to (x, 2t-1)$, where $x \in |K|$. Define now a map $F : |K| \times I \to R^{N+1}$ by setting for any $x \in |K|$,

$$F(x,t) = \begin{cases} (\pi F'_1(x, 2t), t), & 0 \leqslant t \leqslant \dfrac{1}{2}, \\ (\pi F'_2(x, 2t-1), t), & \dfrac{1}{2} \leqslant t \leqslant 1. \end{cases}$$

It is clear that F is a linear map of M' in R^{N+1} which gives a fine joining map of the pair (f, h). Hence we have by Theorem 12,

$$\theta_F^{N-1} \in {}^{\rho_N}\Theta_{f,h}^{N-1}(K),$$

where

$$\theta_F^{N-1}(\sigma \times \tau) = \phi(F(\sigma \times I), F(\tau \times I)),$$

for any $(N-1)$-cell $\sigma \times \tau$ of $\tilde{K}^*$. Now it is evident that

$$\phi(F(\sigma \times I), F(\tau \times I)) = \phi({F'}_1(\sigma \times I), {F'}_1(\tau \times I)) \\ + \phi({F'}_2(\sigma \times I), {F'}_2(\tau \times I)).$$

Hence we get

$$\theta_F^{N-1}(\sigma \times \tau) = \theta_{f,g}^{N-1}(\sigma \times \tau) + \theta_{g,h}^{N-1}(\sigma \times \tau).$$

for any $(N-1)$-cell $\sigma \times \tau$ of $\tilde{K}^*$. Thus

$$\theta_F^{N-1} = \theta_{f,g}^{N-1} + \theta_{g,h}^{N-1}, \text{ or} {}^{\rho_N}\Theta_{f,g}^{N-1}(K) + {}^{\rho_N}\Theta_{g,h}^{N-1}(K) = {}^{\rho_N}\Theta_{f,h}^{N-1}(K),$$

which proves (4).

As the equation (2) is quite simple to verify and the equation (3) follows then from (4) in setting $h = f$, the theorem is completely proved.

§8 A relation between the various isotopy classes

Between the classes ${}^{\rho_N}\Theta_f^{N-1}(K)$, ${}^{\rho_N}\Theta_g^{N-1}(K)$ and ${}^{\rho_N}\Theta_{f,g}^{N-1}(K)$ of any two linear imbeddings f, g of a finite simplicial complex K in an oriented R^N we have a relation as given in the following

Theorem 14 For any pair of linear imbeddings f, g of a finite simplicial complex K in an oriented R^N we have

$$2{}^{\rho_N}\Theta_{f,g}^{N-1}(K) = {}^{\rho_N}\Theta_f^{N-1}(K) - {}^{\rho_N}\Theta_g^{N-1}(K). \tag{1}$$

Proof. For any two singular chains A and B in $R^{N+1} = R^N \times L$ (L is the line $-\infty < t < +\infty$ as before) with their point-sets $|A|$ and $|B|$ disjoint from each other, we shall denote by $j(A, B)$ the singular chain lying on the unit sphere S^N of R^{N+1} carried by the pointset $j(|A| \times |B|)$, where $j(x, y)$ denotes the point of intersection of S^N with the ray from the origin O of R^{N+1} parallel to the directed line joining x to y, for any pair of points $x \neq y$ in R^{N+1}. Let $F : |K| \times I \to R^{N+1}$ be the mapping associatd with the pair of linear imbeddings f and g which may be supposed to be normal, by Theorem 9 of §5, Theorem 10 of §6 and Theorem 4 of §3. For any cell $\sigma \times \tau \in \tilde{K}^*$ with $\dim \sigma + \dim \tau \leqslant N-3$, the set $R(\sigma, \tau)$ of points $j(F(x,t), F(y,t))$

where $x \in |\sigma|$, $y \in |\tau|$, $t \in I$, is of dimension $\leqslant N-2$. Similarly, for any cell $\sigma \times \tau \in \tilde{K}^*$ with $\dim \sigma + \dim \tau \leqslant N-2$, the set $R'(\sigma,\tau)$ of points $j(F(x,t), F(y,t))$, where $x \in |\sigma|$, $y \in |\tau|$, $t = 0$ or 1, is also of dimension $\leqslant N-2$. Hence we may take a point e on the unit sphere S^{N-1} in R_0^N not lying in any of the above sets $R(\sigma,\tau)$ or $R'(\sigma,\tau)$. Take orientations of S^N and S^{N-1} coherent with those of R^{N+1} and R^N. For any singular chain C of dimension N (resp. $N-1$) lying on S^N (resp. on S^{N-1}) with $|\partial C|$ disjoint from e the local covering degree of C at e will be denoted by $\mathrm{Deg}_e^{(N)} C$ (resp. $\mathrm{Deg}_e^{(N-1)} C$).

Define now φ_f, $\varphi_g \in C^{N-1}(\tilde{K}^*)$ and $\psi \in C^{N-2}(\tilde{K}^*)$ by

$$\varphi_f(\sigma \times \tau) = \mathrm{Deg}_e^{(N-1)} j(f(\sigma), f(\tau)),$$
$$\varphi_g(\sigma \times \tau) = \mathrm{Deg}_e^{(N-1)} j(g(\sigma), g(\tau)),$$

and

$$\psi(\xi \times \eta) = (-1)^{\dim \xi} \cdot \mathrm{Deg}_e^{(N)} j(F(\xi \times I), F(\eta \times I)),$$

in which

$$\sigma \times \tau, \xi \times \eta \in \tilde{K}^*, \quad \dim \sigma + \dim \tau = N-1 \text{ and } \dim \xi + \dim \eta = N-2.$$

Note that all these cochains are well defined owing to our choice of the point e. From the definition of the class ${}^{\rho_N}\Theta_f^{N-1}(K)$, we have then (cf. the remark at the end of §1 and also the end of §2)

$$\theta_f = \bar{\rho}_N \varphi_f = (1 + (-1)^N T_K)\varphi_f \in {}^{\rho_N}\Theta_f^{N-1}(K).$$

Similarly,

$$\theta_g = \bar{\rho}_N \varphi_g = (1 + (-1)^N T_K)\varphi_g \in {}^{\rho_N}\Theta_g^{N-1}(K).$$

For any singular chains A and B in R^{N+1} let $s(A,B)$ denote the singular chain carried by the point set $s(|A| \times |B|)$, where $s(x,y) = y - x$, for any points x, $y \in R^{N+1}$, considered as vectors. If $\dim A + \dim B = N+1$, $|\partial A|$ disjoint from $|B|$ and $|A|$ disjoint from $|\partial B|$, then the chain $s(A,B)$ has a local covering degree at the origin O of R^{N+1} (oriented) which will be denoted by $\mathrm{Deg}_O^{(N+1)} s(A,B)$. It is known that

$$\mathrm{Deg}_O^{(N+1)} s(A,B) = (-1)^{\dim A} \cdot \phi(A,B),$$

and it equals also the global covering degree of the chain $j(\partial A, B) + (-1)^{\dim A} j(A, \partial B)$ on the oriented sphere S^N so that

$$\mathrm{Deg}_O^{(N+1)} s(A,B) = \mathrm{Deg}_e^{(N)} j(\partial A, B) + (-1)^{\dim A} \cdot \mathrm{Deg}_e^{(N)} j(A, \partial B),$$

if each term in the right-hand side has a meaning.

Consider now a segment $I' = [-\varepsilon, 1+\varepsilon]$ where $\varepsilon > 0$ is arbitrary. For any $(N-1)$-cell $\sigma \times \tau \in \tilde{K}^*$, the singular chains $A = F(\sigma \times I)$ and $B = F(\tau \times I')$ meet then the requirements about A, B above so that we have

$$\begin{aligned}
&(-1)^{\dim \sigma+1} \cdot \phi(F(\sigma \times I), F(\tau \times I')) \\
=& Deg_O^{(N+1)} s(F(\sigma \times I), F(\tau \times I')) \\
=& \mathrm{Deg}_e^{(N)} j(\partial F(\sigma \times I), F(\tau \times I')) \\
&+ (-1)^{\dim \sigma+1} \cdot \mathrm{Deg}_e^{(N)} j(F(\sigma \times I), \partial F(\tau \times I')) \\
=& \mathrm{Deg}_e^{(N)} j(F(\partial\sigma \times I), F(\tau \times I')) \\
&+ (-1)^{\dim \sigma+1} \cdot \mathrm{Deg}_e^{(N)} j(F(\sigma \times I), F(\partial\tau \times I')) \\
&+ (-1)^{\dim \sigma} \cdot \mathrm{Deg}_e^{(N)} j(F(\sigma \times (1) - \sigma \times (0)), F(\tau \times I')) \\
&+ (-1)^{N} \cdot \mathrm{Deg}_e^{(N)} j(F(\sigma \times I), F(\tau \times (1+\varepsilon) - \tau \times (-\varepsilon))).
\end{aligned}$$

Note that the last term evidently vanishes while the other terms in the last expression have values unchanged if we replace I' by I. Similarly, it is evident that

$$\phi(F(\sigma \times I), F(\tau \times I)) = \phi(F(\sigma \times I), F(\tau \times I')).$$

Furthermore, we have also for $i = 0, 1$,

$$\begin{aligned}
&\mathrm{Deg}_e^{(N)} j(F(\sigma \times (i)), F(\tau \times I)) = \mathrm{Deg}_e^{(N-1)} j(F(\sigma \times (i)), F(\tau \times (i))) \\
=& \begin{cases} \mathrm{Deg}_e^{(N-1)} j(g(\sigma), g(\tau)), & i = 1, \\ \mathrm{Deg}_e^{(N-1)} j(f(\sigma), f(\tau)), & i = 0. \end{cases}
\end{aligned}$$

Hence we get

$$\begin{aligned}
&(-1)^{\dim \sigma+1} \cdot \phi(F(\sigma \times I), F(\tau \times I)) \\
=& \mathrm{Deg}_e^{(N)} j(F(\partial\sigma \times I), F(\tau \times I)) \\
&+ (-1)^{\dim \sigma+1} \cdot \mathrm{Deg}_s^{(N)} j(F(\sigma \times I), F(\partial\tau \times I)) \\
&+ (-1)^{\dim \sigma} \cdot \mathrm{Deg}_e^{(N-1)} j(g(\sigma), g(\tau)) \\
&- (-1)^{\dim \sigma} \cdot \mathrm{Deg}_e^{(N-1)} j(f(\sigma), f(\tau)).
\end{aligned}$$

Using the definitions of the cochains θ_f, θ_g, ψ and $\theta_{f,g}^{N-1}$, we have then

$$\begin{aligned}
\theta_{f,g}^{N-1}(\sigma \times \tau) &= \phi(F(\sigma \times 1)F(\tau \times I)) \\
&= \psi(\partial\sigma \times \tau) + (-1)^{\dim \sigma}\psi(\sigma \times \partial\tau) + \varphi_f(\sigma \times \tau) - \varphi_g(\sigma \times \tau),
\end{aligned}$$

or

$$\theta_{f,g}^{N-1}(\sigma \times \tau) = \delta\psi(\sigma \times \tau) + \varphi_f(\sigma \times \tau) - \varphi_g(\sigma \times \tau). \tag{2}$$

Now

$$\begin{aligned}
\theta_{f,g}^{N-1}(\sigma \times \tau) &= \phi(F(\sigma \times I), F(\tau \times I)) \\
&= (-1)^{(\dim\sigma+1)(\dim\tau+1)} \cdot \phi(F(\tau \times I), F(\sigma \times I)) \\
&= (-1)^{\dim\sigma\dim\tau+N} \cdot \theta_{f,g}^{N-1}(\tau \times \sigma) \\
&= (-1)^N \cdot T_K\theta_{f,g}^{N-1}(\sigma \times \tau).
\end{aligned}$$

Therefore

$$\begin{aligned}
2\theta_{f,g}^{N-1}(\sigma \times \tau) &= (1 + (-1)^N T_K)\theta_{f,g}^{N-1}(\sigma \times \tau) \\
&= \bar{\rho}_N\theta_{f,g}^{N-1}(\sigma \times \tau) \\
&= \delta\bar{\rho}_N\psi(\sigma \times \tau) + \bar{\rho}_N\varphi_f(\sigma \times \tau) - \bar{\rho}_V\varphi_g(\sigma \times \tau),
\end{aligned}$$

or

$$2\theta_{f,g}^{N-1}(\sigma \times \tau) = \delta\bar{\rho}_N\psi(\sigma \times \tau) + \theta_f(\sigma \times \tau) - \theta_g(\sigma \times \tau).$$

As $\sigma \times \tau$ is any $(N\quad 1)$-cell of $\tilde{K}^*$, we have therefore

$$2\theta_{f,g}^{N-1}\tilde{\rho}_N\theta_f - \theta_g,$$

or

$$2^{\rho_N}\Theta_{f,g}^{N-1}(K) = {}^{\rho_N}\Theta_f^{N-1}(K) - {}^{\rho_N}\Theta_g^{N-1}(K), \quad \text{q.e.d.}$$

Theorem 15　If f and g are two linear imbeddings of a finite simplicial complex K in R^N such that $f(|K|)$ and $g(|K|)$ both lie in an $(N-1)$-dimensional linear subspace R^{N-1} of R^N. Then ${}^{\rho_N}\Theta_{f,g}^{N-1}(K) = 0$.

Proof. Let S^{N-1} be the unit sphere in $R_0^N = R^N \times (0) \subset R^{N+1} = R^N \times L$ and e the point on S^{N-1} such that the radius through e is orthogonal to the space $R_0^{N-1} = R^{N-1} \times (0) \subset R_0^N$. Deform now slightly f, g to linear imbeddings f' and g' such that they form a normal pair. The deformation will be chosen to be sufficiently small so that e does not lie in any of the sets $R(\sigma,\tau)$ or $R'(\sigma,\tau)$ as in the proof of the preceding theorem, with (f,g) replaced by (f',g'). Using the same notations as in the preceding proof we have then

$$\varphi_{f'}(\sigma \times \tau) = \mathrm{Deg}_e^{(N-1)} j(f'(\sigma), f'(\tau)) = 0,$$
$$\varphi_{g'}(\sigma \times \tau) = \mathrm{Deg}_e^{(N-1)} j(g'(\sigma), g'(\tau)) = 0,$$

and

$$\psi'(\xi \times \eta) = (-1)^{\dim \xi}\mathrm{Deg}_e^{(N)} j(F'(\xi \times I), F'(\eta \times I)) = 0,$$

in which $\sigma \times \tau$, $\xi \times \eta \in \tilde{K}^*$, $\dim \sigma + \dim \tau = N - 1$, $\dim \xi + \dim \eta = N - 2$, and F' is the map associated with the pair (f', g'). As in the equation (2) before we have then

$$\theta_{f',g'}^{N-1}(\sigma \times \tau) = \delta\psi'(\sigma \times \tau) + \varphi_{f'}(\sigma \times \tau) - \varphi_{g'}(\sigma \times \tau) = 0,$$

for any $(N-1)$-cell $\sigma \times \tau$ of $\tilde{K}^*$. Hence $\theta_{f',g'}^{N-1} = 0$ or ${}^{\rho_N}\Theta_{f,g}^{N-1}(K) = 0$, q.e.d.

References

[1] Wu Wen-tsün. *Deut. Math. Vers.*, 1958, 61: 65-75.

[2] ——. On the isotopy of complexes in a Euclidean space, I, II. *Science Record* (New Ser.), 1959, 3.

[3] ——. *Science Record* (New Ser.), 1957, 1: 377-380.

[4] ——. A theory of imbedding and immersion in Euclidean spaces. Peking (Mimeographed), 1957.

[5] ——. *Acta Math. Sinica*, 1953, 3: 261-290.

31. Topologie Combinatoire Et Invariants Combinatoires*

Un ensemble de points E situédans un espace euclidien est dit *convexe* si, pour deux points quelconques x et y de E, le segment rectiligne xy est contenu dans E.

Etant donnés dans un espace Euclidien $p+1$ points linéairement indépendants a_0, $\cdots$, a_p, le plus petit ensemble convexe contenant tous ces points (c'est-à-dire l'intersection de tous les ensembles convexes qui les contiennent) est un ensemble compact qui s'appelle *simplexe euclidien de dimension* p et les points a_i, où $i = 0$, $\cdots$, p, s'appellent ses *sommets*. Ce simplexe sera désigné par $\overline{(a_0 \ldots a_p)}$. Les simplexes euclidiens de dimension $n \leqslant p$ et dont les sommets a_{i_0}, $\cdots$, a_{i_n} sont des sommets du simplexe $\overline{(a_0 \ldots a_p)}$ portent le nom de *faces* de ce simplexe; il y en a $2^{p+1}-1$ (ou 2^{p+1} si l'on convient de considérer l'ensemble vide comme une face de dimension -1 de tout simplexe). Tout espace homéomorphe à un simplexe euclidien de dimension p s'appelle tout court *simplexe de dimension p*. Ses points et ses sous-ensembles correspondants à ceux du simplexe euclidien par cette homéomorphie sont dits alors ses *sommets* et *faces* respectivement.

On appelle *polyèdre fini* tout espace topologique compact que l'on peut représenter comme réunion des termes d'une suite finie $\bar{K} = \{\bar{\sigma}_m\}$ qui sont ses sous-espaces fermés $\bar{\sigma}_m$, chacun d'eux étant un simplexe et l'intersection de deux simplexes quelconques étant une face de chacun des deux (done en particulier 1'ensemble vide), et toutes les faces d'un simplexe quelconque de $\bar{K}$ figurant aussi parmi les $\bar{\sigma}_m$ de $\bar{K}$. Une telle représentation $\bar{K}$ du polyèdre fini P est dite sa *subdivision simpliciale* et ce polyèdre, muni d'une subdivision simpliciale $\bar{K}$, constitue ce qu'on appelle un *complexe simplicial fini* ou un *complexe* tout court. Il sera désigné par $(P, \bar{K})$ et P s'appellera *l'espace* de ce complexe.

A tout complexe simplicial fini$(P, \bar{K})$ un schéma abstrait peut être associé, composé d'ensemble fini $A = \{a_0, \cdots, a_N\}$ de tout les sommets de ses simplexes et de famille des sous-ensembles $\{a_{i_0}, \cdots, a_{i_p}\}$ de A. tels que $\overline{(a_{i_0} \ldots a_{i_p})} \in \bar{K}$, c'est-à-dire tels que l'ensemble $\{a_{i_0}, \cdots, a_{i_p}\}$ est celui de tous les sommets d'un simplexe $\bar{\sigma}_m$ de

* *Colloq. Math.*, 1959, 7: 1-8.

$\bar{K}$. Alors

(1) $\{a_{j_0}, \cdots, a_{j_q}\} \subset \{a_{i_0}, \cdots, a_{i_p}\}$ *et* $\overline{(a_{i_0} \ldots a_{i_p})} \in \overline{K}$ entraînent $\overline{(a_{j_0} \ldots a_{j_q})} \in \overline{K}$

et tout $a_i \in A$ est élément d'au moins un de ces sous-ensembles de A. Le schéma assujetti à ces conditions s'appelle *complexe simplicial fini abstrait* ou *complexe abstrait* tout court. Il sera désignó par $K = \{\sigma_m\} == \{(a_{i_0} \ldots a_{i_p})\}$ et les sous-ensembles de A de la forme $\{a_{i_0}, \cdots, a_{i_p}\}$ dont K se compose le seront par $(a_{i_0} \ldots a_{i_p})$. Pour un tel schéma abstrait, les éléments $a_{i_0}, \cdots, a_{i_p}$ s'appelleront *sommets* du simplexe abstrait $\sigma_m == (a_{i_0} \ldots a_{i_p})$ de dimension p et tout simplexe abstrait $(a_{j_0} \cdots a_{j_q})$ tel que $\{a_{j_0}, \cdots, a_{j_q}\} \subset \{a_{i_0}, \cdots, a_{i_p}\}$ sera dit une *face* de dimension q de σ_m. On peut ainsi considérer le couple (P, K), où P est un espace compact et $K = \{\sigma_m\}$ est un complexe abstrait, comme correspondant au complexe simplicial fini$(P, \bar{K})$ dont le polyèdre P est l'espace et dont $\bar{K} = \{\bar{\sigma}_m\}$ est une subdivision simpliciale. Ainsi, à tout $\sigma_m = (a_{i_0} \ldots a_{i_p}) \in K$ vient correspondre un simplexe $\bar{\sigma}_m = \overline{(a_{i_0} \ldots a_{i_p})} \in \bar{K}$ étant un sous-ensemble fermé de P.

La partie de la topologie que l'on peut nommer *topologie du polyèdre fini* a pour l'objet précisément l'étude des propriétés topologiques des polyèdres finis en partant des schémas abstraits associés aux subdivisions simpliciales de ces polyèdres.

Soient $K = \{\sigma_n\}$ et $K' = \{\sigma'_n\}$ deux complexes abstraits associés à deux subdivisions simpliciales d' un même polyèdre fini P. Il peut arriver que, pour tout simplexe abstrait $\sigma_m \in K$, la réunion des $\bar{\sigma}_n \in \bar{K}$ tels que $\bar{\sigma}'_n \subset \bar{\sigma}_n$ coïncide avec $\bar{\sigma}_m$. On dit alors que le complexe (P, K') est une *subdivision simpliciale* du complexe (P, K). Deux complexes (P, K_1) et (P, K_2) ayant le même espace P sont dits *équivalents au sens combinatoire* lorsqu'il en existe une subdivision simpliciale commune. Les propriétés d'un complexe (P, K) qui sont invariantes par rapport à l'équivalence combinatoire portent le nom des *invariants combinatoires* de ce complexe et on appelle *topologie combinatoire* la partie de la topologie qui traite des invariants combinatoires des complexes simpliciaux finis.

Une présomption fondamentale de la topologio combinatoire et qui n'est pas démontrée jusqu'à présent (elle est connue sous le nom allemand de "Hauptvermutung") veut que deux complexes simpliciaux finis dont l'espace commun est un même polyèdre soient toujours équivalents an sens combinatoire. S'il en est ainsi, l'étude de la structure topologique des polyèdres finis se ramenerait à la topologie combinatoire. En particulier, les invariants topologiques d'un polyèdre fini P (c'est-à-dire ses propriétés invariantes par rapport aux transformations biunivoques et bicontinues) coïncideraient avec les invariants combinatoires d'un complexe quelconque (P, K) associé à ce polyèdre de la manière décrite plus haut. Quoi qu'il en soit, les invariants topologiques d'un polyèdre fini sont en tout cas des invariants combinatoires de tout complexe qui lui est associé. O'est l'une des raisons qui justifient l'intérêt à la topologie combinatoire.

Or, la notion de complexe se laissant formuler d'une manière alastraite, il en est de même des notions de subdivision simpliciale et d'équivalence combinatoire. Considérons en effet un complexe abstrait $K_1 = \{\sigma_m\}$ dont $A_1 = \{a_0, \cdots, a_N\}$ est l'ensemble de tous les sommets. Soit $(a_j a_k) \in K_1$ un simplexe de dimension 1. Ajoutons à l'ensemble A_1 un nouveau sommet α et considérons la famille K_2 des sous-ensembles de $A_2 = A_1 \cup (a)$ définie par deux conditions suivantes:

(2) les simplexes de K_1 n'ayant pas $(a_j a_k)$ pour face appartiennent à K_2.

(3) si $(a_j a_k a_{i_0} \ldots a_{i_p}) \in K_1$ (ou bien si l'ensemble $\{a_{i_0}, \cdots, a_{i_p}\}$ est vide), les sous-ensembles de A_2 de la forme $(a_j a a_{i_0} \ldots a_{i_p})$, $(a a_k a_{i_0} \ldots a_{i_p})$ et $(a a_{i_0} \ldots a_{i_p})$ appartiennent à K_2.

On vérifie aisément que K_2 satisfait à ia condition (1) et qu'il est par conséquent un complexe abstrait, A_2 étant l'ensemble de ses sommets et les sous-ensembles de A_2 assujettis à (2) et (3) étant tous ses simplexes. K_2 est dit une *subdivision élémentaire* de K_1.

D'après un théorème fondamental de la topologie combinatoire (voir [1] et [4]), si deux complexes (P, K) et (P, K') ayant le même polyèdre pour l'espace sont équivalents au sens combinatoire, les complexes abstraits K et K' qui leurs sont associés se trouvent en relation suivante:

(4) il existe une suite finie $K = K_1, K_2, \cdots, K_n = K'$ de complexes abstraits tels que de deux termes consécutifs de cette suite l'un est toujours une subdivision élémentaire de l'autre.

La réciproque de ce théorème étant évidente, on est conduit aux définitions suivantes:

Deux complexes abstraits K et K^* sont *isomorphes* s'il existe entre les ensembles A et A^* de leurs sommets une correspondance biunivoque et telle qu'un sous-ensemble de A est un simplexe de K lorsque, et soulement lorsque le sous-ensemble correspondant de A^* est un simplexe de K^*.

Deux complexes abstraits K^* et K'^*, sont *équivalents au sens combinatoire* s'il existe deux complexes abstraits K et K' isomorphes à K^* et K'^* respectivement et qui sont en relation (4).

Les propriétés des complexes abstraits invariantes par rapport à cette équivalence combinatoire s'appellent les *invariants combinatoires* des complexes abstraits.

En vertu du théorème fondamental précité, la topologie combinatoire peut donc être ramenée à l'étude des invariants combinatoires des complexes abstraits, donc de l'équivalence combinatoire entre ces complexes, qui est par définition de nature purement algébrique. Or l'étude de leurs invariants combinatoires, en tant que propriétés invariantes par rapport à leurs subdivisions élémentaires, peut nous en fournir trop au point de vue des besoins de la topologie des polyèdres finis, mais-comme il a été noté-nous fournira en tout cas, à côté des propriétés peut-être topologiquement non-invariantes, donc inutiles, toutes celles qui sont en effet des invariants topologiques distincts.

Passons en revue les invariants combinatoires d'un complexe abstrait, les plus connus à l'état actuel de la science. Donnons-nous à ce but un complexe abstrait K composé de simplexes abstraits σ_m et dont l'ensemble des sommets $A = \{a_0, \cdots, a_N\}$ est supposé ordonné par une relation d'une manière arbitraire, mais définitive, à savoir de la manière suivante : $a_0 < \ldots < a_N$. Soit $d(K)$ la dimension maximum des simplexes de K et $n_r(K)$ le nombre de ceux qui sont de dimension r. Alors,

(I) $d(K)$ est un invariant combinatoire de K; il s'appelle la *dimension* de K.

(II) $\chi(K) = \sum\limits_{r=0}^{d(K)} (-1)^r n_r(K)$ est aussi un invariant combinatoire de K ; il s'appelle la *car actéristique d'Euler-Poincaré* de K.

Un peu plus compliquée que ces deux invariants numériques est la notion de groupe d'homologie de K. Soit $\{\sigma_m^r\}$, où $m \in T_r$, l'ensemble de tous les simplexes de K qui sont de dimension r. étant donné un groupe abélien G, soit C_r (K, G) le groupe abélien de toutes les combinaisons linéaires de la forme $\sum\limits_{i \in T_r} g_i \sigma_i^r$, où $g_i \in G$, l'addition étant entendue au sens naturel. Considérons un système d'homomorphismes

$$\partial_r : \begin{cases} C_0(K, G) \to 0, & \text{pour} \quad r = 0, \\ C_r(K, G) \to C_{r-1}(K, G), & \text{pour} \quad r > 0 \end{cases}$$

tels que l'on ait

$$\partial_r(g\sigma^r) = \sum_{k=0}^{r} (-1)^k g(a_{i_0}, \cdots, a_{i_{k-1}}, a_{i_{k+1}}, \cdots, a_{i_r})$$

pour tout $g \in G$ et tout simplexe $\sigma^r = (a_{i_0} \ldots a_{i_r})$ de K, où $r \geqslant 0$ et $i_0 < \ldots < i_r$. On a alors $\partial_r \partial_{r+1} = 0$ pour tout $r \geqslant 0$, d'où Im $\partial_{r+1} \subset$ Noy ∂_r, et on sait que pour $0 \leqslant r \leqslant d(K)$.

(III) $H_r(K, G) = \mathrm{Noy}\partial_r/\mathrm{Im}\partial_{r+1}$ est un invariant combinatoire de K; il s'appelle son *groupe d'homologie* de dimension r, relatif au groupe des coefficients G.

Ainsi, dans les définitions des invariants (I)—(III) n'interviennent que

(i) la dimension des simplexes de K,

(ii) le nombre de ses simplexes de chaque dimension,

(iii) la relation entre deux simplexes de K dont les dimensions ne diffèrent que par l'unité et qui s'exprime par 0 lorsqu'aucun de ces simplexes n'est une face de l'autre et par -1 ou $+1$ lorsqu'il en est une.

Les autres invariants combinatoires connus, par exemple le groupe fondamental, les groupes de cohomologies, les opérations cohomologiques et même les groupes d'homotopie de Hurewicz, peuvent être définis en partant des propriétés analogues, mais parfois plus compliquées, en faisant intervenir encore

(iv) les relations générales entre les faces qui sont simplexes de K.

Il y a done lieu d'envisager le problème suivant:

Quels sont les invariants combinatoires d'un complexea bstrait K qui se laissent définir à l'aid ede sa structure combinatoire, en particulier à l'aide de ses propriétés (i) - (iv)?

Il semble que c'est Mayer [3] qui a été le premier à aborder le problème dans cet ordre d'idées, en établissant le théorème qui peut être formulé comme suit:

Théorème 1 (Mayer) *Tout invariant combinatoire de K ayant la forme*

(IV) $$\vartheta(K) = \sum_{r=0}^{d(K)} a_r n_r(K)$$

est un multiple de la caractéristique (II) *d'Euler-Poincaré.*

Le but de cette communication① est d'envisager le même problème sous un aspect moins général, à savoir limité à ces invariants *enumériques*, et d'indiquer une méthode générale de les définir à l'aide, des propriétés (i)—(iv).

En vue d'appliquer (iv), introduisons les nombres suivauts:

(5) $$n_{rs,k}(K)\,\text{où}\,0 \leqslant \begin{pmatrix} r \\ s \end{pmatrix} \leqslant d(K) \quad \text{et} \quad -1 \leqslant k \leqslant \min(r, s),$$

qui est celui des couples ordonnés composés d'un simplexe de K de dimension r et d'un autre de dimension s qui ont *exactement* une face commune de dimension k;

(6) $$n_{r_1 \ldots r_t}(K)\,\text{où}\,t \geqslant 2,$$

① Elle a été faite à Wroclaw, le 30. XII. 1957, au Séminaire de topologie de l'Institut Mathématique de l'Académie Polonaise des Sciences.

qui est le nombre des t-tuples $(\sigma_1, \cdots, \sigma_i)$ de simplexes de K tels que σ_g est de dimension r_q pour $1 \leqslant q \leqslant t$ et qu'aucun sommet n'est comnun à tous les σ_q d'un même t-tuple.

On a en particulier

$$n_{rr,r}(K) = n_r(K), \tag{7}$$

$$n_{r_1 r_2,-1}(K) = n_{r_1 r_2}(K). \tag{8}$$

Posons

$$\varphi(K) = \sum_{r,s=0}^{d(K)} a_{rs,k} n_{rs,k}(K), \tag{V}$$

$$\psi_i(K) = \sum_{r_1,\cdots,r_t=0}^{d(K)} a_{r_1 \ldots r_t} n_{r_1,\cdot,r_t}(K). \tag{VI}$$

Par analogie an théorème de Mayer, on a les théorèmes de Wu [5]et de Lee [2] permettant de trouver les invariants combinatoires qui sont des combinaisons linéaires de la forme (V) et (VI) des nombres (5) et (6).

Notons d'abord que la caractéristique d'Euler-Poincaré (II) est de la forme (V), car on a d'après (7)

$$\chi(K) = \sum_{r=0}^{d(K)} (-1)^r n_{rr,r}(K),$$

et qu'il en est de même du carré de cette earactéristique

$$\chi^2(K) = \sum_{r,s=0}^{d(K)} (-1)^{r+s} n_{rs,k}(K). \tag{9}$$

Quant à (VI), il s'agit du suivant

Théorème 2 (Wu) *Les nombres*

$$\chi_t^+(K) = \sum_{r_1,\cdots,r_t=0}^{d(K)} (-1)^{r_1+\ldots+r_t} n_{r_1\ldots r_t}(K) \tag{10}$$

sont des invariants combinatoires du complexe abstrait K, et même des invariants topologiques du polyèdre fini correspondant.

Pour la genèse des invariants (10) et la démonstration de leur invariance, on voudra consulter le travail [5].

Le cas particulier do (10) pour $t = 2$, c'est-à-dire le nombre qui est gal d'après (8) à

$$\chi_2^+(K) = \sum_{r,s=0}^{d(K)} (-1)^{r+s} n_{rs,-1}(K),$$

est done en même temps de la forme (V). Or on a le

Théorème 3 (Lee) *Tous les invariants combinatoires de K de la forme* (V) *sont des combinaisons linéaires des invariants* $\chi(K)$, $\chi^2(K)$, et $\chi_2^+(K)$.

Pour illustrer la méthode employée par Lee dans la démonstration de son théorème, je vais citer ici sa démonstration du théorème 1 qui est bien plus simple et naturelle quo la démonstration originale de Mayer. Ce théorème étant trivialement vrai pour $d(K) = 0$, procédons par induction en admettant qu'il est vrai pour tout $d(K) < d$. Par conséquent, pour $d(K) = d$, tout invariant combinatoire de K qui est de la forme (IV) se laisse représenter par la formule

$$\vartheta(K) = a\sum_{r=0}^{d-1} (-1)^r n_r(K) + bn_d(K) = a\chi(K) + [b - (-1)^{n_d} a] n_d(K).$$

K' étant une subdivision élémentaire de K, on a donc la même formule

$$\vartheta(K') = a\chi(K') + [b - (-1)^{n_d} a] n_d(K').$$

On a $\vartheta(K) = \vartheta(K')$ par suite de l'invariance admise de $\vartheta(K)$, et en même temps $\chi(K) = \chi(K')$ par suite de l'invariance de $\chi(K)$. Les deux formules entraînent done

$$[b - (-1)^{n_d} a][n_d(K) - n_d(K')] = 0$$

pour tout complexe abstrait K de dimension $d(K) \leqslant d$ et pour toutes les subdivisions élémentaires K' de K. En particulier, pour K et K' tels que $n_d(K) \neq n_d(K') \neq 0$, on aura par conséquent $b = (-1)^{n_d} a$, d'où $\vartheta(K) = a\chi(K)$, ce qui achève la démonstration du théorème de Mayer.

Aucun théorème analogue à celui de Lee, mais concernant les invariants combinatoires de la forme (VI), ne m'est connu jusqu'à présent. Il semble cependant qu'il existe bien des invariants de cette forme, différents de $\chi_r^+(K)$, et qu'ils peuvent être trouvés dela même manière. Enfin, en introduisant d'autres nombres, analogues à (5) et (6), et en cherchant à établir des invariants combinatoires numériques s'exprimant par leurs combinaisons linéaires, on se voit bienen possession d'une méthode générale qui se prête à les définir systématiquement an moyen des propriétés (i)—(iv) de complexes abstraits.

Ajoutons, pour terminer, deux remarques:

1. Parmi les invariants de nature algébrique (e'est-à-dire qui s'expriment par des nombres, groupes, homomorphismes etc.), les invariants topologiques les plus importants sont, presque tous, en même temps des invariants du type d'homotopie. Tels sont les groupes d'homologie, ceux d'homotopie et autres. Il y a toutefois des exceptions, comme par exemple la dimension-et, plus généralement, les invariants de la forme (10) -qui sont des invariants topologiques des espaces de complexes abstraits sans en être des invariants du type d'homotopie. On a en effet pour le segment rectiligne I et le triangle $\varDelta$, qui sont de même type d'homotopie, $\chi_2^+(I) = 2$ et $\chi_2^+(\varDelta) = 0$. Ainsi, ce sont des invariants topologiques d'un genre nouveau, promettant d'être applicables avec avantage dans des problèmes de nature topologique, mais non homotopique, donc tels que celui de la classification topologique des espaces, de leur isotopie, du plongement d'un espace dans un autre etc. Ces invariants paraissent en outre les plus simples dans leur genre (excepté la dimension).

2. La topologie combinatoire, qui coucentrait sur elle l'attention des topologues pendant la première trentaine d'années de ce siècle, semble cesser peu à peu de susciter leur intérêt, malgré que *tous* les invariants topologiques des polyèdres finis se trouvent parmi les invariants combinatoires des complexes abstraits. I1 est done peut-être désirable de poursuivre les recherches dans ce domaine. Ce qui vient d'être exposé iei est déstiné à montrer qu'*il y a une possibilité de faire avancer ces recherches d'une facon méthodique*, en particulier dans l'ordre d'idées des théorèmes établis par Mayer et Lee pour des invariants numériques particuliers.

Travaux cités

[1] Alexander J W. *The combinatorial theory of oomplexes.* Annals of Mathematics, 1950, 31: 292-320.

[2] Lee K C. *über die Eindeutigkeit von einigen kombinatorischen Invarianten endlichen Komplexes.* Science Record, 1957, 1: 279-281.

[3] Mayer W. *A new homology theory.* Annals of Mathematics, 1942, 43: 594-605.

[4] Reidemeister K. *Einführuny in die kombinatorisohe Topologie.* Braunschweig, 1932.

[5] Wu W T. *Topological invariants of new type of finite polyhedrons.* Acta Mathematica Sinica, 1953, 3: 261-289.

32. On Certain Invariants of Cell-Bundles*

A locally trivial fibre bundle with numerical space R^n of dimension n as fibres will be called an *n-cell-bundle*, or simply a *cell-bundle*. More precisely, an *n-cell-bundle* is a triple (E, f, B) in which f is a continuous mapping of the space E onto the space B such that each point b of B has a neighbourhood U_b with its inverse image homeomorphic under a certain map φ to the product space $U_b \times R^n$ with $\varphi(f^{-1}(b')) \equiv (b') \times R^n$, for any $b' \in U_b$. Remark that we suppose nothing about the structural group of the bundle. In this paper we shall describe a method of introducing, invariants of such bundles which turn out to be Stiefel-Whitney classes or other characteristic, classes in case that the bundle has the linear group as its structural group.

§ 1. Let p be a fixed prime number. Let $\widetilde{X}$ be a Hausdorff space and t a periodic homeomorphism of $\widetilde{X}$ onto itself of period p and without fixed points. Let $X = \widetilde{X}/t$ be the corresponding modular space and π the projection. Then we may associate to the pair $(\widetilde{X},\ t)$ or the triple $(\widetilde{X},\ X,\ \pi)$ a system of *Smith classes* $\Phi^i_{(p)}(\widetilde{X}, t) \in H^i(X, I_{(i)})$, where $I_{(i)}$ is the group of integers I or the group of integers mod $p\, I_p$ according as i is even or odd (cf.e.g. [1]). Denote by ρ_p the reduction mod p, and put $\rho_p \Phi^i_{(p)}(\widetilde{X}, t) = \Phi^i_p(\widetilde{X}, t)$, then we have

$$\Phi^{2i}_p(\widetilde{X}, t) \cup \Phi^{2j}_p(\widetilde{X}, t) = \Phi^{2i+2j}_p(\widetilde{X}, t),$$

$$\Phi^{1}_p(\widetilde{X}, t) \cup \Phi^{2i}_p(\widetilde{X}, t) = \Phi^{2i+1}_p(\widetilde{X}, t),$$

$$\Phi^{1}_p(\widetilde{X}, t) \cup \Phi^{1}_p(\widetilde{X}, t) = \begin{cases} \Phi^{2}_p(\widetilde{X}, t), & p = 2, \\ 0, & p > 2. \end{cases}$$

If $(\widetilde{X}',\ t')$ or $(\widetilde{X}',\ X',\ \pi')$ be a second pair or triple and $\tilde{j} : \widetilde{X}' \to \widetilde{X}$ and $j : X' \to X$ are continuous maps with $t\tilde{j} = \tilde{j}t', \pi\tilde{j} = j\pi'$, then

$$j * \Phi^i_{(p)}(\widetilde{X}, t) = \Phi^i_{(p)}(\widetilde{X}, t').$$

As a particular case, let X be a Hausdorff space and $\widetilde{X}^*_p$ be the subspace of the p-fold topological product $\widetilde{X}_p = \underbrace{X \times \cdots \times X}_{p}$ consisting of all points $(x_1, \cdots, x_p)$ with x_i

* *Sci. Record(N.S.)*, 1959, 3: 137-142.

not all identical. Let t_p be the cyclic transformation $t_p(x_1,\cdots,x_p)=(x_2,\cdots,x_p,x_1)$ and X_p^* the modular space of $\widetilde{X}_p^*$ with respect to t_p. We shall put

$$\Phi^i_{(p)}(\widetilde{X}_p^*,t_p)=\Phi^i_{(p)}(X)\in H^i(X_p^*,I_{(i)}),$$
$$\Phi^i_p(\widetilde{X}_p^*,t_p)=\Phi^i_p(X)\in H^i(X_p^*,\mathrm{mod}p).$$

In particular, when $X=R^n$, it has been shown (cf. [1]) that the cohomology group mod p of X_p^* is generated by the classes $\Phi^i_p(X)$, $0\leqslant i<n(p-1)$, while

$$\Phi_p^{n(p-1)}(R^n)=0.$$

Consider now an n-cell-bundle $(E,\ f,\ B)$ and the subspace E_p^{**} (resp.$\tilde{E}_p^{**}$) of E_p^* (resp. $\tilde{E}_p^*$) which is the union of all $(F_b)_p^*$ (resp.$(\tilde{F}_b)_p^*$), where $F_b=f^{-1}(b)$, $b\in B$. For a fixed $b\in B$, let the inclusion map of $(F_b)_p^*$ in E_p^{**} and $(\tilde{F}_b)_p^*$ in $\tilde{E}_p^{**}$ be j and $\tilde{j}$ respectively. Denote also the cyclic transformations in $\tilde{E}_p^*$ and $(\tilde{F}_b)_p^*$ by t_E and t_F respectively, then $t_F\tilde{j}=\tilde{j}t_E$ and $t_E(\tilde{E}_p^{**})=\tilde{E}_p^{**}$ so that

$$j^*\Phi^i_p(\tilde{E}_p^{**},t_E)=\Phi^i_p(F_b).$$

As $F_b\equiv R^n$, we see by the stated properties about the classes $\Phi^i_p(F_b)$ that the bundle $(E_p^{**},\ \pi,\ B)$ with projection π enjoys a certain property which permits to apply the so-called theorem of Leray-Hirsch (cf. e.g. [2] p. 472). It follows that the cohomology ring mod p of E_p^{**} has an additive basis consisting of classes of the form $\Phi^i_p(\tilde{E}_p^{**},t_E)\cup\pi*U$, where $0\leqslant i<n(p-1)$, and $\{U\}$ is an additive basis of the cohomology ring mod p of B. In particular, we have

$$\Phi_p^{n(p-1)}+\sum_{i=1}^{n(p-1)}\Phi_p^{n(p-1)-i}\cup\pi^*Q_p^1=0, \tag{1}$$

in which $\Phi^i_p=\Phi^i_p(\tilde{E}_p^{**},t_E)$, and

$$Q_p^i\in H^i(B,\mathrm{mod}p),\quad 1\leqslant i\leqslant n(p-1) \tag{2}$$

are uniquely determined by the formula (1). Thus to any n-cellbundle $(E,\ f,\ B)$ for which nothing about structural group is assumed may be associated for each prime p a system of invariants Q_p^i defined by means of (1) (cf. also [5]).

§ 2. Suppose now the n-cell-bundle $(E,\ f,\ B)$ possesses an orthogonal group as its structural group. Then we may define a system of Stiefel-Whitney classes $W^i\in H^i$ $(B,\ \mathrm{mod}\ 2)$, $0\leqslant i\leqslant n$, as well as a system of Pontrjagin classes $P^{4k}\in H^{4k}$ (B),

$0 \leqslant k \leqslant n' = [n/2]$. To (E, f, B) we may associate a projective-space-bundle, say (P, g, B), in an evident manner and also for each odd prime p a lens-spacebundle (L_p, g_p, B) as follows. Form the p-fold Whitney product $(\tilde{E}_p, f_p, B)$ of the bundle (E, f, B) and let $(\tilde{L}_p, \tilde{g}_p, B)$ be the associated sphere bundle of which the fibres are the spheres of fixed radius in the fibres of $(\tilde{E}_p, f_p, B)$. Let t_p be the cyclic transformation in $\tilde{E}_p$ defined by $t_p(x_1, \cdots, x_p) = (x_2, \cdots, x_p, x_1)$, where $x_i \in f_p^{-1}(b)$, $b \in B$. Then $t_p(\tilde{L}_p) = \tilde{L}_p$ and L_p is the modular space of $\tilde{L}_p$ with respect to t_p. As shown in [3] and [4], there exists in P a system of classes $U_2^i \in H^i$ $(P, \bmod 2)$ and in L_p a system of classes $U_p^i \in H^i$ $(L_p, \bmod p)$ such that

$$\sum_{i=0}^{n} U_2^{n-i} \cup g^* W^i = 0 \, in \, P \tag{3}$$

and

$$\sum_{k=0}^{n'q} U_p^{n(p-1)-4k} \cup g_p^* R_p^{4k} = 0 \,\mathrm{in}\, L_p, \tag{4}$$

in which

$$R_p^{4k} = (-1)^k \sum_{k_1+\cdots+k_q=k} 1^{n-2k_1} \cdot 2^{n-2k_2} \cdots q^{n-2k_q} \cdot P_p^{4k_1} \cup P_p^{4k_2} \cup \cdots \cup P_p^{4k_q}, \tag{5}$$

$$q = (p-1)/2, \tag{5'}$$

and P_p^{4k} are the mod p reduced Pontrjagin classes. In particular, we have $R_3^{4k} = P_3^{4k}$.

By definition, we see that P and L_p are subspaces of E_2^{**} and $E_p^{**} (p>2)$ respectively, and, if the inclusion map be denoted by j, we have

$$j^* \Phi_p^i = U_p^i.$$

From (1) we get therefore

$$\sum_{i=0}^{n} U_2^{n-i} \cup g^* Q_2^i = 0 \,\mathrm{in}\, P,$$

and

$$\sum_{k=0}^{n(p-1)} U_p^{n(p-1)-k} \cup g_p^* Q_p^k = 0 \,\mathrm{in}\, L_p.$$

By comparing with (3) and (4), we see that in the case of an n-cell-bundle possessing orthogonal group as structural group:

$$Q_2^i = W^i, \qquad \text{for} \quad p = 2,$$

$$Q_p^i = \begin{cases} 0, & i \not\equiv 0 \bmod 4, \\ R_p^{4k}, & i = 4k, \end{cases} \qquad \text{for} \quad p > 2.$$

Thus the invariants Q_p^i introduced in §1 are reduced in this case to either the usual Stiefel-Whitney classes or combinations of mod p Pontrjagin classes of the bundle.

§ 3. Let us consider an n-dimensional manifold M and the diagonal Δ in $M \times M$. A neighbourhood V of Δ in $M \times M$ will be called *cellic* if for any point $x \in M$, the set of all points $(x, y) \in V$ is homeomorphic to an n-cell E_x, and if each point $x \in M$ possesses a neighbourhood U_x with the union of all $E_{x'}$, $x' \in U_x$, homeomorphic in an evident manner to $E_x \times U_x$. The n-manifold M will be called *cellic* if any neighbourhood V' of the diagonal Δ in $M \times M$ contains a cellic neighbourhood of Δ in $M \times M$. Among the cellic manifolds we may cite differentiable manifolds. The author is ignorant of the existence of non-cellic manifolds.

For any cellic neighbourhood V of the n-dimensional cellic manifold M, we have an n-cell-bundle (V, f, M) where f is defined by $f(x,y) = x$, $x \in M$, $(x, y) \in V$. By (1) of §1, we may then define certain classes $Q_{p,V}^i$ of M such that

$$\Phi_{p,V}^{n(p-1)} + \sum_{i=1}^{n(p-1)} \Phi_{p,V}^{n(p-1)-i} \cup \pi_V^* Q_{p,V}^i = 0, \tag{6}$$

in which $\Phi_{p,V}^i$, π_V, etc. have analogous meaning as before. If V, V' are two cellic neighbourhoods, let V'' be a cellic neighbourhood contained in both V and V'. Denote the inclusion maps of V'' in V and V'' by j and j' respectively, then we have $j{*}\Phi_{p,V}^i = \Phi_{p,V''}^i$, $j' * \Phi_{p,V'}^i = \Phi_{p,V''}^i$, and $\pi_{V''} = \pi_V j = \pi_{V'} j'$. Hence on applying $j*$ and $j'*$ to $(6)_V$ and $(6)_{V'}$ respectively we get

$$\Phi_{p,V''}^{n(p-1)} + \sum_{i=1}^{n(p-1)} \Phi_{p,V''}^{n(p-1)-i} \cup \pi_{V''}^* Q_{p,V}^i = 0,$$

and

$$\Phi_{p,V''}^{n(p-1)} + \sum_{i=1}^{n(p-1)} \Phi_{p,V''}^{n(p-1)-i} \cup \pi_{V''}^* Q_{p,V'}^i = 0.$$

It follows that $Q_{p,V}^i = Q_{p,V'}^i = Q_{p,V''}^i$ so that $Q_{p,V}^i$ is independent of V and may be denoted simply by $Q_p^i = Q_p^i(M)$. Thus, for any cellic manifold M we may define for each prime p a system of classes $Q_p^i \in H^i$ (M, mod p) which are, by their very definition, topological invariants of the manifold M. In particular, if M is differentiable and D any differential structure of M, then cellic neighbourhoods V of M may be defined by means of the differential structure D. The corresponding n-cell-bundle (V, f, M) may be shown to be equivalent to the tangent bundle of M with respect to D and hence

$$Q_p^i = Q_{p,V}^i = W^i(D), \qquad \text{for} \quad p = 2,$$

$$Q_p^i = Q_{p,V}^i = \begin{cases} 0, & i \not\equiv 0 \mod 4, \\ R_p^{4k}(D), & i = 4k, \end{cases} \qquad \text{for} \quad p > 2,$$

in which $W^i(D)$ and $R_p^{4k}(D)$ are the Stiefel-Whitney classes and combinations of mod p Pontrjagin classes $R_p^{4k}(D)$ (cf. (5)) of the differentiable manifold M with respect to the differential structure D. It follows that *the classes $W^i(D)$ and $R_p^{4k}(D)$ are all topological invariants of the manifold M*. Thus the above developments furnish an intrinsic *topological* definition of the classes $W^i(D)$ and $R_p^{4k}(D)$ and give a simple alternative proof of their topological invariance (cf. also [5]).

§ 4. As a further application let us suppose that the cellic manifold M of dimension n is imbeddable in the Euclidean space R^N of dimension N. Let $f: M \to R^N$ be any such imbedding. For any cellic neighbourhood V of M, let $\tilde{f}_p : \tilde{V}_p^{**} \to (\tilde{R}^N)_p^*$ be the map defined by $\tilde{f}_p((x, y_1), \cdots, (x, y_p)) = (f(y_1), \cdots, f(y_p))$, where $x \in M$, $(x, y_i) \in V$ and $((x, y_1), \cdots, (x, y_p)) \in \tilde{V}_p^{**}$.Let t_V and t_R be the cyclic transformations in $\tilde{V}_p^{**}$ and $(\tilde{R}^N)_p^*$ respectively. Then $\tilde{f}_p t_V = t_R \tilde{f}_p$ so that $\tilde{f}_p$ induces a map $f_p : V_p^{**} \to (R^N)_p^*$, and we have $f_p^* \Phi_p^i \ ((\tilde{R}^N)_p^*, t_R) = \Phi_p^i(\tilde{V}_p^{**}, t_V)$. Since $\Phi_p^i((\tilde{R}^N)_p^*, t_R) = 0$ for $i \geqslant N(p-1)$, we get

$$\Phi_p^i = \Phi_p^i(\tilde{V}_p^{**}, t_V) = 0 \quad \text{for} \quad i \geqslant N(p-1). \tag{7}$$

Let

$$\Phi_p^{n(p-1)+j} = \sum_{i=1}^{n(p-1)} \Phi_p^{n(p-1)-i} \cup \pi^* Q_p^{i,j}, \quad j \geqslant 0,$$

in which the classes $Q_p^{i,j} \in H^{i+j}$ (M, mod p) are uniquely determined, since (V_p^{**}, π, M) is a bundle verifying the property of Leray-Hirsch. By (1) the classes $Q_p^{i,j}$ may be determined in terms of Q_p^i recurrently.In particular, we have for $p = 2$, $Q_2^{i,j} = \bar{W}^{i+j}$, where $\bar{W}^i$ are determined by means of the following recurrence relations:

$$\sum_{i+j=k} Q_2^i \cup \bar{W}^i = \begin{cases} 0, & k > 0 \\ \text{mod 2 unit class}, & k = 0. \end{cases}$$

By (7) we get therefore the following theorem:

If the cellic manifold M of dimension n is imbeddable in R^N, then

$$Q_p^{i,j} = 0, \quad for \quad j \geqslant (N-n)(p-1).$$

In particular, for $p = 2$, we get $\bar{W}^i = 0$. for $i \geqslant N - n$, cf. [6].

References

[1] Wu Wen-tsün. *Science Record*, New Ser., 1957, I: 377-380.

[2] Serre J P. *Annals of Math.*, 1951, 54: 425-505.

[3] Wu Wen-tsün. *Scientia Sinica*, 1954, 3: 353-367.

[4] ——. *Acta Mathematica Sinica*, 1955, 5: 37-63.

[5] Thom R. *Ann. Ec. Norm. Sup.* 1952, 3(69): 110-182.

[6] Wu Wen-tsün. *Scientia Sinica*, 1958, 7: 365-387.

33. On the Isotopy of a Finite Complex in a Euclidean Space I*

§ 1. In a previous paper[1] we have proved that any two C^r-imbeddings ($1 \leqslant r \leqslant \infty$) of a compact C^r-manifold of dimension n in a Euclidean $(2n+1)$ -space R^{2n+1} are C^r-isotopic when $n > 1$, in contrast to the case $n = 1$ which forms the subject of the knot theory. In this paper and in the succeeding one we shall study the isotopy of a finite simplicial complex K of dimension n imbedded in a Euclidean space R^N of arbitrary dimension N, and a necessary and sufficient condition for the "linear isotopy" of any two "linear imbeddings" of such an n-complex in a Euclidean $(2n+1)$ -space when $n > 1$ will be given.

§ 2. Let E be a complex with an operator t without fixed cells and having a period 2. Denote by ρ one of the two operators $d = 1 - t$ and $s = 1 + t$ and then the other by $\bar{\rho}$. Considering ρ, $\bar{\rho}$ and t as operators in the integral cochain groups $C^r(E)$ and the integral cocycle groups $Z^r(E)$, we may define *special cohomology groups.* ${}^\rho H^r(E)$ of P. A. Smith as the groups ${}^\rho Z^r(E)/\delta\bar{\rho}C^{r-1}(E) = \mathrm{Ker}\rho \cap Z^r(E)/\delta\bar{\rho}C^{r-1}(E)$. If E is the singular complex $S(X)$ of a space X with a homeomorphism t without fixed points of period 2 which induces an operator in $S(X)$ denoted by the same letter t, then the groups ${}^\rho H^r(S(X))$, etc., will be denoted simply by ${}^\rho H^r(X)$, etc. Using the exact sequence of Smith-Richardson about special groups we may prove the following

Lemma 1 If S^{N-1} is an $(N-1)$ -sphere ($N > 1$) and t the antipodal map, we have

$$ {}^{\rho_N}H^{N-1}(S^{N-1}) \approx Z \quad \text{(the group of integers)}, $$
$$ {}^{\bar{\rho}_N}H^{N-1}(S^{N-1}) \approx Z_2 \quad \text{(the group of integers mod 2)}, $$

in which $\rho_N = 1 + (-1)^{N-1}t$, $\bar{\rho}_N = 1 + (-1)^N t$.

A generator of ${}^{\rho_N}H^{N-1}(S^{N-1})$ is given by $\bar{\rho}_N \sum^{N-1}$, where $\sum^{N-1}$ is a generator of $H^{N-1}(S^{N-1})$, and $\rho : H^r(S^{N-1}) \to {}^{\bar{\rho}}H^r(S^{N-1})$ is the homomorphism induced by the correspondence $u \to \rho u$, for any r-cocycle u of S^{N-1}. If S^{N-1} is oriented, then a definite generator of ${}^{\rho_N}H^{N-1}(S^{N-1})$ corresponding to that orientation may be chosen

* *Sci. Record(N.S.)*, 1958, 3: 342-351.

which will be denoted by ${}^{\rho_N}\sum^{N-1}$ in what follows.

§ 3. For any space X let $\tilde{X}^*$ be the *two-fold reduced product* which is the topological product $X \times X$ with its diagonal removed. Let t_X be the cyclic transformation in $\tilde{X}^*$ defined by $t_X(x, y) = (y, x)$, $x, y \in X$, $x \neq y$. Then the special groups ${}^{\rho}H^r(\tilde{X}^*)$ will be denoted simply by ${}^{\rho}\tilde{H}^r_{(2)}(X)$.

Let f be a topological imbedding of X in an oriented Euclidean space R^N of dimension N. Then f will *induce* a map $\bar{f}$ of $\tilde{X}^*$ in the unit sphere S^{N-1} of R^N, such that $\bar{f}\ (x, y)$ is the point of intersection with S^{N-1} of the ray from the origin of R^N and parallel to the directed line from $f(x)$ to $f(y)$. Let us orient S^{N-1} coherently as R^N, denote by t the antipodal map of S^{N-1}, and let ${}^{\rho_N}\sum^{N-1} \in {}^{\rho_N}H^{N-1}\ (S^{N-1})$ be as in §2. As $\bar{f}t_X = t\bar{f}$, $\bar{f}$ will induce a homomorphism

$$ {}^{\rho_N}\bar{f}^* : {}^{\rho_N}H^{N-1}(S^{N-1}) \to {}^{\rho_N}H^{N-1}(\tilde{X}^*) = {}^{\rho_N}\tilde{H}^{N-1}_{(2)}(X). $$

Definition ${}^{\rho_N}\Theta^{N-1}_f(X) = {}^{\rho_N}\bar{f}^* {}^{\rho_N}\sum^{N-1} \in {}^{\rho_N}\tilde{H}^{N-1}_{(2)}(X)$ will be called the *isotopy class* of the imbedding f of X in the oriented R^N.

Remark. The class $\Theta^{N-1}_f(X)$ of an imbedding f of X in R^N introduced in [2] §7, when multiplied by 2, is the ordinary cohomology class associated with the special class ${}^{\rho_N}\Theta^{N-1}_f(X)$.

Two topological imbeddings f, g of a space X in an R^N are said to be *isotopic* if there exists a continuous mapping F of $X \times I$ in R^N where $I = [0, 1]$ such that, defining $F_t : X \to R^N$ by $F_t(x) = F(x, t)$, $x \in X$, we have $F_0 \equiv f$ and $F_1 \equiv g$, and each F_t is a topological imbedding. The mapping F is called an *isotopy* from f to g. The following theorem is trivial to prove:

Theorem 1 *${}^{\rho_N}\Theta^{N-1}_f(X)$ is an isotopy invariant of topological imbeddings f of X in an oriented R^N, i.e., for two imbeddings f and g of X in the oriented R^N to be isotopic, it is necessary that ${}^{\rho_N}\Theta^{N-1}_f(X) = {}^{\rho_N}\Theta^{N-1}_g(X)$.*

§ 4. Any finite simplicial n-complex K will be considered as a Euclidean complex in a certain Euclidean space of a sufficiently high dimension and the space of K will be denoted by $|K|$. A pair of simplexes σ and τ of K will be said to be *disjoint* if they have no vertices in common. With the complex K we may associate the *two-fold reduced product* $\tilde{K}^*_2 = \tilde{K}^*$ consisting of all cells $\sigma \times r$ of the product complex $K \times K$ for which (σ, τ) are pairs of disjoint simplexes in K. As shown in [3] (cf. also [4]) $|\tilde{K}^*|$ is a deformation retract of $|\tilde{K}|^*$ with a deformation commutative with the cyclic transformation $t_{|K|}$ so that the groups ${}^{\rho}H^r(|\tilde{K}^*|) = {}^{\rho}\tilde{H}^r_{(2)}(|K|)$ and ${}^{\rho}H^r(\tilde{K}^*) = {}^{\rho}\tilde{H}^r_{(2)}(K)$ may be canonically identified. Note that the special groups

${}^{\rho}H^{r}(\tilde{K}^*)$ are defined with respect to transformations t_K of period 2 in $\tilde{K}^*$ induced by $t_{|K|}$ such that $t_K(\sigma\times\tau) == (-1)^{\dim^{\sigma}\dim^{\tau}}(\tau\times\sigma)$.

If f is a topological imbedding of $|K|$ in an oriented R^N, then the isotopy class ${}^{\rho_N}\Theta_f^{N-1}(|K|)$ introduced in §3, when identified to be a class of $\tilde{H}^{\rho_N{}^{N-1}_{(2)}}(K)$, will also be denoted by ${}^{\rho_N}\Theta_f^{N-1}(K)$.

§ 5. A topological imbedding (resp. a continuous mapping) f of $|K|$ in R^N will be said to be a *linear imbedding* (resp. a *linear mapping*) of K in R^N if f is linear on each simplex of K. Two linear imbeddings (resp. linear mappings) f, g of K in R^N will be said to be *normal* with respect to a pair of disjoint simplexes σ, τ of K if (i) $f(\sigma)$ and $f(\tau)$ are disjoint, so are $g(\sigma)$ and $g(\tau)$, and (ii) for any points $x\in|\sigma|$, $y\in|\tau|$, the line joining $f(x)$ and $f(y)$ is not parallel to the line joining $g(x)$ and $g(y)$. The two linear imbeddings (resp. mappings) f, g of K in R^N are said to be *normal* with respect to K if they are normal with respect to any pair of disjoint simplexes σ, τ of K for which $\dim\sigma+\dim\tau\leqslant N-2$. Two linear imbeddings f, g of K in R^N will be said to be *linearly isotopic* if there exists an isotopy F of f, g such that F is a linear mapping of a certain simplicial subdivision of $K\times I$ in R^N.

Let R^{N+1} be the product space $R^N\times L$ where L is a line $-\infty<t<+\infty$. The subspace $R^N\times(t)$ will be denoted by R_t^N. For any two continuous mappings f and g of $|K|$ in R^N, we may *associate* the mapping $F:|K|\times L\to R^{N+1}$ defined in the manner that $F(|K|\times(t))\subset R_t^N, F(x,0)=(f(x),0)$, $F(x,1)=(g(x),1)$, and that F maps $(x)\times L$ to the line joining F $(x,$ $0)$ and F $(x,$ $1)$, where $x\in|K|$. Let R^N be oriented previously and R^{N+1} be oriented as the product space $R^N\times L$, where L is oriented by increasing t. Let f, g be two linear imbeddings of K in R^N which are normal with respect to K and $F:|K|\to R^{N+1}$ be their associated mapping. Then with respect to the above chosen orientation of R^{N+1}, the index of intersection $\phi(F(\sigma\times I),\ F(\tau\times I))$ for any $(n-1)$ -cell $\sigma\times\tau$ of $\tilde{K}^*$ is well-defined owing to the normality of the pair f, g. Hence we may define an $(N-1)$ -dimensional cochain $\theta_{f,g}^{N-1}$ in the complex $\tilde{K}^*$ by setting

$$\theta_{f,g}^{N-1}(\sigma\times\tau)=\phi(F(\sigma\times I),F(\tau\times I)),\tag{1}$$

where $\sigma\times\tau\in\tilde{K}^*$, $\dim\sigma+\dim\tau=N-1$. It may be directly verified that $\rho_N\theta_{f,g}^{N-1}=0$ and $\delta\theta_{f,g}^{N-1}=0$ so that $\theta_{f,g}^{N-1}$ is a ρ_N-cocycle of $\tilde{K}^*$with respect to t_K.

Definition The special cohomology class ${}^{\rho_N}\Theta_{f,g}^{N-1}(K)\in{}^{\rho_N}\tilde{H}_{(2)}^{N-1}(K)$ of $\theta_{f,g}^{N-1}$ will be called the *isotopy class* of the normal pair of linear imbeddings f and g of K in the oriented R^N.

Remark. $\theta_{f,g}^{N-1}$ is still well-defined by (1) even if (f, g) is only a normal pair of linear *mappings*, but with $f(\sigma)$, $f(\tau)$ and also $g(\sigma)$, $g(\tau)$ disjoint for any $(N-1)$-cell $\sigma \times \tau$ of $\tilde{K}^*$.

As a preliminary for our main objective considered in the succeeding paper, we are in need of the following two theorems.

Theorem 2 *Arbitrarily near to two linear imbeddings f, g of a finite simplicial complex K in R^N there exist normal pairs of linear imbeddings f' and g' of K in R^N which are linearly isotopic to f and g respectively.*

Theorem 3 *Given two linear imbeddings f, g of a finite simplicial complex K in R^N. Then for any normal pair of linear imbeddings f', g' of K in R^N linearly isotopic to f and g respectively, the isotopy class $_{\rho_N}\Theta_{f',g'}^{N-1}(K)$ is independent of the normal pair f', g' chosen.*

§ 6. The proof of Theorem 2 is achieved by successive modifications of f, g near each vertex of K. The modifications are based on the following two lemmas, both of which are easily proved by simple dimensionality considerations.

Lemma 2 Let K be the complex consisting of two disjoint simplexes $\sigma = (a_0, \cdots, a_p)$ and $\tau = (b_0, \cdots, b_q)$ as well as their faces. Let f, g be two linear imbeddings (resp. linear mappings) of K in R^N of a dimension $N \geqslant p+q+2$. If f, g are normal with respect to the complex of disjoint simplexes σ and $\xi = (b_1, \cdots, b_q)$, then arbitrarily near to $f(b_0)$ there exist points b_0' such that, defining a linear imbedding (resp. linear mapping) f' of K in R^N by setting $f'(a_i) = f(a_i)$, $f'(b_i) = f(b_i)$, $j \neq 0$ and $f'(b_0) = b_0'$, the pair f', g will be normal with respect to the complex K.

Lemma 3 Let f be a linear imbedding in R^N of a finite simplicial complex K of which the vertices are $a_1, \cdots, a_m$. Then any linear mapping f' of K in R^N for which $f'(a_i)$ are sufficiently near to $f(a_i)$ is also a linear imbedding of K in R^N linearly isotopic to f.

§ 7. Proof of Theorem 3.

Let $J = [0, 1]$ and $M = K \times J$. Let (f', g') and (f'', g'') be two normal pairs of linear imbeddings of K in R^N, each linearly isotopic to f and g respectively. By at most small deformations, we may suppose without loss of generality that the totality of image points of all vertices of K under f', g', f'', g'' are in general position. Then there exists a certain simplicial subdivision M' of M with $K \times (0)$and $K \times (1)$ as subcomplexes and linear mappings h_f and h_g of M' in R^N such that $h_f(x, 0) = f'(x)$, $h_f(x, 1) = f''(x)$, $h_g(x, 0) = g'(x)$, $h_g(x, 1) = g''(x)$ for any $x \in |K|$, that $h_{f,t}$, $h_{g,t}$ are linear imbeddings of a simplicial subdivision M_t' of $K \times (t)$ in R^N, and that the totality of image points of all vertices of M' under h_f and h_g are in general position. Using Lemmas 2, 3, we may suppose without loss of generality that the pair h_f, h_g is normal. The ρ_N-cocycle $\theta^{N-1}_{h_f,h_g} \in^{\rho_N} Z^{N-1}(\tilde{M}'^*)$ is then well-defined. Let $j_i : K \to M'$ be the simplicial map defined by $j_i(a) = (a, i)$, $i = 0$ or 1, and a an arbitrary vertex of K. Let $\tilde{j}_i : \tilde{K}^* \to \tilde{M}'^*$ be the cell map induced by j_i. Then $\tilde{j}_0^{\#}\theta^{N-1}_{h_f,h_g} = \theta^{N-1}_{f',g'}$, $\tilde{j}_1^{\#}\theta^{N-1}_{h_f,h_g} = \theta^{N-1}_{f'',g''}$. It follows that $\theta^{N-1}_{f',g'}$ and $\theta^{N-1}_{f'',g''}$ are ρ_N-cohomologous, which proves our theorem.

§ 8. From Theorems 2 and 3 it follows that to any pair of linear imbeddings f and g of a finite simplicial complex K in R^N a unique class ${}^{\rho_N}\Theta^{N-1}_{f,g}(K) \in {}^{\rho_N}\tilde{H}^{N-1}_{(2)}(K)$ may be defined, namely, the class ${}^{\rho_N}\Theta^{N-1}_{f',g'}(K)$ of any normal pair of linear imbeddings f', g' of K in R^N linearly isotopic to f and g.

Definition The class ${}^{\rho_N}\Theta^{N-1}_{f,g}(K)$ will be called the *isotopy class* of the pair of linear imbeddings f and g.

Among the classes ${}^{\rho_N}\Theta^{N-1}_{f}(K)$, ${}^{\rho_N}\Theta^{N-1}_{g}(K)$ and ${}^{\rho_N}\Theta^{N-1}_{f,g}(K)$, we have a relation given in the following

Theorem 4 *For any pair of linear imbeddings f, g of a finite simplicial complex K in R^N we have*

$$2\,{}^{\rho_N}\Theta^{N-1}_{f,g}(K) = {}^{\rho_N}\Theta^{N-1}_{f}(K) - {}^{\rho_N}\Theta^{N-1}_{g}(K). \tag{2}$$

Proof. For any two singular chains A and B in $R^{N+1} = R^N \times L$ with their point sets $|A|$ and $|B|$ disjoint from each other, we shall denote by $\alpha(A, B)$ the singular chain lying on the unit sphere S^N of R^{N+1} carried by the point set $\alpha(|A| \times |B|)$, where $\alpha(x, y)$ denotes the point of intersection of S^N with the ray from the origin O of R^{N+1} parallel to the directed line joining x to y, for any pair of points $x \neq y$ in R^{N+1}. Let $F : |K| \times I \to R^{N+1}$ be the mapping associated to the pair of linear imbeddings f and g which may be supposed normal, by Theorems 2, 3. For any cell $\sigma \times \tau \in \tilde{K}^*$ with dim $\sigma +$ dim $\tau \leqslant N-3$, the set $R(\sigma, \tau)$ of points $\alpha(F(x,t), F(y,t))$ where $x \in |\sigma|$, $y \in |\tau|$, $t \in I$, is of dimension $\leqslant N-2$. Similarly, for any cell $\sigma \times \tau \in \tilde{K}^*$ with dim $\sigma +$ dim $\tau \leqslant N-2$, the set $R'(\sigma, \tau)$ of points $\alpha(F(x, t), F(y, t))$, where $x \in |\sigma|$, $y \in |\tau|$, $t = 0$ or 1 is also of dimension $\leqslant N-2$. Hence we may take a point e on the unit sphere S^{N-1} in R^N_0 not lying in any of the above sets $R(\sigma, \tau)$ or $R'(\sigma, \tau)$. Take orientations of S^N and S^{N-1} coherent with those of R^{N+1} and R^N. For any singular chain C of dimension N (resp. $N-1$) lying on S^N (resp. S^{N-1}) with $|\dot{C}|$ disjoint from e the local covering degree of C at e will be denoted by $\mathrm{Deg}^{(N)}_e(C)$ (resp. $\mathrm{Deg}^{(N-)}_e(C)$).

Define now $\varphi_f, \varphi_g \in C^{N-1}(\tilde{K}^*)$, $\psi \in C^{N-2}(\tilde{K}^*)$ by

$$\varphi_f(\sigma \times \tau) = \mathrm{Deg}_e^{(N-1)}\alpha(f(\sigma), f(\tau)),$$
$$\varphi_g(\sigma \times \tau) = \mathrm{Deg}_e^{(N-1)}\alpha(g(\sigma), g(\tau)),$$

and

$$\psi(\xi \times \eta) = (-1)^{\dim \xi} Deg_e^{(N)}\alpha(F(\xi \times I), F(\eta \times I)),$$

in which $\sigma \times \tau$, $\xi \times \eta \in \tilde{K}^*$, dim σ+ dim $\tau = N-1$, and dim ξ+ dim $\eta = N-2$. Then $\theta_f = \bar{\rho}_N \varphi_f \in^{\rho_N} \Theta_f^{N-1}$, $\theta_g = \bar{\rho}_N \varphi_g \in^{\rho_N} \Theta_g^{N-1}$. By direct verification we have

$$\theta_{f,g}^{N-1}(\sigma \times \tau) = \delta\psi(\sigma \times \tau) + \varphi_f(\sigma \times \tau) - \varphi_g(\sigma \times \tau).$$

It follows that

$$2\theta_{f,g}^{N-1}(\sigma \times \tau) = \delta \bar{f}_N \psi(\sigma \times \tau) + \theta_f(\sigma \times \tau) - \theta_g(\sigma \times \tau),$$

which proves (2).

References

[1] Wu Wen-tsün. *Science Record, New* Ser., 1958, II: 271-275.

[2] Wu Wen-tsün. *Scientia Sinica*, 1958, 4: 365-387.

[3] Wu Wen-tsün. *Acta Math. Sinica*, 1953, 3: 261-290.

[4] Shapiro A. *Annals of Math.*, 1957, 66: 256-269.

34. On the Isotopy of a Finite Complex in a Euclidean Space II*

§ 1. In what follows $K = K^n$ will denote a Euclidean finite simplicial complex of dimension n in a Euclidean space of sufficiently high dimension. For any linear imbeddings f, g of K in R^N, an invariant ${}^{\rho_N}\Theta_{f,g}^{N-1}(K) \in^{\rho_H} H^{N-1}(\widetilde{K}^*) =^{\rho_N} \widetilde{H}_{(2)}^{N-1}(K)$ with respect to a fixed orientation of R^N has been introduced in Part I (cf. [1]). The aim of this part is to prove the following.

Theorem 1 *Two linear imbeddings f and g of the complex K^n in R^{2n+1} are linearly isotopic if and only if ${}^s\Theta_{f,g}^{2n}(K) = 0$, where $n > 1$, and ${}^s\Theta_{f,g}^{2n}(K)$ is referred to a fixed orientation of R^{2n+1}.*

Remark. For the isotopy of an n-complex in a Euclidean N-space of dimension $N \geqslant 2n+2$, cf. [2].

§ 2. Let us consider any continuous mapping F of $|K| \times L$ in $R^{2n+2} = R^{2n+1} \times L$, where L is the line $-\infty < t < +\infty$ such that:

(i) $F(|K| \times (t)) \subset R_t^{2n+1} = R^{2n+1} \times (t)$.

(ii) F is differentiable and regular on $|\sigma| \times L$ for each $\sigma \in K$.

(iii) For any pair of disjoint simplexes $\sigma_1, \sigma_2 \in K$, with dim σ_1+ dim $\sigma_2 \leqslant 2n-1$, the set $F(|\sigma_1| \times I)$ and $F(|\sigma_2| \times I)$ are disjoint.

(iv) The mappings $f, g : |K] \to R^{2n+1}$ defined by $F(x,0) = (f(x), 0)$, $F(x,1) = (g(x), 1)$ for $x \in |K|$ are both imbeddings of $|K|$ in R^{2n+1}.

Such a mapping F will be said to be *normal.*

Let R^{2n+1} be oriented and R^{2n+2} be oriented as a product space $R^{2n+1} \times L$, L being oriented by increasing t. Then for any normal mapping $F : |K| \times L \to R^{2n+2}$ we may define a $2n$-dimensional s-co-cycle $\theta_F \in^s C^{2n}(\widetilde{K}^*)$ by

$$\theta_F(\sigma_1 \times \sigma_2) = \phi(F(\sigma_1 \times I), F(\sigma_2 \times I)), \tag{1}$$

where $\sigma_1 \times \sigma_2 \in \widetilde{K}^*$, dim σ_1+ dim $\sigma_2 = 2n$.

Let $\sigma, \tau \in K$ be any pair of disjoint simplexes with dim $\sigma = n$, dim $\tau = n-1$, and $\chi_{\sigma,r}$ be the s-cochain of dimension $2n-1$ in $\widetilde{K}^*$ defined by $\chi_{\sigma,r}(\sigma \times \tau) = -\chi_{\sigma,\tau}(\tau \times \sigma) =$

---

* *Sci. Record(N.S.)*, 1959, 3: 342-351.

$+\lambda$, while $\chi_{\sigma,\tau}(\xi \times \eta) = 0$ for any other $(2n-1)$-cell $\xi \times \eta$ of $\widetilde{K}^*$, λ +1 or -1 supposed given.

Lemma We can modify the normal map F to a normal map F' such that

$$\theta_{F'} = \theta_F + \delta\chi_{\sigma,\tau}. \tag{2}$$

Proof. Choose interior points p of σ and q of τ and also t, $0 < t < 1$ such that $\widetilde{\boldsymbol{p}} = F(p,t), \widetilde{q} = F(q,t)$ are not double points of the map F. By at most a further deformation if necessary, we may suppose without loss of generality that F is flat about $\widetilde{\boldsymbol{p}}$. Let S^n be and n-sphere of centre $\widetilde{\boldsymbol{p}}$ and sufficiently small radius $\varepsilon > 0$ in $F(\sigma \times I)$ with the topmost point $\widetilde{\boldsymbol{p}}_+ = F(p_+, t_+)$, where $t_+ > t$. The interior of S^n will be denoted by $E^{n+1} \subset F(\sigma \times I)$.

Let P^{n+2} be an $(n+2)$-plane through $\widetilde{q}$, orthogonal to $F(\tau^{n-1} \times (t))$, transversal to $F(\tau \times I)$ at $\widetilde{q}$, and intersect R_t^{2n+1} in an $(n+1)$-plane P^{n+1}. Let H'^{n+1} be a sufficiently small $(n+1)$-dimensional hemi-sphere in P^{n+2}, with $\widetilde{q}$ in its interior, of which the bottom is an $(n+1)$-cell E'^{n+1} with boundary S'^n, centre $\widetilde{q}$, and radius $\varepsilon' > 0$. The topmost point of S'^n is $\widetilde{q}'_+ = (q'_+, t'_+)$. We shall take H'^{n+1} and $\varepsilon' > 0$ so small that $t'_+ < t_1$ and that S'^n is disjoint from $F(|K| \times I)$. We may join $\widetilde{\boldsymbol{p}}$ and $\widetilde{q}'$ by a simple are c of class C^∞ in R_t^{2n+1} such that $c(0) = \widetilde{\boldsymbol{p}}$, $c(1) = \widetilde{q}'$, c is orthogonal to the plane of $S^{n-1} = S^n \cap R_t^{2n+1}$ and $S'^{n-1} = S'^n \cap P^{n+1}$, and c meets $F(K| \times I)$ only in $\widetilde{p}$ an the solid hemisphere bounded by H'^{n+1} only in $\widetilde{\boldsymbol{q}}'$. Any orientation of S'^{n-1} will induce naturally one of S'^n and also one of H'^{n+1} and P^{n+2} so that $\partial(H'^{n+1} - E'^{n+1}) = S'^n$. We shall orient S'^{n-1} so that

$$\phi(P^{n+2}, F(\tau \times L)) = \lambda. \tag{3}$$

By the tube construction employed in [3], we may construct a C^∞—system of $(n-1)$-spheres $S_s^{n-1}, 0 \leqslant s \leqslant 1$, with centre $c(s)$, radius $\varepsilon(s) > 0$ such that $S_0^{n-1} = S^{n-1}$, $S_1^{n-1} = S'^{n-1}$, $\varepsilon(0) = \varepsilon, \varepsilon(1) = \varepsilon', \varepsilon(s)$ linearly dependent on s and that, having taken $\varepsilon, \varepsilon' > 0$ sufficiently small, $T^n = \Sigma S_s^{n-1}$ will be a *topological* tube about axis c, which, duly oriented, will give $\partial T^n = S_1^{n-1} - S_0^{n-1}$, the S_0^{n-1} being oriented so as to induce an orientation of E^{n+1} which coincides with the given orientation of $F(\sigma \times I)$.

In R^{2n+2} we may then construct a simple are c_+ of class C^∞ joining $\widetilde{\boldsymbol{p}}_+$ and $\widetilde{q}_+$ such that $c_+(0) = \widetilde{\boldsymbol{p}}_+, c_+(1) = \widetilde{q}'_+$, and the t-coordinate of $c_+(s)$ depends linearly on s, and that the segment $\overline{c(s)c_+(s)}$ is orthogonal to the plane of S_s^{n-1} and of length $= \varepsilon(s)$. Let S_s^n be the n-sphere with centre $c(s)$ and radius $\varepsilon(s)$. Set $T^{n+1} = \Sigma\ S_s^n$, $H^{n+1} = E^{n+1} + T^{n+1} + (H'^{n+1} - E'^{n+1})$. We may then, taking $\varepsilon, \varepsilon' > 0$ sufficiently small, define $F' : |K| \times L \to R^{2n+2}$ by $F' \equiv F/|K| \times L - F^{-1}(E^{n+1}) \cap (|\sigma| \times I)$, and $F'(E^{n+1}) \equiv H^{n+1} - E^{n+1}$ in the manner that T^{n+1} meets $F(|K| \times I)$ only in S^n and that F' satisfies all the conditions (i)–(iv) except perhaps the regularity of the map at points of S^n and S'^n. This may be achieved, however, by a further deformation if necessary, cf. [4] Props. 1 and 2. This resulting map F' satisfies then (2) owing to (3).

§ 3. Proof of Theorem 1

Let f, g be two linear imbeddings of K in R^{2n+1} wiht ${}^s\Theta^{2n}_{f,g} = 0$ and we have to prove the linear isotopy of f and g. By Theorems 2 and 3 of Part I we may suppose without loss of generality that the pair (f, g) is normal so that $\theta^{2n}_{f,g} \in^s \Theta^{2n}_{f,g}(K) \in^s H^{2n}(\widetilde{K}^*)$ is well-defined. Let $F : |K| \times L \to R^{2n+2}$ be the map associated with the pair (f, g) as defined in Part I. By hypothesis we have $\theta_F = \theta^{2n}_{f,g} s0$. By successive applications of the Lemma in §2 it follows that we may modify the map F to a normal map $F' : |K| \times L \to R^{2n+2}$ with $F'/|K| \times (0) + |K| \times (1) + |K^{n-1}| \times \boldsymbol{I} \equiv F$ and with $\theta_{F'} = 0$. We may suppose also that $F'/|K| \times \boldsymbol{I}$ has no triple points and that at any double point, the two pieces of $F'(|K| \times I)$ are transversal to each other, cf. again [4]. Remark that, owing to the normality of the pair (f, g), the map F' deduced from the associated map F can have double points only as common points of intersection in the interiors of $F'(|\sigma_1| \times I)$ and $F'(|\sigma_2| \times I)$, where the n-simplexes $\sigma_1, \sigma_2 \in K$ are *disjoint.* For each pair of disjoint n-simplexes $\sigma_1, \sigma_2 \in K$, $F'(|\sigma_1| \times I)$ and $F'(|\sigma| \times I)$ then meet transversally in and even number of points of which half a number has an index of inter-section $+1$ and the other half an index of intersection -1. As $n > 1$, these double points may then be succcessively removed in pairs by a modified construction of Whitney (cf. [5] and [3], also [6]) such that the resulting mapping in each step is always *admissible* in the sense that it maps $|K| \times (t)$ into R^{2n+1}_t, $0 \leqslant t \leqslant 1$. The final mapping F'' gives then an isotopy between f and g such that F'' coincides with F on $|K| \times (0) + |K| \times (1) + |K^{n-1}| \times I$, that F'' is an admissible imbedding of $|K| \times I$ in R^{2n+2} and that F'' is differentiable and regular on each $\sigma \times I$, where σ is any n-simplex of K.

By a theorem of Cairns (cf. [7] and also [8]), for each n-simplex σ of K, we may take a polyhedral approximation F'''_σ to $F''/\sigma \times I$ of $\sigma \times I$ in R^{2n+2} such that the mapping $F''' : |K| \times I \to R^{2n+2}$ defined by $F'''/\sigma \times I \equiv F'''_\sigma$ coincides with F'' on $|K| \times (0) + |K| \times (1) + |K^{n-1}| \times I$ and that F''' is a linear imbedding of a certain simplicial subdivision of $K \times I$ in R^{2n+2} which is admissible. The mapping F''' gives then a linear isotopy between f and g as required. This prover the sufficiency part of our theorem. The necessity part follows easily from the techniques developed in Part I. In fact we have the following more general theorem of which the proof is similar to that of Theorem 3 in Part I.

Theorem 2 *If two imbeddings f, g of a finite simplicial complex K in a Euclidean space R^n are linearly isotopic, then* ${}^{\rho_N}\Theta_{f,g}^{N-1}(K) = 0$.

As a corollary of Theorem 1, we have also the following theorem, which is the combinatorial analogue of the corresponding theorem for the C^r-isotopy of C^r-n-manifolds in $R^{2n+1}, n > 1$. (cf. [4]).

Theorem 3 *Any two linear imbeddings of a combinatorial manifold of dimension $n > 1$ in a Euclidean $(2n+1)$-space are linearly isotopic.*

References

[1] Wu Wen-tsün. *Science Record*, New Ser., 1959, III (8): 342-347.
[2] Van Kampen E R. *Hamb. Abh.*, 1933, 9: 72-78.
[3] Wu Wen-tsün. *Scientia Sinica*, 1959, 8: 133-150.
[4] Wu Wen-tsün. *Science Record*, New Ser., 1958, II: 271-275.
[5] Whithey H. *Annals of Math.*, 1944, 45: 220-246.
[6] Shapiro A. *Annals of Math.*, 1957, 66: 256-259.
[7] Cairns S S. *Annals of Math.*, 1936, 37: 409-415.
[8] Whitehead J H C. *Annals of Math.*, 1940, 41: 809-824.

35. 关于 Leray 的一个定理*

这里所谓 Leray 定理, 是指在适当条件下, 一个空间与它的一个覆盖的神经复合形有相同的同调群而言. Leray 的原证 (以及 Borel, Cartan, Serre 等在各种变化形式的证明), 奠基于他的 Converture 理论 (亦或用及束论与谱序列论). 本文将按照 Eilenberg-Steenrod 的体系给出另一证明. 我们的证明虽只适用于有限覆盖, 但易于推广到基本群的情形, 而已知方法则不适用. 我们也同样讨论了关于同伦群与同伦型的情形.

§1 与覆盖相关的规范映像

设 $\mathscr{F} = \{F_\lambda\}$ 是正常空间 E 的一个有限闭覆盖而 K 是它的神经复合形, 与 $\mathscr{F}$ 中集 F_λ 相当的顶点是 a_λ. 命 $\mathscr{U}$ 是一 E 的有限开覆盖 $\{U_\lambda\}$, 与 $\mathscr{F}$ 相似, 使 $F_\lambda \subset U_\lambda$, $F_{\lambda_1}\cdots F_{\lambda_r} = \varnothing$与$U_{\lambda_1}\cdots U_{\lambda_r} = \varnothing$ 同义. 对每一 U_λ 设 f_λ 是任一 E 上的连续函数, 使在 $E - U_\lambda$ 上 $f_\lambda = 0$, 而在 U_λ 上 $f_\lambda > 0$ (由 Urysolm 引理这样的 f_λ 必存在). 对于这样一组 $\{f_\lambda\}$ 可依下式定义一 E 到 K 的空间 $|K|$ 的一个**规范映像** φ 如下:

$$\varphi(x) = \sum_\lambda \frac{f_\lambda(x)}{\sum\limits_\mu f_\mu(x)} \cdot a_\lambda.$$

设有限开覆盖 $\mathscr{U}' = \{U'_\lambda\}$ 与函数组 f'_λ 具有以上相似的性质而命由此确定的规范映像为 $\varphi' : E \to |K|$, 则有限开覆盖 $\{U_\lambda \cap U'_\lambda\}$ 仍与 $\mathscr{F}$ 相似, 对此如前作一组函数 g_λ 并以 ψ 表所定的规范映像 $E \to |K|$, 则

$$\varphi_t(x) = \sum_\lambda \frac{t \cdot g_\lambda(x) + (1-t) f_\lambda(x)}{t \cdot \sum\limits_\mu g_\mu(x) + (1-t)\cdot \sum\limits_\mu f_\mu(x)} \cdot a_\lambda, \quad 0 \leqslant t \leqslant 1,$$

与

$$\varphi'_t(x) = \sum_\lambda \frac{t \cdot g_\lambda(x) + (1-t) f'_\lambda(x)}{t \cdot \sum\limits_\mu g_\mu(x) + (1-t)\cdot \sum\limits_\mu f'_\mu(x)} \cdot a_\lambda, \quad 0 \leqslant t \leqslant 1$$

---

* *Acta Math. Sinica*, 1961, 11: 348-356; *Chinese Math.*, 1962, 2: 398-410; *Sci. Sinica*, 1961, 10: 793-805.

各为 φ, ψ 间与 φ', ψ 间的伦移. 由此知 φ 的同伦类与覆盖 $\mathscr{U}$. 及函数组 f_λ 的选择无关, 因而引出了一个唯一的准同构

$$\Phi^* : H^*(K) \to H^*(E),$$

这里 H^* 指任一 Eilenberg-Steenrod 意义下的正常空间偶的同调群系统. 我们将称 Φ^* 为覆盖 $\mathscr{F}$ 所定的规范准同构.

今考虑一正常空间 E, 一个 E 的闭子集 E', 与 E 的一个有限闭覆盖 $\mathscr{F}$. $\mathscr{F}$ 中集合与 E' 的非空交集构成一 E' 的一个有限闭覆盖 $\mathscr{F}''$, 以下将称为 $\mathscr{F}$ 在 E' 上的限制. 可能 $\mathscr{F}''$ 中集合的一部分早已构成一 E' 的一个闭覆盖, 例如 $\mathscr{F}'$. 命 K, K'', 与 K' 各为 $\mathscr{F}, \mathscr{F}''$ 与 $\mathscr{F}'$ 的神经复合形. 我们将恒同 K', K'' 中与 $\mathscr{F}'$ 中 $F'_\lambda = F_\lambda \cdot E' \neq \varnothing$ 相当的顶点为 K 中与 $\mathscr{F}$ 的 F_λ 相当的顶点 a_λ, 于是 K' 是 K'' 的一个子复合形, 而 K'' 又是 K 的一个子复合形. 命 $\mathscr{F}$ 与 $\mathscr{F}'$ 所定的规范映像各为

$$\Phi^* : H^*(K) \to H^*(E), \quad 与\ \Phi'^* : H^*(K') \to H^*(E').$$

于是我们有

引理 1 Φ^*, Φ'^* 与上边准同构 δ^* 以及由嵌入 $i : E' \subset E, j : |K'| \subset |K|$ 引起的准同构可交换, 即下面的图形是可交换的:

$$\begin{array}{ccccc} H^*(K) & \xrightarrow{j^*} & H^*(K') & \xrightarrow{\delta^*} & H^*(K) \\ \downarrow{\scriptstyle \Phi^*} & & \downarrow{\scriptstyle \Phi'^*} & & \downarrow{\scriptstyle \Phi^*} \\ H^*(E) & \xrightarrow{i^*} & H^*(E') & \xrightarrow{\delta^*} & H^*(E) \end{array}$$

证 取一 E 的有限开覆盖 $\mathscr{U} = \{U_\lambda\}$ 使与 $\mathscr{F}$ 相似, $\mathscr{U}$ 在 E' 上的限制 $\mathscr{U}''$, 于是也与 $\mathscr{F}$ 在 E' 上的限制 $\mathscr{F}''$ 相似, 且 $\mathscr{U}''$ 中与 $\mathscr{F}'$ 中集合相当的那些集合也构成了一个 E' 的开覆盖 $\mathscr{U}'$, 与 $\mathscr{F}'$ 相似. 今对每一 U_λ 作一连续函数 f_λ 使在 $E - U_\lambda$ 上 $f_\lambda = 0$ 而在 $U_\lambda = 0$上$f_\lambda > 0$. 对覆盖 $\mathscr{U}$ 与函数组 f_λ 作规范映像 $\varphi : E \to |K|$, 又对覆盖 $\mathscr{U}'$(或 $\mathscr{U}''$) 与函数组 f'_λ 作规范映像 $\varphi' : E' \to |K'|$(或 $\varphi'' : E' \to |K''|$), 此处 f'_λ 为 f_λ 在 $U_\lambda \cdot E'$ 上的限制. 于是对 $x \in E'$, 有

$$\varphi(x) = \varphi''(x) = \sum{}' \frac{f_\lambda{}'(x)}{\sum' f_{\mu'}(x) + \sum'' f_{\mu''}(x)} \cdot a_\lambda + \sum{}'' \frac{f_\lambda{}''(x)}{\sum' f_{\mu'}(x) + \sum'' f_{\mu''}(x)} \cdot a_\lambda{}'',$$

$$\varphi'(x) = \sum{}' \frac{f_\lambda{}'(x)}{\sum' f_{\mu'}(x)} \cdot a_\lambda{}'.$$

在诸和中 λ', μ' 跑过使 $F_\lambda \cdot E' \in \mathscr{F}'$ 的一切 λ, 而 λ'', μ'' 跑过使 $F_\lambda \cdot E' \in \mathscr{F}''$ 但 $\bar{\in} \mathscr{F}'$ 的一切 λ. 因对 $x \in E' \sum' f_{\mu'}(x) > 0$, 故

$$\varphi_t(x) = \sum{}' \frac{f_\lambda{}'(x)}{\sum' f_{\mu'}(x) + (1-t)\sum'' f_{\mu''}(x)} \cdot a_\lambda{}'$$
$$+ (1-t) \cdot \sum{}'' \frac{f_\lambda{}''(x)}{\sum' f_{\mu'}(x) + (1-t) \cdot \sum'' f_{\mu''}(x)} \cdot a_\lambda{}''$$

给出了 φ 到 φ' 的一个伦移, 即 $\varphi_i \simeq j\varphi'$. 延拓这一伦移即得 $\varphi \simeq \bar{\varphi}$, 此处 $\bar{\varphi} i = j\varphi'/E'$. 由此即得上图的可交换性.

从上述证明亦可获得

引理 2 存在映像 φ 与 E 到 $|K|$ 的一个规范映像同伦使 $\varphi(E') \subset |K'|$. 假设 E, E' 都弧连通. 若取任一点 $O \in E'$ 为 $\pi_1(E)$ 与 $\pi_1(E')$ 的参考点, 又取 $\varphi(O)$ 为 $\pi_1(|K|)$ 与 $\pi_1(|K'|)$ 的参考点, 则下图

$$\begin{array}{ccc} \pi_1(E') & \xrightarrow{i_*} & \pi_1(E) \\ \Big\downarrow{\scriptstyle \varphi_*} & & \Big\downarrow{\scriptstyle \varphi_*} \\ \pi_1(|K'|) & \xrightarrow{j_*} & \pi_1(|K|) \end{array}$$

可交换其中 φ_*, i_* 与 j_* 各为由 φ 与嵌入 $i : E' \subset E, j : |K'| \subset |K|$ 所引起的准同构.

§2 Leray 定理

所谓 Leray 定理的原来形式可述之如下:

设 E 是一复紧 Hausdorff 空间, $\mathscr{F} = \{F_\lambda\}$ 是 E 的一个有限闭覆盖, 具有以下性质: $\mathscr{F}$ 中集合的每一非空交集 $F_{\lambda_1} \cdots F_{\lambda_s} \neq \varnothing$ 对一系数环 R 而言具有与一个点一样的 Alexander-Cêch 上同调环, 则对 R 而言 $\mathscr{F}$ 的神经复合形具有与空间 E 一样的 Alexander-Cêch 上同调环.

定义在正常空间偶上满足 Eilenberg-Steenrod 公理系统的一个 (上) 同调理论将称为具有强 excision 性质, 如果以下成立: 对任一正常空间偶 (E_1, F_1) 到偶 (E_2, F_2) 的连续映像 φ, 只要 φ 是 $E_1 - F_1$ 到 $E_2 - F_2$ 上的一个同拓变换, 准同构 $\varphi^* : H^*(E_2, F_2) \to \to H^*(E_1, F_1)$ 即是一同构上. 我们将把 Leray 定理加强为如下形式:

定理 1 设 H^* 为定义在正常空间偶上具有强 excision 性质的一个上同调理论, 设 E 是一复紧 Hausdorff 空间, $\mathscr{F}$ 是 E 的一个有限闭覆盖, 使其中集合的任意非空交集在维数 $r-1$ 与 r 有与一个点相同的上同调群, 则 E 在维数 r 有与 $\mathscr{F}$ 的

神经复合形 K 同构的上同调群, 且 $H^r(|K|)$ 与 $H^r(E)$ 间的同构可为对覆盖 $\mathscr{F}$ 的任一 E 到 $|K|$ 中的规范映像所实现.

证 试对 $\mathscr{F}$ 中集合的个数进行归纳证明. 在 $\mathscr{F}$ 中只有一个集合时, 定理是不足道的. 假设对 $\leqslant n$ 个集合的情形已证明而 $\mathscr{F}$ 中有 $n+1$ 个集合 $F_0, F_1, \cdots, F_n$, 记 $F_1, \cdots, F_n$ 的并集为 E_1, 改记 F_0 为 E_2, 又记 $E_1 \cdot E_2$ 为 E_0. 因 $E - E_1$ 即为 $E_2 - E_1 \cdot E_2$ 而 $E - E_2$ 即为 $E_1 - E_1 \cdot E_2$, 由强 excision 性质知 $(E; E_1, E_2)$ 成一正则仨 (proper triad). 由于 $H^*(E_2)$ 的平凡性, 这一正则仨的 Mayer-Vietoris 确列约化成

$$\cdots \xrightarrow{j_1^*} H^{r-1}(E_1) \xrightarrow{j_0^*} H^{r-1}(E_0) \xrightarrow{\delta^*} H^r(E) \xrightarrow{j_1^*} H^r(E_1) \xrightarrow{j_0^*} H^r(E_0) \to \cdots$$

其中 δ^* 是上边准同构而 j_0^*, j_1^* 是由嵌入引出的准同构. 今记 E 的覆盖 $\mathscr{F}$ 在 E_0, E_1 上的限制当除去 F_0 后的覆盖为 $\mathscr{F}_0$ 与 $\mathscr{F}_1$, 而其神经复合形为 K_0 与 K_1, 二者都视作覆盖 $\mathscr{F}$ 的神经复合形 K 的子复合形. 命 K 中相当于 $\mathscr{F}$ 的 F_0 的顶点为 a_0, 并记 a_0 与 K_0 的联合 (join) 为 K_2, 则 K_0 是 K_1, K_2 的交而 K 是其和. 因 K_2 的同调是平凡的故复合形仨 $(K; K_1, K_2)$ 的 Mayer-Vietoris 确列约化为

$$\cdots \xrightarrow{i_1^*} H^{r-1}(K_1) \xrightarrow{i_0^*} H^{r-1}(K_0) \xrightarrow{\delta^*} H^r(K) \xrightarrow{i_1^*} H^r(K_1) \xrightarrow{i_0^*} H^r(K_0) \to \cdots$$

其中 δ^* 为上边准同构, 而 i_0^*, i_1^* 为由嵌入引起的准同构, 试考虑以下图形:

$$\begin{array}{ccccccccccc}
\cdots \to & H^{r-1}(E_1) & \xrightarrow{j_0^*} & H^{r-1}(E_0) & \xrightarrow{\delta^*} & H^r(E) & \xrightarrow{j_1^*} & H^r(E_1) & \xrightarrow{i_0^*} & H^r(E_0) & \to \cdots \\
& \uparrow{\scriptstyle \Phi_1^*} & & \uparrow{\scriptstyle \Phi_0^*} & & \uparrow{\scriptstyle \Phi^*} & & \uparrow{\scriptstyle \Phi_1^*} & & \uparrow{\scriptstyle \Phi_1^*} & \\
\cdots \to & H^{r-1}(K_1) & \xrightarrow{i_0^*} & H^{r-1}(K_0) & \longrightarrow & H^r(K) & \longrightarrow & H^r(K_1) & \longrightarrow & H^r(K_0) & \to \cdots
\end{array}$$

其中 $\Phi^*, \Phi_0^*, \Phi_1^*$ 各为相应诸覆盖所确定的规范准同构. 由§1 的引理 1, 这个图形是可交换的. 因 $\mathscr{F}_0$ 与 $\mathscr{F}_1$ 中的集合个数都 $\leqslant n$, 故由归纳假设 Φ_0^* 与 Φ_1^* 都是同构上, 由所谓 "五" 引理 ("five" lemma) Φ^* 也是一个同构上, 即所欲证.

推论 在关于 H^*, E 与 $\mathscr{F}$ 的相同假设下并设 $\mathscr{F}$ 中集合的任意非空交集在每一维数都有与一个点相同的上同调群, 则在 H^* 有环结构时, $H^*(|K|)$ 与 $H^*(E)$ 环同构.

证 由于 $H^r(|K|) \approx H^r(E)$ 可为一连续映像, 即与 $\mathscr{F}$ 相关的任意规范映像所引出.

§3 对基本群的推广

设 E 是一弧连通正常空间而 C_1, C_2 是 E 的两个弧连通开集 (或闭集) 满足以下条件:

1° C_1 与 C_2 的和集是 E.

2° C_1 与 C_2 的交是有限个互不相遇的开集 (或闭集) B_j 之和, $1 \leqslant j \leqslant m$, 每一个 B_j 都是弧连通的.

在 C_i, B_j 都是闭集的情形, 我们将添入下一

假设 (*H*) 在 E 中有包含 B_j 的互不相遇的弧连通开集 U_j, 使 B_j 是 U_j 的伦移收缩核.

对于这样的空间仨 $(E; C_1, C_2)$ 在每一 B_j 中取一点 O_j, 并设 B_j, C_1, C_2 与 E 的基本群的参考点各为 O_j, O_1, O_1 与 O_1. 对每一 $j \geqslant 1$ 与 $\leqslant m$ 在 C_i 中取一固定道路 l_j^i 从 O_1 至 O_j, 其中 l_1^i 约化为点 $O_1 (i = 1, 2)$, 于是闭道路 $l_j^1 (l_j^2)^{-1}$ 代表了 $\pi_1(E)$ 中的一个元素, 将记之为 $\beta_j, 1 \leqslant j \leqslant m$, 这里 $\beta_1 = 0$, 而 $j > 1$ 时 $\beta_j \neq 0$. 又对 B_j 中的每一两端在 O_j 的闭道路 l, 对应 $l \to l_j^i \cdot l \cdot (l_j^i)^{-1}$ 引出一个准同构 $\lambda_j^i : \pi_1(B_j) \to \pi_1(C_i), i = 1, 2$, 而嵌入 $C_i \subset E$ 所引出的准同构 $\pi_1(C_i) \to \pi_1(E)$ 则将记为 $h_i^*, i = 1, 2$.

下述引理为 Van Kampen 一个定理 [4] 的改变形式, 其证明将从略, 对此亦可参阅 P.Olum 文 [6].

引理 1 在以上假定之下 $\pi_1(E)$ 的构造由 $\pi_1(C_1), \pi_1(C_2)$ 与准同构 λ_j^i 所完全决定, 精确言之, 有

(i) 若 $\{\alpha_k^1\}$ 与 $\{\alpha_l^2\}$ 各为 $\pi_1(C_1)$ 与 $\pi_1(C_2)$ 的母素组, 则 $\pi_1(E)$ 有一母素组 $\{h_1^*(\alpha_k^1), h_2^*(\alpha_l^2), \beta_j\}$.

(ii) 若 $R_r^1(\alpha_k^1) = 1$ 与 $R_s^2(\alpha_l^2) = 1$ 各为以上 $\pi_1(C_1)$ 与 $\pi_1(C_2)$ 的母素组的完全关系组, 则 $\pi_1(E)$ 的以上母素组的一个完全关系组为

$$R_r^1(h_1^*(\alpha_k^1)) = 1,$$
$$R_s^2(h_2^*(\alpha_l^2)) = 1,$$

与

$$\bar{\lambda}_j^1(r_{jt}) \cdot \beta_j = \beta_j \cdot \bar{\lambda}_j^2(r_{jt}), \quad 1 \leqslant j \leqslant m,$$

这里 $\{r_{jt}\}$ 各为 $\pi_1(B_j)$ 的母素组, $1 \leqslant j \leqslant m$, 而 $\bar{\lambda}_j^i = h_j^* \lambda_j^i, i = 1, 2$.

定理 2 若 $\mathscr{U} = \{U_\lambda\}$ 是一连通正常空间 E 的有限开覆盖, 而 $\mathscr{U}$ 中集合的每一非空交集 $U_{\lambda_1} \cdots U_{\lambda_r} \neq \varnothing$ 都是弧连通与单连通的, 则 $\pi_1(E)$ 与 $\pi_1(|K|)$ 同构, 这里 K 是 E 的神经复合形. 且这个同构可如下实现: 试考虑任一与覆盖 $\mathscr{U}$ 以及如§1 中一组连续函数 f_λ 所定的规范映像 $\bar{\varphi} : E \to |K|$, 以及任一同伦于 $\bar{\varphi}$ 的连续映像 φ. 命 O 为 E 的任一点而 $\pi_1(E)$ 与 $\pi_1(|K|)$ 各以 O 与 $\varphi(O)$ 为参考点, 则 φ 引出同构

$$\pi_1(E) \approx \pi_1(|K|).$$

证 $\mathscr{U}$ 中集合的个数 $=1$ 时定理显然. 今设定理对 $\leqslant n-1$ 的情形已证明而 $\mathscr{U}$ 中集合的个数是 $n>1$. 命 C_1, 设为 U_1, 是 $\mathscr{U}$ 中的一个集合, 使 $C_2=\sum\limits_{\lambda=2}^{n}U_\lambda$ 仍是弧连通的, 于是 C_1 与 C_2 的交是有限多个弧连通分支 B_j 的和集. $\mathscr{U}'=\{U_2,\cdots,U_m\}$ 在 B_j 上的限制 $\mathscr{U}_j$ 作为 B_j 的一个开覆盖显然满足定理中的条件. 命 a 是 K 中与 U_1 相当的一个顶点而 L_j,K_2 各为 $\mathscr{U}_j$ 与 C_2 的覆盖 $\mathscr{U}'=\{U_2,\cdots,U_n\}$ 的神经复合形, 都视作 K 的子复合形, 于是 a 与诸 L_j 的和的联合 K_1 可视作 $\mathscr{U}$ 在 $C_1=U_1$ 的限制覆盖的神经复合形. 注意诸子复合形 L_j 都是连通的而且互不相遇, 因为 $\mathscr{U}$ 中集合的任意非空交集都是弧连通的.

与§1 的引理 2 相同, 可作一映像 φ 同伦于 E 到 $|K|$ 的一个规范映像 $\bar{\varphi}$, 使 $\varphi(B_j)\subset|L_j|,\varphi(C_i)\subset|K_i|$. 在 B_j 中取 O_i 并取 O_i,O_1, 与 O_1 为 $\pi_1(B_j),\pi_1(C_i)$ 与 $\pi_1(E)$ 的参考点. 同样基本群 $\pi_1(|L_j|),\pi_1(|K_i|)$ 与 $\pi_1(|K|)$ 将各以 $\varphi(O_j),\varphi(O_1)$ 与 $\varphi(O_1)$ 为参考点. 在 C_i 中将取自 O_1 至 O_j 的固定路线 l_j^i (l_1^i 约化为点 O_1) 并定义准同构 $\lambda_j^i:\pi_1(B_j)\to\pi_1(C_i)$ 如前. 于是 $\varphi(l_j^i)$ 为 $|K_i|$ 中从 $\varphi(O_1)$ 到 $\varphi(O_j)$ 的道路 ($\varphi(l_1^i)$ 约化为点 $\varphi(O_1)$), 由此可定义准同构 $\mu_j^i:\pi_1(|L_j|)\to\pi_1(|K_i|)$. 命 $h_i^*:\pi_1(C_i)\to\pi_1(E)$ 与 $k_i^*:\pi_1(|K_i|)\to\pi_1(|K|)$ 为由嵌入 $h_i:C_i\subset E_i$ 与 $k_i:|K_i|\subset|K|$ 所引出的准同构而 φ^* 为由 φ 所引出相应基本群间的准同构. 于是下面的图形是可交换的, 这在 $i=2$ 时是由于§1 的引理 2, 而在 $i=1$ 时则由于 $\pi_1(C_1)=\pi_1(|K_1|)=0$ 而是显然的

$$
\begin{array}{ccccc}
\pi_1(B_j) & \xrightarrow{\lambda_j^i} & \pi_1(C_i) & \xrightarrow{h_i^*} & \pi_1(E)\\
\uparrow{\scriptstyle\varphi_*} & & \uparrow{\scriptstyle\varphi_*} & & \uparrow{\scriptstyle\varphi_*}\\
\pi_1(|L_j|) & \xrightarrow{\mu_j^i} & \pi_1(|K_i|) & \xrightarrow{k_i^*} & \pi_1(|K|)
\end{array}
$$

由归纳假设有

$$\varphi_*:\pi_1(|L_j|)\approx\pi_1(B_j),$$

$$\varphi_*:\pi_1(|K_2|)\approx\pi_1(C_2),$$

显然亦有

$$\pi_1(|K_1|)\approx\pi_1(C_1).$$

对于 K 的空间 $|K|$ 有 $|K|=|K_1|\cup|K_2|,|K_1|\cdot|K_2|=\sum|L_j|$, 而复合形仨 $(|K|;|K_1|,|K_2|)$ 显然满足假设 (H). 应用引理于 $(|K|;|K_1|,|K_2|)$ 与 $(E;C_1,C_2)$ 并比较二者即得 $\pi_1(|K_1|)\approx\pi_1(E)$. 参考点 O 的选择则显然是无关的.

注 在下节中将叙述另一证明, 可避免使用 Van Kampen 定理, 而在高维同伦群的情形亦有效.

§4 对同伦群的推广

因为对高维同伦群没有像同调群那样的 Mayer-Vietoris 序列或基本群那样的 Van Kampen 定理, 故在将 Leray 定理推广到高维同伦群时, 在技术上须略作改变, 我们将从下述简单情形开始.

引理 1 设 $\{U_1, U_2\}$ 是空间 E 的一个开覆盖. 假设 U_1, U_2 与 $U_1 \cdot U_2$ 都是 r 连通的, 即弧连通且 $\pi_1 = \pi_2 = \cdots = \pi_r = 0$. 则 E 也是 r 连通的.

证 设 f 是一个 s 维球 $S(s \leqslant r)$ 到 E 的一个连续映像, 剖分 S 使对每一单纯形 $\sigma \in S$, 象 $f(|\sigma|)$ 全在三个开集 U_1, U_2 或 $U_1 \cdot U_2$ 之一以内. 取 $U_1 \cdot U_2$ 的一个定点 0 并定义一同伦 $h : S \times [0,1] \to E$ 使 $f \equiv h/S \times (0)$ 伦移为一个定值映像 $h(S \times (1)) = 0$ 如下: 对每一 S 的顶点 $a, h((a) \times [0,1])$ 将为一从 $f(a)$ 到 O 的道路, 在 $f(a) \in U_1 \cdot U_2$ 时, 此道路将全在 $U_1 \cdot U_2$ 中, 而在 $f(a) \in U_1 - U_1 \cdot U_2$ (或 $\in U_2 - U_1 \cdot U_2$) 时, 将全在 $U_1 - U_1 \cdot U_2$(或 $U_2 - U_1 \cdot U_2$ 中). 假设 h 已在 $S^k \times [0,1]$ 上定义, 这里 S^k 指 S 的 k 维骨架, $k < s$, 使对 S 的每一 k 维单纯形 $\sigma^k, h(|\sigma^k| \times [0,1])$ 在 $f(|\sigma^k|) \subset U_1 \cdot U_2$ 时全在 $U_1 \cdot U_2$ 中, 否则全在 U_1 或 U_2 中. 于是对每一 S 的 $k+1$ 维单纯形 σ^{k+1}, h 已在 $\sigma^{k+1} \times [0,1]$ 的边界上定义而其象当 $f(|\sigma^{k+1}|) \subset U_1 \cdot U_2$ 时全在 U_1 或 U_2 中. 因这些开集的 π_{k+1} 都 $= 0$, 故 h 可推广至 $\sigma^{k+1} \times [0,1]$ 使其象全在 $U_1 \cdot U_2$ 中, 或 U_1 中或 U_2 中. 由此即得所需的同伦 $h : S \times [0,1] \to E$ 而有 $\pi_s(E) = 0$.

引理 2 设 $\mathscr{U} = \{U_0, U_1, \cdots, U_n\}$ 是空间 E 的一个开覆盖, 使

(i) $U_i \cdot U_0 \neq \varnothing, i \neq 0$.

(ii) 若 $U_{\lambda_i} \cdot U_{\lambda_j} \neq \varnothing, i, j = 1, \cdots, s$, 则 $U_{\lambda_1} \cdots U_{\lambda_s} \neq \varnothing (\lambda_i = 0, 1, \cdots, n)$, 又

(iii) 任意非空交集 $U_{\lambda_1} \cdots U_{\lambda_s} (\lambda_i = 0, 1, \cdots, n)$ 都是 r 连通的.

则 E 是 r 连通的, 且 $\mathscr{U}$ 的神经复合形 K 也是 r 连通的.

证 在 $n = 1$ 时即上引理 1. 假设引理在 $\leqslant n-1$ 的情形已经证明, 于是 $E_0 = \sum\limits_{i=0}^{n-1} U_i$ 是 r 连通的. 空间 $E' = E_0 \cdot U_n$ 有一开覆盖 $\mathscr{U}'$, 由 $U_0', \cdots, U_{n-1}'$ 中的非空集合所构成, 这里 $U_i' = U_i \cdot U_n$. 这一覆盖显然具有与 (i), (ii), (iii) 相仿的性质, 因而由归纳假定 E' 是 r 连通的. 应用引理 1 到空间 E 与它的覆盖 $\{E_0, U_n\}$ 即知 E 是 r 连通的.

对于神经复合形 K 可设 K' 是覆盖 $\mathscr{U}'$ 的神经复合形, 视之为一 K 的子复合形. 同样 E_0 的覆盖 $\{U_0, U_1, \cdots, U_{n-1}\}$ 的神经复合形 K_0 可视为 K 的子复合形, 而 K_0 又以 K' 为子复合形. 命 a_n 为 K 中与 U_n 相当的顶点, 则 K 为 K_0 与 $a_n K'$ 之和而 K' 为二者之交. 由归纳假设 K' 与 K_0 都是 r 连通的. 对于 $a_n K'$ 这一点更属显然, 今 $|a_n K'|$ 与 $|K_0|$ 在 $|K|$ 中有邻域 V_n 与 V_0 使 V_n, V_0 与 $V_n \cdot V_0$ 各以

$|a_nK'|, |K_0|$, 与 $|K'|$ 为变状收缩核. 应用引理 1 即得空间 $V_0+V_n=|K|$ 是 r 连通的, 即引理的第二部分.

对任意有限单纯复合形 K 命 $\mathrm{Cl}\,\sigma$ 为由 $\sigma\in K$ 所定的 K 的闭子复合形, $\mathrm{St}\,\sigma$ 为由以 σ 为面的一切单纯形所构成的开子复合形, 而 $\overline{\mathrm{St}}\,\sigma$ 为由 $\mathrm{St}\,\sigma$ 中一切单纯形与其面所构成的闭子复合形. 对 K 中任一闭子复合形 L, 将以 $\mathrm{St}\,L$ 表由一切 $\mathrm{St}\,\sigma, \sigma\in L$ 的并集. 对 K 的任一开子复合形 $L, K-L$ 是一闭子复合形而 $|K|-|K-L|$ 将简记为 $|L|$. 于是有

引理 3 对 K 的任意闭子复合形 L, 命 f 为 $|L|$ 到 $|K|$ 中的连续映像使对任意 $\sigma\in L$ 与 $x\in|\sigma|$, 有 $f(x)\in|\mathrm{St}\,\mathrm{Cl}\,\sigma|$. 则 f 同伦于 $|L|$ 的恒同映像.

证 对任意 $\sigma\in L$, 相应于 $\tau\prec\sigma$ 的诸集合 $|\mathrm{St}\tau|$ 构成一个 $|\mathrm{St}\,\mathrm{Cl}\,\sigma|$ 的开覆盖, 显然对任意 r 都满足引理 2 的条件 ($|\mathrm{St}\,\sigma|$ 的作用如 U_0). 因之 $|\mathrm{St}\,\mathrm{Cl}\,\sigma|$ 对任意 r 都是 r 连通的. 今定义一同伦 $h:|L|\times[0,1]\to|K|$, 伦移 $f\equiv h/|L|\times(0)$ 为 $h/|L|\times(1)\equiv|L|$ 的恒同映像如下：假设 h 已定义于 $|L^k|$ 上使对任意 $\tau\in L^k$ 有 $h(|\tau|\times[0,1])\subset|\mathrm{St}\,\mathrm{Cl}\,\tau|$. 考虑任一 $k+1$ 维单纯形 $\sigma\in L$, 则 h 已在 $|\sigma|\times[0,1]$ 的边界上定义而其象 $\subset|\mathrm{St}\,\mathrm{Cl}\,\sigma|$, 因而由于这一集合的 $r-1$ 连通性可推广之至 $|\sigma|\times[0,1]$ 使其象 $\subset|\mathrm{St}\,\mathrm{Cl}\,\sigma|$. 由归纳可得所求同伦 h, 故 $f\simeq$ 恒同映像.

今考虑空间 E 的一个任意有限覆盖 $\mathscr{U}=\{U_\lambda\}$, 以 K 为神经复合形, a_λ 为与 U_λ 相应的顶点, 对 K 的任意单纯形 $\sigma=(a_{\lambda_0}\cdots a_{\lambda_s})$ 集合 $U_{\lambda_0}\cdots U_{\lambda_s}$ 与 $U_{\lambda_0}+\cdots+U_{\lambda_s}$ 将各称为 σ 的支柱与盖集, 并将记之为 $\mathrm{Sup}\,\sigma$ 与 $\mathrm{Cov}\,\sigma$. 对 K 的任意子复合形 L 所有 $\sigma\in L$ 时 $\mathrm{Cov}\,\sigma$ 的并集将记作 $\mathrm{Cov}\,L$. 显然 τ 是 $\sigma\in K$ 的面 (即 $\tau\prec\sigma$) 时有 $\mathrm{Sup}\,\tau\supset\mathrm{Sup}\,\sigma$ 而 $\mathrm{Cov}\,\tau\subset\mathrm{Cov}\,\sigma$. 子复合形 L 到 E 的一个连续映像 f 将称为可允许的 (对 $\mathscr{U}$ 来说), 如果对任意 $\sigma\in L$, 有 $f(|\sigma|)\subset\mathrm{Cov}\,\sigma$. 对任意集合 $A\subset E$, 命 $\mathscr{U}$ 中包含 A 的集合的全体 (设有的话) 为 $U_{\lambda_0},\cdots,U_{\lambda_s}$, 则 K 中的单纯形 $\sigma=(a_{\lambda_0}\cdots a_{\lambda_s})$ 将称为 A 的载元而记之为 $\mathrm{Car}\,A$. 另一面 $\mathscr{U}$ 中所有使 $U_\lambda\cdot U_{\lambda_0}\cdots U_{\lambda_s}\neq\varnothing$ 的集合 U_λ 的并集将称为 A 的膨涨而记作 $\mathrm{Exp}\,A$. 易见对任意 $A\subset B\subset E$, 这里 B 至少全在 $\mathscr{U}$ 的一个集合中时, 有 $\mathrm{Exp}\,A\subset\mathrm{Exp}\,B, \mathrm{Car}\,B\prec\mathrm{Car}\,A, \overline{\mathrm{St}}\,\mathrm{Car}\,A\subset\overline{\mathrm{St}}\,\mathrm{Car}\,B$, 又对任一 $\sigma\in\overline{\mathrm{St}}\,\mathrm{Car}\,A$ 有 $\mathrm{Cov}\,\sigma\subset\subset\mathrm{Exp}\,A$.

我们将称一个覆盖为星状离散的, 如果对 $\mathscr{U}$ 的任意有限多个集合 $U_{\lambda_1},\cdots,U_{\lambda_s}$, 只须 $U_{\lambda_i}\cdot U_{\lambda_j}\neq\varnothing, i,j=1,\cdots,s$, 即有 $U_{\lambda_1}\cdots U_{\lambda_s}\neq\varnothing$. 则易见 $\overline{\mathrm{St}}\,\mathrm{Car}\,A$ 为 $\mathrm{Exp}\,A$ 的一个覆盖的神经复合形, 这个覆盖系由 $\mathscr{U}$ 中与 $\mathrm{Sup}\,\mathrm{Car}\,A$ 有非空交集的集合所构成. 我们又称 $\mathscr{U}$ 为 r 连通的, 如果 $\mathscr{U}$ 中集合的任意非空交集都是 r 连通的.

定理 3 设连通正常空间 E 有一星状离散与 r 连通的有限开覆盖 $\mathscr{U}$, 则 E 在维数 $1,2,\cdots,r$ 中有与神经复合形 K 相同的同伦群, 且这些同伦群的同构关系可为与 $\mathscr{U}$ 相关的任意 E 到 $|K|$ 的规范映像或 $|K^{r+1}|$ 到 E 的任意可允许映像 ψ 所实现, 这种可允许映像必然存在.

证　由引理 2 对任意 $\sigma\in K$ 集合 $\operatorname{Cov}\sigma$ 是 r 连通的. 今定义一 $|K^{r+1}|$ 到 E 的可允许映像如次: 对任一 K 的顶点在 $U_\lambda=\operatorname{Cov}(a_\lambda)$ 中任取一点 $\bar{a}_\lambda$, 记之为 $\psi(a_\lambda)$. 假设 ψ 已在 $|K^k|$ 上有定义, $k\leqslant r$, 使对每一 $\sigma\in K^k$, 有 $\psi(|\sigma|)\subset\operatorname{Cov}\sigma$. 试考任一 K 的 $k+1$ 维单纯形. 映像 ψ 已在 σ 的边界 $\dot{\sigma}$ 上定义使 $\psi(|\dot{\sigma}|)\subset\sum\limits_{t<\sigma}\operatorname{Cov}\tau\subset\operatorname{Cov}\sigma$, 而 $\psi(|\dot{\sigma}|)$ 是 r 连通的. 因 $k\leqslant r$ 故 ψ 可拓广到 $|\sigma|$ 使 $\psi(|\sigma|)\subset\operatorname{Cov}\sigma$. 这证明了 $\psi:|K^{r+1}|\to E$ 的存在.

其次试证 ψ 与覆盖 $\mathscr{U}$ 所定的一个规范映像 φ 引出在维数 $s\leqslant r$ 的同构

$$\psi_*:\pi_s(|K|)\approx\pi_s(E),$$
$$\varphi_*:\pi_s(E)\approx\pi_s(|K|)$$

并互为其逆如下 (参考点可取为 K 的一个顶点与其在 ψ 下的象). 证明将分 $\varphi_*\psi_*$ 在维数 $s\leqslant r$ 时恒同以及 ψ_* 在维数 $s\leqslant r$ 时为同构上二部分以完成之. 试先论前者. 为此试考虑 K 中一 k 维单纯形 $\sigma=(a_{\lambda_0}\cdots a_{\lambda_k})$ 的任一点 $x,k\leqslant r+1$, 于是 $\psi(x)\in U_{\lambda_0}++\cdots+U_{\lambda_k}=\operatorname{Cov}\sigma$. 命 $U_{\lambda_{i_1}},\cdots,U_{\lambda_{i_a}}(a\geqslant 1)$ 或尚有 $U_{\mu_1},\cdots,U_{\mu_b}$, 为 $\mathscr{U}$ 中含有 $\psi(x)$ 的集合全体, 则 $\varphi\psi(x)$ 在单纯形 $a_{\lambda_{i_1}}\cdots a_{\lambda_{i_a}}a_{\mu_1}\cdots a_{\mu_b}$ 中但不在面 $a_{\mu_1}\cdots a_{\mu_b}$ 上, 因而 $\varphi\psi(x)\in|\operatorname{St}\operatorname{Cl}\sigma|$. 由引理 3 故有 $\varphi\psi/K^{r+1}\simeq$ 恒同而有在维数 $s\leqslant r$ 时, $\varphi_*\psi_*=$ 恒同. 为证 ψ_* 在维数 $s\leqslant r$ 时为同构上, 试考虑一个 s 维球 S 到 E 的任一连续映像 f, 这里 $s\leqslant r$. 剖分 S 使对 S 的每一单纯形 $\xi,f(|\xi|)$ 全在某一 U_λ 内. 我们将定义一同伦 $h:S\times[0,1]\to E$ 变 $f\equiv h/S\times(0)$ 为 $g\equiv h/S\times(1)$ 使 $g=\psi g_0,g_0:S\to|K^r|$ 如下: 对 S 的任一顶点 b, 命 U_λ 为含有 $f(b)$ 的 $\mathscr{U}$ 中的某集, 则 $h((b)\times[0,1])$ 将取为全在 U_λ 中联 $f(b)$ 至 $\psi(a_\lambda)$ 的一个道路. 又置 $g_0(b)=a_\lambda$. [在 b 是 S 上的参考点时, a_λ 将取为 $\pi_s(|K|)$ 的参考点而 $h(b\times[0,1])$ 将为约化为点 $f(b)=\psi(a_\lambda)$ 的点道路]. 假设 h 在 $S^k\times[0,1]$ 与 g_0 在 S^k 上都已定义并满足以下性质 $(k<S)$: 对任意 S 的 k 维单纯形 ξ 有 (i) $h(|\xi|\times[0,1])\subset\operatorname{Exp}f(|\xi|)$, 与 (ii) $g_0(|\xi|)\subset\subset|[\overline{\operatorname{St}}\operatorname{Car}f(|\xi|)]^{r+1}|$. 今考 S 的任一 $k+1$ 维单纯形 η, 则 $g_0(|\dot{\eta}|)$ 已定义且 $\subset\sum\limits_{\zeta\in\dot{\eta}}|[\overline{\operatorname{St}}\operatorname{Car}f(|\zeta|)]^{r+1}|\subset|[\overline{\operatorname{St}}\operatorname{Car}f(|\eta|)]^{r+1}|$, 这里 $\overline{\operatorname{St}}\operatorname{Car}f(|\eta|)$ 是 $\operatorname{Exp}f(|\eta|)$ 被 $\mathscr{U}$ 中与 $\operatorname{Sup}\operatorname{Car}f(|\eta|)$ 有非空交集的集合所成覆盖的神经复合形. 由引理 2 复合形 $\overline{\operatorname{St}}\operatorname{Car}f(|\eta|)$ 是 r 连通的, 故 $g_0/|\dot{\eta}|$ 可拓广至 $g_0:|\eta|\to|[\overline{\operatorname{St}}\operatorname{Car}f(|\eta|)]^{r+1}|$. 由此得 $\psi g_0(|\eta|)\subset\psi(|[\overline{\operatorname{St}}\operatorname{Car}f(|\eta|)]^{r+1}|)\subset\operatorname{Cov}[\overline{\operatorname{St}}\operatorname{Car}f(|\eta|)]^{r+1}\subset\operatorname{Exp}f(|\eta|)$. 定义 $h/|\eta|\times\times(1)$ 为 $\psi g_0/|\eta|$, 则 h 在 $|\eta|\times[0,1]$ 的边界上有定义而其象全含于 $\operatorname{Exp}f(|\eta|)$. 仍由引理 2 $\operatorname{Exp}f(|\eta|)$ 是 r 连通的. 故 h 可拓广至 $|\eta|\times[0,1]$ 使其象 $\subset\operatorname{Exp}f(|\eta|)$. 如此依归纳可最后得一映像 $h(S\times[0,1])$ 给出一 f 到映像 ψg_0 的同伦, 这里 $g_0:S\to\to|K^{r+1}|$. 这证明了 ψ_* 在维数 $s\leqslant r$ 时是同构上. 证毕.

附注　关于同伦群不能有像定理 1 关于同调群那样的定理, 例如设 E 是 2 维球, U_1, U_2 是比半球略加扩大的两个开半球, 使 $U_1 \cdot U_2$ 为环绕赤道的一个带形, 则 U_1, U_2 与 $U_1 \cdot U_2$ 都有 $\pi_{r-1} = 0, \pi_r = 0$, 只须 $r > 2$. 但 $\pi_r(E)$ 已知在 $r = 3, 4$ 等都是 $\neq 0$ 的.

§5 对同伦型的推广

J.H.C.Whitehead 称一个连通的紧致统为一 CR 空间, 如果它也是一个 ANR 集. 如 Whitehead 所证明, 一个 CR 空间 X 到 CR 空间 Y 的映像 f 如果对每一维数 $n \geqslant 1$ 都引出一同构上 $\pi_n(X) \approx \pi_n(Y)$, 则 f 必是一同伦等价映像 (参阅 [5]). 与前一定理相结合, 即得

定理 4 若 $\mathscr{U}$ 是一 CR 空间 E 的有限星状离散的开覆盖, 使 $\mathscr{U}$ 中集合的每一非空交集都是可缩成一点的, 则空间 E 有与 $\mathscr{U}$ 的神经复合形 K 相同的同伦型, 且其同伦等价可被任一对 $\mathscr{U}$ 的规范映像 $\varphi: E \to |K|$ 所实现, 同样亦可为任一 $|K|$ 到 E 的可允许映像 (这种映像必然存在) 所实现, 而二者互相为逆.

在以下我们将给出定理 4 的一个变形, 应用闭覆盖而不用开覆盖. 为此试先回忆绝对收缩核的定义: 一个正则可分空间 E 称为绝对收缩核, 如果对任意正则可分空间 A, B, 这里 A 是 B 的闭子集, 任意 A 到 E 的连续映像必可拓广到 B. 下述定理 (引理 1) 是属于 Aronszajn 与 Borsuk 的 (参阅 [1]):

引理 1 若正则可分空间 E 是两个闭子集的并集, 而 F_1, F_2 与 $F_1 \cdot F_2$ 都是绝对收缩核, 则 E 也是绝对收缩核.

下述引理是§4 中引理 2 的类似.

引理 2 设 $\mathscr{F} = \{F_0, F_1, \cdots, F_n\}$ 是正则可分空间 E 的一个闭覆盖, 使

(i) $F_i \cdot F_0 \neq \varnothing, i \neq 0$;

(ii) 若 $F_{\lambda_i} \cdot F_{\lambda_j} \neq \varnothing, i, j = 1, \cdots, r$, 则 $F_{\lambda_1} \cdots F_{\lambda_r} \neq \varnothing$ $(\lambda_i = 0, 1, \cdots, n)$; 又

(iii) $\mathscr{F}$ 中集合的任意非空交集都是绝对收缩核.

于是 E 自身也是绝对收缩核.

证 对 n 行归纳法.

定理 4′ 若 $\mathscr{F} = \{F_\lambda\}$ 是 CR 空间 E 的一个有限星状离散闭覆盖, 使 $\mathscr{F}$ 中集合的任意非空交集都是绝对收缩核, 则空间 E 与 $\mathscr{F}$ 的神经复合形 K 有相同的同伦型, 且其同伦等价可为任一对 $\mathscr{F}$ 的规范映像 $\varphi: E \to |K|$ 所实现, 亦可为任意可允许映像 (这样的映像必然存在) $|K| \to E$ 所实现, 而二者互相为逆.

证 因每一绝对收缩核都是可缩为一点的, 故与定理 3 中的证明相同, 可证使 $\psi(|\sigma|) \subset \operatorname{Cov} \sigma$ 的可允许映像 $\psi: |K| \to E$ 必然存在.

命 φ 为与闭覆盖 $\mathscr{F}$ 相关的任一 E 到 $|K|$ 中的规范映像. 定理的证明将分 $\varphi\psi \simeq$ 恒同与 $\psi\varphi \simeq$ 恒同两部分以完成之.

为证前者试先注意对任一点 $a \in |\sigma|, \sigma = (a_{\lambda_0} \cdots a_{\lambda_r}) \in K$, 有 $\psi(a) = \bar{a} \in \operatorname{Cov}\sigma$, 因而 $\varphi(\bar{a}) = \varphi\psi(a) \in |\operatorname{St}\operatorname{Cl}\sigma|$. 由§4 的引理 3, 故有 $\varphi\psi \simeq$ 恒同.

为证第二部分试注意对 E 的任一点 $\bar{a}$, 有 $\psi\varphi(\bar{a}) \in \operatorname{Exp}(\bar{a})$. 由引理 2 这一集合是一绝对收缩核 ($\mathscr{F}$ 中任一含 $\bar{a}$ 的集合可作为引理 2 中的 F_0). 今作一同伦 $h: E\times[0,1] \to E$ 使 $h/E\times(0) \equiv \psi\varphi, h/E\times(1) \equiv E$ 的恒同映像, 而对任意 $\bar{a} \in E$ 有 $h(\bar{a}\times[0,1]) \subset\subset S(\bar{a})$ 如次: 假设 h 对所有 $\bar{a}\times[0,1], \bar{a} \in \operatorname{Sup}\tau, \tau \in K, \dim\tau > k$, 都已作出并满足上述要求. 试考 K 的任一 k 维单纯形 σ. 命 $\sigma_j, j = 1, \cdots, s$ 为 K 中以 σ 为面的 $k+1$ 维单纯形全体, 则 $\operatorname{Sup}\sigma \supset \sum \operatorname{Sup}\sigma_j$ 而 h 已在 $\sum \operatorname{Sup}\sigma_j \times [0,1] + \operatorname{Sup}\sigma\times(0) + +\operatorname{Sup}\sigma\times(1) \subset \operatorname{Sup}\sigma\times[0,1]$ 上定义, 且其象 $\subset \operatorname{Exp}\operatorname{Sup}\sigma$. 由引理 2 末一集合是绝对收缩核, 故可拓广 h 至 $\operatorname{Sup}\sigma\times[0,1]$, 使其象 $\subset \operatorname{Exp}\operatorname{Sup}\sigma$. 对任意点 $\bar{a} \in \operatorname{Sup}\sigma$ 但不 $\in \sum \operatorname{Sup}\sigma_j$ 即有 $\operatorname{Car}\bar{a} = \sigma$ 致 $h(\bar{a}\times[0,1]) \subset \operatorname{Exp}\operatorname{Sup}\sigma = \operatorname{Exp}(\bar{a})$. 由归纳故可得所求同伦 h 而 $\psi\varphi \simeq$ 恒同得证.

参考文献

[1] Aronszajn N and Borsuk K. Sur la somme et le produit combinatoire des rétraites absolus. *Fund.Math.*, 1932, 18: 193-197.

[2] Eilenberg S and Steenrod N E. Foundations of algebraic topology, Vol.I, 1952.

[3] Leray J. Sur la forme des espaces topologiques et sur les points fixes des représentations. *J. de Math.*, 1945, 24: 95-167.

[4] Van Kampen E R. On the connection between the fundamental groups of some related spaces. *Amer.J.Math.*, 1933, 60: 261-267.

[5] Whitehead J H C. On the homotopy type of ANR's. *Bull.Amer.Math.Soc.*, 1948, 54: 1133-1145.

[6] Olum P. Non-abelian cohomology and Van Kampen's theorem. *Annals of Math.*, 1958, 68: 658-668.

36. 某些实二次曲面的示性类*

吴文俊　李培信

前　　言

对于任意微分流形 M, 可定义 Stiefel-Whitney 示性类 $W^i(M) \in H^i(M, Z_2)$ 与 Понтрягин 示性类 $P^{4k}(M) \in H^{4k}(M)$. 对于任意复流形 M, 则可定义陈省身示性类 $C^{2i}(M)$, 这时视 M 为实微分流形时, $W^i(M)$ 与 $P^{4k}(M)$ 都可自 $C^{2i}(M)$ 定出 (见 [8]). 一些重要流形的示性类的具体计算虽原则上有一般方法, 但并不简单, 其已知者就作者所知犹如下述:

1° 实投影空间, 见 Stiefel [5].

2° 复投影空间, 见陈省身 [2].

3° 实 Stiefel 流形, 见吴文俊 [6], 复 Stiefel 流形亦可同法决定.

4° 非退化复二次曲面, 见 Morie-Hèlèue Schwarte [4] 或吴文俊 [7] 附录.

5° 复 Grassmann 流形, 见 Vesentini [1], 其中只计算出了极少一部分.

6° 在 Borel 与 Hirzebruch [9] 中, 提供了一个一般方法, 应用群表示论, 对 1°, 2°, 4°, 5° 都给出了明显公式.

在本文中, 我们将应用吴文俊在 [8] 中所示的方法以计算投影空间中某些实二次曲面的示性类.

§1　实二次曲面 $Q^n_{p,q}$ 的示性类

1. P^{p+q+1} 表示 $p+q+1$ 维的实投影空间, 在 P^{p+q+1} 里取定了齐次坐标 $(x_0, \cdots, x_p, y_0, \cdots, y_q)$.

$Q^n_{p,q}(n = p+q, p \leqslant q)$ 表示 P^{p+q+1} 里的被方程

$$x_0^2 + \cdots + x_p^2 - y_0^2 - \cdots - y_q^2 = 0 \tag{1}$$

所确定的曲面, 在这一节里我们要计算 $Q^n_{p,q}$ 的 Stiefel-Whitney 示性类.

* *Acta Math. Sinica*, 1962, 12 : 203-215; *Translated in Chinese Math.*, 1963, 3 : 218-231.

2. 覆盖空间 $Q^n_{p,q}$.

P^p, P^q 分别表示 p 维和 q 维的实投影空间, 在它们里面分别取定了齐次坐标 $(x_0,\cdots,x_p)$ 和 $(y_0,\cdots,y_q)$. 定义 $Q^n_{p,q}$ 到 $P^p\times P^q$ 的一映射 π 如下:

$$\pi:\begin{cases} Q^n_{p,q}\to P^p\times P^q,\\ (x_0,\cdots,x_p,y_0,\cdots,y_q)\to(x_0,\cdots,x_p)\times(y_0,\cdots,y_q)=\pi((x_0,\cdots,x_p,y_0,\cdots,y_q)).\end{cases}$$

映射 π 具有下述性质:

1° π 是把 $Q^n_{p,q}$ 映到 $P^p\times P^q$ 上.

证 设 $(x_0,\cdots,x_p)\times(y_0,\cdots,y_q)\in P^p\times P^q$, 令 $k=\sqrt{\dfrac{y_0^2+\cdots+y_q^2}{x_0^2+\cdots+x_p^2}}$, 那么由 $(kx_0)^2+\cdots+(kx_p)^2=k^2(x_0^2+\cdots+x_p^2)=y_0^2+\cdots+y_q^2$ 就推出点 $(kx_0,\cdots,kx_p,y_0,\cdots,y_q)\in Q^n_{p,q}$, 而 $\pi((kx_0,\cdots,kx_p,y_0,\cdots,y_q))=(kx_0,\cdots,kx_p)\times(y_0,\cdots,y_q)=(x_0,\cdots,x_p)\times(y_0,\cdots,y_q)$ 所以 π 是把 $Q^n_{p,q}$ 映到 $P^p\times P^q$ 上.

2° 设 $(x_0,\cdots,x_p,y_0,\cdots,y_q)\in Q^n_{p,q}$, $(x'_0,\cdots,x'_p)\times(y'_0,\cdots,y'_q)\in P^p\times P^q$ 和 $\pi((x_0,\cdots,x_p,y_0,\cdots,y_q))=(x'_0,\cdots,x'_p)\times(y'_0,\cdots,y'_q)$. 由 $\pi((x_0,\cdots,x_p,y_0,\cdots,y_q))=(x_0,\cdots,x_p)\times(y_0,\cdots,y_q)=(x'_0,\cdots,x'_p)\times(y'_0,\cdots,y'_q)$ 得出 $x_i=\lambda x'_i$, $i=0,\cdots,p$, λ 是一非零实数, $y_j=\mu y'_j$, $j=0,1,\cdots,q$, μ 是一非零实数; 又由 $(x_0,\cdots,x_p,y_0,\cdots,y_q)=(\lambda x'_0,\cdots,\lambda x'_p,\mu y'_0,\cdots,\mu y'_q)\in Q^n_{p,q}$ 得出 $\lambda^2x_0'^2+\cdots+\lambda^2x_p'^2=\mu^2y_0'^2+\cdots+\mu^2y_q'^2$ 即 $\dfrac{\lambda}{\mu}=\pm\sqrt{\dfrac{y_0'^2+\cdots+y_q'^2}{x_0'^2+\cdots+x_p'^2}}$, 令 $k=\sqrt{\dfrac{y_0'^2+\cdots+y_q'^2}{x_0'^2+\cdots+x_p'^2}}$, 由上述就有 $\pi^{-1}((x'_0,\cdots,x'_p)\times(y'_0,\cdots,y'_q))$ 是由两个点 $(kx'_0,\cdots,kx'_p,y'_0,\cdots,y'_q)$ 和 $(-kx'_0,\cdots,-kx'_p,y'_0,\cdots,y'_q)$ 组成.

3° $\{Q^n_{p,q},\pi\}$ 是 $P^p\times P^q$ 的一具有两叶的覆盖空间.

证 在 P^p 和 P^q 中分别选取邻域组 $U_i, i=0,1,\cdots,p$ 和 $V_j, j=0,1,\cdots,q$, 如下:

$$\begin{cases} U_i=\{x=(x_0,\cdots,x_p)|x_i\neq 0\}, & i=0,1,\cdots,p,\\ V_j=\{y=(y_0,\cdots,y_q)|y_j\neq 0\}, & j=0,1,\cdots,q.\end{cases}$$

我们知道它们就分别作成 P^p 和 P^q 的覆盖. 所以 $U_i\times V_j, i=0,1,\cdots,p, j=0,1,\cdots,q$ 就作成 $P^p\times P^q$ 的覆盖. $\pi^{-1}(U_i\times V_j)=X_1\cup X_2$ 由两个分量 X_1 和 X_2 组成,

$$X_1=\left\{\left(k\frac{x_0}{x_i},\cdots,k\frac{x_{i-1}}{x_i},k,k\frac{x_{i+1}}{x_i},\cdots,k\frac{x_p}{x_i},\frac{y_0}{y_j},\cdots,\frac{y_{j-1}}{y_j},1,\frac{y_{j+1}}{y_j},\cdots,\frac{y_q}{y_j}\right)\right.$$
$$\left.\middle|\, x_i\neq 0,y_j\neq 0,k-\sqrt{\frac{x_i^2(y_0^2+\cdots+y_q^2)}{y_j^2(x_0^2+\cdots+x_p^2)}}\right\},$$

$$X_2=\left\{\left(-k\frac{x_0}{x_i},\cdots,-k\frac{x_{i-1}}{x_i},-k,-k\frac{x_{i+1}}{x_i},\cdots,-k\frac{x_p}{x_i},\frac{y_0}{y_j},\cdots,\frac{y_{j-1}}{y_j},1,\frac{y_{j+1}}{y_j},\cdots,\frac{y_q}{y_j}\right)\right.$$
$$\left.\left|x_i\neq 0,y_j\neq 0,k=\sqrt{\frac{x_i^2(y_0^2+\cdots+y_q^2)}{y_j^2(x_0^2+\cdots+x_p^2)}}\right.\right\}.$$

显然, 可见在 π 下它们都分别是与 $U_i\times V_j$ 同胚, 这样我们就证明了 $\{Q^n_{p,q},\pi\}$ 是 $P^p\times P^q$ 的一具有两叶的覆盖空间.

3. 纤维从 $\mathscr{B}_1=\{Q^n_{p,q},\pi,P^p\times P^q\}$ 的 Stiefel-Whitney 示性类.

1° 已知投影空间的同调构造如下:

$H^1(P^p;Z_2)\cong Z_2, X$ 表其不为零的元素,

$H^i(P^p;Z_2)\cong Z_2$, 它的不为零的元素就是 X^i, 这里 X^i 表示 X 的 i 次幂, 乘积了解为上积.

$H^1(P^q;Z_2)\cong Z_2, Y$ 表其不为零的元素,

$H^i(P^q;Z_2)\cong Z_2$, 它的不为零的元素就是 Y^i, 由 Künneth 公式就知道 $H^k(P^p\times P^q;Z_2)$ 的生成元如下:

$$\begin{aligned}X^0\times Y^k,\cdots,X^k\times Y^0,&\quad 当\ 0\leqslant k\leqslant p,\\X^0\times Y^k,\cdots,X^p\times Y^{k-p},&\quad 当\ p<k<q,\\X^{k-q}\times Y^q,\cdots,X^p\times Y^{k-p},&\quad 当\ q\leqslant k\leqslant n,\end{aligned}$$

这里 X^0 和 Y^0 分别表 $H^0(P^p;Z_2)$ 和 $H^0(P^q;Z_2)$ 的非零元素.

2° $W(\mathscr{B}_1)=1+W^1(\mathscr{B}_1),W^1(\mathscr{B}_1)\in H^1(P^p\times P^q;Z_2)$ 表丛 $\mathscr{B}_1$ 的 Stiefel-Whitney 示性类, 在这一小节里我们要来计算 $W^1(\mathscr{B}_1)$.

X_0 和 X_1 分别表示 $H_0(P^p;Z)\cong Z$ 和 $H_1(P^p;Z)\cong Z_2$ 的生成元, Y_0 和 Y_1 分别表示 $H_0(P^q;Z)\cong Z$ 和 $H_1(P^q;Z)\cong Z_2$ 的生成元; 那么 $H_1(P^p\times P^q:Z)$ 的生成元就是 $X_0\times Y_1$ 和 $X_1\times Y_0$.

几何地表示 $X_0\times Y_1$ 和 $X_1\times Y_0$ 如下: 在 P^p 中选定一点 $(1,\underbrace{0,\cdots,0}_{p})$ 表示 X_0, 再在 P^p 中另外选定一点 $(0,1,\underbrace{0,\cdots,0}_{p-1})$, 联结这两点的投影直线是 $\{(\lambda,\mu,\underbrace{0,\cdots,0}_{p-1})|$, λ,μ 不同时为零$\}$, 我们就取它几何地来表示 X_1. 同样在 P^q 中选取一点 $(1,\underbrace{0,\cdots,0}_{q})$ 表示 Y_0, 再在 P^q 中另外选取一点 $(0,1,\underbrace{0,\cdots,0}_{q-1})$, 联结这两点的投影直线是 $\{(\alpha,\beta,0,\cdots,0)|\alpha,\beta$ 不同时为零$\}$, 我们就取这条投影直线几何地来表示 Y_1. 由此就可以得到 $X_0\times Y_1$ 和 $X_1\times Y_0$ 的几何表示如下:

$$X_0 \times Y_1 : \{(1, \underbrace{0, \cdots, 0}_{p}) \times (\alpha, \beta, \underbrace{0, \cdots, 0}_{q-1}) | \alpha, \beta \text{ 不同时为零}\},$$

$$X_1 \times Y_0 : \{(\lambda, \mu, \underbrace{0, \cdots, 0}_{p-1}) \times (1, \underbrace{0, \cdots, 0}_{q}) | \lambda, \mu \text{ 不同时为零}\}.$$

为了计算 $W^1(\mathscr{B}_1)$, 只要计算 $W^1(\mathscr{B}_1)$ 在 $H_1(P^p \times P^q; Z)$ 的生成元上取的值, 为此在 $X_1 \times Y_0$ 上造向量场 (即通常所谓的截面) f 如下:

$$f((\lambda, \mu, \underbrace{0, \cdots, 0}_{p-1}) \times (1, \underbrace{0, \cdots, 0}_{q}))$$

$$= \left(\frac{1}{\sqrt{1 + \left(\frac{\mu}{\lambda}\right)^2}}, \frac{\frac{\mu}{\lambda}}{\sqrt{1 + \left(\frac{\mu}{\lambda}\right)^2}}, 0, \cdots, 0, 1, \underbrace{0, \cdots, 0}_{q} \right) \in Q^n_{p,q},$$

当$\lambda \neq 0$.

$$f((0, 1, \underbrace{0, \cdots, 0}_{p-1}) \times (1, \underbrace{0, \cdots, 0}_{q})) = (0, 1, 0, \cdots, 0, 1, \underbrace{0, \cdots, 0}_{q}) \in Q^n_{p,q}.$$

考虑点 $(\lambda, \mu, 0, \cdots, 0) \times (1, 0, \cdots, 0), \lambda \neq 0$; 当 $\dfrac{\mu}{\lambda} \to +\infty$ 时, 这点趋于点 $(0, 1, \underbrace{0, \cdots, 0}_{p-1}) \times (1, \underbrace{0, \cdots, 0}_{q})$; 当 $\frac{\mu}{\lambda} \to -\infty$ 时, 这点也趋于点 $(0, 1, \underbrace{0, \cdots, 0}_{p-1}) \times (1, \underbrace{0, \cdots, 0}_{q})$; 但是当 $\dfrac{\mu}{\lambda} \to +\infty$ 时,

$$f((\lambda; \mu, \underbrace{0, \cdots, 0}_{p}) \times (1, \underbrace{0, \cdots, 0}_{q})) = \left(\frac{1}{\sqrt{1 + \left(\frac{\mu}{\lambda}\right)^2}}, \frac{\frac{\mu}{\lambda}}{\sqrt{1 + \left(\frac{\mu}{\lambda}\right)^2}}, 0, \cdots, 0, 1, \underbrace{0, \cdots, 0}_{q} \right)$$

趋于点 $(0, 1, 0, \cdots, 0, 1, \underbrace{0, \cdots, 0}_{q})$; 而当 $\dfrac{\mu}{\lambda} \to -\infty$ 时,

$$f((\lambda, \mu, \underbrace{0, \cdots, 0}_{p}) \times (1, \underbrace{0, \cdots, 0}_{q})) = \left(\frac{1}{\sqrt{1 + \left(\frac{\mu}{\lambda}\right)^2}}, \frac{\frac{\mu}{\lambda}}{\sqrt{1 + \left(\frac{\mu}{\lambda}\right)^2}}, 0, \cdots, 0, 1, \underbrace{0, \cdots, 0}_{q} \right)$$

趋于点 $(0, -1, 0, \cdots, 0, 1, \underbrace{0, \cdots, 0}_{q})$. 由此可见点 $(0, 1, \underbrace{0, \cdots, 0}_{p-1}) \times (1, \underbrace{0, \cdots, 0}_{q})$ 是向量场 f 的奇点.

类似地可以证明 $(1,\underbrace{0,\cdots,0}_{p})\times(0,1,\underbrace{0,\cdots,0}_{q-1})$ 也是一个奇点, 这样我们就证明了下面的

命题 1 $W^1(\mathscr{B}_1)=X^1\times Y^0+X^0\times Y^1\in H^1(P^p\times P^q;Z_2)$.

4. $Q^n_{p,q}$ 的同调性质.

在这一小节里我们要利用纤维丛的 Gysin 序列来计算 $Q^n_{p,q}$ 的同调性质.

对于球纤维丛 $\mathscr{B}_1=\{Q^n_{p,q},\pi,P^p\times P^q\}$ 有一所谓 Gysin 序列如下:

$$\cdots\to H^{r-1}(P^p\times P^q)\xrightarrow{\Psi_{r-1}}H^r(P^p\times P^q)\xrightarrow{\pi_r}H^r(Q^n_{p,q})\xrightarrow{\Phi_r}H^r(P^p\times P^q)$$
$$\xrightarrow{\Psi_r}H^{r+1}(P^p\times P^q)\to\cdots$$

它是正合的. 这儿的系数群可以是任意的, 但在我们这里只考虑系数群是 Z_2. 考虑下面图形:

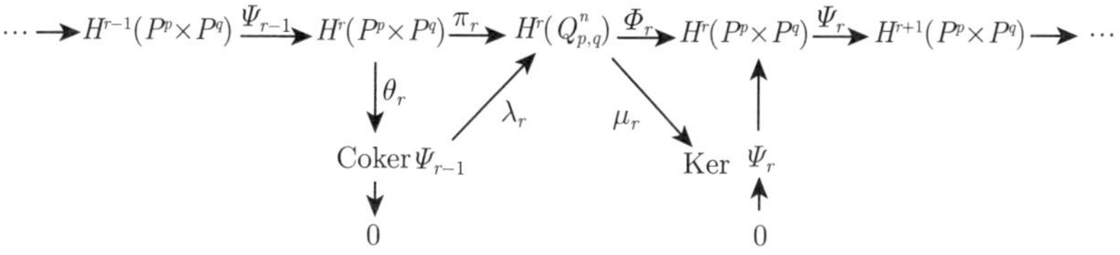

$\cdots\to H^{r-1}(P^p\times P^q)\xrightarrow{\Psi_{r-1}}H^r(P^p\times P^q)\xrightarrow{\pi_r}H^r(Q^n_{p,q})\xrightarrow{\Phi_r}H^r(P^p\times P^q)\xrightarrow{\Psi_r}H^{r+1}(P^p\times P^q)\to\cdots$
θ_r
λ_r
μ_r
Coker Ψ_{r-1}
Ker Ψ_r
0
0

这儿 θ_r 是自然映射, λ_r 是由 π_r 导出的同态, 由 $\operatorname{Im}\Psi_{r-1}=\operatorname{Ker}\pi_r$, 就知道 λ_r 是唯一确定的, 而且 λ_r 是同构映入的映射; μ_r 是由 Φ_r 诱导出的映射, 由 $\operatorname{Ker}\Psi_r=\operatorname{Im}\Phi_r$, 知道 μ_r 是在上的同态; 由 $\operatorname{Ker}\Phi_r=\operatorname{Im}\pi_r$ 就可以推出 $\operatorname{Ker}\mu_r=\operatorname{Im}\lambda_r$, 我们的系数群 Z_2 是一域, 所以 $H^r(Q^n_{p,q})$ 还是 Z_2 上的向量空间, 因此由 $\operatorname{Ker}\mu_r=\operatorname{Im}\lambda_r$ 就可以推出

$$H^r(Q^n_{p,q})\cong\operatorname{Coker}\Psi_{r-1}+\operatorname{Ker}\Psi_r.$$

由一已知的定理 (参阅 [3]) 有: 设 $u\in H^r(P^p\times P^q)$, 则 $\Psi_r(u)=u\cup W^1(\mathscr{B}_1)$. 由此可以求得下面的命题

命题 2

$p<q$ 的情形	$0\leqslant r\leqslant p$	$p<r<q$	$q\leqslant r\leqslant n$
$H^r(P^p\times P^q)$ 的生成元	$X^0\times Y^r,\cdots,$ $X^r\times Y^0$	$X^0\times Y^r,\cdots,$ $X^p\times Y^{r-p}$	$X^{r-q}\times Y^q,\cdots,$ $X^p\times Y^{r-p}$
Ker Ψ_r	0	0	$X^{r-q}\times Y^q+\cdots$ $+X^p\times Y^{r-p}$
Coker Ψ_{r-1}	$\theta_r(X^0\times Y^r)=\cdots$ $=\theta_r(X^r\times Y^0)$	0	0

$p=q$ 的情形	$0\leqslant r<p$	$r=p=q$	$q<r\leqslant n$
$H^r(P^p\times P^q)$ 的生成元	$X^0\times Y^r,\cdots,$ $X^r\times Y^0$	$X^0\times Y^r,\cdots,$ $X^r\times Y^0$	$X^{r-q}\times Y^q,\cdots,$ $X^p\times Y^{r-p}$
Ker $\varPsi_r$	0	$X^0\times Y^r+\cdots$ $+X^r\times Y^0$	$X^{r-q}\times Y^q,\cdots,$ $X^p\times Y^{r-p}$
Coker $\varPsi_{r-1}$	$\theta_r(X^0\times Y^r)=\cdots$ $=\theta_r(X^r\times Y^0)$	$\theta_r(X^0\times Y^r)=\cdots$ $=\theta_r(X^r\times Y^0)$	0

证　$p<q$ 的情形.

1° $0\leqslant r\leqslant p$: $\varPsi_r(X^i\times Y^{r-i})=X^i\times Y^{r-i}\cup(X^1\times Y^0+X^0\times Y^1)=X^{i+1}\times$ $\times Y^{r-i}+X^i\times Y^{r+1-i}$, 而 $X^{i+1}\times Y^{r-i}$ 和 $X^i\times Y^{r+1-i}$ 都是 $H^{r+1}(P^p\times P^q)$ 的生成元, 所以 $\varPsi_r(X^i\times Y^{r-i})\neq 0$, 所以 $\operatorname{Ker}\varPsi_r=0$ 和 $\theta_{r+1}(X^{i+1}\times Y^{r-i})=\theta_{r+1}(X^i\times Y^{r-i+1})$ 即 $\operatorname{Coker}\varPsi_{r-1}$ 有生成元 $\theta_r(X^0\times Y^r)=\cdots=\theta_r(X^r\times Y^0)$.

2° $p<r<q$: 与 1° 的考虑一样可得 $\operatorname{Ker}\varPsi_r=0$, 对 $p<r<q$ 有 $H^{r-1}(P^p\times P^q)$ 的生成元为 $X^0\times Y^{r-1},\cdots,X^p\times Y^{r-p-1}$, $H^r(P^p\times P^q)$ 的生成元为 $X^0\times Y^r,\cdots,X^p\times Y^{r-p}$, 而 $\varPsi_{r-1}(X^0\times Y^{r-1})=X^0\times Y^r+X^1\times Y^{r-1},\cdots,\varPsi_{r-1}(X^{p-1}\times Y^{r-p})=X^{p-1}\times Y^{r-p+1}+X^p\times Y^{r-p},\varPsi_{r-1}(X^p\times Y^{r-p-1})=X^p\times Y^{r-p}$, 这些元素正好作成 $H^r(P^p\times P^q)$ 的生成元, 所以 $\operatorname{Coker}\varPsi_{r-1}=0$.

3° $q\leqslant r\leqslant n$: $\varPsi_r(X^{r-q+i}\times Y^{q-i})=X^{r-q+i}\times Y^{q-i}\cup(X^1\times Y^0+X^0\times Y^1)=X^{r-q+i+1}\times Y^{q-i}+X^{r-q+i}\times Y^{q-i+1}, i=0,\cdots,q-r+p$, 而 $X^{r-q}\times Y^{q+1}=X^{p+1}\times Y^{r-p}=0$, 所以有 $\varPsi_r(X^{r-q}\times Y^q+\cdots+X^p\times Y^{r-p})=0$, 所以得到 $\operatorname{Ker}\varPsi_r$ 的生成元为 $X^{r-q}\times Y^q+\cdots+X^p\times Y^{r-p}$, 而 $\operatorname{Coker}\varPsi_{r-1}=0$.

对 $p=q$ 的情形证明完全类似.

根据 $H^r(Q^n_{p,q})\cong\operatorname{Coker}\varPsi_{r-1}+\operatorname{Ker}\varPsi_r$ 和命题 2, 即得下面的

命题 3

$p<q$	$0\leqslant r\leqslant p$	$p<r<q$	$q\leqslant r\leqslant n$
$H^r(Q^n_{p,q})$ 的生成元	$Z^r=\pi_r(X^0\times Y^r)$ $=\cdots$ $=\pi_r(X^r\times Y^0)$	0	$Z^r=\varPhi_r^{-1}(X^{r-q}$ $\times Y^q+\cdots$ $+X^p\times Y^{r-p})$

$q=p$	$0\leqslant r<p$	$p=q=r$	$q<r\leqslant n$
$H^r(Q^n_{p,q})$ 的生成元	$Z^r=\pi_r(X^0\times Y^r)$ $=\cdots$ $=\pi_r(X^r\times Y^0)$	Z_1^r,Z_2^r $Z_1^r=\pi_r(X^0\times Y^r)$ $=\cdots$ $=\pi_r(X^r\times Y^0)$ $Z_2^r=\varPhi_r^{-1}(X^0\times Y^r$ $+\cdots+X^r\times Y^0)$	$Z^r=\varPhi_r^{-1}(X^{r-q}\times Y^q$ $+\cdots+X^p\times Y^{r-p})$

5. $Q^n_{p,q}$ 的示性类.

$$
\begin{aligned}
W^k(Q^n_{p,q}) &= \pi^* W^k(P^p \times P^q) = \pi^* \sum_{i+j=k} W^i(P^p) \times W^j(P^q) \\
&= \pi^* \sum_{i+j=k} \binom{p+1}{i}\binom{q+1}{j} X^i \times Y^j \\
&= \sum_{i+j=k} \binom{p+1}{i}\binom{q+1}{j} \pi^*(X^i \times Y^j) \\
&= \begin{cases} \displaystyle\sum_{i+j=k} \binom{p+1}{i}\binom{q+1}{j} Z^k, & k \leqslant p \\ 0, & k > p \end{cases} = \begin{cases} \dbinom{n+2}{k} Z^k, & k \leqslant p, \\ 0, & k > p. \end{cases}
\end{aligned}
$$

所以我们就得到下面的

定理 1

$$
W^k(Q^n_{p,q}) = \begin{cases} \dbinom{n+2}{k} Z^k, & k \leqslant p, \\ 0, & k > p. \end{cases}
$$

§2 $H^n_{p,q}$ 的示性类

1. $P_{p+q+1}(C)$ 表示 $p+q+1$ 维的复投影空间, 在它里面取定了齐次坐标 $(z_0, \cdots, z_{p+q+1})$, $H^n_{p,q}(n=p+q, p \leqslant q)$ 表示 $P_{p+q+1}(C)$ 里的被方程

$$
z_0\bar{z}_0 + z_1\bar{z}_1 + \cdots + z_p\bar{z}_p - z_{p+1}\bar{z}_{p+1} - \cdots - z_{p+q+1}\bar{z}_{p+q+1} = 0
$$

所确定的超二次曲面, 在这一节里我们要计算 $H^n_{p,q}$ 的 Stiefel-Whitney 示性类, 陈省身示性类和 Понтрягин 示性类.

2. 空间 $H^n_{p,q}$ 的纤维化.

$P_p(C)$ 和 $P_q(C)$ 分别表示 p 维和 q 维的复投影空间, 定义 $H^n_{p,q}$ 到 $P_p(C) \times P_q(C)$ 的一映射 π 如下:

$$
\pi : \begin{cases} H^n_{p,q} \to P_p(C) \times P_q(C), \\ (z_0, \cdots, z_{p+q+1}) \to (z_0, \cdots, z_p) \times (z_{p+1}, \cdots, z_{p+q+1}) = \pi(z_0, \cdots, z_{p+q+1}). \end{cases}
$$

映射 π 具有下面的性质:

1° π 是把 $H^n_{p,q}$ 映到 $P_p(C) \times P_q(C)$ 上.

证 设 $(z_0, \cdots, z_p) \times (z_{p+1}, \cdots, z_{p+q+1}) \in P_p(C) \times P_q(C)$.

令

$$A=\sqrt{\frac{z_{p+1}\bar{z}_{p+1}+\cdots+z_{p+q+1}\bar{z}_{p+q+1}}{z\bar{z}_0+\cdots+z_p\bar{z}_p}},$$

那么由

$$(Az_0)(\overline{Az_0})+\ldots+(Az_p)(\overline{Az}_p)=A^2(z_0\bar{z}_0+\cdots+z_p\bar{z}_p)$$
$$=z_{p+1}\bar{z}_{p+1}+\cdots+z_{p+q+1}\bar{z}_{p+q+1}$$

就得出 $(Az_0,\cdots,Az_p,z_{p+1},\cdots,z_{p+q+1})\in H^n_{p,q}$, 而

$$\pi(Az_0,\cdots,Az_p,z_{p+1},\cdots,z_{p+q+1})$$
$$=(Az_0,\cdots,Az_p)\times(z_{p+1},\cdots,z_{p+q+1})=(z_0,\cdots,z_p)\times(z_{p+1},\cdots,z_{p+q+1}),$$

所以 π 把 $H^n_{p,q}$ 映到 $P_p(C)\times P_q(C)$ 上.

2° 设 $(z_0,\cdots,z_p)\times(z_{p+1},\cdots,z_{p+q+1})\in P_p(C)\times P_q(C)$, 则 $\pi^{-1}((z_0,\cdots,z_p)\times(z_{p+1},\cdots,z_{p+q+1}))$ 同胚于一维的球.

证 设点 $(z'_0,\cdots,z'_p,z'_{p+1},\cdots,z'_{p+q+1})\in H^n_{p,q}$ 有 $\pi(z'_0,\cdots,z'_{p+q+1})=(z_0,\cdots,z_p)\times(z_{p+1},\cdots,z_{p+q+1})$, 则 $\pi(z'_0,\cdots,z'_{p+q+1})=(z'_0,\cdots,z'_p)\times(z'_{p+1},\cdots,z'_{p+q+1})=(z_0,\cdots,z_p)\times(z_{p+1},\cdots,z_{p+q+1})$, 所以 $z'_i=\lambda z_i,i=0,\cdots,p,\lambda$ 为非零复数, $z'_j=\mu z_j,j=p+1,\cdots,p+q+1,\mu$ 为非零复数. 由 $(z'_0,\cdots,z'_{p+q+1})=(\lambda z_0,\cdots,\lambda z_p,\mu z_{p+1},\cdots,\mu z_{p+q+1})\in H^n_{p,q}$ 得出: $(\lambda z_0)(\overline{\lambda z_0})+\cdots+(\lambda z_p)(\overline{\lambda z_p})=(\mu z_{p+1})(\overline{\mu z_{p+1}})+\cdots+(\mu z_{p+q+1})(\overline{\mu z_{p+q+1}})$, 即 $\dfrac{\lambda}{\mu}\left(\dfrac{\bar{\lambda}}{\mu}\right)=\dfrac{z_{p+1}\bar{z}_{p+1}+\cdots+z_{p+q+1}\bar{z}_{p+q+1}}{z_0\bar{z}_0+\cdots+z_p\bar{z}_p}$, 所以有

$$\pi^{-1}((z_0,\cdots,z_p)\times(z_{p+1},\cdots,z_{p+q+1}))=\Big\{(\lambda z_0,\cdots,\lambda z_p,z_{p+1},\cdots,z_{p+q+1})|\lambda\bar{\lambda}=A,$$
$$A=\frac{z_{p+1}\bar{z}_{p+1}+\cdots+z_{p+q+1}\bar{z}_{p+q+1}}{z_0\bar{z}_0+\cdots+z_p\bar{z}_p}\Big\}.$$

由此可见 $\pi^{-1}((z_0,\cdots,z_p)\times(z_{p+1},\cdots,z_{p+q+1}))$ 同胚于一维球 S^1.

3° $\mathscr{B}_2=\{H^n_{p,q},\pi,P_p(C)\times P_q(C)\}$ 是以 S^1 为纤维的球纤维丛.

证 在 $P_p(C)$ 和 $P_q(C)$ 中分别选取邻域组如下:

$$U_i=\{z=(z_0,\cdots,z_p)|z_i\neq 0\},\qquad i=0,\cdots,p.$$
$$V_j=\{z=(z_{p+1},\cdots,z_{p+q+1})|z_{j+p+1}\neq 0\},\quad j=0,\cdots,q.$$

我们知道它们就分别作成 $P_p(C)$ 和 $P_q(C)$ 的覆盖, 所以 $U_i\times V_j,i=0,\cdots,p;j=0,\cdots,q$ 就作成 $P_p(C)\times P_q(C)$ 的覆盖. S^1 表单位圆 $z\bar{z}=1$, S^1 上的点 z 可以由一个参数 θ, $0\leqslant\theta=\arg z<2\pi$ 唯一地来确定.

$$\pi^{-1}(U_i\times V_j)$$
$$=\left\{\left(\lambda\frac{z_0}{z_i},\cdots,\lambda\frac{z_{i-1}}{z_i},\lambda,\lambda\frac{z_{i+1}}{z_i},\cdots,\lambda\frac{z_p}{z_i},\frac{z_{p+1}}{z_{p+j+1}},\cdots,\frac{z_{p+j}}{z_{p+j+1}},1,\frac{z_{p+j+2}}{z_{p+j+1}},\cdots,\frac{z_{p+q+1}}{z_{p+j+1}}\right)\right.$$
$$\left.\left|z_i\neq 0,z_j\neq 0,\lambda\bar{\lambda}=\frac{z_i\bar{z}_i(z_{p+1}\bar{z}_{p+1}+\cdots+z_{p+q+1}\bar{z}_{p+q+1})}{z_{p+j+1}\bar{z}_{p+j+1}(z_0\bar{z}_0+\cdots+z_p\bar{z}_p)}\right.\right\}.$$

定义映射 φ 如下:

$$\varphi:\begin{cases}\pi^{-1}(U_i\times V_j)\to U_i\times V_j\times S^1\\ \varphi\Big(\Big(\lambda\frac{z_0}{z_i},\cdots,\lambda\frac{z_{i-1}}{z_i},\lambda,\lambda\frac{z_{i+1}}{z_i},\cdots,\lambda\frac{z_p}{z_i},\frac{z_{p+1}}{z_{p+j+1}},\cdots,\\ \quad\frac{z_{p+j}}{z_{p+j+1}},1,\frac{z_{p+j+2}}{z_{p+j+1}},\cdots,\frac{z_{p+q+1}}{z_{p+j+1}}\Big)\Big)\\ \quad=(z_0,\cdots,z_p)\times(x_{p+1},\cdots,z_{p+q+1})\times\theta,\quad \theta=\arg\lambda.\end{cases}$$

容易看到 φ 是一拓扑映射, 所以 $\mathscr{B}_2=\{H^n_{p,q},\pi,P_p(C)\times P_q(C)\}$ 是一以 S^1 为纤维的球纤维丛.

3. 纤维丛 $\mathscr{B}_2=\{H^n_{p,q},\pi,P_p(C)\times P_q(C)\}$ 的示性类.

1° $W(\mathscr{B}_2)=1+W^1(\mathscr{B}_2)+W^2(\mathscr{B})$, 由于复投影空间的奇维数的同调群和上同调群皆为零, 所以 $W^1(\mathscr{B}_2)=0$, 这样就只须计算 $W^2(\mathscr{B})$.

2° $H_2(P_p(C)\times P_q(C))\cong H_2(P_p(C))\otimes H_0(P_q(C))+H_0(P_p(C))\otimes H_2(P_q(C))$, 所以 $H_2(P_p(C)\times P_q(C))$ 的基为 $X_0\times Y_2, X_2\times Y_0$, 这里 X_0 和 X_2 分别表 $P_p(C)$ 的零维和 2 维的同调群的基, Y_0 和 Y_2 分别表 $P_q(C)$ 的零维和 2 维的同调群的基.

3° 计算丛 $\mathscr{B}_2$ 的 Euler 示性类 $X(\mathscr{B}_2)\in H^2(P_p(C)\times P_q(C);Z)$, 我们分下面几步来计算:

首先我们确定 $X_0\times Y_2$ 和 $X_2\times Y_0$ 的几何表示. 为此在 $P_p(C)$ 里选定一点 $(1,\underbrace{0,\cdots,0}_{p})$ 表示 X_0, 再在 $P_p(C)$ 中另选一点 $(0,1,\underbrace{0,\cdots,0}_{p-1})$, 联结这两点的复投影直线是: $\{(\lambda,\mu,\underbrace{0,\cdots,0}_{p-1})|\lambda,\mu$ 是不同时为零的复数$\}$, 我们取这条复投影直线几何地来表示 X_2, 同样在 $P_q(C)$ 中选定点 $(1,\underbrace{0,\cdots,0}_{q})$ 表示 Y_0, 再在 $P_q(C)$ 中另外选取一点 $(0,\beta,\underbrace{0,\cdots,0}_{q-1})$, 联结这两点的复投影直线是 $\{(\alpha,\beta,0,\cdots,0)|\alpha,\beta$ 是不同时为零的复数$\}$, 我们就以这条复投影直线几何地来表示 Y_2. 这样我们就有 $X_2\times Y_0$ 和 $X_0\times Y$ 的几何表示如下:

$$X_2\times Y_0:\{(\lambda,\mu,\underbrace{0,\cdots,0}_{p-1})\times(1,\underbrace{0,\cdots,0}_{q})|\lambda,\mu\text{是不同时为零的复数}\},$$

$$X_0 \times Y_2 : \{(1, \underbrace{0, \cdots, 0}_{p}) \times (\alpha, \beta, \underbrace{0, \cdots, 0}_{q-1}) | \alpha, \beta \text{是不同时为零的复数}\}.$$

在 $X_2 \times Y_0$ 上造一向量场 f (即通常所谓的截面) 如下:

$$f((\lambda, \mu, 0, \cdots, 0) \times (1, 0, \cdots, 0))$$
$$= \left(\frac{1}{\sqrt{1+\frac{\mu}{\lambda}\left(\frac{\mu}{\lambda}\right)}}, \frac{\frac{\mu}{\lambda}}{\sqrt{1+\frac{\mu}{\lambda}\left(\frac{\mu}{\lambda}\right)}}, 0, \cdots, 0, 1, \underbrace{0, \cdots, 0}_{q} \right) \in H^n_{p,q}, \quad \text{当}\lambda \neq 0;$$

$$f((0, 1, 0, \cdots, 0) \times (1, 0, \cdots, 0)) = (0, 1, 0, \cdots, 0, 1, \underbrace{0, \cdots, 0}_{q}) \in H^n_{p,q}, \text{当} \left|\frac{\mu}{\lambda}\right| \to \infty$$

时, 点 $(\lambda, \mu, 0, \cdots, 0) \times (1, 0, \cdots, 0)$ 趋于点 $(0, 1, 0, \cdots, 0) \times (1, 0, \cdots, 0)$, 但当 $\left|\frac{\mu}{\lambda}\right| \to \infty$ 时 $\dfrac{\frac{\mu}{\lambda}}{\sqrt{1+\left(\frac{\mu}{\lambda}\right)\overline{\left(\frac{\mu}{\lambda}\right)}}}$ 的极限是不确定的, 所以 $f((\lambda, \mu, 0, \cdots, 0) \times (1, 0, \cdots, 0)) = \left(\dfrac{1}{\sqrt{1+\frac{\mu}{\lambda}\left(\frac{\mu}{\lambda}\right)}}, \dfrac{\frac{\mu}{\lambda}}{\sqrt{1+\frac{\mu}{\lambda}\overline{\left(\frac{\mu}{\lambda}\right)}}}, 0, \cdots, 0, 1, 0, \cdots, 0 \right)$, 当 $\left|\frac{\mu}{\lambda}\right| \to \infty$ 时, 极限是不确定的, 所以点 $(0, 1, 0, \cdots, 0) \times (1, 0, \cdots, 0)$ 是向量场 f 的奇点. 现在来计算这个奇点的指数, $U = \{(\lambda, \mu, 0, \cdots, 0) \times (1, 0, \cdots, 0) | \mu \neq 0\}$ 是点 $(0, 1, 0, \cdots, 0) \times (1, 0, \cdots, 0)$ 在 $X_2 \times Y_0$ 里的邻域, 考虑 $U' = \{(1, \mu, 0, \cdots, 0) \times (1, 0, \cdots, 0) | \mu\bar{\mu} = r$, r 是大于零的常数$\} \subset U$,

$$f((1, \mu, 0, \cdots, 0) \times (1, 0, \cdots, 0)) = \left(\frac{1}{\sqrt{1+\mu\bar{\mu}}}, \frac{\mu}{\sqrt{1+\mu\bar{\mu}}}, 0, \cdots, 0, 1, 0, \cdots, 0 \right)$$
$$= \left(\frac{1}{\sqrt{1+r^2}}, \frac{\mu}{\sqrt{1+r^2}}, 0, \cdots, 0, 1, 0, \cdots, 0 \right), \text{当}(1, \mu, 0, \cdots, 0) \times (1, 0, \cdots, 0) \in U',$$
$$\varphi f((1, \mu, 0, \cdots, 0) \times (1, 0, \cdots, 0)) = (1, \mu, \underbrace{0, \cdots, 0}_{p-1}) \times (1, \underbrace{0, \cdots, 0}_{q}) \times \arg \mu.$$

当 μ 跑过 1 维球 U' 时, $\arg \mu$ 正好从 0 变到 2π, 所以向量场 f 的奇点 $(0, 1, 0, \cdots, 0) \times (1, 0, \cdots, 0)$ 的指数是 1. 类似地可以得到点 $(1, 0, \cdots, 0) \times (0, 1, 0, \cdots, 0)$ 也是指数为 1 的奇点. 这样我们就得到下面的

命题 4

$$X^2(\mathscr{B}_2) = X^0 \times Y^2 + X^2 \times Y^0 \in H^2(P_p(C) \times P_q(C); Z);$$

$$W^2(\mathscr{B}_2) = X^0 \times Y^2 + X^2 \times Y^0 \in H^2(P_p(C) \times P_q(C); Z_2),$$

这里 $X^0 \times Y^2; X^2 \times Y^0$ 既表整系数的上同调类, 又表示由它法 2 约化后的模 2 系数的上同调类.

4. $H^n_{p,q}$ 的同调性质.

如同§1 中一样, 在这一节里我们利用球纤维丛的 Gysin 序列来计算 $H^n_{p,q}$ 的同调性质.

$\mathscr{B}_2 = \{H^n_{p,q}, \pi, P_p(C) \times P_q(C)\}$的Gysin 序列如下:

$$\cdots \to H^{r-1}(P_p(C) \times P_q(C)) \xrightarrow{\Psi_{r-1}} H^{r+1}(P_p(C) \times P_q(C)) \xrightarrow{\pi_{r+1}} H^{r+1}(H^n_{p,q})$$
$$\xrightarrow{\Phi_{r+1}} H^r(P_p(C) \times P_q(C)) \xrightarrow{\Psi_r} H^{r+2}(P_p(C) \times P_q(C)) \to \cdots$$

我们考虑下面图形

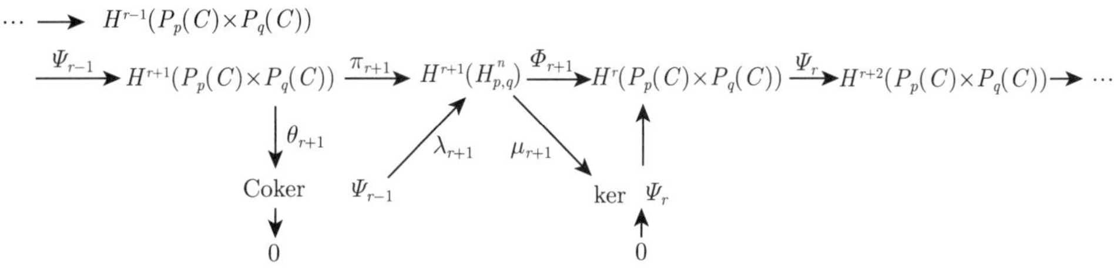

这儿 θ_{r+1} 是自然映射, λ_{r+1} 是 π_{r+1} 所诱导的同态, 因为 Gysin 序列的正合性, 从 $\operatorname{Im}\Psi_{r-1} = \ker\pi_{r+1}$ 就知道 λ_{r+1} 是唯一确定的而且是同构映入的映射, μ_{r+1} 是由 Φ_{r+1} 诱导出的映射, 由 $\ker\Psi_r = \operatorname{Im}\Phi_{r+1}$ 就知道 μ_{r+1} 是在上的同态, 由 $\ker\Phi_{r+1} = \operatorname{Im}\pi_{r+1}$ 就可以得出 $\operatorname{Im}\lambda_{r+1} = \ker\mu_{r+1}$, 所以有

$$H^{r+1}(H^n_{p,q}) \cong \operatorname{Cok}\Psi_{r-1} + \operatorname{Ker}\Psi_r.$$

复投影空间的整系数的上同调群如下:

$$H^0(P_p(C)) \cong H^2(P_p(C)) \cong \cdots \cong H^{2p}(P_p(C)) \cong Z,$$
$$H^1(P_p(C)) \cong H^2(P_p(C)) \cong \cdots \cong H^{2p-1}(P_p(C)) = 0.$$

设 X^2 表 $H^2(P_p(C))$ 的基, 则 $(X^2)^i = X^{2i}$ 就是 $H^{2i}(P_p(C))$ 的基, $(X^2)^i$ 表在上积意义下的 i 次幂, Y^2 表 $H^2(P_q(C))$ 的基. 由 Künneth 公式就可以求得 $P_p(C) \times P_q(C)$ 的同调性质.

由已知的结果知道: 设 $u \in H^r(P_p(C) \times P_q(C))$, 则 $\Psi_r(u) = u \cup X(\mathscr{B}_2)$. 利用这个结果就可以求得下面的

命题 5

$p<q$ 的情形	$0\leqslant 2r\leqslant 2p$	$2p<2r<2q$	$2q\leqslant 2r\leqslant 2(p+q)=2n$
$H^{2r}(P_p(C)\times P_q(c))$ 的基	$X^0\times Y^{2r},\cdots,X^{2r}\times Y^0$	$X^0\times Y^{2r},\cdots,X^{2p}\times Y^{2(r-p)}$	$X^2(r-q)\times Y^{2q},\cdots,X^{2p}\times Y^{2(r-p)}$
$H^{2r-1}(P_p(C)\times P_q(C))$ 的基	0	0	0
ker Ψ_{2r}	0	0	$X^{2(r-q)}\times Y^{2q}+\cdots+X^{2p}\times Y^{2(r-p)}$
ker Ψ_{2r-1}	0	0	0
Coker Ψ_{2r-1}	0	0	0
Coker Ψ_{2r-2}	$\theta_{2r}(X^0\times Y^{2r})=\cdots=(-1)^i\theta_{2r}(X^{2i}\times Y^{2r-2i})=\cdots=(-1)^r\theta_{2r}(X^{2r}\times Y^0)$	0	0
$p=q$ 的情形	$0\leqslant 2r<2p$	$r=p=q$	$2q<2r\leqslant 2(p+q)=2n$
$H^{2r}(P_p(C)\times P_q(C))$ 的基	$X^0\times Y^{2r},\cdots,X^{2r}\times Y^0$	$X^0\times Y^{2r},\cdots,X^{2r}\times Y^0$	$X^{2(r-q)}\times Y^{2q},\cdots,X^{2p}\times Y^{2(r-p)}$
$H^{2r-1}(P_p(C)\times P_q(C))$ 的基	0	0	0
Ker Ψ_{2r}	0	$X^0\times Y^{2r}+\cdots+X^{2r}\times Y^0$	$X^{2(r-q)}\times Y^{2q}+\cdots+X^{2p}\times Y^{2(r-p)}$
Ker Ψ_{2r-1}	0	0	0
Coker Ψ_{2r-1}	0	0	0
Coker Ψ_{2r-2}	$\theta_{2r}(X^0\times Y^{2r})=\cdots=(-1)^i\theta_{2r}(X^{2i}\times Y^{2r-2i})=\cdots=(-1)^r\theta_{2r}(X^{2r}\times Y^0)$	0	0

证 $p<q$ 的情形.

1° $0\leqslant 2r\leqslant 2p$: 设 $X^{2i}\times Y^{2j}\in H^{2r}(P^p(C)\times P^q(C)),i+j=r,\Psi_{2r}(X^{2i}\times Y^{2j})=(X^{2i}\times Y^{2j})\cup X^2(\mathscr{B}_2)=(X^{2i}\times Y^{2j})\cup(X^0\times Y^2+X^2\times Y^0)=X^{2i}\times Y^{2(j+1)}+X^{2(i+1)}\times Y^{2j},i=0,\cdots,\gamma;j=0,\cdots,\gamma$, 而 $X^0\times Y^{2(r+1)}+X^2\times Y^{2r},X^2\times Y^{2r}+X^4\times Y^{2(r-1)},\cdots,X^{2r}\times Y^2+X^{2(r+1)}\times Y^0$ 是 $H^{2r+2}(P_p(C)\times P_q(C))$ 中的无关元素, 所以 $\mathrm{Ker}\,\Psi_{2r}=0$ 和 $\theta_{2r}(X^0\times Y^{2r})=-\theta_{2r}(X^2\times Y^{2r-2})=\cdots=(-1)^r\theta_{2r}(X^{2r}\times Y^0)$, 所以 $\mathrm{Coker}\,\Psi_{2r-2}$ 的基为 $\theta_{2r}(X^0\times Y^{2r})=-\theta_{2r}(X^2\times Y^{2r-2})=\cdots=(-1)^r\theta_{2r}(X^{2r}\times Y^0)$.

2° $2p<2r<2q$: 与 1° 中的考虑一样可得 $\mathrm{Ker}\,\Psi_{2r}=0$, 此时 $H^{2r}(P_p(C)\times P_q(C))$ 的基 $X^0\times Y^{2r},\cdots,X^{2p}\times Y^{2(r-p)}$ 在 Ψ_{2r} 下映成 $X^0\times Y^{2(r+1)}+X^2\times Y^{2r},X^2\times Y^{2r}+X^4\times Y^{2(r-1)},\cdots,X^{2(p-1)}\times Y^{2(r-p+1)}+X^{2p}\times Y^{2(r-p)},X^{2p}\times Y^{2(r-p)}$, 而它们正好就是 $H^{2(r+1)}(P_p(C)\times P_q(C))$ 的基, 所以 $\mathrm{Coker}\,\Psi_{2r}=0$.

3° $2q \leqslant 2r \leqslant 2(p+q) = 2n$：在 Ψ_{2r} 下 $X^{2(r-q)} \times Y^{2q}, \cdots, X^{2p} \times Y^{2(r-p)}$ 映成 $X^{2(r-q+1)} \times Y^{2q}, X^{2(r-q+1)} \times Y^{2q} + X^{2(r-q+2)} \times Y^{2(q-1)}, \cdots, X^{2(p-1)} \times Y^{2(r-p+2)} + X^{2p} \times Y^{2(r-p+1)}, X^{2p} \times Y^{2(r-p+1)}$, 所以 $\operatorname{Ker}\Psi_r$ 的基为 $X^{2(r-q)} \times Y^{2q} + \cdots + X^{2p} \times Y^{2(r-p)}$ 和 $\operatorname{Coker}\Psi_{2r-2} = 0$.

对于 $p = q$ 的情形可以类似地证明.

由这个命题即可推得下面的

命题 6

$p<q$ 的情形	$0 \leqslant 2r \leqslant 2p$	$p<r<q$	$2q \leqslant 2r \leqslant 2n$
$H^{2r}(H^n_{p,q})$ 的基	$Z^{2r} = \pi_{2r}(X^0 \times Y^{2r})$ $= \cdots$ $= \pi_{2r}(X^{2r} \times Y^0)$	0	0
$H^{2r+1}(H^n_{p,q})$ 的基	0	0	$\Phi^{-1}_{2r+1}(X^{2(r-q)} \times$ $Y^{2q} + \cdots$ $+ X^{2p} \times Y^{2(r-p)})$

$p=q$ 的情形	$0 \leqslant r < p$	$p=r=q$	$q<r \leqslant n$
$H^{2r}(H^n_{p,q})$ 的基	$Z^{2r} = \pi_{2r}(X^0 \times Y^{2r})$ $= \cdots$ $= \pi_{2r}(X^{2r} \times Y^0)$	$Z^{2r} = \pi_{2r}(X^0 \times Y^{2r})$ $= \cdots$ $= \pi_{2r}(X^{2r} \times Y^0)$	0
$H^{2r+1}(H^n_{p,q})$ 的基	0	$\Phi^{-1}_{2r+1}(X^0 \times Y^{2r} + \cdots$ $+ X^{2r} \times Y^0) = Z^{2r}$	$\Phi^{-1}_{2r+1}(X^{2(n-q)}$ $\times Y^{2q} + \cdots + X^{3p}$ $\times Y^{2(r-p)}) = Z^{2r}$

5. $H^n_{p,q}$ 的 Stiefel-Whitney 示性类.

根据 [8] 中所证明的一条定理就有

$$W(H^n_{p,q}) = \pi^*(W(P_p(C) \times P_q(C))) \cup \pi^* W(\mathscr{B}_2),$$

$$\pi^* W^{2k}(P_p(C) \times P_q(C)) = \sum_{i+j=k} \pi^* W^{2i}(P_p(C)) \times \pi^* W^{2j}(P_q(C)),$$

$$W^{2j}(P_p(C)) = \begin{pmatrix} p+1 \\ i \end{pmatrix} (X^2)^j, \quad X^2 \text{是} H^2(P_p(C)) \text{的基}.$$

$$\pi^* W^{2k}(P_p(C) \times P_q(C)) = \sum_{i+j=k} \binom{p+1}{i} \binom{q+1}{j} \pi^*(X^{2i} \times X^{2j})$$

$$= \begin{cases} \displaystyle\sum_{i+j=k} \binom{p+1}{i} \binom{q+1}{j} Z^{2k}, & k \leqslant p \\ 0, & k > p \end{cases} = \begin{cases} \dbinom{n+2}{k} Z^{2k}, & k \leqslant p, \\ 0, & k > p, \end{cases}$$

$$\pi^* W(\mathscr{B}_2) = \pi^*(1 + W^1(\mathscr{B}_2) + W^2(\mathscr{B}_2)) = 1 + \pi^* W^2(\mathscr{B}_2) = 1 + 2Z^2 = 1.$$

所以我们得到下面的

定理 2

$$W^{2k}(H^n_{p,q}) = \begin{cases} \dbinom{n+2}{k} Z^{2k}, & k \leqslant p, \\ 0, & k > 0. \end{cases}$$

$$W^{2k+1}(H^n_{p,q}) = 0.$$

6. $H^n_{p,q}$ 的Понтрягин示性类.

$P_0(\mathscr{B}_2)$ 表 $\mathscr{B}_2$ 的实系数的Понтрягин示性类, $P_0(\mathscr{B}_2) = 1 + P_0^4(\mathscr{B}_2)$, 由 [8] 所证明的结果有

$$\begin{aligned} P_0^4(\mathscr{B}_2) &= C^2(\mathscr{B}_2) \cup C^2(\mathscr{B}_2) = X(\mathscr{B}_2) \cup X(\mathscr{B}_2) = (X^0 \times Y^2 + X^2 \times Y^0)^2 \\ &= X^0 \times Y^4 + X^4 \times Y^0 + 2(X^2 \times Y^2), \end{aligned}$$

$$P_0^{4i}(P_p(C)) = \binom{p+1}{i} (X^2)^{2i}, X^2\text{是}H^2(P_p(C))\text{的基},$$

$$P_0(H^n_{p,q}) = \pi^* P_0(\mathscr{B}_2) \cup \pi^*(P_0(P_p(C) \times P_q(C)),$$

$$\begin{aligned} P_0^{4k}(P_p(C) \times P_q(C)) &= \sum_{k_1+k_2=k} P_0^{4k_1}(P_p(C)) \times P_0^{4k}(P_q(C)) \\ &= \sum_{k_1+k_2=k} \binom{p+1}{k_1}\binom{q+1}{k_2} X^{4k_1} \times Y^{4k_2}, \end{aligned}$$

$$\pi^* P_0(\mathscr{B}_2) = \pi^*(1 + X^0 \times Y^4 + X^4 \times Y^0 + 2(X^2 \times Y^2)) = 1 + 4Z^4,$$

$$\pi^* P_0^{4k}(P_p(C) \times P_q(C)) = \sum_{k_1+k_2=k} \binom{p+1}{k_1}\binom{q+1}{k_2} \pi^*(X^{4k_1} \times Y^{4k_2})$$

$$= \begin{cases} \displaystyle\sum_{k_1+k_2=k} \binom{p+1}{k_1}\binom{q+1}{k_2} Z^{4k}, & 2k \leqslant p \\ 0, & 2k > p \end{cases}$$

$$= \begin{cases} \dbinom{n+2}{k} Z^{4k}, & 2k \leqslant p, \\ 0, & 2k > p. \end{cases}$$

这样我们就有下面的

定理 3

$$P_0^{4k}(H^n_{p,q}) = \begin{cases} \left[\dbinom{n+2}{k} + 4\dbinom{n+2}{k-1}\right] Z^{4k}, & 2k \leqslant p, \\ 0, & 2k > p. \end{cases}$$

参考文献

[1] Vesentini E. Construction géometrique des classes de Chern de quelques. variétés de Grassmann Complexes. Colloque de Topologie algebrique, Paris, 1956: 97-120.

[2] Chern S S (陈省身). Characteristio classes of Hermition manifold. *Annals of Math.*, 1946, 47: 85-121.

[3] Chern S S (陈省身)and Spanier E. The homology structure of sphere bundles. *Proc. Nat. Acad. Sci.*, 1950, 36: 248-255.

[4] Marie-H élène Schwartz. Classes de Chern des quadriques complexes. *Bull. des. Sci. Math.*, 1956, 80: 144-155.

[5] Stiefel E. Über Richtungsfelder in den projektiven Räumen und einen Satz aus der reellen Algebra. *Comment. Math. Helv.*, 1941, 13: 201-218.

[6] Wu W T (吴文俊). Sur les classes cáracteristiques d'un espace en fibres spheres. C.R.Paris, 1948, 227: 582-584.

[7] Wu W T (吴文俊). Notes on complex manifold and algebraic varieties I Plücker's formula (即将在中国科学发表).

[8] Wu W T (吴文俊). Les classe caractéristiques des espaces fibrés sphériques Paris. Hermann, 1952.

[9] Borel A and Hirzebruch F. Characteristic classes and homogeneous spaces. I. *Amer. Jour. of Math.*, 1958, 80: 459-538.

37. On the Imbedding of Orientable Manifolds in a Euclidean Space*

§1 Statement of the theorem

Whitney[6-7] has proved that a C^∞-n-manifold M can always be C^∞-imbedded in a Euclidean $2n$-space. He and others have also conjectured that it can be C^∞-imbedded in a Euclidean $(2n-1)$-space if M is *orientable* and $n > 1$. This conjecture, so far as the author knows, has been proved in the affirmative only in the following special cases (R^k = Euclidean space of dimension k):

1°. M is of dimension 2 (trivial).

2°. M is nonclosed; in that case whether M is orientable or not is immaterial, (Hirsch[2]).

3°. M is simply connected (Haefliger[1]).

4°. M is closed, orientable and of dimension $n = 3$ or 6 (Hirsch[3]).

The present paper has the aim to prove the following

Theorem *Any closed orientable C^∞-n-manifold M can be topologically imbedded in an R^{2n-1} for $n \geqslant 4$ and can be C^∞-imbedded in R^{2n-1} for $n \geqslant 5$.*

By combining with the results of Hirsch, it follows that the conjecture of Whitney is true in all cases except when $n = 4$. Even in that case the topological imbedding can be at least assured.

§2 Smoothing of corners in an imbedding or an isotopy

Let M_0^{n-1} be a closed C^∞-$(n-1)$ -submanifold of a C^∞-n-manifold M^n such that M_0^{n-1} divides M^n into two parts M_+^n and M_-^n with $M_+^n \cap M_-^n = M_0^{n-1}$. Let f: $M^n \to R^N$ be a mapping for which both f/M_+^n and f/M_-^n are C^∞-imbeddings, and T_x^+, T_x^-, and T_x^0 be the tangent spaces at $f(x) \in f(M_0^{n-1})$ to $f(M_+^n)$, $f(M_-^n)$,

* *Sci. Sinica*, 1963, 12: 25-33.

and $f(M_0^{n-1})$ respectively. Let L_x^+ and L_x^- be the lines in T_x^+ and T_x^- normal to T_x^0, oriented in such a way that they point to the interior of $f(M_+^n)$ and $f(M_-^n)$ respectively. Then we shall say that f is a piecewise C^∞-imbedding of M^n in R^N with a (possible) *corner* at $f(M_0^{n-1})$ if for each $x \in M_0^{n-1}$, the angle between the oriented lines L_x^+ and L_x^- is > 0 and $\leqslant \pi$. The following theorem shows that such a corner can always be smoothed.

Proposition 1. Any piecewise C^∞-imbedding f of M^n in R^N with a corner at $f(M_0^{n-1})$ can be approximated to arbitrary nearness by a C^∞-imbedding g of M^n in R^N which coincides with f on $M^n - V$ where V is a given neighbourhood of M_0^{n-1} in M^n.

Proof. Let N_x be the normal $(N-n+1)$ - plane to $f(M_0^{n-1})$ at $f(x)$. Then $\{N_x\}$ is a C^∞-family of planes over the C^∞-manifold $f(M_0^{n-1})$. Take $\varepsilon > 0$ sufficiently small such that the sets $\tilde{V}_x = \{y/y \in N_x, ||y-x|| < 2\varepsilon\}$ are mutually disjoint, and, denote the union of all $\tilde{U}_x = \{y/y \in N_x, ||y-x|| < \varepsilon\}$ $x \in M_0^{n-1}$ by $\tilde{U}$, then $\tilde{U}$ is a C^∞-manifold in R^N for which the mapping $\pi : \tilde{U} \to f(M_0^{n-1})$ defined by $\pi(\tilde{U}_x) = f(x)$ is a C^∞-map. Set $U_x^+ = \tilde{U}_x \cap f(M_+^n)$, $U_x^- = \tilde{U}_x \cap f(M_-^n)$, and $U_x = U_x^+ + U_x^-, U^\pm = \sum U_x^\pm, U = U^+ + U^-$. Define $k^\pm : U^\pm \to M_0^{n-1} \times [0, \pm 1]$ by $k^\pm(u) = (\pi(u), \pm||u-x||/\varepsilon), u \in U^\pm$. Then for small ε, $k^\pm$ are C^∞-diffeomorphisms which map each $U_x^\pm$ to the segment $(x) \times [0, \pm 1]$ with $(k^\pm)^{-1} = h^\pm$. Take a continuously differentiable function $\lambda(t)$ with $\lambda(0) = 0$, and $\lambda(t) = 1$ for t near to 1. Define now a map g^* of M^n in R^N as follows. For $z \in M_-^n - f^{-1}(U^+)$ we set $g^*(z) = f(z)$. For $z \in f^{-1}(U^+)$, let $k^+ f(z) = (x, t), x \in M_0^{n-1}, 0 \leqslant t < 1$ and set

$$g^*(z) = [1-\lambda(t)] \cdot \left[t \cdot \frac{\partial h^-(x,0)}{\partial t} + f(x)\right] + \lambda(t) h^+(x,t).$$

It is easy to verify that g^* is a C^1-imbedding. By the method of Whitney[8] g^* can then be approximated by a C^∞-imbedding g as desired.

Two C^∞-imbeddings f, g of a closed C^∞-manifold M in R^N will be said to be *C^∞-isotopic* if there exists a C^∞-imbedding F of the C^∞-manifold $M \times [t_0, t_1]$ in $R^N \times [t_0, t_1], (t_0 < t_1)$ such that

$$\begin{aligned} &F/M \times (t_0) \equiv f/M, \\ &F/M \times (t_1) \equiv g/M, \\ &F/M \times (t) \subset R^N \times (t), \quad t_0 \leqslant t \leqslant t_1. \end{aligned}$$

The map F will then be called a *C^∞-isotopy* between f and g. For $t_0 < \tau_1 < \cdots <$

$\tau_k < t_1$ we shall say that f, g are *piecewise* C^∞*-isotopic* with (possible) *corners* at levels τ_i if there exists an imbedding F of the C^∞-manifold $M\times[t_0,t_1]$ in $R^N\times[t_0,t_1]$ such that $F/M\times[\tau_i,\tau_{i+1}]$ are all C^∞-isotopies between $h_i/M\equiv F/M\times(\tau_i)$ and $h_{i+1}/M\equiv F/M\times(\tau_{i+1}),(\tau_0=t_0,\tau_{k+1}=t_1)$. The mapping F will then be called a *piecewise* C^∞-isotopy between f and g. The systems $(M\times[t_0,t_1],M\times(\tau_i),F)$ are now in a similar situation as the systems (M^n,M_0^{n-1},f) in the above theorem. By applying the same procedure with evident modifications the corners can then be smoothed and we have therefore the following

Proposition 2. If two C^∞-imbeddings f, g of a closed C^∞-manifold in R^N are piecewise C^∞-isotopic under the isotopy F, then they are also C^∞-isotopic under an isotopy G arbitrarily near to F and coinciding with F except in prescribed neighbourhoods of the corners.

The above proposition implies the following corollary which amounts to the transitivity of C^∞-isotopies:

If the C^∞-imbeddings f, g, and h of a closed C^∞-manifold in R^N are such that f and g as well as g and h are both C^∞-isotopic, then so are f and h.

§3 Some lemmas

Let D^k be an ordinary closed k-disc with centre O, and M a closed C^∞-n-manifold. A C^∞-diffeomorphism f of $D^k\times M$ in R^N with $N\geqslant n+k$ will be said to be *normal* with respect to M with *normal length* $\varepsilon>0$ if for each $x\in M$, f maps $D^k\times(x)$ to a k-disc of centre $f(0,x)$, of radius ε, and with plane normal to $f((0)\times M)$.

Lemma 1 *Any* C^∞*-diffeomorphism* f *of* $D^k\times M$ *in* R^N *is* C^∞*-isotopic to one normal with respect to* M *such that during the isotopy* $f/(0)\times M$ *is kept fixed.*

Proof. For $x\in M$ and $u=(u^1,\cdots,u^k)\in D^k$, $||u||=\sqrt{\sum(u^i)^2}\leqslant 1$, the map f can be put in the form:

$$f(u,x)=f(0,x)+\sum_{i=1}^k\frac{\partial f(0,x)}{\partial u^i}\cdot u^i+\sum_{i,j=1}^k a_{ij}(u,x)u^iu^j,$$

in which

$$a_{ij}(u,x)=a_{ji}(u,x)=\int_0^1(1-t)\cdot f_{u^iu^j}(tu,x)dt$$

and

$$a_{ij}(0,x)=\frac{1}{2}f_{u^iu^j}(0,x).$$

Take $\varepsilon > 0$ and < 1 and define for each $0 \leqslant \tau \leqslant 1$ a C^∞-map $f_\tau : D^k \times M \to R^N$ given by

$$f_\tau(u, x) = \begin{cases} f([1-2\tau(1-\varepsilon)]u, x), & 0 \leqslant \tau \leqslant \dfrac{1}{2}, \\ f(0,x) + \displaystyle\sum_{i=1}^{k} \frac{\partial f(0,x)}{\partial u^i} \cdot \varepsilon u^i + 2(1-\tau) & \\ \quad \cdot \displaystyle\sum_{i,j=1}^{k} a_{ij}(\varepsilon u, x) \cdot \varepsilon u^i \cdot \varepsilon u^j, & \dfrac{1}{2} \leqslant \tau \leqslant 1. \end{cases}$$

For $\varepsilon > 0$ sufficiently small f_τ is then a piecewise C^∞-isotopy of $f_0 = f$ into a C^∞-diffeomorphism f_1. The latter is now easily seen to be piecewise C^∞-isotopic to a normal diffeomorphism as desired. The Lemma follows now from Proposition 2 in §2.

Let D^{n-1} be the ordinary $(n-1)$-disc

$$x_1^2 + \cdots + x_{n-1}^2 \leqslant 1,$$
$$x_n = \cdots = x_{2n-2} = 0, \quad x_0 = 0,$$

in the R^{2n-1} with coordinates $x_0, x_1, \cdots, x_{2n-2}$. For any $x \in S^{n-2} = \partial D^{n-1}$, let N_x be the n-plane normal to S^{n-2} at x in $R_0^{2n-2} : x_0 = 0$. Let $u = u(x)$, $v = v(x) \in N_x$ be two C^∞-fields of normal vectors to S^{n-2} with $||u|| = ||v|| = \varepsilon > 0$ sufficiently small and $u(x), v(x)$ mutually orthogonal for each $x \in S^{n-2}$. Let C_x be the circle with centre $x \in S^{n-2}$ and passing through $x + u(x)$ and $x + v(x)$. Then the collection of all such circles $C_x, x \in S^{n-2}$, is naturally a C^∞-manifold M of dimension $n-1$ which is C^∞-diffeomorphic to $S^1 \times S^{n-2}$ in an evident manner.

Lemma 2 *The manifold M described above bounds a C^∞-manifold C^∞-diffeomorphic to $S^1 \times D^{n-1}$ under f with $f(S^1 \times \operatorname{Int} D^{n-1})$ contained in the half-space $R_+^{2n-1} : x_0 > 0$.*

Proof. Since $n \geqslant 4$, we have $\pi_{n-2}(V_{2n-2,2}) = 0$ for the Stiefel manifold $V_{2n-2,2}$ of orthonormal pairs of vectors of length ε in R_0^{2n-2}. It follows that there exists a homotopy

$$h_t : S^{n-2} \to V_{2n-2,2}, \quad 0 \leqslant t \leqslant 1,$$

with

$$h_t(x) = (u_t(x), v_t(x)),$$
$$u_1(x) = u(x), \quad v_1(x) = v(x),$$
$$u_0(x) = \varepsilon e_n, \quad v_0(x) = \varepsilon e_{n+1},$$

$$||u_t(x)|| = ||v_t(x)|| = \varepsilon,$$

for all $x \in S^{n-2}$. Now define a mapping

$$f_0 : S^1 \times D^{n-1} \to R^{2n-1}$$

as follows. For

$$z = (a\cos\theta + b\sin\theta, tx) \in S^1 \times D^{n-1},$$

where $0 \leqslant t \leqslant 1$, $x \in \partial D^{n-1} = S^{n-2}$, set

$$f_0(x) = \begin{cases} x + 2(1-t)e_0 + u_{2t-1}(x)\cos\theta + v_{2t-1}(x)\sin\theta, & \dfrac{1}{2} \leqslant t \leqslant 1, \\ 2tx + e_0 + \varepsilon e_n \cos\theta + \varepsilon e_{n+1}\sin\theta, & 0 \leqslant t \leqslant 1. \end{cases}$$

It is easy to verify that f_0 defines an imbedding of $S^1 \times D^{n-1}$ in R^{2n-1} with

$$f_0(S^1 \times \operatorname{Int} D^{n-1}) \subset R^{2n-1}_+,$$

and $f_0/S^1 \times S^{n-2}$ coincides with the given manifold M. To see this, let

$$\begin{aligned} z &= (a\cos\theta + b\sin\theta, tx), \\ z' &= (a\cos\theta' + b\sin\theta', t'x') \end{aligned}$$

be any two points of $S^1 \times D^{n-1}$, with

$$\begin{aligned} &0 \leqslant \theta, \theta' \leqslant 2\pi, \\ &0 \leqslant t, t' \leqslant 1, \\ &x, x' \in S^{n-2}. \end{aligned}$$

We have to prove that

$$f_0(z) = f_0(z') \text{ implies } z = z'.$$

Case 1. $\dfrac{1}{2} \leqslant t, t' \leqslant 1$.

By comparing the components of e_0 and x, we have

$$\begin{aligned} &2(1-t) = 2(1-t'), \\ &[1 + (u_{2t-1}(x), x)\cos\theta + (v_{2t-1}(x), x)\sin\theta]\cdot x \\ &= [1 + (u_{2t'-1}(x'), x')\cos\theta' + (v_{2t'-1}(x'), x')\sin\theta']\cdot x'. \end{aligned}$$

Hence $t = t'$ and for ε sufficiently small $x = x'$. Whence

$$u_{2t-1}(x)\cos\theta + v_{2t-1}(x)\sin\theta$$
$$= u_{2t-1}(x)\cos\theta' + v_{2t-1}(x)\sin\theta'.$$

As $u_{2t-1}(x), v_{2t-1}(x)$ are mutually orthogonal and non-zero, we have $\theta = \theta' \bmod 2$ so that $z = z'$.

Case 2. $0 \leqslant t \leqslant \frac{1}{2}, \frac{1}{2} \leqslant t' \leqslant 1$.

By comparing the components of e_0 we get

$$2(1-t') = 1.$$

Whence $t' = \frac{1}{2}$ and

$$f_0(z') = x' + e_0 + e_n\cos\theta' + e_{n+1}\sin\theta'.$$

By comparing the components of x, x', e_n, e_{n+1} we get then

$$x = x', \quad \theta = \theta' \bmod 2\pi$$

so that $z = z'$.

Case 3. $0 \leqslant t,\ t' \leqslant \frac{1}{2}$.

We get again

$$t = t', \quad x = x', \quad \theta = \theta' \bmod 2\pi$$

so that $z = z'$ by comparing the various components.

It follows that

$$f_0(z) \neq f_0(z') \text{ whenever} z \neq z',$$

i.e., f_0 is an imbedding. A C^∞-imbedding f as desired may now be obtained by applying the corner-smoothing procedure described in § 2.

§4 Proof of the theorem

Let M be a closed orientable C^∞-manifold of dimension $n \geqslant 4$. Let C_i $(i = 1, \cdots, r)$ be r simple-closed C^∞-curves in M disjoint from each other so that, by taking a fixed point O in M and r paths L_i from O to a point of C_i, the r closed paths $L_iC_iL_i^{-1}$ will form a system of generators of $\pi_1(M)$. Take r C^∞-tubes T_i about C_i disjoint from each other so that, owing to the orientability of M, there exist C^∞-diffeomorphisms h_i of $\partial D_i^2 \times D_i^{n-1}$ onto T_i with $h_i/\partial D_i^2 \times (\tilde{O}_i) \equiv C_i$, where D_i^2, D_i^{n-1}

are ordinary discs of dimensions 2 and $n-1$ respectively, and $\tilde{O}_i$ is the centre of D_i^{n-1}. Let D_0^n be a closed C^∞-disc of dimension n in M disjoint from all T_i. Removing now the interior of D_0^n and the interiors of all $h_i(\partial D_i^2 \times D_i^{n-1})$ and adjoining $D_i^2 \times \partial D_i^{n-1}$ to it with $\partial(D_i^2 \times \partial D_i^{n-1})$ identified to $h_i(\partial D_i^2 \times \partial D_i^{n-1})$ under h_i, we get a manifold M^* simply-connected and bounded by the $(n-1)$-sphere $\partial D_0^n = S_0^{n-1}$ which can be endowed with a differential structure compatible with those of M and $D_i^2 \times \partial D_i^{n-1}$ and is unique up to C^∞-diffeomorphism. By a theorem of Hirsch ([3]Theorem 1), M^* can be C^∞-imbedded in a Euclidean $(2n-2)$-space. Let $R^{2n-1} = R^{2n-2} \times (-\infty < t < +\infty)$ be a Euclidean $(2n-1)$-space with coordinates $(x_1, \cdots, x_{2n-2}, t)$, R_t^{2n-2} the subspace $R^{2n-2} \times (t)$ and let $f^* : M^* \subset R_0^{2n-2}$ be such a C^∞-imbedding. Our task is now to replace the mapping f^* on the interior of $D_i^2 \times \partial D_i^{n-1}$ by one giving a homeomorphism of $h_i(\partial D_i^2 \times D_i^{n-1})$ in R^{2n-1}. For this purpose let us take in R_2^{2n-2} a set of r mutually disjoint ordinary $(n-1)$ -discs

$$E_i^{n-1}: \begin{cases} (x_1 - a_1^{(i)})^2 + \cdots + (x_{n-1} - a_{n-1}^{(i)}) \leqslant \rho_i^2, \\ x_n = \cdots = x_{2n-2} = 0, \\ t = 2, \end{cases}$$

with $\partial E_i^{n-1} = S_i^{n-2}$. Let S^{n-2} be an ordinary $(n-2)$ -sphere with $\alpha_i : S^{n-2} \equiv \partial D_i^{n-1}$ and $\beta_i : S^{n-2} \equiv S_i^{n-2}$ the natural homeomorphisms. According to a wellknown theorem of Whitney[8], any two C^∞-imbeddings of a closed C^∞-m-manifold in an R^{2m+2} are C^∞-isotopic, whether the manifold is connected or not. Let O_i be the centre of D_i^2. Applying the above theorem to the union of manifolds $\sum f^*((O_i) \times \partial D_i^{n-1})$ and $\sum S_i^{n-2}$, we get a set of C^∞-diffeomorphisms g_i of $S^{n-2} \times [0,2]$ in R^{2n-1} such that:

$$\begin{aligned} &g_i/S^{n-2} \times (t) \subset R_t^{2n-2}, \quad 0 \leqslant t \leqslant 2, \\ &g_i(x,t) = (\pi_0 f^*(O_i, \alpha_i(x)), t), \quad x \in S^{n-2}, \quad 0 \leqslant t \leqslant 1, \\ &g_i(x,2) = \beta_i(x), \quad x \in S^{n-2}, \end{aligned}$$

and $g_i/S^{n-2} \times [0,2]$ are disjoint from each other, (π_0 being the projection onto R_0^{2n-2}). The Lemma 1 of §3 gives now a set of C^∞-diffeomorphisms $\tilde{g}_i$ of $D_i^2 \times \partial D_i^{n-1} \times [0,1]$ in R^{2n-1} such that

$$\begin{aligned} &\tilde{g}_i/D_i^2 \times \partial D_i^{n-1} \times (t) \subset R_t^{2n-2}, \quad 0 \leqslant t \leqslant 1, \\ &\tilde{g}_i/O_i \times \partial D_i^{n-1} \times [0,1] \equiv g_i/S^{n-2} \times [0,1], \end{aligned}$$

$$\tilde{g}_i/D_i^2 \times \partial D_i^{n-1} \times (0) \equiv f^*/D_i^2 \times \partial D_i^{n-1}.$$

$\tilde{g}_i/D_i^2 \times \partial D_i^{n-1} \times (1)$ is normal with respect to $\partial D_i^{n-1} \times (1)$, with normal length $\varepsilon > 0$ sufficiently small, and $\tilde{g}_i/D_i^2 \times \partial D_i^{n-1} \times [0,1]$ are disjoint from each other.

Now the isotopy $g_i/S^{n-2} \times (t), 1 \leqslant t \leqslant 2$ induces an equivalence between the normal bundles of $g_i(S^{n-2} \times (t))$ in R_t^{2n-2}, and hence $\tilde{g}_i/D_i^2 \times \partial D_i^{n-1} \times (1)$ can be extended to a C^∞-diffeomorphism $\tilde{g}_i$ of $D_i^2 \times \partial D_i^{n-1} \times [1,2]$ into R^{2n-1} such that for each $1 \leqslant t \leqslant 2$,

$$\begin{aligned} &\tilde{g}_i/O_i \times \partial D_i^{n-1} \times [1,2] \equiv g_i/S^{n-2} \times [1,2], \\ &\tilde{g}_i/D_i^2 \times \partial D_i^{n-1} \times (t) \subset R_t^{2n-2}. \end{aligned}$$

$\tilde{g}_i/D_i^2 \times \partial D_i^{n-1} \times (t)$ is normal with respect to $\partial D_i^{n-1} \times (t)$ with normal length $\varepsilon > 0$ sufficiently small and $\tilde{g}_i/D_i^2 \times \partial D_i^{n-1} \times (t)$ are disjoint from each other.

Now each of the manifold $\tilde{g}_i(\partial D_i^2 \times \partial D_i^{n-1} \times (2)) \subset R_2^{2n-2}$ is one as described in the Lemma 2 of §2. Hence by this Lemma, each of $\tilde{g}_i/\partial D_i^2 \times \partial D_i^{n-1} \times (2)$ may be extended to a C^∞-imbedding $\tilde{g}_i^*$ of $\partial D_i^2 \times D_i^{n-1}$ in R^{2n-1} such that

$$\tilde{g}_i^*/\partial D_i^2 \times \partial D_i^{n-1} \equiv \tilde{g}_i/\partial D_i^2 \times \partial D_i^{n-1} \times (2).$$

$\tilde{g}_i^*/\partial D_i^2 \times \operatorname{Int} D_i^{n-1}$ is contained in the half-space $t > 2$, and $\tilde{g}_i^*/\partial D_i^2 \times D_i^{n-1}$ are mutually disjoint.

Now take a pair of fixed orthogonal points u, v on a circle S^1 and let $\gamma_i : S^1 \equiv \partial D_i^2$ be natural homeomorphisms.Define a mapping f of $M - \operatorname{Int} D_0^n$ into R^{2n-1} as follows. Let f coincide with f^* on

$$M^* - \sum \operatorname{Int}(D_i^2 \times \partial D_i^{n-1}) \equiv M - \sum \operatorname{Int} h_i(\partial D_i^2 \times D_i^{n-1}) - \operatorname{Int} D_0^n.$$

For $\omega^{(i)} = \gamma_i(u)\cos\theta + \gamma_i(v)\sin\theta \in \partial D_i^2, x^{(i)} \in \partial D_i^{n-1}$ and $0 \leqslant \tau \leqslant 1$, we set

$$f(\omega^{(i)}, \tau x^{(i)}) = \begin{cases} \dfrac{1}{\tau} \cdot \tilde{g}_i(\tau\omega^{(i)}, x^{(i)}, 4(1-\tau)), & \dfrac{3}{4} \leqslant \tau \leqslant 1, \\ \tilde{g}_i(\omega^{(i)}, x^{(i)}, 4-4\tau), & \dfrac{1}{2} \leqslant \tau \leqslant \dfrac{3}{4}, \\ \tilde{g}_i^*(\omega^{(i)}, 2\tau x^{(i)}), & 0 \leqslant \tau \leqslant \dfrac{1}{2}. \end{cases}$$

It is clear that f gives an imbedding of $M - \operatorname{Int} D_0^n$ into the half-space $t \geqslant 0$ of R^{2n-1}. Now by a theorem of Haefliger ([1], Theorem 1, Corollary), any two C^∞-imbeddings of an m-sphere in an R^{2m} are C^∞-isotopic if $m \geqslant 4$. Hence, for $n \geqslant 5$, $f/\partial D_0^n$ may be extended to a C^∞-imbedding f' of $\partial D_0^n \times [-1,0]$ in R^{2n-1} such that

$$f'/\partial D_0^n \times (0) \equiv f/\partial D_0^n,$$
$$f'/\partial D_0^n \times (t) \subset R_t^{2n-2}, \quad -1 \leqslant t \leqslant 0.$$

$f'/\partial D_0^n \times (-1)$ is an ordinary $(n-1)$-sphere with centre O in R_{-1}^{2n-2}. For any $x \in \partial D_0^n$, define further

$$f(tx) = \begin{cases} f'(x, -2(1-t)), & \dfrac{1}{2} \leqslant t \leqslant 1, \\ O + 2t[f'(x,-1) - O], & 0 \leqslant t \leqslant \dfrac{1}{2}. \end{cases}$$

Then f gives an imbedding of M in R^{2n-1} for $n \geqslant 5$, which can be modified to give a C^∞-imbedding of M in R^{2n-1} by smoothing the corners. For $n = 4$, we define f/D_0^n to be a cone over the base $f/\partial D_0^n$ with vertex in the halfspace $t < 0$ of R^{2n-1}, which gives a topological imbedding of M in R^{2n-1}. Thus our theorem is proved.

References

[1] Haefliger A. Differentiable imbeddings. *Bulletin Amer. Math. Soc.*, 1961, 67: 109-112.

[2] Hirsch M W. On imbedding differentiable manifolds in Euclidean space. *Annals of Math.*, 1961, 73: 566-571.

[3] Hirsch M W. The imbedding of bounding manifolds in Euclidean space. *Annals of Math.*, 1961, 74: 494-497.

[4] Milnor J. On the Relationship between Differentiable Manifolds and Combinatorial Manifolds (mimeographed), 1956.

[5] Steenrod N E. *The Topology of Fiber Bundles*. Princeton, 1951.

[6] Whitney H. The self-intersections of a smooth n-manifold in $2n$-space. *Annals of Math.*, 1944, 45: 220-246.

[7] Whitney H. The singularities of a smooth n-manifold in $(2n-1)$-space. *Annals of Math.*, 1944, 45: 247-293.

[8] Whitney H. Differentiable manifolds. *Annals of Math.*, 1936, 3: 645-680.

38. 欧氏空间中的旋转*

设 ρ 是 n 维欧氏空间 R^n 绕原点的一个旋转, 我们熟知在 n 是奇数时, ρ 必有一固定的旋转轴, 且一般说来, R^n 中有通过原点而两两相垂直的 m 个平面 $P_1,\cdots,P_m$, 这里 $n-1\leqslant 2m\leqslant n$, 使 ρ 是每一平面 P_i 上的一个旋转. 换言之, 在 R^n 中可取一组正交规范化基, 使对此基而言, 表示旋转 ρ 的矩阵形如

$$M_\rho=\begin{bmatrix} A_1 & & & \\ & \ddots & & \\ & & A_m & \\ & & & E \end{bmatrix},$$

这里 $A_i=\begin{bmatrix}\cos\alpha_i, & -\sin\alpha_i \\ \sin\alpha_i, & \cos\alpha_i\end{bmatrix}$, E 在 n 是偶数时不存在, 在 n 是奇数时 $=[1]$, 除此外 M_ρ 的元素都是 0. 本小品文将对此定理提供一简单的拓扑证明, 它的依据是 Lefschetz 的不动点定理, 参阅 Pontrjagin《组合拓扑学》, 冯康译, 第三章.

为此, 命 $G_{n-2,2}=G$ 是 R^n 中所有过原点的平面自然形成的空间即通称所谓 Grassmann 流形者是. G 有一胞腔剖分, 它的胞腔与所谓 Schubert 符号 $[i,j]$ 者相对应, 这里 $0\leqslant i\leqslant j\leqslant n-2$, 而 $[i,j]$ 的维数是 $i+j$, G 的维数则为 $2(n-2)$. 因之在 $0\leqslant k\leqslant n-2$ 时, k 维的胞腔有 $[0,k],[1,k-1],\cdots,\left[\left[\frac{k}{2}\right],k-\left[\frac{k}{2}\right]\right]$ 等 $\left[\frac{k}{2}\right]+1$ 个, 这里 $[a]$ 表 $\leqslant a$ 的最大整数. 而在 $n-2\leqslant(n-2)+k\leqslant 2(n-2)$ 时, $(n-2)+k$ 维胞腔只有 $[k,n-2],[k+1,n-3],\cdots,\left[k+\left[\frac{n-2-k}{2}\right],n-2-\left[\frac{n-2-k}{2}\right]\right]$ 等 $\left[\frac{n-2-k}{2}\right]+1$ 个. 因之 G 的 Euler-Poincaré 示性数为

$$\begin{aligned}\chi(G)&=\sum_{k=0}^{n-2}(-1)^k\left(\left[\frac{k}{2}\right]+1\right)+\sum_{k=1}^{n-2}(-1)^{n-2+k}\cdot\left(\left[\frac{n-2-k}{2}\right]+1\right)\\&=\sum_{k=0}^{n-2}(-1)^k\left(\left[\frac{k}{2}\right]+1\right)+\sum_{i=0}^{n-3}(-1)^i\cdot\left(\left[\frac{i}{2}\right]+1\right)\\&=\begin{cases}(-1)^{2m-2}\cdot m+0=m, & n=2m\text{时},\\ 0+(-1)^{2m-2}\cdot m=m, & n=2m+1\text{时}.\end{cases}\end{aligned}$$

---

* *Shuxue Jinzhan*, 1963, 6: 96-97.

不论何时均有 $\chi(G) \neq 0$.

命 ρ 在 G 中引起的变换为 $\overline{\rho}$, 则由于旋转群的连通性 $\overline{\rho}$ 显然同伦于恒同映像, 因而 $\overline{\rho}$ 的 Lefschetz 数为 $\Lambda(\bar{\rho}) = \chi(G) \neq 0$, 故由 Lefschetz 定理 $\bar{\rho}$ 有一不动点, 即 ρ 在 R^n 中有一不动平面. 命之为 P_1 而与之垂直的空间为 R^{n-2}, 则同法知 ρ 在 R^{n-2} 中所引起的旋转亦有一不动平面, 命之为 P_2. 依次类推得 ρ 的一组不动平面 $P_1, \cdots, P_m, m = [n/2]$, 由此即得本小品文开首所述的定理.

上法自然亦可用之于酉空间而知任一酉矩阵可在酉变换下化为对角形.

39. A Theorem on Immersion*

Notations. R^N = Euclidean N-space; $V_{N,m}$ = Stiefel manifold of m-frames in R^N; $G_1 = \pi_{n-2}(V_{n+1,3}) \approx Z$ for n = even, and $\approx Z_2$ for n = odd with α_1 = a generator; $G_2 = \pi_1(V_{n+1,n}) \approx Z_2 (n > 1)$ with generator α_2; $\rho_i : G_i \to Z_2$ is the homomorphism with $\rho_i(\alpha_i) = 1 \bmod 2$, $i = 1, 2; i_1 : S^{n-2} \subset V_{n,2}$ and $i_2 : V_{n,2} \subset V_{n+1,3}$ are the inclusion maps in the natural fibrations $V_{n,2}/S^{n-2} \to S^{n-1}$ and $V_{n+1,3}/V_{n,2} \to S^n$ respectively; γ = generator of $i_1 * \pi_{n-1}(S^{n-2}) \subset \pi_{n-1}(V_{n,2}) \approx Z_2$ for n = odd $\geqslant 5$, and $\approx Z_2 + Z$ for n = even $\geqslant 6; G_3 = \pi_{n-1}(V_{n+1,3}) \approx Z_4, 0, Z_2 + Z_2, Z_2$ for $n \equiv 0, 1, 2, 3 \bmod 4$ respectively; $i_2* : \pi_{n-1}(V_{n,2}) \to \pi_{n-1}(V_{n+1,3})$ is an epimorphism with $i_2{*}\gamma \in \pi_{n-1}(V_{n+1,3})$ of order 2 for $n \equiv 0, 2, 3 \bmod 4, n \geqslant 5; \sigma : Z_2 \to G_3$ is the homomorphism with $\sigma(1 \bmod 2) = i_2 * \gamma$.

Let M^n be a closed orientable C^∞-manifold of dimension $n \geqslant 5$. Take a C^∞-imbedding $f : M^n \subset R^{2n+1}$, and consider the normal 3-field bundle $\Re$ over M^n determined by f. The characteristic class of $\Re$ is independent of the imbedding f, and is the *unreduced* normal Whitney class $\overline{W}^{n-1} \in H^{n-1}(M^n, G_1)$.

Lemma $\overline{W}^{n-1} = 0$.

It follows that a cross section exists in $\Re$ over the $(n-1)$-squellette K^{n-1} of a smooth triangulation K of M^n, and the second obstruction of cross sections may be considered. Let $Z^n_s, Z^n_{s_0} \in H^n(M^n, G_3)$ be the obstructions to cross sections s, s_0 over K^{n-1} in $\Re$, $D^{n-2} = D(s, s_0) \in H^{n-2}(M^n, G_1)$ the difference class of the sections s, s_0, and $\overline{W^2} \in H^2(M^n, G_2)$ the normal Whitney class of dimension 2 of M^n. Then the formula of Boltyanski-Liao gives for $n \geqslant 5$[1]:

$$z^n_s = z^n_{s_0} + S^2_q D^{n-2} + D^{n-2} \cup \overline{W}^2, \tag{1}$$

in which the $\cup-$product is determined by the pairing of groups $G_1 \cdot G_2 \subset G_3$ with $\alpha_1 \cdot \alpha_2 = i_2{*}\gamma$, and the square by the pairing $G_1 \cdot G_1 \subset G_3$ with $\alpha_1 \cdot \alpha_1 = i_2{*}\gamma$. (1) may also be written as

$$z^n_s = z^n_{s_0} + \sigma S^2_q \rho_1 D^{n-2} + \sigma(\rho_1 D^{n-2} \cup \rho_2 \overline{W}^2). \tag{2}$$

---

* *Sci. Sinica*, 1964, 13: 160.

Determining from the formulas in Ref. [4] the Stiefel-Whitney classes of a manifold in terms of squares, we have, however, $S_q^2 X + X \cup \rho_2 \overline{W}^2 = 0$ for any $X \in H^{n-2}(M^n, Z_2)$. It follows that (2) becomes $Z_s^n = Z_{s_0}^n$ and $I(M^n) = Z_s^n$. $M^n \in G_3 = \pi_{n-1}(V_{n+1,3})$ is independent of the cross sections s over K^{n-1} and of the imbedding f, and it is a well-defined invariant of M^n.

Theorem 1 *For any C^∞-immersion of M^n in R^N with $N \geqslant 2n-1$ a normal $(N-2n+2)$-field exists if and only if $I(M) = 0$.*

By a theorem of Hirsch[2] we set further

Theorem 2 *M^n is C^∞-immersible in R^{2n-2} if and only if $I(M^n) = 0$.*

References

[1] Boltyanski V G. *Изв. Ак. Наук СССР*. Сер. Матем., 1956, 20: 99-136.

[2] Hirsch G. *Trans. Amer. Math. Soc.*, 1959, 93: 242-276.

[3] Paechter G F. *Quart. J. Math.*, 1958, 9: 8-27.

[4] Wu W T C R, 1950, 230: 508-511.

40. On the Immersion of C^∞-3-Manifolds in a Euclidean Space*

For a C^∞-manifold M, let $T_x(u)$ be its tangent space at a point $x \in M$, and $T(M)$ be its tangent bundle. Any C^∞-map f of a C^∞-manifold M in a Euclidean n-space R^n will induce a map $df : T(M) \to R^n$, which, in case f is an immersion, is a linear map in the sense that, for each $x \in M$, df is a nondegenerate linear map of $T_x(M)$ in R^n. Denote the set of regular homotopy classes of C^∞-maps of M in R^n by $\pi_R^n(M)$ and the set of linear homotopy classes of $T(M)$ in R^n by $\pi_L^n(M)$, then the theory of Hirsch (see Ref.[2]) shows that for $n > m = \dim M$, $\pi_R^n(M) \sim \pi_L^n(M)$ as sets and the 1-1-correspondence is induced by $f \to df$. Suppose now M is parallelizable with a fixed C^∞-m-$field$ $\{v_1(x), \cdots, v_m(x)\}$, $x \in M$, so that $T(M)$ is accordingly homeomorphic to $M \times R^m$. With linear map λ of $T(M)$ in R^n may then be associated canonically a map $\overline{\lambda}$ of M in $V_{n,m}$, the Stiefel manifold of sets of n linearly independent vectors in R^n. This correspondence $\lambda \to \lambda$ gives an equivalence as sets $\pi_L^n(M) \sim \pi(M, V_{n,m})$, the latter being the set of ordinary homotopy classes of M in $V_{n,m}$. With this remark we shall determine in the present note $\pi_{\dot{R}}^n(M)$ in case M is an orientable C^∞-3-manifold, which is necessarily parallelizable.

In what follows let M be thus an oriented C^∞-3-manifold with a fixed C^∞-3-field $\{v_1(x), v_2(x), v_3(x)\}$, $x \in M$, so that $\pi_{\dot{R}}^n(M) \sim \pi(M, V_{n,3})$. For $n \geqslant 6$, the determination of $\pi(M, V_{n,3})$ is immediate. For $n = 4, 5$, the determination of $\pi(M, V_{n,3})$ becomes a problem of second obstruction, which has been treated by Olum in [3] and Whitney, etc.in [1, 4, 5] respectively. The final results are given in Theorems 1 and 2 below.

Consider first the case $n = 4$. Let $(e_0, \cdots, e_3)$ be a basis of R^4. For any vector $e = x_0e_0 + \cdots + x_3e_3$ of R^4, let us put $J_1e = x_1e_0 - \mathrm{x}_0e_1 + x_3e_2 - x_2e_3$, $J_2e = x_2e_0 - x_3e_1 - x_0e_2 + x_1e_3$, $J_3e = J_1J_2e$. For any immersion f of M in R^4, let $u_i(x) = df(v_i(x))$, i =1, 2, 3, and $u_0(x)$ be the unit vector normal to M at $x \in M$ such that $(u_0(x), u_1(x), u_2(x), x_3(x))$ determines the same orientation of R^4 as (e_0, e_1, e_2, e_3).

* *Sci. Sinica*, 1964, 13: 335-336.

Let $u_i(x) = \sum_{j=1}^{3} a_{ij}(x)J_j u_0(x)$, i =1, 2, 3. Then $x \to u_0(x)$ defines a map α_f of M in oriented S^3 with degree a_f. Similarly, $x \to [\alpha_{ij}(x)]$ defines a map β_f of M in $V_{4,3}$, which has a degree b_f and induces a homomorphism $h_f : \pi_1(M) \to \pi_1(V_{4,3}) \approx Z_2$.

Theorem 1 *If M is closed (resp. nonclosed), then to any pair of integers a, b and a homomorphism $h \in Hom(\pi_1(M), Z_2)$(resp. h only) there are immersions f of M in R^4 with $(a_f, b_f, h_f) = (a, b, h)$(resp. $h_f = h$ only). Moreover, two immersions f, g of M in R^4 are regularly homotopic if and only if $(a_f, b_f, h_f) = (a_g, b_g, h_g)$(resp. $h_f = h_g$ only).*

Proof. For any point $(u_1, u_2, u_3) \in V_{4,3}$, let u_0 be the unit vector normal to u_1, u_2, u_3 such that $(u_0,\ u_1, u_2, u_3)$ determines the same orientation of R^4 as (e_0, e_1, e_2, e_3). Let $u_i = \sum_{j=1}^{3} \alpha_{ij} J_j u_0$, i =1, 2, 3, then $(u_1, u_2, u_3) \to (u_0, [\alpha_{ij}])$ defines a homeomorphism h of $V_{4,3}$ onto $S^3 \times V_{3,2}$, in which $\pi_1(V_{3,2}) \approx Z_2$, $\pi_2(V_{3,2}) = 0$, $\pi_3(V_{3,2}) \approx Z$. It follows that $\pi(M,\ V_{4,3}) \sim \pi(M, S^3) \times \pi(M, V_{3,2})$. Now $\pi(M, S^3)$ is determined by the degree and $\pi(M, V_{3,2})$ is determined by the degree and the induced homomorphism of $\pi_1(M)$ in $\pi_1(V_{3,2}) \approx Z_2$ by Olum(Th.IIb in[4]). Hence we get the theorem as desired.

Theorem 2 *The normal Whitney class $\overline{W}_f^2$ with integer coefficients of an immersion f of M in R^5 is of the form $2X$, where X may be any class in $H^2(M)$. Moreover, given X the set of regular homotopy classes of immersions with normal Whitney class $2X$ is in 1-1-correspondence with the cosets in $H^3(M)/4X_\cup H^1(M)$.*

Proof. Let $\pi_i = \pi_i(V_{5,3})$, so that $\pi_1 = 0$, $\pi_2 \approx Z$, $\pi_3 \approx Z$. With fixed generators α of π_2 and β of π_3 they will be canonically identified to Z. Then $H^2(V_{5,3}, \pi_2) = H^2(V_{5,3}) \approx Z$ has a generator Σ^2 such that $2\Sigma^2$ is the characteristic class of the canonical circle bundle $V_{5,4} \to V_{5,3}$. Now to any immersion f of M in R^5 the associated map $\overline{df} : M \to V_{5,3}$ induces the normal bundle over M from the bundle $V_{5,4} \to V_{5,3}$. Hence $\overline{df}^*(2\Sigma^2)\overline{W}_f^2$. For any $X \in H^2(M)$, let $g : M \to V_{5,3}$ be maps with $g^*\Sigma^2 = X$. By Hirsch such maps are homotopic to maps $\overline{df}$ associated with immersions f so that $\overline{W}_f^2 = 2\overline{df}^*\Sigma^2 = 2X$, which proves the first part of the theorem.

Now from the homotopy sequence of the natural fibration $V_{5,3}/S^2 = V_{5,2}$ we see that $j* : \pi_3(S^2) \to \pi_3(V_{5,3})$ induced by the injection $j : S^2 \subset V_{5,3}$ is given by $j*\gamma = \pm 2\beta$, where γ is a generator of $\pi_3(S^2) \approx Z$. The function $\eta : \pi_2 \to \pi_3$ obtained by the composition $S^3 \xrightarrow{h} S^2 \to V_{5,3}(h \in \gamma)$ is thus given by $\eta(\alpha) = \pm 2\beta$, and we have the Whitehead product $[\alpha, \alpha] = \pm 4\beta$. Hence for $X \in H^2(M, \pi_2) = H^2(M)$ and $Y \in H^1(M, \pi_2) = H^1(M)$, we have $X \cup_0 Y = \pm 4X \cup Y \in H^3(M, \pi_3) = H^3(M)$, in which $\cup_0$ means cup products with the Whitehead product as pairing, and $\cup$ with the usual pairing. By the theorem of Whitney-Postnikov the homotopy classes of maps f of M in $V_{5,3}$ with $f^*\Sigma^2 = X$ are in 1-1-correspondence with the cosets $H^3(M)/X \cup_0 H^1(M, \pi_2)$, from which the second part of the theorem follows.

References

[1] Eilenberg S & MacLane S. *Annals of Math.*, 1954, 60: 513-557.
[2] Hirsch W. *Trans.Amer.Math.Soc.*, 1959, 93: 242-276.
[3] Olum P. *Annals of Math.*, 1953, 58: 458-480.
[4] Postnikov M M. *C.R.URSS*, 1949, 64: 461-462.
[5] Whitney H. *Ann.of Math.*, 1949, 50: 270-284.

41. On the Notion of Imbedding Classes*

§ 1. Let $\zeta = \{\tilde{X}, X, \omega\}$ be an S^0-bundle or a 2-sheeted covering having a base-connected polytope X, projection ω, and a structural group T consisting of two elements: 1 and t, with t operating without fixed points in $\tilde{X}$. For any coefficient group G, let $G(\zeta)$ be the system of local groups in X determined by the homomorphism $\psi : \pi_1(X, x_0) \to AutG$ such that for any $\alpha \in \pi_1(X, x_0)$, $\psi(\alpha)(g) = +g$ or $-g$ according as $\alpha \in \omega_* \pi_1(\tilde{X}, \tilde{x}_0)$ or not $(\omega(\tilde{x}_0) = x_0)$. Let $G_{(k)} = G$ or $G(\zeta)$ according as k is even or odd. Then any pairing of groups $G' \otimes G'' \to G$ will induce natural pairings of local systems of groups $G'_{(k)} \otimes G''_{(l)} \to G_{(k+l)}$, which will induce, in turn, cup products $H^q(X, G'_{(k)}) \smile H^r(X, G''_{(l)}) \subset H^{q+r}(X, G_{(k+l)})$. Write $d = 1 - t$, $s = 1 + t$, $\rho = d$ or s, and then $\bar{\rho} = s$ or a, and let $H^q_{(\rho)}(\tilde{X}, G)$ be the Smith special groups of ζ. Then we have canonical isomorphisms $\bar{\omega}^* : H^q_{(d)}(\tilde{X}, G) \approx H^q(X, G)$ and $H^q_{(s)}(\tilde{X}, G) \approx H^q(X, G(\zeta))$. Let $\mu^* : H^q_{(\rho)}(\tilde{X}, G) \to H^{q+1}_{(\bar{\rho})}(\tilde{X}, G)$ be the Smith homomorphism and 1_d be the d-cohomology class associated with the unit class of $\tilde{X}$. Then $A^q(\zeta) = \bar{\omega}^*(\mu^*)^q 1_d \in H^q(X, Z_{(q)})$ will be called the *Smith classes* of ζ, for which we have $A^q(\zeta) \smile A^r(\zeta) = A^{q+r}(\zeta)$.

Let $N(\zeta)$ be the N-fold Whitney sum of ζ with itself and π^N be the bundle of coefficients $\mathscr{B}(\pi_{N-1}(S^{N-1}))$ of $N\zeta$, as defined in § 30 of [3]. Then $\pi^N = Z_{(N)}$. In particular, $N\zeta$ is always orientable for N even. Let $W^N(N\zeta) \in H^N(X, \pi^N)$ be the (unreduced) Whitney class of $N\zeta$. Since for any Whitney sum $\xi \oplus \eta$ of S^{m-1} and S^{n-1} bundles ξ and $\eta (m, n \geqslant 1)$ we have $W^{m+n}(\xi \oplus \eta) = W^m(\xi) \smile W^n(\eta)$ with natural pairing of local coefficients and $W^1(\zeta) = A^1(\zeta)$ for an S^0-bundle ζ, we get

Proposition 1. For the 2-sheeted covering ζ, we have

$$W^N(N\zeta) = A^N(\zeta) = (A^1(\zeta))^N.$$

§ 2. For any integer $N > 0$, let $\mathscr{G}_N^2$ be the set of nonempty sequences of positive integers $I = (i_1, \cdots, i_s)$ with $i_1 \geqslant 2i_2, \cdots, i_{s-1} \geqslant 2i_s$ and $e(I) = 2i_1 - n(I) < N - 1$ or $I = (N-1)$ where $n(I) = \sum_{r=1}^{s} i_r$. Let $\mathscr{G}_N \subset \mathscr{G}_N^2$ be the subset with the further restriction $i_s > 1$. Let ζ be the same as in § 1, and suppose that $A^N(\zeta) = 0$ where N is *even*. Then as in [4] we shall define for each $I \in \mathscr{G}_N$ a *secondary Smith class* of ζ

* *Sci. Sinica*, 1964, 13: 681-682.

as follows:

Let $S^\infty \to P^\infty$ be the universal S^0-bundle η over the infinite dimensional projective space P^∞, and $f : X \to P^\infty$ the map with $f * \eta = \zeta$. Let e_1 and e_2 be the generators of $H^*(P^\infty, Z_2)$ and $H^*(P^\infty)$ respectively. Define for each $I \in \mathscr{G}_N^2$ a class $e_I \in H^{n(Z)}(P^\infty, Z_2)$ by $Sq^I e_N = \rho_2 e_N \cup e_I$ where $\rho_2 =$ reduction mod 2 and $e_N = (e_2)^{N/2}$. Set $e_I(\zeta) = f^* e_I$. Let $K(Z, N-1) \to P(Z, N) \to K(Z, N)$ be the path bundle over the Eilenberg-MacLane space $K(Z, N)$, and $K(Z, N-1) \xrightarrow{i} S_N^\infty \xrightarrow{\pi} P^\infty$ the bundle induced from it by a map $x : P^\infty \to K(Z, N)$ with $x^* t_N = e_N$ where t_N is the canonical class of $K(Z, N)$. For each $I \in \mathscr{G}_N$ take (Cf.[1]) a representative x_I in $(Sq^I + e_I \smile)_\pi(e_N) \in H^{N-1+n(I)}(S_N^\infty, Z_2)/L_N$ where $L_N = (Sq^I + \pi^* e_I \smile) H^{N-1}(S_N^\infty) + \pi^* H^{N-1+n(I)}(P^\infty, Z_2) = (Sq^I + \pi^* e_I \smile) H^{N-1}(S_N^\infty)$. Then $H^*(S_N^\infty, Z_2)$ is a polynomial algebra generated by $\pi^* e_1$, and all x_I, $I \in \mathscr{G}_N$, with the single relation $(\pi^* e_1)^N = 0$. Now $(xf)^* t_N = f^* e_N = A^N(\zeta) = 0$ by hypothesis so that f has liftings $\hat{f}_N : X \to S_N^\infty$ with $\pi \hat{f}_N \sim f$. As in Proposition 2.2 of [2], the set $\{\hat{f}_N^* x_I\}$ corresponding to all liftings $\hat{f}_N$ is then a well-defined coset in $H^{N-1+n(I)}(X, Z_2)/L_I(\zeta)$ with $L_I(\zeta) = (Sq^I + e_I(\zeta) \smile)$. $H^{N-1}(X)$, which will be called a *secondary Smith class* of ζ and will be denoted by $S_I^{N-1+n(I)}(\zeta)$. This coset is moreover independent of the representatives x_I chosen.

Now $N\zeta$ is orientable for N even, and $W^N(N\zeta) = A^N(\zeta) = 0$ by Proposition 1 so that by Peterson-Stein is also defined for each $I \in \mathscr{G}_N$ a secondary characteristic class $\Phi_I^{N-1+n(I)}(N\zeta) \in H^{N-1+n(I)}(X, Z_2)/M_I$ where $M_I = (Sq^I + Q_I(N\zeta) \smile) H^{N-1}(X)$, $Q_I(N\zeta) = h^* Q_I$, $h : X \to G_N$ is the map inducing $N\zeta$ from the universal S^{N-1}-bundle γ_N over the grassmannian G_N of the oriented N-planes in R^∞, $Q_I \in H^{n(I)}(G_N, Z_2)$ is defined by $Sq^I W^N = \rho_2 W^N \smile Q_I$, and $W^N \in H^N(G_N)$ is the universal Euler class in G_N. We have then

Proposition 2. For the 2-sheeted covering ζ with $A^N(\zeta) = 0$, N even, we have for each $I \in \mathscr{G}_N$

$$S_I^{N-1+n(I)}(\zeta) = \Phi_I^{N-1+n(I)}(N\zeta).$$

To prove this, consider the commutative diagram

$$\begin{array}{ccccccc} & & S_N^\infty & \xrightarrow{\tilde{g}} & K_N & \xrightarrow{\tilde{\lambda}} & P(Z,N) \\ & {\scriptstyle \tilde{f}_N} \nearrow & \downarrow & & \downarrow & & \downarrow \\ X & \xrightarrow[f]{} & P^\infty & \xrightarrow[g]{} & G_N & \xrightarrow[\lambda]{} & K(Z,N) \end{array}$$

in which $K_N \to G_N$ is the $K(Z, N-1)$-bundle induced from $P(Z, N) \to K(Z, N)$ by a map λ with $\lambda * t_N = W^N$, $(\tilde{g}, g) : (S_N^\infty, F^\infty) \to (K_N, G_N)$ be the maps with

$g * \gamma_N = N\eta$ and $\hat{f}_N$ a lifting of f as before. For $h = gf$, we have then $h * \gamma_N = N\zeta$ and $g^* W^N = e_N$. Hence a comparison of the definitions of Q_I and e_I shows that $L_I(\zeta) = M_I(\zeta)$ for each $I \in \mathscr{G}_N$. Now $\lambda g = z$ is a map with $x^* t_N = e_N$, and $\tilde{h} = \tilde{g}\hat{f}_N$ is a lifting of h. Let us take $\varphi_I \in H^{n(I)}(K_N, Z_2)$ as in Th.4.2 of [2], and then take $x_I \in H^{n(I)}(S_N^\infty, Z_2)$ to be $x_I = \tilde{g}^* \varphi_I$ in defining $S_I^{N-1+n(I)}(\zeta)$. Then $\tilde{h}^* \varphi_I = \tilde{f}_N^* x_I$ so that $S_I(\zeta) = \Phi_I(N\zeta)$ as to be proved.

For $I = (2)$, let us set $S_I^{N-1+n(I)}(\zeta) = S^{N+1}(\zeta)$, then by Prop.4.4 of [2], we get the following

Corollary *For the 2-sheeted covering ζ with $A^N(\zeta) = 0$, N even $\geqslant 4$, $S^{N+1}(\zeta)$ is the second obstruction of $N\zeta$.*

§ 3. Let K be a simplicial complex and $\tilde{K}_2^*$(resp. K_2^*) its 2-fold deleted (resp. cyclic) product. Denote the natural 2-sheeted covering of $\tilde{K}_2^*$ over K_2^* by ζ_K. Then the Smith classes $A^q(\zeta_K) \in H^q(K_2^*)$ or $H^q(K_2^*, Z(\zeta_K))$ according as q is even or odd have been called the *imbedding classes* of K, and were denoted by $\Phi^q(K)$. If N is *even* and $\Phi^N(K) = 0$ so that $A^N(\zeta_K) = 0$, then the secondary Smith classes of ζ_K will be called the *secondary imbedding classes* of K and will be denoted by

$$Y_I^{N-1+n(I)}(K) = S_I^{N-1+n(I)}(\zeta_K) \in H^{N-1+n(I)}(K_2^*, Z_2)/L_I(\zeta_K)$$

for any $I \in \mathscr{G}_N$. In particular, we set

$$Y^{N+1}(K) = S^{N+1}(\zeta_K)$$

for the sequence $I = (2)$.

For any $I \in \mathscr{G}_N^2$, let $\mathrm{L}_I^2(\xi_K) = (Sq^I + e_I(\zeta_K) \smile)$. $H^{N-1}(K_2^*, Z_2)$ and α : $H^{N-1+n(I)}(K_2^*, Z_2)/L_I(\zeta_K) \to H^{N-1+n(2)}(K_2^*, Z_2)/L_I^2(\zeta_K)$ be the canonical homomorphism, then the secondary imbedding classes $O_2^I(K)$, $I \in \mathscr{G}_N^2$, of Yo as defined in [4] are connected with the classes Y_I by

$$\alpha * Y_I^{N-1+n(I)}(K) = O_2^I(K).$$

The verification is direct.

References

[1] Peterson F P. *Amer. J. Math.*, 1960, 82: 649-652.

[2] Peterson F P & Stein N. *Annals of Math.*, 1962, 76: 510-523.

[3] Steenrod N E. *Topology of Fiber Bundles*, Princeton, 1951.

[4] Yo G T. *Scientia Sinica*, 1963, 12: 1072.

42. On the Imbedding of Manifolds in a Euclidean Space(1)*

In what follows by manifolds and imbeddings we shall mean closed C^∞-manifolds and C^∞-imbeddings. The notation $M^n \subset R^N$ will mean that the n-manifold M^n is imbeddable in the Euclidean N-space R^N. For any integer $q > 0$, let J_q be the set of primes p with $2p \leqslant q+2$, and let $(C_{q,p})$ stands for the condition $H^i(M^n, Z_p) = 0$ for $1 \leqslant i \leqslant q$. We shall prove:

Theorem 1 *Assume* $n \geqslant 2q+3$ *and* $(C_{q,p})$ *for all primes* p. *Then* $M^n \subset R^{2n-q}$ *always.*

Theorem 2 *Assume* $n \geqslant 2q+3$ *and* $(C_{q-1,p})$ *for all* $p \in J_q$. *Then* $M^n \subset R^{2n-q}$ *iff the imbedding class* $\Phi^{2n-q}(M^n) = 0$.

Theorem 3 *Assume* $n \geqslant 2q+3$, q *even, and* $(C_{q-2,p})$ *for all* $p \in J_q$. *Then* $M^n \subset R^{2n-q}$ *iff* $\Phi^{2n-q}(M^n)$, *and Yo's secondary imbedding class* $Y^{2n-q+1}(M^n)$ *both* $= 0$.

Corollary *For* $n \geqslant 5$, $M^n \subset R^{2n-1}$ *iff* $\Phi^{2n-1}(M^n) = 0$. *For* $n \geqslant 7$, $M^n \subset R^{2n-2}$ *iff* $\Phi^{2n-2}(M^n) = 0$ *and* $Y^{2n-1}(M^n) = 0$.

Denote by $\tilde{M}^*$(resp.M^*) the two-fold deleted product (resp.deleted cyclic product) of M^n. Then the proof of the theorems depends on the following

Lemma *For a prime* p *suppose that* $H^1(M^n, Z_p) = \cdots = H^k(M^n, Z_p) = 0 (1 \leqslant k \leqslant n-2)$, *and that* M^n *is orientable in case* $p > 2$. *Then* $H^r_{(s)}(\tilde{M}^*, Z_p) = 0$, *and* $H^r_{(d)}(\tilde{M}^*, Z_p) = 0$ *for* $r \geqslant 2n-k$.

Assuming for the moment the truth of the lemma Theorem 3 will be proved as follows: Only the sufficiency part needs to be considered. Let ζ be the S^0-bundle over M^* with bundle space $\tilde{M}^*$ and projection ω. According to Haefliger [1], for $n \geqslant 2q+3$ or $2N \geqslant 3(n+1)$ where $N = 2n-q$, $M^n \subset R^N$ iff the S^{N-1}-bundle $N\zeta$ over M^* has a cross section. As shown in [4], the first and second obstructions of $N\zeta$ are just the ordinary and secondary imbedding classes $\Phi^N(M^n)$ and $Y^{N+1}(M^n)$. When these are zero, $N\zeta$ has then a cross section over the $(N+1)$-sequelette $K^{(N+1)}$ of a certain subdivision K of M^*. Suppose that this cross section has been extended

* *Sci. Sinica*, 1964, 13: 682-683.

over the squelette $K^{(N+j)}$, $j \geqslant 1$. Then the obstruction to further extension to $K^{(N+j+1)}$ is in $H^{N+j+1}(M^*, \pi^N_{N+j})$ where π^N_{N+j} is the bundle of coefficients of $\pi_{N+j} = \pi_{N+j}(S^{N-1})$ associated with the bundle $N\zeta$. In the notation of [4], π^N_{N+j} is just the system of local coefficients $\pi_{N+j}(\zeta)$ or the simple system of coefficients π_{N+j} according as N is odd or even so that $H^{N+j+1}(M^*, \pi^N_{N+j}) \approx H^{N+j+1}_{(s)}(\tilde{M}^*, \pi_{N+j})$ or $H^{N+j+1}_{(d)}(\tilde{M}^*, \pi_{N+j})$ according as N is odd or even. Now by a theorem of Serre [2], π_{N+j} is finite in the range considered, and has p-primary components only for $p \in J_{j+2}$. The conditions $(C_{q-2,p})$ for $p \in J_q$would imply M^n orientable whenever $q \geqslant 3$, and, in particular, when J_q contains certain odd primes. By the lemma we have then both $H^{N+j+1}_{(s)}(\tilde{M}^*, Z_p) = 0$ and $H^{N+j+1}_{(d)}(\tilde{M}^*, Z_p) = 0$ for $j \geqslant 1$, and $p \in J_q$. Hence the obstruction in $H^{N+j+1}(M^*, \pi^N_{N+j})$ is 0 for $j \geqslant 1$, and a cross section in $N\zeta$ exists over M^* so that $M^n \subset R^N$. This proves Theorem 3 and in a similar manner Theorems 1 and 2.

It remains to prove the lemma. For this let $\tilde{T}$ be a normal tube of the diagonal $\tilde{\Delta}$ in $\tilde{M} = M \times M$ defined by means of a Riemann metric of M. Then $\tilde{T}$ separates $\tilde{M}$ into two bounded manifolds $\tilde{M}^+$ and $\tilde{M}^-$ with $\tilde{M}^- \supset \tilde{\Delta}$. Let $\omega(\tilde{M}\pm) = M\pm$ and $\omega(\tilde{T}) = T$. From the condition $(C_{k,p})$ we get $H^j(\tilde{M}, Z_p) = 0$ for $2n-1 \geqslant j \geqslant 2n-k$. It follows that $H^j(\tilde{M}^+, \tilde{T}; Z_p) \approx H^j(\tilde{M}, \tilde{\Delta}; Z_p) = 0$ for such j. Now $\tilde{T}$ is the tangent bundle of M so that for $2n-k-1 \leqslant j \leqslant 2n-2$, the Gysin sequence gives $H^j(\tilde{T}, Z_p) \approx H^{j-n+1}(M, Z_p) = 0$. From the exact sequence of the pair $(\tilde{M}^+, \tilde{T})$ we get, therefore, $H^j(\tilde{M}^+, Z_p) = 0$ for $2n-k \leqslant j \leqslant 2n-2$. In the exact sequence $0 = H^{2n-1}(\tilde{M}^+, \tilde{T}; Z_p) \rightarrow H^{2n-1}(\tilde{M}^+, Z_p) \rightarrow H^{2n-1}(\tilde{T}, Z_p) \xrightarrow{\delta^*} H^{2n}(\tilde{M}^+, \tilde{T}; Z_p) \rightarrow H^{2n}(\tilde{M}^+, Z_p) = 0$, we see that δ^* is always an isomorphism for $p = 2$ and is one for $p > 2$ if M is orientable. Hence under the hypothesis of the lemma we have $H^j(\tilde{M}^+, Z_p) = 0$ for $j \geqslant 2n-k$. As $\tilde{M}^+$ is a 2-sheeted covering of M^+ and $\tilde{M}^+$, M^+ are deformation retracts of $\tilde{M}^*$, M^* respectively, we get then $H^j(\tilde{M}^*, Z_p) = 0$ for $j \geqslant 2n-k$, $p \geqslant 2$ and $H^j(M^*, Z_p) = 0$ for $j \geqslant 2n-k$, $p > 2$. For $p = 2$ we have now $H^{2n-i}(M^*, Z_2) \approx H^i(M^*, T; Z_2) \approx H^i(M^*M, \Delta\ Z_2) = 0$ for $i \leqslant k$ by a known theorem of Richardson-Smith-Thom(cf.e.g.[2])(M^*M is the quotient space of $M \times M$ with respect to the transformation $t(x, y) = (y, x)$). The lemma now follows from the exact sequence of Richardson-Smith: $\cdots \rightarrow H^r(\tilde{M}^*, Z_p) \rightarrow H^r_{(s)}(\tilde{M}^*, Z_p) \rightarrow H^{r+1}_{(d)}(\tilde{M}^*, Z_p) \rightarrow \cdots$ and the fact that $H^r_{(d)}(\tilde{M}^*, Z_p) \approx H^r(M^*, Z_p)$ for all p.

References

[1] Haefliger A. *Comm. Math. Helv.*, 1962, 37: 155-176.

[2] Nakaoka M. *J. Osaka City Univ.*, 1956, 7: 51-102.

[3] Serre J P. *Annals of Math.*, 1951, 54: 425-505.

[4] Wu W T. On the notion of imbedding classes, *Scientia Sinica*, 1964, 13: 681.

[5] Haefliger A & Hirsch M W. *Topology*, 1963, 2: 129-136.

[6] Levine J P. *Not. Amer.Math.Soc.*, 1962, 9: 220.

43. On Complex Analytic Cycles and Their Real Traces*

Let V be a complex analytic manifold of complex dimension n and V^0 a real submanifold of V of real dimension n. Let X be a complex analytic subvariety of complex dimension r such that X and V^0 determine mod 2 homology classes $\varphi(X)$ and $\varphi(V^0)$ of dimensions $2r$ and n respectively in V. Suppose that $X^0 = X \cap V^0$ is of real dimension r so that the real analytic subvariety X^0 of V^0 determines, according to Borel-Haefliger[1], a mod 2 homology class $\psi(V^0)$ of dimension r in V^0. Then, a recent result of W.L.Chow states that, under the hypothesis that both V and X are algebraic, there is a relation between the intersection products mod 2

$$\varphi(X) \cdot \varphi(V^0) \sim \psi(X^0) \cdot \psi(X^0),$$

in which $\sim$ means numerical equivalence with respect to real algebraic homology classes in $H_*(V^0, Z_2)$. The present paper has the purpose to show that Chow's relation, in the case that X is nonsingular, will remain true without the hypothesis of being algebraic. Moreover, the $\sim$ relation turns out to be an exact equality as homology classes in V. The proof of Chow depends on the fact that in a (projective) algebraic variety, any two algebraic cycles are rationally equivalent, and hence homologous, to algebraic cycles in general position. As this is in general not true for complex analytic cycles in complex analytic varieties, Chow's proof cannot be extended to meet our purpose. On the other hand our proof is based on the following remark, which may be traced back to Ehresmann[4]: V^0 possesses neighbourhoods in V which are homoeomorphic to neighbourhoods of the diagonal in $V^0 \times V^0$. As the intersections are of a local character, everything is then reduced to a study of the diagonal in $V^0 \times V^0$, which can be easily done.

The homology theory used in this paper will be the usual singular homology theory for which the homology groups will be denoted by $H_*(V, V'; G)$ for a pair of spaces $V' \subset V$. The support of a singular chain x in a space V will then be denoted by $|x|(\subset V)$. On the other hand, the homology groups of a locally compact space V

* *Sci. Sinica*, 1965, 14: 831-839.

as introduced by Borel-Moore will be denoted by $H_*^{BM}(V,G)$. For a pair of a finite complex V and a subcomplex V' we have then $H_*(V,V';G)=H_*^{BM}(V-V',G)$, as shown by these authors.

§1 Intersections in a bounded manifold

Let M be a compact bounded manifold of dimension n with boundary $\dot{M}$. Let M' be a second copy of M with boundary $\dot{M}'$ and h be the homeomorphism of M onto M'. Identifying points x and $h(x)$ for any point $x \in \dot{M}$, we get then a closed manifold $\tilde{M}=M\cup M'$ which is called the double of M. The triad $(\tilde{M};M,M')$ is then a proper one in the sense of Eilenberg-Steenrod([5]I § 14). For, since $\tilde{M}$, etc.are all manifolds, the singular homologies of $\tilde{M}$, $(\tilde{M},M)$, etc. coincide with the corresponding Cêch homologies so that the inclusion maps

$$k:\quad (M,\dot{M})\subset(\tilde{M},M')$$

and

$$k':\quad (M',\dot{M}')\subset(\tilde{M},M)$$

will induce isomorphisms of homology groups in all dimensions by the excision property for Cêch groups(Eilenberg-Steenrod[5] X §5). Consequently we have a Mayer-Vietoris exact sequence on coefficients Z_2:

$$\begin{aligned}\cdots\to H_r(M,Z_2)+H_r(M',Z_2)&\xrightarrow{\beta} H_r(\tilde{M},Z_2)\xrightarrow{\tilde{\partial}} H_{r-1}(\dot{M},Z_2)\\ \xrightarrow{\alpha} H_{r-1}(M,Z_2)&+H_{r-1}(M',Z_2)\to\cdots\end{aligned}\tag{1}$$

The homomorphisms in(1)are defined respectively as

$$\alpha=i_*+i'_*,\quad \beta=j_*+j'_*,$$
$$\tilde{\partial}=\partial k_*^{-1}l_*=\partial' k_*'^{-1}l'_*,$$

in which

$$\begin{aligned}&i:\dot{M}\subset M, && i':\dot{M}'\subset M',\\ &j:M\subset\tilde{M}, && j':M'\subset\tilde{M}',\\ &l:\tilde{M}\subset(\tilde{M},M'), && l':\tilde{M}\subset(\tilde{M},M),\\ &k:(M,\dot{M})\subset(\tilde{M},M'), && k':(M',\dot{M}')\subset(M',M)\end{aligned}$$

are all inclusions, while

$$\partial : H_r(M, \dot{M}; Z_2) \to H_{r-1}(\dot{M}, Z_2)$$

and

$$\partial' : H_r(M', \dot{M}'; Z_2) \to H_{r-1}(\dot{M}', Z_2)$$

are boundary homomorphisms.Let $r : \tilde{M} \to M$ be the map defined by $r(x) = x$, $rh(x) = x$ for any $x \in M$, then $rj =$ ident. It follows that $j_* : H_r(M, Z_2) \to H_r(\tilde{M}, Z_2)$ is a monomorphism.Similarly for $j'_* : H_r(M', Z_2) \to H_r(\tilde{M}, Z_2)$.

Now for any $\boldsymbol{X} \in H_r(M, \dot{M}; Z_2)$ let us take a singular relative-cycle $x \in \boldsymbol{X}$ and set $\tilde{x} = x + h(x)$. Then $\tilde{x}$ is a singular cycle mod 2 in $\tilde{M}$ and determines a class $\tilde{\boldsymbol{X}} \in H_r(\tilde{M}, Z_2)$. If we take a second relative-cycle x' in $\boldsymbol{X}$, then $x' = x + \partial a + b$ for some singular chains a, b with supports $|a| \subset M$ and $|b| \subset \dot{M}$. It follows that $\tilde{x}' = \tilde{x} + \partial \tilde{a}$ where $\tilde{a} = a + h(a)$. The class $\tilde{\boldsymbol{X}}$ is thus independent of the relative-cycle x chosen in $\boldsymbol{X}$. The correspondence $\boldsymbol{X} \to \tilde{\boldsymbol{X}}$ induces then a homomorphism

$$e_* : H_r(M, \dot{M}; Z_2) \to H_r(\tilde{M}, Z_2).$$

It is clear from the definition that $k_*^{-1} l_* e_*$ is the identity homomorphism of $H_r(M, \dot{M}; Z_2)$ onto itself. In particular we see that

$$\tilde{\partial} e_* = \partial : H_r(M, \dot{M}; Z_2) \to H_{r-1}(\dot{M}, Z_2)$$

and that e_* is a monomorphism.

Lemma 1 *Let M be a compact bounded manifold with boundary $\dot{M}$ and $\tilde{M} = M \cup M'$ be the double of M. Then we have the decomposition(not direct sum):*

$$H_r(\tilde{M}, Z_2) = j_* H_r(M, Z_2) + j'_* H_r(M', Z_2) + e_* H_r(M, \dot{M}; Z_2).$$

Proof. Let $\tilde{\boldsymbol{X}} \in H_r(\tilde{M}, Z_2)$. Set $k_*^{-1} l_* \boldsymbol{X} = \boldsymbol{X} \in H_r(M, \dot{M}; Z_2)$ in which l_*, k_* are defined as before. Then $\tilde{\partial}(\tilde{\boldsymbol{X}} + e_* \boldsymbol{X}) = \tilde{\partial} \tilde{\boldsymbol{X}} + \partial k_*^{-1} l_* \tilde{\boldsymbol{X}} = 0$. By the Mayer-Vietoris exact sequence(1)we have therefore $\tilde{\boldsymbol{X}} + e_* \boldsymbol{X} = j_* \boldsymbol{Y} + j'_* \boldsymbol{Y}'$ for some $\boldsymbol{Y} \in H_r(M, Z_2)$ and $\boldsymbol{Y}' \in H_r(M', Z_2)$, q.e.d.

With M, M', $\tilde{M}$ etc. as before, it is clear that for any classes $\boldsymbol{X} \in H_r(M, \dot{M}; Z_2)$ and $\boldsymbol{Y} \in H_s(M, Z_2)$ the intersection $e_* \boldsymbol{X} \cdot j_* \boldsymbol{Y} \in H_{r+s-n}(\tilde{M}, Z_2)$ in $\tilde{M}$ contains cycles with supports lying in M so that $e_* \boldsymbol{X} \cdot j_* \boldsymbol{Y}$ is in the image of $j_* : H_{r+s-n}(M, Z_2) \to H_{r+s-n}(\tilde{M}, Z_2)$. As j_* are monomorphisms, $j_*^{-1}(e_* \boldsymbol{X} \cdot j_* \boldsymbol{Y})$ is a well defined class in $H_{r+s-n}(M, Z_2)$. This justifies the following.

Definition. $j_*^{-1}(e_*\boldsymbol{X}\cdot j_*\boldsymbol{Y})$ will be called the intersection in the bounded manifold M of $\boldsymbol{X}\in H_r(M,\dot{M};Z_2)$ and $\boldsymbol{Y}\in H_s(M,Z_2)$. It will be denoted by $(\boldsymbol{X}\cdot\boldsymbol{Y})_M$ or simply $\boldsymbol{X}\cdot\boldsymbol{Y}$ if no confusion can arise.

Lemma 2 *Let M be a compact bounded combinatorial (or differentiable) manifold of dimension n with boundary $\dot{M}$. Then for any*

$$\boldsymbol{X}\in H_r(M,\dot{M};Z_2),\quad \boldsymbol{Y}\in H_{n-r}(M,\dot{M};Z_2)$$

we have

$$(e_*\boldsymbol{X}\cdot e_*\boldsymbol{Y})_{\tilde{M}}=0.$$

Proof. By means of simplicial subdivisions of the combinatorial manifold $\tilde{M}$ we can take. in $e_*\boldsymbol{X}$ and $e_*\boldsymbol{Y}$ simplicial cycles of the form $\tilde{x}=x+h(x)$ and $\tilde{y}=y+h(y)$ such that x, y are simplicial chains in M intersecting in finite number of points in $M-\dot{M}$ and ∂x, ∂y have their supports in $\dot{M}$ disjoint from each other. As $h(x)$ and $h(y)$ will then be simplicial chains in M' intersecting in points which are images under h of those of x and y in M and the indices of intersection are respectively equal mod 2, we have clearly $\tilde{x}\cdot\tilde{y}=0$ so that $e_*\boldsymbol{X}\cdot e_*\boldsymbol{Y}=0$ in $\tilde{M}$. The differentiable case follows now from the combinatorial case by the theorem of Cairns.

Proposition 1 *Let M be a compact bounded combinatorial (or differentiable) manifold of dimension n with boundary $\dot{M}$ and $\boldsymbol{X}\in H_r(M,\dot{M};Z_2)$. Then $\boldsymbol{X}=0$ if and only if $(\boldsymbol{X}\cdot\boldsymbol{Y})_M=0$ for any $\boldsymbol{Y}\in H_{n-r}(M,Z_2)$.*

Proof. By Lemma 1, $H_{n-r}(\tilde{M},Z_2)$ is generated by classes in $j_*H_{n-r}(M,Z_2)$, $j'_*H_{n-r}(M',Z_2)$, and $e_*H_{n-r}(M,\ \dot{M};\ Z_2)$. Set $e_*\boldsymbol{X}=\tilde{\boldsymbol{X}}\in H_r(\tilde{M},Z_2)$.Then $\tilde{\boldsymbol{X}}\cdot\tilde{\boldsymbol{Z}}=0$ for any class $\tilde{\boldsymbol{Z}}\in e_*H_{n-r}(M,\dot{M};Z_2)$ by Lemma 2 and $\tilde{\boldsymbol{X}}\cdot j_*\boldsymbol{Y}=\tilde{\boldsymbol{X}}\cdot j'_*h_*\boldsymbol{Y}$ for any class $\boldsymbol{Y}\in H_{n-r}(M,Z_2)$. Hence $\tilde{\boldsymbol{X}}=0$ if and only if $\tilde{\boldsymbol{X}}\cdot j_*\boldsymbol{Y}=0$ for any $\boldsymbol{Y}\in H_{n-r}(M,Z)$, i.e., if and only if $(\boldsymbol{X}\cdot\boldsymbol{Y})_M=0$ for any $\boldsymbol{Y}\in H_{n-r}(M,Z_2)$. As e_* is a monomorphism, the proposition is proved.

As the intersection is of a local character, we have readily the following

Proposition 2 *Let $\hat{M}$ be a closed combinatorial (or differentiable) manifold of dimension n and M a bounded combinatorial (or differentiable) submanifold of $\hat{M}$ of the same dimension n with boundary $\dot{M}$. Consider the homomorphisms*

$$H_r(\hat{M},Z_2)\xrightarrow{\hat{j}_*}H_r(\hat{M},\overline{\hat{M}-M};Z_2)\xrightarrow{\hat{k}_*^{-1}}H_r(M,\dot{M};Z_2)$$

and

$$\hat{i}_*:\quad H_s(M,Z_2)\to H_s(\hat{M},Z_2),$$

in which $\hat{i}_*$, $\hat{j}_*$, $\hat{k}_*$ *are injections with* $\hat{k}_*^{-1}$ *as the excision isomorphism. Then for any* $\hat{\boldsymbol{X}} \in H_r(\hat{M}, Z_2)$ *and* $\boldsymbol{Y} \in H_s(M, Z_2)$, *we have*

$$\hat{i}_*(\hat{k}_*^{-1}\hat{j}_*\hat{\boldsymbol{X}} \cdot \boldsymbol{Y})_M = (\hat{\boldsymbol{X}} \cdot \hat{i}_*\boldsymbol{Y})_{\tilde{M}}.$$

§2 Real submanifolds of a complex analytic manifold

Let M be a real differential manifold of dimension $2n$. We shall denote by $T_x(M) = T_x$ the tangent space of M at $x \in M$, by $\tau(M) = \tau$ the tangent bundle of M, and by $T(M) = T$ the bundle space of τ. An almost complex structure J in M is then a differential field of linear transformations J_x in each tangent space T_x such that $J_x^2 =$—ident. A differential submanifold M^0 of M is then said to be *real* with respect to J if for each $x \in M^0$, the tangent space to M^0 at x, $T_x(M^0)$, has only 0-vector in common with its image under J_x, $J_xT_x(M^0)$. If M admits a complex analytic structure, then M has a natural almost complex structure J defined in the following manner. Take any coordinate neighbourhood about $x \in M$ with complex coordinates $z_1, \cdots, z_n$. Then the linear space $T_x(M)$ has a complex structure spanned by vectors $\left.\dfrac{\partial}{\partial z_k}\right|_x = \dfrac{\partial}{\partial z_k}$. Set $J_x\left(\dfrac{\partial}{\partial z_k}\right) = i\dfrac{\partial}{\partial z_k}$, then J_x induces a linear transformation in T_x independent of the local coordinate system $z_1, \cdots, z_n$ in U and defines the structure J in question.A submanifold M^0 which is real with respect to J thus induced will then be called simply a *real* submanifold of the complex analytic manifold M. Denote the tangent bundle of M^0 by τ^0 and the image bundle of τ^0 under J by υ^0, which is isomorphic to τ^0 by J. Then, in case the dimension of the *real* submanifold M^0 is n, the restriction of the tangent bundle τ of M to M^0 is clearly isomorphic to the Whitney sum of τ^0 and υ^0. The converse is also true and had been stated long ago by Ehresmann[4]. As a detailed proof of this fact is lacking in the literature and it is important for our purposes, we shall give such a proof below.

Proposition 3 *Let* M *be a real-analytic manifold of dimension* $2n$ *and* M^0 *an analytic submanifold of* M *of dimension* n. *Let* $\{N_x^0/x \in M^0\}$ *be a differential family of* n*-dimensional tangent planes of* M *over* M^0 *such that the tangent space* $T_x(M) = T_x$ *of* M *at* $x \in M^0$ *is spanned by the tangent space* $T_x(M^0) = T_x^0$ *of* M^0 *at* x *and the space* N_x^0.*Suppose that the bundle* υ^0 *over* M^0 *with* $N^0 = \cup N_x^0$ *as bundle space is isomorphic to the tangent bundle* τ^0 *of* M^0 *with* $T^0 = \cup T_x^0$ *as bundle space,* *and the isomorphism is realized by the bundle map $J^0 : T^0 \to N^0$. Then there exists a neighbourhood U of M^0 in M and a complex analytic structure in U for which the induced almost complex structure J coincides with J^0 on M^0.*

Proof. It is known that there exists a complex analytic manifold $\tilde{M}^0$ of complex dimension n unique up to a complex analytic homeomorphism which has M^0 as a *real* submanifold (Schutrick[6], Whitney-Bruhat[7]). Denote the almost complex structure thus induced by $\tilde{J}$. Let us take an arbitrary (differential) Riemannian metric $d\,s_0^2$ in M^0 and denote the corresponding scalar product in T_x^0 for any $x \in M^0$ by $\langle\ \ \rangle$. We can then extend this scalar product to $T_x(\tilde{M}^0) = \tilde{T}_x^0$ by setting $\langle v_x, w_x\rangle = \left\langle \tilde{J}_x^{-1} v_x, \tilde{J}_x^{-1} w_x \right\rangle$ for $v_x,\ w_x \in \tilde{J}_x T_x^0$ and $\langle v_x, w_x\rangle = \langle w_x, v_x\rangle = 0$ for $v_x \in T_x^0$, $w_x \in \tilde{j}_x T_x^0$. Now M^0 has neighbourhoods W in $\tilde{M}^0$ which has M^0 as a deformation-retract. Hence the above field of scalar products over M^0 can be extended throughout W and gives thus a Riemannian metric $d\tilde{s}_0^2$ in W. Let an exponential map be defined with respect to this metric. For any $x \in M^0$ and $\varepsilon > 0$ let $W_{x,\varepsilon}$ be the set of points $\exp\, v_x$ with $v_x \in \tilde{J}_x T_x^0$ and $||v_x|| < \varepsilon$. Then for $\varepsilon > 0$ sufficiently small the sets $W_{x.\varepsilon}$ will be disjoint from each other for different x of M^0 and will fulfil a tubular neighbourhood of M^0 in $\tilde{M}$ which will be denoted by W_ε. In just the same manner the Riemannian metric ds_0^2 in M^0 can be extended to one ds^2 in a neighbourhood U of M^0 in M for which the scalar products induced in T_x for $x \in M^0$ are given by $\langle v_x, \omega_x\rangle = \left\langle (J_x^0)^{-1} v_x, (J_x^0)^{-1} \omega_x \right\rangle$ for $v_x,\ \omega_x \in J_x^0 T_x^0$ and $\langle v_x, \omega_x\rangle = \langle \omega_x, v_x\rangle = 0$ for $v_x \in T_x^0$, $\omega_x \in J_x^0 T_x^0$. For $\varepsilon > 0$ sufficiently small let U_ε be the tubular neighbourhood of M^0 in M formed by all $U_{x,\varepsilon} = \{\exp v_x / v_x \in J_x^0 T_x^0,\ ||v_x|| < \varepsilon\}$ where $x \in M^0$, and the exponential map exp be defined with respect to the metric ds^2. Define now a map h of W_ε into U_ε such that $h \exp\, v_x = \exp(J_x^0 \tilde{J}_x^{-1} v_x)$, for $x \in M^0$, $v_x \in \tilde{J}_x T_x^0$, and $||v_x|| < \varepsilon$. Then h is a diffeomorphism of W_ε onto U_ε which reduces to identity on M^0. The complex analytic structure on W_ε is thus transported by h to one on U_ε, which clearly meets the requirements of the proposition.

§3 Diagonal in the square of a manifold

Let M be a real-analytic closed manifold of dimension n with a Riemannian metric ds^2. The square of M, i.e., the product manifold $M \times M$ will then also be a realanalytic manifold of dimension $2n$ possessing a Riemannian metric of the form $d\tilde{s}^2 = ds'^2 + ds''^2$, where ds'^2, ds''^2 are prototypes of ds^2 in the two factors of $M \times M$. The diagonal in $M \times M$ will be denoted by Δ and the diagonal map

$M \to M \times M$ by d with $d(x) = (x, x) = \tilde{x}$ for any $x \in M$. For each $x \in M$ the normal space at $\tilde{x}$ with respect to $d\tilde{s}^2$ in $M \times M$ to the diagonal Δ will be denoted by N_x, and the set of points $\exp v$, with $v \in N_x$, $||v|| < \varepsilon$ (resp.$||v|| \leqslant \varepsilon$) will be denoted by $W_{\varepsilon,x}$(resp. $\overline{W}_{\varepsilon,x}$). Then for $\varepsilon > 0$ sufficiently small the sets $W_{\varepsilon.x}$(resp. $\overline{W}_{\varepsilon,x}$) for $x \in M$ are mutually disjoint and their union forms an open (resp.closed) neighbourhood W_ε(resp.its closure $\overline{W}_\varepsilon$) of Δ in $M \times M$. The union of all normal spaces N_x, $x \in M$, is the bundle space N of the normal bundle v of the diagonal Δ in $M \times M$. It is well known that v is isomorphic to the tangent bundle τ of M and an explicit isomorphism may be established in the following manner. Denote the tangent space of M at x by T_x and let U be a coordinate neighbourhood containing x in M with coordinates $x_1, \cdots, x_n$. Consider the neighbourhood $U \times U$ of $\tilde{x} = (x, x)$ in $M \times M$ with coordinates $(x'_1, \cdots, x'_n, x''_1, \cdots, x''_n)$ such that $(x'_1, \cdots, x'_n, x''_1, \cdots, x''_n) \to (x'_1, \cdots, x'_n)$ and $\to (x''_1, \cdots, x''_n)$ represent the projections of $U \times U$ on the two factors. Let the union of all T_x with $x \in M$ be denoted by T, then $\tilde{d}_*\left(\dfrac{\partial}{\partial x_i}\right) = \dfrac{\partial}{\partial x''_i} - \dfrac{\partial}{\partial x'_i}$, $i = 1, \cdots, n$, defines an isomorphism of T_x onto N_x. It is easily verified that $\tilde{d}_*$ is independent of the coordinate neighbourhood chosen about x and induces thus an isomorphism $\tilde{d}_*$ of the tangent bundle τ to the normal bundle v.

Lemma 3 *For any real-analytic submanifold A of M and any point $x \in A$ the tangent space to $A \times A$ at $\tilde{x} = d(x)$ is the direct sum $d_* T_x(A) \oplus d_* T_x(A)$.*

Proof. Take coordinate neighbourhood U about x in M with coordinates $x_1, \cdots, x_n$ such that $U \cap A$ is given by $x_1 = \cdots = x_k = 0$, k being the codimension of A in M. Then in the neighbourhood $U \times U$ about $\tilde{x}$ in $M \times M$ with coordinates $x'_1, \cdots, x'_n$, $x''_1, \cdots, x''_n$ as above the set $(U \times U) \cap (A \times A)$ is given by $x'_1 = \cdots = x'_k = 0$, $x''_1 = \cdots = x''_k = 0$. Hence the tangent space to $A \times A$ at $\tilde{x}$ is generated by the vectors $\dfrac{\partial}{\partial x'_i}, \dfrac{\partial}{\partial x''_i}$, $k+1 \leqslant i \leqslant n$, or $\dfrac{\partial}{\partial x''_i} - \dfrac{\partial}{\partial x'_i}$ and $\dfrac{\partial}{\partial x'_i} + \dfrac{\partial}{\partial x''_i}$, $k+1 \leqslant i \leqslant n$. The former set of vectors spans the space $d_* T_x(A)$ and the latter set the space $d_* T_x(A)$. Hence the assertion.

By §2, Prop.3, so far $\varepsilon > 0$ is sufficiently small, $W_{2\varepsilon}$ possesses a complex analytic structure for which Δ is a real submanifold, and the associated J-homomorphism at any point $\tilde{x} = d(x)$, $x \in M$, is such that $J_{\tilde{x}}(v) = \tilde{d}_x d_*^{-1}(v)$ for any $v \in T_{\tilde{x}}^{\Delta}$, the tangent space to Δ at $\tilde{x}$. Let X be any irreducible complex analytic set in $W_{2\varepsilon}$ of complex dimension $d(X) = r$ such that $X^0 = X \cap \Delta$ is of real dimension $\dim X^0 = r$. By Borel-Haefliger[1], X possesses a fundamental class $[X] \in H_{2r}^{BM}(W_{2\varepsilon})$ and the real analytic set X^0 in Δ possesses also a fundamental class $[X^0] \in H_r(\Delta, Z_2)$ or $[d^{-1}X^0] \in H_r(M, Z_2)$. Denote now the composite homomorphisms

$$H_{2r}^{BM}(W_{2^\varepsilon}) = H_{2r}(\overline{W}_{2^\varepsilon}, \overline{W}_{2^\varepsilon} - \overline{W}_{2^\varepsilon}) \xrightarrow{j_*} H_{2r}(\overline{W}_{2^\varepsilon}, \overline{W}_{2^\varepsilon} - W_\varepsilon) \xrightarrow{k_*^{-1}} H_{2r}(\overline{W}_\varepsilon, \overline{W}_\varepsilon - W_\varepsilon)$$

and

$$H_{2r}(M \times M, Z_2) \xrightarrow{j_*} H_{2r}(M \times M, M \times M - W_\varepsilon; Z_2) \xrightarrow{k_*^{-1}} H_{2r}(\overline{W}_\varepsilon \overline{W}_\varepsilon - W_\varepsilon; Z_2)$$

by λ_* and μ_* respectively, in which j_* and k_*^{-1} are injection homomorphisms and excision isomorphisms respectively. Then we have

Proposition 4 *For an irreducible nonsingular complex analytic set X in W_{2^ε} with $d(X) = \dim X^0 = r$ where $X^0 = X \cap \Delta$ we have*

$$\rho_2 \lambda_*[X] = \mu_*([d^{-1}X^0] \otimes [d^{-1}X^0]),$$

in which

$$\rho_2 : H_{2r}(\overline{W}_\varepsilon, \overline{W}_\varepsilon - W_\varepsilon) \to H_{2r}(\overline{W}_\varepsilon, \overline{W}_\varepsilon - W_\varepsilon; Z_2)$$

is the reduction mod 2.

Proof. By Prop.1 of §1, it is enough to prove that for any

$$Y \in H_{2n-2r}(\overline{W}_\varepsilon, Z_2) \approx H_{2n-2r}(\Delta, \cdot Z_2)$$

we have

$$(\rho_2 \lambda_*[X] \cdot Y)_{\overline{W}_\varepsilon} = (\mu_*([d^{-1}X^0] \otimes [d^{-1}X^0) \cdot Y)_{\overline{W}_\varepsilon}. \tag{2}$$

Nov by Lemma 3 and § 2, X and $d^{-1}X^0 \times d^{-1}X^0$ are tangent to each other along points of X^0, with common tangent space to a point $\tilde{x} = (x, x) \in \Delta$ as $d_*T_x \oplus J_{\tilde{x}} d_*\, T_x$, in which T_x is the tangent space to X^0 at x, and $J_{\tilde{x}}$ is the J-homomorphism at $\tilde{x}$ associated to the complex-analytic structure in W_{2^ε}. The metric $d\tilde{s}^2$ in $M \times M$ will induce a metric ds'^2 in X and also one ds''^2 in $d^{-1}X^0 \times d^{-1}X^0$. Let the corresponding exponential maps in X and $d^{-1}X^0 \times d^{-1}X^0$ be denoted by exp′ and exp″ respectively. Take $\eta > 0$ sufficiently small. Then X^0 has neighbourhoods N' and N'' in X and $d^{-1}X^0 \times d^{-1}X^0$consisting of points $\exp' J_{\tilde{x}} v$ and $\exp'' J_{\tilde{x}} v$ respectively, with $v \in d_*T_x$, $x \in d^{-1}X^0$, and $||v|| < \eta$. Let α be a C^∞-function on $[0, \eta]$ with $\alpha = 0$ on $\left[0, \frac{\eta}{3}\right]$, $= 1$ on $\left[\frac{2\eta}{3}, \eta\right]$, and $\geqslant 0$, $\leqslant 1$ on $\left[\frac{\eta}{3}, \frac{2\eta}{3}\right]$. For any $x \in d^{-1}X^0$, $v \in d_*T_x$, and $||v|| < \eta$, let $g'(x,v)$ (resp. $g''(x,v)$) be the unique geodesic with respect to $d\tilde{s}^2$ in $W_{2\varepsilon}$ joining $\exp J_{\tilde{x}}v$ to $\exp' J_{\tilde{x}}v$ (resp. $\exp'' J_{\tilde{x}}v$). Denote by $\psi'(x, v)$ (resp.ψ'', (x,v)) the point on $g'(x,v)$ (resp. on $g''(x,v)$) dividing it in the ratio $\alpha(||v||) : 1-\alpha(||v||)$. Define a map ψ' (resp. ψ'') of X (resp. $d^{-1}X^0 \times d^{-1}X^0$) into W_{2^ε} (resp. $M \times M$) by $\psi'(\exp' J_{\tilde{x}}v) = \psi'(x,v)$ (resp. $\psi''(\exp'' J_{\tilde{x}}v) = \psi''(x,v)$) for $\exp' J_{\tilde{x}}v \in N'$(resp. for $\exp'' J_{\tilde{x}}v \in N''$), while $\psi' \equiv$ ident. (resp. $\psi'' \equiv$ ident.) for points in $X - N'$(resp. in $d^{-1}X^0 \times d^{-1}X^0 - N''$). Then $\psi'(X)$ and $\psi''(d^{-1}X^0 \times d^{-1}X^0)$ are respectively isotopic to X and $d^{-1}X^0 \times d^{-1}X^0$ and coincide with each other in a certain neighbourhood U of Δ in $M \times M$. Now with respect to a sufficiently fine subdivision of $M \times M$, the given class Y will contain simplicial cycle Y with support arbitrarily near to Δ so that it will meet $\psi'(X)$ and $\psi''(d^{-1}X^0 \times d^{-1}X^0)$ only in points in U which may be taken to be isolated ones by further deformations of Y if necessary. These points will contribute same values of indices of intersection for the intersection products, occuring in both sides of (2). It follows that (2) is true and hence the proposition.

§4 The formula of Chow

We are now in a position to prove the complex form of Chow's formula as stated in the introduction. The preceding sections have reduced the situation to the case of a complex analytic neighbourhood of the diagonal in the square $M \times M$ of the manifold M. We begin therefore by the study of the homologies in $M \times M$:

Proposition 5 *Let M be a closed manifold of dimension n and X, $Y \in H_*(M, Z_2)$. Then in $\tilde{M} = M \times M$ we have*

$$(X \otimes Y) \cdot d_*[M] = d_*(X \cdot Y), \tag{3}$$

in which $[M]$ is the fundamental class mod 2 of M and $d : M \to \tilde{M}$ is the diagonal map.

Proof. Let D and $\tilde{D}$ be the duality isomorphisms in M and $\tilde{M}$ respectively, and $G^* = \tilde{D}d_*D : H^i(M,\ Z_2) \to H^{i+n}(\tilde{M}, Z_2)$ the Gysin homomorphisms. Denote the unit class mod 2 of M by $1 = D[M] \in H^0(M, Z_2)$, then for any class $Z \in H^*(\tilde{M}, Z_2)$ we have

$$Z \cup G^*1 = G^*d^*Z.$$

Hence

$$\tilde{D}((X \otimes Y) \cdot d_*[M]) = \tilde{D}(X \otimes Y) \cup \tilde{D}d_*[M]$$

$$
\begin{aligned}
&= (DX \otimes DY) \cup \tilde{D}d_* [M] \\
&= (DX \otimes DY) \cup G^* 1 \\
&= G^* d^* (DX \otimes DY) \\
&= G^* (DX \otimes DY) \\
&= G^* D(X \cdot Y) \\
&= \tilde{D}d_* (X \cdot Y).
\end{aligned}
$$

As $\tilde{D}$ is an isomorphism, we get therefore (3).

Now let V be a complex analytic manifold of complex dimension n and V^0 a real submanifold of real dimension n of V. Let $\mathscr{Z}_r(V)$ be the group of complex analytic cycles of complex dimension r in V and $\tilde{\mathscr{Z}}_r(V)$ the subgroup generated by complex analytic irreducible nonsingular subvarieties X for which $X \cap V^0 = X^0$ has real dimension r. Let

$$\varphi : \mathscr{Z}_r(V) \to H_{2r}(V, Z_2)$$

be the homomorphism such that for any complex analytic irreducible subvariety X of complex dimension r in V, $\varphi(X)$ is the fundamental class reduced mod 2 of X. Let

$$\rho : \mathscr{Z}_r(V) \to H_{2r}(V^0, Z_2)$$

be the homomorphism introduced by Borel-Haefliger in [1], such that for any complex analytic irreducible subvariety X of complex dimension r of V, $\rho(X)$ is the fundamental class mod 2 of $X^0 = X \cap V^0$ if dim $X^0 = r$, and is 0 in the contrary case. Then the complex form of Chow's theorem in question may be stated as follows.

Theorem *Let V be a closed complex analytic manifold of complex dimension n and V^0 a real submanifold of V, closed and of real dimension n. Then for any complex analytic cycle $X \in \hat{\mathscr{Z}}_r(V)$ we have*

$$\varphi(X) \cdot i_*^0[V^0] = i_*^0(\rho(X) \cdot \rho(X)), \tag{4}$$

in which $i^0 : V^0 \subset V$ is the injection and $[V^0]$ is the fundamental class of $H_r(V^0, Z_2)$ represented by V^0 as singular cycle.

Remark. It seems that the theorem remains true for any $X \in \mathscr{Z}_r(V)$, but we have not been able to prove it.

Proof. It is sufficient to prove the formula(4)in case that X is a complex analytic irreducible nonsingular subvariety of complex dimension r in V for which f $X^0 = X \cap V^0$ is of real dimension r. Now by Prop.3 there exist neighbourhoods W, W_ε of

V^0 in V with $\overline{W}_\varepsilon \subset W$, neighbourhoods U, U_ε of the diagonal $d(V^0)$ in $V^0 \times V^0$ with $\bar{U}_\varepsilon \subset U$, a complex analytic structure of U, and a complex analytic homeomorphism h of W onto U with $h(\overline{W}_\varepsilon) = \bar{U}_\varepsilon$ such that the restriction of h on V^0 reduces to the diagonal map d of V^0 in $V^0 \times V^0$. Moreover, by §3 this complex analytic structure of U and the complex analytic homeomorphism h can be chosen such that $X^0 \times X^0$ will be tangent to $h(X)$ at each point of $d(X^0)$. Consider now the homeomorphisms

$$H_{2r}(V, Z_2) \xrightarrow{j_*^0} H_{2r}(V, V - W_\varepsilon; Z_2) \xrightarrow{k_*^{0-1}} H_{2r}(\overline{W}_\varepsilon, \overline{W}_\varepsilon - W_\varepsilon; Z_2)$$

and

$$H_{2r}(V^0 \times V^0, Z_2) \xrightarrow{j_*} H_{2r}(V^0 \times V^0, V^0 \times V^0 - U_\varepsilon; Z_2) \xrightarrow{k_*^{-1}} H_{2r}(\bar{U}_\varepsilon, U_\varepsilon; Z_2),$$

in which j_*, j_*^0 are injections while k_*^{-1} and k_*^{0-1} are excision isomorphisms. By Props.3 and 4 we have then

$$h_* k_*^{0-1} j_*^0 \varphi(X) = k_*^{-1} j_*(\rho(X) \otimes \rho(X)) \in H_{2r}(\bar{U}_\varepsilon, U_\varepsilon; Z_2).$$

Hence for intersections in the bounded manifold $\overline{W}_\varepsilon$, we have

$$k_*^{0-1} j_*^0 \varphi(X) \cdot i_{\varepsilon *}^0[V^0] = h_{\varepsilon *}^{-1}[k_*^{-1} j_*(\rho(X) \otimes \rho(X)) \cdot d_{\varepsilon *}[V^0]], \tag{5}$$

in which $i_\varepsilon^0 : V^0 \subset \overline{W}_\varepsilon$ is the injection, $d_\varepsilon : V^0 \to \bar{U}_\varepsilon$ is the diagonal map, and $h_\varepsilon : \overline{W}_\varepsilon \to \bar{U}_\varepsilon$ is the restriction of h. Let $\bar{i}_* : H_*(\bar{U}_\varepsilon, Z_2) \to H_*(V^0 \times V^0, Z_2)$ be the injection, then by Props.2 and 5 we have

$$\begin{aligned}\bar{i}_*(k_*^{-1} j_*(\rho(X) \otimes \rho(X)) \cdot d_{\varepsilon *}[V^0]) &= (\rho(X) \otimes \rho(X)) \cdot d_{\varepsilon *}[V^0] \\ &= d_*(\rho(X) \cdot \rho(X)) \\ &= \bar{i}_{*d_{\varepsilon *}}(\rho(X) \cdot \rho(x)).\end{aligned}$$

As $\bar{i}_*$ is well known to be a homomorphism, we get

$$k_*^{-1} j_*(\rho(X) \otimes \rho(X)) \cdot d_{\varepsilon *}[V^0] = d_{\varepsilon *}(\rho(X) \cdot \rho(X)). \tag{6}$$

Consider now the injection homomorphism

$$\bar{i}_*^0 : H_*(\overline{W}_\varepsilon, Z_2) \to H_*(V, Z_2)$$

such that

$$\bar{i}_*^0 i_{\varepsilon *}^0 = i_*^0.$$

By Props.2 and 5 again we get

$$\bar{i}_*^0[k_*^{0-1}j_*^0\varphi(X)\cdot i_{\varepsilon*}^0[V^0]] = \varphi(X)\cdot \bar{i}_*^0 i_{\varepsilon*}^0[V^0] = \varphi(X)\cdot i_*^0[V^0].$$

From (5) and (6) we get therefore

$$\begin{aligned}\varphi(X)\ \cdot\ i_*^0\left[V^0\right] &= \bar{i}_*^0 h_{\varepsilon*}^{-1}(k_*^{-1}j_*(\rho(X)\otimes\rho(X))\cdot d_{\varepsilon*}\left[V^0\right])\\ &= \bar{i}_*^0 h_{\varepsilon*}^{-1} d_{\varepsilon*}(\rho(X)\cdot\rho(X))\\ &= \bar{i}_*^0 i_{\varepsilon*}^0(\rho(X)\cdot\rho(X))\\ &= i_*^0(\rho(X)\cdot\rho(X)).\end{aligned}$$

This proves the theorem.

References

[1] Borel A et Haefliger A. La classe d'homologie fondamentale d'un espace analytique. *Bull.Soc.Math.France*, 1961, 89: 461-513.

[2] Cartan H. Variétès analytiques-réelles et variétès analytiques complexes. *Bull.Soc.Math. France*, 1957, 85: 77-100.

[3] Chow W L. On the real traces of analytic varieties. *Amer.J.Math.*, 1963, 85: 723-733.

[4] Ehresmann C. Sur les variétès presque complexes. *Proc.Intern.Cong.Math.*, Cambridge, 1950: 412-419.

[5] Eilenberg S et Steenrod N E. *Foundations of algebraic topology.* Princeton, 1952.

[6] Schutrick H B. Complex extension. *Quart.J.*, 1958, 9: 189-201.

[7] Whitney H et Bruhat F. Quelques propriétès fondamentales des ensembles analytiques réels. *Comm.Math.Helv.*, 1959, 33: 132-160.

44. On Critical Sections of Convex Bodies*

Abstract The study of critical chords of a manifold imbedded in a Euclidean space constitutes one of the earliest applications of the celebrated critical point theory of Morse (cf.[3], p.183). As an extension we shall study in the present paper the critical sections of any dimension of a convex body imbedded in a Euclidean space. For this purpose we state first in §I the Morse theory for bounded manifolds which is essentially contained in [4]. The particular case of an ellipsoid is then studied in §III, which permits to determine the Betti numbers of complex grassmanian manifolds via this theory of Morse.

§1 The Morse inequalities under boundary conditions

Let M be a compact bounded C^∞-manifold with boundary $\dot{M}$. Suppose that f is a C^∞-function on M verifying the following boundary conditions:

B1. f is constant on $\dot{M}$.

B2. The value of f at any point interior to M is greater than the constant value of f on $\dot{M}$.

B3. There is a neighbourhood of $\dot{M}$ in M, which contains no critical points of f.

The usual Morse inequalities can then be generalized to take the following form:

Theorem 1 *Let f be a C^∞-function on M verifying conditions* B1—3 *and having only nondegenerate critical points. Let C_λ be the number of critical points of f of index λ and let $P_\lambda =$ rank over F of $H_\lambda(M, \dot{M}; F)$, in which F is any field. Then for any λ we have*

$$C_\lambda - C_{\lambda-1} + \cdots \pm C_0 \geqslant P_\lambda - P_{\lambda-1} + \cdots \pm P_0, \tag{1}$$

and for $\lambda = n =$ dim M,

$$C_n - C_{n-1} + \cdots \pm C_o = P_n - P_{n-1} + \cdots \pm P_0. \tag{2}$$

* *Sci. Sinica*, 1965, 14: 1721-1728.

In particular, we have

$$C_\lambda \geqslant p_\lambda. \tag{3}$$

We have also

$$C_{2\lambda} = p_{2\lambda} \tag{4}$$

if n is even and all $C_\lambda = 0$ for λ odd.

Proof. The usual proof of the Morse inequalities in the case of closed manifolds requires only slight modification to bring out the above theorem. For example, in the proof given in §5 of [2], it is only necessary to replace the sequence

$$\phi \subset M^{a_1} \subset \cdots \subset M^{a_k}$$

given on p.30 of that book by the sequence

$$\dot{M} \subset M^{a_1} \subset \cdots \subset M^{a_k},$$

and then apply the same reasoning.

For applications of the Morse inequalities the following proposition seems often to simplify the matter.

Theorem 2 *Let M and f be as in Theorem 1 with* dim $M = n$. *Let M_0 be a C^∞-submanifold of dimension n_0 of M lying wholly in its interior and f^0 be the restriction of f on M_0. Suppose that there exists a C^∞-field $\{N_x/x \in M_0\}$ of $(n-n_0)$-dimensional C^∞-submanifolds of M over M_0, which are disjoint from each other, transverse everywhere to M_0, fill a certain neighbourhood of M_0 in M, and satisfy the following condition: For any point $x \in M_0$ the restriction f_x of f on N_x has x as a nondegenerate critical point of index μ_x. Then any (nondegenerate) critical point c of f^0 on M_0 is also a (non-degenerate) critical point of f on M and the indices λ^0 resp. λ of c considered as critical point of f^0 resp. f are connected by the relation*

$$\lambda = \lambda^0 + \mu_c. \tag{5}$$

Proof. We may take a coordinate neighbourhood U about c in M with coordinates $x_1, \cdots, x_{n_0}, y_1, \cdots, y_{n-n_0}$ such that in this coordinate system $c = (0, \cdots, 0, 0, \cdots, 0)$, $M_0 \bigcap U$ is given by $y_1 = \cdots = y_{n-n_0} = 0$ and for any $a = (a_1, \cdots, a_{n_0}, 0, \cdots, 0)$ on $M_0 \bigcap U$, N_a is given by $x_1 = a_1, \cdots, x_{n_0} = a_{n_0}$. The hypothesis shows that $\dfrac{\partial f}{\partial y_j} \equiv 0.$ $1 \leqslant j \leqslant n - n_0$ on $M_0 \bigcap U$ so that $\dfrac{\partial^2 f}{\partial x_i \partial y_j} = 0$ at c for any $1 \leqslant i \leqslant n_0$,

$1 \leqslant j \leqslant n-n_0$. Moreover, the $(n-n_0)$ – rowed square matrix $\left(\dfrac{\partial^2 f(0)}{\partial y_j \partial y_{j'}}\right)$ is nondegenerate and of index μ_c. Hence the n-rowed square matrix

$$\begin{pmatrix} \dfrac{\partial^2 f(0)}{\partial x_i \partial x_{i'}} & \dfrac{\partial^2 f(0)}{\partial x_i \partial y_{j'}} \\ \dfrac{\partial^2 f(0)}{\partial x_{i'} \partial y_j} & \dfrac{\partial^2 f(0)}{\partial y_j \partial y_{j'}} \end{pmatrix} = \begin{pmatrix} \dfrac{\partial^2 f(0)}{\partial x_i \partial x_{i'}} & 0 \\ 0 & \dfrac{\partial^2 f(0)}{\partial y_j \partial y_{j'}} \end{pmatrix}$$

is nondegenerate if and only if $\left(\dfrac{\partial^2 f(0)}{\partial x_i \partial x_{i'}}\right)$ is so. In that case its index will be equal to the sum of the indices of the matrices $\left(\dfrac{\partial^2 f(0)}{\partial x_i \partial x_{i'}}\right)$ and $\left(\dfrac{\partial^2 f(0)}{\partial y_j \partial y_{j'}}\right)$, which is just the equality (5).

§2 Critical m-sections of convex bodies

Let E be a bounded convex body with C^∞-smooth boundary in the n-dimensional Euclidean space R^n. Let $\tilde{G}_m$(resp.G_m) be the grassmannian of all m-planes (resp.all m-planes through the origin 0 of R^n) in R^n The subspace of $\tilde{G}_m$ consisting of all m-planes meeting E will be denoted by $\hat{G}_m$ which is a bounded manifold of dimension $(n-m)(m+1)$ with boundary $\hat{G}_m$. For any $X \in \hat{G}_m$, let $A(X)$ be the area of the cross-section $X \bigcap E$ of X with E. Then A is a C^∞-function on $\hat{G}_m$ which takes value 0 on its boundary $\hat{G}_m$ and is positive in its interior.

Definition. For an m-plane X in R^n, $X \bigcap E$ will be called a *critical m-section* with *critical value c* if X is in the interior of $\hat{G}_m$ and is a critical point with critical value c of the function A. The section $X \bigcap E$ is then said to be *nondegenerate* of *index* λ if X is so for the function f.

Theorem 3 *If all critical m-sections of the convex body E are nondegenerate, then the number C_λ of those of index λ is at least equal to the number p_λ of partitions of $\lambda-n+m$ into m numbers $\geqslant 0$ and $\leqslant n-m$. In particular, the total number of critical m-sections is at least equal to* $\Sigma_{P\lambda} = \dbinom{n}{m}$.

Proof. By the Morse inequality under boundary conditions we have

$$\begin{aligned} C_\lambda &\geqslant \text{rank over } Z_2 \text{ of } H_\lambda(\hat{G}_m, \dot{\hat{G}}_m; Z_2) \\ &= \text{rank over } Z_2 \text{ of } H^{(n-m)(m+1)-\lambda}(\hat{G}_m, Z_2). \end{aligned}$$

Hence the theorem is a consequence of the following

Lemma *Rank over* Z_2 of $H^{(n-m)(m+1)-\lambda}(\hat{G}_m, Z_2) = p_\lambda$.

Proof. Let us take the origin O to be in the interior of E. For any $X_0 \in G_m$, let $X_0^\perp$ be the orthogonal complement through O of x_0. The orthogonal projection of E onto $X_0^\perp$ is then a convex body $E(X_0)$ in $X_0^\perp$. For any $x \in E(X_0)$, let (X_0, x) be the m-plane through x and parallel to X_0. Then $\hat{G}_m$ is the space of all pairs (X_0, x) with $X_0 \in G_m$ and $x \in E(X_0)$. For $(X_0, x) \in \hat{G}_m$, set $h_t(X_0, x) = (X_0, x(t))$ where $x(t)$ is the point on the segment Ox dividing it in the ratio $1-t:t$. Then h_t gives a deformation retraction of $\hat{G}_m$ into G_m. It follows that

$$H^{(n-m)(m+1)-\lambda}(\hat{G}_m, Z_2) \approx H^{(n-m)(m+1)-\lambda}(G_m, Z_2) \approx H_{\lambda-n+m}(G_m, Z_2).$$

The lemma follows now from the known fact[1] that Rank over Z_2 of $H_{\lambda-n+m}(G_m, Z_2) = p_\lambda$.

§3 Critical sections of an ellipsoid

In R^n with rectangular coordinates $x_1, \cdots, x_n$, let E be the ellipsoid

$$\sum_{k=1}^{n} \frac{x_k^2}{\alpha_k^2} \leqslant 1, \quad 0 < \alpha_1 < \cdots < \alpha_n, \tag{1}$$

and $\dot{E}$ be the boundary surface of E.

For any subset I of m indices of the index set $K = \{1, \cdots, n\}$, let X_I be the m-plane defined by

$$x_j = 0, \quad j \in K - I = J. \tag{2}$$

For $I = \{i_1, \cdots, i_m\}$ with $1 \leqslant i_1 < \cdots < i_m \leqslant n$, set also

$$\lambda_I = \sum_{s=1}^{m}(i_s - s) + (n-m). \tag{3}$$

Theorem 4 *There are just* $\binom{n}{m}$ *critical m-sections $X_I \bigcap E$ of E where I runs over all subsets of m indices of K. All such critical sections are nondegenerate and the index of $X_I \bigcap E$ is λ_I.*

Proof. Let X_0 be an m-plane for which $X_0 \bigcap E$ is a critical section. We shall show first that X_0 passes necessarily through the origin O of R^n. For if not, the centre C of the section $X_0 \bigcap E$ will be different from O. Let X' be the m-plane through O parallel to X_0. For any pair of points $x' \in X' \bigcap E$ and $x \in X_0 \bigcap E$ with $O_{x'}$

parallel to C_x, it is clear by convexity of E that $\overline{C_x} < \overline{O_{x'}}$ so that $A(X_0) < A(X')$ and $\overline{O_{x'}}/\overline{C_x} > a$ where $a > 1$ is independent of x, x'. For t with $|t|$ sufficiently small, let C_t be the point on the ray OC with $\overline{CC_t} = |t|$ and C_t between O, C resp. C between O, C_t according as $t < 0$ resp. $t > 0$. Let X_t be the m-space through C_t and parallel to X_0. With x, x' as before let x_t be the point on $X_t \bigcap E$ such that $C_t x_t$ is parallel to C_x. Set $\overline{OC} = c$. Then the convexity of E shows that

$$\overline{C_t x_t} < \frac{c+t}{c} \cdot \overline{Cx} - \frac{t}{c} \cdot \overline{O'x} < \left[1 - \frac{t}{c} \cdot (a-1)\right] \cdot \overline{Cx} \text{ for } t > 0,$$

resp.

$$\overline{C_t x_t} > \left[1 - \frac{t}{c} \cdot (a-1)\right] \cdot \overline{Cx} \text{ for } t < 0.$$

It follows that

$$A(x_t) < \left[1 - \frac{t}{c} \cdot (a-1)\right]^m \cdot A(X_0) \text{ for } t > 0,$$

resp.

$$A(X_t) > \left[1 - \frac{t}{c} \cdot (a-1)\right]^m \cdot A(X_0) \text{ for } t < 0.$$

Let $\tau > 0$ be sufficiently small and $g : (-\tau, +\tau) \to \hat{G}_m$ be the curve defined by $g(t) = X_t$. Then $\dfrac{dA(X_t)}{dt}$ at $X = X_0$ is $\leqslant -\dfrac{m}{c} \cdot (a-1)$. $A(X) < 0$ so that X_0 cannot be a critical section, as to be proved.

Now G_m is a C^∞-submanifold of $\hat{G}_m$ lying wholly in its interior. For any $X \in G_m$, let N_X be the submanifold in $\tilde{G}_m$ consisting of all m-planes X_0 in $\tilde{G}_m$ parallel to X. Then the field of submanifolds $\{N_x/X \in G_m\}$, with similar calculations as above, is easily seen to be one verifying the conditions in Theorem 2 with each X as a nondegenerate maximum of the restriction A_X of the function A on N_X. By Theorem 2 it is therefore sufficient to consider the restriction A^0 of the function A on G_m.

Let X_0 be therefore a critical m-section of E through the origin O. We can then change the coordinates $(x_1, \cdots, x_n)$ to $(y_1, \cdots, y_m, z_1, \cdots, z_{n-m})$ such that $\dot{E}$ will be given by

$$\sum_{i=1}^{m} \frac{y_i^2}{\beta_i^2} + \sum_{\substack{1 \leqslant i \leqslant m \\ 1 \leqslant j \leqslant n-m}} 2\varepsilon_{ij} y_i z_j + \sum_{j=1}^{n-m} \frac{z_j^2}{\gamma_j^2} = 1$$

and X_0 is given by the equations $z_1 = \cdots = z_{n-m} = 0$. For fixed k and l with $1 \leqslant k \leqslant m$ and $1 \leqslant l \leqslant n-m$, let us consider the m-plane X_t given by

$$X_t : \begin{cases} z_j = 0, & j \neq l, \\ z_l = t y_k. \end{cases}$$

The square of the area of the section $X_t \bigcap E$ is then of the form

$$B(t) = c \cdot \prod_{i \neq k} \beta_i^2 / \left(\frac{1}{\beta_k^2} + 2\varepsilon_{kl} t + \frac{t^2}{\gamma_l^2} \right) \left(1 + \sum_{i \neq k} \beta_i^2 \varepsilon_{il}^2 t^2 \right),$$

in which c is a universal constant. That X_0 is a critical section implies that $\dfrac{dB(t)}{dt} = 0$ for $t = 0$. It follows that all $\varepsilon_{kl} = 0$ and the axis of the section $X_0 \bigcap E$ is also the axis of the ellipsoid E. There exists therefore a subset $I = \{i_1, \cdots, i_m\}$ with $1 \leqslant i_1 < \cdots < i_m \leqslant m$ of the index set $K = \{1, \cdots, n\}$ such that $X_0 = X_I$ is given by the system of equations (2).

It remains now to determine the index of the critical section X_I. For this let U_I be the neighbourhood of X_I in G_m consisting of the m-planes X_a represented by the systems of equations

$$X_a : x_j = \sum_{i \in I} a_{ij} x_i, \quad j \in J, \tag{4}$$

in which a is the matrix (a_{ij}).

Consider the function

$$F(x) \equiv \sum_{k \in K} x_k^2 + \lambda_0 \cdot \sum_{k \in K} \left(\frac{x_k^2}{\alpha_k^2} - 1 \right) + 2 \sum_{j \in J} \cdot \lambda_j \left(x_j - \sum_{i \in I} a_{ij} x_i \right), \tag{5}$$

in which λ_0, λ_j are all parameters. The extremities (x) of an axis of the section $X_a \bigcap E$ are then determined by the systems of equations

$$\sum_{k \in K} \frac{x_k^2}{\alpha_k^2} = 1, \tag{6}$$

$$x_j = \sum_{i \in I} a_{ij} x_i, \quad j \in J, \tag{7}$$

and

$$\frac{1}{2} \frac{\partial F}{\partial x_i} \equiv x_i + \lambda_0 \cdot \frac{x_i}{\alpha_i^2} - \sum_{j \in J} \lambda_j a_{ij} = 0, \quad i \in I, \tag{8}$$

$$\frac{1}{2} \frac{\partial F}{\partial x_j} \equiv x_j + \lambda_0 \cdot \frac{x_j}{\alpha_j^2} + \lambda_j = 0, \quad j \in J. \tag{9}$$

Eliminating λ_j and x_j, we get then

$$\sum_{s \in I} \left[\left(\delta_{rs} + \sum_{j \in J} a_{rj} a_{sj} \right) + \lambda_0 \left(\frac{\delta_{rs}}{\alpha_s^2} + \sum_{j \in J} \frac{a_{rj} a_{sj}}{\alpha_j^2} \right) \right] x_s = 0, \quad r \in I. \tag{10}$$

As x_i, $i \in I$, cannot all be 0 by (6) and (7), we have

$$\left| \delta_{rs} + \sum_{j \in J} a_{rj} a_{sj} + \lambda_0 \left(\frac{\delta_{rs}}{\alpha_s^2} + \sum_{j \in J} \frac{a_{rj} a_{sj}}{\alpha_j^2} \right) \right| = 0. \tag{11}$$

Multiplying (8) by x_i, (9) by x_j, and adding all these equations, we get by (6) and (7)

$$\sum_{k \in K} x_k^2 + \lambda_0 = 0.$$

Hence λ_0 is the negative square of the length of a demi-axis of the section $X_a \bigcap E$. Denoting the m roots of the equation (11) in λ_0 by $\lambda_0^{(t)}$, $1 \leqslant t \leqslant m$, we have therefore

$$[A(X_a)]^2 = c_m \cdot (-1)^m \cdot \lambda_0^{(1)} \cdots \lambda_0^{(m)},$$

in which c_m is a positive numerical constant independent of $a = (a_{ij})$.

By (11) we have then ($\cdots$ means terms involving a_{ij} of degrees > 2),

$$\begin{aligned} [A(X_a)]^2 &= c_m \cdot \frac{\left| \delta_{rs} + \sum\limits_{j \in J} a_{rj} a_{sj} \right|}{\left| \delta_{rs}/\alpha_s^2 + \sum\limits_{j \in J} a_{rj} a_{sj}/\alpha_j^2 \right|} \\ &= c_m \cdot \prod_{i \in I} \alpha_i^2 \cdot \frac{1 + \sum\limits_{i \in I} \sum\limits_{j \in J} a_{ij}^2 + \cdots}{1 + \sum\limits_{i \in I} \sum\limits_{j \in J} a_{ij}^2 \alpha_i^2/\alpha_j^2 + \cdots} \\ &= c_m \cdot \prod_{i \in I} \alpha_i^2 \cdot \left[1 + \sum_{i \in I} \sum_{j \in J} a_{ij}^2 \left(1 - \frac{\alpha_i^2}{\alpha_j^2} \right) + \cdots \right], \end{aligned}$$

or

$$[A(X_a)]^2 = [A(X_I)]^2 \cdot \left[1 + \sum_{i \in I, j \in J} a_{ij}^2 \left(1 - \frac{\alpha_i^2}{\alpha_j^2} \right) + \cdots \right]. \tag{12}$$

The matrix $\left(\dfrac{\partial^2 [A(X_a)]^2}{\partial a_{ij} \partial a_{i'j'}} \right)$ at $X_a = X_I$ is therefore a diagonal matrix of $m(n-m)$ rows and columns with elements in the diagonal equal to

$$2c_m \cdot \prod_{i \in I} \alpha_i^2 \cdot \left(1 - \frac{\alpha_i^2}{\alpha_j^2} \right), \quad i \in 1, j \in J. \tag{13}$$

As all elements in the set (13) are nonzero and the number of negative ones is $\lambda_I^0 = \sum_{s=1}^{m} (i_s - s)$, the index of X_I of the function $[A(X)]^2$ on G_m is equal to λ_I^0. The theorem follows now from Theorem 2 in §I and the Lemma below.

Lemma *Let f be a C^∞-function on a compact C^∞-manifold M, supposed to be everywhere positive if M is closed and satisfy the conditions* B 1-3 *in* §I *in case M is bounded. Then a point x of M is a (nondegenerate) critical point of f if and only if it is so for f^2, and at a nondegenerate critical point the indices of f and f^2 are equal to each other.*

Proof. This will be clear by writing down the partial derivatives of f^2 at x in terms of those of any f in local coordinates.

§4 Determination of Betti numbers of Grassmannian manifolds via Morse theory

In the proof of the theorem in §II the Betti numbers mod 2 of a grassmannian manifold is assumed to be known. On the other hand, the calculations as in §III permit to determine the Betti numbers of a complex grassmannian manifold directly by the general theory of Morse as indicated in §I in the following manner.

Let C^n be the unitary space of complex dimension n with coordinates $z_k = x_k + i_{y_k}$, $1 \leqslant k \leqslant n$, and R^{2n} the associated Euclidean space of coordinates $x_1, \cdots, x_n, y_1, \cdots, y_n$. Let E be the ellisoid

$$E : \sum_{k=1}^{n} \frac{x_k^2 + y_k^2}{\alpha_k^2} \leqslant 1 \tag{1}$$

with $0 < \alpha_1 < \cdots < \alpha_n$. Let G_m^c be the complex grassmannian of all unitary m-space through the origin O in C^n. For any $Z \in G_m^c$, the area $A(Z)$ of $Z \bigcap E$ is then a C^∞-function on G_m^c, whose critical points consist of unitary m-spaces Z_I defined by

$$Z_I : z_j = 0, \quad j \in J = K - I, \tag{2}$$

with I any subset of m indices of the index set $K = \{1, \cdots, n\}$. Consider the neighbourhood U_I of all unitary m-spaces Z_c defined by the systems of equations

$$Z_c : z_j = \sum_{i \in I} c_{ij} z_i, \quad j \in J, \tag{3}$$

with $C_{ij} = a_{ij} + ib_{ij}$, a_{ij}, b_{ij} all real, $c = \text{matrix}(c_{ij})$. Now (3) can also be written in the form

$$Z_c : \begin{cases} x_j = \sum_{i\in I} a_{ij}x_i - \sum_{i\in I} b_{ij}y_i, \\ y_j = \sum_{i\in I} b_{ij}x_i + \sum_{i\in I} a_{ij}y_i, \end{cases} \quad j \in J. \tag{4}$$

Equation (12) of §III gives therefore

$$[\mathrm{A}(Z_c)]^2 = [A(Z_I)]^2 \cdot \left[1 + 2 \sum_{i\in I, j\in J} (a_{ij}^2 + b_{ij}^2)\left(1 - \frac{\alpha_i^2}{\alpha_j^2}\right) + \cdots \right],$$

in which $\cdots$ means the terms involving a_{ij}, b_{ij} of degrees > 2. The index of Z_I is thus even and equal to $2\lambda_I^0 = 2\sum_{s=1}^{m}(i_s - s)$. The Morse inequalities in Theorem 1 give therefore the following result of Ehresmam[1]: The complex grassmannian G_m^c has nonzero Betti numbers only in even dimensions. The Betti number of dimension 2d is then given by the number of subsets $I = (i_1, \cdots, i_m)$ with $1 \leqslant i_1 < \cdots < i_m \leqslant n$, for which

$$\lambda_1^0 = \sum_{s=1}^{m}(i_s - s) = d.$$

References

[1] Ehresmann Ch. Sur la topologie de certains espaces homogènes. *Annals of Math.*, 1934, 35: 396-443.

[2] Milnor J. *Morse Theory.* Princeton, 1963.

[3] Morse M. *The Calculus of Variations in the Large.* New York, 1934.

[4] Morse M & Van Schaak G B. The critical point theory under general boundary conditions. *Annals of Math.*, 1934, 35: 545-571.

45. S_k 型奇点所属的同调类*

§1 符号说明

在以下所用同调系统依 [1], 系数环取模 2 整数域 Z_2, 在记号中一概略去.

记 $\Omega_d(m)$ 为一切定义在整数集 $1,2,\cdots,m$ 上面满足条件

$$0 \leqslant \omega(1) \leqslant \cdots \leqslant \omega(m),$$
$$d(\omega)=\sum_{i=1}^{m}\omega(i)=d$$

的一切整数值函数 ω 的集合. 又记 $\Omega_d(N,m)$ 为 $\Omega_d(m)$ 中满足 $\omega(m)\leqslant N-m$ 的一切 ω 的子集.

命 R^N 为一固定的 N 维欧氏空间, G_m^N 为过 R^N 中的一切 m 面 (即过原点的 m 维子平面) 所成的格拉斯曼流形, 已知 $H^d(G_m^N)$ 与集合 $\Omega_d(N,m)$ 一一对应. 与 $\omega\in\Omega_d(N,m)$ 对应的类将记为 W_ω.

记 γ_m^N 为 G_m^N 上的标准 m 面丛. 设 ξ 为一有限维底空间上的一个 m 面丛, 由一映像 $f: M\to G_m^N$ 所引起. 则 f^*W_ω 只须 N 充分大即与 N 无关而将记为 $W_\omega(\xi)$. 若 M 为一 C^∞-m 维流形而 ξ 为 M 的切丛 $\tau(M)$, 则 $W_\omega(\tau(M))$ 将径记为 $W_\omega(M)$.

定义映像 $D: G_m^N\to G_{N-m}^N$ 如下, 设 $R^m\in G_m^N$ 为 R^N 中的一个 m 面, 则 $D(R^m)$ 为 R^N 与 R^m 正交的 $(N-m)$ 面. 于是对每一 $\omega\in\Omega_d(N,m)$ 有一 $D\omega\in\Omega_d(N,N-m)$ 使 $D^*W_{D\omega}=W_\omega$. 若函数 $\omega\in\Omega_d(N,m)$ 为

$$\omega(i)=\begin{cases}0, & 1\leqslant i\leqslant i_0,\\ \beta_1, & i_0+1\leqslant i\leqslant i_1,\\ \vdots & \vdots\\ \beta_s, & i_{s-1}+1\leqslant i\leqslant i_s,\\ N-m, & i_s+1\leqslant i\leqslant m,\end{cases}$$

---

* *Acta Math. Sinica*, 1974, 17: 28-37. 本文结果系 1966 年得到. Porteous 与 Ronga 亦有类似结果, 见 [5, 6], 但与本文所用方法不同, 结果的表达式也不一样.

其中 $0<\beta_1<\cdots<\beta_s<N-m$(可有 $i_0=0$ 或 $i_s=m$), 则

$$D\omega(i)=\begin{cases} m-i_s, & 1\leqslant i\leqslant N-m-\beta_s,\\ m-i_{s-1}, & N-m-\beta_s+1\leqslant i\leqslant N-m-\beta_{s-1},\\ \quad\vdots & \qquad\vdots\\ m-i_0, & N-m-\beta_1+1\leqslant i\leqslant N-m.\end{cases}$$

简记 $W_{D\omega}$ 为 $\overline{W}_\omega\in H^d(G_m^N)$. 特别在 $\omega=\omega_j$, 这里 $\omega_j(i)=0,\ 1\leqslant i\leqslant m-j$, $\omega_j(i)=1,\ m-j<i\leqslant m$ 时, 有 $D\omega_j(i)=0,\ 1\leqslant i\leqslant N-m-1, D\omega_j(N-m)=j$, 此时将简记 $W_{\omega j}=W^j$, $\overline{W}_{\omega j}=D^*W_{D\omega j}=\overline{W}^j$. 前者为 γ_m^N 的 Stiefel-Whitney 示性类, 而后者为 γ_m^N 的对偶 Stiefel-Whitney 示性类. 由陈省身的一个定理, 知 $H^*(G_m^N)$ 可由 W^j, $1\leqslant j\leqslant m$ 在上积下生成, 亦可由诸 $\overline{W}^j$, $1\leqslant j\leqslant m$ 所生成.

G_m^N 中任一类可由所谓 Schubert 循环表示, 详见 [4].

设 M, N 为维数各是 m, n 的 C^∞ 仿紧流形, 而 f 为 M 到 N 的 C^∞ 映像. 命 $J^1(M,N)$ 为 $M\times N$ 上的节丛 (节 = jet), $s=\min(m,n)$, 而 j_k 为 $J^1(M,N)$ 中由秩 $\leqslant s-k$ 的一切节所成的子空间. 已知 J_k 为一实代数簇且可表成流形堆

$$J_k=(J_k-J_{k+1})\cup\cdots\cup(J_{s-1}-J_s)\cup J_s,$$

其补维数 (在 $J^1(M,N)$ 中) 为

$$\mathrm{codim}J_k=(m-s+k)(n-s+k).$$

依 Borel-Haefliger[1], 此时 J_k 具有一基本类

$$J_k\in H_{\rho-(m-s+k)(n-s+k)}(J^1(M,N)),$$

此处 $\rho=\dim J^1(M,N)=mn+m+n$.

设 $f^1:M\to .J^1(M,N)$ 为 f 的诱导节映像. 今设 f^1 横截于 J_k, 即 f^1 横截于 $J_k-J_{k+1},\cdots,J_{s-1}-J_s,J_s$, 简记作 $f'\pitchfork J_k$, 则 $(f^1)^{-1}J_k$ 亦为一流形堆, 具有基本类

$$T_k(f)\in H_{m-(m-s+k)(n-s+k)}(M).$$

若 $i:J_k\to J^1(M,N)$, $j:(f^1)^{-1}J_k\to M$ 为恒同映像, 而

$$\Delta:H_p(x)\to H^{\gamma-p}(x)$$

指 r 维仿紧流形的 Veblen-Poincaré 对偶, 则

$$\Delta j_*T_k(f)=(f^1)^*\Delta i_*T_k\in H^{(m-s+k)(n-s+k)}(M).$$

记作 $T^k(f)$.

§2 定理及其证明

定理 设 $s=\min(m,n)$, $t=\max(m,n)$, N 充分大, 则当 $f:M^m\to N^n$ 而 $f^1:M\to J^1(M,N)$ 横截于 J_k 时,

$$T^k(f)=\sum W_\omega(M)\cup f^*\bar{W}_{\theta(\omega)}(N),$$

其中 $\sum$ 展开于一切 $\omega\in\Omega(N,m)$ 上, 这里 ω 满足

$$\begin{aligned}&\omega(i)=0,\quad i\leqslant s-k,\\&\omega(m)\leqslant n-s+k,\\&0\leqslant d(\omega)\leqslant(m-s+k)(n-s+k),\end{aligned}$$

又 $\theta(\omega)\in\Omega(N,N-n)$ 满足

$$\begin{aligned}&\theta(\omega)(i)=0,\quad i\leqslant N-t-k,\\&\theta(\omega)(N-t-k+1+i)=n-s+k-\omega(m-i),\\&\qquad 0\leqslant i\leqslant m-s+k-1,\end{aligned}$$

特别在 $k=1$ 时有

$$T^1(f)=\begin{cases}\displaystyle\sum_{i=0}^{m-n+1}W^i(M)\cup f^*\bar{W}^{m-n+1-i}(N), & m\geqslant n,\\ \displaystyle\sum_{i=0}^{n-m+1}\bar{W}^i(M)\cup f^*W^{n-m+1-i}(N), & m\leqslant n.\end{cases}$$

注 在 $m\geqslant n$, $k=1$ 时, 上述 $T^1(f)$ 的公式见 [3].

为证明这一定理, 在 $G_m^N\times G_n^N=B$ 上作一 R^{mn} 丛 (E,B,π), 使 $(R^m,R^n)\in G_m^N\times G_n^N$ 上的纤维为 $\mathrm{Hom}(R^m,R^n)$, 命 Σ_k 为 E 的下述子集:

$$\Sigma_k=\left\{(R^m,R^n,f)\Big/\begin{array}{c}(R^m,R^n)\in G_m^N\times G_n^Nf\in\mathrm{Hom}(R^m,R^n),\\ \mathrm{Runk}\ f\leqslant\min(m,n)-k\end{array}\right\}$$

定义

$$i:\begin{cases}G_m^N\times G_n^N=B\to E,\\(R^m,R^n)\to(R^m,R^n,f_0),\end{cases}$$

使 f_0 为 $R^m(\subset R^N)$ 到 $R^n(\subset R^N)$ 上的垂直投影. 又记

$$S_k=(1\times D)i^{-1}\Sigma_k\subset G_m^N\times G_{N-n}^N,$$

$[S_k] = S_k$ 的基本类,

$\{S_k\}$ 为 $[S_k]$ 在 $G_m^N \times G_{N-n}^N$ 中的对偶类.

则定理的证明依赖于下面二个引理：

引理 A 视 Σ_k 为一流形堆

$$\Sigma_k = (\Sigma_k - \Sigma_{k+1}) \bigcup (\Sigma_{k+1} - \Sigma_{k+2}) \bigcup \cdots \bigcup (\Sigma_{m-1} - \Sigma_m) \bigcup \Sigma_m,$$

则 $i: B \to E$ 横截于 Σ_k.

引理 B 在 $G_m^N \times G_{N-n}^N$ 中有

$$\{S_k\} = \Sigma W_\omega \otimes W_{\theta(\omega)},$$

其中 $\Sigma, \omega, \theta(\omega)$ 各如定理叙述中所示.

引理 A, B 的证明见§3—4.

定理的证明 从 C^∞-映像 $f: M \to N$ 引出一可交换图像：

$$\begin{array}{ccccccc}
& & J^1(M,N) & \xrightarrow{g} & E & & \\
& \nearrow^{f^1} & \downarrow \pi_0 & & \pi \downarrow \uparrow i & \nwarrow^{j} & \\
M & \xrightarrow{1 \times j} & M \times N & \xrightarrow{g_m \times g_n} & G_m^N \times G_n^N & \xrightarrow{1 \times D} & G_m^N \times G_{N-n}^N
\end{array}$$

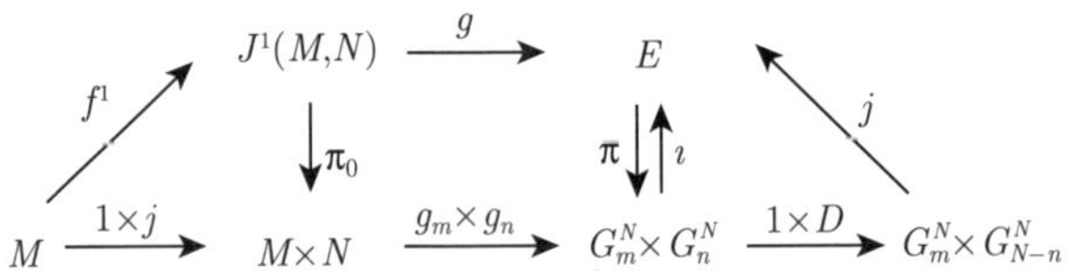

其中, f^1 为 f 所引出的节映像, g_m, g_n 各为引出 M, N 切丛的映像, 可视为嵌入, π_0, π 为投影, $j = i(1 \times D)^{-1}$, 而 g 为视节为相应切面间线性同态时所得的映像.

命 E' 为 E 中在 $(g_m \times g_n)(M \times N)$ 上的部分, 则易见在 E 中 $E' \pitchfork \Sigma_k$. 故 $f^1 \pitchfork J_k$ (在 $J^1(M,N)$ 中) 或 $gf^1 \pitchfork g(J_k)$ (在 E' 中) 等价于 $gf^1 \pitchfork \Sigma_k$ (在 E 中). 由引理 A 即得 ($\{\Sigma_k\}$ 指 Σ_k 所定基本类在 E 中的对偶类)

$$T^k(f) = (f^1)^* g^* \{\Sigma_k\} = (f^1)^* g^* (j*)^{-1} \{S_k\},$$

由引理 B 即得定理.

§3 引理 A 的证明

设 $n \geqslant m$ ($m \geqslant n$ 的情形与此类似).

试考虑任一点

$$p = (R_0^m, R_0^n, f_0) \in (\Sigma_k - \Sigma_{k+1}) \cap V,$$

此处

$$V = i(B).$$

于是 R_0^m 到 R_0^n 上的正交投影 f_0 恰有维数 $m-k$. 置

$$\operatorname{Ker} f_0 = R_0^k \subset R_0^m,$$

$$\operatorname{Im} f_0 = \bar{R}_0^{m-k} \subset R_0^n.$$

在 R_0^m 与 R_0^n 中各取 R_0^k 与 $\overline{R}_0^{m-k}$ 的正交补空间为 R_0^{m-k}, R_0^{n-m+k}, 即 ($\dot{+}$ 表直和):

$$\begin{aligned} R_0^m &= R_0^k \dot{+} R_0^{m-k}, \\ R_0^n &= \bar{R}_0^{m-k} \dot{+} R_0^{n-m+k}. \end{aligned}$$

于是 $R_0^k, \overline{R}_0^{m-k}, R_0^{n-m+k}$ 互相垂直, 又 $R_0^k, R_0^{m-k}, R_0^{n-m+k}$ 互相垂直. 置

$$\begin{aligned} R_0^{n+k} &= R_0^k \dot{+} R_0^{m-k} \dot{+} R_0^{n-m+k}, \\ \bar{R}_0^{n+k} &= R_0^k \dot{+} \bar{R}_0^{m-k} \dot{+} R_0^{n-m+k}, \\ \bar{R}_0^m &= R_0^k \dot{+} \bar{R}_0^{m-k}, \\ R_0^{n-m+2k} &= R_0^k \dot{+} R_0^{n-m+k}, \\ R_0^{N-n-k} &= R^N \text{ 中 } \bar{R}_0^{n+k} \text{ 的正交补空间}. \end{aligned}$$

于是

$$R_0^{m-k} \subset R_0^{N-n-k} \dot{+} \bar{R}_0^{m-k}.$$

定义

$$T = \left\{ (R^m, R^n, f) \middle/ \begin{array}{c} R^m = R^k \dot{+} R_0^{m-k}, \quad R^k \subset R_0^{n-m+2k} \\ R^n = R_0^{n-m+k} \dot{+} \bar{R}^{m-k}, \quad \bar{R}^{m-k} \subset \bar{R}_0^m \\ f : R^m \to R^n \text{线性} \end{array} \right\}$$

$$Q = \{(R^m, R^n, f)^{\varepsilon T} / f(R^k) = (0)\} \subset T \cap \textstyle\sum_k.$$

则

$$\begin{aligned} &\dim T = \dim G_k^{n-m+2k} + \dim G_{m-k}^m + mn = kn + mn, \\ &\dim Q = \dim G_k^{n-m+2k} + \dim G_{m-k}^m + (m-k)n = mn, \\ &\dim V = \dim B = m(N-m) + n(N-n), \\ &\dim T \cap V = \dim G_k^{n-m+2k} + \dim G_{m-k}^m = kn, \end{aligned}$$

$$\begin{aligned}\dim E &= \dim G_m^N + \dim G_n^N + mn\\ &= m(N-m) + n(N-n) + mn.\end{aligned}$$

故

$$\dim T \cap V = \dim T + \dim V - \dim E.$$

显然有

1°　　在 p 处有 $T \pitchfork V$　(在 E 中).

在以下并将证明

2°　　在 p 处 $Q \pitchfork T \cap V$　(在 T 中).

由 1°, 2° 得

在 p 处 $V \pitchfork Q$　(在 E 中).

因之更有

在 p 处 $V \pitchfork \Sigma_k$　(在 E 中).

故

$$i(B) \pitchfork \Sigma_k - \Sigma_{k+1} \quad (\text{在 } E \text{ 中}).$$

同样又有

$$i(B) \pitchfork \Sigma_{k+1} - \Sigma_{k+2}, \cdots, \Sigma_m.$$

故

$$i(B) \pitchfork \Sigma_k, \quad \text{q. e. d.}$$

2° 的证明:

在以下指针范围为

$$1 \leqslant i, j \leqslant k; \qquad 1 \leqslant g, h, l \leqslant m-k;$$
$$1 \leqslant p \leqslant n-m+k, \quad 1 \leqslant r \leqslant N-n-k.$$

在 R^N 中取有法 (orthonormal) 向量基 (a_i, b_h, c_p, d_r) 使 [1] 中表相应向量空间中的向量基):

$$\begin{aligned}&R_0^k = [a_1, \cdots a_k] = [a_i],\\ &\bar{R}_0^{m-k} = [b_1, \cdots, b_{m-k}] = [b_h],\\ &R_0^{n-m+k} = [c_1, \cdots, c_{n-m+k}] = [c_p],\\ &R_0^{N-n-k} = [d_1, \cdots, d_{N-n-k}] = [d_r],\end{aligned}$$

$$\begin{aligned}R_0^{m-k}&=[b_1+e_1,\cdots,b_{m-k}+e_{m-k}]\\&=[b_h+e_h],e_h\in R_0^{N-n-k}.\end{aligned}$$

考虑 T 的邻域 $\cup$, 使 $\cup$ 中任一 (R^m,R^n,f) 有下列性质:

$$\begin{aligned}&R^m=R^k\dot{+}R_0^{m-k},\quad R^k\subset R_0^{n-m+2k},\\&R^n=R_0^{n-m+k}\dot{+}\bar{R}^{m-k},\quad \bar{R}^{m-k}\subset\bar{R}_0^m,\\&R^k\text{ 在 }R_0^k\text{上的正交投影不退化},\\&\bar{R}^{m-k}\text{在}\bar{R}_0^{m-k}\text{上的正交投影不退化}.\end{aligned}$$

于是有

$$\begin{aligned}R^k&=\left[a_t+\sum_p\alpha_{ip}c_p\right],\\\bar{R}^{m-k}&=\left[b_h+\sum_i\beta_{hi}a_t\right],\\R^m&=\left[a_t+\sum_p\alpha_{tp}c_p,b_h+e_h\right],\\R^n&=\left[b_h+\sum_h\beta_{hi}a_i,c_p\right],\\f&:R^m\to R^n.\end{aligned}$$

由下二式所定:

$$f\left(a_i+\sum_p\alpha_{ip}c_p\right)=\sum_p\xi_{ip}c_p+\sum_h\eta_{ih}\left(b_h+\sum_j\beta_{hj}a_j\right),$$

$$f(b_h+e_h)=\sum_p\zeta_{ip}c_p+\sum_l\tau'_{hl}\left(b_l+\sum_i\beta_{li}a_i\right).$$

在 U 中可取坐标系统为

$$\alpha_{ip},\beta_{hi},\xi_{ip},\eta_{ih},\zeta_{hp},\tau_{hl},$$

此处

$$\tau_{hl}=1-\tau'_{hl}.$$

在此坐标系统中 $p=(R_0^m, R_0^n, f_0)$ 相当于原点. 又 $q=(R^m, R^n, f)\in Q\cap U$ 时应有 $f(R^k)=(0)$, 即

$$f(a_i+\sum_p \alpha_{ip}c_p)=0.$$

因而 $Q\cap U$ 由以下方程组所定:

$$Q\cap U: \xi_{ip}=0, \quad \eta_{ih}=0.$$

设 $(R^m, R^n, f)\in U$ 如前而命

$$R_\perp^k = R^n \text{ 在 } \bar{R}_0^{n+k} \text{ 中的正交补空间},$$

则

$$R_\perp^k=\left[a_i-\sum_h \beta_{hi}b_h\right].$$

若 $(R^m, R^n, f)\in U\cap V$ 因而 f 为

$$R^m=\left[a_i+\sum_p \alpha_{ip}c_p, b_h+e_h\right]$$

到

$$R^n=\left[b_h+\sum_h \beta_{hi}a_i, c_p\right]$$

上的垂直投影, 则应有 λ_{ij}, μ_{hj} 使

$$\begin{aligned}f\left(a_i+\sum_p \alpha_{ip}c_p\right)&=a_i+\sum_p \alpha_{ip}c_p-\sum_j \lambda_{ij}\left(a_i-\sum_h \beta_{hi}b_h\right)\\&=\sum_p \xi_{ip}c_p+\sum_h \eta_{ih}\left(b_h+\sum_j \beta_{hi}a_j\right),\end{aligned}$$

$$\begin{aligned}f(b_h+e_h)&=b_h-\sum_j \mu_{h1}\left(a_i-\sum_l \beta_{lj}b_l\right)\\&=\sum_p \zeta_{hp}c_p+\sum_l \tau'_{hl}\left(b_l+\sum_i \beta_{li}a_i\right).\end{aligned}$$

从第一方程比较 a_t 得

$$\delta_{ij}-\lambda_{ij}=\sum_h \eta_{ih}\beta_{hj},$$

比较 c_p 得

$$\alpha_{ip} = \xi_{ip},$$

比较 b_h 得

$$\sum_j \lambda_{ij}\beta_{hj} = \eta_{jh}.$$

又从第二方程比较 b_l 得

$$\delta_{hl} + \sum_j \mu_{hj}\beta_{l_j} = \tau'_{hl},$$

比较 c_p 得

$$0 = \zeta_{hp},$$

比较 a_i 得

$$-u_{hi} = \sum_l \tau'_{hl}\beta_{li}.$$

消去 λ_{ij}, μ_{hi} 即得 $V \cap U$ 的方程组为

$$V \cap U : \begin{cases} a_{ip} = \xi_{ip}, \quad \zeta_{hp} = 0, \\ \eta_{ih} = \sum\limits_j \beta_{h_j}\left(\delta_{ij} - \sum\limits_l \eta_{il}\beta_{l_j}\right) = \beta_{hi} - \sum\limits_{l,j} \eta_{il}\beta_{li}\beta_{hj}, \\ \tau'_{hl} = \delta_{hl} - \sum\limits_j \left(\beta_{lj} \sum\limits_g \tau'_{hg}\beta_{gj}\right). \end{cases}$$

记 ξ_{ip} 等的微分为 ξ_{ip}, 余类推, 则 $Q \cap U$ 与 $V \cap U$ 在 p 处的切面各由下方程组所定:

$$\mathrm{Q} \cap \mathrm{U} : \dot{\xi}_{ip} = 0, \quad \dot{\eta}_{ih} = 0,$$

$$\mathrm{V} \cap \mathrm{U} : \begin{cases} \dot{a}_{ip} = \dot{\xi}_{ip}, \quad \dot{\zeta}_{hp} = 0, \\ \dot{\eta}_{ih} = \dot{\beta}_{hi}, \quad \dot{\tau}_{hl} = 0. \end{cases}$$

显然 $Q \cap U, V \cap U$ 在 p 处的切面横截, 因而 Q 与 V 在 T 中 p 处横截. 证毕.

§4 引理 B 的证明

先设 $n \geqslant m$. 此时

$$S_k = \{(R^m, R^{N-n})/\dim(R^m \cap' R^{N-n}) \geqslant k\} \subset G_m^N \times G_{N-n}^N.$$

试考任意

$$\alpha \in \Omega_d(N, N-n), \quad \beta \in \Omega_{d'}(N, m), \quad d + d' = k(n-m+k).$$

在 R^N 中取两组在一般位置的平面序列:

$$\begin{aligned}&(\mathrm{A}) : A_1 \subset \cdots \subset A_{N-n}, \dim A_i = a(i) + i,\\ &(\mathrm{B}) : B_1 \subset \cdots \subset B_m, \dim B_j = \beta(j) + j,\end{aligned}$$

对此两组平面序列定义的 Schubert 循环 $[\alpha]^*$, $[\beta]^*$ 各由满足

$$\begin{aligned}&\dim\,(R^{N-n} \cap A_i) \geqslant i, \quad i = 1, \cdots, N-n,\\ &\dim\,(R^m \cap B_j) \geqslant j, \quad j = 1, \cdots, m\end{aligned}$$

的一切 R^{N-n} 与 R^m 的簇所代表. 于是引理 B 与下等价 ($[\alpha]$, $[\beta]$ 指相应的 Schubert 下同调类):

$$([\beta] \times [a]) \cdot [S_k] = \begin{cases} 0, & \alpha, \beta\text{不满足下关系 } (C) \text{ 时},\\ 1, & \alpha, \beta\text{满足下关系 } (C) \text{ 时}.\end{cases}$$

$$(C)\begin{cases} \alpha(i) = 0, & i \leqslant N-n-k,\\ \beta(j) = 0, & j \leqslant m-k,\\ \alpha(N-n-k+i) + \beta(m-i+1) = n-m+k, & 1 \leqslant i \leqslant k.\end{cases}$$

为此, 先设

$$(R^m, R^{N-n}) \in ([\beta]^* \times [\alpha]^*) \cap S_k \neq \varnothing.$$

取满足

$$i + j \leqslant k-1, \quad i, j \geqslant 0$$

的 i, j 而暂置

$$A_{N-n-i} = A, \quad B_{m-j} = B,$$

则

$$\begin{aligned}&\dim\,(R^m \cap R^{N-n}) = h \geqslant k,\\ &\dim\,(R^{N-n} \cap A) \geqslant N-n-i,\\ &\dim\,(R^m \cap B) \geqslant m-j.\end{aligned}$$

由此得

$$\dim\,(R^m \cap R^{N-n} \cap A) \geqslant h-i,$$

$$\dim\left(R^m\cap R^{N-n}\cap B\right)\geqslant h-j,$$
$$\dim\left(R^m\cap R^{N-n}\cap A\cap B\right)\geqslant h-i-j.$$

故

$$\dim\left(A\cap B\right)\geqslant k-i-j>0,$$
$$\dim A+\dim B\geqslant N+k-i-j,$$

或

$$\alpha(N-n-i)+\beta(m-j)\geqslant n-m+k,(i+j\leqslant k-1).$$

故有

$$\sum_{\substack{i+j=k-1\\ i,j\geqslant 0}}(a(N-n-i)+\beta(m-j))\geqslant k(n-m+k).$$

但

$$\sum_{\substack{0\leqslant i\leqslant N-n-1\\ 0\leqslant j\leqslant m-1}}(\alpha(N-n-i)+\beta(m-j))=k(n-m+k).$$

故得 (C) 式. 因之在 (C) 不满足时, 有

$$([\beta]\times[\alpha])\cdot[S_k]=0.$$

其次设 α,β 满足 (C), 于是有

$$\alpha(N-n-i)+\beta(m-j)\begin{cases}\geqslant n-m+k, & i+j<k-1\text{时},\\ =\pi-m+k, & i+j=k-1\text{时};\end{cases}$$

$$\dim(A_{N-n-i}\cap B_{m-j})\begin{cases}\geqslant k-i-j, & i+j<k-1\text{时},\\ =1, & i+j=k-1\text{时},\\ =0, & i+j>k-1\text{时}.\end{cases}$$

今取 $v_i\neq 0, v_i\in A_{N-n-k+i}\cap B_{m+1-i}, 1\leqslant i\leqslant k$. 可证诸 v_i 线性无关. 否则设 $v_1,\cdots,v_{i-1}$ 线性无关而 v_i 线性依赖于 $v_1,\cdots,v_{i-1}$ 时, 将有

$$A_{N-n-k+i}\cap B_{m+1-i}=A_{N-n-k+i-1}\cap B_{m+1-i},$$

而二者维数一为 1, 一为 0. 同样诸 v_i 与 A_{N-n-k},B_{m-k} 线性无关 (即诸 v_i 的任意线性组合都不在 A_{N-n-k} 中或 B_{m-k} 中).

今设

$$(R_0^m,R_0^{N-n})\in([\beta]^*\times[\alpha]^*)\cap S_k,$$

则如前有

$$\dim(R_0^m \cap R_0^{N-n} \cap A_{N-n-k+i} \cap B_{m+1-i}) \geqslant k-(k-i)-(i-1)=1.$$

故 $R_0^m \cap R_0^{N-n}$ 应含有 $v_1, \cdots, v_k$. 又

$$R_0^{N-n} \supset A_{N-n-k},$$
$$R_0^m \supset B_{m-k}.$$

故 R_0^{N-n} 恰由 A_{N-n-k} 与诸 v_i 所张成, 而 R_0^m 恰由 B_{m-k} 与诸 v_i 所张成. 换言之, $([\beta]^* \times [\alpha]^*) \cap S_k$ 恰含有一个元素 (R_0^m, R_0^{N-n}).

在 (R_0^m, R_0^n) 附近选取局部坐标时, 可验证 $[\beta] \times [\alpha]$ 与 $[S_k]$ 在 (R_0^m, R_0^{N-n}) 处相交指数为 1. 由此即得 $n \geqslant m$ 时的引理.

其次设 $m \geqslant n$, 则

$$S_k = \{(R^m, R^{N-n})/\dim(R^m \cap R^{N-n}) \geqslant m-n+k\}.$$

故在以上论证中易 k 为 $m-n+k$ 即得. 证毕.

附录　格拉斯曼流形的切丛

设 C^N 为复数域上的 N 维向量空间, $G_n^N(C)$ 为 C^N 中一切复 n 维子向量空间所成的格拉斯曼流形, $\tau = \tau_{n,N}$ 为其切丛, $\xi = \xi_{n,N}$ 与 $\eta = \eta_{n,N}$ 各为 $G_n^N(C)$ 上的标准 C^n 丛与 C^{N-n} 丛. 项武忠与 Szczarba (*Amer, J.Math.*, 1964, 86: 698—704) 最早给出了 τ 与 ξ, η 间的某些关系式, Porteous 则在 [5] 中证明了下面的公式 ($*$ 表丛的对偶)

$$\tau \approx \xi^* \otimes \eta, \tag{1}$$

并用之以证明 S_k 型奇点的同调类公式. 我们将指出, 上述公式 (I) 实质上早已隐含于陈省身 1946 的工作中 (*Annals of Math.*, 47: 85—121), 说明如下.

对 $G_n^N(C)$ 的一个坐标邻域 U 在 C^N 中取一向量组 $\{e_A, A=1, \cdots, N\}$ 的复解析族使

1° $e_A \cdot \bar{e}_B = \delta_{AB}, 1 \leqslant A, B \leqslant N$;

2° $e_1, \cdots, e_n$ 所定子向量空间 $\in U \subset G_n^N(C)$;

3° e_A 复解析依赖于 U 中的复坐标.

依陈文, 置 $(1 \leqslant A, B \leqslant N)$

$$\theta_{AB} = de_A \cdot \bar{e}_B \text{ 或 } de_A = \sum_B \theta_{AB} e_B,$$

则 nN 个外微分式

$$\theta_{ir}, \quad 1 \leqslant i \leqslant n, \quad n+1 \leqslant r \leqslant N,$$

在 U 中构成 $G_n^N(C)$ 在各点对偶切空间中的基. 考虑两个坐标邻域间向量组族的变换

$$e_i' = \sum_{j=1}^{n} a_{ij} e_j, \quad 1 \leqslant i \leqslant n,$$

$$e_r' = \sum_{s=n+1}^{N} b_{rs} e_s, \quad n+1 \leqslant r \leqslant N,$$

其中 $\alpha = (a_{ij})$, $\beta = (b_{rs})$ 都是酉矩阵, 代表从 ξ, η 的转换函数. 由陈文公式 (9), (10) 得 θ_{ir} 的相应变换为

$$\theta_{ir}' = \Sigma a_{ij} \bar{b}_{rs} \theta_{js},$$

即 $G_n^N(C)$ 的对偶切丛 τ^* 的转换函数为

$$\gamma = (a_{ij} \bar{b}_{rs}) = \alpha \otimes \bar{\beta} = \alpha \otimes^{tr} \beta^{-1}.$$

因而 $G_n^N(c)$ 的切丛的转换函数为

$$\gamma^* = \alpha^* \otimes \bar{\beta}^* = \alpha^* \otimes \beta,$$

即 (I) 式 (参阅 F.Hirzelruch. *Topological Methods in Algelraic Geometry* 一书).

参考文献

[1] Borel A, Haefliger A. La classe d'homologie fondamermtale d'un espace analytique. *Bull. Soc. Math.France*, 1961, 89: 461-513.

[2] Haefliger A, Kosinski A. Un théorème de Thom sur les singularités des applications différentiables. *Sem.Cartan*, 1956/7.

[3] Thom R. Les singularitès des applications différentiables. *Ann.Inst.Fouruèr*, 1955/6, 6: 43-87.

[4] Wu Wen-tsün. Sur les classes caractéristiques des structures fibrées sphériques. *Act.Sci. Ind.*, 1952: 1183.

[5] Porteous I R. Simple singularities of maps. *Proc. Liverpool Singularities-Symposinm*, 1970, 1: 286-307.

[6] Ronga F. Le calcul des classes duales aux singularitès de Boardman d'ordre 2. *Comm. Math.Helv.*, 1972, 47: 15-35.

46. On Universal Invariant Forms*

§1 Introduction

The concept of integral invariant or invariant form has already had a fruitful effect on the theory and application of mechanical system. In 1974, Mr, H. C. Lee (李华宗) ([6]) in our country introduced also the concept of universal integral invariant or universal invariant form for the hamilton system, and proved that besides those already discussed by poincaré and F. Cartan, there is no other such invariant forms. Those universal invariant forms under Lee's meaning, can be generalized to following more general understanding. Let M be a space of n variables, G be an infinite transformation group on M under E. Cartan's meaning, a vector field X on M will be called belonging to G. If the transformation in the local single parametric group produced by x only needs to be sufficiently close to the identity transformation then it belongs to G. We denote the collection of these vector fields as $\mathscr{L}G$. Accordingly an exterior differential form θ on M will be defined as an universal invariant form of G (In the following we abbreviate it as the universal invariant form). If for any vector field $X \in \mathscr{L}G$, the $\mathscr{L}ic$ derivative of θ along x : $\mathscr{L}_x\theta = 0$. When M is a symplectic manifold and g is an infinite transformation group formed by all symplectic transformatioms on M, the universal invariant form of G is similar to that defined by H. C. Lee.

E. Cartan has pointed out that there are six classes of primitive infinite transformation groups, where four classes are single (see 1°, 2°, 4°, 6° in the following). Cartan's results up to now still **have not** been proved, we list them as follows:

1° The group G_n^{I} formed by all transformations on n variables.

2° The group G_n^{II} formed by all transformations preserving the volume element

$$\theta = dx_1 \Lambda \cdots \Lambda dx_n.$$

Invariant on n variables $x_1, \cdots, x_n$.

---

* *Acta Math. Sinica*, 1975, 18: 263-273.

3° The group G_n^{III} formed by all transformations which varies only a non-zero constant factor of the above mentioned volume element θ on n variables $x_1, \cdots, x_n$.

4° The group G_n^{IV} formed by all regular transformations preserving the form

$$Q = dp_i \Lambda dq_i \, (= dp_1 \Lambda dq_1 + \cdots + dp_n \Lambda dq_n) \, .$$

Invariant on $2n$ variables $p_1, \cdots, p_n, q_1, \cdots, q_n (n > 2)$.

5° The group G_n^{V} formed by all transformations on $2n$ variables $p_i, q_i (i=1, \cdots, n; n \geqslant 2)$ which vary only a non-zero constant factor of the above mentioned form Q.

6° The group G_n^{VI} formed by all transformations preserving the form

$$\omega = dt + p_i dq_i - q_i dp_i$$
$$(= dt + p_1 dq_1 + \cdots + p_n dq_n - q_1 dp_1 - \cdots - q_n dp_n).$$

Invariant on $2n + 1$ variables $t, p_i, q_i (i = 1, \cdots, n)$.

This paper will determine the universal invariant form of these above mentioned infinite groups. The group G_n^{IV} in 4° is what H. C. Lee exploring. The case of 1° is insignificant, the groups in 3° and 5° are subgroups of the groups in 2° and 4° respectively, therefore its exploration may be concluded trivially to the latter. Hence we need to explore only G_n^{II} in 2° and G_n^{VI} in 6° (see §2 and §4).

Many conservative laws in mechanics reflect a certain symmetry of the mechanical system. They can be expressed by using the concept of universal invariant forms of certain subgroups in G_n^{IV}, these subgroups are formed by all transformations preserving a group of function that is the so-called "moved constants". The complete determination of the universal invariant forms of these subgroups is equivalent to the determination of the corresponding conservative law of the system (see §5).

The function, vector field and form etc. Mentioned in this paper belong to C^∞. More accurately, they should be treated as sprout bundle section of function, vector field, the form. Similarly, the socalled transformation also implies the local homeomorphism C^∞ transformation, the transformation group is the so-called pseudo-group. But since we consider only the problems of local property on the whole, hence sprout bundle, pseudo-group such vocabularies are not very necessary and have not been used in this paper.

In many formulas, in accordance with the custom of differential geometry overlapping exponent indicate to take sum, we only write the sigma sign Σ and its range of taking sum clearly when the range of indices may cause confusions, otherwise neglected.

§2 Inifinite group of volumn-preserving transformations

Let G_n^{II} be the single infinite group of type II formed by all transformations preserving the volume element

$$\theta = dx_1 \Lambda \cdots \Lambda dx_n.$$

Invariant on n variables $x_1, \cdots, x_n$. For any vector field $X = X^i \dfrac{\partial}{\partial x_i}$, we have

$$\mathscr{L}_X \theta = \frac{\partial X^i}{\partial x_i} \theta.$$

Therefore the necessary and sufficient condition for $X \in \mathscr{L} G_n^{\mathrm{II}}$ is

$$\operatorname{div} X = \frac{\partial X^i}{\partial x_i} = 0.$$

Hence $\mathscr{L}\, G_n^{\mathrm{II}}$ contains the following $\dfrac{1}{2}\, n\,(n+1)$ special vector fields :

$$A_i = \frac{\partial}{\partial x_i}, \quad i = 1, \cdots, n.$$

$$A_{ij} = x_j \frac{\partial}{\partial x_i}, \quad i, j = 1, \cdots, n;\ i \neq j.$$

Now let

$$\theta = a_{i_1 \cdots i_r} dx_{i_1} \Lambda \cdots \Lambda dx_{i_r}$$

Be an universal invariant form of G_n^{II}, where $a_{i_1} \cdots {}_{i_r}$ are functions of $x_1, \cdots, x_n$, anti-symmftric for the lower index $i_1, \cdots, i_r$. Accordingly for any $X \in G_n^{\mathrm{II}}$, particularly for $X = A_i$ or A_{ij} we have $\mathscr{L}_X \theta = 0$. We have computed

$$\mathscr{L}_{A_i} \theta = \frac{\partial a_{i_1 \cdots i_r}}{\partial x_i} dx_{i_r} \Lambda \cdots \Lambda dx_{i_r},$$

$$\mathscr{L}_{Aij} \theta = x_i \frac{\partial a_{i_1 \cdots i_r}}{\partial x_i} dx_{i_1} \Lambda \cdots \Lambda dx_{i_r} + f a_{i i_1 \cdots i_{r-1}} dx_i \Lambda dx_{i_1} \Lambda \cdots \Lambda dx_{i_{r-1}}.$$

By $\mathscr{L}_{A_i} \theta = 0$ we obtain $\dfrac{\partial a_{i_1 \cdots i_r}}{\partial x_i} = 0$, hence $a_{i_1 \cdots i_r}$ are constants. If $r = n$ then $\theta = n! a_{1 \cdots n} dx_1 \Lambda \cdots \Lambda dx_n$ is a constant multiple of θ. Let $r < n$. Then for any r indices $(i,\ i_1, \cdots,\ i_{r-1})$ such that $i < i_1 < \cdots < i_{r-1}$ we can take any index j

not equal to i and $i_1, \cdots, i_{r-1}$, for this pair (i, j) the condition $\mathscr{L}_{Aij}\theta = 0$ gives $a_{ii_1\cdots i_{r-1}} = 0$. by this we obtain $\theta = 0$ and have the following

Theorem The unique universal invariant form of the infinite group $G_n^{\rm II}$ of type ii is a constant multiple of the volume variable θ.

§3 Regular transformation infinite group ——H. C. Lee's theorem

Let $G_n^{\rm IV}$ be the single infinite group of type IV formed by all regular transformations preserving the symplectic form

$$Q = dp_i \Lambda dq_i \tag{1}$$

invariant on $2n$ variables p_i, $q_i (i = 1, \cdots, n)$. H.C.Lee has proved the following described

Theorem ([6], 1947) The unique universal invariant form of the infinite group $G_n^{\rm iv}$ is a constant multiple of Q and its outer power (外乘幂) $Q^2 = Q\Lambda Q$, $Q^3 = Q^2\Lambda Q, \cdots, Q^n = Q^{n-1}\Lambda Q$.

Since this theorem and the computation in its proof is needed to use in future, therefore we repeat it according to the form a bit different from the original paper as follows.

Showing the indices of range $1, \cdots, n$ with the latin letters $i, j, k, \cdots$ and showing the indices of range $1, 2, \cdots, 2n$ with the greek letters $\alpha, \beta, \lambda, \mu, \cdots$. We introduce the new variable as follows

$$x^i = q_i, \quad x^{n+i} = p_i. \tag{2}$$

Also we show the anti-symmetric matrix which are the inverses of each others with $\varepsilon^*, \varepsilon_*$:

$$\varepsilon^* = (\varepsilon^{\alpha\beta}) = \begin{pmatrix} 0 & I \\ -I & 0 \end{pmatrix}, \tag*{(3)*}$$

$$\varepsilon_* = (\varepsilon_{\alpha\beta}) = \begin{pmatrix} 0 & -I \\ I & 0 \end{pmatrix}, \tag*{$(3)_*$}$$

where o and I are the zero matrix and identity matrices of order n respectively, accordingly Q recomes

$$Q = \frac{1}{2}\,\varepsilon_{\alpha\beta} dx^\alpha \Lambda dx^\beta. \tag{4}$$

With respect to any vector field

$$X = X^\lambda \frac{\partial}{\partial x^\lambda},$$

where X^λ is a function of x^a, there is

$$\mathscr{L}_X Q = \frac{1}{2}\varepsilon_{\alpha\beta}(dX^a \Lambda dx^\beta + dx^a \Lambda dX^\beta) = -d\left(\varepsilon_{a\beta} X^\beta dx^\alpha\right).$$

Therefore the necessary and sufficient condition of $x \in \mathscr{L} G_n^{\text{iv}}$ is

$$d(\varepsilon_{\alpha\beta} X^\beta dx^\alpha) = 0.$$

Since we only consider within a local range, therefore from poincaré's lemma the condition becomes that there is a function H determined to a constant such that

$$\varepsilon_{\alpha\beta} X^\beta dx^\alpha = dH,$$

or $\varepsilon_{\alpha\beta} X^\beta = \dfrac{\partial H}{\partial x^\alpha}$, $X^\lambda = \varepsilon^{\lambda\mu}\dfrac{\partial H}{\partial x^\mu}$. By this we obtain the following

Lemma The necessary and sufficient condition for $X \in \mathscr{L} G_n^{\text{IV}}$ is that there is a function H such that

$$X = \varepsilon^{\lambda\mu} \frac{\partial H}{\partial_x^\mu} \frac{\partial}{\partial x^\lambda}. \tag{5}$$

X is determined in this lemma by H uniquely, in the future we will denote it as X_H. Conversely, H determines a constant by $X \in \mathscr{L} G_n^{\text{IV}}$ uniquely, in future we will denote a difference of -constants as H_X.

Now let any form of -degree r $(1 \leqslant r \leqslant 2n)$:

$$\theta = A_{\alpha_1 \cdots \alpha_r} dx^{\alpha_1} \Lambda \cdots \Lambda dx^{a_r}, \tag{6}$$

where $A_{\alpha_1 \cdots \alpha_r}$ are functions of x^λ and anti-symmetric for the lower indices $\alpha_1, \cdots, \alpha_r$. For any

$$X = \varepsilon^{\lambda\mu} \frac{\partial H}{\partial x^\mu} \frac{\partial}{\partial x^\lambda} \in \mathscr{L} G_n^{\text{IV}},$$

there is

$$\mathscr{L}_X \theta = \left[\varepsilon^{\lambda\mu} \frac{\partial A_{\alpha_1 \cdots \alpha_r}}{\partial x^\lambda} \frac{\partial H}{\partial x^\mu} + \delta^\nu_{\alpha i} \varepsilon^{\lambda\mu} J^i_\lambda A_{\alpha_1 \cdots \alpha_r} \frac{\partial^2 H}{\partial x^\mu \partial x^\nu}\right] dx^{\alpha_1} \Lambda \cdots \Lambda dx^{\alpha_r},$$

where J^i_λ means to change the i-th index α_i in $\alpha_{\alpha_1} \cdots_{\alpha_r}$ to the operator of λ:

$$J^i_\lambda A_{\alpha_1 \cdots \alpha_i \cdots \alpha_r} = A_{\alpha_1 \cdots \alpha_{i-1} \lambda \alpha_{i+1} \cdots \alpha_r}, \tag{7}$$

the necessary and sufficient condition for the form θ is to an universal invariant form is that for any H hence for any $\frac{\partial H}{\partial x^{\mu}}$ and $\frac{\partial^2 H}{\partial x^{\mu}\partial x^{v}}$ we should have $\mathscr{L}_X\theta = 0$. By this we obtain the following H.C.Lee's system of equations:

$$\begin{cases} \varepsilon^{\lambda\mu}\dfrac{\partial A_{\alpha_1\cdots\alpha_r}}{\partial x^{\lambda}} = 0, \\ \left(\delta^{v}_{ai}\varepsilon^{\lambda\mu} + \delta^{\mu}_{ai}\varepsilon^{\lambda v}\right) J^{i}_{\lambda} A_{\alpha_1\cdots\alpha_r} = 0. \end{cases} \tag{8}$$

By the front part of the system of equations, all $A_{\alpha_1}\cdots_{\alpha_r}$ are constants. By the rear part (where $\mu,\ v, \alpha_1,\ \cdots,\ \alpha_r$ are arbitrary). H. C. Lee uses pure algebraic method to obtain that when r is an even number 2_s,

$$\theta = c\cdot \underbrace{\Omega\Lambda\cdots\Lambda\Omega}_{s},$$

where c is a constant, and when $r =$ odd number

$$\theta = 0.$$

By this Lee has proved his theorem.

§4 Trangential transformation infinite group

Let G_n^{VI} be the single infinite group of type vi formed by all tangential transformations preserving the form

$$\omega = dt + p_i dq_i - q_i dp_i.$$

Invariant on $2n+1$ variarbes $t, p_i, q_i (i = 1,\ \cdots,\ n)$. Applying the similar symbols in §3 we may write ω as

$$\omega = dt + \varepsilon_{\alpha\beta}x^{\alpha}dx^{\beta}, \tag{1}$$

hence

$$d\omega = \varepsilon_{\alpha\beta}dx^{\alpha}\Lambda dx^{\beta} = 2\Omega,$$

here Ω is the same as (4) of §3.

Consider any vector field

$$X = T\frac{\partial}{\partial t} + X^{\lambda}\frac{\partial}{\partial x^{\lambda}},$$

where T, X^{λ} are functions of x^{a} and t, setting

$$K = \frac{1}{2}\left(T + \varepsilon_{\mu\beta}x^{a}X^{\beta}\right), \tag{2}$$

$$K_\lambda = \frac{\partial K}{\partial x^\lambda}, \quad K_{\lambda\mu} = K_{\mu\lambda} = \frac{\partial^2 K}{\partial x^\lambda \partial x^\mu}, \tag{3}$$

then we have

$$\mathscr{L}_{x}\omega = 2\frac{\partial K}{\partial t}dt + 2\left(\frac{\partial K}{\partial x^\lambda} + \varepsilon_{\alpha\lambda}X^\alpha\right)dx^\lambda.$$

By $\mathscr{L}_X\omega = 0$ we obtain $\frac{\partial K}{\partial t} = 0$ or K is independent of t and $X^a = \varepsilon^{\alpha\lambda}K_\lambda$, hence

$$X = (2K - x^\alpha K_\alpha)\frac{\partial}{\partial t} + \varepsilon^{\alpha\lambda}K_\lambda\frac{\partial}{\partial x^\alpha}. \tag{4}$$

Its inverse is obviously true therefore we obtain the following

Lemma　The necessary and sufficient condition for $x \in \mathscr{L}G_n^{\mathrm{IV}}$ is that X possesses the representation of (4), where K is any function of x^λ but independent of t.

Now we write any form of degree r as

$$\theta = \varphi + \psi \Lambda dt, \tag{5}$$

$$\varphi = A_{\alpha_1}\cdots_{\alpha_r}dx^{\alpha_1}\Lambda\cdots\Lambda dx^{\alpha_r}, \tag{5$'$}$$

$$\psi = B_{\beta_1}\cdots_{\beta_{r-1}}dx^{\beta_1}\Lambda\cdots\Lambda dx^{\beta_r}, \tag{5$''$}$$

where all A and B are functions of x^j and t, and anti-symmeyric for the lower index, for the X determined by (4), we directly compute and obtain

$$\mathscr{L}_X\theta = \tilde{A}_{\alpha_i\cdots\alpha_r}dx^{\alpha_1}\Lambda\cdots\Lambda dx^{\alpha_r} + \tilde{b}_{\beta_i\cdots\beta_{r-1}}dx^{\beta_1}\Lambda\cdots\Lambda dx^{\beta_{r-1}}\Lambda dt, \tag{6}$$

$$\begin{aligned}\tilde{A}_{\alpha_i\cdots\alpha_r} = {}& 2K\frac{\partial A_{\alpha_i\cdots\alpha_r}}{\partial t} \\ & - \left(x^\lambda\frac{\partial A_{\alpha_i\cdots\alpha_r}}{\partial t} + \varepsilon^{\lambda\mu}\frac{\partial A_{\alpha_i\cdots\alpha_r}}{\partial x^\mu} - (-1)^{r-j}\delta^\lambda_{\alpha j}B_{\alpha_1\cdots\alpha_j\cdots\alpha_r}\right)K_\lambda \\ & + \left(\delta^v_{\alpha i}\varepsilon^{\lambda\mu}J^i_\lambda A_{\alpha_i\cdots\alpha_r} - (-1)^{r-j}\delta^v_{\alpha_j}x^\mu B_{\alpha_1\cdots\hat{\alpha}j\cdots\alpha_r}\right)K_{\mu v},\end{aligned} \tag{6$'$}$$

$$\begin{aligned}\tilde{B}_{\beta_1\cdots\beta_{r-1}} = {}& 2K\frac{\partial B_{\beta_1\cdots\beta_{r-1}}}{\partial t} \\ & - \left(x^\lambda\frac{\partial B_{\beta_1\cdots\beta_{r-1}}}{\partial t} + \varepsilon^{\lambda\mu}\frac{\partial B_{\beta_1\cdots\beta_{r-1}}}{\partial x^\mu}\right)K_\lambda \\ & + \delta^v_{\beta_j}\varepsilon^{\lambda\mu}j^j_\lambda B_{\beta_1\cdots\beta_{r-1}}K_{\mu v}.\end{aligned} \tag{6$''$}$$

Because of K, K_λ and $K_{\mu v}$ can be selected arbitrarily, therefore the necessary and sufficient condition for θ to be the universal invariant form of group G_n^{vi} is that the following equalities hold:

$$\frac{\partial B_{\beta_1\cdots\beta_{r-1}}}{\partial t}=0,\quad \varepsilon^{\lambda\mu}\frac{\partial B_{\beta_1\cdots\beta_{r-1}}}{\partial x^\mu}=0, \tag{7}$$

$$\left(\delta^v_{\beta_j}\varepsilon^{\lambda\mu}+\delta^\mu_{\beta_j}\varepsilon^{\lambda v}\right)J^j_\lambda B_{\beta_1\cdots\beta_{r-1}}=0, \tag{7$'$}$$

$$\frac{\partial A_{\alpha_1\cdots\alpha_r}}{\partial t}=0, \tag{8}$$

$$\varepsilon^{\lambda\mu}\frac{\partial A_{\alpha_1\cdots\alpha_r}}{\partial x^\mu}-(-1)^{r-j}\delta^\lambda_{\alpha_j}b_{\alpha_1\cdots\hat{\alpha}_j\cdots\alpha_r}=0, \tag{8$'$}$$

$$\begin{aligned}&\left(\delta^v_{\alpha j}\varepsilon^{\lambda\mu}+\delta^\mu_{\alpha j}\varepsilon^{\lambda\mu}\right)j^i_\lambda A_{\alpha_1\cdots\alpha_r}\\&-(-1)^{r-j}\left(\delta^v_{\alpha_j}x^\mu+\delta^\mu_{\alpha_j}x^v\right)B_{\alpha_1\cdots\hat{\alpha}_j\cdots\alpha_r}=0.\end{aligned} \tag{8$''$}$$

By (7) we know all B are constants and by (7)$'$ such as the proof of Lee's theorem in §3 we know (b = constant)

$$\psi=\begin{cases} b\cdot(d\omega)^s, & r=\text{ odd } 2s+1,\\ 0, & r=\text{even}.\end{cases} \tag{9}$$

Now set

$$A'_{\alpha_1\cdots\alpha_r}=A_{\alpha_1\cdots\alpha_r}-(-1)^{r-j}\varepsilon_{r\alpha_j}x^r B_{\alpha_1\cdots\alpha_j\cdots\alpha_r}, \tag{10}$$

by direct computation we know that (8), (8)$'$, (8)$''$ can be changed to the following equalities respectively:

$$\frac{\partial A'_{\alpha_1\cdots\alpha_r}}{\partial t}=0, \tag{11}$$

$$\varepsilon^{\lambda\mu}\frac{\partial A'_{\alpha_1\cdots\alpha_r}}{\partial x^\mu}=0, \tag{11$'$}$$

$$\left(\delta^v_{\alpha_i}\varepsilon^{\lambda\mu}+\delta^\mu_{\alpha_i}\varepsilon^{\lambda v}\right)J^i_\lambda A'_{\alpha_1\cdots\alpha_r}=0. \tag{11$''$}$$

Still similar to the proof of Lee's theorem in §3 we know that there is a constant a such that:

$$\varphi'=A'_{\alpha_1\cdots\alpha_r}dx^{\alpha_1}\Lambda\cdots\Lambda dx^{\alpha_r}=\begin{cases} a\cdot(d\omega)^s, & r=\text{even } 2s,\\ 0, & r=\text{odd}.\end{cases} \tag{12}$$

First let $r = 2s$ be an even number, then by (0) we obtain $\psi = 0$ and (10) becomes $A'_{\alpha_1}\cdots_{\alpha_r} = A_{\alpha_1}\cdots_{\alpha_r}$. Hence by (12) we obtain

$$\theta = \varphi' = a \cdot (d\omega)^s, \quad r = 2s. \tag{13}$$

Next let $r = 2s + 1$ be an odd number. Tnen $\varphi' = 0$ and (10) gives

$$A_{\alpha_1}\cdots_{\alpha_r} = (-1)^{j+1}\varepsilon_{r\alpha j}x^r B_{\alpha_1}\cdots_{\alpha_j}\cdots_{\alpha_r}.$$

By (5)′, (5)″ and (1) we obtain

$$\begin{aligned}\varphi &= A_{\alpha_1}\cdots_{\alpha_r}dx^{\alpha_1}\Lambda\cdots\Lambda dx^{\alpha_r}\\ &= (-1)^{j+1}\varepsilon_{r\alpha j}x^r B_{\alpha_1}\cdots_{\alpha_j}\cdots_{\alpha_r}dx^{\alpha_1}\Lambda\cdots\Lambda dx^{\alpha_r}\\ &= \varepsilon_{x\alpha j}x^r dx^{\alpha}i\Lambda B_{\alpha_1}\cdots_{\alpha_j}\cdots_{\alpha_r}dx^{\alpha_1}\Lambda\cdots\widehat{\Lambda dx^{\alpha}}j\Lambda\cdots\Lambda dx^{\alpha_r}\\ &= (\omega - dt)\,\Lambda\phi.\end{aligned}$$

By (5) and (9) we obtain

$$\begin{aligned}\theta &= (\omega - dt)\,\Lambda\psi \;+\; \psi\Lambda dt = \omega\Lambda\psi\\ &= b\cdot\omega\Lambda(d\omega)^s, \quad r = 2s + 1.\end{aligned}$$

To sum up, we have the following

Theorem The unique universal invariant form of the infinite group G_n^{VI} is a constant multiple of $\omega\Lambda(d\omega)^s$ and $(d\omega)^s(s = 0,\ 1,\ 2,\ \cdots,\ n)$.

§5 Regular transformation infinite group possessing dffinite symmetry

Still using the notations in §3 we consider the infinite group G_n^{IV} formed by all regular transformations oresfrving the form

$$\Omega = dp_i\Lambda dq_i = \frac{1}{2}\varepsilon_{\alpha\beta}dx^{\alpha}\Lambda dx^{\beta} \tag{1}$$

invariant on $2n$ variables $p_i,\ q_i(i = 1,\ \cdots,\ n)$. Let $\mathscr{G}$ be a given Lie's group acting on the phase space M on the right, that there is a mapping $\Phi : M \times \mathscr{G} \to M$ such that for any $x \in M,\ g \in \mathscr{G}$, setting $\Phi\,(x,\ g) = \Phi_x(g) = \Phi_g(x) \in M$, we have $\Phi_g \in G_n^{\mathrm{IV}}$ and $\Phi_{gg'} = \Phi_{g'}\Phi_g(g,\ g\prime \in \mathscr{G})$. We also let Φ_g be non-degenerate when g is not an identity element e in $\mathscr{G}$, denote the Lie algebra of $\mathscr{G}$ as g, then for any $a \in g$, regarding α

as a left invariant vector field on $\mathscr{G}$, $Y_\alpha(x) = \Phi_{x*}a(c)$ defines a vector field Y_α on m. By $\Phi_g \in G_n^{\mathrm{IV}}(g \in \mathscr{G})$ we know $Y_a \in \mathscr{G}\, G_n^{\mathrm{iv}}$ or $\mathscr{L}_{Y\dot{a}}\Omega = 0$. Similarly it is easy to prove that corresponding $a \to y_a$ is a Lie's homomorphism from g to $\mathscr{L}\, G_n^{\mathrm{IV}}$. We call the collection denoting $Y_a(a \in g)$ as the Lie algebra $\mathscr{G}$. It is easily known that for any function H, if H is invariant under $\mathscr{G}$, that is for any $g \in \mathscr{G}$, there is $\Phi_g^* H = H$, then for any $Y \in \mathscr{G}$, we have $\mathscr{L}_Y H = 0$, or H is the constant of the motion produced by Y, or H takes similar values on each integral curve produced by Y (refer to, for example, [1]).

For any two functions H, K define the poisson bracket to be

$$(H,K) = \frac{\partial\,(H,K)}{\partial\,(p_i,q_i)} = \frac{1}{2}\varepsilon_{\alpha\beta}\frac{\partial\,(H,K)}{\partial\,(x^\alpha,x^\beta)} = \varepsilon_{\alpha\beta}H_\alpha K_\beta, \tag{2}$$

here $H_a = \dfrac{\partial H}{\partial x^a}$, similar for the others. According to the lemma in §3 from H, K we can determine two vector fields in G_n^{IV}

$$X_H = \varepsilon^{\lambda\mu}H_\mu\frac{\partial}{\partial x^\lambda},\quad X_K = \varepsilon^{\alpha\beta}K_\beta\frac{\partial}{\partial x^\alpha}.$$

Accordingly

$$\begin{aligned}[X_H,X_K] &= \left[\varepsilon^{\lambda\mu}H_\mu\frac{\partial}{\partial x^\lambda},\varepsilon^{\alpha\beta}K_\beta\frac{\partial}{\partial x^\alpha}\right]\\ &= \varepsilon^{\alpha\beta}\varepsilon^{\lambda\mu}\left(H_\mu K_{\beta\lambda}\frac{\partial}{\partial x^\alpha} - K_\beta H_{\alpha\mu}\frac{\partial}{\partial x^\lambda}\right).\end{aligned}$$

Interchanging $(\alpha,\ \beta)$ and $(\lambda,\ \mu)$ in the first term on the right hand side, and α and β in the second term, we then ortain

$$\begin{aligned}[X_H,X_K] =& \varepsilon^{\lambda\mu}\left(\varepsilon^{\alpha\beta}H_\beta K_{\mu\alpha} - \varepsilon^{\beta\alpha}K_\alpha H_{\beta\mu}\right)\frac{\partial}{\partial x^\lambda}\\ =& \varepsilon^{\lambda\mu}(\varepsilon_{\beta\alpha}H_\beta K_\alpha)_\mu\frac{\partial}{\partial x^\lambda},\end{aligned}$$

or

$$[X_{H'},X_K] = X_{(H,K)}.$$

Hence the functiom under the poisson bracket and the vector field under the Lee bracket possess certain dualities.

Now we denotf the collection of invariant functions under $\mathscr{G}$ as $\mathscr{H}$. Since the poisson bracket is invariant under regular transformations. Therefore for arbitrary $g \in \mathscr{L}, \phi_g \in G_n^{\mathrm{IV}}$, we have $(\Phi_g^* H,\ \Phi_g^* K) = \phi_g^*(H,K)$, hence when H, $K \in \mathscr{H}$, we also have $(H, K) \in \mathscr{H}$. Also by the known properties of the poisson bracket. We know that $\mathscr{H}$ becomes a Lie algebra under this bracket. If according to the lemma in §3 we denote the collection of all vector fields X_H corresponding to $H \in \mathscr{H}$ as $\mathscr{K} \subset \mathscr{L}G_n^{\mathrm{IV}}$. Then the Lie algebra formed by $\mathscr{K}$ under the Lie's bracket and the lie algebra formed by $\mathscr{H}$ under the poisson bracket possess the previous described dual property. We will denote the infinite group generated by the regular transformations produced by the vector field X in $\mathscr{K}$ with G_*^{IV}, and denote $\mathscr{K}$ as $\mathscr{L}G_*^{\mathrm{IV}}$.

Our purpose is to determine those forms which are invariant forms under the usual meaning for all H in $\mathscr{H}$, or that is the universal invariant form of G_*^{IV}. Every such universal invariant form corresponds to a consfevative law possessing relative symmetry with $\mathscr{G}$ in physics.

Because of this for any vector field $Y_a, a \in g$ in $\mathscr{G}$, according to the lemma in §3 take a relative function f_a (determine to a constant) such that $Y_a = X_{f_a}$. For $a, b \in g$ by $[Y_a,\ Y_b] = Y_{(a,b)}$ and $[X_{f_a}, X_{f_b}] = X_{(f_a, f_b)}$ we obtain $(f_a,\ f_b) = f_{[a,b]}$ (differs by a constant). Let the whole group of f_a $(a \in g)$ be $\mathscr{C}$, also let the collection after adding all arbitrary functions $F(f_{a_1},\ \cdots,\ f_{a_k}),\ (a_i \in g)$ in $\mathscr{C}$ be $\bar{\mathscr{C}}$, then it is easy to see that $\bar{\mathscr{C}}$ is the Smallest function set possessing the following two properties and $\supset \mathscr{C}$:

1° There is a function basis $f_1,\ \cdots,\ f_m$, the rank of its Jacobi expression $= m$.

2° For any $f', f'' \in \bar{\mathscr{C}}$ we also have $(f', f'') \in \bar{\mathscr{C}}$.

Proof. Take a basis $a_i,\ i = 1,\ \cdots,\ m\,(m = \dim\ \mathscr{G})$ for g, and set $f_{a_i} = f_i$, then f_i satisfies 1° and $(f_i, f_j) = c_{ij}^k f_k$ (differ by a constant). Here c_{ij}^k is the structure constant of g. Accordingly for any $f', f'' \in \bar{\mathscr{C}}, f', f''$ are functions of f_i and has $(f', f'') = \dfrac{\partial f'}{\partial_{fi}} \cdot \dfrac{\partial f''}{\partial_{fj}} \cdot (f_i, f_j) \in \bar{\mathscr{C}}$, that is what we want to prove.

The function set possessing the two properties of 1°, 2° is called a function group (see [2] chapter 9 or [5]§69). By the theory of function group we know that we can take a standard function basis $p_1,\ \cdots,\ \bar{p}_{r+s},\ \bar{q}_1,\ \cdots,\ q_r$ in $\bar{\mathscr{C}}$ and it can be spanned into a function basis $\bar{p}_1,\ \cdots,\ p_n,\ q_1, \cdots, q_{m.}$ on m such that it possesses the following relations (as above, particulariy see [5] theorfm 69.6) :

$$\begin{cases} (\bar{q}_i, \bar{q}_j) = 0, \\ (\bar{p}_i, \bar{p}_j) = 0, \quad (i, j = 1, \cdots,\ n), \\ (\bar{p}_i, \bar{q}_j) = \delta_{ij}. \end{cases} \tag{3}$$

Accordingly $(\bar{p}_i,\ \bar{q}_i)$ can be treated as a group of new coordinate on the $(p_i;\ q_i)$.

Phase space and the transformations from $(p_i,\ q_i)$ to $(\bar{p}_i,\ \bar{q}_i)$; $\in G_n^{\mathrm{iv}}$ and (in the expressiom i is from 1 to n):

$$\Omega = d\bar{p}_i \Lambda d\bar{q}_i.$$

for any $H \in \mathscr{H}$ and $F = F(f_1, \cdots, f_m) \in \bar{\mathscr{C}}$, we have

$$(H, F) = \frac{\partial F}{\partial f_i}(H, f_i) = -\frac{\partial F}{\partial f_i}\mathscr{L}_{Y ai} H = 0. \tag{4}$$

Conversely, if the function H such that for arbitrary $F \in \bar{\mathscr{C}}$ we have $(H, F) = 0$, then for arbitrary $a \in g$, we have $\mathscr{L}_{ya} H = \mathscr{L}_{X_{fa}} H = (f_a, H) = 0$, hence H is invariant under $\mathscr{G}$ or $H \in \mathscr{H}$. By this we know that $\mathscr{H}$ and $\bar{\mathscr{C}}$ form two function groups which are the inverses of each others, and $\mathscr{H}$ has a standard function basis $\bar{p}_{s+1}, \cdots, \bar{p}_a, \bar{q}_{r+s+1}, \cdots, \bar{q}_n$ refer to [2] and [5].

If we write any vector field x in $\mathscr{L}G_n^{\mathrm{IV}}$ as

$$X = \sum_{i=1}^{n} \frac{\partial h}{\partial \bar{p}_i}\frac{\partial}{\partial \bar{q}_i} - \sum_{i=1}^{n} \frac{\partial H}{\partial \bar{q}_i}\frac{\partial}{\partial \bar{p}_i},$$

where H is a certain function of $\bar{p}_i$, $\bar{q}_i$, then when $X \in \mathscr{H} = \mathscr{L}G_*^{\mathrm{IV}}$, we should have

$$\mathscr{L}_X \bar{p}_i = -\frac{\partial H}{\partial \bar{q}_i} = 0, \quad i = 1, \cdots, r+s,$$
$$\mathscr{L}_X \bar{q}_i = \frac{\partial H}{\partial \bar{p}_i} = 0, \quad i = 1, \cdots, r.$$

Summarying the above mentioned, we obtain:

Lemma The necessary and sufficient condition for $X \in \mathscr{H} = \mathscr{L}G_*^{\mathrm{IV}}$ is to have a function H which only depends on

$$\bar{p}_{r+1}, \cdots, \bar{p}_n, \bar{q}_{r+s+1}, \cdots, \bar{q}_n \tag{5}$$

such that

$$X = \sum_{i=r+1}^{n} \frac{\partial H}{\partial \bar{p}_i}\frac{\partial}{\partial \bar{q}_i} - \sum_{i=r+i+1}^{n} \frac{\partial H}{\partial \bar{q}_i}\frac{\partial}{\partial \bar{p}_i}. \tag{6}$$

Introduce the notations

$$\begin{cases} t = n - r - s, \\ y^i = \bar{q}_{r+i}, \\ y^{s+i} = \bar{p}_{r+i}, \qquad (i = 1, \cdots, s) \\ z^i = \bar{q}_{r+s+i}, \\ z^{t+i} = \bar{p}_{r+s+i}. \quad (i = 1, \cdots, t) \end{cases} \tag{7}$$

Take the following range of indices:

$$\begin{cases} a, b, c, \cdots = 1,\ 2,\ \cdots,\ s \\ A, B, C, \cdots = 1,\ 2,\ \cdots,\ 2t. \end{cases} \tag{8}$$

Also take matrix of order $2s$ which are the inverses of each others

$$\eta^* = (\eta^{AB}) = \begin{pmatrix} 0 & I \\ -I & 0 \end{pmatrix},$$

$$\eta_* = (\eta_{AB}) = \begin{pmatrix} 0 & -I \\ I & 0 \end{pmatrix}, \tag{9}$$

where 0 and I are the zero matrix and identity matrices of order t.

Accordingly any $X \in \mathscr{K}$ of expression (6) can be rewritten

$$X = \eta^{AB} \frac{\partial H}{\partial z^b} \frac{\partial}{\partial z^A} + \frac{\partial H}{\partial y^{s+a}} \frac{\partial}{\partial y^a}, \tag{10}$$

where H is an arbitrary function only depending on z^a and y^{s+a}.

Now let θ be any universal invariant form of degree m of G_*^{iv}. Write θ as

$$\theta = \sum_{k \geqslant 0} \sum_{(i)} du_{i_1} \Lambda \cdots \Lambda du_{i_k} \Lambda \theta_{i_1} \cdots {}_{i_k},$$

where all u_i represent one of $\bar{p}_1, \cdots, \bar{p}_r$, $\bar{q}_1, \cdots, \bar{q}_r$ and $\theta_{i_1} \cdots {}_{i_k}$ is a form of dz^a and dy^a, dy^{s+a}, their coefficients are all functions of the variables $\bar{p}_i$, $\bar{q}_i$, for X in expression (10) we obviously have

$$\mathscr{L}_X \theta = \sum_{k \geqslant 0} \sum_{(i)} du_{i_1} \Lambda \cdots \Lambda du_{i_k} \Lambda \mathscr{L}_X \theta_{i_1 \cdots i_k},$$

And $\mathscr{L}_X \theta_{i_1} \cdots {}_{i_k}$ does not contain any differential du, hence by $\mathscr{L}_X \theta = 0$ we obtain $\mathscr{L}_X \theta_{i_1} \cdots {}_{i_k} = 0$ that is all $\theta_{i_1} \cdots {}_{i_k}$ are universal invariant forms.

By this the problem is concluded as determining the universal invariant forms of the following shape (let it be of degree $\bar{m}$)

$$\bar{\theta} = \sum_{k \geqslant 0} \sum_{(i)} d\bar{u}_{i_1} \Lambda \cdots \Lambda d\bar{u}_{i_k} \Lambda \bar{\theta}_{i_1} \cdots {}_{i_k},$$

where $\bar{u}_i$ shows $\bar{p}_{r+1}, \cdots, \bar{p}_{r+s}, \bar{q}_{r+1}, \cdots, \bar{q}_{r+s}$ that is one of y^a, y^{s+a} and $\bar{\theta}_{i_1} \cdots {}_{i_k}$ is a form of degree $\bar{m} - k$ of dz^A, its coefficients are all functions of the variables $\bar{p}_i$, $\bar{q}_i$.

Now we first consider a special case such as $X = X' \in \mathscr{K}$ of expression (10), where H is A_n arbitrary function of z^A, but independent of y^{s+a} :

$$X' = \eta^{AB} \frac{\partial H}{\partial z^B} \frac{\partial}{\partial z^A}. \tag{11}$$

Accordingly

$$\mathscr{L}_{X'}\bar{\theta} = \sum_{k\geqslant 0}\sum_{(i)} d\bar{u}_{i_1}\Lambda\cdots\Lambda d\bar{u}_{i_k}\Lambda\mathscr{L}_{X'}\bar{\theta}_{i_1}\cdots_{i_k},$$

and from $\mathscr{L}_{X'}\bar{\theta} = 0$, similar to the above, we obtain $\mathscr{L}_{X'}\bar{\theta}_{i_1}\cdots_{i_k} = 0$, here X' is to show as in expression (11), besides it is arbitrary. By this we know that $\bar{\theta}_{i_1\cdots i_k}$ is the universal invariant form of all regular transformation infinite groups preserving the forms

$$\Omega_x = \frac{1}{2}\eta_{AB}dz^A\Lambda dz^B = \sum_{i=r+s+1}^{n} d\bar{p}^i\Lambda d\bar{q}^i \tag{12}$$

invariant on $2t$ variables z^A (The variarles $\bar{p}_1,\cdots, \bar{p}_{r+s}, \bar{q}_1, \cdots, \bar{q}_{r+s}$ are regarded as constant parameters here). By H. C. Lee's theorem we obtain

$$\bar{\theta}_{i_1}\cdots_{i_k} = \begin{cases} 0, & \bar{m}-k = \text{odd}, \\ f_{i_1}\cdots_{i_k}(\Omega_x)^{(m-k)/2}, & \bar{m}-k = \text{even}, \end{cases}$$

where $f_{i_1}\cdots_{i_k}$ is a function only depending on $\bar{p}_1, \cdots, \bar{p}_{r+s}, \bar{q}_1, \cdots, \bar{q}_{r+s}$.

Up to now the original problem is concluded further as the problem of determining the universal invariant form of the following shape

$$\varphi = \sum_{h\geqslant 0}\sum_{(a,{}^1b)} g_{a_1}\cdots_{a_k b_1}\cdots_{b_l} dy^{a_1}\Lambda\cdots\Lambda dy^{a_k}\Lambda dy^{s+b_1}\Lambda\cdots\Lambda dy^{s+b_l}\Lambda(\Omega_x)^h,$$

where all g are functions of $\bar{p}_1, \cdots, \bar{p}_{r+s}, \bar{q}_1, \cdots, \bar{q}_{r+s}$, anti-symmetric for the lower index a, also anti-symmferic for the lower index b, and $2h + k + l$ is a fixed integer, that is the degree of φ.

Now take any fixed c among the indices $l, \cdots, s$ also take a special $X = X_c \in \mathscr{K}$ shape as $X_c = \dfrac{\partial H}{\partial y^{s+c}}\dfrac{\partial}{\partial y^c}$. Where H is an arbitrary function depending only on y^{s+c}, accordingly $\mathscr{L}_{X_c}\Omega_z = 0, \mathscr{L}_{X_c}(dy^{s+b}) = 0$, also when $a \neq c$ $\mathscr{L}_{X_c}(dy^a) = 0$, and $\mathscr{L}_{X_c}(dy^c) = \dfrac{\partial^2 H}{\partial(y^{s+c})^2}dy^{s+c}$. By $\mathscr{L}_{X_c}\varphi = 0$ (H is arbitrary), it is easy to know that in φ any term whenever contains dy^c, it must at the same tine contains dy^{s+c} and has the factor $dy^c\Lambda dy^{s+c}$, also g is a function independemt of y^c. Since this holds for any exponential c in $1, \cdots, s$, therefore we can write φ as the following shape:

$$\varphi = \sum_{l\geqslant 0}\sum_{(b)} \dot{\psi}_{b_1}\cdots_{b_l},$$

where every $\psi_{b_1\cdots b_l}$ has the following shape ($a_i \neq b_i$):

$$\psi = dy^{s+b_1}\Lambda\cdots\Lambda dy^{s+b_l}\Lambda\psi',$$

$$\psi' = \sum_{k\geqslant 0} g_{a_1\cdots a_k} dy^{a_1} \Lambda dy^{s+a_1} \Lambda \cdots \Lambda dy^{a_k} \Lambda dy^{s+a_k} \Lambda (\Omega_z)^{c-k}.$$

Here all g are functions depending only on $\bar{p}_1, \cdots, \bar{p}_{r+s}, \bar{q}_1, \cdots, \bar{q}_r$, also symmetric for the lower index α, and the degree of φ' is then an even number, let it be $2e$. When $k=0$, the relative coefficient g in the expression of φ was denoted as g_0.

Now take a general $X \in \mathscr{K}$ given by expression (10), where H are all arbitrary functions of y^{s+a} and z^A, by computation we obtain

$$\mathscr{L}_X \psi = dy^{s+b_1} \Lambda \cdots \Lambda dy^{s+b_c} \Lambda \mathscr{L}_X \psi',$$

$$\mathscr{L}_X \psi' = \sum \tilde{g}_{a_1\cdots a_k, c, A} dy^{a_1} \Lambda dy^{s+a_1} \Lambda \cdots \Lambda dy^{a_k} \Lambda dy^{s+a_k} \Lambda dy^{s+c} \Lambda dz^A \Lambda (\Omega_z)^{c-k-1} + \cdots,$$

$$\tilde{g}_{a_1\cdots a_k, c, A} = -\left[(k+1)\, g_{ca_1\cdots a_k} + (c-k)\, g_{a_1\cdots a_k}\right] \frac{\partial^2 H}{\partial z^A \partial y^{s+c}}.$$

Since for any H we should have $\mathscr{L}_X \varphi = 0$, therefore we obtain ($c, a_i, b_j$ are mutually not equal)

$$g_{ca_1\cdots a_k} = -\frac{c-k}{k+1} \cdot g_{a_1\cdots a_k}.$$

Hence

$$g_{a_1\cdots a_k} = (-1)^k \binom{c}{k} \cdot g_0.$$

By this we obtain

$$\begin{aligned}\psi &= g_0 \cdot dy^B \Lambda \sum_{a \neq b} (-1)^k \binom{c}{k} dy^{a_1} \Lambda dy^{s+a_1} \Lambda \cdots \Lambda dy^{ak} \Lambda dy^{s+a_k} \Lambda (\Omega_x)^{c-k} \\ &= g_0 \cdot dy^B \Lambda \left(\sum_{a \neq b} dy^{s+a} \Lambda dy^a + \Omega_x \right)^c \\ &= g_0 \cdot dy^B \Lambda \left(\Omega - \sum_{i=1}^{r} d\bar{p}_i \Lambda d\bar{q}_i \right)^c.\end{aligned}$$

In the above expression $dy^b = dy^{s+b_1} \Lambda \cdots \Lambda dy^{s+b_l}$.

From the above we know that any universal invariant form θ of G_*^{IV} must be a sum formula, where each term is the outer product of some $d\bar{p}_1, \cdots, d\bar{p}_{r+s}, d\bar{q}_1, \cdots, d\bar{q}_r$ and the quadratic form Ω, and with an arbitrary function depending only on $\bar{p}_1, \cdots, \bar{p}_{r+s}, \bar{q}_1, \cdots, \bar{q}_r$ as its coefficient, in other words, since $\bar{p}_1, \cdots, \bar{p}_{r+s}, \bar{q}_1, \cdots, \bar{q}_r$ is a function basis of $\bar{\mathscr{C}}$, we have already proved the following

Theorem　The universal invariant form of the infinite group G_*^{IV} forms a differential ring produced by functions in $\bar{\mathscr{C}}$ and the form Ω.

Notice that the function in $\bar{\mathscr{C}}$ can completely be determined through simple operations and integrations from $\mathscr{G}$ of the effect to phase space $(p_i,\ q_i)$, hence the conservative law corresponding to a known symmetric group $\mathscr{G}$ can completely be determined through computations.

Bibliography

[1] Abraham R, Marsden J E. Foundations of Mechanics, 1967.

[2] Caratheodony C. Variationsrechnung und partielle differential gleichungen erster ordnung, bd. I, 1956.

[3] Certan E. Lecons sur les Invariants Intégraux, 1958.

[4] Chern S S (陈省身). Pseudo-groupes continus infinis. géomètrie différentielle (colloque cnrs 1953), 119-136.

[5] Eisenhart L P. Continuous groups of Transformations, 1961.

[6] Lee H C (李华宗). The universal integral invariants of hamiltonian systems and application to the theory of canonical transformations. Proc. Roy. Soc. Edinburgh, Sect. A, 1947, 72: 237-246.

47. 代数拓扑的一个新函子*

数学研究现实世界中的空间形式与数量关系, 就代数拓扑而论, 它的主要对象是拓扑空间这一类空间形式. 它的主要工具与方法是使空间在通常所称函子 F 与以数、群、环、代数等表达的数量关系相对应, 并通过这些代数结构来探讨空间的特性与变化. 在已知的函子中, 主要有同伦函子 π_*、同调函子 H_* 与 H^*、K 理论中的 K 函子, 以及 Adams 的 J 函子等. 但一般说来, 这些已知函子的计算都不是容易的, 甚至原则上不可能计算, 即使 H 函子也是如此, 但若限制于实域 $\underline{R}$, 则这些函子就有可能计算出来, 突出的例子如 Hopf 关于 H 空间与 Cartan 关于齐性空间实域上 H^* 函子的确定. 近年来, Sullivan[1] 提出了一种实域 (或一般指针为 0 的域) 上 DGA 代数的极小模型理论, 据此可导出一种包括 π_* 与 H^* 在实域上部分在内的新函子. 在以下, 我们称这一新函子为 I^* 函子, 它的计算比之已知函子的计算要有办法得多, 在不少实域 $\underline{R}$ 上 π 与 H 不能计算的场合, I^* 函子却能计算出来, 本文将所得初步结果作一简单报道.

设 A, B 是实域 $\underline{R}$ 上的 DGA 代数, 即 $\underline{R}$ 上有微分运算的分次反称代数, 简称 A_x 到 B 保持这些代数结构的代数同态为 DGA 同态, 我们只考虑 A, B 等都是连通、单连通与有限型, 即 $H_0 \approx \underline{R}, H_1 = 0, \dim_{\underline{R}} H_r$ 有限的情形. 所有这些代数与同态构成一个范畴, 记作 $\mathscr{A}$. 在 $\mathscr{A}$ 中的代数 A 称为自由的, 如果 A 作为代数除通常的模以及反称结合等关系外并无其他关系. A 称为可分解的, 如果对任一 $a \in A$, 有 $da \in A^+ \cdot A^+ (A^+$ 指 A 中次数> 0 的理想). 自由又可分解的 $A \in \mathscr{A}$ 称为极小的, 这些极小 DGA 代数构成 $\mathscr{A}$ 的一个子范畴, 记作 $\mathscr{M}$. 任一 $M \in \mathscr{M}$ 可记作 $M \approx \Lambda(x_\alpha, d)$, 其中 x_α 为代数产生 M 的基, d 为微分运算. 据 Sullivan, 对任一 $A \in \mathscr{A}$, 有 (确定至同构的)Min $A \in \mathscr{M}$ 与 (确定至同伦的)DGA 同态 $\rho_A : \text{Min } A \to A$ 使 $\rho_{A*} : H(\text{Min } A) \approx H(A)$.

设 K 是可数局部有限且属有限型的连通、单连通单纯复形, 则据 Sullivan 与 De Rham 理论相当, 从 K 上的片段外微分形式可定一 $\underline{R}$ 上 DGA 代数 $A^*_{PL}(K) \in \mathscr{A}$, 且从 Min $A^*_{PL}(K)$ 可定出 $\pi_*(K) \otimes_z \underline{R}$ 与 $H^*(K, \underline{R})$, 特别有 $H(\text{Min } A^*_{PL}(K)) \approx H^*(K, \underline{R})$.

今设 $\mathscr{W}$ 是与上述类型复形具有相同伦型的空间以及连续映像所构成的空间范畴, 而 $\mathscr{W}_0$ 是与 $\mathscr{W}$ 中有限复形具有相同伦型的空间所成的子范畴, 从 Sullivan 的一般理论容易得出下述.

* *Kexue Tongbao*, 1975, 20: 311-312.

基本引理 对任一 $X \in \mathscr{W}$, 与任一同伦于 X 的复形 K, $\mathrm{Min}\ A_{PL}^*(K)$ 的同构型与 K 无关, 记之为 $I^*(K)$, 则有 $H(I^*(X)) \approx H^*(X, \underline{R})$, 且对任意 $X, Y \in \mathscr{W}$ 与连续映像 $f: X \to Y$, 有确定至同伦的 DGA 同态 $f^I: I^*(Y) \to I^*(X)$, 使引出的 $\underline{R}$ 上分次代数同态 $H(I^*(Y)) \to H(I^*(X))$ 即 $f^*: H^*(Y, \underline{R}) \to H^*(X, \underline{R})$.

以下是所得的初步结果综述.

定理 X 设 $X, Y \in \mathscr{W}$, 则

$$I^*(X \times Y) \approx I^*(X) \underset{\underline{R}}{\otimes} I^*(Y).$$

定理 V 设 $X, Y \in \mathscr{W}$, 则

$$I^*(X \vee Y) \approx \mathrm{Min}(I^*(X) \underset{\underline{R}}{\oplus} I^*(Y))$$

($\underset{\underline{R}}{\oplus}$ 指次数>0 处取直和, 次数$=0$ 处即取作 $\underline{R}$).

设 $X \in \mathscr{W}$, 记 X 在一点的闭路空间的通用覆叠空间为 $\tilde{\Omega}X$, 则按 Milnor 有 $\tilde{\Omega}X \in W$. 对 $M = \Lambda(x_\alpha, y_\beta, d) \in \mathscr{M}$, 其中 $\deg x_\alpha > 2$, $\deg y_\beta = 2$, 作 $\tilde{\Omega}M = \Lambda(\xi_\alpha, d)$, 其 $\deg \xi_\alpha = \deg x_\alpha - 1$, 而 d 恒$=0$. 于是有

定理 $\tilde{\Omega}$ 设 $X \in \mathscr{W}$, 则

$$I^*(\tilde{\Omega}X) \approx \tilde{\Omega}(I^*(X)).$$

设 $M = \Lambda(x_\alpha, y_\beta, d) \in \mathscr{M}$, 其中 $\deg x_\alpha =$ 奇数, $\deg y_\beta =$ 偶数, 记形如 $x_{\alpha_1} \cdots x_{\alpha_r} y_{\beta_1}^{\varepsilon_1} \cdots y_{\beta_s}^{\varepsilon_s} (\varepsilon_k \geqslant 1, r+s \geqslant 1)$ 的一切元素为 $\{a_i\}$ 并设 $da_i = \sum \alpha_{ij} a_j$ ($\deg a_j \neq \deg a_i + 1$) 时 $\alpha_{ij} = 0$). 作由 $a_i^{\sum}$ 产生的 $\sum^0 M \in \mathscr{A}$, 使 $\sum^0 M$ 的 0 次部分为 $\underline{R}$, 次数>0 的分次模部分以 $a_i^{\sum}$ 为基, 这里 $\deg a_i^{\sum} = \deg a_i + 1$, $da_i^{\sum} = \sum \alpha_{ij} a_j^{\sum}$, 又基中任两元素相乘都等于 0, 于是有

定理 $\sum$ 若 $X \in \mathscr{W}_0$, 则

$$I^*\left(\sum X\right) \approx \mathrm{Min} \overset{0}{\sum} I^*(X).$$

设纤维丛 $F \subset E \xrightarrow{p} B$, 其中 $F, E, B \in W$, E, B 有剖分且对此 p 是单纯映像, 则丛将称为单式的. 此时有

定理 E 对单式纤维丛 $F \subset E \xrightarrow{p} B$ 在 $I^*(B) \underset{\underline{R}}{\otimes} I^*(F)$ 中可引入扭曲微分运算 d_τ 使所得 DGA 代数记为 $I^*(B) \underset{\tau}{\otimes} I^*(F)$ 时有

$$I^*(E) \approx \mathrm{Min}(I^*(B) \underset{\tau}{\otimes} I^*(F)).$$

设有纤维方

$$\begin{array}{ccc} F\subset Z & \longrightarrow & Y \\ \downarrow & & \downarrow g \\ X & \xrightarrow{f} & B \end{array}$$

其中 $F\subset Y\xrightarrow{g}B$ 是单式纤维丛, $F\subset Z\to X$ 是 f 下的诱导丛, 这时纤维方将称为单式的, 在 f 与 g 下, $f^I:I^*(B)\to I^*(X)$ 与 $g^I:I^*(B)\to I^*(Y)$, 使 $I^*(X)$ 与 $I^*(Y)$ 各成为 $I^*(B)$ 右与左作用的右与左 DGA 代数 $\in\mathscr{A}$, 命 $P(X)\in\mathscr{A}$ 是 $I^*(X)$ 作为 $I^*(B)$ 右 DGA 代数的 Bar 分解, $DP(X)$ 是 $P(X)$ 的全复形. 由 $I^*(B)$ 的作用可定一 $DP(X)\underset{\underline{R}}{\otimes}I^*(Y)$ 的商代数 $DP(X)\underset{I^*(B)}{\otimes}I^*(Y)\in\mathscr{A}$, 则有

定理 □ 对上述单式纤维方有

$$I^*(Z)\approx\mathrm{Min}(DP(X)\underset{I^*(B)}{\otimes}I^*(Y)),$$

特别有

$$H^*(Z)\approx\mathrm{Tor}_{I^*(B)}(I^*(X),I^*(Y)),$$

定理 C 设 G 是紧致李群, U 是 G 的闭子群, 而齐性空间 $G/U=M\in\mathscr{W}$, 若 $C_M\in\mathscr{A}$ 是 M 的 Cartan 代数, 则

$$I^*(M)\approx\mathrm{Min}\,C_M.$$

以上诸定理包括了 Baum, Borel, Brown, Cartan, Dold, Eilenberg-Moore, Hirsch, Kahn, Oniscêk, Smith 等许多有关实域上拓扑的已知定理, 而结果较明确.

以上仅就范畴 $\mathscr{W}$ 或 $\mathscr{W}_0$ 而言, 对于其他空间范畴, 例如仿紧空间, 也可通过 Cartan-Leray 关于甲, 盖, 束的理论引进相应的 I^* 函子并进行类似的讨论.

参考资料

[1] Sullivan D. Differential forms and Topology of Manifolds. *Symp. Tokyo on Topology*, 1973.

48. 代数拓扑 I^* 函子论 —— 齐性空间的实拓扑*

摘要 自古以来, 人们往往以数来刻画形的某些特征, 即所谓形的度量, 如道路的长短、田地的面积、器物的容量, 以致点集的测度等.

在代数拓扑中, 拓扑空间、复形、流形这类复杂的"形"难以用通常的数来描述, 近代数学则引用群、环、代数等"代数系统"来刻画这些复杂的"形". 代数拓扑中, 陆续出现了一些用以度量"形"的"数". 这种度量称为"函子". 现已引进的函子主要有 H, π, J, K 等, 我们最近引进的函子, 记作 I^*. 与传统函子相比较优越之处是"能计算". 经过检验, I^* 函子对主要十几种几何作法都是能计算的, 而传统的函子并非如此.

由于 I^* 函子"能计算", 因此它比传统函子更易于驾驭, 估计会有较广泛的应用. 但是, I^* 函子目前须把"形"局限于单连通, 而"数"只保留了无限部分, 这是不足之处, 也是今后要解决的重要问题.

代数拓扑的主要工具是一些函子, 把某种空间范畴对应为某种代数的范畴, 并通过这些代数结构来考察空间的几何特性与变化, 这些已知函子包括了同伦函子 π_*、同调函子 H_* 或 H^*、K 理论中的 K 函子, 以及 Adams 的 J 函子等. 依据最近 Sullivan 关于极小模型的理论 ([3, 8]), 可以提出一种新函子, 称之为 I^* 函子, 它包括了 π_* 的一部分 $\pi_* \underset{z}{\otimes} R$, 也包括了 H^* 的一部分 $H^*(,R)$, 这里 R 指实域, 下同 (实质上 R 可代以任一指针为 0 的域, 又 $H^*(,R)$ 中的 R 以后将略去不写).

一般说来, 已知函子的计算都不是容易的, 甚至无法计算, 不仅是 π_*, J, K, 即使 H^* 函子也是如此. 但 I^* 函子的计算却有办法得多, 在不少 (实域 R 上) π_* 与 $H*$ 不能计算的场合, I^* 函子却能计算出来. 在本文以及有关论著中, 我们将就各方面阐明这一点.

本文除引入 I^* 函子外, 并对闭路空间与齐性空间作 I^* 的具体计算, 由此以获得 Cartan 与 Oniscêk 的已知定理 ([2, 5, 6]). 后者关于闭路空间 $H^*(,R)$ 以及齐性空间 $\pi_* \underset{z}{\otimes} R$ 的结果, 只通过原来的函子 H^* 与 π_* 来进行, 都得来不易, 但这里用 I^* 函子则一切计算都变成平凡, 且某些假设还可减轻.

---

* *Acta Math. Sinica*, 1975, 18: 162-172.

§1　单连通空间的 I^* 函子

我们将考虑以下一些空间与代数的范畴:

$\mathscr{K}$—— 连通与单连通可数局部有限且为有限型的单纯复形及单纯映像所成的范畴.

$\mathscr{W}$ —— 与 K 中复形同伦等价的空间及连续映像所成的范畴.

$\mathscr{G}$ —— 实域上分次模与保次模同态所成的范畴.

$\mathscr{H}$—— 实域上分次反称代数及相应同态所成的范畴.

设 A 是一个分次反称带有微分运算 d (符合通常规律) 的实域上代数 (简称为 DGA 代数). 若有 $H_0(A,d)=R, H_1(A,d)=0$, 则 A 称为连通与单连通的. 若 A 中的代数运算除通常规律外不再有其他关系, 则 A 称为自由的, 这时 A 必是一若干奇数次生成元 $x_1,x_2,\cdots$ 上的外代数 (记作 $E(x_1,x_2,\cdots)$) 与一偶数次生成元 $y_1,y_2,\cdots$ 上多项式代数 (记作 $R[y_1,y_2,\cdots]$) 的张量积, 即

$$A=E(x_1,x_2,\cdots)\otimes R[y_1,y_2,\cdots].$$

($\otimes$ 指 $\underset{R}{\otimes}$, 下同) 有时也将记作

$$A=F(x_1,x_2,\cdots,y_1,y_2,\cdots).$$

此时以诸 x_i 与 y_j 为基的实域上分次模将记为 G_rA.

设 A 中的微分运算 d 具有下述性质:

$$da\in A^+\cdot A^+,\quad (a\in A)$$

这里 A^+ 是 A 中一切次数 >0 的元素所成的子代数, 则 A 称为可分解的.

一切实域上连通单连通具有有限型的 DGA 代数及保持 DGA 结构的代数同态 (简称 DGA 同态) 所成范畴将记为 $\mathscr{A}$. 由 $\mathscr{A}$ 中一切自由且可分解的代数 (称为极小 DGA 代数) 所成的子范畴将记作 $\mathscr{M}$.

通常实域上同调环的作法给出了范畴间的同调函子

$$H:\mathscr{W}\to\mathscr{H},$$

与

$$H:\mathscr{K}\to\mathscr{H}.$$

对空间 X 的同伦群 $\pi_i(X)$, 记 $\pi^i(X)=Hom(\pi_i(X),R),\pi^*(X)=\Sigma\pi^j(X)$, 则有函子

$$\pi^*:\mathscr{W}\to\mathscr{G},$$

与

$$\pi^* : \mathscr{K} \to \mathscr{G}.$$

依据 Sullivan([8]), 对任一 $A \in a$ 可作一 $\min A \in m$ 与 DGA 同态

$$\rho : \min A \to A$$

使 ρ 引出

$$\rho_* : H(\min A) \approx H(A).$$

这样的 $\min A$ 由 $A \in \mathscr{A}$ 所唯一确定至一同构, $\min A$ 称 A 的极小模型, ρ 称标准同态. 又对 $A, B \in \mathscr{A}$ 以及 DGA 同态 $f : A \to B$ 可作 DGA 同态 $g : \min A \to \min B$ 使下图像 (ρ_A, ρ_B 为标准同态)

(I)
$$\begin{array}{ccc} A & \xrightarrow{f} & B \\ {\scriptstyle\rho_A}\uparrow & & \uparrow{\scriptstyle\rho_B} \\ \min A & \xrightarrow{g} & \min B \end{array} \tag{I}$$

引出可交换图像

(II)
$$\begin{array}{ccc} H(A) & \xrightarrow{f_*} & H(B) \\ {\scriptstyle\rho_{A^*}}\uparrow\approx & & \approx\uparrow{\scriptstyle\rho_{B^*}} \\ H(\min A) & \xrightarrow{g_*} & H(\min B) \end{array} \tag{II}$$

同态 g 由 f, ρ_A, ρ_B 唯一确定至一同伦 (详见 [3]§ 15).

Sullivan 证明对任一 $K \in \mathscr{K}$, 可从 K 上 PL 外形式作一 DGA 代数 $A_{PL}^*(K) \in a$ 使以积分运算引出代数同构

$$h : H(A_{P_L}^*(K)) \approx H^*(K).$$

h 将称为标准同构. 记

$$\min A_{P_L}^*(K) = I^*(K),$$

则 Sullivan 理论指出除

$$H(I^*(K)) \approx H^*(K)$$

外并有

$$G_r(I^*(K)) \approx \pi^*(K).$$

因而有以下函子关系图:

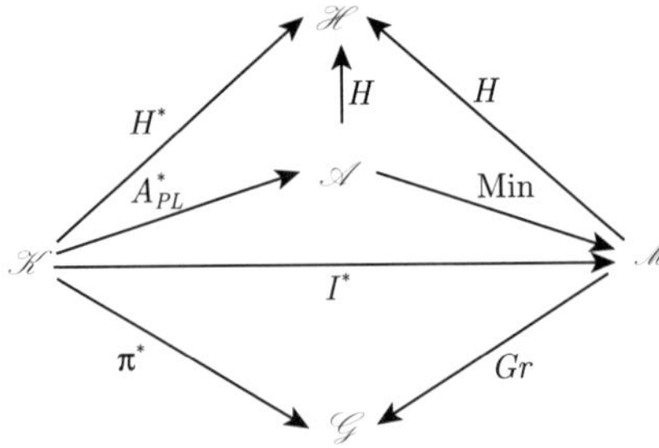

今设 $X \in \mathscr{W}$ 而 x 同伦等价于 $K, L \in K, g : |K| \to |L|$ 使 $g : |K| \simeq |L|$. 作 K 的单纯剖分 K', 恒同映像与 g 的单纯逼近 $s : K' \to K, s' : K' \to L$. 由下可交换图 ($\rho_K, \rho_L$ 是标准同态, s^A, s'^A 各由 s, s' 引出)

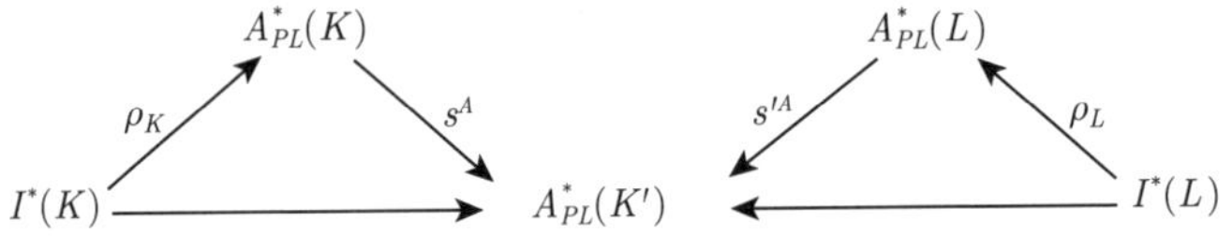

作 H 函子得下交换图

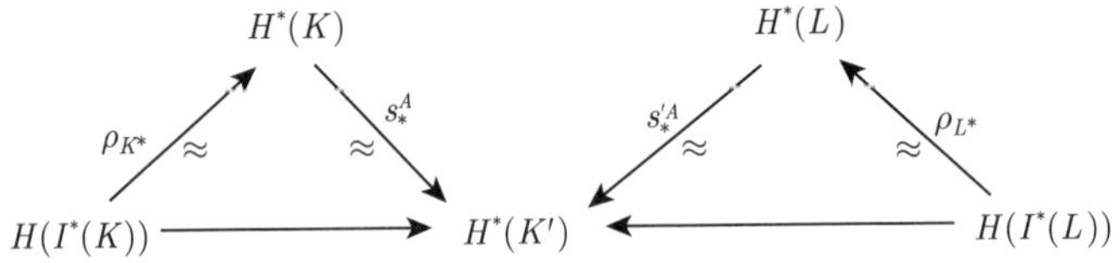

由 $A^*_{PL}(K') \in \mathscr{A}$ 的极小模型的唯一性得

$$I^*(K) \approx I^*(L).$$

因之 $I^*(K)$(差一 DGA 同构) 只依赖于 $X \in \mathscr{W}$ 而可记作 $I^*(X)$. 于是又有以下诸范畴间的函子关系图:

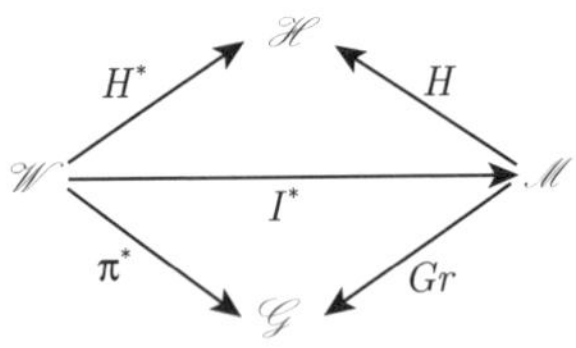

§2　闭路空间的 I^* 函子

设 $x(\in \mathscr{W})$ 是一有限型的空间. 记 X 对一固定参考点的闭路空间为 ΩX, ΩX 的通用覆叠空间为 $\tilde{\Omega} X$, 则据 Milnor 也有 $\tilde{\Omega} X \in \mathscr{W}$. 本节目的在证明 $\tilde{\Omega} X$ 的 I^* 函

子由 X 的 I^* 函子所完全确定.

对任一 $A\in\mathscr{M}$ 定义 $\tilde{\Omega}A\in\mathscr{M}$ 如下. 设

$$A=E(x_i)\otimes R[y_j,z_k],$$

其中 $\deg x_i=$ 奇数, $\deg y_1=$ 偶数 $>2,\deg z_k=2$, 而微升运算 d 除符合 $\mathscr{M}$ 中代数一般的要求外任意. 定义

$$\tilde{\Omega}A=E(\tilde{y}_1)\otimes R[\tilde{x}_i]\in\mathscr{M},$$

其中 $\deg\tilde{y}_j=\deg y_j-1,\deg\tilde{x}_i=\deg x_i-1$, 而微分运算 $d=0$.

于是本节的结果可概述为:

定理 A 设 $X\in\mathscr{W}$, 则

$$I^*(\tilde{\Omega}X)=\tilde{\Omega}(I^*X). \tag{1}$$

我们需要下面的引理, 其证明甚易从略:

引理 设 $Z\in\mathscr{W}$ 的实域上上同调代数为

$$H^*(Z)\approx E(v_1,v_2,\cdots)\otimes R[u_1,u_2,\cdots], \tag{2}$$

其中 $\deg v_j=$ 奇数(>1), $\deg u_i=$ 偶数, 则

$$I^*(Z)\approx H^*(Z)$$

(代数同构), 即

$$I^*(Z)\approx E(\eta_1,\eta_2,\cdots)\otimes R[\xi_1,\xi_2,\cdots], \tag{3}$$

其中 $\deg\eta_j=\deg v_j,\deg\xi_i=\deg u_i$, 而微分运算 $d=0$.

定理 A 的证明 在 X 是有限型空间时, $Z=\tilde{\Omega}X$ 是有限型 H 空间. 在实数域系数上, 据 Hopf-Samelson 定理, $H^*(Z)$ 形如 (2) 式, 由引理即得 (3) 式. 由 Sullivan 定理得

$$\pi^*(\tilde{\Omega}X)\approx Gr(\eta_j,\xi_i),$$

此式右边指以 η_j,ξ_i 为基的 R 上分次模. 因而

$$\pi^*(X)\approx Gr(x_i,y_j,z_k),$$

其中 $\deg x_i=\deg\xi_i+1=$ 奇数, $\deg \mathrm{y}_j=\deg\eta_j+1=$ 偶数 $>2,\deg z_k=2$. 仍由 Sullivan 定理得

$$I^*(X)\approx E(x_i)\otimes R[y_j,z_k], \tag{4}$$

带有适当的微分运算 d. 比较 (3) 与 (4) 即得 (1) 式, 证毕.

作为定理 A 的一个特例, 设 $X \in \mathscr{W}$ 的实域上上同调代数 $H^*(X)$ 为以下三种代数的张量积:

1) 以 x_i 为基的外代数 $H_1, \deg x_i = 2s_i + 1 > 1$.

2) 以 y_j 为基的多项式代数 $H_2, \deg y_j = 2l_j$.

3) 以 u_k 为基的截头多项式代数 $H_3, \deg u_k = 2r_k, (u_k)^{m_k} = 0, (u_k)^{m_k - 1} \neq 0$.

由此假定易知

$$I^*(X) \approx E(x_i, \upsilon_k) \otimes R[y_j, u_k],$$

其中 $\deg \upsilon_k = m_k \cdot \deg u_k - 1$, 微分运算为 $d\upsilon_k = (u_k)^{mk}$, 而 $dx_i = 0, dy_j = 0, du_k = 0$. 由本节定理即得

$$H^*(\tilde{\Omega}X) \approx I^*(\tilde{\Omega}X) \approx \tilde{\Omega}(I^*(X)) \approx E(\eta_{j'}, \mu_{k'}) \otimes R[\xi_i, \upsilon_k],$$

其中 $\deg \xi_i = 2s_i, \deg \eta_{j'} = 2l_{j'} - 1 > 1, \deg \mu_{k'} = 2r_{k'} - 1 > 1$($k'$ 对应于 $r_k > 1$ 的 k, j' 对应于 $l_j > 1$ 的 j), $\deg \upsilon_k = 2r_k m_k - 2$. 参阅 Oniscek[5] 一文中的定理 3.3.

§3　齐性空间的 I^* 函子与 Cartan 定理的推广

设 G 是紧致连通李群, U 是 G 的连通闭子群, M 是齐性空间 $G/U, \dim G = r, \dim U = r - n, \dim M = n$.

据 Hopf 定理, $H^*(G)$ 中有一组次数为奇数的本原元素 $x_1, \cdots, x_N$ 构成一线性空间 P_G, 使 $H^*(G)$ 即 P_G 上的外代数:

$$H^*(G) \approx E(P_G) = E(x_1, \cdots, x_N).$$

记 G 的分类空间为 $B_G, H^*(B_G)$ 中维数 > 0 的子代数为 $H^+(B_G)$, 可分解类所成子代数为 D_G, 又记 $Q_G = H^+(B_G)/D_G, H^*(B_G)$ 到 Q_G 的投影为 p_G. 则有模同构

$$\lambda_G : P_G \approx Q_G$$

使 $\deg \lambda_G(x_i) = \deg x_i + 1$. 又在 $H^*(B_G)$ 中可取 $y_i \in p_G^{-1} \lambda_G x_i$ 使 $H^*(B_G)$ 是 y_i 上的多项式代数:

$$H^*(B_G) \approx R[y_1, \cdots, y_N].$$

模同态

$$\tau_G^* : \begin{cases} P_G \to H * (B_G), \\ x_i \to y_i, \end{cases}$$

使 $p_G\tau_G^* = \lambda_G$, 称为 G 的一个转置.

对 U 同样有 P_U, B_U, Q_U 等, 使 P_U 以 $u_1, \cdots, u_L$ 为基,

$$H^*(U) \approx E(P_U) = E(u_1, \cdots, u_L).$$

又可取 $v_j \in p_U^{-1}\lambda_U u_j$ 与转置

$$\tau_U^* : \begin{cases} P_U \to H^*(B_U), \\ u_j \to v_j, \end{cases}$$

使

$$H^*(B_U) \approx R[v_1, \cdots, v_L].$$

据 Hopf-Samelson, 嵌入 $i : U \subset G$ 引出 $i^* : H^*(G) \to H^*(U)$ 使 $i^*(P_G) \subset P_U$. 此外有自然映像 $\rho : B_U \to B_G$. 于是有可交换图像:

$$\begin{array}{ccccccc}
E(x_1,\cdots,x_N) \approx H^*(G) \supset & P_G & \xrightarrow[\approx]{\lambda_G} & Q_G & \xleftarrow{P_G} & H^*(B_G) & \approx R[y_1,\cdots,y_N] \\
\quad\quad\quad\quad\quad\quad\quad {\scriptstyle i^*}\downarrow & {\scriptstyle i^*}\downarrow & & {\scriptstyle \rho^*}\downarrow & & {\scriptstyle \rho^*}\downarrow & \\
E(u_1,\cdots,u_L) \approx H^*(U) \supset & P_U & \xrightarrow[\lambda_U]{\approx} & Q_U & \xleftarrow[p_U]{} & H^*(B_U) & \approx R[v_1,\cdots,v_L]
\end{array}$$

在 ρ^* 下将有

$$\rho^* y_i = p_i(v_1, \cdots, v_L) \in R[v_1, \cdots, v_L].$$

今对取定的转置 τ_G^* 作 DGA 代数

$$\begin{aligned} C_M^* &= H^*(G) \otimes H * (B_U) \\ &\approx E(x_1, \cdots, x_N) \otimes R[v_1, \cdots v_L], \end{aligned}$$

其中微分运算 d 由下二式确定:

$$\begin{aligned} &dx_i = \rho^* \tau_G^* x_j = P_i(v_1, \cdots, v_L) \in H^*(B_U), \\ &dv_j = 0. \end{aligned}$$

C_M^* 称齐性空间 $M = G/U$ 的 Cartan 代数, 所谓 Cartan 定理, 是指 M 在实域上的上同调代数由 Cartan 代数如下完全决定:

$$H^*(M) \approx H(C_M^*),$$

本节目的在证明 M 单连通时, M 的 I^* 函子也同样由 Cartan 代数 C_M^* 所完全决定:

定理 B　若 $M=G/U$ 单连通, 则

$$I^*(M)\approx \min C_M^*.$$

显然这一定理包括了 Cartan 定理:

$$H^*(M)\approx H(I^*(M))\approx H(\min C_M^*)\approx H(C_M^*).$$

为证明定理 B, 先对原来 Cartan 定理的证明依 [6] 作一概括介绍如下 (符号略有变动).

取以下指数范围:

$$\begin{aligned}&\alpha,\beta,\gamma,\cdots=n+1,\cdots,r,\\&a,b,c,\cdots=1,2,\cdots,n,\\&i,j,k,\cdots=1,2,\cdots,r.\end{aligned}$$

记 G 上一切左不变外形式所成 DGA 代数为 $\Lambda(G)$. 由于 G 是紧致的, 在 G 的李代数中可取在伴随表示下不变的内积与归一正交基 e_i 使 e_α 构成 U 的李代数的归一正交基, 记 G 的结构常数为 c_{ij}^k, G 上与 e_i 对偶的左不变外形式为 ω^i. 则 c_{ij}^k 对诸指针都反称, 且

$$\begin{aligned}&c_{a\beta}^a=0,\\&[e_i,e_j]=c_{ij}^k e_k,\\&d\omega^i=-\frac{1}{2}c_{jk}^i\omega^j\Lambda\omega^k.\end{aligned}$$

对任一 $g\in G$, 由 $x\to g^{-1}xg(x\in G)$ 所定 G 的内自同构引出 $\Lambda(G)$ 的自同态

$$Ad_g:\begin{cases}\Lambda(G)\to\Lambda(G),\\ \omega^i\to A_j^i(g)\omega^j,\end{cases}$$

则 $g\to A(g)=(A_j^i(g))$ 即 G 的伴随表示.

今除次数为 1 的 ω^i 又引进次数为 2 的生成元 Ω^i 而作 G 的所谓 Weil 代数

$$W=E(\omega^1,\cdots,\omega^r)\otimes R[\Omega^1,\cdots,\Omega^r].$$

命 $Ad_g(g\in G)$ 作用于 W 上使 Ad_g 为 W 的代数同态而

$$\begin{aligned}&Ad_g\omega^i=A_j^i(g)\omega^j,\\&Ad_g\Omega^i=A_j^i(g)\Omega^j.\end{aligned}$$

W 中在一切 Ad_g 下不变的元素 $a(Ad_g a = a, g \in G$任意) 构成一 W 的子代数 W^I. 在 W^I 中可引入微分运算 d 使 W^I 成一 DGA 代数. 在 W^I 中由 ω^i(与Ω^i). 生成的子代数则各记为 Λ^I(与S^I).

依 Hopf 定理,

$$H^*(G) \approx \Lambda^I \approx E(p_1, \cdots, p_N),$$

此处 p_μ 为次数 = 奇数的本原外形式并构成线性空间 P_G 的一组基. 依 Cartan-Weil-Chevalley 定理, 本原外形式都是可转置的 (反之亦然), 换言之, 对于每一 p_μ 有一 $\bar{p}_\mu \in W^I$, 使 $\bar{p}_\mu$ 中置 $\Omega^i = 0$ 后得 p_μ, 且

$$d\bar{p}_\mu = \bar{s}_\mu \in S^I.$$

于是 S^I 是一多项式代数, 以 $\bar{s}_\mu$ 为基, 且与 $H^*(B_G)$ 同构:

$$H^*(B_G) \approx S^I \approx R[\bar{s}_1, \cdots, \bar{s}_r],$$

若置 $\tau_G^*(p_\mu) = \bar{s}_\mu$, 则 $\tau_G^* : P_G \to H^*(B_G)$ 即为一转置.

记 W 中由 $\omega^\alpha, \omega^a, \Omega^a$ 所产生而在一切 $Ad_h(h \in U)$ 下不变的元素所成的子代数为 W_U^I(W_U^I 与 W^I 同在 W 中但无包含关系), 在 W_U^I 中可引入微分运算 d 使成一 DGA 代数. 今对任一 $a \in W^I$, 命 $\sigma(a)$ 为在 a 中置 $\Omega^a = 0$ 后所得元素, 则 $\sigma(a) \in W_U^I$, 且

$$\sigma : W^I \to W_U^I$$

为一 DGA 同态.

记 W_U^I 中由只含 Ω^a 的元素所成子代数为 S_U^I, 则 S_U^I 之于 U, 恰如 S^I 之于 G, 而知 S_U^I 为以若干 $\bar{s}_{U1}, \cdots, \bar{s}_{UL}$ 为基的多项式代数, 且

$$H^*(B_U) \approx s_U^I \approx R[\bar{s}_{U1}, \cdots, \bar{s}_{UL}].$$

在 σ 下有 $\sigma p_\mu = p_\mu$. 命 $\sigma\bar{p}_\mu = \bar{p}_\mu^0, \sigma\bar{s}_\mu = \bar{s}_\mu^0$, 则有

$$d\bar{p}_\mu^0 = \bar{s}_\mu^0 \approx P_\mu(\bar{s}_{U1}, \cdots, \bar{s}_{UL}),$$

这里 p_μ 是一些确定的多项式, 而 $\sigma\bar{s}_\mu = \bar{s}_\mu^0$ 所定同态 $\sigma : S^I \to S_U^I$ 即本节初由 $\rho : B_U \to B_G$ 所引起的同态 $\rho^* : H^*(B_G) \to H^*(B_U)$.

在 W_U^I 中有两 DGA 子代数. 其一为由 W_U^I 中一切形如

$$\mathrm{C} = \sum_{k_p} \sum_{(p)} s_{u_1 \cdots u_k} \bar{p}_{\mu_1}^0 \cdots \bar{p}_{\mu_k}^0, \quad s\mu_1 \cdots \mu_k \in s_U^I$$

的元素所成代数. 由前所述可见与本节初所说 Cartan 代数 C_M^* 在对应 $x_i \to \bar{p}_i^0$, $v_\lambda \leftrightarrow \bar{s}_{U\lambda}(i=1,\cdots,N;\lambda=1,\cdots,L)$ 下同构, 故即可记为 C_M^*, 另一为由 W_U^I 中一切只由 ω^a 产生而在 $Ad_h(h\in U)$ 下不变的元素所成的子代数 W_B^I. 若 $j_C: C_M^*\subset W_U^I$, $j_B: W_B^I\subset W_U^I$ 表恒同同态, 则有

$$j_{C_*}: H(C_M^*)\approx H(W_U^I),$$
$$j_{B_*}: H(W_B^I)\approx H(W_U^I).$$

记 G 到 $G/U=M$ 的投影为 π, M 的 De Rham 代数为 $D^*(M)$. 命 $D^*(M)$ 中在 G 的自然作用下不变的外形式所成 DGA 子代数为 Φ. 因 M 紧致, 故嵌入 $i:\Phi\subset D^*(M)$ 引出同构

$$i_*: H(\Phi)\approx H(D*(M))\approx H*(M),$$

对 $\varphi\in\Phi$, 易知 $\pi^*\varphi$ 为 G 上左不变且在 $Ad_h(h\in U)$ 下不变的外形式, 因而可视为 W_B^I 的元素, 且 W_B^I 的元素都可由此获得, 由此并可得一 DGA 同构

$$\gamma:\Phi\approx W_B^I.$$

至此易证定理 B 如下:

作 DGA 同态 ρ,ρ_U 等所成图像

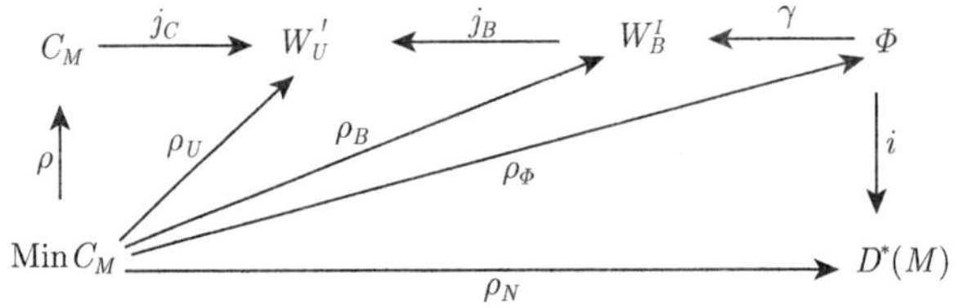

其中 j_C, j_B, γ, i 已知引出 H 函子的同构, ρ 为标准同态, $\rho_U=j_C\rho, \rho_B$ 与 ρ_Φ 各因 j_B^*,γ^* 为同构而可依次自 ρ_U,ρ_B 作出使 $j_B\rho_B,\gamma\rho_\Phi$ 各同伦于 ρ_U,ρ_B(参见 [3]§§ 15—16). 又 $\rho_M=i\rho_\Phi$. 由此可得

$$\rho_{M_*}: H(\min C_M)\approx H(D^*(M))\approx H^*(M).$$

因而 $\min C_M\approx I^*(M)$, 如所欲证.

§4 Cartan 代数的简化与 Oniscek 定理

从上节所证定理 B 可得出关于齐性空间上同调代数的 Cartan 定理. 同样可得出关于齐性空间同伦群的 Oniscek 定理 ([6]). 为此将先对 Cartan 代数作一些简化如下:

设连通单连通 DGA 代数

$$C = E(x_1, \cdots, x_s) \otimes R[v_1, \cdots, v_t] \in \mathscr{A} \tag{1}$$

中的 x_i 是奇次数而 v_j 是偶次数, 又其微分运算 d 满足

$$\begin{cases} dx_i = P_i(v_1, \cdots, v_t) \in R[v_1, \cdots, v_t], \\ dv_j = 0, \end{cases} \tag{2}$$

则 C 称为一 C 型代数. 记 $v_1, \cdots, v_t$ 二次以上多项式所成子代数为 $Q[v_1, \cdots, v_t]$, 若对 C 有

$$dx_i \in Q[v_1, \cdots, v_t],$$

则 C 称为已规范化. 显然规范化的 C 极小而有

$$\min C \approx C.$$

若 $s = 1, t = 1$ 即

$$C = E(x) \otimes R[y],$$

又有 $dx = y$ 则 C 称为一初等 C 型代数并记作 $C_0(x, y)$, 若对两 C 型代数 (1) 与

$$C' = E(x_1', \cdots, x_{s'}') \otimes R[v_1', \cdots, v_{t'}']$$

存在 DGA 同态

$$f : C \to C', g : C' \to C$$

使

$$fg = 1, \quad gf = 1,$$

$$f(v_t) \in R[v_1', \cdots, v_{t'}'], \quad g(v_j') \in R[v_1, \cdots, v_t],$$

则称 C, C' 等价, 记作 $C \sim C'$. 此时显然有 $g_*^{-1} = f_* : H(C) \approx H(C')$ 以及 $\min C \approx \min C'$.

对于任一 C 型代数 (1) 设

$$dx_i = \Sigma a_{ij} y_j + Q_i, \quad Q_i \in Q[v_1, \cdots, v_t],$$

若矩阵 $[a_{ij}]$ 的秩为 r, 则称 C 的秩为 r (在 $\deg y_j \neq \deg x_i + 1$ 时, $a_{ij} = 0$). 用简单的变量替换易证下述:

规范化引理 若 C 型代数 (1) 的秩为 r, 则

$$C \sim C' = C_0(x_1', y_1') \otimes \cdots \otimes C_0(x_r', y_r') \otimes C^R,$$

其中

$$C^R = E(x'_{r+1}, \cdots, x'_s) \otimes R[v'_{r+1}, \cdots, v'_t]$$

为一规范化 C 型代数, C^R 极小且有

$$\min C \approx \min C' \approx \min C^R \approx C^R.$$

今设 G 为紧致连通李群, U 为 G 的连通闭子群, 并如前节考虑可交换图

$$\begin{array}{ccccc}
P_G & \xrightarrow[\approx]{\lambda_G} & Q_G & \xleftarrow{p_G} & H^*(B_G) \\
{\scriptstyle i^*}\downarrow & & {\scriptstyle \rho^*}\downarrow & & \downarrow{\scriptstyle \rho^*} \\
P_U & \xrightarrow[\lambda_U]{\approx} & Q_U & \xleftarrow[p_U]{} & H^*(B_U)
\end{array}$$

已知在 i^* 下有分解

$$\begin{aligned}
P_G &= P_1 \dot{+} P_2, \\
P_U &= R_1 \dot{+} R_2, \\
i^* &: P_1 \approx R_1, \\
P_2 = \operatorname{Ker} i^*&, R_2 \approx \operatorname{Coker} i^*.
\end{aligned}$$

对 P_1, P_2, R_1, R_2 各取基

$$\begin{aligned}
P_1 &= [x'_1, \cdots, x'_m], \quad P_2 = [x''_1, \cdots, x''_p], \\
R_1 &= [u'_1, \cdots, u'_m], \quad R_2 = [u''_1, \cdots, u''_q],
\end{aligned}$$

使

$$\begin{aligned}
i^* x'_\lambda &= u'_\lambda, \quad \lambda = 1, \cdots, m, \\
i^* x''_\mu &= 0, \quad \mu = 1, \cdots, p.
\end{aligned}$$

又取转置

$$\tau_G^* : \begin{cases}
P_G \to H*(B_G), \\
x'_\lambda \to y'_\lambda, & \lambda = 1, \cdots, m, \\
x''_\mu \to y''_\mu, & \mu = 1, \cdots, p,
\end{cases}$$

$$\tau_U^* : \begin{cases}
P_U \to H^*(B_U), \\
u'_\lambda \to v'_\lambda, & \lambda = 1, \cdots, m, \\
u''_v \to v''_v, & v = 1, \cdots, q,
\end{cases}$$

使

$$H^*(B_G) \approx R[y'_\lambda, y''_\mu],$$
$$H^*(B_U) \approx R[\upsilon'_\lambda, \upsilon''_v].$$

在 $M = G/U$ 单连通时, 其 Cartan 代数可取为

$$C^*_M = E(x'_\lambda, x''_\mu) \otimes R[\upsilon'_\lambda, \upsilon''_v],$$

其中微分运算 d 为

$$dx'_\lambda = \rho^* \tau^*_G x'_\lambda \in R[\upsilon'_\lambda, \upsilon''_v],$$
$$dx''_\mu = \rho^* \tau^*_G x''_\mu \in R[\upsilon'_\lambda, \upsilon''_v].$$

故 C^*_M 为一 C 型代数. 由于

$$p_U dx'_\lambda = \rho^* \lambda_G x'_\lambda = \lambda_U i^* x'_\lambda = \lambda_U u'_\lambda = \upsilon'_\lambda \bmod D_U,$$

故有

$$dx'_\lambda = \upsilon'_\lambda + Q'_\lambda, \quad Q'_\lambda \in Q[\upsilon'_{\lambda'}, \upsilon''_v].$$

同样有

$$dx'_\mu = Q''_\mu, \qquad Q''_\mu \in Q[\upsilon'_\lambda, \upsilon''_v],$$

由此可知 C^*_M 的秩为 m.

应用规范化引理, 可得

$$\min C^*_M \approx E(x''_1, \cdots, x''_p) \otimes R[\upsilon''_1, \cdots, \upsilon''_q].$$

由 Sullivan 理论即得

$$\pi^s(M) \approx \begin{cases} \mathrm{Ker}^s i^*, & s = \text{奇数时}, \\ \mathrm{Coker}^{s-1} i^*, & s = \text{偶数时}. \end{cases}$$

这里 $\mathrm{Ker}^s, \mathrm{Coker}^{s-1}$ 各指 Ker 与 Coker 中次数为 s 与 $s-1$ 的部分.

上式给出了在 $M = G/U$ 单连通时, 由 $i^*: P_G \to P_U$ 以确定 $\pi^*(M)$ 的秩的明显公式. 在 $M = G/U$ 连通而不单连通时, 可考虑 G 的通用覆叠空间而归结为上述情形. 此即 Oniscek 定理, 重述如次.

定理 C (Oniscek[6]) 设 G 为紧致连通李群, U 为连通闭子群, 则齐性空间 G/U 的 π_* 的秩为

$$\mathrm{Rank}\, \pi_{*\text{奇}}(G/U) = \dim \mathrm{Ker}^* [P_G \xrightarrow{i^*} P_U],$$

$$\operatorname{Rank} \pi_{*\text{偶}}(G/U) = \dim \operatorname{Coker}^{*-1} [P_G \xrightarrow{i^*} P_U].$$

附注　上述 Oniscek 定理也可直接地证明如下.

由于 $I^*(G) \approx H^*(G), I^*(U) \approx H^*(U)$ 都是奇数次元素上的外代数, 故依 Sullivan 定理有

$$i = \text{偶数时}, \quad \pi^i(G) = 0, \pi^i(U) = 0.$$

从 $M = G/U$ 的同伦正合序列

$$\pi_r(U) \to \pi_r(G) \to \pi_r(M) \to \pi_{r-1}(U) \to \pi_{r-1}(G)$$

即得正合序列

$$\pi^{r-1}(G) \to \pi^{r-1}(U) \to \pi^r(M) \to \pi^r(G) \to \pi^r(U),$$

$$\pi^r(M) \approx \begin{cases} \operatorname{Coker} [\pi^{r-1}(G) \to \pi^{r-1}(U)], & r = \text{偶数时}, \\ \operatorname{Ker} [\pi^r(G) \to \pi(U)], & r = \text{奇数时}. \end{cases}$$

由此仍依 Sullivan 定理即得 Oniscek 公式. 这一证明不必假定 G 是李群或紧致性, 只须 G 是 $\mathscr{W}$ 中有限型拓扑群即可.

附记 (1975 年 7 月 30 日) §2 定理 A 已见 D.Sullivan 在 Manifolds Tokyo 1973(Univ of Tokyo Press, 1975) 中一文. 又 Sullivan 已得出圆象空间的 I^* 函子, 而这一计算要难得多.

参考文献

[1] Borel A. Sur la Cohomologie des espaces fibrés principaux et des espaces homogènes de groupes de Lie Compacts. *Annals of Math.*, 1953, 57: 115-207.

[2] Cartan H. Cohomologie réelle d'un espace fibré principal différentiable. Sém. Cartan, 1949/50, Exp.19, 20.

[3] Friedlander E, Griffiths P A, Morgan J. Homotopy theory and differential forms. *mimeographed*, 1972.

[4] Kahn D W. The existence and applications of anticommutative cochain algebras, [1]. *J. Math.*, 1963, 7: 376-395.

[5] Oniscêk A L. On cohomologies of spaces of paths. *Mat.Sbor.* 1958, 44: 3-52.

[6] Oniscék A L. On transitive compact groups of transformations. *Mat.Sbor.*, 1963, 60: 447-485.

[7] Rashevskii P K, Real cohomology of homogeneous spaces. *Usp.Mat.Nauk*, 1969, 24(3): 23-90.

[8] Sullivan D. Differential forms and topology of manifolds. *Symp.Tokyo on Topology*, 1973.

[9] 吴文俊. 代数拓扑的一个新函子. 科学通报, 1975, 9: 311-312.

49. 代数拓扑 I^* 函子论 —— 纤维方的实拓扑*

摘要 数学是研究现实世界中空间形式与数量关系的科学. 自古以来, 数学就往往用数量来刻画空间形式的某些特征, 即所谓形的量度. 例如, 道路的长度、田地的面积、器物的容量以及近代数学中点集的测度等.

在代数拓扑中, 则通常用所谓代数量如群、环等来量度或刻画拓扑空间, 这种量度通称函子, 已知的如同调函子 H、同伦函子 π 以及较新出现的 J, K 函子等, 是代数拓扑的主要工具. 在资料 [1] 中, 我们提出了一种新函子 I^*, 它比已知经典的函子 π, H, J, K 等更易于计算与使用. 本文详细证明了资料 [1] 中一部分结果, 着重讨论纤维方的理论. 就 H 函子来说, 有所谓 Eilenberg-Moore 谱序列 [2, 3], 但只能给出 H 函子间的某些复杂"关系", 通过 I^* 函子, 却能给出确切的等式, 这说明 I^* 函子比 H 函子更好.

§1 空间的 I^* 函子

首先, 对 Sullivan 的极小模型理论作一简介 (参见资料 [4, 5]), 由此引入 I^* 函子.

设 A 是一个分次反称带有微分运算 d (符合通常规律) 的实域 $\underline{R}$ 上的代数 (简称为 DGA 代数). 若有 $H_0(A, d) = \underline{R}, H_1(A, d) = 0$, 则 A 称为**连通与单连通的**. 若 $\dim H_r(A, d) =$ 有限, 则 A 称为**有限型的**, 若 A 中的代数运算除通常规律外不再有其他关系, 则 A 称为**自由的**, 若 A 中的微分运算 d 具有下述性质: $da \in A^+ \cdot A^+ (a \in A)$, 这里 A^+ 是 A 中一切次数 > 0 的元素所成的子代数, 则 A 称为**可分解的**. 一切实域上连通、单连通、有限型的 DGA 代数及保持 DGA 结构的代数同态 (简称 DGA同态) 所成范畴记作 $\mathscr{A}$, 由 $\mathscr{A}$ 中一切自由且可分解的代数 (简称**极小 DGA 代数**) 所成的子范畴记作 $\mathscr{M}$.

对任一 DGA 代数 $A \in \mathscr{A}$, 可作 $\mathrm{Min} A \in \mathscr{M}$ 与 DGA 同态 $\rho : \mathrm{Min}\, A \to A$ 引出 DGA 同构 $\rho_* : H(\mathrm{Min}\, A) \approx H(A)$, 这样的 $\mathrm{Min}\, A$ 唯一确定至一 DGA 同构, 称为 A 的**极小模型**, ρ 称**标准同态**. 又对 $A, B \in \mathscr{A}$ 以及 DGA 同态 $f : A \to B$ 作 DGA 同态 $g_0 : \mathrm{Min}\, A \to \mathrm{Min}\, B$, 使左下图像**同伦可交换** ($\rho_A, \rho_B$ 为标准同态), 而 g 也由 f 唯一确定至一同伦, 并引出右下可交换图像:

---

* *Sci. Sci. Sinica*, 1975, 18: 464-482.

$$
\begin{array}{ccc}
A & \xrightarrow{f} & B \\
\rho_A \uparrow & & \uparrow \rho_B \\
\mathrm{Min}A & \xrightarrow{g} & \mathrm{Min}B
\end{array}
\qquad
\begin{array}{ccc}
H(A) & \xrightarrow{f_*} & H(B) \\
\rho_{A*} \uparrow \approx & & \approx \uparrow \rho_{B*} \\
H(\mathrm{Min}A) & \xrightarrow{g_*} & H(\mathrm{Min}B)
\end{array}
$$

图 1

设 $\mathscr{K}$ 是由连通、单连通、可数、局部有限且为有限型的单纯复形及单纯映像所成的范畴. 对任一 $K \in \mathscr{K}$, 可定义一 DGA 代数 $A^*(K) \in \mathscr{A}$ 如下：对任一 $\sigma \in K$, 命 $A^*(\sigma) = \sum A^q(\sigma)$ 为在 σ 上一切外微分形式在外乘积与外微分下所成的 $\underline{R}$ 上 DGA 代数. 设函数

$$\omega : \sigma(\in K) \to A^q(\sigma)$$

满足下述协合关系：对任意 $\sigma' \subset \sigma \in K$, 记 i 为恒同映像 $\sigma' \to \sigma$ 引出 DGA 同态

$$i^A : A^*(\sigma) \to A^*(\sigma')$$

时, 有 $i^A\omega(\sigma) = \omega(\sigma')$, 这样的 q 次协合函数 ω 成一 $\underline{R}$ 模, 记作 $A^q(K)$. 在外乘积与外微分下, $\sum A^q(K)$ 自然成一 $\underline{R}$ 上的 DGA 代数, 记作 $A^*(K) \in \mathscr{A}$. 令

$$\mathrm{Min}\, A^*(K) = I^*(K),$$

则 $I^*(K)$ 给出了 K 在实域上的 H^* 函子以及 $\pi_*(K)\underset{z}{\otimes}\underline{R}$, 例如 ($H^*$ 中的 $\underline{R}$ 略去不写, 下同)：

$$H(I^*(K)) \approx H^*(K).$$

记通常 K 的 $\underline{R}$ 上上链代数为 $C^*(K)$. 将 $\omega \in A^*(K)$ 对每一 $\sigma \in K$ 积分, 得一保持微分运算的模同态

$$\int : A^*(K) \to C^*(K),$$

由此引出模同构

$$\int_* : H(A^*(K)) \approx H(C^*(K)) \approx H^*(K).$$

设 $L \in K$, 而 $f : L \to K$ 是单纯映像, 则有自然 DGA 同态 $f^A : A^*(K) \to A^*(L)$, 使下图可交换：

$$
\begin{array}{ccc}
A^*(K) & \xrightarrow{f^A} & A^*(L) \\
\int \downarrow & & \downarrow \int \\
C^*(K) & \xrightarrow{f^*} & C^*(L)
\end{array}
$$

图 2

今设 L 是 K 的子复形, f 是恒同映像, 定义上链复形 $A^*(K,L)=\sum A^q(K,L)$, 使下图可交换

$$\begin{array}{ccccccccc}
0 & \longrightarrow & A^q(K,L) & \longrightarrow & A^q(K) & \xrightarrow{f^A} & A^q(L) & \longrightarrow & 0\\
 & & \downarrow d & & \downarrow d & & \downarrow d & & \\
0 & \longrightarrow & A^{q+1}(K,L) & \longrightarrow & A^{q+1}(K) & \xrightarrow{f^A} & A^{q+1}(L) & \longrightarrow & 0
\end{array}$$

图 3

其中各横行都是正合序列. 同样, 定义 $C^*(K,L)=\sum C^q(K,L)$, 则由积分同态

$$\int : A^*\to C^*$$

可引出模同态

$$\int : A^*(K,L)\to C^*(K,L).$$

由于

$$\int_* : H(A^*(K))\approx H(C^*(K)),$$
$$\int_* : H(A^*(L))\approx H(C^*(L)),$$

因而也引出模同构:

$$\int_* : H(A^*(K,L))\approx H(C^*(K,L)).$$

按 Sullivan 理论引入空间的 I^* 函子如下.

设 $\mathscr{W}$ 是与 $\mathscr{K}$ 中复形有相同伦型的空间以及连续映像所成的范畴. 设 $X\in\mathscr{W}$ 而 X 同伦等价于 $K,L\in\mathscr{K},g:|K|\to|L|$ 使 $g:|K|\simeq|L|$. 作 K 的单纯剖分 K', 恒同映像与 g 的单纯逼近 $s:K'\to K$ 与 $s':K'\to L$. 由下面可交换图 (ρ_K,ρ_L 是标准同态)

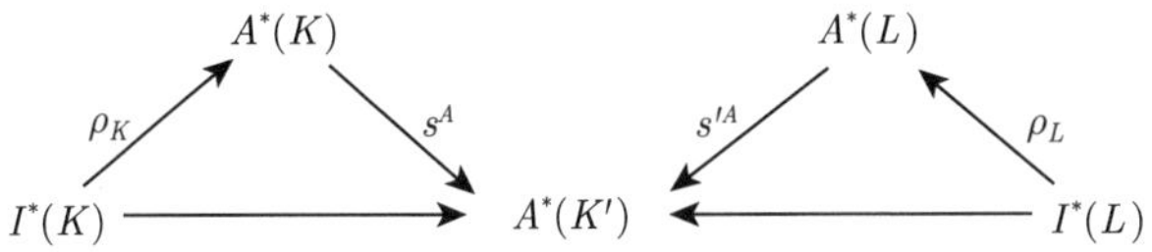

图 4

作 H 函子得下面可交换图:

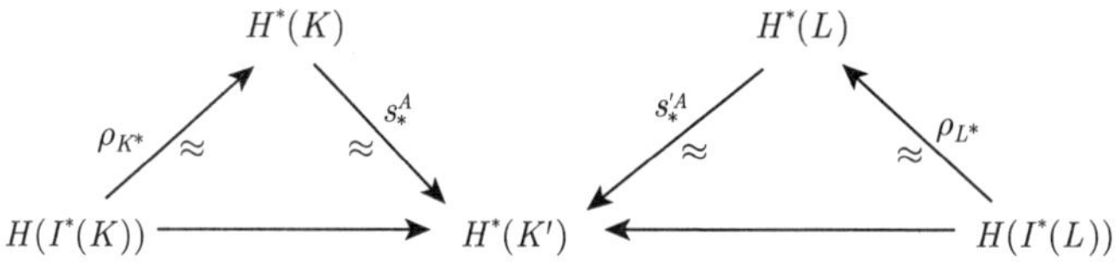

图 5

由 $A^*(K') \in \mathscr{A}$ 的极小模型的唯一性得

$$I^*(K) \approx I^*(L).$$

因之 $I^*(K)$(差一 DGA 同构) 只依赖于 $X \in \mathscr{W}$ 而可记作 $I^*(X)$, 即 $X \in \mathscr{W}$ 的 I^* 函子.

§2 单式纤维丛的 Leray 谱序列

设纤维丛

$$F \subset Y \xrightarrow{g} B, \tag{1}$$

其中 F 是纤维, Y 是丛空间, B 是底空间. 假设 F, Y, B 都属于 $\mathscr{K}$, 特别有 $\pi_1(B) = 0, g$ 是单纯映像, 则 (1) 式称为单式纤维丛. 此时命 B_q 为 B 的 q 维骨架, $Y_q = g^{-1}(B_q), i_q : Y_q \to Y$ 为恒同映像.

今将 $A^*(Y)$ 与 $C^*(Y)$ 分滤如下:

$$\begin{aligned} &F^p A^q(Y) = \ker[A^q(Y) \overset{i_{p-1}^A}{\to} A^q(Y_{p-1})] \\ =& \{\omega^{\in A^q(Y)} / \omega(\tilde{\sigma}) = 0, \tilde{\sigma} \in Y_{p-1}\}, \end{aligned} \tag{2}$$

$$\begin{aligned} &F^p C^q(Y) = \ker[C^q(Y) \overset{i_{p-1}^*}{\to} C^q(Y_{p-1})] \\ =& \{f^{\in C^q(Y)} / f(\tilde{\sigma}) = 0, \tilde{\sigma} \in Y_{p-1}, \dim \tilde{\sigma} = q\}. \end{aligned} \tag{3}$$

显然积分同态 $\int : A^*(Y) \to C^*(Y)$ 使

$$\int F^p A^q(Y) \subset F^p C^q(Y),$$

因而引出谱序列各项的同态 $\int_r : E_r^{p,q} A^*(Y) \to E_r^{p,q} C^*(Y)$.

引理 1 同态 $\int_r$ 在 $r \geqslant 1$ 时都是同构, 特别有 ($\otimes$ 指 $\underset{R}{\otimes}$, 下同)

$$\int_2 : E_2^{p,q} A^*(Y) \approx E_2^{p,q} C^*(Y) \approx H^p(B) \otimes H^q(F). \tag{4}$$

证 作可交换图

$$
\begin{array}{ccccccccc}
 & & 0 & & 0 & & & & \\
 & & \downarrow & & \downarrow & & & & \\
0 & \longrightarrow & F^{p+1}A^{p+q}(Y) & = & F^{p+1}A^{p+q}(Y) & & & & \\
 & & \downarrow & & \downarrow & & & & \\
0 & \longrightarrow & F^{p}A^{p+q}(Y) & \longrightarrow & A^{p+q}(Y) & \longrightarrow & A^{p+q}(Y_{p-1}) & \longrightarrow & 0 \\
 & & \downarrow & & \downarrow & & \| & & \\
0 & \longrightarrow & A^{p+q}(Y_p, Y_{p-1}) & \longrightarrow & A^{p+q}(Y_p) & \longrightarrow & A^{p+q}(Y_{p-1}) & \longrightarrow & 0 \\
 & & \downarrow & & \downarrow & & & & \\
 & & 0 & & 0 & & & &
\end{array}
$$

图 6

其中两横行与中间纵行都依定义为正合序列. 左边纵行的虚线箭头即由此唯一补出. 由此图得

$$
E_0^{p,q}A^*(Y) \approx F^pA^{p+q}(Y)/F^{p+1}A^{p+q}(Y) \approx A^{p+q}(Y_p, Y_{p-1}),
$$

且 E_0 的微分运算由下面可交换图定出:

$$
\begin{array}{ccc}
E_0^{p,q}A^*(Y) & \xrightarrow{d_0} & E_0^{p,q+1}A^*(Y) \\
\wr\wr & & \wr\wr \\
A^{p+q}(Y_p, Y_{p-1}) & \xrightarrow{d} & A^{p+q+1}(Y_p, Y_{p-1}).
\end{array}
$$

图 7

由此得 $E_1^{p,q}A^*(Y) \approx H_{p+q}A^*(Y_p, Y_{p-1})$.

对于 $C^*(Y)$ 的分滤也可作与图 6 相仿的图 (即将图 6 中的 A 换成 C), 而得 $E_1^{p,q}C^*(Y) \approx H_{p+q}C^*(Y_p, Y_{p-1})$. 由于 $\int$ 同态的函子性得

$$
\int_1 : E_1^{p,q}A^*(Y) \approx E_1^{p,q}C^*(Y).
$$

由谱序列同构定理即得

$$
\int_r : E_r^{p,q}A^*(Y) \approx E_r^{p,q}C^*(Y), \quad r \geqslant 1,
$$

因已知 $E_2^{p,q}C^*(Y) \approx H^p(B) \otimes H^q(F)$, 故得引理 1.

对单式纤维丛 (1) 式除 $A^*(Y)$ 外, 引入另一 DGA 代数 $\bar{A}^*(Y)$ 如下.

考虑 B 上的函数

$$
\eta : \sigma(\in B) \to A^r(\sigma) \otimes A^s(g^{-1}\sigma), \tag{5}
$$

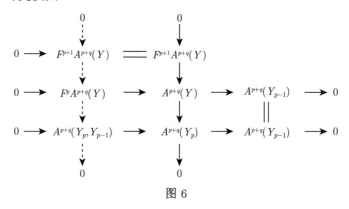

使满足下述协合条件:

对任意 $\sigma' \subset \sigma(\in B)$, 恒同映像 $i:\sigma'\subset\sigma, \tilde{i}:g^{-1}\sigma'\subset g^{-1}\sigma$, 有

$$\eta(\sigma')=(i^A\otimes\tilde{i}^A)\eta(\sigma).$$

这样的函数 η 自然构成一 $\underline{R}$ 上的模, 记作 $\bar{A}^{r,s}(Y)$, 置

$$\begin{cases}\bar{A}^q(Y)=\sum\limits_{r+s=q}\bar{A}^{r,s}(Y),\\ \bar{A}^*(Y)=\sum\limits_{q\geqslant 0}\bar{A}^q(Y).\end{cases}\tag{6}$$

在外微分形式的外乘积与外微分下, $\bar{A}^*(Y)$ 自然地成一 DGA 代数.

对 $\bar{A}^*(Y)$ 依下式分滤:

$$F^p\bar{A}^*(Y)=\sum_{r\geqslant p}\bar{A}^{r,*}(Y)=\left\{\eta^{\in\bar{A}^*(Y)}/\eta(\sigma)\in\sum_{r\geqslant p}A^r(\sigma)\otimes A^*(g^{-1}\sigma),\sigma\in B\right\}.\tag{7}$$

引理 2 对分滤 (7) 式的 Leray 谱序列有

$$E_2^p\bar{A}^*(Y)\approx H^p(B)\otimes H^*(F).\tag{8}$$

证 由 (7) 式易得 (参阅 (28) 式)

$$\begin{aligned}&E_0^p\bar{A}^*(Y)\approx\bar{A}^{p,*}(Y)\approx\{\eta^{\in\bar{A}^*(Y)}/\eta(\sigma)\in A^p(\sigma)\otimes A^*(g^{-1}\sigma),\sigma\in B\},\\&\qquad d_0=d|A^*(g^{-1}\sigma),\\&E_1^p\bar{A}^*(Y)\approx\{\tilde{\eta}:\sigma(\in B)\to A^p(\sigma)\otimes H^*(g^{-1}\sigma)/(C)\}.\end{aligned}$$

式中 (C) 指下述协合条件:

对任意 $\sigma'\subset\sigma(\in B), i, \tilde{i}$ 如前, 有

$$\tilde{\eta}(\sigma')=(i^A\otimes\tilde{i}^*)\tilde{\eta}(\sigma).$$

由于 $\pi_1(B)=0$, 得

$$E_1^p\bar{A}^*(Y)\approx A^p(B)\otimes H^*(F),$$
$$d_1=d|A^*(B).$$

由此即得引理 2.

今定义一 DGA 同态

$$h:\bar{A}^*(Y)\to A^*(Y)\tag{9}$$

如下: 设 $\eta \in \bar{A}^{r,s}(Y) \subset \bar{A}^*(Y)$. 对任一 $\tilde{\sigma} \in Y$, 命 $g\tilde{\sigma} = \sigma \in B$, 恒同映像 $\tilde{\sigma} \subset g^{-1}\sigma$ 为 $i_{\tilde{\sigma}}$, 投影 $\tilde{\sigma} \to \sigma$ 为 $g_{\tilde{\sigma}}$. 若

$$\eta(\sigma) = \sum \alpha_j \otimes \beta_j \in A^r(\sigma) \otimes A^s(g^{-1}\sigma),$$

命

$$(h\eta)(\tilde{\sigma}) = \sum g_{\tilde{\sigma}}^A \alpha_j \cdot i_{\tilde{\sigma}}^A \beta_j \in A^{r+s}(\tilde{\sigma}),$$

则 $(h\eta)(\tilde{\sigma}), \tilde{\sigma} \in Y$, 符合协合条件而确定一 $h\eta \in A^{r+s}(Y)$. 由 $\eta \to h\eta$ 所定的同态 (9) 式, 明显地为一 DGA 同态,

引理 3 由 (9) 式所定 DGA 同态 h 引出代数同构

$$h_*: H(\bar{A}^*(Y)) \approx H(A^*(Y)), \tag{10}$$

因而有

$$I^*(Y) \approx \operatorname{Min} \bar{A}^*(Y). \tag{11}$$

证 易见 h 对分滤 (7) 与 (2) 式有

$$hF^p\bar{A}^*(Y) \subset F^p A^*(Y).$$

因而 h 引出 Leray 谱序列的同态 $h_r: E_r^p\bar{A}^*(Y) \to E_r^p A^*(Y)$.

由引理 1 和引理 2 得

$$h_2: E_2^p\bar{A}^*(Y) \approx E_2^p A^*(Y).$$

故由谱序列同构定理得模同构 (10) 式. 由于 h 保持乘法, 故 h_* 为代数同构. 由 (9) 与 (10) 式即得 (11) 式:

$$I^*(Y) \approx \operatorname{Min} A^*(Y) \approx \operatorname{Min} \bar{A}^*(Y).$$

§3 纤维方的 I^* 函子

设有复形 $F, B, X, Y, Z \in \mathscr{K}$ 与可交换的单纯映像图:

$$\begin{array}{ccc} F \subset Z & \xrightarrow{\bar{f}} & Y \\ \bar{g} \downarrow & & \downarrow g \\ X & \xrightarrow{f} & B \end{array}$$

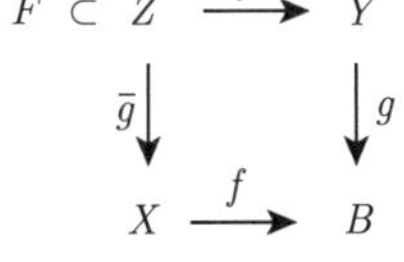

图 8

其中 $F \subset Y \xrightarrow{g} B$ 为单式纤维丛, $F \subset Z \xrightarrow{\bar{g}} X$ 是由 $f: X \to B$ 诱出的单式纤维丛. 称这些复形与映像的集体为一单式纤维方. 我们的目的, 在从 B, X, Y 的 I^* 函子及其在 f, g 下的诱导同态

$$\begin{cases} f^I: I^*(B) \to I^*(X), \\ g^I: I^*(B) \to I^*(Y), \end{cases} \tag{12}$$

确定 Z 的 I^* 函子.

依第二节, 我们已定义

$$\begin{cases} \bar{A}^*(Y) = \sum \bar{A}^q(Y), \bar{A}^q(Y) = \sum\limits_{r+s=q} \bar{A}^{r,s}(Y), \\ \bar{A}^*(Z) = \sum \bar{A}^q(Z), \bar{A}^q(Z) = \sum\limits_{r+s=q} \bar{A}^{r,s}(Z), \end{cases} \tag{13}$$

使

$$\begin{cases} \mathrm{Min}\, \bar{A}^*(Y) \approx I^*(Y), \\ \mathrm{Min}\, \bar{A}^*(Z) \approx I^*(Z). \end{cases} \tag{14}$$

在 f 下, $A^*(B)$ 依 (15) 式右作用于 $A^*(X)$ 使成为 $A^*(B)$ 右 DGA 代数 ($\xi \in A^*(X)$, $\beta \in A^*(B)$):

$$R(f^A): \begin{cases} A^*(X) \otimes A^*(B) \to A^*(X), \\ \xi \otimes \beta \to \xi \cdot f^A \beta. \end{cases} \tag{15}$$

同样, $A^*(B)$ 可左作用于 $\bar{A}^*(Y)$, 使成一 $A^*(B)$ 左 DGA 代数

$$L(g^A): A^*(B) \otimes \bar{A}^*(Y) \to \bar{A}^*(Y). \tag{16}$$

其定义如下: 设 $\beta \in A^*(B)$ 与 $\eta \in \bar{A}^*(Y)$. 对任一 $\sigma \in B$, 又设

$$\eta(\sigma) = \sum \alpha_j \otimes \beta_j \in A^*(\sigma) \otimes A^*(g^{-1}\sigma),$$

则 $L(g^A)(\beta \otimes \eta) \in \bar{A}^*(Y)$ 由下式所决定:

$$L(g^A)(\beta \otimes \eta)(\sigma) = \sum \beta(\sigma) \cdot \alpha_j \otimes \beta_j \in A^*(\sigma) \otimes A^*(g^{-1}\sigma).$$

其次定义一 DGA 同态

$$\lambda: A^*(X) \otimes \bar{A}^*(Y) \to \bar{A}^*(Z) \tag{17}$$

如下: 设 $\xi \in A^*(X), \eta \in \bar{A}^*(Y)$. 对任一 $\tau \in X$, 命 $f\tau = \sigma \in B$. 设

$$\eta(\sigma) = \sum \alpha_j \otimes \beta_j \in A^*(\sigma) \otimes A^*(g^{-1}\sigma).$$

于是 $\lambda(\xi\otimes\eta)\in\bar{A}^*(Z)$ 由下式所决定:

$$\lambda(\xi\otimes\eta)(\tau)=\sum\xi(\tau)\cdot f_\tau^A\alpha_j\otimes\bar{f}_\tau^A\beta_j\in A^*(\tau)\otimes A^*(\bar{g}^{-1}\tau),$$

这里 $f_\tau:\tau\to\sigma,\bar{f}_\tau:\bar{g}^{-1}\tau\to g^{-1}\sigma$ 为自然映像.

记 $A^*(X)\otimes\bar{A}^*(Y)$ 中由一切形如

$$R(f^A)(\xi\otimes\beta)\otimes\eta-\xi\otimes L(g^A)(\beta\otimes\eta)$$

的元素所产生的理想 $(\xi\in A^*(X),\beta\in A^*(B),\eta\in\bar{A}^*(Y))$ 为 W, 则

$$A^*(X)\otimes\bar{A}^*(Y)/W$$

为一 DGA 代数, 记作

$$A^*(X)\underset{A^*(B)}{\otimes}\bar{A}^*(Y).$$

其自然投影为

$$\pi:A^*(X)\otimes\bar{A}^*(Y)\to A^*(X)\underset{A^*(B)}{\otimes}\bar{A}^*(Y).$$

容易验证 $\lambda(W)=0$, 因而 λ 引出一 DGA 同态

$$\mu:A^*(X)\underset{A^*(B)}{\otimes}\bar{A}^*(Y)\to\bar{A}^*(Z)\tag{18}$$

使下图可交换:

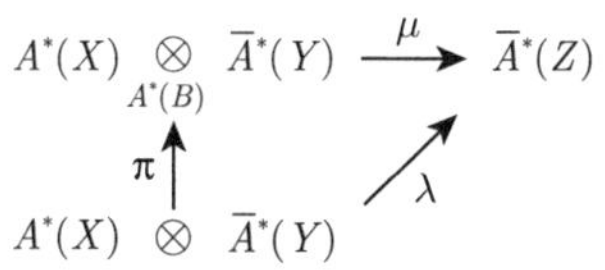

图 9

引入

$$A^+(B)=\{b^{\in A^*(B)}/\deg b>0\},$$

$$P_{-n}(X)=A^*(X)\otimes\underbrace{A^+(B)\otimes\cdots\otimes A^+(B)}_{n}\otimes A^*(B),\quad n\geqslant 0,$$

$$P(X)=\sum_{n\geqslant 0}P_{-n}(X),$$

$$\varepsilon:P(X)\to A^*(X),$$

$$\varepsilon=\begin{cases}R(f^A)|P_0(X),\\ 0\quad|P_{-n}(X),\quad n>0.\end{cases}$$

上述 $P(X)$ 即 $A^*(X)$ 作为 $A^*(B)$ 右 DGA 代数的 Bar 作法. $P(X)$ 在拌和乘积以及通常 "外" 微分 d_E 下成一 DGA 代数, 而

$$\cdots \to P_{-n}(X) \xrightarrow{d_E} P_{-n+1}(X) \xrightarrow{d_E} \cdots \to P_0(X) \xrightarrow{\varepsilon} A^*(X) \to 0$$

构成 $A^*(X)$ 的 $A^*(B)$ 的正常投影分解.

记 $P(X)$ 在全微分下的全复形为 $DP(X)$, 则有

$$H(DP(X)) \approx H(A^*(X)) \approx H^*(X). \tag{19}$$

又 $A^*(B)$ 自然地右作用于 $DP(X)$ 而可定义 DGA 代数

$$P^* = DP(X) \underset{A^*(B)}{\otimes} \bar{A}^*(Y). \tag{20}$$

易见 μ 引出一 DGA 同态

$$\theta : P^* \to \bar{A}^*(Z) \tag{21}$$

使下图可交换:

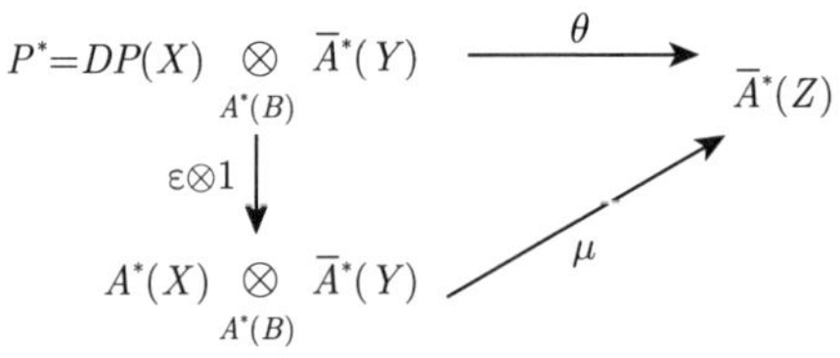

图 10

作为确定 $I^*(Z)$ 的第一步, 将先证明

引理 4 DGA 同态 θ 引出代数同构

$$\theta_* : H(P^*) \approx H(\bar{A}^*(Z)) \approx H^*(Z). \tag{22}$$

由引理 4 即得

推论 1 Z 的 I^* 函子为

$$I^*(Z) \approx \operatorname{Min} P^* \approx \operatorname{Min}(DP(X) \underset{A^*(B)}{\otimes} \bar{A}^*(Y)), \tag{23}$$

Z 的上同调代数则为

$$H^*(Z) \approx H(P^*) \approx \operatorname{Tor}_{A^*(B)}(A^*(X), \bar{A}^*(Y)). \tag{24}$$

引理 4 的证明 依第二节引理 2 分滤 $\bar{A}^*(Y)$ 为

$$F^p \bar{A}^*(Y) = \sum_{r \geqslant p} \bar{A}^{r,*}(Y).$$

由此得到

$$\begin{aligned}&E_0^p\bar{A}^*(Y)\approx\bar{A}^{p,*}(Y),\\&E_1^p\bar{A}^*(Y)\approx A^p(B)\otimes H^*(F),\quad d_1=d|A^*(B),\\&E_2^p\bar{A}^*(Y)\approx H^p(B)\otimes H^*(F).\end{aligned}$$

同样分滤 $\bar{A}^*(Z)$ 为

$$F^p\bar{A}^*(Z)=\sum_{r\geqslant p}\bar{A}^{r,*}(Z),$$

得到

$$\begin{aligned}&E_0^p\bar{A}^*(Z)\approx\bar{A}^{p,*}(Z),\\&E_1^p\bar{A}^*(Z)\approx A^p(X)\otimes H^*(F),d_1=d|A^*(X),\\&E_2^p\bar{A}^*(Z)\approx H^p(X)\otimes H^*(F).\end{aligned}\tag{25}$$

依 $DP(X)$ 的总次数与 $\bar{A}^*(Y)$ 的分滤次数之和来分滤 $P^*=DP(X)\underset{A^*(B)}{\otimes}\bar{A}^*(Y)$:

$$F^pP^*=\sum_i[DP(X)]^i\underset{A^*(B)}{\otimes}F^{p-i}\bar{A}^*(Y).$$

依次可得

$$\begin{aligned}E_0^pP^*&\approx\sum_i[DP(X)]^i\underset{A^*(B)}{\otimes}E_0^{p-i}\bar{A}^*(Y),d_0=d_0|E_0\bar{A}^*(Y),\\E_1^pP^*&\approx\sum_i[DP(X)]^i\underset{A^*(B)}{\otimes}E_1^{p-i}\bar{A}^*(Y)\\&\approx\sum_i[DP(X)]^i\underset{A^*(B)}{\otimes}A^{p-i}(B)\otimes H^*(F)\\&\approx[DP(X)]^p\otimes H^*(F),d_1=d|DP(X),\\E_2^pP^*&\approx H^p(DP(X))\otimes H^*(F)\approx H^p(X)\otimes H^*(F).\end{aligned}\tag{26}$$

对于由 (17), (18) 式和图 9, (20), (21) 式和图 10 所定义的 DGA 同态

$$\theta:P^*=DP(X)\underset{A^*(B)}{\otimes}\bar{A}^*(Y)\to\bar{A}^*(Z).$$

易见有

$$\theta F^pP^*\subset F^p\bar{A}^*(Z),$$

因而 θ 引出 Leray 谱序列各项的 DGA 同态 $\theta_r:E_rP^*\to E_r\bar{A}^*(Z)$, 由 (25), (26) 式得

$$\theta_2:E_2P^*\approx E_2\bar{A}^*(Z),$$

故由谱序列同构定理得代数同构 (22) 式, 如所欲证.

§4　主要定理

设纤维方 (如第三节):

$$\begin{array}{ccc} F \subset Z & \xrightarrow{\bar{f}} & Y \\ \bar{g}\downarrow & & \downarrow g \\ X & \xrightarrow{f} & B \end{array}$$

图 11

按第三节引理 4 的推论, 将 Z 的 I^* 函子与 H^* 函子用 B, X, Y 的 A^* 以及 f, g 表示出:

$$I^*(Z) \approx \operatorname{Min}(DP(X) \underset{A^*(B)}{\otimes} \bar{A}^*(\mathrm{Y})),$$

$$H^*(Z) \approx \operatorname{Tor}_{A^*(B)}(A^*(X), \bar{A}^*(Y)),$$

但 $A^*(B)$ 等并无内在意义. 本节目的在于将 $I^*(Z)$ 与 $H^*(Z)$ 用具有内在意义的 $I^*(B)$ 等以及在 f, g 下的关系来具体表出.

为此, 考虑 I^* 函子与标准 DGA 同态

$$\rho_X : I^*(X) \to A^*(X),$$

$$\rho_Y : I^*(Y) \to \bar{A}^*(Y),$$

$$\rho_B : I^*(B) \to A^*(B).$$

又设与 f, g 相应 I^* 函子间的 DGA 同态为

$$f^I : I^*(B) \to I^*(X),$$

$$g^I : I^*(B) \to I^*(Y),$$

在 f^I 与 g^I 下, $I^*(X)$ 与 $I^*(Y)$ 各可视为 $I^*(B)$ 右 DGA 与 $I^*(B)$ 左 DGA, 与第三节一样, 作 $I^*(X)$ 的 $I^*(B)$ 正常投影分解

$$\varepsilon : P^I(X) \to I^*(X),$$

或

$$\cdots \to P^I_{-n}(X) \xrightarrow{d_E} \cdots \xrightarrow{d_E} P^I_0(X) \xrightarrow{\varepsilon} I^*(X) \to 0,$$

其中

$$P^I_{-n}(X) = I^*(X) \otimes \underbrace{I^+(B) \otimes \cdots \otimes I^+(B)}_{n} \otimes I^*(B), \quad n \geqslant 0,$$

$$I^+(B) = \{\lambda^{\in I^*(B)}/\deg \lambda > 0\},$$

$$P^I(X) = \sum_{n \geqslant 0} P^I_{-n}(X),$$

$$\varepsilon = \begin{cases} R(f^I)|P^I_0(X), \\ 0|P^I_{-n}(X), n > 0, \end{cases}$$

$$R(f^I): \begin{cases} I^*(X) \otimes I^*(B) \to I^*(X), \\ \xi \otimes \beta \to \xi \cdot f^I(\beta). \end{cases}$$

记 $P^I(X)$ 在全微分下的全复形为 $DP^I(X)$. 于是 $\rho_X : I^*(X) \to A^*(X)$ 可导出一保持 $I^*(B)$ 右作用的 DGA 同态

$$\tilde{\rho}_X : DP^I(X) \to DP'(X) = \sum_{n \geqslant 0} A^*(X) \otimes \underbrace{I^+(\sigma) \otimes \cdots \otimes I^+(B)}_{n} \otimes I^*(B),$$

使

$$\rho_X \varepsilon = \varepsilon \tilde{\rho}_X, \quad \text{右面的 } \varepsilon \text{ 由 } R(\rho_X f^I) \text{ 所定.}$$

与第三节推论相当有下述

定理 1 Z 的 I^* 函子为

$$I^*(Z) \approx \mathrm{Min}(DP^I(X) \underset{I^*(B)}{\otimes} I^*(Y)).$$

定理 2 Z 的上同调代数为

$$H^*(Z) \approx \mathrm{Tor}_{I^*(B)}(I^*(X), I^*(Y)).$$

在 f^I, g^I 与 $f^A, \bar{g}^A, \rho_B, \rho_X, \rho_Y$ 可交换, 即 $\rho_X f^I = f^A \rho_B, \rho_Y g^I = g^A \rho_B$ 时, 定理 1, 2 容易由第三节的推论得出. 但一般说来上述同态只是同伦可交换的. 在此情形, 可作 DGA 代数 $J^*(X), J^*(Y)$ 以及 DGA 同态 $h, h_i, k, k_i (i = 1, 2)$, 使在图 12 中有

$$h_0 h = f^A \rho_B, \quad h_1 h = \rho_X f^I,$$

$$k_0 k = g^A \rho_B, \quad k_1 k = \rho_Y g^I,$$

$$h_i^* : H(J^*(X)) \approx H(A^*(X)),$$

$$k_i^* : H(J^*(Y)) \approx H(\bar{A}^*(Y)).$$

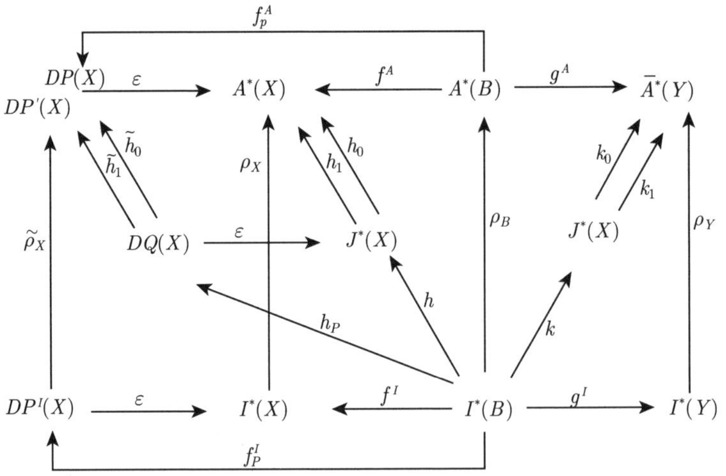

图 12

在 h 下视 $J^*(X)$ 为 $I^*(B)$ 右 DGA, 则与前同样可作 $J^*(X)$ 的正常投影分解:

$$\varepsilon : Q(X) = \sum Q_{-n}(X) \to J^*(X),$$

$$Q_{-n}(X) = J^*(X) \otimes \underbrace{I^+(B) \otimes \cdots \otimes I^+(B)}_{n} \otimes I^*(B), \quad n \geqslant 0.$$

记 $Q(X)$ 在全微分下的全复形为 $DQ(X)$, 则可补出保持 $I^*(B)$ 与 $A^*(B)$ 作用的 DGA 同态

$$\tilde{h}_0 : DQ(X) \to DP(X),$$
$$\tilde{h}_1 : DQ(X) \to DP'(X)$$

使在图 12 中有

$$f_P^A \rho_B = \tilde{h}_0 h_P, \quad \tilde{\rho}_X f_P^I = \tilde{h}_1 h_P,$$

其中 f_P^A, f_P^I, h_P 各为由 f^A, f^I 与 h 所引出的右作用.

此时有以下可交换的 DGA 同态与作用图 13, 其中 $R(\quad), L(\quad)$ 各表示 $(\quad)$ 中同态所引出的右与左作用:

$$
\begin{array}{ccccccc}
\text{I} & DP(X) & \xleftarrow{R(f_P^A)} & A^*(B) & \xrightarrow{L(g^A)} & \bar{A}^*(Y) \\
& \| & & \uparrow{\rho_B} & & \| \\
\text{II} & DP(X) & \xleftarrow{R(f_P^A\rho_B)} & I^*(B) & \xrightarrow{L(g^A\rho_B)} & \bar{A}^*(Y) \\
& \uparrow{\tilde{h}_0} & & \| & & \uparrow{k_0} \\
\text{III} & DQ(X) & \xleftarrow{R(h_P)} & I^*(B) & \xrightarrow{L(k)} & \bar{J}^*(Y) \\
& \uparrow{\tilde{h}_1} & & \| & & \uparrow{k_1} \\
\text{IV} & DP'(X) & \xleftarrow{R(\tilde{h}_1 h_P)} & I^*(B) & \xrightarrow{L(k_1 k)} & \bar{A}^*(Y) \\
& \uparrow{\tilde{\rho}_X} & & \| & & \uparrow{\rho_Y} \\
\text{V} & DP^I(X) & \xleftarrow{R(f_P^I)} & I^*(B) & \xrightarrow{L(g^I)} & I^*(Y)
\end{array}
$$

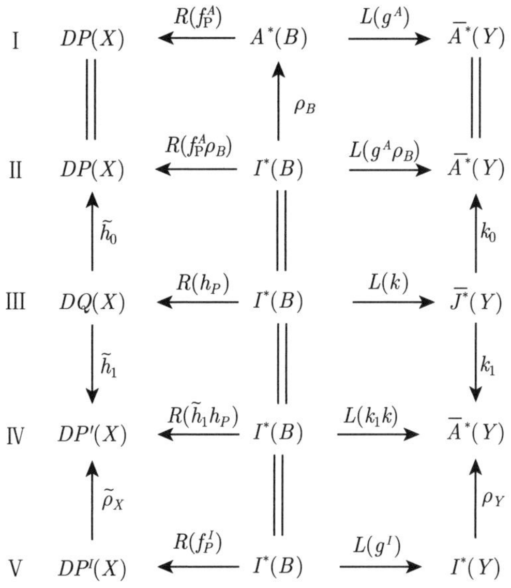

图 13

图 13 中每一行确定了一张量积, 第 i 行者记作 $T^{(i)}$, 例如

$$T^{(1)} = [DP(X) \underset{A^*(B)}{\otimes} \bar{A}^*(Y)]^{(1)}.$$

[ ]$^{(1)}$ 指 $A^*(B)$ 各依第 1 行 f_P^A 与 g^A 右与左作用于 $DP(X)$ 与 $\bar{A}^*(Y)$, 余类推. 各纵行 DGA 同态引出了这些张量积的 DGA 同态

$$T^{(1)} \leftarrow T^{(2)} \leftarrow T^{(3)} \rightarrow T^{(4)} \leftarrow T^{(5)}.$$

由于诸纵向 DGA 同态都引出 H 函子的同构, 因而也有同构

$$H(T^{(1)}) \approx H(T^{(2)}) \approx H(T^{(3)}) \approx H(T^{(4)}) \approx H(T^{(5)}).$$

由此知

$$\text{Min}\, T^{(1)} \approx \text{Min}\, T^{(2)} \approx \cdots \approx \text{Min}\, T^{(5)},$$

而从 $\text{Min}\, T^{(1)} \approx \text{Min}\, T^{(5)}$ 与引理 4 即得定理 1. 定理 2 则为定理 1 的简单推论.

下面是主要定理 1 与 2 的一些特例与简单推论:

1. 若 X 为一点, 则 $Z = F$. 记 $P^I(\underline{R})$ 为 $\underline{R}$ 作为 $I^*(B)$ 平凡作用下的正常投影分解, 则由定理 1, 2 有

$$I^*(F) \approx \text{Min}(DP^I(\underline{R}) \underset{I^*(B)}{\otimes} I^*(Y)),$$

$$H^*(F) \approx \mathrm{Tor}_{I^*(B)}(\underline{R}, I^*(Y)). \tag{27}$$

在某些情形下, 例如 B 或 Y 是 $H-$ 空间或紧致李群的分类空间时有 $I^* \approx H^*$, 于是 (27) 式变为

$$H^*(F) \approx \mathrm{Tor}_{H^*(B)}(\underline{R}, I^*(Y)) \text{ 或 } \mathrm{Tor}_{I^*(B)}(\underline{R}, H^*(Y)) \text{ 等}.$$

2. 设 G 是紧致李群, U 是 G 的闭子群, $G/U = M$ 的 $\pi_1 = 0$. 设 $Y = B_U, B = B_G(\in \mathscr{K})$ 是 U 与 G 的分类空间, $B_U \to B_G$ 为 B_G 上主丛以 G/U 为纤维的相配丛, 考虑纤维方

$$\begin{array}{ccc} G/U \subset Z & \longrightarrow & B_U \\ \downarrow & & \downarrow \\ X & \xrightarrow{f} & B_G \end{array}$$

图 14

则 $Z \to X$ 是以 G/U 为纤维从 f 诱导而来 X 上主丛的相配丛, 此时 $f^*H^*(B_G)$ 构成丛的示性环. 我们称 $f^I H^*(B_G) \subset I^*(X)$ 为丛的 I^*-示性环.

由于此时 $I^*(B_G) \approx H^*(B_G), I^*(B_U) \approx H^*(B_U)$, 故定理 1, 2 给出

$$I^*(Z) \approx \mathrm{Min}(DP^I(\mathrm{X}) \underset{H^*(B_G)}{\otimes} \mathrm{H}^*(\mathrm{B}_U)),$$

$$\mathrm{H}^*(Z) \approx \mathrm{Tor}_{H^*(B_G)}(I^*(X), H^*(B_U)),$$

故 $I^*(Z)$ 与 $H^*(Z)$ 由 $I^*(X)$ 以及丛的 I^*-示性环 (以及带有通用性质的 DGA 同态 $H^*(B_G) \to H^*(B_U)$) 所完全确定, 特别在 X 具有 $I^*(X) \approx H^*(X)$ 的性质, 例如 $X =$ 紧致对称空间时, 上式中 $I^*(X)$ 可易为 $H^*(X)$, 此即 Baum-Smith 在资料 [6] 中所考虑的情形.

3. 结合前面 1 与 2 段使图 14 的 $X =$ 一点, 则得

$$I^*(G/U) \approx \mathrm{Min}(DP^I(\underline{R}) \underset{H^*(B_G)}{\otimes} H^*(B_U)),$$

与

$$H^*(G/U) \approx \mathrm{Tor}_{H^*(B_G)}(\underline{R}, H^*(B_U)).$$

具体计算给出关于 $H^*(G/U)$ 的 Cartan 定理及其对 $I^*(G/U)$ 的推广 (见资料 [7, 8]).

§5 由 Cartan-Leray 同调理论引出的 I^* 函子

前面各节定义于 PL 范畴 $\mathscr{K}$ 或 $\mathscr{W}$ 的 I^* 函子, 也可定义于其他空间范畴而与适当的 H 函子相应. 本节给出与 Cartan-Leray 同调理论相应的 I^* 函子, 并证明与第四节中相当的纤维方定理也同样成立, 所用词汇符号参阅 Sem.Cartan, 1950/1, Exp.14—21, 基环恒假定是实域.

设空间 X, C 是 X 上 DGA 代数所代表的甲, 单位元素记作 $1_C, S$ 是 X 上 DGA 代数所成的束, C 的支柱记作 σ, 对 C 可确定一束记作 $\mathscr{F}$ 而从 S 可定一截面所成的甲, 记作 ΓS. 若 C 到 $\Gamma\mathscr{F}C$ 的自然嵌入是一同构, 则 C 是完全的.

设空间 X, Y 与映像 $f: X\to Y$, 对 X 上的甲 A 可作 Y 上的甲 fA, 其支柱为 $\sigma(fa)=\overline{f(\sigma(a))}$, $a\in A$, 在 A 精致时, fA 也精致, 而 A 精致且完全时, fA 也完全.

设 X 连通与单连通, X 上的甲 C 将称为一良甲, 如果以下诸条件满足:

C1. $C\in\mathscr{A}$, 即 C 为连通单连通的 DGA 代数.

C2. C 的单位元素 1_C 的支柱为 X.

C3. C 精致且完全.

C4. $\mathscr{F}C$ 是 X 上的基束.

命 $\mathscr{H}$ 为具有以下诸性质的空间 X 所成的范畴:

X1. X 为仿紧的连通单连通空间.

X2. X 为有限型空间.

X3. X 上有良甲存在.

作为 $\mathscr{H}$ 中的空间有满足 $X1,2$ 的有限维紧致有距可分空间, 局部有限单纯复形与微分流形等.

设　$X\in\mathscr{H}, C$ 是 X 上的良甲, 则

$$H(C)=H(\Gamma\mathscr{F}C)\approx H^*(X),$$

即 X 在实域上的 H^* 函子.

今设 C_1, C_2 是 $X\in\mathscr{H}$ 上的任两良甲, 则有自然 DGA 同态

$$C_1=\Gamma\mathscr{F}C_1\xrightarrow{f_1}\Gamma\mathscr{F}(C_1\circ C_2)\xleftarrow{f_2}\Gamma\mathscr{F}C_2=C_2,$$

其中 f_1, f_2 各由对应 $\alpha_1\to\alpha_1\circ 1_{C_2}$ 与 $\alpha_2\to 1_{C_1}\circ\alpha_2$ 引出 $(\alpha_1\in C_1, \alpha_2\in C_2)$. 由于 $C_1\circ C_2$ 精致, 且

$$H(\mathscr{F}(C_1\circ C_2))\approx H(\mathscr{F}C_1\otimes\mathscr{F}C_2)\approx H(\mathscr{F}C_1)\otimes H(\mathscr{F}C_2)$$

平凡, 故 $\mathscr{F}(C_1 \circ C_2)$ 为 X 上基束, 而 f_1, f_2 各引出同构

$$H(C_1) \underset{\approx}{\overset{f_{1*}}{\to}} H(\Gamma\mathscr{F}(C_1 \circ C_2)) \underset{\approx}{\overset{f_{2*}}{\leftarrow}} H(C_2) (\approx H^*(X)).$$

因而 $\operatorname{Min} C_1$ 与 $\operatorname{Min} C_2 \in \mathscr{M}$ 在

$$\operatorname{Min} C_i \overset{\rho_i}{\to} C_i \overset{f_i}{\to} \Gamma\mathscr{F}(C_1 \circ C_2), \quad i = 1, 2$$

下面都是 $\Gamma\mathscr{F}(C_1 \circ C_2)$ 的极小模型, 而有

$$\operatorname{Min} C_1 \approx \operatorname{Min} C_2.$$

据此可作

定义 对 $X \in \mathscr{H}$ 由任一 X 上良甲 C 所定的 $\operatorname{Min} C \in \mathscr{M}$, 将称为 X 的 I^* 函子, 记作 $I^*(X)$.

今设空间 $F, B, X, Y, Z \in \mathscr{H}$ 与可交换的映像图

$$\begin{array}{ccc} F \subset Z & \overset{\bar{f}}{\longrightarrow} & Y \\ \bar{g} \downarrow & & \downarrow g \\ X & \overset{f}{\longrightarrow} & B \end{array}$$

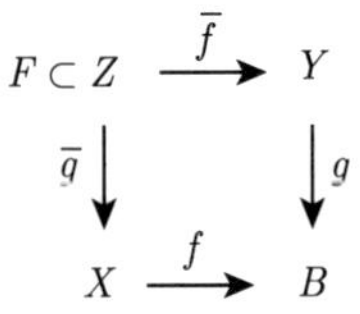

图 15

其中 $g: Y \to B$ 是以 F 为纤维的局部平凡丛, $\bar{g}: Z \to X$ 为 f 下的诱导丛. 称这些空间与映像的集体为 $\mathscr{H}$ 中的纤维方.

在 B, X, Y, Z 上可取良甲 $A^*(B), A^*(X), A^*(Y), A^*(Z)$ 以及 DGA 甲同态

$$\bar{f}^A : A^*(Y) \to A^*(Z),$$
$$g^A : A^*(B) \to A^*(Y),$$
$$f^A : A^*(B) \to A^*(X),$$

各引出 H^* 函子的同态, 由此又有诱导同态

$$g^I : I^*(B) \to I^*(Y),$$
$$f^I : I^*(B) \to I^*(X).$$

我们的目的, 在证明当 F 为紧致时, 与第四节的相应定理也同样成立, 即 $I^*(Z)$ 与 $H^*(Z)$ 由 B, X, Y 的 I^* 函子及 f^I, g^I 所完全确定.

由于 $A^*(Y), A^*(Z)$ 是良甲, $gA^*(Y), \bar{g}A^*(Z)$ 都精致且完全. 故置①

$$\begin{cases} \bar{A}^*(Z) = \varGamma\mathscr{F}(A^*(X) \circ \bar{g}A^*(Z)), \\ \bar{A}^*(Y) = \varGamma\mathscr{F}(A^*(B) \circ gA^*(Y)), \end{cases} \tag{28}$$

则由对应 ($w \in A^*(Z), v \in A^*(Y), 1_B$ 为 $A^*(B)$ 的单位元素, 余类推)

$$w \to 1_X \circ \bar{g}w,$$
$$v \to 1_B \circ gv.$$

所引出 DGA 同态

$$A^*(Z) \stackrel{\approx}{\to} \varGamma\mathscr{F}\bar{g}A^*(Z) \to \bar{A}^*(Z),$$
$$A^*(Y) \stackrel{\approx}{\to} \varGamma\mathscr{F}gA^*(Y) \to \bar{A}^*(Y),$$

依 Cartan 有

$$H^*(Z) \approx H(A^*(Z)) \approx H(\bar{A}^*(Z)),$$
$$H^*(Y) \approx H(A^*(Y)) \approx H(\bar{A}^*(Y)).$$

因而

$$\mathrm{Min}\,\bar{A}^*(Z) \approx \mathrm{Min}\,A^*(Z) \approx I^*(Z),$$
$$\mathrm{Min}\,\bar{A}^*(Y) \approx \mathrm{Min}\,A^*(Y) \approx I^*(Y).$$

对任意 $b \in A^*(B), u \in A^*(X)$, 对应

$$b \to b \circ g1_Y,$$
$$u \to u \circ g1_Z,$$

各引出 DGA 同态

$$\overset{+}{g}{}^* : A^*(B)(\approx \varGamma\mathscr{F}A^*(B)) \to \bar{A}^*(Y),$$
$$\overset{-}{g}{}^* : A^*(X)(\approx \varGamma\mathscr{F}A^*(X)) \to \bar{A}^*(Z).$$

又对任意 $b \in A^*(B), v \in A^*(Y)$, 易证对应

$$j^* : \begin{cases} A^*(B) \otimes gA^*(Y) \to A^*(X) \otimes \bar{g}A^*(Z), \\ b \otimes gv \to f^A b \otimes \bar{g}\bar{f}^A v, \end{cases}$$

① 第二节中 $\bar{A}^*(Y)$ 的引入系仿照这里 (28) 式而得, 再者, 第二节引理 2 证明中的 $E_1^p\bar{A}^*(Y)$ 一式也和这里同样证明 (参阅 Sém.Cartan, 1950/1, Exp.15), 但此时应用局部有限复形 B 中的单位分解.

引出 DGA 同态

$$j^*: \bar{A}^*(Y) \to \bar{A}^*(Z).$$

在 f^A 下, $A^*(B)$ 如下右作用于 $A^*(X)$, 使成为 $A^*(B)$ 右 DGA 代数 ($u \in A^*(X), b \in A^*(B)$):

$$R(f^A): \begin{cases} A^*(X) \otimes A^*(B) \to A^*(X), \\ u \otimes b \to u \cdot f^A b. \end{cases}$$

同样, 在 $\overset{+}{g}{}^*$ 下, $A^*(B)$ 左作用于 $\bar{A}^*(Y)$, 使成为 $A^*(B)$ 左 DGA 代数 ($\eta \in \bar{A}^*(Y), b \in A^*(B)$):

$$L(\overset{+}{g}{}^*): \begin{cases} A^*(B) \otimes \bar{A}^*(Y) \to \bar{A}^*(Y), \\ b \otimes \eta \to g^* b \cdot \eta. \end{cases}$$

与第三节相同引入以下诸 DGA 代数与 DGA 同态的可交换图

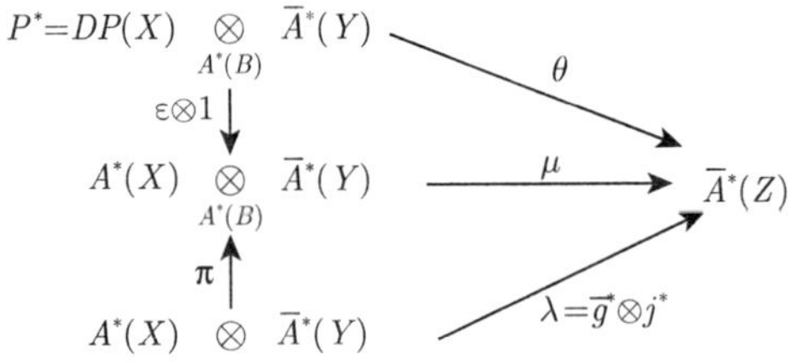
$P^*=DP(X) \underset{A^*(B)}{\otimes} \bar{A}^*(Y)$
θ
$\varepsilon\otimes 1$
$A^*(X) \underset{A^*(B)}{\otimes} \bar{A}^*(Y)$
μ
$\bar{A}^*(Z)$
π
$A^*(X) \otimes \bar{A}^*(Y)$
$\lambda=\bar{g}^*\otimes j^*$

图 16

其中 $\varepsilon: P(X) \to A^*(X)$ 为 $A^*(X)$ 作为 $A^*(B)$ 右 DGA 代数的 Bar 作法, $DP(X)$ 为相应全复形, 余类推.

今在 $\bar{A}^*(Y), \bar{A}^*(Z)$ 以及 P^* 中引入以下诸分滤:

$$F^p\bar{A}^*(Y) = \sum_{i\geqslant p} \Gamma\mathscr{F}(A^i(B) \circ gA^*(Y)),$$

$$F^p\bar{A}^*(Z) = \sum_{i\geqslant p} \Gamma\mathscr{F}(A^i(X) \circ \bar{g}A^*(Z)),$$

$$F^pP^* = \sum_i [DP(X)]^i \underset{A^*(B)}{\otimes} F^{p-i}\bar{A}^*(Y).$$

于是与引理 4 相同, 在 F 紧致的假设下, 有代数同构

$$\theta_*: H(P^*) \approx H(\bar{A}^*(Z)) \approx H^*(Z).$$

由此又得下述

定理 1′ 设 F 紧致则 Z 的 I^* 函子为

$$I^*(Z) \approx \operatorname{Min}(DP^I(X) \underset{I*(B)}{\otimes} I^*(Y)),$$

其中 $P^I(X) \to I^*(X)$ 为 $I^*(X)$, 作为 f^I 下 $I^*(B)$ 右 DGA 代数的 Bar 作法, 而 $DP^I(X)$ 为相应全复形.

定理 2′ 设 F 紧致则 Z 的上同调代数为

$$H^*(Z) \approx \operatorname{Tor}_{I*(B)}(I^*(X), I^*(Y)).$$

证明与第三、四节中相应引理 4 与定理 1, 2 的证明完全平行, 故从略.

参考资料

[1] 吴文俊. 代数拓扑的一个新函子. 科学通报, 1975, 20(7): 311.

[2] Moore J C. *Séminaire Cartan*, 1959-1960, Exp.7.

[3] Smith L. *Trans.Amer.Math.Soc.*, 1967, 129: 58.

[4] Friedlander E, Griffiths P A & Morgan J. *Mimeog.Notes*, 1972.

[5] Sullivan D. *Symp.Tokyo on Topology*, 1973.

[6] Baum P F & Smith L. *Comm.Math.Helo.*, 1967, 42: 171.

[7] Cartan H. *Séminaire Cartan*, 1949-1950, Exp.19-20.

[8] 吴文俊. 代数拓扑 I^* 函子论 —— 齐性空间的实拓扑. 数学学报, 1975, 18(3): 162.

50. Theory of I^*-Functor in Algebraic Topology——Effective Calculation and Axiomatization of I^*-Functor on Complexes*

Abstract According to a classical definition due to Engels, the pure mathematics has space forms and quantitative relations in the exterior world as its objects of study. These two fundamental notions of mathematics are, however, not to be considered as unrelated, but are often interconnected by " measures". Previously we have introduced the concept of I^* which serves as a measure of space forms by means of quantitative relations. This measure is called a "functor" to follow the current terminology in algebraic topology. This I^*-functor or I^*-measure has the advantage over other known functors of being in general "calculable" to be understood roughly in the following sense: If a new space form is constructed geometrically from some given space forms, the I^*-functor of this new space form is completely determined by the I^*-functors of the given space forms. We have given illustrations of this point in various papers. The aims of the present paper are twofold. First, we not only show the calculability of this functor in principle, but also give a method of effective calculations for practical purposes in the case of finite complexes. Secondly, we have listed a set of representative properties of I^* which are sufficient to characterize it completely, forming thus a so-called axiomatic system in the current terminology. The case of infinite complexes is also considered.

In papers [5, 6, 7], the author basing himself on Sullivan's theory of minimal models[2, 4], has introduced the notion of I^*-functor of spaces and has. pointed out that in many cases the I^*-functor is "calculable" while the usual H-and π-functors are often "non-calculable", even restricted to the real field domain. The present paper makes further studies to explain this point. Moreover, we give for the category $\mathscr{K}_0$ of connected, simply-connected finite complexes methods to calculate effectively the I^*-

* *Sci. Sinica*, 1976, 19: 647-664.

functor from the combinatorial structure of the complex and establish also axiomatic system for this functor.

The notations in the present paper as those in the preceding papers[5, 6, 7], are to be understood here.

I I^*-functor of K/L and $K' \cup K''$

Let $K \in \mathscr{K}$, $L \in \mathscr{K}_0$, and $f: L \subset K$. Let $C_L$be the cone over L. Denote the union $K \cup C_L$ by $K/L \in \mathscr{K}$. Then we have a commutative diagram of simplicial maps:

$$\begin{array}{ccc} C_L & \xrightarrow{\tilde{j}} & K/L \\ {\scriptstyle j}\uparrow & & \uparrow{\scriptstyle \tilde{f}} \\ L & \xrightarrow{f} & K \end{array}$$

Set

$$C_f = \sum_{\pi \geqslant 0} C_n,$$

$$C_0 = \{x^{\varepsilon A^*(K)}/x \text{ takes constant value } \varepsilon R \text{ on the simplexes of } L\},$$

$$C_n = \operatorname{Ker}[f_n^A : A^n(K) \to A^n(L)], \quad n > 0.$$

It is easy to see that C_f is a DGA-algebra $\varepsilon\mathscr{A}$ and there is a natural DGA-morphism $i:\ C_f \to A^*(K)$ with the following sequence exact for $n > 0$:

$$0 \to C_n \xrightarrow{i} A^n(K) \xrightarrow{f^A} A^n(L) \to 0.$$

Proposition 1 Min $C_f \approx I^*(K/L)$.

Proof. Let $x \in C_f$ and define $\tau x = \tilde{x} \in A^*(K/L)$ as follows. If $\deg x = 0$ with x taking on the constant value $c \in R$ by all simplexes in L, then set $\tilde{x} \in A^0(K/L)$ to take on the same constant value c on all simplexes in K/L. If deg $x > 0$, then $f^A ix = 0$. Hence we can take $\tilde{x} \in A^*(K/L)$ with $\tilde{f}^A\tilde{x} = ix$ and $\tilde{j}^A\tilde{x} = 0$. Clearly $x \to \tilde{x}$ is a DGA-morphism $\tau : C_f \to A^*(K/L)$ and the diagram below is commutative:

$$\begin{array}{ccccc} & \overset{\tau}{\nearrow} & A^*(K/L) & \xrightarrow{\tilde{j}^A} & A^*(C_L) \\ & & \downarrow{\scriptstyle \tilde{f}^A} & & \downarrow{\scriptstyle j^A} \\ C_f & \xrightarrow{i} & A^*(K) & \xrightarrow{f^A} & A^*(L) \end{array}$$

We prove now

$$\tau_* : H(C_f) \approx H(A * (K/L)) = H * (K/L), \tag{1}$$

from which the proposition follows immediately.

To see this, let $\tilde{x} \in A^*(K/L)$ with $d\tilde{x} = 0$. In case deg $\tilde{x} = 0$ with $\tilde{x}$ taking constant value $c \in R$ on all simplexes of K/L, let us set $x \in A^0(K)$ to take the same constant value c on all simplexes of K. Then $x \in C_0$, $dx = 0$ and $\tau x = \tilde{x}$. If deg $\tilde{x} > 0$, then in C_L we have $d\tilde{j}^A\tilde{x} = 0$ so that there exists $y \in A^*(C_L)$ with $dy = j^A\tilde{x}$. As j^A is an epimorphism, we can take $\tilde{y} \in A^*(K/L)$ with $\tilde{j}^A\tilde{y} = y$. Set $\tilde{z} = \tilde{x} - d\tilde{y}$, then $\tilde{j}^A\tilde{z} = 0$. Hence $f^A\tilde{f}^A\tilde{z} = j^A\tilde{j}^A\tilde{z} = 0$ or $\tilde{f}^A\tilde{z} \in C_f$. Now $\tau\tilde{f}^A\tilde{z} = \tilde{z} = \tilde{x} - d\tilde{y} \sim \tilde{x}$ and $d\tilde{f}^A\tilde{z} = 0$. From these we see that τ_* is an epimorphism.

Next suppose $z \in C_f$ with $dz = 0$ and $\tau z = d\tilde{a}$, $\tilde{a} \in A^*(K/L)$. As $d\tilde{j}^A\tilde{a} = 0$, C_L is contractible and $\tilde{j}^A$ is an epimorphism, there exists $\tilde{b} \in A^*(K/L)$ with $\tilde{j}^A\tilde{a} = d\tilde{j}^A\tilde{b}$. Then $\tilde{f}^A(\tilde{a} - d\tilde{b}) \in C_f$ and $z = d\tilde{f}^A(\tilde{a} - d\tilde{b}) \sim 0$. Hence τ_* is a monomorphism.

The isomorphism (1) is thus proved. Hence the Proposition 1.

Theorem 1 *$I^*(K/L)$ is completely determined by the natural DGA-morphism*

$$g = f^I : I^*(K) \to I^*(L). \tag{2}$$

Proof. Set $I^*(K) = M$, $I^*(L) = N$. From §§15—16 of [2], we see easily that there exist $K' \in \mathscr{K}$, $L' \in \mathscr{K}_0$, $f' : L' \subset K'$ with $M \approx I^*(K')$, $N \approx I^*(L')$ and the following diagram is homotopically commutative (ρ being the canonical homomorphisms) :

$$\begin{array}{ccc} M & \xrightarrow{\quad g \quad} & N \\ {\scriptstyle\rho}\downarrow & & \downarrow{\scriptstyle\dot{\rho}} \\ A^*(K') & \xrightarrow{\quad f'^A \quad} & A^*(L') \end{array}$$

Moreover, we have $K' \simeq K$, $L' \simeq L$, $f' \simeq f$, $K'/L' \simeq K/L$ so that Proposition 1 gives

$$I^*(K/L) \approx I^*(K'/L') \approx \text{Min } C_{f'}.$$

As K', L', f' are constructed from (2) , so $I^*(K/L)$ is completely determined by (2) , as to be proved.

The determination of $I^*(K/L)$ (to be denoted by J_g) and the natural DGA-morphisms $I^*(K/L) \to I^*(K)$ (to be denoted by j_g) from (2) or

$$g : M \to N$$

will be called the *J-construction*. The above gives only an existence proof of such J-construction. In the next section, we shall give some explicit constructions of J_g and j_g in the special case $L = S^n$.

Entirely analogous to Proposition 1 and Theorem 1, the same method can be applied to the study of union of complexes as in the Mayer-Vietoris sequence. Thus, let K, K', K'', $L \in \mathscr{K}$, $K = K' \cup K''$, $L = K' \cap K''$ and $f' : L \subset K'$, $f'' : L \subset K''$. Set $D_{f', f''} = \sum_{n \geqslant 0} D_n$, where

$$\begin{aligned} D_0 =& \{(a', a'')^{\varepsilon A^0(K') \times A^0(K'')} / a', a'' \text{ take the same constant} \\ & \text{values } \varepsilon R \text{ on simplexes of } L\}, \\ D_n =& \{(a', a'')^{\varepsilon A^n(K') \times A^n(K'')} / f'^A a' = f''^A a''\}, \quad n > 0. \end{aligned}$$

Then $D_{f', f''}$ forms naturally a DGA-algebra easily seen to be $\varepsilon \mathscr{A}$. We have then (proof omitted) :

Proposition 2 Min $D_{f', f''} \approx I^*(K' \cup K'')$.

Theorem 2 *$I^*(K' \cup K'')$ is completely determined by*

$$f'^I : I^*(K') \to I^*(L),$$

and

$$f''^I : I^*(K'') \to I^*(L).$$

II *J*-construction of DGA-morphism $g : M \to N$ IN case $N \approx I^*(S^n)$

Let M, $N \in \mathscr{M}$ and $N \approx I^*(S^n)$, $n \geqslant 2$. The purpose of this section is to construct explicitly $J_g \in \mathscr{M}$ and the DGA-morphism

$$j_g : J_g \to M$$

from a given DGA-morphism $g : M \to N$. This is the J-construction of I. For this, let

$$N \approx I^*(S^n) = \begin{cases} \Lambda(s), & n = \text{odd}, \\ \Lambda(s, t), & n = \text{even}, \end{cases}$$

in which

$$\deg s = n, \quad ds = 0,$$

and also

$$\deg t = 2n - 1, \quad dt = s^2,$$

for even n. Let $\bar{N} \in \mathscr{A}$ be the DGA-algebra as follows:

$$\bar{N} = R \oplus R\bar{s},$$

the degrees, multiplications and differentiations in $\bar{N}$ being given by

$$\deg \bar{s} = n, \quad \bar{s}^2 = 0, \quad d\bar{s} = 0.$$

Let $\tilde{S} \in \mathscr{A}$ be the DGA-algebra

$$\tilde{S} = R \oplus R\tilde{s},$$

with degrees, multiplications and differentiations as follows:

$$\deg \tilde{s} = n + 1, \quad \tilde{s}^2 = 0, \quad d\tilde{s} = 0.$$

Denote by γ the natural DGA-morphism from N to $\overline{N}$ which maps s to $\bar{s}$, and also t to 0 in case n is even. Set

$$\gamma g = \bar{g} : M \to \overline{N}.$$

Define now $K_{\bar{g}}$ according as $\bar{s} \in \operatorname{Im} \bar{g}$ or not:

$$K_{\bar{g}} = \begin{cases} R \oplus \sum_{r>0} \bar{g}_r^{-1}(0), & \text{for } \bar{s} \in \operatorname{Im} \bar{g}, \\ M \otimes \tilde{S}, & \text{for } \bar{s} \notin \operatorname{Im} \bar{g}. \end{cases}$$

Clearly $K_{\bar{g}}$ under operations in M, is a DGA-algebra of $\mathscr{A}$. Define also a DGA-morphism

$$k_{\bar{g}} : K_{\bar{g}} \to M$$

as follows. In case $\bar{s} \in \operatorname{Im} \bar{g}$, $k_{\bar{g}}$ is the natural inclusion, while in case $\bar{s} \notin \operatorname{Im} \bar{g}$, we set

$$k_{\bar{g}}(a \otimes 1) = a, k_{\bar{g}}(a \otimes \tilde{s}) = 0 \quad (a \in M).$$

In what follows we shall construct $J_{\bar{g}} \in \mathscr{M}$ and DGA-morphism $j_{\bar{g}} : J_{\bar{g}} \to M$ from $K_{\bar{g}}$ and $k_{\bar{g}}$, and $J_{\bar{g}}, j_{\bar{g}}$ will be taken to be the $J_{\bar{g}}$ and $j_{\bar{g}}$ in the beginning of this section, which will be discussed in two seperate cases.

Case I. $\bar{s} \in \operatorname{Im} \bar{g}$.

Define now

$$J_{\bar{g}} = \text{Min } K_{\bar{g}} \in \mathscr{M},$$

and also

$$j_{\bar{g}} = k_{\bar{g}}\rho_{\bar{g}} : J_{\bar{g}} \to M,$$

in which $\rho_{\bar{g}} : J_{\bar{g}} \to K_{\bar{g}}$ is any canonical homomorphism.

Case II. $\bar{s} \notin \text{Im } \bar{g}$.

Take any set $x_i \in Z(M)$ forming an additive homology basis of $H(M)$. Then $H(K_{\bar{g}})$ has an additive homology basis:

$$x_i, \tilde{s}, x_i \otimes \tilde{s}.$$

In the set $x_i \otimes \tilde{s}$, let those of the lowest degree (say $m_{\bar{g}} \geqslant n+3$) be denoted by y_j, while the others with degree $> m_{\bar{g}}$ be denoted by z_k. Introduce η_j and construct the DGA-algebra

$$K_{\bar{g}}^1 = K_{\bar{g}} \otimes \Lambda(\eta_j),$$

in which

$$\deg \eta_j = m_{\bar{g}} - 1,$$
$$d\eta_j = y_j.$$

As $k_{\bar{g}}(y_j) \sim 0$ (in fact $= 0$) , we may take $a_j \in M$ with $k_{\bar{g}}(y_j) = da_j$ (in fact a_j may be taken to be 0) . Define the DGA-morphism

$$k_{\bar{g}}^1 : K_{\bar{g}}^1 \to M,$$

by

$$k_{\bar{g}}^1(a) = k_{\bar{g}}(a), \quad a \in K_{\bar{g}},$$
$$k_{\bar{g}}^1(\eta_j) = a_j.$$

It is easy to see that $K_{\bar{g}}^1$ has an additive homology basis:

$$x_i, \tilde{s}, y_j^1, z_k^1,$$

in which

$$\text{deg} y_j^1 = m_{\bar{g}}^1 > m_{\bar{g}},$$
$$\deg z_k^1 > m_{\bar{g}}^1,$$

and under $k^1_{\bar{g}}$,

$$k^1_{\bar{g}}(y^1_j) \sim 0, \quad k^1_{\bar{g}}(z^1_k) \sim 0.$$

Now introduce η^1_j and construct the DGA-algebra

$$K^2_{\bar{g}} = K^1_{\bar{g}} \otimes \Lambda(\eta^1_j),$$

in which

$$\deg \eta^1_j = m^1_{\bar{g}} - 1, \quad d\eta^1_j = y^1_j.$$

Define also a DGA-morphism

$$k^2_{\bar{g}} : K^2_{\bar{g}} \to M,$$

by

$$k^2_{\bar{g}}(a) = k^1_{\bar{g}}(a), \quad a \in K^1_{\bar{g}},$$
$$k^2_{\bar{g}}(\eta^1_j) = a^1_j,$$

in which

$$a^1_j \in M, \quad \deg a^1_j = m^1_{\bar{g}} - 1, \quad da^1_j = k^1_{\bar{g}}(y^1_j).$$

In $K^2_{\bar{g}}$,there is additive homology basis as before and the preceding constructions can be continued to get

$$K_{\bar{g}} \subset K^1_{\bar{g}} \subset K^2_{\bar{g}} \subset \cdots .$$

In this manner, we get a DGA-algebra

$$K^\infty_{\bar{g}} = K_{\bar{g}} \otimes \Lambda \quad (\eta_j, \eta^1_j, \eta^2_j, \cdots),$$

and a DGA-morphism

$$k^\infty_{\bar{g}} : K^\infty_{\bar{g}} \to M,$$

with

$$k^\infty_{\bar{g}}(a) = k_{\bar{g}}(a), \quad a \in K_{\bar{g}},$$
$$k^\infty_{\bar{g}}(\eta_j) = a_j, \quad k^\infty_{\bar{g}}(\eta^r_j) = a^r_j, \quad r = 1, 2, \cdots .$$

Moreover $H(K^\infty_{\bar{g}}) \approx H(M) \oplus H(\tilde{S})$ has an additive homology basis

$$x_i, \bar{s}.$$

Define now the DGA-algebra

$$J_{\bar{g}} = \mathrm{Min}\, K^\infty_{\bar{g}} \in \mathscr{M},$$

and the DGA-morphism

$$j_{\bar{g}} : J_{\bar{g}} \to M$$

by

$$j_{\bar{g}} = k_{\bar{g}}^{\infty} \rho_{\bar{g}}^{\infty},$$

in which

$$\rho_{\bar{g}}^{\infty} : J_{\bar{g}} \to K_{g}^{\infty}$$

is any canonical homomorphism. Then we have

$$H_q(J_{\bar{g}}) \approx H_q(M) \oplus H_q(\tilde{S}), \quad q > 0,$$

and

$$j_{\bar{g}*} : H(J_{\bar{g}}) \to H(M)$$

is an isomorphism on $H(M)$, and is 0 on $H(\tilde{S})$.

III Privileged morphisms of minimal models

Let $A,\ B \in \mathscr{A}$, $f : A \to B$ be a DGA-morphism, M =Min$A$, N = Min$B$, and $\rho_A : M \to A$, $\rho_B : N \to B$ be canonical homomorphisms. The collection of all DGA-morphisms $g : M \to N$ induced from f will be denoted by $G(f)$. In general, the diagram

$$\begin{array}{ccc} A & \xrightarrow{\ f\ } & B \\ {\scriptstyle \rho_A}\uparrow & & \uparrow{\scriptstyle \rho_B} \\ M & \xrightarrow{\ g\ } & N \end{array} \tag{1}$$

is only homotopically commutative. It is easy to give examples with f given for which no ρ_A, ρ_B and g can be chosen to make (1) commutative. However, when f is an epimorphism, e.g. when in the case $A = A^*(K)$, $B = A^*(L)$ and the DGA-morphism $f = f^A : A \to B$ induced from a simplicial map $f : L \subset K$, then we have the following:

Theorem 3 *Let $A, B \in \mathscr{A}$ and $f : A \to B$ be a DGA-epimorphism. Then for $M =$ Min A, $N =$ Min B and a given canonical homomorphism $\rho_B : N \to B$, there are a canonical homomorphism ρ_A: $M \to A$ and a DGA-morphism $g : M \to N$ such that (1) is commutative.*

Definition and notation The morphisms $g : M \to N$ in the theorem will be called *privileged morphisms* associated with $f : A \to B$ whose collection will be denoted by $G^0(f) \subset G(f)$. In case $f : L \subset K$ and $f = f^A : A^*(K) \to A^*(L)$, $G^0(f)$ and $G(f)$ will also be denoted by $G^0(K, L)$ and $G(K, L)$ respectively.

Proof of Theorem 3. We shall go into detail of the proof only in the case $N \approx I^*(S^n)$ and $B_n = 0$ for $m > n$. The proof of the general case is similar, but more complicate.

For this, let us take s as the generator of degree n in N, and set $\rho_B s = c$. As f is an epimorphism, we have $a \in A$ with $fa = c$. The choice of such an a will be explained below.

Denote by $M^{(m)}$ the minimal DGA-algebra generated by generators of M of degree $\leqslant m$, with $M^{(0)} = M^{(1)} = R$. We shall extend $M^{(m)}$ successively as

$$M^{(0)} = M^{(1)}(= R) \subset M^{(2)} \subset \cdots \subset M^{(m)} \subset M^{(m+1)} \subset \cdots \subset M,$$

and define DGA-morphisms

$$\rho_A^{(m)} : M^{(m)} \to A,$$
$$g^{(m)} : M^{(m)} \to N,$$

such that the following induction hypothesis is observed:

$H1_m^0$. $\rho_A^{(m)}/M^{(m-1)} = \rho_A^{(m-1)}, \quad g^{(m)}/M^{(m-1)} = g^{(m-1)}$.

$H2_m^0$. $\rho_{A*}^{(m)} : \begin{cases} H_q(M^{(m)}) \approx H_q(A), & q \leqslant m, \\ H_{m+1}(M^{(m)}) \subset H_{m+1}(A). \end{cases}$

$H3_m^0$. The diagram below is commutative :

$$\begin{array}{ccc} A & \xrightarrow{\ f\ } & B \\ {\scriptstyle \rho_A^{(m)}}\uparrow & & \uparrow{\scriptstyle \rho_B} \\ M^{(m)} & \xrightarrow{\ g^{(m)}\ } & N \end{array}$$

We shall construct successively from

$$G^{(m)} = \{M^{(m)}, \rho_A^{(m)}, g^{(m)}\},$$

which satisfies $H1_m^0 - 3_m^0$ to the set

$$G^{(m+1)} = \{M^{(m+1)}, \rho_A^{(m+1)}, g^{(m+1)}\},$$

satisfying $H1^0_{m+1} - 3^0_{m+1}$ as follows.

The construction of $G^{(2)}$ is easy. Suppose that $G^{(m)}$ has been constructed with $m \leqslant n-2$.

From the induction hypothesis $H1^0_m - 2^0_m$, we have exact sequences

$$\begin{cases} 0 \to H_{m+1}(M^{(m)}) \overset{\rho^{(m)}_{A*}}{\to} H_{m+1}(A) \to \mathrm{Coker}_{m+1}\rho^{(m)}_{A*} \to 0, \\ 0 \to \mathrm{Ker}_{m+2}\rho^{(m)}_{A*} \to H_{m+2}(M^{(m)}) \overset{\rho^{(m)}_{A*}}{\to} H_{m+2}(A). \end{cases} \tag{2$_m$}$$

Take now

$$\begin{cases} e^{(m)}_i \in A_{m+1}, de^{(m)}_i = 0, \\ \xi^{(m)}_j \in M^{(m)}_{m+2}, d\xi^{(m)}_j = 0, \\ x^{(m)}_j \in A_{m+1}, \rho^{(m)}_A \xi^{(m)}_j = dx^{(m)}_j, \end{cases} \tag{3$_m$}$$

such that the $e^{(m)}_i \in Z_{m+1}(A)$ form an additive homology basis of $\mathrm{Coker}_{m+1}\rho^{(m)}_{A*}$, and the $\xi^{(m)}_j \in Z_{m+2}(M^{(m)})$ form one of $\mathrm{Ker}_{m+2}\rho^{(m)}_{A*}$.

Since $H(B) \approx H(N) \approx H(S^n)$ and $m \leqslant n-2$, we have $fe^{(m)}_i \sim 0$. As f is an epimorphism, we have $h^{(m)}_i \in A_m$ with

$$fe^{(m)}_i = dfh^{(m)}_i. \tag{4$_m$}$$

From $m \leqslant n-2$ we have further

$$\rho_B g^{(m)} \xi^{(m)}_j = 0. \tag{5$_m$}$$

From $H3^0_m$ we get

$$dfx^{(m)}_j = fdx^{(m)}_j = f\rho^{(m)}_A \xi^{(m)}_j = \rho_B g^{(m)} \xi^{(m)}_j = 0,$$

so that $fx^{(m)}_j \in Z_{m+1}(B)$. As before, we have $y^{(m)}_j \in A_m$ with

$$fx^{(m)}_j = dfy^{(m)}_j. \tag{6$_m$}$$

Define now

$$M^{(m+1)} = M^{(m)} \otimes \Lambda(\varepsilon^{(m+1)}_i, \zeta^{(m+1)}_j), \tag{7$_m$}$$

in which

$$\begin{cases} \deg \varepsilon^{(m+1)}_i = \deg \xi^{(m+1)}_j = m+1, \\ d\varepsilon^{(m+1)}_i = 0, \\ d\zeta^{(m+1)}_i = \xi^{(m)}_j \in M^{(m)}_{m+2}. \end{cases} \tag{8$_m$}$$

Then $M^{(m+1)} \in \mathscr{M}$. Define also the morphism

$$\rho_A^{(m+1)} : M^{(m+1)} \to A$$

by

$$\begin{cases} \rho_A^{(m+1)}/M^{(m)} = \rho_A^{(m)}, \\ \rho_A^{(m+1)}\varepsilon_i^{(m+1)} = e_i^{(m)} - dh_i^{(m)}, \\ \rho_A^{(m+1)}\zeta_j^{(m+1)} = x_j^{(m)} - dy_j^{(m)}. \end{cases} \tag{9$_m$}$$

Clearly $d\rho_A^{(m+1)}\varepsilon_i^{(m+1)} = \rho_A^{(m+1)}d\varepsilon_i^{(m+1)} = 0$ and

$$d\rho_A^{(m+1)}\zeta_j^{(m+1)} = dx_j^{(m)} = \rho_A^{(m)}\xi_j^{(m)} = \rho_A^{(m+1)}\xi_j^{(m)} = \rho_A^{(m+1)}d\zeta_j^{(m+1)}.$$

Hence $\rho_A^{(m+1)}$ can be uniquely extended to a DGA-morphism from $M^{(m+1)}$ to A which will still be denoted by $\rho_A^{(m+1)}$. If we define

$$g^{(m+1)} = 0, \tag{10$_m$}$$

then from $(9)_m$, $(10)_m$ we see that $G^{(m+1)}$ so obtained will satisfy $H1^0_{m+1}$ and $H3^0_{m+1}$. Besides, $H2^0_{m+1}$ is also easily verified.

Suppose now $m = n-1$ and construct $G^{(n)}$ from $G^{(n-1)}$ as follows.

Take as before $e_i^{(n-1)} \in A_n$, $\xi_j^{(n-1)} \in M_{n+1}^{(n-1)}$, $x_j^{(n-1)} \in A_n$ to satisfy $(3)_{n-1}$ such that $e_i^{(n-1)}$ form an additive homology basis of $\mathrm{Coker}_n\rho_{A*}^{(n-1)}$, and $\xi_j^{(n-1)}$, one of $\mathrm{Ker}_{n+1}\rho_{A*}^{(n-1)}$.

As $f\,e_i^{(n-1)} \in Z_n(B)$ and $H_n(B) \approx R$ is generated by c, there are $r_i \in R$ and $h_i^{(n-1)} \in A_{n-1}$ such that

$$fe_i^{(n-1)} = r_i c + dfh_i^{(n-1)}. \tag{4$_{n-1}$}$$

If some $r_i \neq 0$, then $f_* : H_n(A) \to H_n(B)$ is an epimorphism and a will be chosen with $da = 0$. Otherwise a will be chosen arbitrarily. Furthermore, as $g^{(n-1)} = 0$, we have as before $fx_j^{(n-1)} \in Z_n(B)$ so that there are $y_j^{(n-1)} \in A_{n-1}$ and $r'_j \in R$ such that

$$fx_j^{(n-1)} = r'_j c + dfy_j^{(n-1)}. \tag{6$_{n-1}$}$$

Now we define $G^{(n)}$ by $(7)_{n-1} - (10)_{n-1}$, where $(9)_{n-1}$ and $(10)_{n-1}$ in view of $(4)_{n-1}$ and $(6)_{n-1}$ are, however, replaced by formulas below:

$$\begin{cases} \rho_A^{(n)}/M^{(n-1)} = \rho_A^{(n-1)}, \\ \rho_A^{(n)}\varepsilon_i^{(n)} = e_i^{(n-1)} - dh_i^{(n-1)} - r_i a, \\ \rho_A^{(n)}\zeta_j^{(n)} = x_j^{(n-1)} - dy_j^{(n-1)}, \end{cases} \tag{9$_{n-1}$}$$

$$\begin{cases} g^{(n)}/M^{(n-1)} = g^{(n-1)} = 0, \\ g^{(n)}\varepsilon_i^{(n)} = 0, \\ g^{(n)}\zeta_j^{(n)} = r'_j s. \end{cases} \qquad (10)_{n-1}$$

It is easy to see that $\rho_A^{(n)}$ can be extended to a DGA-morphism with the obtained $G^{(n)}$ satisfying $H1_n^0 - 3_n^0$.

Let $m \geqslant n$. Then $(5)_n$ holds still since $B_{m+2} = 0$. Hence we can construct $G^{(m+1)}$ from $G^{(m)}$ as before.

Set now

$$M = \bigcup_m M^{(m)},$$
$$\rho_A : M \to A,$$
$$g : M \to N,$$

such that

$$\rho_A/M^{(m)} = \rho_A^{(m)},$$
$$g/M^{(m)} = g^{(m)}.$$

Then $\{M, \rho_A, g\}$ thus obtained, meets the requirement of the theorem.

Ⅳ *J*-construction determined by combinatorial spheres in *K*

Let $K \in \mathscr{K}$, $f : L \subset K$, and L be a combinatorial sphere of dimension $n \geqslant 2$, C_L be a combinatorial cell with L as boundary, and K/L be the complex $K \cup C_L \in \mathscr{K}$. Denote for simplicity

$$A = A^*(K), \quad B = A^*(L), \quad \tilde{A} = A^*(K/L),$$

and let $f = f^A : A \to B$, $C = C_f$, $i : C \subset A$, $\tau : C \to \tilde{A}$, $\tilde{f} = \tilde{f}^A : \tilde{A} \to A$ as in I. Set also $M \approx \text{Min}\, A \approx I^*(K)$, $N \approx \text{Min}\, B \approx I^*(L) \approx I^*(S^n)$, then by III there are DGA-morphisms $g : M \to N$, $g \in G^0(K, L)$ with the diagram below commutative (ρ being the canonical homomorphisms) :

$$\begin{array}{ccc} A & \xrightarrow{f} & B \\ \rho \uparrow & & \uparrow \rho \\ M & \xrightarrow{g} & N \end{array}$$

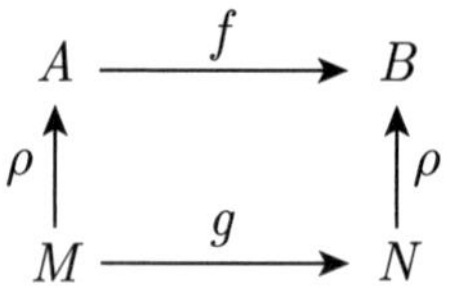

Let $\bar{N} = R \oplus R\bar{s}$, $\gamma : N \to \bar{N}$, and $\bar{g} = \gamma g : M \to \bar{N}$ as in II. As $\bar{g}_n : M_n \to \bar{N}_n$ is nothing but the homomorphism $f^* : \pi_n^*(K) \to \pi_n^*(L)$ induced by $f : L \to K$, the morphism $\bar{g}$ is completely determined by $L \subset K$.

Definition. The unique DGA-morphism

$$\bar{g} : M \to \bar{N}$$

determined by the combinatorial sphere $L \subset K(L, K \in \mathscr{K})$ will be called the *characteristic homomorphism* of L w. r. t. K.

The present section then aims at proving the following:

Theorem 4 *From the J-construction (II) of the characteristic homomorphism* $\bar{g} : M \to \bar{N}$*, we get*

$$J_{\bar{g}} \approx \operatorname{Min} \tilde{A} \approx I^*(K/L),$$

$$j_{\bar{g}} \in G^0(K/L, K),$$

which make the diagram below commutative ($\rho, \tilde{\rho}$ being canonical homomorphisms):

$$\begin{array}{ccc} \tilde{A} & \xrightarrow{\tilde{f}} & A \\ \tilde{\rho} \uparrow & & \uparrow \rho \\ J_{\bar{g}} & \xrightarrow{j_{\bar{g}}} & M \end{array}$$

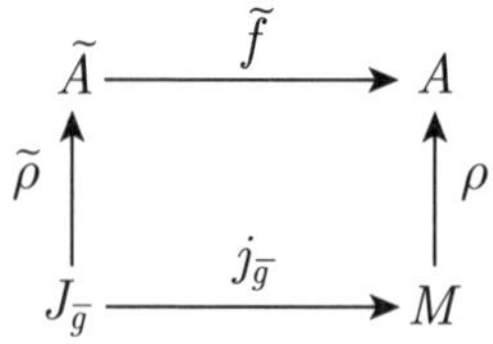

Proof. We shall distinguish two cases whether $\bar{s} \in \operatorname{Im} \bar{g}$ or not.

Case 1. $\bar{s} \in \operatorname{Im} \bar{g}$.

In this case we have $J_{\bar{g}} = \operatorname{Min} K_{\bar{g}}$, $K_{\bar{g}} = R \oplus \sum_{r>0} \bar{g}_r^{-1}(0)$, $k_{\bar{g}} : K_{\bar{g}} \subset M$, $j_{\bar{g}} = k_{\bar{g}} \rho_{\bar{g}} : J_{\bar{g}} \to M$, and $\rho_{\bar{g}} : J_{\bar{g}} \to K_{\bar{g}}$ is some canonical homomorphism. Consider now the following diagram

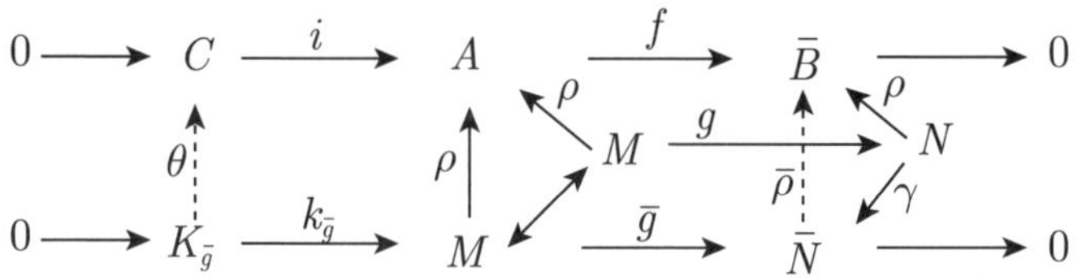

in which the two rows are. exact in each positive degree, while the morphism $\bar{\rho} : \bar{N} \to B$ is determined by $\bar{\rho}(\bar{s}) = \rho s$. Since the degrees of elements of $B \approx A^*(L)$ are all$\leqslant n$, $\bar{\rho}$ is naturally a DGA-morphism. Then $\theta : K_{\bar{g}} \to C$ is the DGA-morphism uniquely determined to make the above diagram commutative. From $\rho_* : H(M) \approx H(A)$, $\bar{\rho}_* : H(\bar{N}) \approx H(B)$ and the 5-Lemma we get

$$\theta_* : H_r(K_{\bar{g}}) \approx H_r(C), \quad r > 0.$$

Moreover, θ_* is clearly an isomorphism also for $r = 0$.

Define now

$$\tilde{\rho} : J_{\bar{g}} \to \tilde{A}$$

by

$$\tilde{\rho} = \tau\theta\rho_{\bar{g}}.$$

Then the following diagram is commutative:

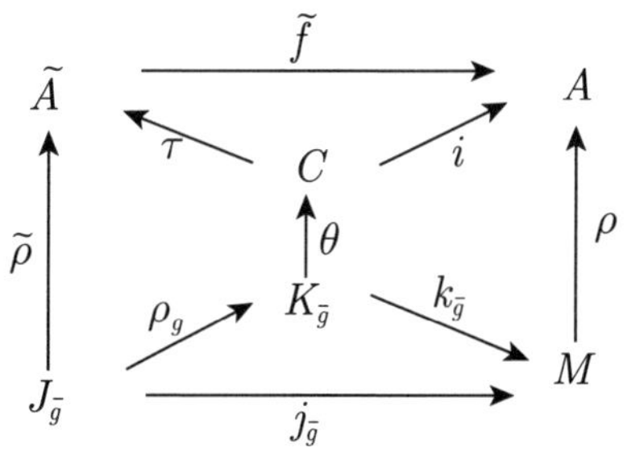

By Proposition 1 of I, $\tau_* : H(C) \approx H(\tilde{A})$. Hence we get

$$\tilde{\rho}_* : H(J_{\bar{g}}) \approx H(\tilde{A}).$$

It follows that $J_{\bar{g}} \approx \text{Min}\tilde{A} \approx I^*(K/L)$, $\tilde{\rho}$ is a canonical homomorphism, and $j_{\bar{g}} \in G^0(K/L, K)$.

Case II. $\bar{s} \notin \text{Im}\, \bar{g}$.

This time $\rho M \subset C$ so that $\rho : M \to A$ determines a DGA-morphism $\rho^c : M \to C$ with $i\rho^c = \rho$. We have then

$$\begin{aligned} H_q(C) &= \rho^c_* H_q(M), \quad q \neq n+1, \\ H_{n+1}(C) &= \rho^c_* H_{n+1}(M) \oplus \delta_* H_n(B), \\ \rho^c_* &: H(M) \subset H(C), \\ \delta_* &: H_n(B) \subset H_{n+1}(C). \end{aligned}$$

The generator of $\delta_* H_n(B)$ is given as follows. Take $a \in A$ with $fa = \rho s$ owing to the epimorphism $f: A \to B$. Then $da \in C$, and $da \in Z(C)$ forms an additive homology basis of $\delta_* H_n(B) \subset H_{n+1}(C)$. Take now any additive homology basis $x_i \in Z(M)$ of $H(M)$, then $H(C)$ has an additive homology basis, viz.

$$\rho^C x_i, \quad da.$$

As $\tilde{f}: \tilde{A} \to A$ is an epimorphism, there exist $\tilde{a} \in \tilde{A}$ with $\tilde{f}\tilde{a} = a$. In what follows we shall suppose that a, $\tilde{a}$ have been so chosen and taken to be fixed.

Define now
$$\varphi: M \otimes \tilde{S} \to \tilde{A},$$
such that
$$\varphi(x) = \tau \rho^c(x), \quad x \in M,$$
$$\varphi(\tilde{s}) = \tau da - d\tilde{a}.$$

As C_L is a combinatorial cell of dimension $n+1$, $(\tau da - d\tilde{a})^2$ is 0 on C_L and a *fortiori* also 0 on K, so $(\tau da - d\tilde{a})^2 = 0$. Hence the above two expressions determine φ to be a multiplicative homomorphism. It is easy to see that φ is a DGA-morphism and the diagram below is commutative:

$$\begin{array}{ccc} \tilde{A} & \xrightarrow{\tilde{f}} & A \\ {\scriptsize \varphi}\uparrow & & \uparrow{\scriptsize \rho} \\ M \otimes \tilde{S} & \xrightarrow{k_{\tilde{g}}} & M \end{array}$$

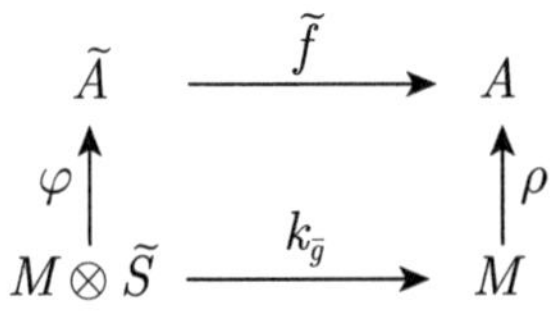

Since $\tau da - d\tilde{a} \sim \tau da$ (in $\tilde{A}$) and $\tau_*: H(C) \approx H(\tilde{A})$, $H(\tilde{A})$ has an additive homology basis, viz.

$$\tau \rho^c x_i, \quad \tau da - d\tilde{a},$$

or

$$\varphi(x_i), \quad \varphi(\tilde{s}),$$

in which the x_i form an additive homology basis of $H(M)$ as before.

Starting from the additive homology basis

$$x_i, \quad \tilde{s}, \quad x_i \otimes \tilde{s}$$

of $H(M \otimes \tilde{S})$, let us now construct successively according to II

$$M \otimes \tilde{S} = K_{\bar{g}} \subset K_{\bar{g}}^1 \subset K_{\bar{g}}^2 \subset \cdots,$$
$$K_{\bar{g}}^\infty = K_{\bar{g}} \otimes \Lambda \ (\eta_j, \eta_j^1, \eta_j^2, \cdots),$$
$$J_{\bar{g}} = \text{Min } K_{\bar{g}}^\infty,$$

and

$$k_{\bar{g}}^i : K_{\bar{g}}^i \to M, \quad k_{\bar{g}}^\infty : K_{\bar{g}}^\infty \to M,$$
$$\rho_{\bar{g}}^\infty : J_{\bar{g}} \to K_{\bar{g}}^\infty, \quad j_{\bar{g}} = k_{\bar{g}}^\infty \rho_{\bar{g}}^\infty : J_{\bar{g}} \to M.$$

Prove now $\varphi : K_{\bar{g}} \to \tilde{A}$ can be successively extended to DGA-morphisms

$$\varphi^i : K_{\bar{g}}^i \to \tilde{A},$$

with following diagram commutative

$$\begin{array}{ccc} \tilde{A} & \xrightarrow{\tilde{f}} & A \\ \varphi^i \uparrow & & \uparrow \rho \\ K_{\bar{g}}^i & \xrightarrow{k_{\bar{g}}^i} & M \end{array}$$

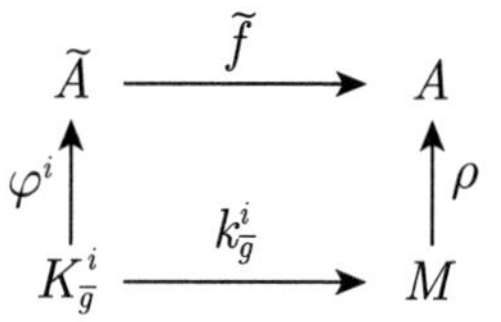

To see this, suppose that $\varphi^{i-1} : K_{\bar{g}}^{i-1} \to \tilde{A}$ has been already defined ($\varphi^0 = \varphi$, $K_{\bar{g}}^0 = K_{\bar{g}}$), and define φ^i as follows.

According to the construction of II, we have

$$K_{\bar{g}}^i = K_{\bar{g}}^{i-1} \otimes \Lambda(\eta_j^{i-1}),$$
$$d\eta_j^{i-1} = y_j^{i-1} \in K_{\bar{g}}^{i-1},$$
$$k_{\bar{g}}^i(\eta_j^{i-1}) = a_j^{i-1} \in M,$$
$$da_j^{i-1} = k_{\bar{g}}^{i-1}(y_j^{i-1}).$$

Then $\varphi^i : K_{\bar{g}}^i \to \tilde{A}$ will be defined as the DGA-morphism determined by the following expressions:

$$\varphi^i(x) = \varphi^{i-1}(x), \quad x \in K_{\bar{g}}^{i-1},$$
$$\varphi^i(\eta_j^{i-1}) = \tau \rho^c(a_j^{i-1}).$$

From φ^i we get then a DGA-morphism

$$\varphi^\infty : K_{\bar{g}}^\infty \to \tilde{A},$$

with

$$\varphi^\infty / K_{\bar{g}}^i = \varphi^i.$$

Define now

$$\tilde{\rho}: J_{\bar{g}} \to \tilde{A}$$

by

$$\tilde{\rho} = \varphi^{\infty} \rho_{\bar{g}}^{\infty},$$

then we have a commutative diagram:

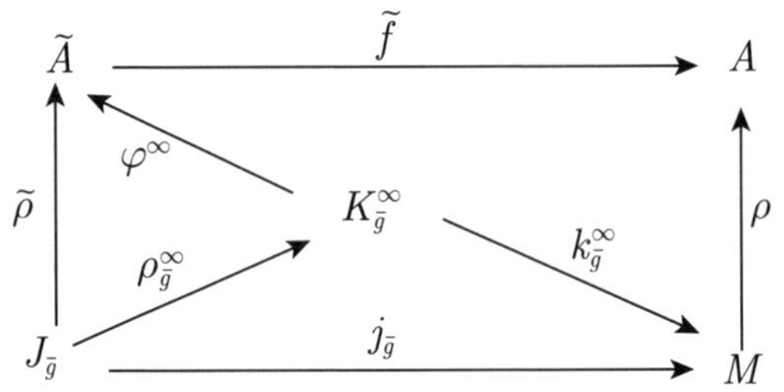

From the construction we know that $\tilde{\rho}_*: H(J_{\bar{g}}) \approx H(\tilde{A})$. Consequently $J_{\bar{g}} \approx \text{Min } \tilde{A} \approx I^*(K/L)$, $\tilde{\rho}$ is a canonical homomorphism, and $j_{\bar{g}} \in G^0(K/L, K)$.

The theorem is now completely proved.

Remark. The construction of $J_{\bar{g}}$ depends on the choice of $\rho: M \to A$, the additive homology basis x^i, and $a \in A$, $\tilde{a} \in \tilde{A}$, $a_j^i \in M$, etc. However, the theorem shows that $J_{\bar{g}}$ is independent of such choice and is completely determined by the characteristic homomorphism $\overline{g}: M \to \overline{N}$.

From the J-construction, we get also easily the following:

Corollary 1°. *Let $L' \in \mathscr{K}$ be any combinatorial sphere in K. Denote the characteristic homomorphism of K w. r. t. L' by*

$$\bar{g}': M \to \overline{I^*(L')},$$

then

$$\bar{g}' j_{\bar{g}}: J_{\bar{g}} \to \overline{I^*(L')}$$

is the characteristic homomorphism of K/L w.r.t. L'.

2°. *If in K/L there exists combinatorial sphere L' of dimension $n+1$ which contains C_L, then the characteristic homomorphism of K w.r.t. L*

$$\bar{g}: I^*(K) \to \overline{I^*(L)}$$

is $\overline{g} = 0^{(*)}$. Denote the combinatorial cell $L' \cap K$ of dimension $n+1$ in K by K', or $L' = K' \cup C_L = K'/L$, and the characteristic homomorphism $M' = I^*(K') = R \to \bar{N}$

(*) If $M, N \in \mathscr{M}$, then $h = 0: M \to N$ will denote the DGA-morphism with $h_0: M_0 \approx N_0 (\approx R)$ and $h_r = 0: M_r \to N_r$ for $r > 0$.

by $\bar{g}' = 0$. *Then the characteristic homomorphism* $h_{\bar{g}} : J_{\bar{g}} \to J_{\bar{g}'}$ *is completely determined by* $\bar{g}$ *which makes the following diagram commutative:*

$$
\begin{array}{ccccc}
J_{\bar{g}} & \xrightarrow{\dot{j}_{\bar{g}}} & M & \xrightarrow{\bar{g}} & \bar{N} \\
{\scriptstyle h_{\bar{g}}}\downarrow & & \downarrow{\scriptstyle 0} & & \updownarrow \\
J_{\bar{g}'} & \xrightarrow{\dot{j}_{\bar{g}'}} & M' & \xrightarrow{\bar{g}'=0} & \bar{N}
\end{array}
$$

V Effective calculations and axiomatic system of *I*-functor on $\boldsymbol{K}^0$

Any $K \in \mathscr{K}_0$ can be represented as

$$K = K_m \supset K_{m-1} \supset \cdots \supset K_1 \supset K_0, \tag{1}$$

in which K_0 is the 2-dimensional squelette, and K_r the union of K_{r-1} with an additional simplex Δ_r, the boundary of Δ_r being

$$\dot{\Delta}_r = L_{r-1} \subset K_{r-1}, \tag{2}$$

so that

$$K_r = K_{r-1} \cup \Delta_r = K_{r-1}/L_{r-1}. \tag{3}$$

Let

$$f_{r-1} : L_{r-1} \subset K_{r-1},$$

and

$$g_{r-1} = f_{r-1}^I : I^*(K_{r-1}) \to I^*(L_{r-1})$$

be any DGA-morphism of $G(K_{r-1}, L_{r-1})$. By I, $I^*(K_r)$ is determined from g_{r-1} by J-construction, and from this it is easy to establish an axiomatic system of I^*-functor on $\mathscr{K}_0$ by induction *w.r.t.* (1).

However, the J-construction of I is not explicit, while all L_{r-1} are combinatorial spheres, so that we shall rather establish axiomatic system of I^*-functor by means of II – IV which permits to furnish at the same time an effective method of calculations of $I^*(K)$ for $K \in \mathscr{K}_0$.

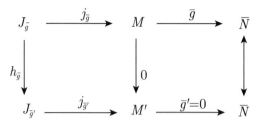

First of all, to any $K \in \mathscr{K}_0$ we have

$$I^*(K) \in \mathscr{M}$$

unique up to DGA-isomorphism, and to any pair of $K \in \mathscr{K}_0$ and combinatorial subsphere $L \in \mathscr{K}_0$ of K, a characteristic DGA-morphism

$$\bar{g} : I^*(K) \to \overline{I^*(L)}.$$

We know also that $I^*$and $\bar{g}$ possess the following properties:

1°. I^* is a homotopic functor, or more precisely, for $K,\ K' \in \mathscr{K}_0$ with $K \simeq K'$, we have $I^*(K) \approx I^*(K')$.

2°. Let $K \in \mathscr{K}_0$ and $L \in \mathscr{K}_0$ be a combinatorial subsphere of K; $\bar{g} : I^*(K) \to \overline{I^*(L)}$ be the characteristic morphism. Construct now by the J-construction of II

$$J_{\bar{g}} \in \mathscr{M}$$

and DGA-morphism

$$j_{\bar{g}} : J_{\bar{g}} \to I^*(K),$$

then

$$J_{\bar{g}} \approx I^*(K/L).$$

3°. Let $K,\ L,\ \bar{g}$ be as in 2°. If $L' \in \mathscr{K}_0$ is any combinatorial subsphere of K and let $\bar{g}' : I^*(K) \to \overline{I^*(L')}$ be the corresponding characteristic morphism, then

$$\bar{g}' j_{\bar{g}} : I^*(K/L) \to \overline{I^*(L')}$$

is the corresponding characteristic morphism.

4°. Let $K \in \mathscr{K}_0$ and $L \in \mathscr{K}_0$ be a combinatorial subsphere of K. If L is the boundary of some combinatorial cell of K, then the characteristic morphis $\bar{g} : I^*(K) \to \overline{I^*(L)}$ is given by $\bar{g} = 0$.

5°. Let $K,\ L,\ \bar{g}$ be as in 2°. If L is the boundary of some combinatorial cell K' of K, then the DGA-morphism

$$h_{\bar{g}} : I^*(K/L) \to \overline{I^*(K'/L)}$$

constructed by IV from the characteristic morphism $\bar{g} = 0 :\ I^*(K) \to I^*(L)$ of 4° is the corresponding characteristic morphism.

6°. For $K', K'' \in \mathscr{K}_0$ we have

$$I^*(K' \vee K'') \approx I^*(K') \vee I^*(K'').$$

7°. If $K \in \mathscr{K}_0$ is an n-dimensional combinatorial sphere, then

$$I^*(K) \approx \begin{cases} E(x), & n = \text{odd}, \\ E(y) \otimes R[x], & n = \text{even}, \end{cases}$$

in which

$$\deg x = n, \quad dx = 0,$$

and also

$$\deg y = 2n - 1, \quad dy = x^2,$$

for even n.

Remark. Owing to 7°, $\overline{I^*(L)}$ and characteristic morphism $\bar{g} : I^*(K) \to \overline{I^*(L)}$ are both meaningful for any combinatorial subsphere $L \in \mathscr{K}_0$ of K.

It is now easy to prove that the above properties 1°–7° form an axiomatic system for the I^*-functor over the category $\mathscr{K}_0$. In other words, we have the following

Theorem 5 *Let* ${}^0I^*$ *be any functor from* $\mathscr{K}_0$ to $\mathscr{M}$ *such that to any* $K \in \mathscr{K}_0$ *we have a* ${}^0I^*(K) \in \mathscr{M}$ *and to any pair of* $K \in \mathscr{K}_0$ *and a combinatorial subsphere* $L \in \mathscr{K}_0$ *of* K *we have a characteristic morphism*

$${}^0\bar{g} : {}^0I^*(K) \to \overline{{}^0I^*(L)},$$

which satisfies the axioms corresponding to 1° – 7°. *Then to any* $K \in \mathscr{K}_0$ *we have*

$${}^0I^*(K) \approx I^*(K), \tag{I}$$

and to any combinatorial subsphere $L' \in \mathscr{K}_0$ *of* K, *there exists a commutative diagram between the above isomorphisms and the various characteristic morphisms, viz.*

$$\begin{array}{ccc} {}^0I^*(K) & \xrightarrow{\;{}^0\bar{g}\;} & \overline{{}^0I^*(L')} \\ \updownarrow & & \updownarrow \\ I^*(K) & \xrightarrow{\;\bar{g}\;} & \overline{I^*(L')} \end{array} \tag{II}$$

Proof. Let $K \in \mathscr{K}_0$ and K be represented by (1)-(3) . Denote the expression(I) by $(\mathrm{I})_r$ in case $K = K_r$, and the expression (II) by $(\mathrm{II})_r$ in case $K = K_r$, and $L' = L'_r \in \mathscr{K}_0$ being any combinatorial subsphere of K_r. As $K_0 \simeq S^2 \vee \cdots \vee S^2$, we know by 1°, 6° ,7° that $(\mathrm{I})_0$ and $(\mathrm{II})_0$ hold true. Suppose that $(\mathrm{I})_{r-1}$ and $(\mathrm{II})_{r-1}$ have been proved and proceed to prove $(\mathrm{I})_r$ and $(\mathrm{II})_r$ as follows.

Let us first prove $(\mathrm{I})_r$. Consider the diagram below

$$\begin{array}{ccccc} {}^0I^*(K_r) & \xrightarrow{\;{}^0\bar{g}_r\;} & {}^0I^*(K_{r-1}) & \xrightarrow{\;{}^0\bar{g}_{r-1}\;} & \overline{{}^0I^*(L_{r-1})} \\ \Updownarrow & & \updownarrow & & \updownarrow \\ I^*(K_r) & \xrightarrow{\;\bar{g}_r\;} & I^*(K_{r-1}) & \xrightarrow{\;\bar{g}_{r-1}\;} & \overline{I^*(L_{r-1})} \end{array}$$

By induction hypothesis $(\mathrm{I})_{r-1}$ and $(\mathrm{II})_{r-1}$, the two vertical arrows on the right are both DGA-isomorphisms, $\bar{g}_{r-1}$, ${}^0\bar{g}_{r-1}$ are both characteristic homomorphisms, and the right square is commutative. By Theorem 4 of IV we have $I^*(K_r) = I^*(K_{r-1}/L_{r-1}) \approx J_{\bar{g}_{r-1}}$ and $\bar{g}_r = j_{\bar{g}_{r-1}} : J_{\bar{g}_{r-1}} \to I^*(K_{r-1})$. By Axiom 2° we have also ${}^0I^*(K_r) = {}^0I^*(K_{r-1}/\ L_{r-1}) \approx J_{{}^0\bar{g}_{r-1}}$ and ${}^0\bar{g}_r = j_{{}^0\bar{g}_{r-1}} : J_{{}^0\bar{g}_{r-1}} \to {}^0I^*(K_{r-1})$. From the construction of J and j, we see then ${}^0I^*(K_r) \approx I^*(K_r)$, i. e., $(I)_r$. Moreover, under these isomorphlsms, the left-hand square in the above diagram is also commutative.

Prove next $(\mathrm{II})_r$. For this, let $L'_r \in \mathscr{K}_0$ be a combinatorial subsphere of K_r and consider two cases separately.

Case I. $\Delta_r \notin L'_r$.

This time $L'_r \subset K_{r-1}$.

Case II. $\Delta_r \in L'_r$.

This time

$$\dim L'_r = \dim \Delta_r,$$

$$L'_r = K'_{r-1} \cup \Delta_r,$$

$$K'_{r-1} = \text{combinatorial cell in } K_{r-1}, \text{ with boundary } L_{r-1}.$$

Consider first the Case I. We have then the diagram below:

$$\begin{array}{ccccc} {}^0I^*(K_r) & \xrightarrow{\;{}^0\bar{g}_r\;} & {}^0I^*(K_{r-1}) & \xrightarrow{\;{}^0\bar{g}'_{r-1}\;} & \overline{{}^0I^*(L'_r)} \\ \Updownarrow & & \updownarrow & & \updownarrow \\ I^*(K_r) & \xrightarrow{\;\bar{g}_r\;} & I^*(K_{r-1}) & \xrightarrow{\;\bar{g}'_{r-1}\;} & \overline{I^*(L'_r)} \end{array}$$

By the induction hypothesis $(\mathrm{II})_{r-1}$, the two right vertical homomorphisms are both DGA-isomorphisms, ${}^0\bar{g}'_{r-1}$, $\bar{g}'_{r-1}$ are characteristic homomorphisms, and the left square is also commutative (with the same symbols as before). Now by Corollary 1° at the end of IV, $\bar{g}' = \bar{g}_r \bar{g}'_{r-1}$ is the characteristic homomorphism, so is ${}^0\bar{g}' = {}^0\bar{g}_r\, {}^0\bar{g}'_{r-1}$ by Axiom 3°. It follows that the diagram is commutative or we have $(\mathrm{II})_r$.

For the Case II, we have also $(\mathrm{II})_r$ by both Corollary 2° of IV and Axioms 4°, 5°. The theorem is now proved.

Remark. The Axiom 7° can also be slightly weakened to '7°. If $K \in \mathscr{K}_0$ is a 2-sphere, then

$$I^*(K) \approx E(y) \otimes R[x],$$

in which

$$\deg x = 2, \quad dx = 0,$$
$$\deg y = 3, \quad dy = x^2.$$

Ⅵ I^*-Functor of countably infinite complexes

Any countably infinite complex $\tilde{K} \in \mathscr{K}$ is the union of finite subcomplexes $K_i \in \mathscr{K}_0$:

$$K_1 \subset K_2 \subset \cdots (\subset \tilde{K}). \tag{1}$$

Write for the inclusion $f_i : K_i \subset K_{i+1}$, then we have a sequence of DGA-morphisms:

$$A^*(K_1) \overset{f_1^A}{\leftarrow} A^*(K_2) \leftarrow \cdots \leftarrow A^*(K_i) \overset{f_i^A}{\leftarrow} A^*(K_{i+1}) \leftarrow \cdots.$$

Clearly $A^*(\tilde{K}) = \varprojlim A^*(K_i)$. As each f_i^A is an epimorphism, we can construct successively according to Theorem 3 of III $\{I^*(K_i),\ \rho_i,\ g_{i-1}\}$ starting from $\{I^*(K_1),\ \rho_1\}$, such that the following diagram is commutative in which the ρ_i are canonical homomorphisms, and $g_i \in G^0(f_i^A)$ are privileged homomorphisms:

$$\begin{array}{ccccccccc}
A^*(K_1) & \overset{f_1^A}{\longleftarrow} & A^*(K_2) & \longleftarrow & \cdots & \longleftarrow & A^*(K_i) & \overset{f_i^A}{\longleftarrow} & A^*(K_{i+1}) \longleftarrow \cdots \\
\rho_1\uparrow & & \rho_2\uparrow & & & & \rho_i\uparrow & & \rho_{i+1}\uparrow \\
I^*(K_1) & \overset{g_1}{\longleftarrow} & I^*(K_2) & \longleftarrow & \cdots & \longleftarrow & I^*(K_i) & \overset{g_i}{\longleftarrow} & I^*(K_{i+1}) \longleftarrow \cdots
\end{array} \tag{2}$$

We shall call (g_i) a *sequence of privileged morphisms.*

Theorem 6 *Represent arbitrarily a countably infinite complex $\tilde{K} \in \mathscr{K}$ as* (1) *and construct a sequence of privileged morphisms*

$$I^*(K_1) \overset{g_1}{\leftarrow} I^*(K_2) \leftarrow \cdots \leftarrow I^*(K_i) \overset{g_i}{\leftarrow} I^*(K_{i+1}) \leftarrow \cdots. \tag{3}$$

Then

$$I^*(\tilde{K}) \approx \text{Min} \varprojlim I^*(K_i). \tag{4}$$

Proof. To any $(\xi_i) \in \varprojlim I^*(K_i)$ for which $\xi_i \in I^*(K_i)$, $g_{i-1}\xi_i = \xi_{i-1}$, let us set

$$\tilde{\rho}(\xi_i) = (\rho_i \xi_i) \in \varprojlim A^*(K_i) = A^*(\tilde{K}).$$

Then we have the DGA-morphism

$$\tilde{\rho} : \varprojlim I^*(K_i) \to A^*(\tilde{K}).$$

From this we get a commutative diagram:

$$\begin{array}{ccc} H(\varprojlim I^*(K_i)) & \xrightarrow{\tilde{\rho}_*} & H(\varprojlim A^*(K_i)) = H^*(\tilde{K}) \\ \downarrow F^I & & \downarrow F^A \\ \varprojlim H(I^*(K_i)) & \xrightarrow{(\rho_{i*})} & \varprojlim H(A^*(K_i)) \end{array}$$

In the diagram F^I and F^A are both natural morphisms. Now F^I is an isomorphism by [1] Chap. VIII and F^A is one by [3]. Moreover $\rho_{i*}: H(I^*(K_i)) \approx H(A^*(K_i)) = H^*(K_i)$ for all i so that $\tilde{\rho}_*$ is also an isomorphism. It follows that $\tilde{\rho}$ is a canonical homomorphism and we have (4) as to be proved.

References

[1] Eilenberg S & Steenrod N E. *Foundations of Algebraic Topology I*, 1952.

[2] Friedlander E, Griffiths P A & Morgan J. *Homotopy Theory and Differential Forms, Mimeog. Notes*, 1972.

[3] Kahn D W. The existence and applications of anticommutative cochain algebras, *ILL. J. Math.*, 1963, 7: 376-395.

[4] Sullivan D. Differential forms and topology of manifolds. *Symp. Tokyo on Topology*, 1973: 37-39.

[5] 吴文俊. 代数拓扑的一个新函子. 科学通报, 1975, 7: 311-312.

[6] 吴文俊. 代数拓扑 I^* 函子论——齐性空间的实拓扑. 数学学报, 1975, 18: 162-172.

[7] Wu Wen-tsün. Theory of I^*-functors in algebraic topology——Real topology of fibre squares. *Scientia Sinica*, 1975, 18: 464-482.

51. Theory of I^*-Functor in Algebraic Topology——I^*-Functor of a Fiber Space*

Wu Wen-Tsün(吴文俊) and Wang Qi-Ming(王启明)

Abstract The present paper is a succession to the previous ones (e.g., [1]) concerning I^*-functor of a topological space. Its aim is twofold. First, we recast our theory about fiber square[1] in a form under much more general conditions sufficient for all practical purposes. Secondly, we prove that the I^*-functor of a fiber space can be completely determined algebraically in terms of some twisted product of the I^*-functors of the base and the fiber, a theorem already announced under more stringent conditions in [2]. As this cannot be achieved for H^*-functor, our theorem shows once more the superiority of the I^*-functor over the H^*-functor, so far as real coefficient domain is concerned.

Ⅰ. De Rham-Sullivan algebra on paracompact spaces

Let X be an arbitrary space. An open covering $\mathscr{U}$ of X is called a *canonical* one ([3]) if it consists of open sets U_x containing x indexed on $x \in X$. The family of all such canonical coverings will be denoted by $\mathscr{R}(X)$ which is partially ordered in relation to $\ll$ such that $\mathscr{U} = \{U_x\} \ll \mathscr{V} = \{V_x\}$ if $U_x \subset V_x$ for all $x \in X$. Let $N_{\mathscr{U}}$ be the nerve complex of $\mathscr{U} \in \mathscr{R}(X)$ and $A^*(\mathscr{U})$ be the de Rham-Sullivan algebra of differential forms on $N_{\mathscr{U}}$ for which we have $H(A^*(\mathscr{U})) \approx H^*(N_{\mathscr{U}})$ on real coefficients by the de Rham-Sullivan theorem (cf. [4—7]) . For $\mathscr{U} \ll \mathscr{V}$, the natural simplicial map $N_{\mathscr{U}} \to N_{\mathscr{V}}$ defined by $U_x \to V_x$ for $x \in X$ is called *canonical* and induces a DGA-morphism

$$A^*(\mathscr{V}) \to A^*(\mathscr{U}).$$

* *Sci. Sinica*, 1978, 21: 1-18. The present paper is a combination of two papers received respectively on February 25, 1976 and December 29, 1975, of which the latter was somewhat revised in October 1977.

The direct limit of $A^*(\mathscr{U})$ with $\mathscr{U} \in \mathscr{R}(X)$ will then be denoted by

$$\check{A}^*(X) = \varinjlim A^*(\mathscr{U}).$$

Now direct limit commutes with the II^*-functor and $\mathscr{R}(X)$ is cofinal in the family of all open coverings so we get easily

Proposition 1 $H(\check{A}^*(X)) \approx \check{H}^*(X)$ *as algebra, in which* $\check{H}^*$ *means the Cêch real cohomology algebra based on family of all open coverings.*

Definition The DGA-algebra on real field $\check{A}^*(X)$ will be called the *de Rham-Sullivan algebra of X*. In case $\check{A}^*(X)$ is of finite type so that its minimal model is well-defined, we shall call the latter the *I^*-functor* (or better the *I^*-measure*) of X and denote it by

$$I^*(X) = \mathrm{Min}\check{A}^*(X).$$

Proposition 2 *For a continuous map $f : X' \to X$, there is a uniquely defined* DGA-*morphism*

$$\check{f}^A : \check{A}^*(X) \to \check{A}^*(X'),$$

which is an epimorphism if $X' \subset X$ and f is the inclusion map.

Proof. For any $\alpha \in \check{A}^*(X)$, let $\alpha_{\mathscr{U}}$ be a representative in $A^*(\mathscr{U})$ for $\mathscr{U} = \{U_x\} \in \mathscr{R}(X)$. Let $\mathscr{U}'$ be the canonical covering of X' consisting of open sets $U'_{x'} = f^{-1}U_x$, where $x = f(x'), x' \in X'$. For $f : X' \subset X$, we have simply $U'_{x'} = U_{x'} \cap X'$ for $x' \in X'$. Now $U'_{x'} \to U_x$ defines a simplicial map $f_{\mathscr{U}}$ of $N_{\mathscr{U}'}$ to $N_{\mathscr{U}}$, which induces a DGA-morphism $f^A_{\mathscr{U}} : A^*(\mathscr{U}) \to A^*(\mathscr{U}')$. The correspondence of $\alpha_{\mathscr{U}}$ to $f^A_{\mathscr{U}}\alpha_{\mathscr{U}}$ will clearly induce a DGA-morphism of $\check{A}^*(X)$ to $\check{A}^*(X')$, which proves the first part of the proposition.

To see that $\check{f}^A$ is an epimorphism for $f : X' \subset X$, let us remark that in this case $\mathscr{U}'$ can be chosen arbitrarily in advance and that $N'_{\mathscr{U}}$ can then be considered as a subcomplex of $N_{\mathscr{U}}$. For any element $\alpha' \in \check{A}^*(X')$ with a representative $\alpha'_{\mathscr{U}'}$ in $N_{\mathscr{U}'}$, we can extend $\alpha'_{\mathscr{U}'}$ to a differential form $\alpha_{\mathscr{U}}$ in $N_{\mathscr{U}}$ by some extension lemma in Sullivan theory ([5]§10) . For the element $\alpha \in \check{A}^*(X)$ represented by $\alpha_{\mathscr{U}}$ we have then $\check{f}^A\alpha = \alpha'$ as to be asserted.

Notations. For $f : X' \subset X$, the image $\alpha' \in \check{A}^*(X')$ of $\alpha \in \check{A}^*(X)$ under $\check{f}^A$ occurring in the above proof will be called the *restriction* of α to X' and will sometimes be written as α/X'.

In case X, X' are both of finite type, so that $I^*(X)$ and $I^*(X')$ are both defined, any morphism (unique up to homotopy) of I^* induced by a continuous map $f : X' \to$

X will be denoted by

$$f^I : I^*(X) \to I^*(X').$$

Let $\alpha \in \check{A}^*(X)$ and f be an arbitrary function on X, continuous or not. Then we can define an element $f\alpha \in \check{A}^*(X)$ as follows.

Take a representative $\alpha_{\mathscr{U}} \in A^*(\mathscr{U})$ of α for some $\mathscr{U} \in \mathscr{R}(X)$. For any simplex $s = U_{x_0} \cdots U_{x_n} \in N_{\mathscr{U}}(U_{x_i} \in \mathscr{U})$, let f_s be the function on s defined by

$$f_s\left(\sum \lambda_i U_{x_i}\right) = \sum \lambda_i f(x_i),$$

in which $\lambda_i \geqslant 0, \sum \lambda_i = 1$. The set of differential forms $f_s \cdot \alpha_{\mathscr{U}}(s)$ for all $s \in N_{\mathscr{U}}$ is then coherent and defines a differential form of $A^*(\mathscr{U})$ to be denoted by $f\alpha_{\mathscr{U}}$. It is clear that $f\alpha_{\mathscr{U}}$ will be the representative of a well-defined element of $\check{A}^*(X)$, which is the $f\alpha$ in question.

Let us restrict henceforth to the case of X being a paracompact space. Then we have:

Proposition 3 (Dimension Lemma) *If X is paracompact with corering dimension $\leqslant n$, then $\check{A}^N(X) = 0$ for any $N > n$.*

Proof. Given any $\alpha \in \check{A}^N(X)$, let us take a canonical covering $\mathscr{U}$ of X with a representative $\alpha_{\mathscr{U}} \in A^N(\mathscr{U})$ of α. As dim $X \leqslant n$ and X is paracompact, there exists a locally finite covering $\mathscr{V} \ll \mathscr{U}$ with dim $N_{\mathscr{V}} \leqslant n$. Take any admissible map $f : N_{\mathscr{V}} \to N_{\mathscr{U}}$. Then it is clear that $f^A\alpha_{\mathscr{U}} = 0 \in A^N(\mathscr{V})$ for any $N > n$. It follows that $\alpha = 0$ as to be proved.

Let $\mathscr{U} = \{U_i\}$ be a locally finite open covering of a paracompact space X. A set of elements $\alpha_i \in \check{A}^*(U_i)$ will be said to be *coherent* if $\alpha_i|U_i \cap U_j = \alpha_j|U_i \cap U_j$ for any pair i, j with $U_i \cap U_j \neq \varnothing$. We have then

Proposition 4 (Combination Principle) *If X is a paracompact space and the family of elements $\alpha_i \in \check{A}^*(U_i)$ is coherent w. r. t. a locally finite open covering $\mathscr{U} = \{U_i\}$ of X, then there exists a unique element $\alpha \in \check{A}^*(X)$ with $\alpha/U_i = \alpha_i$ for each i.*

Proof. Let us take a shrinking $\mathscr{U}' = \{U_i'\} \ll \mathscr{U}$ of $\mathscr{U}$ and also take canonical open coverings $\mathscr{U}_i = \{U_x^{(i)}/x \in U_i\}$ on U_i, and $\mathscr{U}_{ij} = \{U_x^{(ij)}/x \in U_i \cap U_j\}$ on $U_i \cap U_j$, such that we have

1) $U_i' \subset \bar{U}_i' \subset U_i$;

2) $\mathscr{U}_{ij} \ll \mathscr{U}_i/U_i \cap U_j, \mathscr{U}_{ij} \ll \mathscr{U}_j/U_i \cap U_j$;

3) α_i has a representative a_i in $N_{\mathscr{U}_i}$;

4) The restrictions of α_i, α_j on $U_i \cap U_j$ have the same representative in $N_{\mathscr{U}_{ij}}$.

We can take a canonical covering $\mathscr{V} = \{V_x\}$ of X such that:

5) For any U_i (and U_j, only finite in number) containing x,V_x is contained in U_i(and U_j) and also in the open set $U_x^{(i)} \in \mathscr{U}_i$ (and $U_x^{(j)} \in \mathscr{U}_j, U_x^{(ij)} \in \mathscr{U}_{ij}$) indeked at x.

6) If V_x meets some $\bar{U}_i'$, then $V_x \subset U_i$.

Consider now any simplex $s = V_{x_0} \cdots V_{x_n} \in N_{\mathscr{V}}$ with $V_{x_0} \cap \cdots \cap V_{x_n} \neq \varnothing$. By 1) and 6) , there is at least one $U_i \in \mathscr{U}$ with $V_{x_k} \subset U_i, k = 1, \cdots, n$. By 5) $s^{(i)} = U_{x_0}^{(i)} \cdots U_{x_n}^{(i)}$ is then a simplex of $N_{\mathscr{U}_i}$ and we shall take the differential form on s induced from $a_i(s^{(i)})$ by natural map $s \to s^{(i)}$ as $a(s)$ which is independent of U_i by 2) , 3) and 4) . The family of forms $a(s)$ thus defined is clearly coherent so that it defines a differential form $a \in A^*(\mathscr{V})$. The unique element $\alpha \in \check{A}^*(X)$ represented by a meets then the purpose of the proposition.

Remark. For the applications in II, the proposition is to be slightly generalized to include the case of coherent family of elements $\alpha_i \in \check{A}^*(U_i) \otimes M$ for some module M on R. The proof is still valid in this case as the tensor product is taken on the real field.

We can define the de Rham-Sullivan algebra $\check{A}^*(X)$ of a paracompact space X as a carapace in the sense of Cartan ([8]) by introducing supports as follows:

Let $\alpha \in \check{A}^*(X)$ and $x \in X$, then $x \notin$ Supp α if there exists some neighborhood U of x such that $\alpha | U = 0$.

Note that the definition is justified owing to the combination principle. By using the same principle it is also easy to see that the carapace thus defined is both fine and complete of which we shall not enter into discussion.

II. Some spectral sequences of a fibration

Consider a fibration

$$F \subset Y \xrightarrow{g} B$$

for which we make the following assumptions:

1°. B is a connected locally finite simplicial complex.

2°. F and Y are both connected and paracompact with $\pi_1(B)$ operating trivially on $\check{H}^*(F)$.

We shall denote the q-squelette of B by B_q, and set $g^{-1}B_q = Y_q$.

The aim of this section is to define two spectral sequences associated with the above fibration each converging to $\check{H}^*(Y)$. In passing, it will settle a question raised after 1950 in Sem. Cartan ([8] Exp. 21) , at least in the present case. (See Prop. 5 below.)

Let the de Rham-Sullivan algebra $\check{A}^*(Y)$ of Y be filtrated as:

$$\check{F}^p = F^p\check{A}^*(Y) = \mathrm{Ker}[\check{A}^*(Y) \to \check{A}^*(Y_{p-1})].$$

This will give rise to a filtration

(I) $$\check{A}^*(Y) = \check{F}_0 \supset \check{F}^1 \supset \cdots \supset \check{F}^p \supset \check{F}^{p+1} \supset \cdots.$$

Proposition 5 *The filtration* (I) *gives rise to a spectral sequence for which*

$$\check{E}_2^p \approx \check{H}^p(B) \otimes \check{H}^*(F) \Rightarrow \check{H}^*(Y).$$

Proof. Let us define DGA-algebras $\check{A}^*(Y_p, Y_{p-1})$ by means of exact sequences

$$0 \to \check{A}^*(Y_p, Y_{p-1}) \to \check{A}^*(Y_p) \xrightarrow{\check{i}_p^A} \check{A}^*(Y_{p-1}) \to 0,$$

in which $\check{i}_p^A$ is induced by the inclusion $i_p : Y_{p-1} \subset Y_p$ and is an epimorphism by Prop.2 of I. As in the proof of Lemma 1 in §2 of [2] we have

$$\begin{aligned} &\check{E}_0^p \approx \check{A}^*(Y_p, Y_{p-1}), && d_0 = d/\check{A}^*(Y_p, Y_{p-1}), \\ &E_1^p \approx C^p(B) \otimes \check{H}^*(F), && d_1 = d/C^*(B), \\ &E_2^p \approx \check{H}^p(B) \otimes H^*(F). \end{aligned}$$

Since π_1 acts trivially on $H^*(F)$, this proves the proposition.

To introduce the second spectral sequence, let us define

$$\begin{aligned} &\bar{A}^*(Y) = \sum_{r\geqslant 0} \bar{A}^r(Y), \\ &\bar{A}^r(Y) = \sum_{s\geqslant 0} \bar{A}^{r,s}(Y), \\ &\bar{A}^{r,s}(Y) = \{\eta : \mathrm{St}\ \sigma \to \check{A}^r(\mathrm{St}\ \sigma) \otimes \check{A}^s(g^{-1}\mathrm{St}\ \sigma)/(C)\}, \end{aligned}$$

in which (C) here as well as (C) below means some natural compatibility conditions.Under natural operations on $\check{A}$, the set $\bar{A}^*(Y)$ becomes a DGA-algebra. Filtrate now $\bar{A}^*(Y)$ by

(II) $$\bar{F}^p = F^p\bar{A}^*(Y) = \sum_{r\geqslant p} \bar{A}^{r,*}(Y).$$

Then we have

Proposition 6 *The filtration* (II) *gives rise to a spectral sequence for which*

$$\bar{E}_2^p \approx \check{H}^p(B) \otimes \check{H}^*(F) \Rightarrow \check{H}^*(Y).$$

For the proof we shall make some preliminaries. For any subspace $B' \subset B$ and $Y' = g^{-1}B'$, let us put as definition the DGA-algebra

$$\tilde{A}^*(B') = \check{A}^*(B') \otimes \check{A}^*(g^{-1}B'),$$

with a differential $\tilde{d} = d$ on $\tilde{A}^*(g^{-1}B')$ while $\hat{d} = 0$ on $\check{A}^*(B')$. For any function f on B' and

$$\tilde{a} = \sum b_i \otimes y_i \in \tilde{A}^*(B')$$

we set by definition

$$f\tilde{a} = \sum (fb_i) \otimes y_i \in \tilde{A}^*(B'),$$

in which $fb_i \in \check{A}^*(B')$ is defined as in I. Then

$$\tilde{d}(f\tilde{a}) = \sum \pm (fb_i) \otimes dy_i = f \cdot \sum \pm b_i \otimes dy_i = f(\tilde{d}\tilde{a}).$$

Hence we have

Lemma 1 *For any function f on B' and $\tilde{a} \in \tilde{A}^*(B')$,*

$$\tilde{d}(f\tilde{a}) = f\tilde{d}\tilde{a}.$$

For any pair of subspaces $B'' \subset B'$ of B there exists by Prop.2 of I the natural DGA-morphism

$$\tilde{A}^*(B') \to \tilde{A}^*(B'').$$

The image of an element $\tilde{a}' = \sum b_i \otimes y_i \in \tilde{A}^*(B')$ is given by

$$\tilde{a}'' = \sum b_i/B'' \otimes y_i/g^{-1}B''.$$

It will be denoted by

$$\tilde{a}'' = \tilde{a}'/B'',$$

and called the *restriction* of $\tilde{a}'$ to B''. Similarly for the restriction $h'' = h'/B'' \in H(\tilde{A}^*(B''))$, where $h' \in H(\tilde{A}^*(B'))$.

Consider now any locally finite open covering $\mathscr{U} = \{U_i\}$ of B. Define a DGA-algebra

$$E = E_{\mathscr{U}} = \sum E^r$$

consisting of elements $(\tilde{a}_i) \in E^r$ with

$$\tilde{a}_i \in \tilde{A}^r(U_i),$$
$$\tilde{a}_i/U_i \cap U_j = \tilde{a}_j/U_i \cap U_j.$$

The DGA-structure will be defined in E by $\tilde{d}(\tilde{a}_i) = (\tilde{d}\tilde{a}_i)$, etc. Let $Z(E)$ be the subalgebra of E consisting of elements $(\tilde{z}_i)$ with

$$\tilde{z}_i \in Z(\tilde{A}^*(U_i)),$$

i.e., with $\tilde{d}\tilde{z}_i = 0$. Let $B(E) = \tilde{d}E$. Define also $H = H_{\mathscr{U}} = \sum H^r$ as the graded algebra consisting of elements $(h_i) \in H^r$ with

$$h_i \in H_r(\tilde{A}^*(U_i)),$$
$$h_i/U_i \cap U_j = h_j/U_i \cap U_j.$$

We have then

Lemma 2 *With natural morphisms i, π, the sequence*

$$0 \to B(E) \xrightarrow{i} Z(E) \xrightarrow{\pi} H \to 0$$

is exact. In other words,

$$H(E_{\mathscr{U}}) \approx H_{\mathscr{U}}.$$

Proof. Let $\{f_i\}$ be a partition of unity subordinate to $\mathscr{U}$ such that

$$\sum f_i = 1,$$

and

$$f_i = 0 \text{ on } X - \bar{U}_i',$$

where $\bar{U}_i'$ =Supp $f_i \subset U_i$.

To see that π is an epimorphism, let us consider any element $(h_i) \in H$. Take $\tilde{z}_i \in h_i, \tilde{z}_i \in Z(\tilde{A}^*(U_i))$, such that restricted on $U_i \cap U_j$ we have $\tilde{z}_i - \tilde{z}_j = \tilde{d}\varphi_{ij}$ for some $\varphi_{ij} \in \tilde{A}^*(U_i \cap U_j)$. Define $\psi_{i,k} \in \tilde{A}^*(U_i)$by

$$\psi_{i,k} = \begin{cases} f_k\varphi_{ik}/U_i \cap U_k, \\ 0/U_i - U_i \cap \bar{U}_k'. \end{cases}$$

By Lemma 1, we have on $U_i \cap U_j \cap U_k$,

$$\begin{aligned}\tilde{d}\psi_{i,k} - \tilde{d}\psi_{j,k} &= f_k\tilde{d}\varphi_{ik} - f_k\tilde{d}\varphi_{jk} \\ &= f_k(\tilde{z}_i - \tilde{z}_k) - f_k(\tilde{z}_j - \tilde{z}_k) \\ &= f_k(\tilde{z}_i - \tilde{z}_j).\end{aligned}$$

As this is also trivially true on $U_i \cap U_j - U_i \cap U_j \cap U_k$, we have

$$\tilde{d}\psi_{i,k} - \tilde{d}\psi_{j,k} = f_k(\tilde{z}_i - \tilde{z}_j)$$

on $U_i \cap U_j$.

Set

$$\tilde{z}'_i = \tilde{z}_i - \tilde{d}\left(\sum_k \psi_{i,k}\right) \in Z(\tilde{A}^*(U_i)).$$

Then on $U_i \cap U_j$ we have

$$\begin{aligned}\tilde{z}'_i - \tilde{z}'_j &= \tilde{z}_i - \tilde{z}_j - \sum_k (\tilde{d}\psi_{i,k} - \tilde{d}\psi_{j,k}) \\ &= \tilde{z}_i - \tilde{z}_j - \sum_k f_k(\tilde{z}_i - \tilde{z}_j) \\ &= 0.\end{aligned}$$

Hence $(\tilde{z}'_i) \in Z(E)$ with

$$\pi(\tilde{z}'_i) = (h_i),$$

i.e.,π is an epimorphism.

To see that $B(E) \xrightarrow{i} Z(E) \xrightarrow{\pi} H$ is exact, let us consider any element $(\tilde{z}_i) \in Z(E)$ with $\pi(\tilde{z}_i) = 0$. There exist then elements $\alpha_i \in \tilde{A}^*(U_i)$ with

$$\tilde{d}\alpha_i = \tilde{z}_i / U_i.$$

Define now $\beta_{i,k} \in \tilde{A}^*(U_i)$ by

$$\beta_{i,k} = \begin{cases} f_k(\alpha_i - \alpha_k)/U_i \cap U_k, \\ 0/U_i - U_i \cap \bar{U}'_k. \end{cases}$$

By Lemma 1, it is easily verified that

$$\tilde{d}\beta_{i,k} = 0/U_i$$

and that

$$\beta_{i,k} - \beta_{j,k} = f_k(\alpha_i - \alpha_j)/U_i \cap U_j.$$

Define now $\alpha_i' \in \tilde{A}^*(U_i)$ by

$$\alpha_i' = \alpha_i - \sum_k \beta_{i,k}.$$

Then on U_i we have

$$\tilde{d}\alpha_i' = \tilde{d}\alpha_i = \tilde{z}_i,$$

and on $U_i \cap U_j$ we have

$$\alpha'_i - \alpha'_j = \alpha_i - \alpha_j - \sum_k f_k(\alpha_i - \alpha_j) = 0.$$

Hence $(\alpha_i') \in E, (\tilde{z}_i) = \tilde{d}(\alpha_i') \in B(E)$, and the sequence (i, π) is exact. As i is clearly a monomorphism, Lemma 2 is proved.

We come now to the

Proof of Proposition 6. From the filtration (II), we get

$$\bar{E}_0^p = \bar{A}^{p,*}(Y) = \{\eta : \mathrm{St}\ \sigma \to \check{A}^p(\mathrm{St}\ \sigma) \otimes \check{A}^*(g^{-1}\mathrm{St}\ \sigma)/(C)\},$$

with

$$\bar{d}_0 = d/\check{A}^*(g^{-1}\mathrm{St}\ \sigma).$$

In other words, $\overline{E}_0$ is just the DGA-algebra E constructed in Lemma 2 with $\mathscr{U} = \{\mathrm{St}\ \sigma/\sigma \ \in\ B\}$. By Lemma 2, we have then

$$\bar{E}_1^p = \{\tilde{\eta} : \mathrm{St}\ \sigma \to \check{A}^p(\mathrm{St}\ \sigma) \otimes H\check{A}^*(g^{-1}\mathrm{St}\ \sigma)/(C)\},$$
$$\bar{d}_1 = d/\check{A}^p(\mathrm{St}\ \sigma).$$

As $\pi_1(B)$ acts trivially on $\check{H}^*(F) \approx H\check{A}^*(g^{-1}\mathrm{St}\sigma)$, we have by combination principle (Prop. 4 of I) ,

$$\begin{aligned}\bar{E}_1^p &\approx \{\tilde{\eta} : \mathrm{St}\ \sigma \to \check{A}^p(\mathrm{St}\ \sigma) \otimes \check{H}^*(F)/(C)\} \\ &\approx \check{A}^p(B) \otimes \check{H}^*(F).\end{aligned}$$

It follows that

$$\bar{E}_2^p \approx \check{H}^p(B) \otimes \check{H}^*(F).$$

Define now a morphism

$$h : \bar{A}^*(Y) \to \check{A}^*(Y)$$

as follows. Consider $\eta \in \bar{A}^{r,*}(Y)$. For any $\sigma \in B$, suppose that

$$\eta(\text{St }\sigma) = \sum \alpha_i \otimes \beta_i \in \check{A}^r(\text{St }\sigma) \otimes \check{A}^*(g^{-1}\text{St }\sigma).$$

Set

$$(h\eta)(g^{-1}\text{St }\sigma) = \sum \check{g}^A \alpha_i \cdot \beta_i \in \check{A}^{r+*}(g^{-1}\text{St }\sigma).$$

By the combination principle (Prop. 4 of I) the compatibility condition on η shows that $h\eta$ gives rise to a well-defined element of $\check{A}^*(Y)$. It is clear that h is a DGA-morphism and moreover

$$hF^p\bar{A}^*(Y) \subset F^p\check{A}^*(Y).$$

Consequently h induces morphisms of spectral sequence with

$$h_{2*} : \bar{E}_2^p \approx \check{E}_2^p (\approx \check{H}^p(B) \otimes \check{H}^*(F))$$

by Prop. 5 and the formula about $\bar{E}_2^p$ above. By isomorphism theorem of spectral sequences h will induce a module-isomorphism

$$h_* : H\bar{A}^*(Y) \approx H\check{A}^*(Y) \approx \check{H}^*(Y).$$

As h is multiplicative, h_* is also an algebraic isomorphism. In particular, we have

$$\bar{E}_2 \Rightarrow \check{H}^*(Y),$$

and Prop. 6 is completely proved.

As in the above proof $h : \bar{A}^*(Y) \to \check{A}^*(Y)$ induces algebraic isomorphism h_*, we have also the following corollary:

Proposition 7 $I^*(Y) \approx \text{Min}\bar{A}^*(Y)$.

III. I^*-functor in a fiber square

Consider a fiber square

$$\begin{array}{ccc} F \subset Z & \xrightarrow{\bar{f}} & Y \\ \bar{g}\downarrow & & \downarrow g \\ X & \xrightarrow{f} & B \end{array}$$

with the following conditions:

1°. All the spaces involved are the connected simply-connected paracompact spaces and all maps are continuous.

2°. B, X have furthermore the same homotopy type as some connected locally-finite simplicial complexes of finite type.

3°. $g : Y \to B$ is a fibration with fiber F and $\bar{g} : Z \to X$ a fibration induced from it by the map f.

In order to prove the fiber-square theorem (Th.1 below) we shall consider the special case with condition 2° replaced by $\bar{2}°$ to which the general case is easily reduced by using induced bundles:

$\bar{2}°$. B, X are furthermore the connected locally-finite simplicial complexes of finite type with f as a simplicial map.

As in II, we have then for the fibrations $F \subset Y \xrightarrow{g} B$ and $F \subset Z \xrightarrow{\bar{g}} X$ the following DGA-algebras ($\sigma \in B, \tau \in X$),

$$\bar{A}^*(Y) = \{\eta : \mathrm{St}\ \sigma \to \check{A}^*(\mathrm{St}\ \sigma) \otimes \check{A}^*(g^{-1}\mathrm{St}\ \sigma)/(C)\},$$
$$\bar{A}^*(Z) = \{\zeta : \mathrm{St}\ \tau \to \check{A}^*(\mathrm{St}\ \tau) \otimes \check{A}^*(g^{-1}\mathrm{St}\ \tau)/(C)\}.$$

Introduce now DGA-morphisms $g^*, \cdots, \lambda$ below:

$$g^* : \check{A}^*(B) \to \bar{A}^*(Y),$$
$$\bar{g}^* : \check{A}^*(X) \to \bar{A}^*(Z),$$
$$j^* : \bar{A}^*(Y) \to \bar{A}^*(Z),$$
$$R(\check{f}^A) : \check{A}^*(X) \otimes \check{A}^*(B) \to \check{A}^*(X),$$
$$L(g^*) : \check{A}^*(B) \otimes \bar{A}^*(Y) \to \bar{A}^*(Y),$$
$$\lambda : \check{A}^*(X) \otimes \bar{A}^*(Y) \to \bar{A}^*(Z).$$

The definitions of these morphisms are as follows:

The morphism g^* is defined by

$$(g * \beta)(\mathrm{St}\ \sigma) = \beta/\mathrm{St}\ \sigma \otimes 1 \in \check{A}^k(\mathrm{St}\ \sigma) \otimes \check{A}^0(g^{-1}\mathrm{St}\ \sigma),$$

for $\beta \in \check{A}^k(B)$. Similarly for $\bar{g}^*$.

Let $\tau \in X$ and $f(\tau) = \sigma \in B$. Let $\eta \in \bar{A}^*(Y)$ and $\eta(\mathrm{St}\sigma) = \sum \alpha_i \otimes \beta_i \in \check{A}^*\cdot$ $(\mathrm{St}\sigma) \otimes \check{A}^*(g^{-1}\mathrm{St}\sigma)$.

Set

$$(j * \eta)(\mathrm{St}\ \tau) = \check{f}_\tau^A \alpha_i \otimes \bar{f}_\tau^A \beta_i \in \check{A}^*(\mathrm{St}\ \tau) \otimes \check{A}^*(\bar{g}^{-1}\mathrm{St}\ \tau),$$

in which

$$\check{f}_\tau^A : \check{A}^*(\mathrm{St}\,\sigma) \to \check{A}^*(\mathrm{St}\,\tau)$$

and

$$\bar{f}_\tau^A : \check{A}^*(g^{-1}\mathrm{St}\,\sigma) \to \check{A}^*(\bar{g}^{-1}\mathrm{St}\,\tau)$$

are DGA-morphisms induced by the continuous maps $f:\mathrm{St}\,\tau \to \mathrm{St}\,\sigma$ and $\bar{f}:\bar{g}^{-1}\mathrm{St}\tau \to g^{-1}\mathrm{St}\sigma$ respectively. Then $j^*\eta$ is a well-defined element of $\bar{A}^*(Z)$ and this defines the morphism j^*.

Finally, the morphisms $R(\check{f}^A)$, etc. are respectively defined by

$$\begin{aligned} R(\check{f}^A)(\xi\otimes\beta) &= \xi\cdot\check{f}^A\beta, \\ L(g^*)(\beta\otimes\eta) &= g^*\beta\cdot\eta, \\ \lambda(\xi\otimes\eta) &= \bar{g}^*\xi\cdot j^*\eta, \end{aligned}$$

in which $\xi\in\check{A}^*(X), \eta\in\bar{A}^*(Y)$ and $\beta\in\check{A}^*(B)$.

Under $R(\check{f}^A)$ the $\check{A}^*(X)$ can be considered as a right $\check{A}^*(B)$-DGA algebra. Construct the corresponding proper projective resolution:

$$\begin{aligned} P(X) &= \sum P_{-n}(X), \\ P_{-n}(X) &= \check{A}^*(X)\otimes\underbrace{\check{A}^+(B)\otimes\cdots\otimes\check{A}^+(B)}_{n}\otimes\check{A}^*(B), \\ \check{A}^+(B) &= \{\beta\in\check{A}^*(B)/\deg\beta>0\}. \end{aligned}$$

Let

$$\varepsilon : P(X)\to\check{A}^*(X)$$

be the augmentation and $DP(X)$ the total complex associated to the resolution so that

$$H(DP(X))\approx H(\check{A}^*(X))\approx\check{H}^*(X).$$

Consider $\bar{A}^*(Y)$ as left $\check{A}^*(B)$-DGA-algebra under $L(g^*)$ and also $DP(X)$ as right $\check{A}^*(B)$-DGA-algebra under natural operations. We can then form quotient algebras

$$A^*(X)\underset{\check{A}*(B)}{\otimes}\bar{A}^*(Y),$$

and

$$P^* = DP(X)\underset{\check{A}^*(B)}{\otimes}\bar{A}^*(Y),$$

and complete the following commutative diagram by introducing natural DGA-morphisms μ and θ, in which π is the natural projection:

$$\begin{array}{ccc}
P^*=DP(X) \underset{\check{A}^*(B)}{\otimes} \bar{A}^*=(Y) & \xrightarrow{\quad\theta\quad} & \\
\Big\downarrow{\scriptstyle \Sigma\otimes 1} & & \\
\check{A}^*=(X) \underset{\check{A}^*(B)}{\otimes} \bar{A}^*=(Y) & \xrightarrow{\quad\mu\quad} & \bar{A}^*=(Z). \\
\Big\uparrow{\scriptstyle \pi} & & \\
\check{A}^*=(X) \otimes \bar{A}^*=(Y) & \xrightarrow{\quad\lambda\quad} &
\end{array}$$

Filtrate now $\bar{A}^*(Y)$ and $\bar{A}^*(Z)$ as in II:

$$F^p\bar{A}^*(Y) = \sum_{r\geqslant p} \bar{A}^{r,*}(Y),$$
$$F^p\bar{A}^*(Z) = \sum_{r\geqslant p} \bar{A}^{r,*}(Z).$$

By Prop. 6 of II, we have then

$$E_0^p\bar{A}^*(Y) \approx \bar{A}^{p,*}(Y) \text{ with some } d_0,$$
$$E_1^p\bar{A}^*(Y) \approx \check{A}^p(B) \otimes \check{H}^*(F), \quad d_1 = d/\check{A}^*(B),$$
$$E_2^p\bar{A}^*(Y) \approx \check{H}^p(B) \otimes \check{H}^*(F).$$

Similarly for $\bar{A}^*(Z)$ we have

$$E_2^p\bar{A}^*(Z) \approx \check{H}^p(X) \otimes \check{H}^*(F).$$

Now filtrate P^* by

$$F^pP^* = \sum_i [DP(X)]^i \underset{\check{A}*(B)}{\otimes} F^{p-i}\bar{A}^*(Y),$$

then we have successively

$$E_0^pP^* \approx \sum_i [DP(X)]^i \underset{\check{A}*(B)}{\otimes} E_0^{p-i}\bar{A}^*(Y), \quad d_0 = d_0/E_0\bar{A}^*(Y),$$
$$E_1^pP^* \approx [DP(X)]^p \otimes \check{H}^*(F), \quad d_1 = d/DP(X),$$
$$E_2^pP^* \approx \check{H}^p(X) \otimes \check{H}^*(F).$$

For the DGA-morphism $\theta : P^* \to \bar{A}^*(Z)$, we have

$$\theta F^p P^* \subset F^p \bar{A}^*(Z),$$

so that θ induces morphisms of spectral sequences

$$\theta_r : E_r P^* \to E_r \bar{A}^*(Z).$$

As

$$\theta_2 : E_2 P^* \approx E_2 \bar{A}^*(Z) [\approx \check{H}^*(X) \otimes \check{H}^*(F)],$$

we see that

$$\theta_* : H(P^*) \approx H(\bar{A}^*(Z)) \approx \check{H}^*(Z).$$

As θ is multiplicative, θ_* is also a multiplicative isomorphism. Hence we have the following

Proposition 8 *The* DGA-*morphism* θ *will induce an algebraic isomorphism* $\theta_* : H(P^*) \approx \check{H}^*(Z)$.

It follows from Prop. 7 of II that

$$I^*(Z) \approx \mathrm{Min}\ \bar{A}^*(Z) \approx \mathrm{Min}\ P^* \approx \mathrm{Min}(DP(X) \underset{\check{A}*(B)}{\otimes} \bar{A}^*(Y)).$$

Using the same method described in IV of [1], we have then the following theorem which generalizes the corresponding theorems in [1]:

Theorem 1 *For the fiber square with conditions* $1^\circ - 3^\circ$ *we have*

$$I^*(Z) \approx \mathrm{Min}(DP^I(X) \underset{I^*(B)}{\otimes} I^*(Y)),$$

in which $DP^I(X)$ *is the total complex of a proper projective resolution of* $I^*(X)$ *considered as a right* $I^*(B)$-DGA-*algebra via* $f^I : I^*(B) \to I^*(X)$,*and* $I^*(Y)$ *is considered as a left* $I^*(B)$-DGA-*algebra via* $g^I : I^*(B) \to I^*(Y)$.

Since any fibration

$$F \subset Y \to B$$

can be considered as a fiber square

$$\begin{array}{ccc} F & \longrightarrow & Y \\ \downarrow & & \downarrow \\ \text{point} & \longrightarrow & B \end{array}$$

we have as a corollary to Theorem 1:

Theorem 2 *For a fibration the I^*-functor of the fiber F is given by*

$$I^*(F) \approx \text{Min } A^*(F),$$

in which

$$A^*(F) = DP^I(\text{point}) \underset{I^*(B)}{\otimes} I^*(Y),$$

with DP^I (point) the total complex of any proper projective resolution of I^*(point) $\approx$ R over $I^*(B)$, which operates trivially on the right on R.

Ⅳ. I^*-functor of a fiber space

Let

$$F \subset Y \xrightarrow{g} B$$

be any fibration under conditions (e.g. Cond. $1° - 3°$ in III) for which Theorem 2 in III holds true, viz. one for which

$$I^*(F) \approx \text{Min } A^*(F)$$

with

$$A^*(F) = DP^I(\text{point}) \underset{I*(B)}{\otimes} I^*(Y).$$

For such fibrations we shall prove in this section that the I^*-functor of the fiber space Y can be determined completely in terms of those of F and B (cf. Theorem E in [2]), viz.

Theorem 3 *In the tensor product $I^*(B) \otimes I^*(F)$, it can be introduced a twisted differential d_τ to be turned into a* DGA-*algebra denoted by $I^*(B) \underset{\tau}{\otimes} I^*(F)$ such that the* DGA-*algebra$I^*(Y)$ is given by*:

$$I^*(Y) \approx \text{Min}(I^*(B) \underset{\tau}{\otimes} I^*(F)).$$

For the proof, we shall start from the previous formula for $I^*(F)$ by choosing an explicit DP^I (point) which can be furnished by the bar construction as follows.

Denote the subalgebra of elements of positive degree of $I^*(B)$ by $I^+(B)$ and set

$$\Omega_B^* = \sum_{n \geqslant 0} [I^+(B)]^n,$$

with generators $[\ ] \in [I^+(B)]^0$ and $[b_1|\cdots|b_n] \in [I^+(B)]^n$. It is a DGA-algebra under shuffle multiplication with degree and differential d_Ω defined by (cf., e. g. [9,10]):

$$\begin{aligned}
\deg[b_1|\cdots|b_n] &= \sum_{i=1}^{n} \deg b_i - n, \\
d_\Omega[b_1|\cdots|b_n] &= \sum_{i=0}^{n-1} (-1)^{s(i)} [b_1|\cdots|b_i b_{i+1}|\cdots|b_n] \\
&\quad + \sum_{i=1}^{n} (-1)^{s(i-1)} \cdot [b_1|\cdots|db_i|\cdots|b_n], \\
s(i) &= \sum_{j\leqslant i} \deg b_j - i = \deg[b_1|\cdots|b_i].
\end{aligned}$$

The total complex DP^I (point) will then be

$$DF^I(\text{point}) = \Omega_B^* \otimes I^*(B).$$

with differential d_P given by

$$\begin{aligned}
d_P([b_1|\cdots|b_n]b) =& (d_\Omega[b_1|\cdots|b_n])b \\
&+ (-1)^{s(n)} \cdot [b_1|\cdots|b_n] d_B b \\
&- (-1)^{s(n-1)} \cdot [b_1|\cdots|b_{n-1}] b_n b,
\end{aligned}$$

in which d_B is the differential of $I^*(B)$.

It follows that

$$\begin{aligned}
A^*(F) &\approx (\Omega_B^* \otimes I^*(B)) \otimes_{I^*(B)} I^*(Y) \\
&\approx \Omega_B^* \otimes I^*(Y),
\end{aligned}$$

with a differential d_F given by

$$\begin{aligned}
d_F([b_1|\cdots|b_n] \otimes y) =& d_\Omega[b_1|\cdots|b_n] \otimes y \\
&- (-1)^{s(n-1)} \cdot [b_1|\cdots|b_{n-1}] \otimes b_n y \\
&+ (-1)^{s(n)} \cdot [b_1|\cdots|b_n] \otimes d_Y y,
\end{aligned}$$

in which d_Y is the differential of $I^*(Y)$, and $b_n y$ is to be understood as the left action of b_n on y via $g^I : I^*(B) \to I^*(Y)$, i.e., $b_n y = g^I(b_n) \cdot y$.

Define now a twisted product

$$I^*(B) \underset{\tau}{\otimes} A^*(F),$$

which, as an algebra, is the same as the tensor product

$$I^*(B) \otimes A^*(F) = I^*(B) \otimes \Omega_B^* \otimes I^*(Y),$$

but with a differential d_τ given by

$$d_\tau = d_B \otimes 1 \pm 1 \otimes d_F + d'_\tau,$$

in which d'_τ is the twisting part given by

$$d'_\tau(b \otimes ([b_1|\cdots|b_n] \otimes y)) = bb_1 \otimes ([b_2|\cdots|b_n] \otimes y).$$

Now it is easy to see that $I^*(B) \underset{\tau}{\otimes} A^*(F)$ with this differential d_τ is just the total complex of the bar resolution of $I^*(Y)$ considered as left $I^*(B)$ – DGA with $I^*(B)$ acting on the left via

$$g^I : I^*(B) \to I^*(Y).$$

It follows that the natural inclusion

$$I^*(Y) \to I^*(B) \underset{\tau}{\otimes} A^*(F) = I^*(B) \otimes \Omega_B^* \otimes I^*(Y)$$

will induce algebraic isomorphism of homology and hence

$$I^*(Y) \approx \mathrm{Min}(I^*(B) \underset{\tau}{\otimes} A^*(F)).$$

The next step of the proof consists then of replacing $A^*(F)$ by $I^*(F)$ in the twisted product as follows. Starting from the natural inclusion

$$I^*(B) \to I^*(B) \underset{\tau}{\otimes} A^*(F),$$

we can construct by the method of Sullivan[4,5] a minimal DGA J^*, a twisted product $I^*(B) \underset{\tau}{\otimes} J^*$ with some twisted differential d_τ and a DGA-morphism

$$\rho : I^*(B) \underset{\tau}{\otimes} J^* \to I^*(B) \underset{\tau}{\otimes} A^*(F),$$

which induces algebraic isomorphism of homologies. Let

$$\bar{\rho} : J^* \to A^*(F)$$

be the composition

$$J^* \xrightarrow{i} I^*(B) \underset{\tau}{\otimes} J^* \xrightarrow{\rho} I^*(B) \underset{\tau}{\otimes} A^*(F) \xrightarrow{p} A^*(F),$$

in which i is the natural inclusion and p is the projection induced by the augmentation $\varepsilon : I^*(B) \to R$. By the theorem of Moore (cf. [9]) , we see that $\overline{\rho}$ will induce algebraic isomorphism of homology and hence $J^* \approx \text{Min } A^*(F) \approx I^*(F)$ with $\overline{\rho}$ as canonical homomorphism. It follows that

$$\text{Min}(I^*(B) \underset{\tau}{\otimes} I^*(F)) \approx \text{Min}(I^*(B) \underset{\tau}{\otimes} A^*(F)) \approx I^*(Y),$$

and the theorem is proved.

V. The case of singular cohomology

The above Theorems 1 and 2 about I^*-funetor of a fiber square hold true also in the case of singular cohomology and may be deduced in the following sketched manner. We extend at first the PL-differential forms of D. Sullivan[4, 5] to the semi-simplicial complexes, giving rise thus to commutative cochains of topological spaces through singular complexes. We prove finally that these cochains and the singular ones can be pieced together by means of a sequence of H-isomorphisms (forward or backward) of DGA's. Whence we get these theorems.

Let K be an s.s. (semi-simplicial) complex. Construct according to D. Sullivan[4, 5] the de Rham complex $A(K) = \underset{q \geqslant 0}{\oplus} A^q(K)$ of K. The elements of A^q are functions which associate to each simplex σ a differential form of degree q in $\Delta[\dim \sigma]$ with polynomials in rational field Q as coefficients such that some "compatibility conditions" are to be observed. The difference of the present case from the case of simplicial complexes lies in the fact that we have to add conditions about degeneracy operators besides those about face operators. But as is the same in the case of simplicial complexes, integration will give rise to a natural chain map ρ of A to C. Here C is the simplicial cochain functor of semi-simplicial complexes. If L is a subcomplex of K, then we have the following exact sequence:

$$0 \to A(K, L) \to A(K) \to A(L) \to 0.$$

The PL de Rham theorem of D. Sullivan can then be extended to the following

Theorem 4 $\rho_* : HA(K) \to HC(K)$ *is an algebraic isomorphism.*

Proof. 1°. ρ_* is an additive isomorphism. To $A(K)$, let us take the filtration

$$F^p A = \text{Ker}\{A(K) \to A(K^{(p-1)})\},$$

in which $K^{(p)}$ is the p-skelette of K. For the corresponding spectral sequence, we find

$$E_1^{p,q}A = H^{p+q}A(K^{(p)}, K^{(p-1)}).$$

Filtrate $C(K)$ in the same manner we get

$$E_1^{p,q}C = H^{p+q}C(K^{(p)}, K^{(p-1)}).$$

Now $A(K^{(p)}, K^{(p-1)}) = \oplus A(\Delta[p], \dot{\Delta}[p])$, in which $\oplus$ is the direct sum with respect to all non-degenerate p-simplexes. We deduce therefore by Lemma 6 in §10 of [5] that ρ induces an isomorphism $E_1A \to E_1C$. Consequently, $\rho_* : HA \to HC$ is an additive isomorphism.

2°. ρ_* is a multiplicative isomorphism: Consider the category $\mathscr{J}$ of all s. s. complexes. In $\mathscr{J}$, let us take as models the set of simplexes $\Delta[n], n = 1, 2, \cdots$. Now A and C are both cochain functors on $\mathscr{J}$. It is easy to prove that C is representable and A is acyclic on the models. As ρ is a natural transformation preserving augmentation, we see that the following diagram is homotopically commutative by the acyclic model theorem:

$$\begin{array}{ccc} A \otimes A & \longrightarrow & A \\ \rho \otimes \rho \downarrow & & \downarrow \rho \\ C \otimes C & \longrightarrow & C \end{array}$$

In the diagram the two horizontal arrows represent exterior product and cup product respectively.

In the case of *simplicial* complexes, the above theorem can be strengthened to the following

Theorem 5 *There exists a functor D from the category $\mathscr{P}$ of all simplicial complexes to the category of all DGA as well as natural transformations (i.e. DGA-morphisms) $\alpha : A \to D$ and $\beta : C \to D$, such that for all $K \in \mathscr{P}, \alpha_*, \beta_*$ are both isomorphisms (in other words, both α and β are H-isomorphisms of DGA's).*

Proof. The method of proof is the same as that of C^∞ de Rham theorem by Bott in [10]. The only modifications are that the $K^{p,q}$ there have to be replaced by the set of coehains which is associated with each p-simplex σ an element in $A^q(\mathrm{St}\ \sigma)$, and that partitions of unity have to be replaced by barycentric coordinates.

Remark. In fact, constructing a bigraded *chain* complex as in A. Weil[11], we can prove that $\rho_* = \beta_*^{-1} \circ \alpha_*$. But this result, though more precise, is not used in what follows.

The above Theorem 5 is not true in the case of s.s.complexes. To avoid this difficulty, we are in need of the following

Lemma 3 (Barratt [12]) *There exist a (subdivision) functor b of $\mathscr{J}$ to $\mathscr{P}$ and a natural transformation $\lambda : b \to 1$ such that*

$$|\lambda| : |bK| \to |K|$$

is a homotopy equivalence for all $K \in \mathscr{J}$.

It is clear that for all $K, \lambda^* : C(K) \to C(bK)$ is an H-isomorphism of DGA's. Consequently by the de Rham theorem on s. s. complexes, we see that $\lambda^* : A(K) \to A(bK)$ is also an H-isomorphism of DGA's.

Suppose that we have a commutative diagram in DGA category below:

$$\begin{array}{ccccc} B & \longleftarrow & A & \longrightarrow & C \\ \downarrow g & & \downarrow f & & \downarrow h \\ B' & \longleftarrow & A' & \longrightarrow & C' \end{array}$$

Eilenberg and Moore have defined the notion of graded groups $\mathrm{Tor}_A\ (B', C')$ and $\mathrm{Tor}_{A'}\ (B', C')$ as well as homomorphism

$$\mathrm{Tor}_f(g, h) : \mathrm{Tor}_A(B, C) \to \mathrm{Tor}_{A'}(B', C').$$

Besides, as described in [9], they have proved the following theorem, viz.

Lemma 4 *If f, g, h are all H-isomorphisms, then $\mathrm{Tor}_f(g, h)$ is an isomorphism.*

Apply now the above results to topological spaces. Let X be a topological space and SX the singular complex of X. Consider SX as an s. s. complex and denote $A(SX)$ and $C(SX)$ by $A(X)$ and $C(X)$. Suppose now given a fiber square

$$\begin{array}{ccc} Z & \longrightarrow & Y \\ \downarrow & & \downarrow g \\ X & \xrightarrow{f} & B \end{array}$$

It gives rise to a commutative diagram below in the DGA category:

$$\begin{array}{ccccccccc} A(X) & \xrightarrow{\lambda^*} & A(bSX) & \xrightarrow{\alpha} & D(bSX) & \xleftarrow{\beta} & C(bSX) & \xleftarrow{\lambda^*} & C(X) \\ f^* \uparrow & & \uparrow f^* & & \uparrow f^* & & \uparrow f^* & & \uparrow f^* \\ A(B) & \xrightarrow{\lambda^*} & A(bSB) & \xrightarrow{\alpha} & D(bSB) & \xleftarrow{\beta} & C(bSB) & \xleftarrow{\lambda^*} & C(B) \\ g^* \downarrow & & \downarrow g^* & & \downarrow g^* & & \downarrow g^* & & \downarrow g^* \\ A(Y) & \xrightarrow{\lambda^*} & A(bSY) & \xrightarrow{\alpha} & D(bSY) & \xleftarrow{\beta} & C(bSY) & \xleftarrow{\lambda^*} & C(Y) \end{array}$$

From the above discussions, each horizontal arrow in the diagram is an H-isomorphism of DGA. Hence, applying Lemma 4 to the diagram successively from right to left, we shall get

$$\mathrm{Tor}_{C(B)}(C(X), C(Y)) \approx \mathrm{Tor}_{A(B)}(A(X), A(Y)).$$

Now Eilenberg-Moore have proved that (cf. [9])

$$H^*(Z) \approx \mathrm{Tor}_{C(B)}(C(X), C(Y)).$$

Therefore we have

$$H^*(Z) \approx \mathrm{Tor}_{A(B)}(A(X), A(Y)).$$

Using methods in [1], we get then the Theorems 1 and 2 for the case of singular cohomology.

References

[1] Wu Wen-tsün. Theory of I^*-functors in algebraic topology-Real topology of fiber squares. *Scientia Sinica*. 1975, 18: 464-482.

[2] 吴文俊. 科学通报, 1975, 20: 311.

[3] Godement R. Théorie des faisceaux. Paris, 1958.

[4] Sullivan D. Differential forms and the topology of manifolds, in *Manifolds Tokyo*, 1975, 1973: 37-49.

[5] Friedlander E, Griffiths P A & Morgan J. Homotopy theory and differential forms. *Mimeog.Notes*, 1972.

[6] Swan R G. Thom's theory of differential forms on simplicial sets. *Topology*, 1975, 14: 271-273.

[7] Cartan H. *Seminaire Cartan*, 1950/51.

[8] Cartan H. *Seminaire Cartan*, 1954/55.

[9] Smith L. Homological algebra and the Eilenberg-Moore spectral sequence. *Trans. Amer. Math. Soc.*, 1967, 129: 58-93.

[10] Bott R. Lectures on characteristic classes and foliations, in *Lecture Notes in Math.*, v. 279. Springer, 1972.

[11] Weil A. Sur les théorèmes de de Rham. *Comm. Math. Helv.*, 1952, 26: 119-145.

[12] Barratt M. Simplicial and semi-simplicial complexes. *Mimeog. Notes.* Princeton Univ., 1956.

52. On Calculability of I^*-Measure with Respect to Complex-Union and Other Related Constructions*

In what follows a connected complex with $H_R^1 = 0$ will be called simply a *c*-complex. For such complexes we may introduce the notion of I^*-measure which was called previously I^*-functor. We have proved that I^*-measure is calculable with respect to most of the geometrical constructions usually met in algebraic topology. In this paper we shall give simple alternative proofs for its calculability with respect to space-union and some relative constructions and give explicit expressions for the calculation. We shall consider the case of complexes only, the generalization to the case of more general spaces is evident.

Ⅰ. Introduction of some relevant notions and symbols

Let N be a minimal DGA, with free generators x_i, $i \in I$. Construct now a free DGA with generators u_i, v_i, $i \in I$, for which deg $u_i =$ deg x_i, deg $v_i =$ deg $x_i + 1$ and $du_i = v_i$, $dv_i = 0$. Denote this DGA by N^{tr} and define a DGA-morphism

$$\tau^N : N^{tr} \to N,$$

by $\tau^N(u_i) = x_i$, $\tau^N(v_i) = dx_i$. It is clear that $H(N^{tr}) = 0$ and τ^N is an epimorphism. We shall call accordingly N^{tr} as the *trivialized* DGA of N and τ^N the corresponding *trivialization*-morphism. To any DGA-morphism of DGA's $\varphi : M \to N$, we shall denote $M \otimes N^{tr}$ by M^N, and $\varphi \otimes \tau^N$ by $\varphi^N : M^N \to N$.

To any diagram of DGA-morphisms between DGA's

$$A_1 \xrightarrow{f_1} B \xleftarrow{f_2} A_2,$$

let us construct a DGA $C = J(f_1, f_2)$ as follows. The elements of C will be of the form (a_1, a_2) with $a_1 \in A_1$, $a_2 \in A_2$, deg $a_1 =$ deg a_2, and $f_1 a_1 = f_2 a_2 \in B$. The DGA-

* *Kexue Tongbao*, 1980, 25: 196-198; *Kexue Tongbao*, 1980, 25: 185-188.

structure in C is defined by $\deg(a_1,\ a_2) = \deg\ a_1$, $(a_1,\ a_2) + (a_1',\ a_2') = (a_1 + a_1',\ a_2 + a_2')$, $(a_1,\ a_2) \cdot (a_1',\ a_2') = (a_1 a_1',\ a_2 a_2')$, $\lambda(a_1, a_2) = (\lambda a_1,\ \lambda a_2)$ for $\lambda \in R$, and $d(a_1,a_2) = (da_1, da_2)$.

II. Proof of the calculability of the I^*-measure with respect to complex-union construction

It is known that for any DGA-mnorphism of DGA's $f : A \to B$ there exist homotopically-commutative diagrams of DGA-morphisms

$$\begin{array}{ccc} A & \xrightarrow{f} & B \\ {\scriptstyle \rho_A}\uparrow & & \uparrow{\scriptstyle \rho_B} \\ M & \xrightarrow{\tilde{f}} & N \end{array} \tag{1}$$

in which $M = \min A$, $N = \min B$, and ρ_A, ρ_B induce H-isomorphisms (i. e., isomorphisms in homology) . Simple examples show that in general such diagrams cannot be made strictly commutative. However, if f is an epimorphism, then we can prove with the usual methods the following.

Lemma *If the DGA-morphism of DGA's $f : A \to B$ is an epimorphism, and $M = \min A$, $N = \min B$ with DGA-morphism $\rho_B : N \to B$ inducing an H-isomorphism, then there exist DGA-morphisms ρ_A and $\tilde{f}$ to make the diagram* (1) *to be a strictly commutative one with ρ_A inducing an H-isomorphism.*

Consider now two c-complexes K_1, K_2 having a c-complex L as their common subcomplex. Denote by K the union of K_1 and K_2. Set also $I^*(K_i) = M_i, I^*(L) = N$. The inclusions $L \subset K_i$ will induce a diagram of DGA-morphisms

$$M_1 \overset{\varphi_1}{\to} N \overset{\varphi_2}{\leftarrow} M_2.$$

We have then

Theorem 1 *The I^*-measure is calculable with respect to the complex-union constrtiction, with an explicit expression*

$$I^*(K) \approx \min J(\varphi_1^N, \varphi_2^N).$$

Proof. Denote the de Rham-Sullivan algebras of K_i and L by $A_i = A^*(K)$ and $B = A^*(L)$ respectively. Then the inclusions $L \subset K_i$ will induce a diagram of

DGA-morphisms:

$$A_1 \xrightarrow{f_1} B \xleftarrow{f_2} A_2,$$

in which f_1, f_2 are known to be both epimorphisms. By the Lemma we can complete from $\rho_B : N \to B$ and the above diagram to a strictly commutative diagram of DGA-morphisms below (with ρ_i, ρ_B all inducing H-isomorphisms):

$$\begin{array}{ccccc} A_1 & \xrightarrow{f_1} & B & \xleftarrow{f_2} & A_2 \\ \rho_1 \uparrow & & \uparrow \rho_B & & \uparrow \rho_2 \\ M_1 & \xrightarrow{\psi_1} & N & \xleftarrow{\psi_2} & M_2 \end{array}$$

As f_i are epimorphisms, we may extend the above diagram to a strictly-commutative diagram below (with ρ_i^N inducing H-isomorphisms):

$$\begin{array}{ccccc} A_1 & \xrightarrow{f_1} & B & \xleftarrow{f_2} & A_2 \\ \rho_1^N \uparrow & & \uparrow \rho_B & & \uparrow \rho_2^N \\ M_1^N & \xrightarrow{\psi_1^N} & N & \xleftarrow{\psi_2^N} & M_2^N \end{array}$$

By diagram-chasing we verify easily that the DGA-morphism

$$\rho = (\rho_1^N, \rho_2^N) : J(\psi_1^N, \psi_2^N) \to J(f_1, f_2)$$

naturally defined from ρ_i^N will also induce H-isomorphism so that we have

$$\min\ J(\psi_1^N, \psi_2^N) \approx \min I(f_1, f_2). \tag{2}$$

From $\varphi_1 \simeq \psi_1, \quad \varphi_2 \simeq \psi_2$ we know also

$$\min J(\varphi_1^N, \varphi_2^N) \approx \min J(\psi_1^N, {\psi_2}^N). \tag{3}$$

On the other hand we see directly from definition

$$A^*(K) \approx J(f_1, f_2). \tag{4}$$

Combining (2) , (3) and (4) we get then the theorem.

III. Other related constructions

Let L be a c-subcomplex of a c-complex K. Construct a cone C_L over L and denote $C_L \cup K$ by $\Delta = \Delta_L(K)$. As a particular case of Theorem 1, we know that the I^*-measure is calculable with respect to such a cone-construction and an explicit formula can be given for such a calculation. On the other hand, we may also prove directly the following theorem as an alternative method.

Theorem 2 *Denote* $I^*(K) = M$, $I^*(L) = N$ *and let the inclusion* $L \subset K$ *induce the DGA-morphisms* $\varphi : M \to N$. *Then the* I^**-measure is calculable with respect to the cone-construction of* Δ *from* $L \subset K$ *with the explicit expression for calculation*:

$$I^*(\Delta) \approx \min\ \mathrm{ker}\varphi^N.$$

Proof. The inclusion $L \subset K$ will induce a morphism of the de Rham-Sullivan algebra $f : A^*(K) \to A^*(L)$. As f is an epimorphism we get by the Lemma the following strictly-commutative diagram of DGA-morphisms

$$\begin{array}{ccccccccc} 0 & \to & \mathrm{Ker}\, f & \to & A^*(K) & \xrightarrow{f} & A^*(L) & \to & 0 \\ & & & & \rho_1 \uparrow & & \uparrow \rho_2 & & \\ & & & & M & \xrightarrow{\psi} & N & & \end{array}$$

In the diagram the upper line is exact and the ρ_i will induce H-isomorphisms. Again as f is an epimorphism, the above diagram can be extended to the following strictly commutative diagram with both upper and lower lines exact:

$$\begin{array}{ccccccccc} 0 & \to & \mathrm{Ker}\, f & \to & A^*(K) & \xrightarrow{f} & A^*(L) & \to & 0 \\ & & \rho \uparrow & & \uparrow \rho_1^N & & \uparrow \rho_2 & & \\ 0 & \to & \mathrm{Ker}\, \psi^N & \to & M^N & \xrightarrow{\psi^N} & N & \to & 0 \end{array}$$

As ρ_1^N and ρ_2 both induce H-isomorphisms so it will be the same for ρ. From the diagram we get

$$\min \ker \psi^N \approx \min\ \ker f.$$

As $\varphi \simeq \psi$ we have

$$\min\ \ker \varphi^N \approx \min\ \ker \psi^N.$$

Directly from the definition we prove further (cf. the proof given in [1]):

$$I^*(\Delta) \approx \min \ker f.$$

Combining the above formulae we get the theorem.

Similarly we get also as a particular case the following

Theorem 3 *The I^*-measure is calculable with respect to the suspension construction of getting ΣK from a finite c-complex K. The explicit expression for such a calculation is given by*

$$I^*(\Sigma K) \approx \min \ker \tau^N,$$

in which $N \approx I^*(K)$ and τ^N is the corresponding trivialization-morphism.

Now any complex may be obtained from its 2-dimensional squelette by sucessive cone constructions in adding each time a simplex. Moreover, the 2-dimensional squelette of a simply-connected complex is homotopically equivalent to a budget of 2-spheres. Hence the above theorems will furnish an algorithmic method for the actual computation of I^*-measure of any simply connected finite complex. It is also easy to establish an axiomatic system for the I^*-measure of such complexes, cf. also [1].

References

[1] Wu Wen-tsün. Theory of I^*-functors in algebraic topology-effective calculation and axiomatization of I^*-functors on complexes. *Scientia Sinica*, 1976, 10: 647-664.

53. de Rham-Sullivan Measure of Spaces and Its Calculability*

1. de Rham-Sullivan theorem for complexes

In the first paper on *L'Analysis Situs*, dated 1895, Poincaré introduced fundamental notions which are nowadays called differential manifolds, complexes, Betti numbers, fundamental groups, etc., thus laying down the foundations of modern algebraic topology. In addition Poincaré posed the problem of determining the Betti numbers of differential manifolds by means of exterior differential forms; see Section 9 of that paper. The problem was clarified by E. Cartan, and only in 1931 was it completely solved by de Rham. The result, now known as the de Rham theorem, may be stated as follows.

Let $A^*(V)$ be the differential graded anticommutative algebra (abbreviated DGA) of exterior differential forms on a differential manifold V. Then the homology of $A^*(V)$ is algebraically (i.e., additively as well as multiplicatively) isomorphic to the real cohomology ring of V as a topological space, viz.

$$H(A^*(V)) \underset{\text{alg}}{\approx} H_R^{**}(V). \tag{1.1}$$

To avoid the difficulties arising from the nonmanipulable notion of "homologous" in the case of differential manifolds, Poincaré developed in the second and third supplements to *L'Analysis Situs* the combinatorial topology with *complex* instead of differential manifolds as the basic subject to be studied. It is natural to ask whether the theory of Cartan and de Rham may be carried over to the much more general case of complexes. However, it was only in the early 1970s that the question was completely settled by Sullivan in the following manner [5, 6].

For a simplicial complex K let us associate to each simplex σ an exterior differential form (abbreviated EDF) $\omega(\sigma)$ on $|\sigma|$ such that for any face τ of σ the restriction of $\omega(\sigma)$ to τ is just $\omega(\tau)$. Call such a compatible set of EDFs $\omega(\sigma)$ simply an EDF on K. Then under natural operations such EDFs on K become a DGA, called the *de*

* *Proc. Chern Symposium*, 1980: 229-245.

Rham-Sullivan algebra of K and denoted by $A^*(K)$. The Sullivan extension of the de Rham theorem then states that the homology of $A^*(K)$ is algebraically isomorphic to the real cohomology ring of the space of K, viz.

$$H(A^*(K)) \underset{\text{alg}}{\approx} H_R^{**}(K). \tag{1.2}$$

The original de Rham theorem has been much studied and proved in various ways, and quite recently several different proofs of the extended de Rham-Sullivan theorem have also appeared. We remark here that one of the proofs of the original de Rham theorem, due to A. Weil in 1952, is extremely instructive in that it is simple in principle, is naturally extendable to the general case of complexes, and moreover is constructive in the sense that it gives an explicit determination of the isomorphisms (1.1) and (1.2) in question. For this reason we give the following

Sketch of Weil's proof (for differential manifolds V). Construct a locally finite simple open covering $\mathscr{U} = (U_i)_{i\in I}$ of V with associated partition of unity (f_i) and nerve N. Define a differential coelement of degree (m,p) as any system $\Omega = (\omega_H)$= $(\omega_{i_0,\cdots,i_p})$ of EDFs of degree m in $U_{|H|} = \cap_{0\leqslant v\leqslant p} U_{i_0} \neq \varnothing$ attached to the sequences $H = (i_0,\cdots,i_p) \subset I$, which depend alternatively on the indices $i_0,\cdots,i_p$. To Ω we have naturally two operators d and δ with $d\Omega$ and $\delta\Omega$ coelements of bidegrees $(m+1,p)$ and $(m,p+1)$ respectively. The retraction of each $U_H \neq \varnothing$ defines an operator I_H in U_H such that $I\Omega = (I_H\omega_H)$ is a coelement of bidegree $(m-1,p)$ satisfying the relation $\Omega = Id\Omega + dI\Omega$ for $m>0$, with a similar relation for $m=0$. Again, for any set $J' = J\cup\{i\} \subset I$ with $U_{J'} \neq \varnothing$ and ω_J an EDF in U_J, let us denote by $f_i\omega$ the EDF in U_J equal to $f_i\omega$ in $U_{J'}$, and to 0 in $U_J - U_J$. The partition of unity $\{f_i\}$ defines now an operator K which associates to each coelement $\Omega = (\omega_H)$ of bidegree (m,p) with $p>0$ the coelement $K\Omega = (\zeta_{i_0\cdots i_{p-1}})$ of bidegree $(m,p-1)$ with

$$\zeta_{i_0\cdots i_{p-1}} = \sum f_k \omega_{ki_0\cdots i_{p-1}}, \tag{1.3}$$

the $\sum$ being extended over such $k\in I$ with $U_{(ki_0\cdots i_{p-1})} \neq \varnothing$. Similarly $K\Omega$ is defined for Ω of bidegree $(m,0)$. We have then $\Omega = K\delta\Omega + \delta K\Omega$.

It may then easily be seen that for a closed EDF ω on V of degree m, $\Xi = \delta(I\delta)^m\omega$ will be a cocycle of dimension m in N, and conversely for a cocycle $\Xi = (\xi_{i_0\cdots i_m})$ of dimension m in $N, \omega = K(dK)^m\Xi$ will be a closed EDF of degree m on V. The correspondence between ω and Ξ will establish additive isomorphism between $H(A^*(V))$ and $H_R^{**}(V)$. It is easy to verify that the isomorphism is also multiplicative as asserted. (See also Bott [2].) □

The above proof can be easily modified to the general case of complexes:

Proof of extended de Rham-Sullivan theorem for complexes. The complex K has a natural simple covering $\mathscr{U}$ consisting of open stars U_i of vertices v_i of K, so that the nerve of $\mathscr{U}$ coincides with K. We may then define coelements of bidegree (m,p) as well as operator I as before. To define the operator K, we may replace the partition of unity $\{f_i\}$ in the following manner. For each vertex v_i let t_i be the function in U_i which takes on the value $t_i = x_i$ for any point in barycentric coordinates $x_i v_i + \sum_{k=1}^{p} x_{jk} v_{jk}$ in a p-simplex of vertices $v_i, v_{j_1}, \cdots, v_{j_p}$. K is then again defined by (1.3) with f_k replaced by t_k. The proof then runs as before. □

Remark. The above proof shows that the de Rham-Sullivan theorem remains true if we consider only EDFs $\sum \alpha_{i_0\cdots i_p} dx^{i_0}\cdots dx^{i_p}$ in a simplex for which $\alpha_{i_0\cdots i_p}$ are polynomials in the barycentric coordinates x^i with *rational* coefficients.

2. I^*-measure of spaces

In what follows we shall consider only DGAs A on R with

$$H_0(A) \approx R, \qquad H_1(A) = 0.$$

For such DGAs Sullivan has shown how to attach a *minimal model* $M = \min A$, unique up to DGA isomorphism, which is characterized by the following two conditions

(M_1) M is free as an algebra and is decomposable, i.e., $dx \in M^+ \cdot M^+$ for any $x \in M(M^+ =$ set of elements of positive degree in $M)$.

(M_2) There exist *canonical* DGA-morphisms $\rho : M \to A$, unique up to DGA homotopy, with induced algebraic isomorphism

$$\rho_* : H(M) \approx H(A).$$

For a DGA morphism

$$f : A \to B$$

it is also proved that there will be induced morphisms $\tilde{f}$ of minimal models to make the following diagram homotopically commutative ($M = \min A$, $N = \min B, \rho_A, \rho_B$) the corresponding canonical morphisms:

$$\begin{array}{ccc} A & \xrightarrow{f} & B \\ \rho_A \uparrow & & \uparrow \rho_B \\ M & \xrightarrow{\tilde{f}} & N_B \end{array}$$

The following complement is sometimes very useful:

Lemma *If* $f : A \to B$ *is an epimorphism, then with* ρ_B *given,* $\tilde{f}$ *and* ρ_A *can be chosen to make the above diagram* strictly *commutative.*

Proof. Straightforward, following the usual steps for the proof of existence of minimal model and the corresponding canonical DGA morphism. □

Consider now connected simplicial complex K for which $H_R^1(K) = 0$ (to be called *c-complexes* for simplicity) so that the minimal model of the de Rham-Sullivan algebra $A^*(K)$ is well defined. Denote it by $I^*(K)$; then it is easily seen that $I^*(K)$ depends only on the space of K. By the very definition there exist canonical morphisms

$$\rho : I^*(K) \to A^*(K)$$

inducing algebraic isomorphisms

$$\rho_* : H(I^*(K)) \approx H(A^*(K)) \approx H_R^{**}(K),$$

which shows that $I^*(K)$ contains complete information about real cohomology of K. In addition, Sullivan has also proved that the graded module associated to $I^*(K)$ is isomorphic to $\pi_*(K) \otimes R$, so that $I^*(K)$ contains also at least partial information about homotopy groups of K.

It goes without saying that the definition of $I^*(X)$ may be easily extended to arbitrary spaces X that are connected and such that $H_R^1(X) = 0$, to be called *c-spaces* in what follows.

The above connection of I^* with H_R^{**} permits us to determine in many cases the I^* of simple c-spaces from the knowledge of H_R^{**}. We have thus

Theorem 2.1 *Consider* $H_R^{**}(X)$ *as a* DGA *with trivial differential; then we have*

$$I^*(X) \approx \min H_R^{**}(X)$$

for c-spaces X *in the following cases*: *compact Lie groups* G, *their classifying spaces* $B_G, G/U$ *with* U *a closed connected subgroup of maximum rank in* G, *and Riemannian symmetric spaces. It is also true for c-spaces* X *with* H_R^{**} *free or* H_R^{**} *a subalgebra in* $A^*(X)$.

Let $M = G/U$ be a homogeneous space with G a compact connected Lie group and U a closed connected subgroup (always with $H_R^1 = 0$) . It is known that $H_R^{**}(G)$ is an exterior algebra on transgressive elements $x_1, \cdots, x_N$ of odd degree and that $H_R^{**}(B_G)$ is a polynomial algebra on generators y_i of even degree $= \deg x_i + 1$, which are transgressives of x_i. Similarly we have $H_R^{**}(U)$ as an exterior algebra on generators $u_1, \cdots, u_L$ and $H_R^{**}(B_U)$ a polynomial algebra on generators $v_1, \cdots, v_L$ with v_j as transgressives of u_j. Let the canonical homomorphism

$$\rho^* : H_R^{**}(B_G) \to H_R^{**}(B_U)$$

be given by

$$\rho^* y_i = P_i(v_1, \cdots, v_L)$$

with P_i some polynomials. Introduce the following Cartan algebra:

$$C_M^* = H_R^{**}(G) \otimes H_R^{**}(B_U)$$

with twisted differential

$$\begin{aligned} dx_i &= P_i(v_1, \cdots, v_L), \\ dv_i &= 0. \end{aligned}$$

Then we have

Theorem 2.2 *$I^*(M) \approx \min C_M^*$, so that on taking homology on both sides, we get Cartan's theorem*

$$H_R^{**}(M) \approx H(C_M^*).$$

Proof. See Wu [9].

3. Determination of H^{**} of a space by means of I^*

The determination of cohomology (or homology) groups and rings is one of the oldest problems in algebraic topology and is far from trivial in its appearance.Besides the direct cell-subdivision method based on the very definition of topology, there are developed since methods of differential forms (for manifolds only) due to E. Cartan, of spectral sequences due to the French school, and of twisted products due to E. H. Brown.

The introduction of I^*-measure furnishes a new method which seems to be a much more powerful one for the determination of cohomology ring in the case of *real* coefficients. It is based on the following principle. The I^*-measures of relatively simple spaces can usually be determined from the very definition as shown in the last section. Now I^*-measure contains information on the real cohomology as well as the real homotopy of a space, so that the knowledge of the I^*-measure of simpler constitutents of a complicated space may be sufficient to determine completely the real cohomology of that complicated space, while the mere knowledge of the real cohomologies of the various constituents is insufficient to do so. It turns out that this is usually in fact the case, and we shall illustrate the applications of this principle in some concrete cases below.

EXAMPLE 3.1 (Cone Construction). Let L be a c-subcomplex of a c-complex K with injection $j : L \subset K$. Let C_L be a cone over L and $\Delta = \Delta_L(K) = K \cup C_L$ which is homotopically the space K/L in shrinking L to a point. Simple examples show that $H_R^{**}(\Delta)$ as a ring is by no means determined by the knowledge of

$$j^* : H_R^{**}(K) \to H_R^{**}(L).$$

However, as $I^*(K), I^*(L)$ contain more information than mere $H_R^{**}(K), H_R^{**}(L)$, it turns out that $H_R^{**}(\Delta)$ can be completely determined in the present case by the knowledge of

$$j^I : I^*(K) \to I^*(L) \tag{3.1}$$

induced by $j : L \subset K$ in the following manner.

For any morphism of DGA algebras

$$j : A \to B,$$

let C be the algebraic map cone of j considered as morphism of modules. For the induced morphism of homology

$$j_H : H(A) \to H(B),$$

we have then the exact sequence

$$0 \to \text{Coker}\, j_H \overset{\delta_H}{\to} H(C) \overset{i_H}{\to} \text{Ker}\, j_H \to 0.$$

We may take then module basis

$$\begin{aligned}
H(A) &= (X_1, \cdots, X_n, Y_1, \cdots, Y_s),\\
H(B) &= (j_H Y_1, \cdots, j_H Y_s, Z_1, \cdots, Z_t),\\
H(C) &= (X_1^c, \cdots, X_n^c, Z_1^c, \cdots, Z_t^c)
\end{aligned}$$

with $i_H X_i^c = X_i, \delta_H Z_j = Z_j^c$.

Take now $\xi_i \in X_i$ and $(\xi_i, \beta_i) \in X_i^c$ with

$$d_A \xi_i = 0,$$
$$d_B \beta_i = -j\xi_i.$$

Suppose that

$$X_i X_j = \sum \lambda_{ij}^k X_k$$

(λ_{ij}^k redundant in incorrect degrees) , so that

$$\xi_i \xi_j = \sum \lambda_{ij}^k \xi_k + d_A \alpha_{ij}$$

for some $\alpha_{ij} \in A$. Set

$$\zeta_{ij} = -\sum \lambda_{ij}^k \beta_k - \beta_i d_B \beta_j + j\alpha_{ij} \in B.$$

Then it is easily verified that $(0, \zeta_{ij})$ is a cycle of C with class Z_{ij}^c depending only on the X's but not on the choice of the ξ's, β's and α's.

Introduce now multiplications and a trivial differential in $H(C)$ to turn it into a DGA, to be denoted by $J(j)$, as follows:

$$X_i^c X_j^c = \sum \lambda_{ij}^k X_k^c + Z_j^c,$$
$$X_i^c Z_j^c = 0, \quad Z_i^c Z_j^c = 0.$$

We have then

Theorem 3.1 $H_R^{**}(\Delta) = J(j^I)$.

In the same manner, for the union K of two c-complexes K_1, K_2 along a common c-subcomplex L we can determine $H_R^{**}(K)$ in terms of the DGA morphisms

$$I^*(K_1) \xrightarrow{j_1^I} I^*(L) \xleftarrow{j_2^I} I^*(K_2), \tag{3.2}$$

but not so in terms of the morphisms

$$H_R^{**}(K_1) \xrightarrow{j_1^H} H_R^{**}(L) \xleftarrow{j_2^H} H_R^{**}(K_2).$$

The construction, however, is too complicated to give here.

EXAMPLE 3.2 (Eilenberg-Moore spectral sequence) . The cohomology of a space is quite often determined by means of spectral sequences of fibrations. For example, suppose that $X \times_B Y$ is the fibre space induced by a map $f : X \to B$ from a fibration $F \subset Y \xrightarrow{j} B$ with B simply connected, so that we have a fibre square

$$\begin{array}{ccc} F\subset X\underset{B}{\times}Y & \longrightarrow & Y \\ \downarrow & & \downarrow j \\ X & \xrightarrow{f} & B \end{array} \tag{3.3}$$

Then there exists a spectral sequence convergent to $H_k^{**}(X\underset{B}{\times}Y)$ over a coefficient field k for which the E_2 term is determined by the morphisms

$$H_k^{**}(X)\xleftarrow{f^*}H_k^{**}(B)\xrightarrow{j^*}M_k^{**}(Y) \tag{3.4}$$

in the following manner:

$$E_2\approx \mathrm{Tor}_{H_k^{**}(B)}(H_k^{**}(X),H_k^{**}(Y))\Rightarrow H_k^{**}\left(X\underset{B}{\times}Y\right). \tag{3.5}$$

In particular we have for the fibration $F\subset Y\xrightarrow{j}B$ the spectral sequence

$$E_2\approx \mathrm{Tor}_{H_k^{**}(B)}(k,H_k^{**}(Y))\Rightarrow H_k^{**}(F). \tag{3.6}$$

These spectral sequences due to Eilenberg and Moore furnish some information about cohomology of $X\times_B Y$ or F from the knowledge of these of X, Y, B. However, only in very rare cases will these spectral sequences collapse to give a complete determination of those of $X\times_B Y$ or F with exact isomorphisms

$$\mathrm{Tor}_{H_k^{**}(B)}(H_k^{**}(X),H_k^{**}(Y))\approx H_k^{**}\left(X\underset{B}{\times}Y\right) \tag{3.7}$$

or

$$\mathrm{Tor}_{H_k^{**}(B)}(k,H_k^{**}(Y))\approx H_k^{**}(F). \tag{3.8}$$

Such cases do arise, but only under very strong conditions, say with $Y\to B$ the canonical fibration $B_U\to B_G$ and $X=$ Riemannian symmetric space, as considered by Baum and Smith[1], or $X=G'/U'$ with U' of maximum rank in G', as considered by Wolf [8] ($k=$ real field, G,G' compact connected Lie groups, U,U' closed connected subgroups) .

Now in case $k=R$, if instead of the cohomology H_R^{**} we use the I^*-measures, which contain much more information than $H_R^{**'}$s, then it turns out that the induced morphisms

$$I^*(X)\xleftarrow{f^I}I^*(B)\xrightarrow{j^I}I^*(Y) \tag{3.9}$$

are enough for the complete determination of $H_R^{**}(X\times_B Y)$(or $H_R^{**}(F)$), viz.

Theorem 3.2 *For the fibre square* (3.3) *of c-spaces we have*

$$H_R^{**}\left(X \underset{B}{\times} Y\right) \approx \mathrm{Tor}_{I^*(B)}(I^*(X), I^*(Y)). \tag{3.10}$$

In particular we have for the fibration $F \subset Y \to B$ *of c-spaces*

$$H_R^{**}(F) \approx \mathrm{Tor}_{I^*(B)}(R, I^*(Y)). \tag{3.11}$$

For proofs see Wu[10]. See also the corresponding theorems in Section 4. We remark only that in the cases considered by Baum and Wolf I^* coincides with H_R^{**} for the various spaces X, Y, B involved, and the above isomorphisms become those in (3.7) , (3.8) .

EXAMPLE 3.3 (Twisted product for fibre space). For a fibration $F \subset Y \to B$ Brown has shown that

$$H^{**}(Y) \approx H(S^*(B) \otimes S^*(F)),$$

in which S^* denotes singular cohomology and H is the homology with respect to a certain twisted differential d_τ in the tensor product. As S^* is transfinite, it is desirable to replace if possible S^* by say H^{**}, which is of course rarely possible. However, if we consider in the case of a real coefficient field I^* instead of H_R^{**}, then it turns out again that I^* will give enough information to determine $H_R^{**}(Y)$ completely, as in the following

Theorem 3.3 *For a fibration* $F \subset Y \to B$ *of c-spaces we have*

$$H_R^{**}(Y) \approx H(I^*(B) \otimes I^*(F)),$$

with some twisted differential d_τ *in the tensor product* $I^*(B) \otimes I^*(F)$.

For a proof see Wu and Wang [12]. See also the corresponding theorem in Section 4.

In certain cases the twisted differential d_τ may be explicitly determined. Thus for F a sphere $S^n (n > 1)$ we have

Theorem 3.4 *For the sphere bundle*

$$S^n \subset Y \to B$$

the twisted differential d_τ *in* $I^*(B) \otimes I^*(S^n)$ *is given as follows*:

For n odd, so that $I^*(S^n) = \Lambda(x)$ *with deg* $x = n, dx = 0$, *we have*

$$d_\tau x = e,$$

where $e \in I^*(B)$ *corresponds to the Euler class of the bundle on passing to the homology.*

For n even, so that

$$I^*(S^n) = \Lambda(x, y)$$

with $\deg x = n$, $\deg y = 2n - 1, dx = 0, dy = x^2$, *we have*

$$d_\tau x = 0,$$
$$d_\tau y = x^2 - p_n,$$

in which p_n *corresponds to the* nth *real Pontrjagin class of the bundle on passing to homology.*

Sketch of Proof. d_τ is determined from consideration of universal bundles over Grassmannian manifolds whose I^*-measures are easily determined. □

4. Calculability of I^*-measure

In algebraic topology one usually associates various algebraic structures (or numbers) to spaces reflecting their topological properties. We may mention thus dimensions, Betti numbers, homology and cohomology groups on rings, homotopy groups, etc. In comparison with the notions length, area, volume, or content associated to elementary geometrical figures reflecting their metrical properties, we shall call such algebraic structures associated to spaces also their "measures". Accordingly, we shall denote by $H_A^*(X)[H_A^{**}(X)]$ the cohomology group [ring] of a space X on a coefficient group [resp. ring] A, instead of the usual notation $H^*(X, A)$, to make clear the character of H_A^* or H_A^{**} being a "measure". Now for a certain measure of a topological space to be fruitful, the measure in question should satisfy, besides being invariant and of finite type in character, the further condition of being "constructive" or "calculable" in the following sense.

In algebraic topology we frequently have to construct new spaces from given ones (as in forming space products, shrinking part of a space to a point, forming a suspension or loop space, etc.) and then determine or "calculate" for the new space the measure in question from those of given ones. In view of this, we shall introduce the notion of "calculability" of a measure M on certain category of spaces as follows:

Definition Let X_α be a set of spaces and G a certain geometric procedure producing some space X from X_α. Then we shall say that M is *calculable* vis-à-vis the geometrical construction G if $M(X)$ is completely determined by means of some algebraic construction from $M(X_\alpha)$ together with the inherent interrelations of the latter ones arising from their mutual geometrical relations and the geometric construction G.

In this sense, the measures H_Z^* and H_R^{**} are both calculable vis-à-vis the space-product construction, and the Künneth formulae give the precise manner of the corresponding algebraic determination. However, it turns out that even the simplest measures in algebraic topology are often noncalculable in the above sense vis-à-vis quite simple geometrical constructions. Thus simple examples show that:

(a) The integral cohomology ring measure H_Z^{**} is not calculable vis-à-vis the space-product construction, though the integral group measure H_Z^* is.

(b) The real cohomology ring measure H_R^{**} is not calculable vis-à-vis the geometrical cone construction, and neither is the integral cohomology group measure H_Z^*, though the real cohomology group measure H_R^* is.

The above cases show that even so simple a measure as H_A^* is not an appropriate one from the point of view of calculability. On the other hand, the I^*-measure introduced in Section 2, based on theory of Sullivan, besides being invariantive and of finite type, is calculable vis-à-vis practically all geometrical constructions usually met in algebraic topology. We shall cite below a few examples.

EXAMPLE 4.1 (Space-union construction). For any diagram of DGA morphisms

$$A_1 \xrightarrow{f_1} B \xleftarrow{f_2} A_2,$$

let $J(f_1, f_2)$ be the DGA consisting of all elements (a_1, a_2) such that $a_i \in A_i$,$\deg a_1 = \deg a_2$, $f_1 a_1 = f_2 a_2$, with an evident DGA structure. For any minimal DGA N with free generators x_i, let N^{tr} be the DGA with free generators u_i, v_i such that

$$\deg u_i = \deg x_i, \quad \deg v_i = \deg x_i + 1,$$
$$du_i = v_i, \quad dv_i = 0.$$

We shall also note by

$$\tau_N : N^{\text{tr}} \to N \tag{4.1}$$

the DGA morphism defined by

$$\tau_N(u_i) = x_i, \quad \tau_N(v_i) = dx_i. \tag{4.2}$$

It is clear that $H(N^{\text{tr}}) = 0$ and τ_N is onto, and accordingly Λ^{tr} will be called the *trivialization* of N.

With these notions we are now in a position to prove that I^*-measure is calculable vis-à-vis the space-union construction with an explicit algebraic determination:

Theorem 4.1 *Let K be the union of two c-complexes* K_1, K_2 *with a c-subcomplex* L *in common. Let the injection of* L *in* K_1, K_2 *induce the* DGA *morphisms*

$$M_1 \xrightarrow{\varphi_1} N \xleftarrow{\varphi_2} M_2, \tag{4.3}$$

in which $M_i = I^*(K_i), N = I^*(L)$. *Set* $\tilde{M}_i = M_i \otimes N^{\mathrm{tr}}, \tilde{\varphi}_i = \varphi_i \otimes \tau_N$, *and form the diagram*

$$\tilde{M}_1 \xrightarrow{\tilde{\varphi}_1} N \xleftarrow{\tilde{\varphi}_2} \tilde{M}_2. \tag{4.4}$$

Then we have

$$I^*(K) \approx \min J(\tilde{\varphi}_1, \tilde{\varphi}_2). \tag{4.5}$$

In particular, for $L =$ a *point, so that* $K = K_1 \vee K_2$, *the* J *on the right-hand side of* (4.5) *is simply the direct sum of* $I^*(K_1)$ *and* $I^*(K_2)$ *with evident* DGA *structure, so that* (4.5) *becomes*

$$I^*(K) \approx \min(I^*(K_1) \oplus I^*(K_2)). \tag{4.6}$$

Sketch of proof. Let $A_i = A^*(K_i), B = A^*(L)$ with DGA morphisms induced by injections

$$A_1 \xrightarrow{f_1} B \xleftarrow{f_2} A_2.$$

The morphisms f_i are epimorphisms and it is clear that

$$A^*(K) \approx J(f_1, f_2)$$

so that

$$I^*(K) \approx \min J(f_1, f_2). \tag{4.7}$$

Now by the Lemma of Section 2 we see that the canonical morphism $\rho_B : N \to B$ and the epimorphisms f_i above can be completed to a *strictly* commutative diagram of DGA morphisms with ρ_i canonical ones:

$$\begin{array}{ccccc} A_1 & \xrightarrow{f_1} & B & \xleftarrow{f_2} & A_2 \\ {\scriptstyle\rho_1}\uparrow & & \uparrow{\scriptstyle\rho_B} & & \uparrow{\scriptstyle\rho_2} \\ M_1 & \xrightarrow{\psi_1} & N & \xleftarrow{\psi_2} & M_2 \end{array}$$

Setting $\tilde{M}_i = M_i \otimes N^{\text{tr}}, \tilde{\psi}_i = \psi_i \otimes \tau_N$, then, as f_i are epimorphisms and N^{tr} is free, we can extend the above diagram further into a strictly commutative one as shown below:

$$\begin{array}{ccccc} A_1 & \xrightarrow{f_1} & B & \xleftarrow{f_2} & A_2 \\ {\scriptstyle \tilde{\rho}_1}\uparrow & & \uparrow{\scriptstyle \rho_B} & & \uparrow{\scriptstyle \tilde{\rho}_2} \\ \tilde{M}_1 & \xrightarrow{\tilde{\psi}_1} & N & \xleftarrow{\tilde{\psi}_2} & \tilde{M}_2 \end{array} \tag{4.8}$$

By diagram-chasing arguments we prove easily that the natural DGA morphism

$$J(\tilde{\psi}_1, \tilde{\psi}_2) \to J(f_1, f_2)$$

defined by the pair $(\tilde{\rho}_1, \tilde{\rho}_2)$ will induce an algebraic isomorphism in homology:

$$H(J(\tilde{\psi}_1, \tilde{\psi}_2)) \approx H(J(f_1, f_2)).$$

It follows that

$$\min J(\tilde{\psi}_1, \tilde{\psi}_2) \approx \min J(f_1, f_2).$$

As it is easy to verify that min $J(\tilde{\psi}_1, \tilde{\psi}_2) \approx$ min $J(\tilde{\varphi}_1, \tilde{\varphi}_2)$, we get the theorem from (4.7) . □

EXAMPLE 4.2 (Cone construction) . As a special case of the preceding one we see that the I^*-measure is also calculable vis-à-vis the cone-construction. As this case is rather of particular importance (cf. Section 5) we give an alternative explicit algebraic construction as follows:

Theorem 4.2 *Let $\Delta = \Delta_L(K)$ be the cone construction over a c-subcomplex L of a c-complex K. The injection $L \subset K$ will induce a* DGA *morphism*

$$\varphi : M \to N$$

in which $M = I^*(K), N = I^*(L)$. Setting $\tilde{M} = M \otimes N^{\text{tr}}, \tilde{\varphi} = \varphi \otimes \tau_N$, then we have

$$I^*(\Delta) \approx \min \text{Ker}\tilde{\varphi}. \tag{4.9}$$

Moreover, we have a canonical DGA *morphism*

$$\gamma(\varphi) : I^*(\Delta) \to I^*(K) = M \tag{4.10}$$

corresponding to the injection $K \subset \Delta$ which is defined as the composition

$$I^*(\Delta) \to \text{Ker}\tilde{\varphi} \to \tilde{M} \to M,$$

with the last DGA *morphism the natural projection.*

Sketch of Proof. Form the exact sequence

$$0 \to \operatorname{Ker} f \to A^*(K) \xrightarrow{f} A^*(L) \to 0, \tag{4.11}$$

in which the DGA morphism f is induced from the injection $L \subset K$. It is easy to verify that

$$I^*(\Delta) \approx \min \operatorname{Ker} f. \tag{4.12}$$

As f is onto, we can complete (4.11) to a strictly commutative diagram

$$\begin{array}{ccccccccc}
0 \longrightarrow & \operatorname{Ker} f & \longrightarrow & A^*(K) & \xrightarrow{\quad f \quad} & A^*(L) & \longrightarrow 0 \\
& \rho\uparrow & & \rho_1\uparrow & & \uparrow\rho_2 & \\
0 \longrightarrow & \operatorname{Ker} \tilde{\psi} & \longrightarrow & \tilde{M} & \xrightarrow{\quad \tilde{\psi} \quad} & N & \longrightarrow 0
\end{array}$$

in which ρ_1, ρ_2 induce algebraic isomorphisms in homology. It follows that ρ also induces algebraic isomorphism in homology, so that the theorem follows immediately from (4.12) and the relation min Ker $\tilde{\psi} \approx$ min Ker $\tilde{\varphi}$, easily verified. □

As a consequence of Theorem 4.2 we get for the suspension construction:

Theorem 4.3 *Let* ΣK *be the suspension of a finite c-complex* K. *Let* $N = I^*(K)$, *then*

$$I^*(\Sigma K) \approx \min \operatorname{Ker} \tau_N .$$

EXAMPLE 4.3 (Holing construction). As the reverse operation for a particular cone construction, we have the following holing construction: Let K be a complex of dimension n and σ a simplex of highest dimension n of K. Let K^σ be the complex obtained from K by removing the interior of σ. Then the I^*-measure is again calculable vis-à-vis such a holing construction of K^σ from K. For this let us first explain some terminology.

For any minimal DGA M we can take generators in each dimension $m \geqslant 2$ to be of the form

$$x_1^m, \cdots, x_{h_m}^m, y_1^m, \cdots, y_{k_m}^m,$$

in which $dx_i^m = 0$ while $dy_1^m, \cdots, dy_{k_m}^m$ are linearly independent. For any cycle $z = z_n$ of degree n of M, let us form a DGA M^z in the following manner. The M^z will possess generators

$$\begin{array}{ll} x_1^m, \cdots, x_{h_m}^m, y_1^m, \cdots, y_{k_m}^m, & \text{in degree } m \leqslant n-2, \\ x_1^{n-1}, \cdots, x_{h_{n-1}}^{n-1}, y_1^{n-1}, \cdots, y_{k_{n-1}}^{n-1}, u_{n-1}, & \text{indegree } n-1, \\ x_1^n, \cdots, x_{h_n}^n, & \text{in degree } n, \end{array}$$

while M^z has no generators in degrees $> n$. The differential will be the same as in M with the further relation $du_{n-1} = z_n$. The multipications in M^z are subject to the sole condition that any product in degree $> n$ of generators of M^z is 0. We have now the following:

Theorem 4.4 *Let K^σ be the holing construction of a complex K of finite dimension n with an n-simplex σ removed. Consider the exact sequence*

$$H_R^{n-1}(K^\sigma) \xrightarrow{j} H_R^{n-1}(\dot{\sigma}) \xrightarrow{\delta} H_R^n(K).$$

If j is onto, then let us take $z = z_n$ to be 0 in $M = I^*(K)$. If $j = 0$, then take z to be a cycle of degree n in M whose class corresponds to the δ-image of a generator of $H_R^{n-1}(\dot{\sigma})$. Then we have

$$I^*(K^\sigma) \approx \min M^z. \tag{4.13}$$

The holing construction was suggested by Professor W. Y. Hsiang, and the theorem above will be used in his study of fixed points under group actions. The proof of the theorem is based on the direct consideration of $A^*(K) \to A^*(K^\sigma)$ and is elementary in character.

EXAMPLE 4.4 (Fibre-product construction). For a fibre square of c-spaces

$$\begin{array}{ccc} F \subset X \underset{B}{\times} Y & \longrightarrow & Y \\ \downarrow & & \downarrow g \\ X & \xrightarrow{f} & B \end{array} \tag{4.14}$$

we have

Theorem 4.5 *Consider $I^*(X)$ and $I^*(Y)$ respectively as a right and a left $I^*(B)$-DGA via induced morphisms f^I and g^I. Take the corresponding bar construction $\mathrm{Bar} I^*(X)$. Then $I^*(X \times_B Y)$ is completely determined from the diagram*

$$I^*(X) \xleftarrow{f^I} I^*(B) \xrightarrow{g^I} I^*(Y) \tag{4.15}$$

as

$$I^*\left(X \underset{B}{\times} Y\right) \approx \min\left(\mathrm{Bar}\, I^*(X) \underset{I^*(B)}{\otimes} I^*(Y)\right). \tag{4.16}$$

In particular we have

$$I^*(F) \approx \min\left(\text{Bar } R \underset{I^*(B)}{\otimes} I^*(Y)\right), \tag{4.17}$$

in which Bar R *is the bar construction of* R *considered as a trivial right* $I^*(B)$-DGA.

Sketch of proof. Let

$$A^*(X) \overset{f^A}{\leftarrow} A^*(B) \overset{g^A}{\rightarrow} A^*(Y)$$

be the diagram of induced DGA morphisms of the corresponding de Rham-Sullivan algebras. Using the usual spectral-sequence arguments, we prove first that the homology of $X \times_B Y$ is isomorphic to the homology of Bar$A^*(X) \otimes_{A^*(B)} A^*(Y)$ or Tor $_{A^*(B)}(A^*(X), A^*(Y))$, so that

$$I^*\left(X \underset{B}{\times} Y\right) \approx \min\left(\text{Bar } A^*(X) \underset{A^*(B)}{\otimes} A^*(Y)\right).$$

The next step consists in replacing all A^* by the corresponding I^*. For details see Wu [10]. □

EXAMPLE 4.5 (Fibre-space construction) . From Theorem 4.5 we deduce the following theorem. For the details of the proof we refer again to Wu and Wang [12]. We remark that Theorems 3.2 and 3.3 are just consequences of Theorems 4.5 and 4.6 on merely taking the homology of both sides in the respective DGA isomorphisms.

Theorem 4.6 *For a fibration* $F \subset Y \to B$ *of c-spaces there exists some twisted differential* d_τ *in* $I^*(B) \otimes I^*(F)$ *such that*

$$I^*(Y) \approx \min(I^*(B) \otimes I^*(F)). \tag{4.18}$$

5. Effective computation and axiomatic system of I^*-measure on the category of simply connected finite polytopes

As any finite complex can be obtained from its 2-dimensional skeleton by adding successively higher-dimensional simplexes each step of which is equivalent to a cone construction, Theorem 4.2 furnishes us a means to compute effectively the I^*-measure of any simply connected finite complex once we know how to compute the I^*-measure of a simply connected 2-dimensional complex. Now any simply connected 2-dimensional complex has the same homotopy type as a budget of 2-spheres joined at a single point. Hence the I^*-measure of a simply connected finite complex can be effectively computed on account of Theorems 4.1−4.3 and the well-known fact that I^*-measure is a homotopy invariant.

Now consider any connected and simply connected finite complex K. Let us represent such a complex by

$$K = K_m \supset K_{m-1} \supset \cdots \supset K_1 \supset K_0,$$

in which K_0 is the 2-dimensional squelette, and K_r is the union of K_{r-1} with an additional simplex Δ_r, the boundary of which is

$$\dot{\Delta}_r = L_{r-1} \subset K_{r-1},$$

so that

$$K_r = K_{r-1} \cup \Delta_r \simeq K_{r-1}/L_{r-1}.$$

Let

$$j_{r-1} : L_{r-1} \subset K_{r-1}$$

be inclusion map, and

$$j^I_{r-1} : I^*(K_{r-1}) \to I^*(L_{r-1})$$

any of the associated morphisms of I^*-measures.

Now we have the following

Lemma *If we know how to compute* $I^*(L'_{r-1}), I^*(L''_{r-1})$ *and the associated* DGA *morphisms*

$$k^I_{r-1} : I^*(L'_{r-1}) \to I^*(L''_{r-1})$$

for any c-subcomplexes $L''_{r-1} \subset L'_{r-1}$ *of* K_{r-1}, *then we know also how to compute* $I^*(L'_r), I^*(L''_r)$, , *and the associated* DGA *morphisms*

$$k^I_r : I^*(L'_r) \to I^*(L''_r)$$

for any c-subcomplexes $L''_r \subset L'_r$ *of* K_r.

Proof. We consider three cases separately.

*Case*1. $L''_r \subset L'_r \subset K_{r-1}$. In this case $I^*(L'_r), I^*(L''_r)$, and $k^I_r : I^*(L'_r) \to I^*(L''_r)$ are already known by hypothesis.

*Case*2. $L''_r \subset K_{r-1}$ but $L'_r \not\subset K_{r-1}$. In that case L'_r contains Δ_r as a subsimplex and may be considered as the cone construction over L_{r-1} of some subcomplex $\bar{L}'_r$ of

$K_{r-1}(L'_r = \Delta_r \cup \bar{L}'_r)$, while L''_r is a subcomplex of $\bar{L}'_r$. It follows from Theorem 4.2 that $I^*(L'_r)$ is determined by the DGA morphism

$$j_r^I : I^*(\bar{L}'_r) \to I^*(L_{r-1}),$$

already known by hypothesis. The associated DGA morphisms

$$I^*(L'_r) \to I^*(L''_r)$$

are then the composition

$$I^*(L'_r) \stackrel{\gamma(j_r^I)}{\to} I^*(\bar{L}'_r) \to I^*(L''_r),$$

of which the morphism on the right is again known by hypothesis.

Case 3. $L''_r \not\subset K_{r-1}$. In this case both L'_r, L''_r contain Δ_r as a subsimplex and can be considered as cone constructions over L_{r-1}:

$$L''_r = \Delta_r \cup \bar{L}''_r, \quad L'_r = \Delta_r \cup \bar{L}'_r.$$

Now $\bar{L}''_r \subset \bar{L}'_r \subset K_{r-1}$, so that $I^*(\bar{L}'_r)I^*(\bar{L}''_r)$, and $j_r^I : I^*(\bar{L}'_r) \to I^*(\bar{L}''_r)$ are known by hypothesis; thus

$$I^*(L'_r), \quad I^*(L''_r), \quad \text{and} \quad k_r^I : I^*(L'_r) \to I^*(L''_r)$$

can be determined by Theorem 4.2. □

In view of Theorems 4.1—4.3 and the above Lemma we thus arrive at the following

Theorem 5.1 *There is an algorithmic procedure permitting us to compute effectively the I^*-measure of any connected and simply connected finite complex up to any prescribed degree.*

The same considerations also furnish us immediately an axiomatic system for the I^*-measure over the category of connected and simply connected finite complexes (or polytopes) as follows.

To each complex K in the category is associated a minimal $I^*(K)$, and to each pair of complex K and subcomplex K' is associated a set of DGA morphisms determined up to homotopy:

$$k^I : I^*(K) \to I^*(K').$$

These DGA algebras and morphisms are characterized by the following axiomatic system (Wu [11]) :

Axiom 1 I^* is a homotopy measure, in other words, for K, K' having same homotopy type in the category, we have $I^*(K) \approx I^*(K')$.

Axiom 2 If $K'' \subset K' \subset K$ are complexes all in the above category and k^I : $I^*(K) \rightarrow I^*(K'), k'^I : I^*(K') \rightarrow I^*(K'')$ are associated DGA morphisms for the pairs $K' \subset K$ and $K'' \subset K'$, then $k'^I k^I : I^*(K) \rightarrow I^*(K'')$ are also associated DGA morphisms for the pair $K'' \subset K$.

Axiom 3 If $\Delta = \Delta_L(K)$ is the cone construction over a subcomplex L of K, both being in the above category, then $I^*(\Delta)$ is given by

$$I^*(\Delta) \approx \min \operatorname{Ker} \tilde{j}^I,$$

where

$$j^I : I^*(K) \rightarrow I^*(L)$$

in any of the associated DGA morphisms of the pair $L \subset K$, and $\tilde{j}^I : I^*(K) \otimes I^*(L)^{\mathrm{tr}} \rightarrow I^*(L)$ is determined from j^I .

Axiom 4 For the cone construction $\Delta = \Delta_L(K)$ as in Axiom 3, the associated DGA morphisms of the pair $K \subset \Delta$ are given by the natural DGA morphisms

$$\gamma(j^I) : I^*(\Delta) \rightarrow I^*(K).$$

Axiom 5 The I^*-measure $I^*(\Delta)$ and the morphism $\gamma(j^I)$ in Axioms 3 and 4 are functorial in character. In other words, if we have pairs $L \subset K$ and $L' \subset K'$ is the category with L', K' as subcomplexes of L and K respectively, then we have the following homotopically commutative diagram with evident DGA morphisms:

$$\begin{array}{ccc} I^*(\Delta) & \rightarrow & I^*(K) \\ \downarrow & & \downarrow \\ I^*(\Delta') & \rightarrow & I^*(K') \end{array}$$

Axiom 6 For any two complexes K', K'' in the category we have

$$I^*(K' \vee K'') \approx \min(I^*(K') \oplus I'(K'')),$$

and the associated DGA morphisms

$$I^*(K' \vee K'') \rightarrow I^*(K')$$

are determined by the composition

$$I^*(K' \vee K'') \rightarrow I^*(K') \oplus I^*(K'') \rightarrow I^*(K').$$

Axiom 7 For a 2-sphere s^2 the I^*-measure is determined as in Theorem 4.5 : $I^*(s) = \Lambda(x, y)$ with deg $x = 2$, deg $y = 3, dx = 0, dy = x^2$.

References

[1] Baum P, Smith L. The real cohomology of differentiable fiber bundles. *Comm. Math. Helv.*, 1967, 42: 171-179.

[2] Bott R. Lectures on characteristic classes and foliations. In *Lectures on Algebraic and Differential Topology* (Bott R, Gitler S, and James I M, Eds.). Springer, Berlin, Heidelberg, New York, 1972.

[3] Friedlander E, Griffiths P A and Morgan J. Homotopy theory and differential forms. *Mimeo. Notes*, 1972.

[4] Poincaré H. *Analysis Situs.* t.1 J. de L'Ecole Polytechnique, 1895.

[5] Sullivan D. Differential forms and topology of manifolds. *Symp. Tokyo on Topology*, 1973.

[6] ——. *Infinitesimal Computations in Topology.* Publ. Math. No. 47, Inst. de Hautes Etudes Scientifiques, 1978.

[7] Weil A. Sur les theoremes de de Rham. *Comm. Math. Helv.*, 1952, 20: 119-145.

[8] Wolf J. The real and rational cohomology of differential fiber bundles. *Trans. Amer. Math.* Soc., 1975, 245: 211-220.

[9] Wu Wen-tsün. Theory of I^*-functions in algebraic topology-Real topology of homogeneous space (in Chinese). *Acta Math, Sinica* , 1975, 18: 162-172.

[10] ——. Real topology of fiber squares. *Scientia Sinica* , 1975, 18: 464-482.

[11] ——. Effective calculation and axiomatization of I^*-functors on complexes. *Scientia Sinica*,1976, 10: 647-664.

[12] Wu Wen-tsün and Wang Qi-ming. I^*-functor of a fibre space. *Scientia Sinica*, 1978, 21: 1-18.

[13] Wu Wen-tsün. Calculability of I^*-Measure vis-ă-vis Space-Union and Related Constructions. *Kuxue Tong Pao*, 1979.

54. A Constructive Theory of Algebraic Topology——Part I. Notions of Measure and Calculability*

In algebraic topology we usually associate with topological spaces various algebraic structures, groups, rings, algebras, etc. which reflect topological properties of the spaces involved. For such algebraic structures we may mention in particular the homology or cohomology groups or rings, the homotopy groups, the K-rings, etc., usually called functors. But we shall call them *measures* and to cope with them we shall adopt the following notations to make clearer the measure-character of these associated algebraic structures: $H_i^G(X) = H_i(X, G), H_G^i(X) = H^i(X, G), H_{\oplus}^G(X) = \sum H_i^G(X), H_G^{\oplus}(X) = \sum H_G^i(X), H_A^*(X) = \sum H_A^i(X)$, etc. Here we use $H_G^{\oplus}$ to indicate that the *measure* in question is an additive group, and H_A^* a multiplicative ring in case A is a ring.

The methods of algebraic topology consist mainly of these steps: translating the topological situations in a problem or a theorem into algebraic languages via convenient measures, manipulating with various algebraic structures and translating back the algebraic relations thus obtained into topological languages to get the final results required, just as in the case of elementary analytic geometry. As a simple illustrative example let us consider the proof of Brouwer's fixed point theorem in the following form:

Let f be a continuous map of an n-disc D with boundary sphere $S = S^{n-1}$ into itself. Let us assume that f has no fixed points. We have to prove that this will lead to a contradiction. Now for any $x \in D$ the ray from $f(x)$ directed toward x will meet S in a point $g(x)$ and we are in the geometrical situation of a commutative diagram of continuous maps in which i is the inclusion and 1 is the identity map: ①

Suppose that we are in possession of a measure M verifying the following axioms:

Ax.1 For any continuous map $h : X \to Y$ there is an induced morphism

---

* *J. Systems Science and Math. Science.*, 1981, 1: 53-68.

① Received March 2. 1981. Part of this work was done at the University of Caiflornia at Berkeley.

$h_M : M(X) \to M(Y)$.

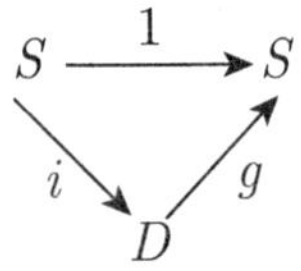

Ax.2 For any continuous maps $h' : X \to Y$ and $h'' : Y \to Z$ the induced morphisms will observe the relation

$$(h''h')_M = h''_M h'_M.$$

Ax.3 For any identity map $1 : X \to X$ the induced morphism $1_M : M(X) \to M(X)$ is an identity.

Besides, let us suppose that measure M can be computed for such simple spaces as S, D, etc. with results:

$$M(S) \neq 0, \quad M(D) = 0.$$

It is now evident that these will lead to a contradiction as desired. We see that in order to prove Brouwer's fixed point theorem, any measure enjoying some formal properies formulated above may be applied equally well. For such measures we may cite $H^G_{n-1}, H^G_{\oplus}, \pi_{n-1}, \pi_n, \pi_{n+1}$, or any homotopy-group-measure π_k for which $\pi_k(S^{n-1}) \neq 0$.

The measures used above are covariantive in view of Axiom 1. We may equally use a measure which is contravariant in the sense that the induced morphism of a map $h : X \to Y$ will be of one of the form $h^M : M(Y) \to M(X)$. With formal properties analogous to those above they may be applied equally well to the proof of Brouwer theorem. Examples of such measures are: $H^{n-1}_G, H^{\oplus}_G, H^*_A, K$, etc.

Algebraic topology has accordingly flourished by means of such measures. However, in order that the algebraic measures will really be utilizable and efficient in applications, some subsidiary requirements should be observed of which we shall cite the following ones:

Requirement 1 Invariance, i.e., the measure in question should be invariant in character. Thus, the measure should be combinatorically invariant in case of complexes, diffeomorphically invariant in case of differential manifolds, and homotopically or topologically invariant in case of general topological spaces.

Requirement 2 Finiteness, i.e. the measure should be of certain finite type, possessing finite basis in each degree in case the measure is some graded algebraic structure, for example.

The groups of chains and cochains of a finite complex observe requirement 2 but not requirement 1, the groups of singular chains of topological spaces and the algebra of exterior differential forms of differential manifolds requirement 1 but not requirement 2. In short, it is tacitly assumed that theorems in algebraic topology should, in their final form, be better stated in terms of measures which are both invariant and of finite type.

Besides these two well adapted requirements we shall introduce a third optional one, *viz.*

Requirement 3 Calculability.

The meaning of the calculability of a certain measure M with respect to a geometrical construction is explained in the following

Definition Let M be a measure on a certain category of spaces. Let X_α be a set of spaces and G a certain geometric construction producing some space X from X_α. Let X_α and X be all in the category so that $M(X_\alpha)$ and $M(X)$ are all well-defined. The measure M will then said to be **calculable** with respect to the geometrical construction G if there exists some algebraic construction A which permits to determine completely $M(X)$ from $M(X_\alpha)$ as well as the algebraic interrelations among $M(X_\alpha)$ arising from the mutual geometrical relations among X_α and the geometrical construction G.

Let us consider the most important measures $M = H_A^*$ and the simplest geometrical constructions $\times$ and $\cup$ where $\times$ is the space-product construction: $(X, Y) \to X \times Y$ and $\cup$ is the space-union construction: $(X, Y, P) \to X \underset{P}{\cup} Y$, in which P is a subspace both of X and of Y. The problem of calculability consists of deciding whether $M(X \times Y)$ can be completely determined by $M(X)$ and $M(Y)$, and whether $M\left(X \underset{P}{\cup} Y\right)$ is from $M(X) \to M(P) \leftarrow M(Y)$ or not.

The answer in case $A = Z$ or R can be seen from the following table:

G \ M	$H_R^{\oplus}$	$H_Z^{\oplus}$	H_R^*	H_Z^*
$\times$	Yes	Yes	Yes	No
$\cup$	Yes	No	No	No

The answer Yes is simply due to the Künneth or Mayer-Vietoris exact sequence which splits in these cases. The answer No may be seen from the following counter-examples.

Ex. 1 Let K be the CW-complex consisting of a 0-cell e^0, two 2-cells e_1^2 and

e_2^2 attached to e^0 to form two 2-spheres S_1^2 and S_2^2, a 3-cell e^3 with an attaching map $\dot{e}^3 \to S_2^2$ of degree 2, and a 4-cell e^4 with an attaching map $\dot{e}^4 \to S_1^2 \vee S_2^2$ which represents the Whitehead product. Let K' be the complex consisting of the same cells as K with the only difference that e^4 will be attached to e^0 to form a 4-sphere S^4. Let L be a complex consisting of a 0-cell e_0^0, a 2-cell e_0^2 and a 3-cell such that $e_0^0 \vee e_0^2$ will be a 2-sphere S_0^2 and e_0^3 will be attached with a map $\dot{e}_0^3 \to S_0^2$ of degree 2.

Let e^3 stand also for the elementary cochain taking value 1 on the cell e^3 and 0 on other cells. Similarly for the others. Then we see that

$$H_Z^2(K) \approx H_Z^2(K') \approx Z \text{ with generator } [e_1^2],$$

$$H_Z^3(K) \approx H_Z^3(K') \approx Z_2 \text{ with generator } [e^3] \text{ of order 2},$$

$$H_Z^4(K) \approx H_Z^4(K') \approx Z \text{ with generator } [e^4],$$

$$H_Z^3(L) \approx Z_2 \text{ with generator } [e_0^3] \text{ of order 2}.$$

All other groups $H_Z^i (i>0)$ are 0. From Künneth formulae we have $H_Z^\oplus (K \times L) \approx H_Z^\oplus (K' \times L)$ as listed in the following table:

	$H_Z^k(K \times L) \approx H_Z^k(K' \times L)$	Generators
k=2	Z	$[e_1^2 \times e_0^0]$
k=3	$Z_2 + Z_2$	$[e^3 \times e_0^0], [e^0 \times e_0^3]$
k=4	Z	$[e^4 \times e_0^0]$
k=5	$Z_2 + Z_2$	$[e_1^2 \times e_0^3], [e_2^2 \times e_0^3 + e^3 \times e_0^2]$
k=6	Z_2	$[e^3 \times e_0^3]$
k=7	Z_2	$[e^4 \times e_0^3]$

Now for the ring measure H_z^* we see that both $H_z^*(K)$ and $H_z^*(K')$ have the trivial multiplicative structure. However, on the cochain level we have

$$e_1^2 \cup e_2^2 = e^4 \text{ in } K \text{ while}$$

$$e_1^2 \cup e_2^2 = 0 \text{ in } K'.$$

Consequently $[e_1^2 \times e_0^0] \cup [e_2^2 \times e_0^3 + e_2^3 \times e_0^2] = 0$ in $K' \times L$ and $= [e^4 \times e_0^3] \neq 0$ in $K \times L$ so that

$$H_Z^*(K \times L) \not\approx H_Z^*(K' \times L).$$

In conclusion, we have

Proposition 1 The H_z^*- measure is non-calculable with respect to the space-product construction.

Ex.2 Let K be the CW-complex consisting of a 0-cell e^0, a 2-cell e^2, two 3-cells e_1^3 and e_2^3, and a 4-cell e^4, with e^0, e^2 forming a 2-sphere S^2, and attaching maps

$\dot{e}_1^3 \to S^2$, $\dot{e}_2^3 \to S^2$ of degrees 2 and 1 respectively. For the complex $e^0 \vee e^2 \vee e_1^3 \vee e_2^3$ we have by Hurewicz theorem $\pi_3 \approx H_3^z \approx Z$ with a generator corresponding to the cycle $e_1^3 - 2e_2^3$. Hence we can attach e^4 by a map $\dot{e}^4 \to e^0 \vee e^2 \vee e_1^3 \vee e_2^3$ such that $\partial e^4 = 2e_1^3 - 4e_2^3$. Let K' be the complex consisting of a 0-cell e^0, a 2-cell e^2, two 3-cells e_1^3 and e_2^3 as in K, a 4-cell e^4 with an attaching map such that $\partial e^4 = e_1^3 - 2e_2^3$, and another 3-cell e_0^3 and 4-cell e_0^4 such that e_0^3 and e^0 form a 3-sphere S_0^3 while e_0^4 is attached with a map $\dot{e}_0^4 \to S_0^3$ of degree 2. Let L be the subcomplex $e^0 \vee e^2 \vee e_1^3$ of K (and K'). Let C_L be the cone over L with a new vertex v and $\Delta = \Delta_L(K)$ (resp. $\Delta' = \Delta_L(K')$)be the union of K(resp. K') and C_L along the common subcomplex L. The Mayer-Vietoris sequence becomes now

$$0 \to \mathrm{Coker}^{k-1} i^H \to H_Z^k(\Delta) \to \mathrm{Ker}^k i^H \to 0$$

in which

$$i^H : H_Z^{\oplus}(K) \to H_Z^{\oplus}(L)$$

is induced by the inclusion map $i : L \subset K$. Similarly for Δ'. Now the only non-trivial homologies of K, K', L are:

$$H_Z^4(K) \approx Z_2 \text{ with generator } [e^4],$$
$$H_Z^4(K') \approx Z_2 \text{ with generator } [e_0^4],$$
$$H_Z^3(L) \approx Z_2 \text{ with generator } [e_1^3].$$

Hence $H_z^k(\Delta) \approx H_z^k(\Delta')$ for $k \neq 4$ and the only case in doubt is for $k = 4$:

$$0 \to Z_2 \to H_Z^4(\Delta)(\text{resp.}\, H_Z^4(\Delta')) \to Z_2 \to 0.$$

It is easy to verify directly that

$$H_Z^4(\Delta) \approx Z_4 \text{ with generator } [e^4] \text{ of order } 4,$$
$$H_Z^4(\Delta') \approx Z_2 + Z_2 \text{ with generators } [e^4] \text{ and } [e_0^4] \text{ both of order } 2.$$

Let us call the construction from (K, L) to Δ the cone-construction over the subcomplex L, then we have

Proposition 2 The $H_z^{\oplus}$-measure is non-calculable with respect to the space-union construction. In fact, it is already non-calculable with respect to the cone-construction over a subspace.

Ex.3 Let L be a 3-sphere and K(resp. K')be the mapping cylinder of L to a 2-sphere with a Hopf map(resp. a constant map). Let C_L be the cone over L and

$\Delta = \Delta_L(K)$ (resp. $\Delta' = \Delta_L(K')$). Now

$$i^* : H_R^*(K) \to H_R^*(L)$$

and

$$i'^* : H_R^*(K') \to H_R^*(L)$$

are the same for the inclusion map $i : L \subset K$ (resp. $i' : L \subset K'$). However $H_R^*(\Delta)$ and $H_R^*(\Delta')$ are non-isomorphic in that Δ is homotopically CP, while Δ' is $S^2 \vee S^4$. This proves the following

Proposition 3 The H_R^*-measure is non-calculable with respect to the space-union construction. In fact it is already non-calculable with respect to the cone-construction over a subspace.

The above proposition shows that for the cone-construction of $\Delta = \Delta_L(K)$ from a pair of complexes $i : L \subset K$ with L finite, the H_R^*-measure of Δ cannot be determined algebraically from the mere knowledge of induced morphism

$$i^H : H_R^*(K) \to H_R^*(L).$$

However if we consider the de Rham-Sullivan measure A^* of exterior differential forms, then the H_R^*-measure of $\Delta = \Delta_L(K)$ is completely determined by the morphism

$$i^A : A^*(K) \to A^*(L).$$

To see this let us consider any DG-morphism of DG-groups (differential graded groups) over R

$$\lambda : A \to B$$

and define the algebraic map cone as the DG-group (B is augmented by $B_{-1} \approx R$ with a generator ε):

$$C^\lambda = \sum C_q^\lambda, \quad C_q^\lambda = A_q \oplus B_{q-1},$$

and

$$d_C(a, b) = (d_A a, -\lambda a - d_B b).$$

Define DG-group $B^+ = \sum B_q^+$ by

$$B_q^+ = B_{q-1}, \quad d_{B+} = -d_B.$$

Then we have a short exact sequence of DG-morphisms

$$0 \to B^+ \xrightarrow{i} C^\lambda \xrightarrow{j} A \to 0$$

with i, j given by

$$i(b) = (0, -b), \quad b \in B^+,$$
$$j(a, b) = a, \quad a \in A, b \in B.$$

Replacing $H_q(B^+)$ by $H_{q-1}(B)$ we get the set of exact sequences

$$0 \to \mathrm{Coker}_{q-1}\lambda_H \xrightarrow{\delta_H} H_q(C_\lambda) \xrightarrow{j_H} \mathrm{Ker}_q\lambda_H \to 0$$

in which

$$\lambda_H : H_\oplus(A) \to H_\oplus(B)$$

is induced by λ, while δ_H, j_H are induced by i and j respectively. It follows that $H(C^\lambda)$, as an additive graded group, will be completely determined from λ as

$$H(C^\lambda) \approx \mathrm{Coker}\,\lambda_H \oplus \mathrm{Ker}\,\lambda_H.$$

Suppose now A, B are DGA-algebras and λ is a DGA-morphism respecting multiplicative structures. We shall see that in this case multiplicative structure may then be introduced in $H(C^\lambda)$ to turn it into a graded ring.

For this purpose let us take a basis in $H(A)$ and $H(B)$:

$$H(A) = R[X_1, \cdots, X_r, Y_1, \cdots, Y_s],$$
$$H(B) = R[Z_1, \cdots, Z_i, \lambda_H Y_1, \cdots, \lambda_H Y_s]$$

such that

$$\mathrm{Ker}\,\lambda_H = R[X_1, \cdots, X_r],$$

and

$$\mathrm{Coker}\,\lambda_H = R[\bar{Z}_1, \cdots, \bar{Z}_t],$$

in which $\bar{Z}_j = Z_j \bmod \mathrm{Im}\lambda_H$. Take any $X_i^\lambda \in H(C^\lambda)$ with $j_H X_i^\lambda = X_i$, and set $Z_j^\lambda = \delta_H \bar{Z}_j$. Then

$$H(C^\lambda) = R[X_1^\lambda, \cdots, X_r^\lambda, Z_1^\lambda, \cdots, Z_t^\lambda].$$

Let us take arbitrarily some $(x_i, b_i) \in X_i^\lambda$ so that $x_{iB} = -db_i$ and $x_i \in X_i$. (We write x_{iB} for λx_i and similarly for the others). Take also some $z_j \in Z_j$ so that $(0 - z_j) \in Z_j^\lambda$.

As $X_i X_j \in \mathrm{Ker}\lambda_H$, we have

$$X_i X_j = \sum \lambda_{ij}^k X_k$$

with eventually redundant $\lambda_{ij}^{k'}s$. Consequently we have for some $a_{ij} \in A$,

$$x_i x_j = \sum \lambda_{ij}^k x_k + d a_{ij}.$$

Set

$$z_{ij} = \sum \lambda_{ij}^k b_k + b_i \cdot d b_j - a_{ij} B.$$

We verify readily that $dz_{ij} = 0$ in B so that $z_{ij} \in Z_{ij} \in H(B)$ and

$$(0, -z_{ij}) \in Z_{ij}^\lambda \in \delta_H \text{ Coker } \lambda_H \subset H(C^\lambda)\ .$$

It is also readily verified that Z_{ij}^λ, as a homology class, is independent of the various x_i, b_i, z_j, a_{ij} chosen. We set now as definition:

Definition The J-construction of DGA-morphism $\lambda : A \to B$ is the DGA-algebra which has an underlying group structure as $H(C^\lambda)$ and, with additive basis X_i^λ, Z_j^λ arbitrarily chosen, has a multiplicative structure given by:

$$\begin{aligned} &X_i^\lambda \cdot X_j^\lambda = \sum \lambda_{ij}^k X_k^\lambda + Z_{ij}^\lambda, \\ &X_i^\lambda \cdot Z_j^\lambda = 0, \\ &Z_i^\lambda \cdot Z_j^\lambda = 0. \end{aligned}$$

The degrees and differentials will be given by

$$\begin{aligned} &\deg X_i^\lambda = \deg X_i, \quad \deg Z_j^\lambda = \deg Z_j + 1, \\ &dX_i^\lambda = 0, \quad dZ_j^\lambda = 0. \end{aligned}$$

Remark. It is easy to verify that this definition of multiplication is intrinsic and is independent of the basis chosen. We can also define it in an intrinsic manner without introducing any basis.

The DGA-algebra thus defined is independant of the various additive basis chosen and hence may be legitimately denoted as $J(\lambda)$. We have now:

Proposition 4 The H_R^*-measure of $\Delta = \Delta_L(K)$ of a pair of complexes $L \subset K$ by the cone-construction with L finite is completely determined by the DGA-morphism

$$i^A : A^*(K) \to A^*(L)$$

induced by the inclusion $i : L \subset K$, *viz.*,

$$H_R^*(\Delta) \approx J(i^A).$$

Proof. Let C^λ be the algebraic map-cone of $\lambda = i^A$. By definition $J(i^A) \approx H(C^\lambda)$ with some multiplicative structure. Let us first define an additive group-homomorphism

$$\tau : H(C^\lambda) \to H_R^{\oplus}(\Delta)$$

as follows. Consider any q-cycle (α, β) of C^λ so that $\alpha \in A^q(K), \beta \in A^{q-1}(L), d_K\alpha = 0$, and $\alpha_L = i^A\alpha = -d_L\beta$. Let $j : L \subset C_L$ be the inclusion and $j^A : A^*(C_L) \to A^*(L)$ be the induced morphism. As j^A is onto we can extend $\beta \in A^*(L)$ to one in $A^*(C_L)$ which will be denoted by β_c. Then the element in $A^*(\Delta)$ which $= \alpha$ on K and $-d_c\beta_c$ on C_L is easily seen to be a cycle in $A^*(\Delta)$ whose class depends only on the class of (α, β) in $H(C^\lambda)$ but is independent of α, β and β_c chosen. This class will then be defined as τ of the class of (α, β).

The τ just defined is only an additive graded-group morphism, and is easily proved to be an additive group isomorphism. With multiplication introduced in $H(C^\lambda)$ as before to turn it into a ring $J(\lambda) = J(i^A)$, we shall show that

$$\tau : J(i^A) \to H_R^*(\Delta)$$

is in fact a graded-ring isomorphism as follows.

For $\lambda = i^A : A^*(K) \to A^*(L)$ let us take an additive basis in $H(A^*(K)) \approx H_R^*(K)$ and $H(A^*(L)) \approx H_R^*(L)$ such that Ker $\lambda_H = R[X_1, \cdots, X_r]$, Coker $\lambda_H = R[\bar{Z}_1, \cdots \bar{Z}_t]$, and

$$H(C^\lambda) = R[X_1^\lambda, \cdots, X_r^\lambda, Z_1^\lambda, \cdots, Z_t^\lambda]$$

as before. Let us take arbitrarily cycles $(\xi_i, \beta_i) \in X_i^\lambda, (0, -\zeta_j) \in Z_j^\lambda$ so that $\xi_i \in A^*(K)$,$\beta_i, \zeta_j \in A^*(L)$ with $d_K\xi_i = 0, \xi_{iL} = -d_L\beta_i, d_L\zeta_j = 0$. Take any $\beta_{iC}, \zeta_{jC} \in A^*(C_L)$ which extend β_i, ζ_j in $A^*(L)$. Define $\tilde{\xi}_i, \tilde{\xi}_j \in A^*(\Delta)$ by

$$\tilde{\xi}_i = \begin{cases} \xi_i \text{ on } K, \\ -d_C\beta_{iC} \text{ on } C_L, \end{cases}$$

$$\tilde{\zeta}_j = \begin{cases} 0 \text{ on K}, \\ +d_C\zeta_{jC} \text{ on } C_L. \end{cases}$$

Then $\tilde{\xi}_i, \tilde{\xi}_j$ are cycles in $A^*(\Delta)$ whose classes $\tilde{X}_i, \tilde{Z}_j$ form an additive basis of $H_R^*(\Delta)$ corresponding to the basis X_i^λ, Z_j^λ of $J(i^A)$ under the morphism τ. Now $\xi_i\xi_j \in X_iX_j \in$ Ker λ_H so that

$$\xi_i\xi_j = \sum \lambda_{ij}^k \xi_k + d_K\alpha_{ij}, \alpha_{ij} \in A^*(K).$$

Set

$$\zeta_{ij} = \sum \lambda_{ij}^k \beta_k + \beta_i \cdot d_L\beta_j - \alpha_{ijL} \in A^*(L).$$

Then $d_L\zeta_{ij} = 0$ so that $\zeta_{ij} \in Z_{ij} \in H_R^*(L)$ and $(0, -\zeta_{ij}) \in Z_{ij}^\lambda = \delta_H Z_{ij} \in H(C^\lambda)$. Take any $\alpha_{ijC} \in A^*(C_L)$ which extends α_{ijL} so that $\zeta_{ijC} = \beta_{iC} \cdot d_C\beta_{jC} + \Sigma\lambda_{ij}^k\beta_{kC} - \alpha_{ijC}$ is an extension of ζ_{ij} to C_L. Define $\tilde{\xi}_{ij} \in A^*(\Delta)$ by

$$\tilde{\zeta}_{ij} = \begin{cases} 0 \text{ on } K, \\ d_C\zeta_{ijC} \text{ on } C_L. \end{cases}$$

Then $\tilde{\xi}_{ij}$ is a cycle in $A^*(\Delta)$ whose class $\tilde{Z}_{ij}$ is such that

$$\tilde{\zeta}_{ij} \in \tilde{Z}_{ij} = \tau(Z_{ij}^\lambda).$$

Define $\tilde{\alpha}_{ij} \in A^*(\Delta)$ by

$$\tilde{\alpha}_{ij} = \begin{cases} \alpha_{ij} \text{ on } K, \\ \alpha_{ijC} \text{ on } C_L. \end{cases}$$

Then we readily verify that

$$\tilde{\xi}_i\tilde{\xi}_j = \sum \lambda_{ij}^k \tilde{\xi}_k + \tilde{\zeta}_{ij} + d_C\tilde{\alpha}_{ij}.$$

As $\tilde{\xi}_i \in \tau X_i^\lambda, \tilde{\xi}_{ij} \in \tau Z_{ij}^\lambda$ and by definition of multiplication in $J(\lambda)$,

$$X_i^\lambda \cdot X_j^\lambda = \sum \lambda_{ij}^k X_k^\lambda + Z_{ij}^\lambda,$$

we see that

$$\tau(X_i^\lambda \cdot X_j^\lambda) = \tau X_i^\lambda \cdot \tau X_j^\lambda.$$

Similarly we see easily that

$$\tau(X_i^\lambda \cdot Z_j^\lambda) = \tau X_i^\lambda \cdot \tau Z_j^\lambda = 0,$$
$$\tau(Z_i^\lambda \cdot Z_j^\lambda) = \tau Z_i^\lambda \cdot \tau Z_j^\lambda = 0.$$

Hence τ respects also the multiplication, which completes the proof.

It is thus seen that by means of A^*-measures $H_R^*(\Delta)$ can be calculated for the cone-construction over a finite subcomplex. However, since A^*-measure is neither invariant nor of finite type, it is desirable to replace A^* by some measure meeting the requirements 1 and 2. Now according to Sullivan, for DGA's A with $H_0(A) \approx R, H_1(A) = 0$ we may define the minimal model $\min A$ of A which is characterized by being, besides trivial in degrees 0 and 1, free and decomposable as well as the existence of canonical morphisms

$$\rho : \min A \to A$$

inducing algebraic isomorphism $\rho_H : H(\min A) \approx H(A)$. We shall restrict ourselves to this case though Sullivan has defined it more generally. For a connected complex K with $H_1^R(K) = 0$ let us set as definition

$$I^*(K) \approx H(A^*(K)).$$

Then $I^*(K)$ is a homotopy invariant of $|K|$ and may be denoted also by $I^*(|K|)$. Then we have:

Theorem 1 *In the category of connected complexes with* $H_1^R = 0$ *the* H_R^*-*measure of the cone-construction over a finite subcomplex is completely determined by the* I^*-*measures involved. More precisely, let* K, L *be a pair of connected complexes with* $H_1^R = 0$ *and* L *a finite subcomplex of* K. *Let* $i : L \subset K$ *be the inclusion inducing a* DGA-*morphism*

$$i^I : I^*(K) \to I^*(L).$$

Then for the union $\Delta = \Delta_L(K) = K \bigcup_L C_L$ *we have*

$$H_R^*(\Delta) \approx J(i^I).$$

Proof. By the general theory of Sullivan we have a homotopically commutative diagram

$$\begin{array}{ccc} A^*(K) & \xrightarrow{i^A} & A^*(L) = B \\ \rho\uparrow & & \uparrow\rho \\ I^*(K) & \xrightarrow{i^I} & I^*(L) \end{array}$$

in which the $\rho's$ are canonical morphisms inducing isomorphisms of H. Let $B^T = A^*(L) \otimes T$ where $T = \{t, dt\}$ is the trivial DGA with deg $t = 0$, deg $dt = 1$. Let

$\tau_s : B^T \to A^*(L) = B$ be the morphism in setting $t = s, dt = 0 (s = 0$ or $1)$. Then the homotopy-commutativity means the existence of a DGA-morphism $h : I^*(K) \to B^T$ such that the following diagram is strictly commutative:

$$\begin{array}{ccc} A^*(K) & \xrightarrow{i^A} & A^*(L) \\ \uparrow{\scriptstyle\rho} & \nearrow_{\tau_0}\ \nearrow^{\tau_1} & \uparrow{\scriptstyle\rho} \\ & B^T & \\ & \nearrow^{h} & \\ I^*(K) & \xrightarrow{i^I} & I^*(L) \end{array}$$

We have then a strictly commutative diagram:

$$\begin{array}{ccc} A^*(K) & \xrightarrow{i^A} & A^*(L) \\ \rho\uparrow & & \uparrow\tau_0 \\ I^*(K) & \xrightarrow{h} & B^T \\ id\downarrow & & \downarrow\tau_1 \\ I^*(K) & \xrightarrow{\tau_1 h} & A^*(L) \\ id\uparrow & & \uparrow\rho \\ I^*(K) & \xrightarrow{i^l} & I^*(L) \end{array}$$

in which all vertical morphisms will induce isomorphisms in H. As the J-construction is clearly functorial, we get isomorphisms

$$J(i^I) \underset{\approx}{\to} J(\tau_1 h) \underset{\approx}{\leftarrow} J(h) \underset{\approx}{\to} J(i^A).$$

Hence by the last Proposition we have

$$J(i^I) \approx H_R^*(\Delta).$$

The theorem is proved.

The same may be said about the union $\tilde{K}$ of any two complexes K', K'' along a common subcomplex L (all with $H_0^R = R$ and $H_1^R = 0$). In fact, let us consider DGA-algebras and DGA-morphisms

$$A' \xrightarrow{\lambda'} B \xleftarrow{\lambda''} A''$$

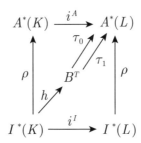

which induce morphisms

$$H(A') \overset{\lambda'_H}{\to} H(B) \overset{\lambda'_H}{\leftarrow} H(A'').$$

Define a DG-group (B augmented to B_{-1} generated by ε):

$$C = \sum C_q,$$

with

$$C_q = A'_q \oplus A''_q \oplus B_{q-1},$$
$$d(a', a'', b) = (da', da'', \lambda' a' - \lambda'' a'' - db).$$

Let

$$\psi_H : H_q(A') \oplus H_q(A'') \to H_q(B)$$

be defined by $\psi_H(v', v'') = \lambda'_H v' - \lambda''_H v''$. Then we have short exact sequences

$$0 \to \text{Coker}_{q-1}\psi_H \overset{\delta_H}{\to} H_q(C) \overset{j_H}{\to} \text{Ker}_q\psi_H \to 0$$

so that as additive graded groups we have

$$H_\oplus(C) = \varphi_H \text{Ker}\,\psi_H \oplus \delta_H\,\text{Coker}\,\psi_H,$$

where φ_H : Ker $\psi_H \to H(C)$ is any right inverse of j_H. By means of multiplications in A',A'' and B we can then introduce multiplications in $H_\oplus(C)$ to turn it into a DGA-algebra with trivial differentials which will be denoted by $J(\lambda', \lambda'')$. Suppose that $A' \to B \leftarrow A''$ is just the same as

$$A^*(K') \overset{i'^A}{\to} A^*(L) \overset{i''^A}{\leftarrow} A^*(K'')$$

with i'^A, i''^A as the morphisms induced by the inclusions $i' : L \subset K', i'' : L \subset K''$. For any cycle $\xi = (\alpha', \alpha'', \beta) \in X \in H_\oplus(C)$ let us extend $\beta \in A^*(L)$ to $\beta'' \in A^*(K'')$ and define $\tilde{\alpha} \in A^*(\tilde{K})$ by

$$\tilde{\alpha} = \begin{cases} \alpha' \text{on } K', \\ \alpha'' + d\beta'' \text{on} K''. \end{cases}$$

Then $\tilde{\alpha}$ is a cycle in $A^*(\tilde{K})$ whose homology class depends on X but not the cycle ξ and extension β'' chosen. Denote this class by τX ; then $\tau : H(C) \to H(A^*(\tilde{K})) \approx H^*_R(\tilde{K})$ is easily seen to be an additive isomorphism. Now the multiplication introduced in

$H(C)$ is such that it will make the isomorphism τ as a multiplicative one. We have then

$$J(i'^A, i''^A) \approx H_R^*(\tilde{K}).$$

We can also replace i'^A, i''^A by morphisms i'^I, i''^I below:

$$I^*(K') \xrightarrow{i'^I} I^*(L) \xleftarrow{i''^I} I^*(K'').$$

In fact, we have the following

Theorem 2 *In the category of complexes with* $H_0^R = R$ *and* $H_1^R = 0$ *the* H_R^**-measure of the union* $\tilde{K}$ *of two complexes* K', K'' *along a common subcomplex* L *is completely determined by the* I^**-measures of* K', K'', L *and the interrelated morphisms, viz.*

$$H_R^*(\tilde{K}) \approx J(i'^I, i''^I).$$

The proof of this theorem parallels that of Theorem 1, but much more complicated. Moreover, theorems 1 and 2 can be generalized to the effect that I^*-measure is calculable with respect to the complex-union construction, in particular with respect to the cone-construction over a finite subcomplex, of which a sketch of the proof of this more general theorem has been given in [4]. For this reason we shall satisfy ourselves in only stating the above theorem 2 with detaits of the construction $J(\lambda', \lambda'')$ and proofs omitted.

For the study of I^*-measure of a complex-union let us first make some preparations.

For DGA's

$$A' \xrightarrow{f'} B \xleftarrow{f''} A''$$

let us define a DGA $C = \sum C_q$ with C_q as the submodule of $A'_q \oplus A''_q$ consisting of elements (a', a'') with $f'a' = f''a''$. The algebraic operations and differentials in C will be defined in a natural manner. We shall denote this DGA C by $\cup(f', f'')$.

Given a commutative diagram of DGA's and their morphisms:

$$\begin{array}{ccccc}
A' & \xrightarrow{f'} & B & \xleftarrow{f''} & A'' \\
{\scriptstyle g'}\uparrow & & \uparrow{\scriptstyle g} & & \uparrow{\scriptstyle g''} \\
M' & \xrightarrow{\varphi'} & N & \xleftarrow{\varphi''} & M''
\end{array}$$

(g', g'') will induce naturally a morphism

$$\tilde{g} : \cup(\varphi', \varphi'') \to \cup(f', f'').$$

We have now the following

Lemma 1 *If in the above commutative diagram* f', f'' *and* φ', φ'' *are all onto and* g', g'', g *all induce isomorphisms in* H, *then* $\tilde{g}$ *will also induce an isomorphism in* H:

$$\tilde{g}_H : H(\cup(\varphi', \varphi'')) \approx H(\cup(f', f'')).$$

Consequently $\min \cup(\varphi', \varphi'') \approx \min \cup(f', f'')$.

Proof. Let us first prove that $\tilde{g}_H$ is onto. Consider any cycle (x', x'') in $C = \cup(f', f'')$ with $dx' = 0, dx'' = 0, f'x' = f''x''$. As g' induces an isomorphism in H we have $x' = g'\xi' + da'$ for some cycle $\xi' \in M'$ and some $a' \in A'$ with $d\xi' = 0$. Similarly $x'' = g''\xi'' + da'', d\xi'' = 0$ with $\xi'' \in M'', a'' \in A''$. Now $\varphi'\xi' - \varphi''\xi''$ is a cycle in N and $g(\varphi'\xi' - \varphi''\xi'') = f'g'\xi' - f''g''\xi'' = f'(x' - da') - f''(x'' - da'') = d(-f'a' + f''a'') \sim 0$ in B. As g induces an isomorphism in H we have therefore $\varphi'\xi' - \varphi''\xi'' \sim 0$ in N, or $\varphi'\xi' - \varphi''\xi'' = d\eta$ for some $\eta \in N$. As φ' is onto we can take $\eta' \in M'$ with $\varphi'\eta' = \eta$. It follows that $\varphi'(\xi' - d\eta') = \varphi''\xi''$ so that $(\xi' - d\eta', \xi'') \in \Gamma = U(\varphi', \varphi'')$. Applying g to $\varphi'(\xi' - d\eta') = \varphi''\xi''$ we get $d(f'a' + f'g'\eta' - f''a'') = 0$ so that $f'a' + f'g'\eta' - f''a''$ is a cycle in B. As g induces an isomorphism in H we have $f'a' + f'g'\eta' - f''a'' = g\zeta + db$ for some $\zeta \in N, b \in B$ with $d\zeta = 0$. As f', φ' are onto we can take $b' \in A', \zeta' \in M'$ with $f'b' = b, \varphi'\zeta' = \zeta$. We have then $f'(a' + g'\eta' - g'\zeta' - db') = f''a''$ so that $(a' + g'\eta' - g'\zeta' - ab', a'') \in C$. As $\varphi'd\zeta' = d\varphi'\zeta' = d\zeta = 0$ we have $\varphi'(\xi' - d\eta' + d\zeta') = \varphi''\xi''$ so that $(\xi' - d\eta' + d\zeta', \xi'') \in \Gamma$. As clearly $\tilde{g}(\xi' - d\eta' + d\zeta', \xi'') = (x', x'') - d(a' + g'\eta' - g'\zeta' - db', a'')$, we see that the class of (x', x'') is in the image of $\tilde{g}_H$ or $\tilde{g}_H$ is onto.

Let us prove that $\tilde{g}_H$ is a monomorphism. Consider for this a cycle $(\xi', \xi'') \in \Gamma$ with $\tilde{g}(\xi', \xi'') = d(a', a''), (a', a'') \in C$ so that $d\xi' = 0, d\xi'' = 0, \xi' = da', \xi'' = da'', \varphi'\xi' = \varphi''\xi''$ and $f'a' = f''a''$. As g', g'' induce isomorphisms in H we have $\xi' = d\alpha', \xi'' = d\alpha''$ for some $\alpha' \in M', \alpha'' \in M''$. As $d(a' - g'\alpha') = da' - g'\xi' = 0$ and g' induces isomorphism in H we have $a' - g'\alpha' = g'\zeta' + dc', d\zeta' = 0$ for some $\zeta' \in M', c' \in A'$. Similarly we have $a'' - g''\alpha'' = g''\zeta'' + dc'', d\zeta'' = 0$ with $\zeta'' \in M'', c'' \in A''$. It follows that

$$g(\varphi'(\alpha' + \zeta') - \varphi''(\alpha'' + \zeta'')) = d(-f'c' + f''c'') \sim 0 \text{ in } B.$$

As g induces an isomorphism in H we have $\varphi'(\alpha'+\zeta')-\varphi''(\alpha''+\zeta'')=d\eta$ for some $\eta\in N$. Take $\eta'\in M'$ with $\varphi'\eta'=\eta$. Then $\varphi'(\alpha'+\zeta'-d\eta')=\varphi''(\alpha''+\zeta'')$ so that $(\alpha'+\zeta'-d\eta',\alpha''+\zeta'')\in\Gamma$. As$(\xi',\xi'')=(d\alpha',d\alpha'')=d(\alpha'+\zeta'-d\eta',\alpha''+\zeta'')$ in Γ we see that $\tilde g_H$ is a monomorphism. This completes the proof of the Lemma.

Remark. Professor L. Taylor of University of Notre Dame has pointed out that the proof may be much simplified in considering the commutative diagram of exact sequences below:

$$\begin{array}{ccccccccc} 0 & \to & \cup(f',f'') & \to & A'\oplus A'' & \to & B & \to & 0 \\ & & \uparrow{\scriptstyle \tilde g} & & \uparrow{\scriptstyle (g',g'')} & & \uparrow{\scriptstyle g} & & \\ 0 & \to & \cup(\varphi',\varphi'') & \to & M'\oplus M'' & \to & N & \to & 0 \end{array}$$

Lemma 2 *Let A, B be* DGA's *with* $H_0\approx R$ and $H_1=0$. *Let the* DGA-*morphism $f:A\to B$ be onto and M, N be minimal models of A and B with $\sigma:N\to B$ as a canonical morphism. Then we can complete f and σ to a strictly commutative diagram below with ρ being a canonical morphism*:

$$\begin{array}{ccc} A & \xrightarrow{f} & B \\ {\scriptstyle\rho}\uparrow & & \uparrow{\scriptstyle\sigma} \\ M & \xrightarrow{\tilde f} & N \end{array}$$

Remark. A simple example shows that the lemma will not be true if f is not onto.

Proof. Let us construct step by step $\{M_n,\rho_n,\tilde f_n\}$ with the diagram (D_n) below commutative:

$$(D_n):\quad \begin{array}{ccc} A & \xrightarrow{f} & B \\ {\scriptstyle\rho_n}\uparrow & & \uparrow{\scriptstyle\sigma} \\ M_n & \xrightarrow{\tilde f_n} & N \end{array}$$

Consider first the case $n=2$. Let $X_i^{(2)}$ be a basis of $H_2(A)$ and $x_i^{(2)}\in X_i^{(2)}$. As σ induces an isomorphism in H we have $fx_i^{(2)}=\sigma y_i^{(2)}+db_i^{(2)}$, and $dy_i^{(2)}=0$ for some $y_i^{(2)}\in N, b_i^{(2)}\in B$. As f is onto we can take $a_i^{(2)}\in A$ with $fa_i^{(2)}=b_i^{(2)}$. Define M_2 as the free DGA of generators $\xi_i^{(2)}$ with $d\xi_i^{(2)}=0$, deg $\xi_i^{(2)}=2$. Define ρ_2 and $\tilde f_2$ by $\rho_2\xi_i^{(2)}=x_i^{(2)}-da_i^{(2)}$ and $\tilde f_2\xi_i^{(2)}=y_i^{(2)}$. This completes the diagram (D_2).

Suppose that (D_n) has been completed and extend (D_n) to (D_{n+1}) as follows. Assume inductively for the construction of (D_n) that $\rho_{nH}:H_q(M_n)\to H_q(A)$ is an isomorphism for $q \leqslant n$ and a monomorphism for $q = n+1$. Consider the diagrams below:

$$\begin{array}{ccccccccc}
0 & \longrightarrow & H_{n+1}(M_n) & \xrightarrow[\rho_{nH}]{} & H_{n+1}(A) & \longrightarrow & \mathrm{Coker}_{n+1}\rho_{nH} & \longrightarrow & 0 \\
 & & \downarrow \tilde{f}_{nH} & & \downarrow f_H & & \downarrow & & \\
0 & \longrightarrow & H_{n+1}(N) & \xrightarrow[\approx]{\sigma_H} & H_{n+1}(B) & \longrightarrow & \mathrm{Coker}_{n+1}\sigma_H(=0) & \longrightarrow & 0
\end{array}$$

$$\begin{array}{ccccccc}
0 & \longrightarrow & \mathrm{Ker}_{n+2}\rho_{nH} & \longrightarrow & H_{n+2}(M_n) & \xrightarrow[\rho_{nH}]{} & H_{n+2}(A) \\
 & & \downarrow & & \downarrow \tilde{f}_{nH} & & \downarrow f_H \\
0 & \longrightarrow & \mathrm{Ker}_{n+2}\sigma_H(=0) & \longrightarrow & H_{n+2}(N) & \xrightarrow[\approx]{\sigma_H} & H_{n+2}(B)
\end{array}$$

Our aim is to introduce new free generators to M_n in forming M_{n+1} to kill $\mathrm{Coker}_{n+1}\rho_{nH}$ and $\mathrm{Ker}_{n+2}\rho_{nH}$.

First consider a basis $X_i^{(n+1)}$ mod Im ρ_{n+1} of $\mathrm{Coker}_{n+1}\rho_{nH}$ and take $x_i^{(n+1)} \in X_i^{(n+1)}$. As $f_H X_i^{(n+1)} \in \mathrm{Im}\,\sigma_H$ we have $fx_i^{(n+1)} = \sigma y_i^{(n+1)} + db_i^{(n+1)}$ with $dy_i^{(n+1)} = 0$. Take $a_i^{(n+1)} \in A$ with $fa_i^{(n+1)} = b_i^{(n+1)}$. Introduce now new free generators $\xi_i^{(n+1)}$ to M_n with $d\xi_i^{(n+1)} = 0$, $\deg \xi_i^{(n+1)} = n+1$ and extend $\rho_n, \tilde{f}_n$ to ρ_{n+1} and $\tilde{f}_{n+1}$ by $\rho_{n+1}\xi_i^{(n+1)} = x_i^{(n+1)} - da_i^{(n+1)}, \tilde{f}_{n+1}\xi_i^{(n+1)} = y_i^{(n+1)}$.

Consider next a basis $Z_i^{(n+1)}$ of $\mathrm{Ker}_{n+2}\rho_{nH}$ and take $z_i^{(n+1)} \in Z_i^{(n+1)}$ so that $\rho_n z_i^{(n+1)} = de_i^{(n+1)}$ for some $e_i^{(n+1)} \in A$. As $\sigma \tilde{f}_n z_i^{(n+1)} = f\rho_n z_i^{(n+1)} = dfe_i^{(n+1)} \sim 0$ in B and σ induces an isomorphism in H we have $\tilde{f}_n z_i^{(n+1)} \sim 0$ in N so that $\tilde{f}_n z_i^{(n+1)} = du_i^{(n+1)}$ for some $u_i^{(n+1)} \in N$. We have then $d(fe_i^{(n+1)} - \sigma u_i^{(n+1)}) = 0$ so that $fe_i^{(n+1)} - \sigma u_i^{(n+1)} = \sigma v_i^{(n+1)} + dc_i^{(n+1)}$ and $dv_i^{(n+1)} = 0$ for some $v_i^{(n+1)} \in N$ and $c_i^{(n+1)} \in B$. As f is onto we have $c_i^{(n+1)} = fh_i^{(n+1)}$ for some $h_i^{(n+1)} \in A$. Introduce new free generators $\zeta_i^{(n+1)}$ of degree $n+1$ to M_n with $d\zeta_i^{(n+1)} = z_i^{(n+1)}$ and extend $\rho_n, \tilde{f}_n$ to ρ_{n+1} and $\tilde{f}_{n+1}$ by $\rho_{n+1}\zeta_i^{(n+1)} = e_i^{(n+1)} - dh_i^{(n+1)}, \tilde{f}_{n+1}\zeta_i^{(n+1)} = u_i^{(n+1)} + v_i^{(n+1)}$.

The DGA $M_{n+1} = M_n \otimes \Lambda(\xi_i^{(n+1)}, \zeta_i^{(n+1)})$ with differentials and $\rho_{n+1}, \tilde{f}_{n+1}$ defined above will then meet the requirements and the induction is completed.

For any free DGA F with free generators u_i let us define F^{tr} as a free DGA with free generators u_i^0, u_i^+ and differentials $du_i^0 = u_i^+, du_i^+ = 0$, the degrees being given by $\deg u_i^0 = \deg u_i, \deg u_i^+ = \deg u_i + 1$. Define a DGA-morphism $\mathrm{tr} : F^{\mathrm{tr}} \to F$ by $\mathrm{tr}(u_i^0) = u_i, \mathrm{tr}(u_i^+) = du_i$. For any DGA G and a morphism

$$g : G \to F$$

let us set $G \otimes F^{\mathrm{tr}} = G_F$ and $g \otimes \mathrm{tr}$ to be $g_F : G_F \to F$. Then we have the following

Theorem 3 *Let $\tilde{K}$ be the union of complexes K', K'' along a common subcomplex L with K', K'', L all satisfying $H_0^R = R$ and $H_1^R = 0$. Then $I^*(\tilde{K})$ is calculable with respect to the complex-union construction. More precisely, let $i' : L \subset K'$ and $i'' : L \subset K''$ be the inclusions; then $I^*(K)$ is completely determined from*

$$I^*(K') \overset{i'^I}{\to} I^*(L) \overset{i''^I}{\to} I^*(K'')$$

as

$$I^*(\tilde{K}) = \min \cup(\tilde{i}', \tilde{i}''),$$

in which

$$\tilde{i}' = (i'^I)_{I^*(L)}, \ \tilde{i}'' = (i''^I)_{I^*(L)}.$$

Proof. From the very definition we have

$$A^*(\tilde{K}) \approx \cup(i'^A, i''^A)$$

in which i'^A, i''^A are induced by the inclusions:

$$A^*(K') \overset{i'^A}{\to} A^*(L) \overset{i''^A}{\leftarrow} A^*(K'').$$

Let us set for simplicity

$$\begin{aligned} A^*(K') &= A', \quad A^*(K'') = A'', \quad A^*(L) = B, \\ I^*(K') &= M', \quad I^*(K'') = M'', \quad I(L) = N. \end{aligned}$$

As i'^A and i''^A are both onto, we can complete the morphisms i'^A, i''^A and g to a strictly commutative diagram below:

$$\begin{array}{ccccc} A' & \xrightarrow{i'^A} & B & \xleftarrow{i''^A} & A'' \\ {\scriptstyle g'}\uparrow & & \uparrow{\scriptstyle g} & & \uparrow{\scriptstyle g''} \\ M' & \xrightarrow{i'^I} & N & \xleftarrow{i''^I} & M'' \end{array}$$

This may be extended naturally to a commutative diagram

$$\begin{array}{ccccc} A' & \xrightarrow{i'^A} & B & \xleftarrow{i''^A} & A'' \\ {\scriptstyle g'_N}\uparrow & & \uparrow{\scriptstyle g} & & \uparrow{\scriptstyle g''_N} \\ M'_N & \xrightarrow{\tilde{i}'} & N & \xleftarrow{\tilde{i}''} & M''_N \end{array}$$

Only the definition of g'_N and g''_N requires explanation. Let the free generators of $N = I^*(L)$ be u_i. As i'^A is onto we can take $a'_i \in A'$ such that $i'^A a'_i = gu_i$. The morphism $g'_N : M'_N = M' \otimes N^{\mathrm{tr}} \to A'$ is then the extension of g' with $g'_N(u_i^0) = a'_i$ and $g'_N(u_i^+) = da'_i$. Similarly for g''_N.

Applying now the Lemma 1 to the above commutative diagram we have then

$$\min \cup(\tilde{i}', \tilde{i}'') \approx \min \cup(i'^A, i''^A) \approx \min A^*(\tilde{K}) \approx I^*(\tilde{K})$$

and the theorem is proved.

In the same manner we have the following

Theorem 4 *The I^*-measure is calculable with respect to the cone-construction. More precisely, let $\Delta = \Delta_L(K)$ be the cone-construction over a finite subcomplex L of K, with K and L both satisfying $H_0^R \approx R$ and $H_1^R = 0$. For the morphism*

$$i^I : I^*(K) \to I^*(L)$$

induced by the inclusion $i : L \subset K$, let

$$\tilde{i} : M_N \to N$$

be the natural extension $\tilde{i} = i^I \otimes \mathrm{tr}$ with $M = I^*(K), N = I^*(L)$. Then we have

$$I^*(\Delta) \approx \min \mathrm{Ker}\, \tilde{i}.$$

In taking the H-functor of I^*-measure, we have as corollary that the H_R^*-measure of the complex-union or the cone-construction is completely determined by the I^*-measures of the original complexes as well as their interrelated morphisms. This gives an alternative formulation of theorems 1 and 2, though in a less precise manner.

References

[1] Friedlander E, Griffiths P A and Morgan J. Homotopy theory and differential forms. *Mimeo-Notes*, 1970.

[2] Sullivan D. Differential forms and topology of manifolds. *Symp. Tokyo on Topology*, 1973.

[3] ——. Infinitesimal computations in topology. *Publ. Math.* No. 47, Inst. de Hautes Etudes Scientifiques, 1978.

[4] Wu Wen-tsün. Calculability of I^*-measure vis-à-vis space-union and related constructions. *Kexue Tong Pao*, 1980, 25: 196-198.

[5] ——. de Rham-Sullivan measure of spaces and its calculability, in *Proc. of Chern Symposium.* Springer, 1980: 229-249.

代数拓扑的构造性理论——I. 量度与能计算性概念

摘要 对某类拓扑空间对应某类代数结构，称之为**量度**. 拓扑中常用的量度有同调群 (或环)H、同伦群 π 等. 通过这些量度的代数探讨，以得出有关拓扑空间的种种结论，乃是代数拓扑的基本方法，这与初等解析几何的方法是类似的.

设 M 是一量度, G 是一几何作法, 从空间 $X_1, X_2, \cdots$作出一新空间 Z. 代数拓扑中经常须从 $M(X_i)$ 获得关于 $M(Z)$ 的知识. 为此引入下面的基本

定义 若 $M(Z)$ 可从 $M(X_i)$ 以及其间相互关系所代数地完全确定, 则称 M 对 G 是**能计算的.**

本文作出实例, 说明在代数拓扑中常用的那些量度, 即使对最简单的作法, 也往往是不能计算的, 例如：

(1) 整系数上同调环 H_Z^* 对空间积作法是不能计算的.

(2) 整系数上同调群 $H_Z^{\oplus}$ 对空间并作法是不能计算的, 甚至对锥形作法也是不能计算的.

(3) 实系数上同调环 H_R^* 对空间并作法是不能计算的, 甚至对锥形作法也是不能计算的.

或许还是这种常用量度的不能计算性, 造成了代数拓扑推理论证的巨大困难. 与之相反, 依据 Sulliven 有理同伦型与极小模型理论引进的 I^* 量度, 则对拓扑中常用的作法却大都是能计算的. 在本文中, 我们给出了 I^* 量度对空间并作法能计算的具体表达式. 详言之, 设复形 K', K'' 有子复形 L 公共, 并以 $\tilde{K}$ 为其并. 又设 $K', K'', L, \tilde{K}$ 都使 I^* 有定义, 而嵌入 $i', L \subset K', i'' : L \subset K''$ 引出 DGA 同态

$$(*)_I \qquad I^*(K') \overset{i'^I}{\to} I^*(L) \overset{i''^I}{\leftarrow} I^*(K''),$$

则从 $(*)_I$ 可定出代数作法 $\cup(\tilde{i}', \tilde{i}'')$ 使

$$I^*(\tilde{K}) = \min \cup(\tilde{i}', \tilde{i}'')$$

(符号解释详见正文).

若在上式两边取 H 函子, 则可见 $H_R^*(\tilde{K})$ 可由 $(*)_I$ 所代数地完全定出, 但从实例可知 $H_R^*(\tilde{K})$ 不可能由下面的 $(*)_H$ 所定出：

$$(*)_H \qquad H_R^*(K') \overset{i'^H}{\to} H_R^*(L) \overset{i''^H}{\leftarrow} H_R^*(K'').$$

这部分说明了 I^* 量度对 H_R^* 量度的优越性, 前者 (对空间并作法) 是能计算的, 而后者则否. 对许多其他作法也有类似情况, 将在以后诸文中陆续说明.

55. De Rham Theorem from Constructive Point of View*

Ⅰ. Introduction

De Rham theorem may be considered as the most important bridge between differential geometry and (algebraic and differential) topology. It may be expressed in the following concise form

$$H(A^*(V)) \approx H_R^*(V), \tag{I}$$

in which V is a differential manifold, $A^*(V)$ the DGA (differential graded algebra) of differential forms on V, $H_R^*(V)$ the usual real cohomology ring of V, and $\approx$ means algebraic isomorphism. Since its first appearance in 1931, various generalizations and proofs have been published even up to quite recent time (e.g.[1, 2, 5, 6, 8, 9]). We shall add in what follows a further one with the special feature of being "constructive" in character.

First let us sketch out the notion of "constructive". For a mathematical logician "constructive" may be understood in various different stages, each of which can be described in precise mathematical terms. However, we shall adopt the attitude of a working mathematician without attaching too seriously a precise meaning to this notion. In the present work we shall understand "constructive" even in a naive sense which may be expressed roughly as follows.

A concept, a definition or a procedure will be considered as "constructive" if it can be described and verified in a finite number of definite steps. A proof of a theorem or a solution to a problem will be considered as "constructive " if it can be achieved in a finite number of steps in a definite manner. Thus, most of the usual definitions occuring in current mathematics such as function (continuous, differentiable, C^∞, or analytic), manifold (topological or differential), differential form, apace, singular

* *Procecdings of the 1981. Shanghai Symposium on Differential Geometry and Differential Equations.* Beijing: Science Press, 1984: 497-528.

chain and singular homology, sheaf, etc., can hardly be considered as constructive in the above sense. Similarly for most of the proofs of theorems in current mathematics which assert mere existence without precise constructions.

From this point of view the above de Rham theorem is not constructive even in its very statement, since it involves the notions of differential manifolds, differential (or C^∞-) forms, singular homologies (or cohomology), etc., which are all non-constructive. The aim of the present paper is to reformulate, or transform, the statement of de Rham theorem into a constructive one and then give a proof which is constructive in our sense.

For this purpose let us first briefly review the early history of algebraic topology. In the first of a series of papers on Analysis Situs Poincaré introduced the notion of Betti numbers (different from present ones by 1) in two different ways. In one way Poincaré defined betti numbers for what we now call differentail manifolds via differential forms and exterior differentiations. In the other way Poincaré introduced the notion of complexes and then defined betti numbers via chains and boundary operations. These two different ways of treatment contradict each other in that one is non-constructive while the other is constructive in the above sense.

It was along this second constructive line of development that topology flourished for about thirty years from Poincaré until E. Cartan revived the study of differential manifolds and differential forms. The two lines of thought were then combined and unified through the theorem bearing the name of de Rham. Following the idea of Poincaré, we shall accourdingly replace the differential manifoids in the statement of de Rham theorem by finite dimensional simplicial complexes which was perhaps first suggested by R. Thom in 1956 [9]. In order to replace the non-constructive notion of differential forms (C^∞- for example) on differential manifolds by a constructive one we shall follow the suggestions of Sullivan [8] and Quillen as follows.

We arrange the vertices of a (locally finite) simplicial complex K of finite dimension n in a definite order $v_0, v_1, \cdots, v_i, \cdots$ and attach to each vertex v_i a variable t_i and a differential dt_i. For each p-simplex $\sigma_J = v_{j_0}\cdots v_{j_p}$ with $J=(j_0,\cdots,j_p), j_0 < \cdots < j_p$, let $\bar{A}^s(\sigma_J)$ be the set of all differential forms of degree s $(0 \leqslant s \leqslant n)$, of the form

$$\sum P_H dt_H = \sum P_H dt_{h_1}\cdots dt_{h_s},$$

in which $\sum$ runs over all ordered sub s-tuples $H=(h_1,\cdots,h_s)$ from $J=(j_0,\cdots,j_p)$, $dt_H = dt_{h_1}\cdots dt_{h_s}$, and P_H are polynomials in t_j, $j\in J$, the coefficients being in a definite field Q, say, the field of rationals. The set $\bar{A}^*(\sigma_J) = \sum \bar{A}^s(\sigma_J)$ will form a DGA over Q under the usual algebraic operations and exterior differentiations.

Let $\bar{B}^*(\sigma_J)$ be the ideal of $\bar{A}^*(\sigma_J)$ generated by the forms

$$t_{j_0} + t_{j_1} + \cdots + t_{j_p} - 1$$

and

$$dt_{j_0} + dt_{j_1} + \cdots + dt_{j_p}.$$

The DGA $\bar{A}^*(\sigma_J)/\bar{B}^*(\sigma_J) = A^*(\sigma_J)$ will be called the algebra of differential forms on σ_J. For abuse of notation, an element expressed as

$$\omega = \sum P_H dt_H$$

in $\bar{A}^*(\sigma_J)$ will also be used sometimes to denote its class in $\bar{A}^*(\sigma_J)$. Let J' be a subset of J and σ_J, the face of σ_J spanned by vertices $v_{j'}$, with $j' \in J' \subset J$. Then the relations

$$t_{j''} = 0, \quad dt_{j''} = 0, \quad j'' \notin J'$$

will induce a morphism of $\bar{A}^*(\sigma_J)$ to $\bar{A}^*(\sigma_{J'})$, which passes to a morphism of $A^*(\sigma_J)$ to $A^*(\sigma_{J'})$, to be called the *restriction* from σ_J to $\sigma_{J'}$. A collection of differential forms $\omega_J \in A^*(\sigma_J)$ associated with simplexes $\sigma_J \in K$ will be said to be *compatible* if for each pair of simplexes with $\sigma_{J'}$, a face of $\sigma_J, \omega_{J'}$ is just the restriction of ω_J from σ_J to $\sigma_{J'}$. Call such a compatible collection $\boldsymbol{\omega} = (\omega_J)$ simply a *differential form* on K. The usual algebraic operations and exterior differentiations can be carried over to the set of all such differential forms to turn it into a DGA over Q, called the *de Rham-Sullivan algebra* of K and denoted by $A^*(K)$ in what follows. On the other hand, the usual groups of cochains of K on the coefficients in Q also form a (noncommutative) DGA over Q under the usual Alexander-Whitney product and the coboundary operator δ, to be denoted by $C^*_Q(K)$ or $C^*(K)$. The DGA's $A^*(K)$ with differential d and $C^*(K)$ with differential δ both possess a homology ring and the reformulated de Rham theorem is just the (algebraic) isomorphism between these rings:

$$H(A^*(K)) \approx H\left(C^*(K)\right). \tag{II}$$

Unlike the original de Rham theorem, all items occuring in the above formula are constructive from their very definitions. We shall accordingly call formula (II) the *Constructive de Rham Theorem.*

In his first paper on Analysis Situs Poincaré already proved that analytic manifolds possess smooth subdivisions. This was later generalized by Cairns and then by J. H. C. Whitehead to arbitrary differential manifolds. With K as any smooth simplicial triangulation of differential manifold V, formula (II) will reduce to (I) of the original de Rham theorem. The constructive de Rham theorem formulated above is thus indeed a generalization of the original one, which is stated in non-constructive terminology.

Ⅱ. Constructive proof of the theorem

We are now in a position to give a constructive proof of the above constructive de Rham theorem. Our method consist of modifying some known proofs, usually non-constructive, of de Rham theorem, generalized or not, to turn them into constructive ones. To fix the idea, we shall consider the proof of the original de Rham theorem due to A. Weil [10] and its modification due to Bott [1]. We shall outline this proof, analyse it into steps, point out the constructive and non-constructive arguments involved, and then show how the modifications should be made.

The proof of Weil consists mainly of the following steps:

A1. The given differential manifold V, supposed to be paracompact, is shown to possess (differentiably) *simple* locally finite open coverings $\mathscr{U} = \{U_i | i \in I\}$. Here differentiably "simple" means that any non-empty intersection $U_J = \bigcap_{i \in J} U_i, J \subset I$, is differentiably contractible with a contraction φ_J of U_J to a point $a_J \in U_J$. The nerve of $\mathscr{U}$ will then be denoted by N and the simplex of N corresponding to J will be denoted by σ_J.

A2. Prove the existence of a partition of unity $\{f_i\}$ subordinate to the covering $\mathscr{U}$ such that f_i's are C^∞-functions over V with support $\subset U_i$ and with

$$f_i \geqslant 0, \quad \sum f_i = 1.$$

A3. Let the index set $I = \{i\}$ be arranged in a definite order. For any ordered $(p+1)$-tuple $H = (h_0, \cdots, h_p)$ with $h_k \in I$, put $U_H = U_{h_0} \cap \cdots \cap U_{h_p}$. Let $A^m(U_H)$ be the set of all C^∞-differential forms of degree m on $U_H (= 0$ for $U_H = \varnothing)$. Then the set of all collections of forms $\omega_H \in A^m(U_H)$

$$\omega = (\omega_H)$$

will be denoted by $W^{m,p}$. It is an R-module under natural operations.

A4. The direct sum

$$W^{**} = \sum W^{m,p}$$

becomes a GA (graded algebra) over R under the multiplication

$$W^{m_1,p_1} \cdot W^{m_2,p_2} \subset W^{m_1+m_2,p_1+p_2}$$

defined by

$$(\eta_{i_0\cdots i_{p_1}})(\zeta_{j_0\cdots j_{p_2}}) = (\omega_{h_0\cdots h_{p_1+p_2}})$$

with

$$\omega_{h_0\cdots h_{p_1+p_2}} = \eta_{h_0\cdots h_{p_1}}\zeta_{h_{p_1}\cdots h_{p_1+p_2}}.$$

It should be understood that $\eta_{h_0\cdots h_{p_1}}$ on the right-hand side of the last formula is the restriction of $\eta_{h_0\cdots h_{p_1}}$ in $U_{h_0\cdots h_{p_1}}$ to $U_{h_0\cdots h_{p_1+p_2}}$. Similarly for $\zeta_{h_{p_1}\cdots h_{p_1+p_2}}$.

A5. Two differentials d and δ are introduced in W^{**}:

$$d : W^{m,p} \to W^{m+1,p}$$

$$\delta : W^{m,p} \to W^{m,p+1}$$

defined as

$$d(\omega_{h_0\cdots h_p}) = (d\omega_{h_0\cdots h_p})$$

$$\delta(\omega_{h_0\cdots h_p}) = (\eta_{h_0\cdots h_{p+1}})$$

$$\eta_{h_0\cdots h_{p+1}} = \sum(-1)^\nu \omega_{h_0\cdots\hat{h}_\nu\cdots h_{p+1}}$$

where $\omega_{h_0\cdots\hat{h}_\nu\cdots h_{p+1}}$ again should be understood as the restriction of $\omega_{h_0\cdots\hat{h}_\nu\cdots h_{p+1}}$ in $U_{h_0\cdots\hat{h}_\nu\cdots h_{p+1}}$ to $U_{h_0\cdots h_{p+1}}$. We then have $d\delta = \delta d$ so that $D = d + (-1)^p\delta|W^{m,p}$ will be a differential of the total complex $W^* = \sum W^q, W^q = \sum W^{q-p,p}$, to be called the *Weil complex* in what follows.

A6. With the aid of the contraction φ_J in each U_J, morphisms

$$I : \begin{cases} W^{m+1,p} \to W^{m,p} \\ (\omega_J) \to (I_J\omega_J) \end{cases}$$

are defined such that

$$dI + Id = \text{ident.}\ |W^{m+1,p}.$$

For the definition of I_J, integration of C^∞-functions has to be used.

A7. With the aid of partition of unity $\{f_i\}$ whose existence is asserted in A2, morphisms

$$K : W^{m,p+1} \to W^{m,p}$$

are defined such that

$$\delta K + K\delta = \text{ident.}|W^{m,p+1}.$$

A8. For the nerve complex N of the simple covering $\mathscr{U}$ define morphisms

$$d : C^p(N) \to W^{0,p}$$

and

$$I : W^{0,p} \to C^p(N)$$

as follows:

For $\gamma \in C^p(N)$ with $\gamma(\sigma_H) = \text{const. } c_H, d\gamma \in W^{0,p}$ is defined as (ω_H) where ω_H is the constant function c_H in U_H. For $(\omega_H) \in W^{0,p}$ so that ω_H are C^∞-functions in $U_H, I(\omega_H)$ is defined as the cochain $\gamma \in C^P(N)$ with $\gamma(\sigma_H) = \omega_H(a_H)$, where a_H is the point in U_H to which U_H is contracted by φ_H. We then have

$$dI + Id = \text{ident.}|W^{0,p},$$

$$Id = \text{ident.}|C^P(N).$$

A9. Define morphisms

$$\delta : A^m(V) \to W^{m,0}$$

and

$$K : W^{m,0} \to A^m(V)$$

as follows. For $\omega \in A^m(V), \sigma\omega$ is defined by

$$\delta\omega = (\omega_i),$$

where $\omega_i \in A^m(U_i)$ is the restriction of ω to U_i. For $(\omega_i) \in W^{m,0}, K(\omega_i)$ is defined with the aid of partition of unity $\{f_i\}$ as the form

$$\omega = \sum f_i\omega_i,$$

in which $f_i\omega_i$ is to be understood as a form on the whole of V, vanishing outside U_i, which is well-defined since the support of f_i lies in U_i. The summation $\sum$ is to be taken over all $i \in I$, also well-defined since $\mathscr{U}$ is locally finite. We then have

$$\delta K + K\delta = \text{ident.}|W^{m,0},$$

$$K\delta = \text{ident.}|A^m(V).$$

A10. The morphisms

$$d : C^*(N) \to W^*$$

and

$$\delta : A^*(V) \to W^*$$

defined in A8, A9 commute with d in $A^*(V), \delta$ in $C^*(N)$, and D in W^*. They induce therefore natural morphisms

$$d_H : H(C^*(N)) \to H(W^*)$$

and

$$\delta_H : H(A^*(V)) \to H(W^*).$$

With the Whitney product in $C^*(N)$, exterior multiplication in $A^*(V)$, and the multiplication in W^* defined as in A4, d_H and δ_H are both ring morphisms preserving products. Moreover, d_H and δ_H are both isomorphisms and this proves the algebraic isomorphism

$$H(A^*(V)) \approx H(C^*(N)).$$

This is the proof in the Bott formulation [1].

A11. In $W^{m,p}$ let $Z^{m,p}$ and $B^{m,p}$ be the submodules defined by

$$Z^{m,p} = \text{Ker}[d\delta : W^{m,p} \to W^{m,p}]$$

$$B^{m,p} = \text{Ker}[d : W^{m,p} \to W^{m+1,p}] \oplus \text{Ker}[\delta : W^{m,p} \to W^{m,p+1}].$$

Then we have the following diagram with evident meaning of the notations Z and B:

$$\begin{array}{ccccccccc}
C^m(N) & \underset{I\delta}{\overset{Kd}{\rightleftarrows}} & W^{0,m-1} & \rightleftarrows \cdots \rightleftarrows & W^{m-p-1,p} & \rightleftarrows \cdots \rightleftarrows & W^{m-1,0} & \underset{I\delta}{\overset{Kd}{\rightleftarrows}} & A^m(V) \\
\cup & & \cup & & \cup & & \cup & & \cup \\
Z^m(N) & \rightleftarrows & Z^{0,m-1} & \rightleftarrows \cdots \rightleftarrows & Z^{m-p-1,p} & \rightleftarrows \cdots \rightleftarrows & Z^{m-1,0} & \rightleftarrows & ZA^m(V) \\
\cup & & \cup & & \cup & & \cup & & \cup \\
B^m(N) & \rightleftarrows & B^{0,m-1} & \rightleftarrows \cdots \rightleftarrows & B^{m-p-1,p} & \rightleftarrows \cdots \rightleftarrows & B^{m-1,0} & \rightleftarrows & BA^m(V).
\end{array}$$

Set

$$H^{m,p} = Z^{m,p}/B^{m,p}.$$

The induced morphisms below are then all (additive) isomorphisms:

$$H_R^m(N) \underset{(I\delta)_H}{\overset{(Kd)_H}{\rightleftarrows}} H^{0,m-1} \rightleftarrows \cdots \rightleftarrows H^{m-p-1,p} \rightleftarrows \cdots \rightleftarrows H^{m-1,0} \underset{(I\delta)_H}{\overset{(Kd)_H}{\rightleftarrows}} H_m(A^*(V)).$$

In particular there are morphisms

$$A^m(V) \underset{\beta=(I\delta)^{m+1}}{\overset{\alpha=(\mathrm{Kd})^{m+1}}{\rightleftarrows}} C^m(N)$$

which induce (additive) isomorphisms

$$H(A^*(V)) \underset{\beta_H}{\overset{\alpha_H}{\rightleftarrows}} H(C^*(N)).$$

These morphisms are also seen to be multiplicative. This is the proof in Weil's original formulation, cf. [10].

A12. The nerve N is proved to have the same homotopy type as V, so that in particular

$$H(C^*(N)) \approx H_R^*(V).$$

Therefore the isomorphisms proved in A10 or A11 become the original de Rham theorem (I):

$$H(A^*(V)) \approx H_R^*(V).$$

Many of the arguments and even the notions themselves above , notably A1, A2, A6, A12, are non-constructive, while A11 is essentially constructive in character. Let us therefore come to the modification of the above arguments to get a constructive proof of the constructive de Rham theorem formulated as (II). As before K will be a locally finite simplicial complex and $A^*(K)$ the de Rham-Sullivan algebra of K. The vertices $v_0, v_1, v_2, \ldots$ and the corresponding index set $I = \{0, 1, 2, \ldots\}$ will be supposed to be arranged in the definite order as indicated. A subset $J = (j_0, j_1, \ldots, j_p)$ of I and the corresponding ordered simplex $\sigma_J = (v_{j_0} \cdots v_{j_p})$ supposed in K will be said to be *normally ordered* if $j_0 < \cdots < j_p$. The modified arguments corresponding to A_i will be denoted respectively as B_i.

B1. For each normally ordered simplex $\sigma_J = (v_{j_0} \cdots v_{j_p})$ of K let K_J be the closed complex determined by the star of σ_J. Thus K_J consists of all simplexes of K which will span with σ_J some simplex belonging to K. The set $\mathscr{K}$ of all such $K_J's$ will take the place of the simple open covering $\mathscr{U}$ in A1. For each K_J in $\mathscr{K}$ corresponding to the normally ordered simplex $\sigma_J = (v_{j_0} \cdots v_{j_p})$ of K. We shall take, to fix the ideas, the first vertex v_{j_0} of σ_J as the point a_J to which the space of K_J will be contracted along linear paths in simplexes of K_J. Remark that the notion of nerve will not be considered here.

B2. For each vertex v_i of K, $i \in I$, let us consider the corresponding barycentric coordinate function t_i defined over the whole space of K as follows. On each simplex of K with v_i as one of its vertex, t_i is the barycentric coordinate in that simplex cooresponding to that vertex. On simplexes of K disjoint from v_i, t_i is taken to be 0. The functions t_i are all well-defined, continuous over the whole space of K and linear on each simplex of K. We also have

$$t_i(v_i) = 1, 0 \leqslant t_i \leqslant 1, \quad \sum t_i = 1,$$

in which $\sum$ is to be taken over all i and is well-defined since K is locally-finite. This set of functions $\{t_i\}$ will take the place of the partition of unity $\{f_i\}$ of A2 .

B3-4. Let $p \geqslant 0, m \geqslant 0$ be fixed integers. For each normally ordered p-simplex σ_J of K let us take a differential form of degree m on $K_J : \omega_J \in A^m(K_J)$. The set of all such collections

$$\omega = (\omega_J)$$

will form some Q-module $W^{m,p}$. Remark that ω_J is itself a (compatible) collection of differential forms on each simplex in the complex K_J.

Convention. For a permutation J' of the normally ordered subset J with $\sigma_J \in K$, we shall set for convenience

$$\omega_{J'} = \varepsilon \omega_J \in A^m(K_J),$$

in which $\varepsilon = +1$ or -1 according as the permutation is even or odd.

The direct sum

$$W^{**} = \sum W^{m,p}$$

will become a GA on Q with a multiplication

$$W^{m_1,p_1} \cdot W^{m_2,p_2} \subset W^{m_1+m_2,p_1+p_2}$$

defined as follows.

Let $\omega' \in W^{m_1,p_1}$, $\omega'' \in W^{m_2,p2}$. For each normally ordered (p_1+p_2)-simplex σ_J of $K, J = (j_0, \cdots, j_{p_1+p_2})$, consider the normally ordered p_1-simplex $\sigma_{J'}, J' = (j_0, \cdots, j_{p_1})$, and the normally ordered p_2-simplex $\sigma_{J''}, J'' = (j_{p_1}, \cdots, j_{p_1+p_2})$. Let

$\tilde{\omega}'_{J'} \in A^{m_1}(K_{J'})$ be the form on $K_{J'}$ in the collection ω'. Similarly for $\omega''_{J''} \in A^{m_2}(K_{J''})$ in ω''. As K_J is clearly a subcomplex of both $K_{J'}$ and $K_{J''}, \omega'_{J'}$ on $K_{J'}$ and $\omega''_{J''}$ on $K_{J''}$ will give rise to differential forms on K_J by simply taking the partial collections which will still be denoted by $\omega'_{J'}, \omega''_{J''}$. The exterior multiplication of these forms will then be a form $\omega_J = \omega'_{J'} \cdot \omega''_{J''} \in A^{m_1+m_2}(K_J)$. The collection of all such forms ω_J is an element of $W^{m_1+m_2, p_1+p_2}$ and will be defined as the product of ω' and $\omega'' : \omega = \omega'\omega''$.

B5. The differentials d and δ will be introduced in W^{**} as follows.

Definition of $d : W^{m,p} \to W^{m+1,p}$.

For $\omega = (\omega_J) \in W^{m,p}$ with $\omega_J \in A^m(K_J)$, dim $\sigma_J = p$, define d simply as $(d\omega_J)$ with $d\omega_J \in A^{m+1}(K_J)$.

Definition of $\delta : W^{m,p} \to W^{m,p+1}$.

Let $\omega = (\omega_J) \in W^{m,p}$ as above, For each normally ordered $(p+1)$-simplex σ_H of K with normally ordered $H = (h_0, \cdots, h_{p+1})$, let $H_\nu = (h_0, \cdots, \hat{h}_\nu, \cdots, h_{p+1})$, Then K_H is a subcomplex of K_{H_ν} so that ω_{H_ν} may be considred as a differential form in K_H for all ν by taking partial collections. We may therefore define

$$\eta_H = \sum (-1)^\nu \cdot \omega_{H_\nu} \in A^m(K_H)$$

and then

$$\delta\omega = (\eta_H) \in W^{m,p+1}.$$

Remark. If H' is a permutation of the normally ordered H, then from $\eta_H = \sum(-1)^\nu \omega_{H_\nu}$ we get also $\eta_{H'} = \sum(-1)^\nu \omega_{H'_\nu}$, cf. the convention in B3-4.

As in A5, we verify that $d\delta = \delta d$ and that

$$D = d + \bar{\delta},$$

where $\bar{\delta} = (-1)^P \delta | W^{m,p}$ is a differential in the total complex

$$W^* = \sum W^q; W^q = \sum W^{q-p,p},$$

which will again be called the *Weil complex* associated with the given complex K.

It is also easy to verify that D verifies the usual formulae about products.

B6. We shall introduce in W^{**} morphisms

$$I : \begin{cases} W^{m+1,p} \to W^{m,p}, \\ (\omega_J) \to (I_J \omega_J), \end{cases}$$

so that

$$dI + Id = \text{ident.}|W^{m+1,p}.$$

For this purpose let us consider any normally ordered p-simplex σ_J of $K, J = (j_0, \cdots, j_p)$, and any simples τ in the complex K_J. Suppose first that τ has v_{j_0} as a vertex with vertices $v_{k_1}, \cdots, v_{k_q}$ besides $v_{j_0}, k_1 < \cdots < k_q$. The barycentric coordinates in τ will satisfy the relations

$$t_{j_0} = 1 - \sum_{s=1}^{q} t_{k_s},$$

$$dt_{j_0} = -\sum_{s=1}^{q} dt_{k_s}.$$

With these relations any form $\alpha \in A^{m+1}(\tau)$ may be uniquely expressed in the form

$$\alpha = \sum P_{h_0 \cdots h_m} dt_{h_0} \cdots dt_{h_m} = \sum P_H dt_H,$$

in which $\sum$ runs over all normally ordered $H = (h_0, \cdots, h_m) \subset (k_1, \cdots, k_q)$ and $P_H = P_{h_0 \cdots h_m}$ are polynomial in $t_{k_1}, \cdots, t_{k_q}$ with coefficients in Q and $dt_H = dt_{h_0} \cdots dt_{h_m}$. Now for any polynomial P in $t_{k_1}, \cdots, t_{k_q}$ let us white it in the form

$$P = \sum_{i \geqslant 0} P^{(i)},$$

in which $P^{(i)}$ is the homogeneous part of degree i. Set

$$I_J \alpha = \sum_{H, \nu, i} \left[(-1)^{\nu-1} \cdot \frac{1}{m+i+1} \cdot P_H^{(i)} \cdot t_{h_\nu} \cdot dt_{H_\nu} \right],$$

in which $H_\nu = (h_0, \cdots, \hat{h}_\nu, \cdots, h_m)$. It is then easy to verify that

$$(dI_J + I_J d)\alpha = \alpha.$$

Consider now an element

$$\omega = (\omega_J) \in W^{m+1,p}.$$

For each $\sigma_J \in K$ as before, $\omega_J \in A^{m+1}(K_J)$ is a compatible collection of differential forms $\omega_\tau \in A^{m+1}(\tau)$ with $\tau \in K_J$. For τ having v_{j_0} as a vertex let us set $I_J \omega_\tau \in A^m(\tau)$ as defined above. For $\tau \in K_J$ disjoint from v_{j_0} let us take τ' to be the simplex spanned by v_{j_0} and τ and define $I_J \omega_\tau$ to be the restriction of $I_J \omega_{\tau'}$ from τ' to τ. It is clear that the collection of all such $I_J\omega_\tau$ for $\tau \in K_J$ is a compatible one and will be defined as $I_J\omega_J \in A^m(K_J)$. From the above relation on I_J over ω_τ we have

$$(dI_J + I_J d)\omega_J = \omega_J.$$

Hence $I\omega = (I_J\omega_J)$ is defined and meets the requirement as stated.

B7. Instead of partition of unity, barycentric coordinate functions $\{t_i\}$ will be used in defining as in A7 morphisms

$$K : W^{m,p+1} \to W^{m,p}$$

as follows. Let

$$\omega = (\omega_J) \in W^{m,p+1}$$

be given with $\omega_J \in A^m(K_J)$ for all normally ordered σ_J of dimension $p+1$ in K. Consider any p-simplex σ_H of K with $H = (h_0, \cdots, h_p)$ normally ordered. Let v_k be a vertex of K which spans with σ_H a simplex σ_{kH} of dimension $p+1$ in K, where $kH = (k, h_0, \cdots, h_p)$. Now the complex K_H contains K_{kH} as a subcomplex and $\omega_{kH} \in \omega$ is a differential form in K_{kH}, cf. the convention in B3-4 . The barycentric coordinate function t_k corresponding to v_k may be considered as a differential form of degree 0 defined in K_{kH} and so by multiplication $t_k\omega_{kH}$ is also a differential form in K_{kH}. On such simplex in K_H not in K_{kH} let us take simply the vanishing form 0. Then these vanishing forms as well as the forms in $t_k\omega_{kH}$ on K_{kH} together form a compatible collection of differential forms on various simplexes in K_H and so define a differetial form in K_H still to be denoted by $t_k\omega_{kH}$. For k such that v_k does not span with σ_H any simplex in K, let us simply understand by the $t_k\omega_{kH}$ the form 0 in K_H. Define now in K_H a differential form ζ_H by

$$\zeta_H = \sum t_k\omega_{kH} \in A^m(K_H),$$

in which the summation $\sum$ is to be extended over all k not in H. The collection (ζ_H) for all normally ordered p-simplexes σ_H of K is then an elemet of $W^{m,p}$ and the morphism K is defined by

$$K\omega = (\zeta_H).$$

It is easy to verify that

$$\delta K + K\delta = \text{ident.}|W^{m,p+1}.$$

B8. We shall define now morphisms

$$d : C^p(K) \to W^{0,p}$$

and

$$I : W^{0,p} \to C^p(K)$$

as follows.

Given $\gamma \in C^p(K)$ let us consider any normally ordered p-simplex σ_H of $K, H = (h_0, \cdots, h_p)$. Define ω_H as the differential form of degree 0 in K_H which is the constant-valued polynomial $\gamma(\sigma_H)$ on each simplex in K_H. The element $d_\gamma \in W^{0,p}$ is then simply defined as the collection (ω_H).

Conversely, let $(\omega_H) \in W^{0,p}$ be given. Then ω_H as a compatible collection of polynomials on simplexes in K_H will take a difinite value c_H on the first vertex v_{h_0} of σ_H. We define $I(\omega_H)$ as the cochain $\gamma \in C^P(K)$ given by

$$\gamma(\sigma_H) = c_H.$$

With d and I thus defined it is easy to verify that

$$dI + Id = \text{ident.}|W^{0,p},$$

$$Id = \text{ident.}|C^p(K).$$

B9. Define the morphisms

$$\sigma : A^m(K) \to W^{m,0}$$

and

$$K : W^{m,0} \to A^m(K)$$

as follows.

Let $\omega \in A^m(K)$ be given. For any vertex v_i of K the complex $K_i = K_{(i)}$ is a subcomplex of K and by taking a partial collection ω will give a differential form $\omega_i \in A^m(K_i)$. The collection (ω_i) is then defined as the element $\delta\omega \in W^{m,0}$.

Conversely, let $(\omega_i) \in W^{m,0}$ be given with $\omega_i \in A^m(K_i)$ corresponding to vertices v_i of K. Consider any p-simplex σ_J of K, with $J = (j_0, \cdots, j_p)$ normally ordered. For each j_s the simplex σ_J belongs to K_{j_s} so that $\omega_{j_s} \in A^m(K_{j_s})$ as a collection of forms on simplexes in K_{j_s} gives a form $\zeta_{j_s} \in A^m(\sigma_J)$. Set

$$\zeta_J = \sum t_s \zeta_{j_s} \in A^m(\sigma_J).$$

Then the collection of ζ_J for all simplexes $\sigma_J \in K$ is clearly a compatible one and (ζ_J) will be defined as $K(\omega_i)$.

By direct verification we have

$$\delta K + K\delta = \text{ident.}|W^{m,0},$$

$$K\delta = \text{ident.}|A^m(K).$$

B10-11. As in A10, the morphisms

$$d : C^*(K) \to W^*$$

and

$$\delta : A^*(K) \to W^*$$

defined in B8, B9 are multiplicative with respect to the Whitney product in $C^*(K)$, exterior multiplication in $A^*(K)$, and the product in W^* defined in B4 . Moreover d and δ above commute with δ in $C^*(K)$, d in $A^*(K)$ and D in W^*, so that they induce ring morphisms

$$d_H : H(C^*(K)) \to H(W^*)$$

and

$$\delta_H : H(A^*(K)) \to H(W^*).$$

Now form the definition of d and δ introduced in B5 the following sequences are exact:

$$0 \to C^p(K) \xrightarrow{d} W^{0,p} \xrightarrow{d} W^{1,p} \longrightarrow \cdots \longrightarrow W^{m,p} \longrightarrow \cdots,$$

$$0 \to A^m(K) \xrightarrow{\delta} W^{m,0} \xrightarrow{\delta} W^{m,1} \longrightarrow \cdots \longrightarrow W^{m,P} \longrightarrow \cdots.$$

From the exactness of these sequences we prove as in [1] that both d, δ induce algebraic isomorphisms:

$$H(A^*(K)) \underset{\approx}{\xrightarrow{\delta_H}} H(W^*) \underset{\approx}{\xleftarrow{d_H}} H(C^*(K)).$$

Moreover, the isomorphisms may be made precise with the aid of I and K introduced in B6-9 as follows.

Set for simplicity

$$\bar{\delta}|W^{m,p} = (-1)^p\delta,$$

$$\bar{\delta}|C^m(K) = (-1)^{m+1}\delta.$$

Then we have

$$d\bar{\delta} = -\bar{\delta}d|W^*,$$

$$d\bar{\delta} = -\bar{\delta}d|C^*(K),$$

$$D = d + \bar{\delta}|W^*.$$

We have by induction,

$$(I\bar{\delta})^{q+1}d = I\bar{\delta}d(I\bar{\delta})^q.$$

Consider now in W^m an element

$$w = \sum_{q=0}^{m} w_q$$

with $w_q \in W^{q,m-q}$. Define a morphism

$$d' : W^* \to C^*(K)$$

by

$$d'w = \sum I(-\bar{\delta}I)^q w_q, \quad (w_{-1} = 0, w_{m+1} = 0).$$

We shall prove that $d'D = (-1)^m \delta d'$, so that d' will induce a morphism

$$d'_H : H(W^*) \to H(C^*(K)).$$

In fact for w given above we have

$$Dw = \sum (d + \bar{\delta})w_q = \sum v_q$$

with $v_q = \bar{\delta}w_q + dw_{q-1} \in W^{q,m+1-q}$. Hence

$$\begin{aligned} d'Dw &= \sum I(-\bar{\delta}I)^q v_q = \sum I(-\bar{\delta}I)^q(\bar{\delta}w_q + dw_{q-1}) \\ &= \sum I(-\bar{\delta}I)^q(\bar{\delta} - \bar{\delta}Id)w_q = \sum I(-\bar{\delta}I)^q \bar{\delta}dIw_q \\ &= \sum (-1)^q (I\bar{\delta})^{q+1} dIw_q = \sum (-1)^q I\bar{\delta}d(I\bar{\delta})^q Iw_q \\ &= Id \sum (-\bar{\delta}I)^{q+1} w_q = -\delta \sum I(-\bar{\delta}I)^q w_q \\ &= (-1)^m \delta d'w. \end{aligned}$$

The proof is completed.

From definition we have

$$d'd = \text{ident.} : C^*(K) \to W^* \to C^*(K),$$

so that

$$d'_H d_H = \text{ident.}: H(C^*(K)) \to H(W^*) \to H(C^*(K)).$$

In particular, d_H is a monomorphism. Moreover, d_H is also an epimorphism. To see this, let $z = \sum z_q, z_q \in W^{q,m-q}$ with $Dz = 0$. Then

$$\bar{\delta} z_q + d z_{q-1} = 0.$$

It is readily verified that

$$z = dIz'_0 + DI \sum_{q \geqslant 1} z'_q$$

with

$$z'_q = \sum_{i=0}^{m-q} (-\bar{\delta} I)^i z_{q+1} \in W^{q,m-q},$$

and

$$\delta(Iz'_0) = 0.$$

Hence the class of z is in the d_H-image of the class of Iz'_0 or d_H is an epimorphism as to be proved.

It follows that

$$d_H : H(C^*(K)) \approx H(W^*).$$

Similarly we have

$$\delta_H : H(A^*(K)) \approx H(W^*).$$

This completes the proof of the generalized de Rham theorem. We see also that the isomorphism in (II) is induced by the morphisms

$$d'\delta = I(-\bar{\delta} I)^m \delta = (-1)^{m(m+1)/2} (I\delta)^{m+1} : A^m(K) \to C^m(K).$$

The modification of arguments in A12 is no more necessary.

III. Realization of the isomoprhism by integration

In the original de Rham theorem (I) the cohomology group $H^*_R(V)$ is the dual of the singular homology group $H^R_*(V)$ so that (I), as an additive isomorphism, may be written in the form

$$H(A^*(V)) \approx \text{Hom}_R(H^R_*(V), R). \tag{I$'$}$$

In such a form the isomorphism may be realized by the integration of differential forms of certain degree m on singular chains of the same dimension m. The concepts involved are again nonconstructive, the same for the usual proofs.

We may again turn the above realization of isomorphisms into a constructive form as follows. Let complex K, etc. be as before. Let $C_m = C_m^Q$ be the group of *finite* chains on coefficients Q in dimension m and $C_* = C_*^Q = \sum C_m$ as usual. Then (II) implies the additive isomorphisms

$$H_m(A^*(K)) \approx H_m(\mathrm{Hom}(C_*(K), Q)). \tag{II}'$$

This isomorphism is realized by some "integration" procedure

$$\int : A^m(K) \to \mathrm{Hom}(C_m(K), Q)$$

to be defined later. The proof is again based on a modification of Weil's original one and will be sketched as follows.

For $m \geqslant 0, p \geqslant 0$ let $W_{m,p}$ be the Q-module of collections $u = (u_J)$, with $u_J \in C_m(K_J)$ running over normally ordered p-simplexes $\sigma_J \in K$ for which only *finite* number is non-zero. Introduce now a set of morphisms b, p, ∂, L as indicated below:

Definition of $b : W_{m+1,p} \to W_{m,p}$.

For $u = (u_J) \in W_{m+1,p}, u_J \in C_{m+1}(K_J)$, let b_J be the usual boundary operator $C_{m+1}(K_J) \to C_m(K_J)$, then by difinition

$$bu = (b_J u_J) \in W_{m,p}.$$

Definition of $b : W_{0,p} \to C_p(K)$.

For $u = (u_J) \in W_{0,p}, u_J \in C_0(K_J)$, let Ind u_J be the sum of coefficients in u_J, then by definition

$$bu = \sum \mathrm{Ind}\ u_J \cdot \sigma_J \in C_p(K).$$

The definition is legitimate since only a finite number of u_J is non-zero.

Definition of $P : W_{m,p} \to W_{m+1,p}$.

For any normally ordered subset $J = (j_0, \cdots, j_p)$ with $\sigma_J \in K$ and any normally ordered m-simplex $\tau = (v_{k_0}, \cdots v_{k_m}) \in K_J$, let $P_J \tau = 0$ if some $k_i = j_0$ while

$$P_J \tau = (-1)^\nu \cdot (v_{k_0} \cdots v_{k_{\nu-1}} v_{j_0} v_{k_\nu} \cdots v_{k_m})$$

if no $k_i = j_0$ and $k_{\nu-1} < j_o < k_\nu$. Let

$$u = (u_J) \in W_{m,p}, \quad u_J = \sum u_{J,\tau}\tau \in C_m(K_J), \quad u_{J,\tau} \in Q.$$

Then by definition

$$P_J u_J = \sum u_{J,\tau} P_J \tau \in C_{m+1}(K_J),$$

$$Pu = (P_J u_J) \in W_{m+1,p}.$$

Definition of $P : C_p(K) \to W_{0,p}$.

For $u = \sum u_J \sigma_J \in C_p(K)$ with $J = (j_0, \cdots, j_p), \sigma_J \in K$ normally ordered and $u_J \in Q$, we put by definition

$$P_J u = u_J v_{j_0} \in C_0(K_J),$$

$$Pu = (P_J u) \in W_{0,p}.$$

Definition of $\partial : W_{m,p+1} \to W_{m,p}$.

Let $x = (x_H) \in W_{m,p+1}, x_H = \sum x_{H,\tau}\tau \in C_m(K_H), x_{H,\tau} \in Q$ be given. Consider any normally ordered subset $J = (j_0, \cdots, j_p)$ with $\sigma_J \in K$. For any vertex $v_k \in K_J$ with $k \notin J$, let $[kJ]$ be the normally ordered subset consisting on indices $k, j_0, \cdots, j_p$ and ε_{kJ} be $+1$ or -1 according as the permutation to bring $(k, j_0, \cdots, j_p)$ into the normal order is even or odd. Set by definition

$$z_J = \sum \varepsilon_{kJ} x_{[kJ],\tau}\tau \in C_m(K_J),$$

in which $\sum$ runs over all indices k as above and all normally ordered m-simplexes $\tau \in k_{[kJ]}$. Then ∂x is defined as

$$\partial x = (z_J) \in W_{m,p}.$$

Definition of $\partial : W_{m,0} \to C_m(K)$.

For $x = (x_j) \in W_{m,0}, x_j \in C_m(K_j)$, we define simply

$$\partial x = \sum x_j \in C_m(K).$$

The definition is legitimate since only a finite number of $x_{j'}s$ is non-zero.

Definition of $L : W_{m,p} \to W_{m,p+1}$.

Let $x = (x_J) \in W_{m,p}, x_J = \sum x_{J,\tau}\tau \in C_m(K_J), x_J\tau \in Q$ be given. For any normally ordered subset $H = (h_0, \cdots, h_{p+1})$, let $H_\nu = (h_0, \cdots, \hat{h}_\nu, \cdots, h_{p+1})$. Any m-simplex τ of K_H belongs also to all K_{H_ν}. Hence we may put

$$y_{H,\tau} = (-1)^\nu \cdot x_{H_\nu,\tau},$$

if the first vertex of τ is some v_{h_v} and

$$y_{H,\tau} = 0$$

if otherwise. We define then

$$y_H = \sum y_{H,\tau}\tau \in C_m(K_H),$$

$$\partial x = (y_H) \in W_{m,p+1}.$$

Definition of $L : C_m(K) \to W_{m,0}$.

For any normally ordered m-simplex τ of K let the first vertex of τ be $v_{i(\tau)}$. For $z = \sum z_\tau \tau \in C_m(K), z_\tau \in Q$, we define then

$$x_j = \sum_{i(\tau)=j} z_\tau \tau \in C_m(K_j),$$

$$Lz = (x_j) \in W_{m,0}.$$

Let us denote the usual boundary operators $C_{r+1}(K) \to C_r(K)$ by either ∂ or b. Then between the various morphisms introduced we have the following relations ($m \geqslant 0, p \geqslant 0$):

$$b^2 = 0 : W_{m+2,p} \to W_{m,p},$$

$$\partial^2 = 0 : W_{m,p+2} \to W_{m,p},$$

$$\partial b = b\partial : W_{m+1,p+1} \to W_{m,p},$$

$$\partial b = b\partial : W_{0,p+1} \to C_p(K),$$

$$\partial b + b\partial : W_{m+1,0} \to C_m(K),$$

$$Pb + bP = \text{ident.}|W_{m,p},$$

$$bP = \text{ident.}|C_p(K),$$

$$L\partial + \partial L = \text{ident.}|W_{m,p},$$

$$\partial L = \text{ident.}|C_m(K).$$

Owing to the above relations we can define a comples W_* with differential D by setting

$$W_* = \sum W_m, W_m = \sum W_{q,m-q},$$

$$D = b + \bar{\partial} : W_m \to W_{m-1},$$

in which

$$\bar{\partial} = (-1)^p \partial | W_{m,p}.$$

Define now morphisms $\partial' : W_m \to C_m(K)$ by

$$\partial' w = \partial \sum (-\bar{\partial} P)^{m-q} w_q,$$

in which $w = \sum w_q \in W_m, w_q \in W_{q,m-q}$. Then

$$\partial' D = b\partial' : W_* \to C_*(K),$$

so that ∂' induces morphisms

$$\partial'_H : H(W_*) \to H(C_*(K)).$$

Similarly, using L and b morphisms $b' : W_m \to C_m(K)$ can be defined with $b'D = \partial b'$, thus inducing morphisms

$$b'_H : H(W_*) \to H(C_*(K)).$$

As in B10-11, both ∂'_H and b'_H are (additive) isomorphisms. It can also be proved that

$$(\partial P)^{m+1} = (-1)^{m(m+1)/2} : C_m(K) \to C_m(K).$$

Define now an "integration" or scalar product

$$\int : (W^{m,p}, W_{m,p}) \to Q$$

as follows.

For the normally ordered set $H = (h_0, \cdots, h_m)$ with $\sigma_H \in K$ and $s_i \geqslant 0$ let us set as definition

$$\begin{aligned} &\int_{\sigma_H} t_{h_0}^{s0} \cdots t_{h_m}^{s_m} \cdot dt_{h_0} \cdots d\hat{t}_{h_\nu} \cdots dt_{h_m} \\ =&(t_{h_0}^{s0} \cdots t_{h_m}^{s_m} \cdot dt_{h_0} \cdots d\hat{t}_{h_\nu} \cdots dt_{h_m}, \sigma_H) \\ =&(-1)^\nu \cdot \frac{s_0! \cdots s_m!}{(\sum s_i + m)!}. \end{aligned}$$

For any differential form $\alpha \in A^m(\sigma_H)$ we then extend the definition of

$$\int_{\sigma_H} \alpha = (\alpha, \sigma_H)$$

by linearity. The definition is legitimate since it is readily verfied that

$$\int_{\sigma_H} \alpha = (\alpha, \sigma_H) = 0$$

whenever α is in the ideal generated by $t_{h_0} + \cdots + t_{h_m} - 1$ and $dt_{h_0} + \cdots + dt_{h_m}$.

Suppose given

$$\omega = (\omega_J) \in W^{m,p}, \quad \omega_J = (\omega_{J,\tau}) \in A^m(K_J), \quad \omega_{J,\tau} \in A^m(\tau), \tau \in K_J$$

and

$$x = (x_J) \in W_{m,p'}, x_J = \sum x_{J,\tau}\tau \in C_m(K_J),$$
$$\tau \in K_J, \dim\tau = m, x_{J,\tau} \in Q.$$

We shall set as definition

$$\int_x \omega = \sum \int_{x_J} \omega_J, \quad \int_{x_J} \omega_J = \sum x_{J,\tau} \int_\tau \omega_{J,\tau},$$

or

$$(\omega, x) = \sum (\omega_J, x_J), \quad (\omega_J, x_J) = \sum x_{J,\tau}(\omega_{J,\tau}, \tau).$$

For $\omega \in W^{m,p}, x \in W_{m+1,p}, y \in W_{m,p+1}, u \in C^p(K)$, $\alpha \in W_{0,p}, \xi \in A^m(K), z \in W_{m,0}$, we readily verify that

$$(d\omega, x) = (\omega, bx), \quad (\delta\omega, y) = (\omega, \partial y),$$

$$(du, \alpha) = (u, b\alpha), \quad (\delta\xi, z) = (\xi, \partial z).$$

From these relations we have the following

Theorem The isomorphism in (II)$'$ is realized by integration.

Proof. In fact, let $\omega \in A^m(K)$ with $d\omega = 0$ and $x \in C_m(K)$ with $\partial x = bx = 0$ be given. Then by induction we have for $q \geqslant 0$:

$$d\delta(I\delta)^q\omega = 0, \quad b(\partial P)^q x = 0,$$

$$(-1)^{m(m+1)/2} \cdot (\omega, x) = ((I\delta)^q\omega, (\partial P)^{m+1-q}x).$$

Hence

$$(\omega, x) = (-1)^{m(m+1)/2} \cdot ((I\delta)^{m+1}\omega, x) = (d'\delta\omega, x)$$

in which

$$(d'\delta)_H : H(A^*(K)) \approx H(C^*(K))$$

is given in B10-11. This proves the theorem.

References

[1] Bott R. Lectures on characteristic classes and of foliations. Lectures on Algebraic and Differential Topology, 1972.

[2] Bousfield A K and Guggenheim V K A M. On PL de Rham theory and rational homotopy type. Mem. AMS, 1976, 8.

[3] De Rham G. Sur l′analysis situs des variétés à n dimensions. J. de Math., 1931, 10: 115-200.

[4] De Rham G. Variétés différentielles, 1955.

[5] Friedlander E, Griffiths P A and Morgan J. Homotopy theory and differential forms. Mimeo. Notes, 1972.

[6] Miller E Y. De Rham cohomology with arbitrary coefficients. Topology, 1978, 17: 193-203.

[7] Poincaré H. Analysis Situs. J. de l′ Ecole Polytechnique, 1985, 1: 1-121.

[8] Sullivan D. Differential forms and topology of manifolds. Symposium Tokyo on Topology, 1973: 37-49.

[9] Swan R D. Thom's theory of differential forms on simplicial set. Topology, 1975, 14: 271-273.

[10] Weil A. Sur les théorèmes de de Rham. Comm. Math. Helv., 1952, 20: 119-145.

[11] Wu Wen-tsün. De Rham-Sullivan measure of spaces and its calculability, in the Chern Symposium 1979, 1980: 229-245.

56. Some Remarks on Jet-Transformations*

I Introduction

The notion of *JET* was introduced by the late Professor Ehresmann and has proved to be of great usefulness in the study of various branches of modern mathematics. The present note gives some remarks about transformations of jets induced by maps and the fibrations connected therewith. We remark that all functions and maps in what follows are understood to be of class infinity.

Let us begin by first making precise some notations. Any k-tuple of integers, different or not, arranged in some definite order, say i, j, $\cdots$, m (k in number) will be denoted by $[i, j, \cdots, m]$, while the UNORDERED k-tuple of the same integers will be denoted by $[[i, j, \cdots, m]]$. Thus $[[1, 2, 1, 3]]$ is the same as $[[2, 3, 1, 1]]$ while $[1, 2, 1, 3]$ is different from $[2, 3, 1, 1]$. For $u = [i, j, \cdots, m]$, we write also the associated un-ordered tuple $[[i, j, \cdots, m]]$ as $[u]$.

Let n be a positive integer fixed in what follows. The set of all ordered (resp. un-ordered) k-tuples of integers chosen from 1, 2, $\cdots$, n will then be denoted by [INDk] (resp. [[INDk]]) and will be arranged in certain definite order.

Let V be a vector space of dimension n over R with a fixed basis and corresponding coordinate system $X = (X_1, X_2, \cdots, X_n)$. Let $V[k]$ be the k-fold tensor product of V by itself. With respect to the chosen basis of V any element of $V[k]$ can be represented in coordinates as (Xu) with u running over the ordered set [INDk]. The subspace of $V[k]$ consisting of all elements (Xu) for which $Xu = Xv$ whenever $[u] = [v]$ will be denoted by $V[[k]]$. The direct sum of all $V[k]$ (resp. all $V[[k]]$) for $k = 1, 2, \cdots, r$ will be denoted by $SV[r]$ (resp. $SV[[r]]$).

Let O be the origin of V. Consider now any fuction (or rather germ of function) F about O on V which takes value 0 at O. For each u in [INDk] let $DXu\ F$ be the value at O of the partial derivative of F with respect to variables Xi with i running over integers in u. Then $(DXu\ F)$ with u running over [INDk] for $k = 1, 2, \cdots, r$ defines an element of $SV[r]$. This is also an element of $SV[[r]]$ and will be called the r-JET of F, to be denoted by JETr F in what follows.

* *Bull. Soc. Math. Belgique*, 1986, 38: 409-414. In Honor of professor Guy HIRSCH

Let us denote by Map(V) the set of all maps (or rather germs of maps) of V into itself preserving the origin of V. For F in Map(V) with $F(X_1, X_2, \cdots, X_n) = (Y_1, Y_2, \cdots, Y_n)$ the set (JETr Y_1, JETr $Y_2, \cdots$, JETr Y_n) will be called the r-JET defined by F and will be simply denoted by JETr F. A local diffeomorphism F in Map(V) will be said to be INVERTIBLE and the associated r-jet too. Two maps F, G in Map(V) are said to be r-EQUIVALENT if JETr F = JETr G. For abwse of notations we shall denote the r-equivalence class of JETr F simply by JETr F too. The set of all these JETr F for F invertible in Map(V) will be denoted by Jr(V) or simply Jr and the subset consisting of all JETr F for which $DXu\ Yi = 0$ whenever u is in [INDk] with $k >= 2$ will be denoted by Jr0(V) or simply Jr0. The aim of the present note is then to make a little more precise the structure of Jr and Jr0, the latter being isomorphic to the general linear group $GL(V)$.

Let H be the composite GF of maps F, G in Map(V) Then the multiplication in Jr defined by JETr $G *$ JETr F = JETr H will turn Jr into a group with Jr0 a subgroup. By a study of the precise structure of Jr in next sections it will follow easily the following

Theorem 1 *With due topology introdluced in* Jr *the subspace* Jr0 *is a strong deformation retract of* Jr *and the quotient space* Jr/Jr0 is *a contractible space.*

As a corollary we have

Theorem 2 *Any vector bundle with V as fiber type and* Jr *as structural group is reducible to one with* Jr0 *as structural group.*

II Standard tables

As a preparation we shall introduce the notion of STANDARD TABLE somewhat alike to Young's table. Thus, for fixed integers $r > 0$ and $k > 0$ and $<= r$ let us fill in the integers 1, 2, ···, r in a table of k rows of lengths which may be different one from the other but all start from the first column. The table of integers will be said to be STANDARD if the integers in each row and also those in the first column are ever increasing. The totality of all such standard tables of r integers in k rows will be denoted by STrk and the totality of STrk for r fixed and $k = 1, 2, \cdots, r$ by STr. For example, the table

$$\begin{array}{lll} 1 & 5 & \\ 2 & & \\ 3 & 4 & 6 \end{array}$$

is a standard one in ST63 while ST4 is consisting of the 15 standard tables below:

ST41: $\begin{array}{llll} 1 & 2 & 3 & 4 \end{array}$

ST42: $\begin{array}{lll} 1 & 2 & 3 \\ 4 & & \end{array}$ $\quad \begin{array}{lll} 1 & 2 & 4 \\ 3 & & \end{array}$ $\quad \begin{array}{lll} 1 & 3 & 4 \\ 2 & & \end{array}$ $\quad \begin{array}{lll} 1 & & \\ 2 & 3 & 4 \end{array}$

ST42: $\begin{array}{ll} 1 & 2 \\ 3 & 4 \end{array}$ $\quad \begin{array}{ll} 1 & 3 \\ 2 & 4 \end{array}$ $\quad \begin{array}{ll} 1 & 4 \\ 2 & 3 \end{array}$

ST43: $\begin{array}{ll} 1 & 2 \\ 3 & \\ 4 & \end{array}$ $\quad \begin{array}{ll} 1 & 3 \\ 2 & \\ 4 & \end{array}$ $\quad \begin{array}{ll} 1 & 4 \\ 2 & \\ 3 & \end{array}$ $\quad \begin{array}{ll} 1 & \\ 2 & 3 \\ 4 & \end{array}$ $\quad \begin{array}{ll} 1 & \\ 2 & 4 \\ 3 & \end{array}$ $\quad \begin{array}{ll} 1 & \\ 2 & \\ 3 & 4 \end{array}$

ST44: $\begin{array}{l} 1 \\ 2 \\ 3 \\ 4 \end{array}$

It is now clear how to form the set of standard tables STs from those in STr where $r = s - 1$ by adjoining the new integer s either at the end of a certain row or in a new row of a single integer at the bottom.

Let T be a standard table in STrk with lengths $L1$, $L2$, $\cdots$, Lk for the successive rows. Then to a set υ in [INDr] and each $i = 1, 2, \cdots, k$ will correspond a set Ti υ in [INDj] with $j = Li$ according to the following rule: Let the t-th integer in the i-th row of T be m, then the t-th element of Ti υ will be the m-th element of υ.

III A formula about composite functions

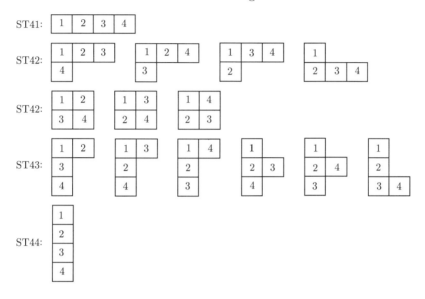

Let $Y = (Y_1, Y_2, \cdots, Y_n)$ now be a set of functions Yi in variables $X = (X_1, X_2, \cdots, X_n)$ which all take value 0 for $X_1 = X_2 = \cdots = X_n = 0$. For υ in [IND$r$] we shall write $DXu\ Yi$ the value at $X_1 = X_2 = \cdots = X_n = 0$ of the partial derivative of Yi with respect to the variables Xj with j running over the integers in υ. Let a standard table T in STrk and an ordered k-tuple w in [INDk] be given. We shall then write

$$Ywi = Yj \text{ for } j = i\text{-th integer in } w,$$

$$Yw = (Yw1, Yw2, \cdots, Ywk).$$

Write also

$$TiDX\upsilon = DXu \; for \; u = Ti\upsilon.$$

Then we shall set as a simplified notation

$$TDX\upsilon Yw = T\text{-PROD}\upsilon(TiDX\upsilon\, Ywi),$$

in which T-PRODυ means that the product is to be taken over $i = 1, 2, \cdots, k$ where k is the number of rows in T.

Let Z now be a function in $Y = (Y_1, Y_2, \cdots, Y_n)$ which in turn is a function of $X = (X_1, X_2, \cdots, X_n)$. We suppose again that $Z = 0$ for $Y_1 = Y_2 = \cdots = Y_n = 0$ and each $Yi = 0$ for $X_1 = X_2 = \cdots = X_n = 0$. Write now DIF1, DIF2, etc for the successive differentials *d, dd,* etc. For υ in [INDk] we set then DIFυ Y the (commutative) product of differentials DIF1 Yi with i running over integers in υ. Similarly for DIFυ X. Under the convention DIF2 $Xi = 0$ for $i = 1, 2, \cdots, n$ we shall have then

$$\text{DIF}r\ Z = \text{SUM}\upsilon r(DX\upsilon\ Z \; * \; \text{DIF}\upsilon X),$$

in which SUMυr means that the summation is to be taken over the range of υ in [INDr]. The aim of this section is to express the partial derivatives $DX\upsilon\ Z$ of Z as a composite function in X in terms of those $DY\upsilon\ Z$ of Z as a function of Y and those $DX\upsilon\ Yi$ of Yi as functions of X. In certain sense it may be considered as a generalization of Faa De Bruno formula in the case of a single variable. For this purpose we have first the following

Lemma 1 *The r-th differential of Z considered as a composite function in the independent variables* $X = (X_1, X_2, \cdots, X_n)$ *is given by*

$$(*)\ \text{DIF}r\ Z = \text{SUM}k\ \text{SUM}wk\ \text{SUM}\upsilon r\ \text{TSUM}rk\ (DYw\ Z * TDX\upsilon\ Yw * DIF\upsilon\ X),$$

in which the various summations are to be taken over respective ranges as follows.

SUMk :	over $k = 1, 2, \cdots, r$
SUMwk :	over all ordered k-tuples w in [INDk]
SUMυr :	over all ordered r-tuples υ in [INDr]
TSUMrk :	over all standard tables T in STrk.

Proof. Forming the first differential of Z we get

$$\begin{aligned}\text{DIF1}\ Z &= \text{SUM}i(DYiZ * \text{DIF1}Yi)\\ &= \text{SUM}i\ \text{SUM}j(DYiZ * DXj\ Yi * \text{DIF1}\ Xj).\end{aligned}$$

This is trivially the formula $(*)$ corresponding to the case $r = 1$. By forming successive differentials and taking into account the rule of formation of standard tables we get easily the formula $(*)$ by induction.

From $(*)$ we get for the partial derivatives of the composite function Z the following formula:

$(**)$ $C\upsilon * DX\upsilon Z = \text{SUM}k\ \text{SUM}wk\ \text{SUM}u\upsilon\ \text{TSUM}rk(DYw\ Z * TDXu\ Yw)$
for any υ in $[\text{IND}r]$, in which SUMu υ means summation over all u in $[\text{IND}r]$ for which $[u] = [\upsilon]$, and $C\upsilon$ is certain integer constant depending on υ, being the number of such u's.

Ⅳ Proof of theorems

We are now in a position to prove Theorem 1 and hence also Theorem 2 stated in Section I. as follows.

For any map F in Map(V) with $F(X_1, X_2, \cdots, X_n) = (Y_1, Y_2, \cdots, Y_n)$ and any t real let us define Ft in Map(V) by

$$DX\upsilon\ Ft = t \wedge (r-1) * DX\upsilon\ F,$$

for any υ in $[\text{IND}r]$. The definition is legitimate according to a known theorem of R. Hermann and passes clearly to germ maps which is invertible if F is so. Hence

$$F \to Ft$$

will induce a jet-map say

$$Rt : \text{J}r \to \text{J}r.$$

Lemma 2 *Rt is a group endomorphism of the group* Jr *with*

$$Rt = \text{ident.}/\text{J}r0,$$

$$R0(\text{J}r) \prec \text{J}r0,$$

in which $\prec$ *stands for "contained in".*

Proof. Consider the composite function $H = GF$ with $F(X) = Y$, $G(Y) = Z$, where $X = (X_1, X_2, \cdots, X_n)$, $Y = (Y_1, Y_2, \cdots, Y_n)$, and $Z = (Z_1, Z_2, \cdots, Z_n)$. For t fixed let $Ft(X) = U$, $Gt(U) = V$, where $U = (U_1, U_2, \cdots, U_n)$, $V = (V_1, V_2, \cdots, V_n)$. Let K be the composite map $K = GtFt$ so that $K(X) = V$. Consider now any table T in STrk with lengths of rows of T supposed to be $L1$, $L2, \cdots$, Lk. For u, υ in [INDr] with $[u] = [\upsilon]$ and w in [INDk] we have clearly for values at origin

$\mathrm{TD}Xu\, Uw = t \wedge ((L1 - 1) + \cdots + (Lk - 1)) * \mathrm{TD}Xu\, Yw$, and

$$\mathrm{DU}w\ Vi = t \wedge (k-1) * DYw\ Zi.$$

As $L1 + \cdots + Lk = r$, we see from the formula (**) in Section III that

$$DX\upsilon\ Vi = t \wedge (r-1) * DX\upsilon Zi.$$

It follows that $K = Ht$ or

$GtFt = Ht = (GF)t.$

This proves that Rt preserves the group structure of Jr. The other assertions are evident.

Theorem 1 now follows immediately from Lemma 2 since Rt gives a deformation-retraction of Jr into Jr0 preserving group structure and keeping Jr0 pointwise fixed which induces thus a contraction of the quotient space Jr/Jr0 into a point, as was to be proved.